AF676505

Feature Papers in Eng 2022

Feature Papers in Eng 2022

Editor

Antonio Gil Bravo

MDPI • Basel • Beijing • Wuhan • Barcelona • Belgrade • Manchester • Tokyo • Cluj • Tianjin

Editor
Antonio Gil Bravo
Universidad Pública de Navarra
Spain

Editorial Office
MDPI
St. Alban-Anlage 66
4052 Basel, Switzerland

This is a reprint of articles from the Special Issue published online in the open access journal *Eng-Advances in Engineering* (ISSN 2673-4117) (available at: https://www.mdpi.com/journal/eng/special_issues/FP_in_Eng_2022).

For citation purposes, cite each article independently as indicated on the article page online and as indicated below:

LastName, A.A.; LastName, B.B.; LastName, C.C. Article Title. *Journal Name* **Year**, *Volume Number*, Page Range.

ISBN 978-3-0365-7530-8 (Hbk)
ISBN 978-3-0365-7531-5 (PDF)

© 2023 by the authors. Articles in this book are Open Access and distributed under the Creative Commons Attribution (CC BY) license, which allows users to download, copy and build upon published articles, as long as the author and publisher are properly credited, which ensures maximum dissemination and a wider impact of our publications.
The book as a whole is distributed by MDPI under the terms and conditions of the Creative Commons license CC BY-NC-ND.

Contents

About the Editor

Antonio Gil Bravo

Antonio Gil Bravo (Full Professor of Chemical Engineering, Universidad Pública de Navarra, Spain): Professor Gil earned his BS and MS in Chemistry at the University of Basque Country (San Sebastián), before receiving his PhD in Chemical Engineering at University of Basque Country (San Sebastián). He undertook postdoctoral research at the Université catholique de Louvain (Belgium), working on Spillover and Mobility of Species on Catalyst Surfaces. The research interests of Professor Gil can be summarized as covering the following topics: the evaluation of the porous and surface properties of solids; pillared claysl gas adsorption; energy and CO2 storage; pollutant adsorption; environmental technologies; environmental management; preparation, characterization and catalytic performance of metal supported nanocatalysts; and industrial waste valorization.

eng

Editorial

Special Issue: Feature Papers in *Eng* 2022

Antonio Gil Bravo

INAMAT2, Science Department, Public University of Navarra, Campus of Arrosadia, Building Los Acebos, E-31006 Pamplona, Spain; andoni@unavarra.es

The aim of this second *Eng* Special Issue is to collect experimental and theoretical re-search relating to engineering science and technology. The general topics published in *Eng* are as follows: electrical, electronic and information engineering; chemical and materials engineering; energy engineering; mechanical and automotive engineering; industrial and manufacturing engineering; civil and structural engineering; aerospace engineering; biomedical engineering; geotechnical engineering and engineering geology; and ocean and environmental engineering. This editorial is an overview of the selected representative studies on these topics.

This book contains 33 papers, including 2 *Review* papers and 1 *Communication*, published by several authors interested in new cutting-edge developments in the field of engineering. Recently, a subcategory of nanotechnology—nano- and microcontainers—has developed rapidly, with unexpected results. Nano- and microcontainers refer to hollow spherical structures in which the shells can be organic or inorganic. These containers can be filled with substances released when excited and can be used in corrosion healing, cancer therapy, cement healing, antifouling, etc. In the first review, the author summarizes the various innovative technologies that have beneficial effects on improving people's lives [1].

Jombo and Zhang [2] report that traditional means of monitoring the health of industrial systems involve the use of vibration and performance monitoring techniques, among others. In these approaches, contact-type sensors, such as accelerometers, proximity probes, pressure transducers and temperature transducers, are installed on the machine to monitor its operational health parameters. However, these methods fall short when additional sensors cannot be installed on the machine due to cost, space constraint or sensor reliability concerns. On the other hand, the use of an acoustic-based monitoring technique provides an improved alternative, as acoustic sensors (e.g., microphones) can be implemented quickly and cheaply in various scenarios and do not require physical contact with the machine. The collected acoustic signals contain relevant operating health information about the machine, yet they can be sensitive to background noise and changes in machine operating condition. These challenges are being addressed from the industrial applicability perspective for acoustic-based machine condition monitoring.

Solar generation has increased rapidly worldwide in recent years, and it is projected to continue to grow exponentially. A problem exists in that the increase in solar energy generation will increase the probability of grid disturbances. The study presented by Soto et al. [3] focuses on analyzing the grid disturbances caused by the massive integration into the transmission line of utility-scale solar energy loaded onto the balancing authority high-voltage transmission lines in four regions of the United States electrical system: (1) California, (2) Southwest, (3) New England, and (4) New York. A statistical analysis of the equality of means was carried out to detect changes in the energy balance and peak power. The results show that, when comparing the difference between hourly net generation and demand, energy imbalance occurs in the regions with the highest solar generation: California and Southwest. No significant difference was found in any of the four regions in relation to the energy peaks. The results imply that regions with greater utility-level solar energy adoption must conduct greater energy exchanges with other

Citation: Gil Bravo, A. Special Issue: Feature Papers in *Eng* 2022. *Eng* **2023**, 4, 1156–1166. https://doi.org/10.3390/eng4020067

Received: 11 April 2023
Accepted: 12 April 2023
Published: 14 April 2023

Copyright: © 2023 by the author. Licensee MDPI, Basel, Switzerland. This article is an open access article distributed under the terms and conditions of the Creative Commons Attribution (CC BY) license (https://creativecommons.org/licenses/by/4.0/).

regions to reduce potential disturbances to the grid. It is essential to bear in mind that, as the installed solar generation capacity increases, the potential energy imbalances created in the grid increase.

The Jiles–Atherton model is commonly used in the hysteresis description of ferromagnetic, ferroelectric, magneto strictive and piezoelectric materials. However, the determination of model parameters is not straightforward because the model involves numerical integration and the solving of ordinary differential equations, both of which are error-prone. As a result, stochastic optimization techniques have been used to explore the vast ranges of these parameters in an effort to identify the parameter values that minimize the error differential between experimental and modelled hysteresis curves. Because of the time-consuming nature of these optimization techniques, Khemani et al. [4] explored the design space of the parameters using a space-filling design. This design provides a narrower range of parameters to look at with optimization algorithms, thereby reducing the time required to identify the optimal Jiles–Atherton model parameters. The authors also indicate that this procedure can be carried out without using expensive hysteresis measurement devices, provided that the desired transformer's secondary voltage is known.

Nuclear energy is currently in the spotlight as a future energy source all over the world amidst the global warming crisis. In the current state of miniaturization, through the development of advanced reactors, such as small modular reactors (SMRs) and micro-reactors, a fission battery was created, inspired by the idea that nuclear energy can be used by ordinary people using the "plug-and-play" concept, such as chemical batteries. As for the design requirements, fission batteries must be economical, standardized, installed, unattended and reliable. Furthermore, the commercialization of reactors is regulated by national bodies, such as the United States (U.S.) Nuclear Regulatory Commission (NRC). At the international level, the International Atomic Energy Agency (IAEA) oversees the safe and peaceful use of nuclear power. However, regulations currently face a significant gap in terms of their applicability to advanced non-light water reactors (non-LWRs). Therefore, Lee and Diaconeasa [5] investigated the regulatory gaps in the licensing of fission batteries concerning safety in terms of siting, autonomous operation and transportation and suggested response strategies to supplement them. To determine the applicability of the current licensing framework to fission batteries, the authors reviewed the U.S. NRC Title 10, Code of Federal Regulations (CFR) and IAEA INSAG-12. To address siting issues, the authors also explored the non-power reactor (NPR) approach for site restrictions and the permit-by-rule (PBR) approach for excessive time burdens. In addition, they discussed how the development of an advanced human–system interface augmented with artificial intelligence and monitored by personnel for fission batteries may enable successful exemptions from the current regulatory operation staffing requirements. Finally, they also indicated that no transportation regulatory challenge exists.

Sharafeldin et al. [6] present in an interesting study that intersections are commonly recognized as crash hot spots on roadway networks. Therefore, intersection safety is a major concern for transportation professionals. Identifying and quantifying the impact of crash-contributing factors are crucial to planning and implementing the appropriate countermeasures. This study covered an analysis of nine years of intersection crash records in the State of Wyoming to identify the contributing factors to crash injury severity at intersections. The study involved an investigation of the influence of roadway (intersection) and environmental characteristics on crash injury severity. The results demonstrated that several parameters related to intersection attributes (pavement friction, urban location, roadway functional classification, guardrails and right shoulder width) and two environmental conditions (road surface condition and lighting) influence the injury severity of intersection crashes. This study also identified the significant roadway characteristics influencing crash severity and explored the key role of pavement friction, which is a commonly omitted variable.

In Ref. [7], Andersen et al. present the use of a high-fidelity neural network surrogate model within a Modular Optimization Framework for the treatment of crud deposition as a constraint while optimizing the light-water reactor core loading pattern. A neural network was utilized for the treatment of crud constraints within the context of an advanced genetic algorithm applied to the core design problem. This proof-of-concept study shows that loading pattern optimization aided by a neural network surrogate model can optimize the manner in which crud distributes within a nuclear reactor without impacting operational parameters such as enrichment or cycle length. Several analysis methods were investigated by the authors. The analysis showed that the surrogate model and genetic algorithm successfully minimized the deviation from a uniform crud distribution against a population of solutions from a reference optimization in which the crud distribution was not optimized. Strong evidence shows that boron deposition in crud can be optimized through the loading pattern. This proof-of-concept study shows that the employed methods provide a powerful tool for mitigating the effects of crud deposition in nuclear reactors.

For the first time, Zayed et al. [8] study the Fokas–Lenells equation in polarization-preserving fibers with multiplicative white noise in the Itô sense. Four integration algorithms were applied by the authors, namely, the method of modified simple equation (MMSE), the method of sine-cosine (MSC), the method of Jacobi elliptic equation (MJEE) and ansatze involving hyperbolic functions.

The next study evaluated unsupervised anomaly detection methods in multispectral images obtained with a wavelength-independent synthetic aperture sensing technique called Airborne Optical Sectioning (AOS) [9]. With a focus on search-and-rescue missions that apply drones to locate missing or injured persons in a dense forest and require real-time operation, the authors evaluated the runtime vs. quality of these methods. Furthermore, they also showed that color anomaly detection methods that normally operate in the visual range always benefit from an additional far infrared (thermal) channel.

Tebuthiuron is a selective herbicide for woody species and is commonly manufactured and sold as a granular formulation. In an interesting study, the authors of [10] investigated the use of infrared spectroscopy for a quality analysis of tebuthiuron granules, specifically the prediction of moisture content and tebuthiuron content. A comparison of different methods showed that near-infrared spectroscopy showed better results than mid-infrared spectroscopy, while a handheld NIR instrument (MicroNIR) showed slightly improved results over a benchtop NIR instrument (Antaris II FT-NIR Analyzer). The best-performing models gave an R2CV of 0.92 and an RMSECV of 0.83% w/w for moisture content, and an R2CV of 0.50 and an RMSECV of 7.5 mg/g for tebuthiuron content. This analytical technique could be used to optimize the manufacturing process and to reduce the costs of post-manufacturing quality assurance.

Thixotropic behavior describes a time-dependent rheological behavior characterized by reversible changes. Fresh cementitious materials often require thixotropic behavior to ensure sufficient workability and proper casting without vibration. Non-thixotropic behavior induces a workability loss. Cementitious materials cannot be considered as an ideal thixotropic material due to cement hydration, which leads to irreversible changes. However, in some cases, cement paste may demonstrate thixotropic behavior during the dormant period of cement hydration. The aim of the work presented by El Bitouri and Azêma [11] was to propose an approach able to quantify the contribution of cement hydration during the dormant period and to examine the conditions under which the cement paste may display thixotropic behavior. The proposed approach consists of a succession of stress growth procedures that allow the static yield stress to be measured. For an inert material, such as a calcite suspension, the structural build-up is due to the flocculation induced by attractive Van der Waals forces. This structural build-up is reversible. For cement paste, there is a significant increase in the static yield stress due to cement hydration. The addition of superplasticizer allows the thixotropic behavior to be maintained during the first hours due to its retarding effect. However, an increase in the superplasticizer dosage leads to a decrease in the magnitude of the Van der Waals forces, which can erase the thixotropic behavior.

Biometrics deals with the recognition of humans based on their unique physical characteristics. It can be based on facial, iris, fingerprint or DNA identification. In Ref. [12], Hafeez et al. considered the iris as a source of biometric verification as it is a unique part of the eye that can never be altered, and it remains the same throughout an individual's life. The authors proposed an improved iris-recognition system including image registration as a main step, as well as an edge-detection method for feature extraction. This PCA-based method was also proposed as an independent iris-recognition method based on a similarity score. The experiments conducted using the developed database demonstrate that the first proposed system reduced the computation time to 6.56 s, and it improved the accuracy to 99.73, while the PCA-based method has less accuracy than this system.

The increasing implementation of distributed renewable generation lead to the need for Citizen Energy Communities. Citizen Energy Communities may be able to be active market players and to solve local imbalances. The liberalization of the electricity sector caused wholesale and retail competition, which is a natural evolution of electricity markets. In retail competition, retailers and communities compete to sign bilateral contracts with consumers. In wholesale competition, producers, retailers and communities can submit bids to spot markets, where the prices are volatile, or can sign bilateral contracts to hedge against spot price volatility. To participate in those markets, communities have to rely on risky consumption forecasts, hours ahead of real-time operation. So, as Balance Responsible Parties, they may pay penalties for real-time imbalances. This paper proposed and tested a new strategic bidding process in spot markets for communities of consumers. The strategic bidding process is composed of a forced forecast methodology for day-ahead and short-run trends for intraday forecasts of consumption. This paper developed by Algarvio [13] also presents a case study where energy communities submit bids to spot markets to satisfy their members using the strategic bidding process. The results show that bidding at short-term markets leads to lower forecast errors than bidding at long and medium-term markets. Better forecast accuracy leads to better fulfillment of a community's programmed dispatch, resulting in lower imbalances and control reserve needs for power system balance. Furthermore, by being active market players, energy communities may save around 35% in their electrical energy costs when compared with retail tariffs.

Corn is an example of an agricultural grain with a specific combustibility level and can promote smoldering fires during storage. The interesting contribution of the study in Ref. [14] conducted an experimental design to numerically evaluate how three parameters, namely particle size, moisture, and air ventilation, influence the smoldering velocity. The work methodology was based on Minitab's experimental design, which defined the number of experiments. First, a pile of corn was heated by a hot plate, and a set of thermocouples registered all temperature variations. Then, a full-factorial experiment was implemented in Minitab to analyze the smoldering, which provided a mathematical equation to represent the smoldering velocity. The results indicate that particle size is the most influential factor in the reaction, with 35% and 45% variation between the dried and wet samples. Moreover, comparing the effect of moisture between corn flour and corn powder samples, variations of 19% and 31% were observed; additionally, analyzing the ventilation as the only variant, the authors noticed variations of 15% and 17% for dried and wet corn flour, respectively, and 27% and 10% for dried and wet corn powder, respectively.

Currently, tissue product producers try to meet consumers' requirements to retain their loyalty. In perforated products, such as toilet paper, these requirements involve the paper being portioned along the perforation line and not outside of it. Thus, it becomes necessary to enhance the behavior of the perforation line in perforated tissue papers. The study presented by Costa Vieira et al. [15] aimed to verify if the perforation line for 0° (the solution found in commercial perforated products) is the best solution to maximize the perforation efficiency. A finite element (FE) simulation was used by the authors to validate the experimental data, where the deviations from the experiments were 5.2% for the case with a 4 mm perforation length and 8.8% for a perforation of 2 mm, and to optimize the perforation efficiency using the genetic algorithm while considering two different cases. In the first case, the blank distance and the perforation line angle were varied, with the best configuration being achieved with a blank distance of 0.1 mm and an inclination angle of 0.56°. For the second case, the blank distance was fixed to 1.0 mm and the only variable to be optimized was the inclination angle of the perforation line. It was found that the best angle inclination was 0.67°. In both cases, it was verified that a slight inclination in the perforation line will favor partitioning and, therefore, the perforation efficiency.

Telecommunication companies collect a deluge of subscriber data without retrieving substantial information. An exploratory analysis of these types of data will facilitate the prediction of varied information that can be geographical, demographic, financial or other. Predictions can therefore be an asset in the decision-making process of telecommunications companies, but only if the information retrieved follows a plan with strategic actions. An exploratory analysis of subscriber data was implemented in this research to predict subscriber usage trends based on historical time-stamped data [16]. The predictive outcome was unknown but approximated using the data at hand. The author used 730 data points selected from Insights Data Storage (IDS). These data points were collected from the hourly statistic traffic table and subjected to exploratory data analysis to predict the growth in subscriber data usage. The Auto-Regressive Integrated Moving Average (ARI-MA) model was used for the forecasting. In addition, the author used the normal Q-Q, correlogram and standardized residual metrics to evaluate the model. This model showed a p-value of 0.007. This result supports the hypothesis predicting an increase in subscriber data growth. The ARIMA model predicted a growth of 3 Mbps, with a maximum data usage growth of 14 Gbps. In the experiment, ARIMA was compared with the Convolutional Neural Network (CNN) and achieved the best results with the UGRansome data. The ARIMA model performed better, with an execution speed that was faster by a factor of 43 for more than 80,000 rows. On average, it takes 0.0016 s for the ARIMA model to execute one row and 0.069 s for the CNN to execute the same row, thus making the ARIMA 43× (0.0690.0016) faster than the CNN model. These results provide a road map for predicting subscriber data usage so that telecommunication companies can be more productive in improving their Quality of Experience (QoE). This study provides a better understanding of the seasonality and stationarity involved in subscriber data usage's growth, exposing new network concerns and facilitating the development of novel predictive models.

Barbosa et al. [17] performed 2D micrometric mapping of different elements in different grain size fractions of the soil of a sample using the X-ray microfluorescence (μ-XRF) technique. The sample was collected in the vicinity of São Domingos, an old mine of massive sulphide minerals located in the Portuguese Iberian Pyrite Belt. As expected, elemental high-grade concentrations of distinct metals and metalloids dependent on the existing natural geochemical anomaly were detected. The authors developed a clustering and k-means statistical analysis considering red–green–blue (RGB) pixel proportions in the produced 2D micrometric image maps, allowing the authors to identify elementary spatial distributions in 2D. The results evidence how elemental composition varies significantly at the micrometric scale per grain-size class and how chemical elements present irregular spatial distributions due to direct dependence on the distinct mineral spatial distributions. Due to this fact, the elemental compositions are more different in coarser grain-size classes, whereas the grinding-milled fraction does not always represent the average of all partial grain-size fractions. Despite the complexity of the performed analysis, the achieved results evidence the suitability of μ-XRF in characterizing natural, heterogeneous, granular soils samples at the micrometric scale, being a very promising high-resolution investigation technique.

In Ref. [18], the author proposed an efficient method of identifying important neurons that are related to an object's concepts by mainly considering the relationship between these neurons and their object concept or class. He first quantified the activation values among neurons, based on which histograms of each neuron were generated. Then, the obtained histograms were clustered to identify the neurons' importance. A network-wide holistic approach was also introduced to efficiently identify important neurons and their influential connections to reveal the pathway of a given class. The influential connections, as well as their important neurons, were carefully evaluated to reveal the sub-network of each object's concepts. The experimental results on the MNIST and Fashion MNIST datasets show the effectiveness of the proposed method.

Safety reporting has long been recognized as critical to reducing safety occurrences by identifying issues early enough such that they can be remedied before an adverse outcome. The study in Ref. [19] examines safety occurrence reporting amongst a sample of 92 New Zealand civilian uncrewed aircraft users. An online survey was created to obtain the types of occurrences that these users have had, how (if at all) these are reported, and why participants did or did not report using particular systems. This work focused on seven types of occurrences that have been highlighted by the Civil Aviation Authority of New Zealand as being reportable using the CA005RPAS form, the template for reporting un-crewed aircraft occurrences to authorities. The number of each type of occurrence was recorded, as well as what percentage of occurrences were reported using the CA005RPAS form, reported using an internal reporting system or not reported. Qualitative questions were used by the authors to understand why participants did or did not report using particular systems. The categorical and numerical data were analyzed using Chi-Squared Tests of Independence, Kruskal–Wallis H Tests and Mann–Whitney U Tests. The qualitative data were analyzed using thematic analysis. The findings reveal that 85.72% of reportable safety occurrences went unreported by pilots, with only 2.74% of occurrences being self-reported by pilots using the CA005RPAS form. The biggest reason for not reporting was that the user did not perceive the occurrence as being serious enough, with not being aware of reporting systems and not being legally required to report also being major themes. Significant differences were also observed by the authors between user groups, thus leading to suggestions on policy changes to improve safety occurrence reporting, such as making reporting compulsory, setting minimum training standards, having an anonymous and non-punitive reporting system, and working with member-based organizations.

Using surrogate safety measures is a common method to assess safety on roadways. Surrogate safety measures allow for a proactive safety analysis; the analysis is performed prior to crashes occurring. This allows for safety improvements to be implemented proactively to prevent crashes, and the associated injuries and property damage. Existing surrogate safety measures primarily rely on data generated by microsimulations, but the advent of connected vehicles has allowed for the incorporation of data from actual cars into safety analyses with surrogate safety measures. In the study by Khanal and Edelmann [20], commercially available connected vehicle data were used to develop crash-prediction models for crashes at intersections and segments in Salt Lake City, Utah. Harsh braking events were identified and counted within the area of influence, inclusive of sixty intersections and thirty segments, and then used to develop crash-prediction models. Other intersection characteristics were considered as regressor variables in the models, such as the intersection's geometric characteristics, connected vehicle volumes, and the presence of schools and bus stops in the vicinity. Statistically significant models were developed by the authors, and these models may be used as a surrogate safety measure to analyze intersection safety proactively. The findings are applicable to Salt Lake City, but similar research methods may be employed by other researchers to determine whether these models are applicable in other cities and to determine how the effectiveness of this method endures through time.

Buried charges pose a serious threat to both civilians and military personnel. It is well established that soil properties have a large influence on the magnitude and variability of loading from explosive blasts in buried conditions. In Ref. [21], work was undertaken to improve techniques for processing pressure data from discrete measurement apparatuses; this was performed by testing truncation methodologies and the area integration of impulses, accounting for the particle size distribution (PSD) of the soils used in testing. Two experimental techniques were investigated by Waddoups et al. to allow for a comparison between a global impulse capture method and an area-integration procedure from a Hopkinson Pressure Bar array. This paper explores an area-limiting approach, based on particle size distribution, as a possible approach to derive a better representation of the loading on the plate, thus demonstrating that the spatial distribution of a loading over a target can be related to the PSD of the confining material.

The rapidly increasing number of drones in the national airspace, including those for recreational and commercial applications, has raised concerns regarding misuse. Autonomous drone-detection systems offer a probable solution to overcoming the issue of potential drone misuse, such as drug smuggling, violating people's privacy, etc. However, detecting drones can be difficult, due to similar objects being in the sky, such as airplanes and birds. In addition, automated drone detection systems need to be trained with ample amounts of data to provide high accuracy. Real-time detection is also necessary, but this requires highly configured devices such as a graphical processing unit (GPU). The work in Ref. [22] sought to overcome these challenges by proposing a one-shot detector called You Only Look Once version 5 (YOLOv5), which can train the proposed model using pre-trained weights and data augmentation. The trained model was evaluated using mean average precision (mAP) and recall measures. The model achieved a 90.40% mAP, a 21.57% improvement over our previous model that used You Only Look Once version 4 (YOLOv4), and was tested on the same dataset.

The paper in Ref. [23] introduces a novel approach to leveraging features learned from both supervised and self-supervised paradigms, to improve image classification tasks, specifically for vehicle classification. Two state-of-the-art self-supervised learning methods, DINO and data2vec, were evaluated and compared by the authors for their representation learning of vehicle images. The former contrasts local and global views, while the latter uses masked prediction on multiple layered representations. In the latter case, supervised learning is employed to finetune a pretrained YOLOR object detector for detecting vehicle wheels, from which definitive wheel positional features are retrieved. The representations learned from these self-supervised learning methods were combined with the wheel positional features for the vehicle classification task. Particularly, a random wheel masking strategy was utilized to finetune the previously learned representations in harmony with the wheel positional features during training of the classifier. The experiments made by the authors show that the data2vec-distilled representations, which are consistent with our wheel masking strategy, outperformed the DINO counterpart, resulting in a celebrated Top-1 classification accuracy of 97.2% for classifying the 13 vehicle classes defined by the Federal Highway Administration.

Many current bioinformatics algorithms have been implemented in parallel programming codes. Some of them have already reached the limits imposed by Amdahl's law, but many can still be improved. Blaszynski and Bielecki [24] presented an approach that allows for the generation of a high-performance code for calculating the number of RNA pairs. The approach allows for the generation of a parallel tiled code with maximum-dimension tiles, which for the discussed algorithm, is in 3D. The experiments carried out on two modern multi-core computers, an Intel(R) Xeon(R) Gold 6326 (2.90 GHz, 2 physical units, 32 cores, 64 threads and 24 MB Cache) and Intel(R) i7(11700KF (3.6 GHz, 8 cores, 16 threads and 16 MB Cache), demonstrate a significant increase in performance and scalability of the generated parallel tiled code. For the Intel(R) Xeon(R) Gold 6326 and Intel(R) i7, target code speedup increased linearly with an increase in the number of threads. The approach presented in this paper to generate a target code can be used by programmers to generate target parallel tiled codes for other bioinformatics codes for which the dependence patterns are similar to those of the code implementing the counting algorithm.

Malware classification is a well-known problem in computer security. Hyperparameter optimization (HPO) using covering arrays (CAs) is a novel approach that can enhance machine learning classifier accuracy. The tuning of machine learning (ML) classifiers to increase classification accuracy is needed nowadays, especially with newly evolving malware. Four machine learning techniques were tuned using cAgen, a tool for generating covering arrays. The results included in Ref. [25] show that cAgen is an efficient approach to achieving the optimal parameter choices for ML techniques. Moreover, the covering array shows significant promise, especially cAgen with regard to the ML hyperparameter optimization community, malware detector community and overall security testing.

Musical timbre is a phenomenon of auditory perception that allows for the recognition of musical sounds. The recognition of musical timbre is a challenging task because the timbre of a musical instrument or sound source is a complex and multifaceted phenomenon that is affected by a variety of factors, including the physical properties of the instrument or sound source, the way it is played or produced, and the recording and processing techniques used. Gonzalez and Prati [26] explored an abstract space with 7 dimensions formed by the fundamental frequency and FFT-Acoustic Descriptors in 240 monophonic sounds from the Tinysol and Good-Sounds databases, corresponding to the 4th octave of the transverse flute and clarinet. This approach allowed the authors to unequivocally define a collection of points and, therefore, a timbral space (Category Theory) that allows for different sounds of any type of musical instrument with its respective dynamics to be represented as a single characteristic vector. The geometric distance allows for studying the timbral similarity between audios of different sounds and instruments or between different musical dynamics and datasets. Additionally, a machine learning algorithm that evaluates timbral similarities through Euclidean distances in the abstract space of seven dimensions was proposed by them. The authors conclude that the study of timbral similarity through geometric distances allowed us to distinguish between audio categories of different sounds and musical instruments, between the same type of sound and an instrument with different relative dynamics, and between different datasets.

When studying horizontally inhomogeneous media, it is necessary to apply tensor modifications of electromagnetic soundings. The use of tensor measurements is of particular relevance in near-surface electrical prospecting because the upper part of the geological section is usually more heterogeneously than the deep strata. In the Enviro-MT system designed for the controlled-source radiomagnetotelluric (CSRMT) sounding method, two mutually perpendicular horizontal magnetic dipoles (two vertical loops) are used for tensor measurements. In Ref. [27], a variant of the CSRMT method with two horizontal electrical dipole sources (two transmitter lines) was proposed. The advantage of such sources is an extended frequency range of 1–1000 kHz in comparison with a frequency range of 1–12 kHz for the Enviro-MT system, the greater operational distance (up to 3–4 km compared to 600–800 m), and the ability to measure the signal at the fundamental frequency and its subharmonics. To implement tensor measurements with the equipment of the CSRMT method described in this work, a technique inducing time-varying polarization of the electromagnetic field (rotating field) was developed by the authors based on the use of two transmitters with slightly different current frequencies and two mutually perpendicular transmitter lines grounded at the ends. In this way, the authors made it possible to change the direction of the electrical and magnetic field polarization continuously. This approach allows for the realization of a technique for tensor measurements using a new modified CSRMT method. In permafrost areas, hydrogenic taliks are widespread. These local objects are important in the context of the study of environmental changes in the Arctic and can be successfully explored using the tensor CSRMT method. For numerical modeling, a 2D model of the talik was used. The results of the interpretation of the synthetic data showed the advantage of bimodal inversion using the CSRMT curves of both TM and TE modes compared with separate inversion of the TM and TE curves. These new data demonstrate the prospects of the tensor CSRMT method in the study of permafrost regions. The problems that can be solved using the CSRMT method in the Arctic permafrost regions are also presented and discussed.

The sugar and alcohol sectors are dynamic as a result of climate alterations, the introduction of sugarcane varieties and new technologies. Despite these factors, Brazil stands out as the main producer of sugarcane worldwide, being responsible for 45% of the production of fuel ethanol. Several varieties of sugarcane have been developed in the past few years to improve features of the plant. This, however, led to the challenge of which variety producers should choose to plant on their property. In order to support this process, the research in Ref. [28] aims to test the application of the analytic hierarchy process (AHP) method to support producers in selecting which sugarcane variety to plant on their property. To achieve this goal, the authors relied on a single case study performed on a rural property located inland of São Paulo state, the main producer state in Brazil. The results demonstrate the feasibility of the used approach, specifically owing to the adaptability of the AHP method.

With the rapid development of modern technologies, autonomous or robotic construction sites are becoming a new reality in civil engineering. Despite various potential benefits of the automation of construction sites, there is still a lack of understanding of their complex nature when combining physical and cyber components in one system. A typical approach to describing complex system structures is to use tools of abstract mathematics, which provide a high level of abstraction, allowing for a formal description of the entire system while omitting non-essential details. Therefore, in Ref. [29], autonomous construction is formalized using categorical ontology logs enhanced by abstract definitions of individual components of an autonomous construction system. In this context, followed by a brief introduction to category theory and ologs, exemplary algebraic definitions were given as a basis for the olog-based conceptual modelling of autonomous construction systems. As a result, any automated construction system can be described without providing exhausting detailed definitions of the system components. Existing ologs can be extended, contracted or revised to fit the given system or situation. To illustrate the descriptive capacity of ologs, a lattice of representations was presented by the authors. The main advantage of using the conceptual modelling approach presented in this paper is that any given real-world or engineering problem could be modelled with a mathematically sound background.

Hypotrochoidal profile contours have been produced in industrial applications in recent years using two-spindle processes, and they are considered effective high-quality solutions for form-fit shaft and hub connections. This study presented by Ziaei [30] mainly concerns analytical approaches to determining the stresses and deformations in hypotrochoidal profile shafts due to pure bending loads. The formulation was developed according to bending principles using the mathematical theory of elasticity and conformal mappings. The loading was further used to investigate the rotating bending behavior. The stress factors for the classical calculation of maximum bending stresses were also determined for all those profiles presented and compiled into the German standard DIN3689-1 for practical applications. The results were compared with the corresponding numerical and experimental results, and very good agreement was found. This study contributes to further refinement of the current DIN3689 standard.

Shell structures have a rich family of boundary layers including internal layers. Each layer has its own characteristic length scale, which depends on the thickness of the shell. Some of these length scales are long, something that is not commonly considered in the literature. In Ref. [31], three types of long-range layers are demonstrated over an extensive set of simulations. The author indicates that the observed asymptotic behavior is consistent with theoretical predictions. These layers are shown to also appear on perforated structures underlying the fact these features are properties of the elasticity equations and not dependent on effective material parameters. The simulations were performed using a high-order finite element method implementation of the Naghdi-type dimensionally reduced shell model. Additionally, the effect of the perforations on the first eigenmodes is discussed. Finally, one possible model for buckling analysis is outlined.

Any stretch of coastline requires protection when the rate of erosion exceeds a certain threshold and seasonal coastal drift fluctuations fail to restore balance. Coast erosion can be caused by natural, synthetic or a combination of events. Severe storm occurrences, onshore interventions liable for sedimentation, wave action on the coastlines and rising sea levels caused by climate change are instances of natural factors. The protective methods used to counteract or prevent coastal flooding are categorized as hard and soft engineering techniques. The paper in Ref. [32] is based on extensive reviews and analyses of scientific publications. In order to establish a foundation for the selection of appropriate adaptation measures for coastal protection, this study compiled the literature on a combination of both natural and artificial models using mangrove trees and polymer-based models' configurations and their efficiency in coastal flooding. Mangrove roots occur naturally and cannot be manipulated, unlike artificial model configuration, which can be structurally configured with different hydrodynamic properties. Artificial models may lack the real structural features and hydrodynamic resistance of the mangrove root that it depicts, and this can reduce its real-life application and accuracy.

In the final manuscript [33], presented as a communication, the author indicates that unmanned aircraft systems (UASs), commonly referred to as drones, are an emerging technology that has changed the way that many industries conduct business. Precision agriculture is one industry that has consistently been predicted to be a major locus of innovation for UASs. However, this has not been the case globally. The agricultural aircraft sector in the United States was used as a case study to consider different metrics in evaluating UAS adoption, including a proposed metric, the normalized UAS adoption index. In aggregate, UAS operators only make up 5% of the number of agricultural aircraft operators. However, the annual number of new UAS operators exceeded that of manned aircraft operators in 2022. When used on a state-by-state basis, the normalized UAS adoption index shows that there are regional differences in UAS adoption, with western and eastern states having higher UAS adoption rates and central states having significantly lower UAS adoption rates. This has implications for UAS operators, manufacturers and regulators as this industry continues to develop at a rapid pace.

Conflicts of Interest: The author declares no conflict of interest.

References

1. Kordas, G. All-Purpose Nano- and Microcontainers: A Review of the New Engineering Possibilities. *Eng* **2022**, *3*, 554–572. [CrossRef]
2. Jombo, G.; Zhang, Y. Acoustic-Based Machine Condition Monitoring—Methods and Challenges. *Eng* **2023**, *4*, 47–79. [CrossRef]
3. Soto, E.A.; Borman, L.B.; Wollega, E.; Leon-Salas, W.O. Analysis of Grid Disturbances Caused by Massive Integration of Utility Level Solar Power Systems. *Eng* **2022**, *3*, 236–253. [CrossRef]
4. Khemani, V.; Azarian, M.H.; Pecht, M.G. Efficient Identification of Jiles-Atherton Model Parameters Using Space-Filling Designs and Genetic Algorithms. *Eng* **2022**, *3*, 364–372. [CrossRef]
5. Lee, D.; Diaconeasa, M.A. Preliminary Siting, Operations, and Transportation Considerations for Licensing Fission Batteries in the United States. *Eng* **2022**, *3*, 373–386. [CrossRef]
6. Sharafeldin, M.; Farid, A.; Ksaibati, K. Investigating the Impact of Roadway Characteristics on Intersection Crash Severity. *Eng* **2022**, *3*, 412–423. [CrossRef]
7. Andersen, B.; Hou, J.; Godfrey, A.T.; Kropaczek, D. A Novel Method for Controlling Crud Deposition in Nuclear Reactors Using Optimization Algorithms and Deep Neural Network Based Surrogate Models. *Eng* **2022**, *3*, 504–522. [CrossRef]
8. Zayed, E.; El-Horbaly, M.; Alngar, M.E.M.; El-Shater, M. Dispersive Optical Solitons for Stochastic Fokas-Lenells Equation with Multiplicative White Noise. *Eng* **2022**, *3*, 523–540. [CrossRef]
9. Seits, F.; Kurmi, I.; Bimber, O. Evaluation of Color Anomaly Detection in Multispectral Images for Synthetic Aperture Sensing. *Eng* **2022**, *3*, 541–553. [CrossRef]
10. Johnson, J.B.; Farquhar, H.; Ismay, M.; Naiker, M. Infrared Spectroscopy for the Quality Control of a Granular Tebuthiuron Formulation. *Eng* **2022**, *3*, 596–619. [CrossRef]
11. El Bitouri, Y.; Azéma, N. On the "Thixotropic" Behavior of Fresh Cement Pastes. *Eng* **2022**, *3*, 677–692. [CrossRef]
12. Hafeez, H.; Zafar, N.; Asad Abbas, C.; Elahi, H.; Osama Ali, M. Real-Time Human Authentication System Based on Iris Recognition. *Eng* **2022**, *3*, 693–708. [CrossRef]
13. Algarvio, H. Strategic Participation of Active Citizen Energy Communities in Spot Electricity Markets Using Hybrid Forecast Methodologies. *Eng* **2023**, *4*, 1–14. [CrossRef]
14. Rosa, A.C.; Teixeira, I.; Lacasta, A.M.; Haurie, L.; Soares, C.A.P.; Tam, V.W.Y.; Haddad, A. Experimental Design for the Propagation of Smoldering Fires in Corn Powder and Cornflour. *Eng* **2023**, *4*, 15–30. [CrossRef]
15. Costa Vieira, J.; Costa Vieira, A.; Ribeiro, M.L.; Fladeiro, P.T.; Costa, A.P. Angle of the Perforation Line to Optimize Partitioning Efficiency on Toilet Papers. *Eng* **2023**, *4*, 80–91. [CrossRef]
16. Nkongolo, M. Using ARIMA to Predict the Growth in the Subscriber Data Usage. *Eng* **2023**, *4*, 92–120. [CrossRef]
17. Barbosa, S.; Dias, A.; Pacheco, M.; Pessanha, S.; Almeida, J.A. Investigating Metals and Metalloids in Soil at Micrometric Scale Using μ-XRF Spectroscopy—A Case Study. *Eng* **2023**, *4*, 136–150. [CrossRef]
18. Alqahtani, A.M. Network Pathway Extraction Focusing on Object Level. *Eng* **2023**, *4*, 151–158. [CrossRef]

19. Walton, C.N.; Henderson, I.L. Safety Occurrence Reporting amongst New Zealand Uncrewed Aircraft Users. *Eng* **2023**, *4*, 236–258. [CrossRef]
20. Khanal, M.; Edelmann, N. Application of Connected Vehicle Data to Assess Safety on Roadways. *Eng* **2023**, *4*, 259–275. [CrossRef]
21. Waddoups, R.; Clarke, S.; Tyas, A.; Rigby, S.; Gant, M.; Elgy, I. An Approach to Quantifying the Influence of Particle Size Distribution on Buried Blast Loading. *Eng* **2023**, *4*, 319–340. [CrossRef]
22. Aydin, B.; Singha, S. Drone Detection Using YOLOv5. *Eng* **2023**, *4*, 416–433. [CrossRef]
23. Ma, S.; Yang, J.J. Image-Based Vehicle Classification by Synergizing Features from Supervised and Self-Supervised Learning Paradigms. *Eng* **2023**, *4*, 444–456. [CrossRef]
24. Blaszynski, P.; Bielecki, W. High-Performance Computation of the Number of Nested RNA Structures with 3D Parallel Tiled Co. *Eng* **2023**, *4*, 507–525. [CrossRef]
25. ALGorain, F.T.; Clark, J.A. Covering Arrays ML HPO for Static Malware Detection. *Eng* **2023**, *4*, 543–554. [CrossRef]
26. Gonzalez, Y.; Prati, R.C. Similarity of Musical Timbres Using FFT-Acoustic Descriptor Analysis and Machine Learning. *Eng* **2023**, *4*, 555–568. [CrossRef]
27. Saraev, A.K.; Shlykov, A.A.; Bobrov, N.Y. Tensor CSRMT System with Horizontal Electrical Dipole Sources and Prospects of Its Application in Arctic Permafrost Regions. *Eng* **2023**, *4*, 569–580. [CrossRef]
28. Schiavon, L.L.P.; Lima, P.A.B.; Crepaldi, A.F.; Mariano, E.B. Use of the Analytic Hierarchy Process Method in the Variety Selection Process for Sugarcane Planting. *Eng* **2023**, *4*, 602–614. [CrossRef]
29. Legatiuk, D.; Luckey, D. Formalising Autonomous Construction Sites with the Help of Abstract Mathematics. *Eng* **2023**, *4*, 799–815. [CrossRef]
30. Ziaei, M. Bending and Torsional Stress Factors in Hypotrochoidal H-Profiled Shafts Standardised According to DIN 3689-1. *Eng* **2023**, *4*, 829–842. [CrossRef]
31. Hakula, H. On Long-Range Characteristic Length Scales of Shell Structures. *Eng* **2023**, *4*, 884–902. [CrossRef]
32. Amos, D.; Akib, S. A Review of Coastal Protection Using Artificial and Natural Countermeasures—Mangrove Vegetation and Polymers. *Eng* **2023**, *4*, 941–953. [CrossRef]
33. Rodriguez, R., III. Measuring the Adoption of Drones: A Case Study of the United States Agricultural Aircraft Sector. *Eng* **2023**, *4*, 977–983. [CrossRef]

Disclaimer/Publisher's Note: The statements, opinions and data contained in all publications are solely those of the individual author(s) and contributor(s) and not of MDPI and/or the editor(s). MDPI and/or the editor(s) disclaim responsibility for any injury to people or property resulting from any ideas, methods, instructions or products referred to in the content.

eng

Review

All-Purpose Nano- and Microcontainers: A Review of the New Engineering Possibilities

George Kordas

Self-Healing Structural Materials Laboratory, World-Class Scientific Center of the Federal State Autonomous Educational Institution of Higher Education, Peter the Great St. Petersburg Polytechnic University, 195251 St. Petersburg, Russia; gckordas@gmail.com

Abstract: Recently, a subcategory of nanotechnology—nano-, and microcontainers—has developed rapidly, with unexpected results. By nano- and microcontainers, we mean hollow spherical structures whose shells can be organic or inorganic. These containers can be filled with substances released when given an excitation, and fulfill their missions of corrosion healing, cancer therapy, cement healing, antifouling, etc. This review summarizes the scattered innovative technology that has beneficial effects on improving people's lives.

Keywords: nanocontainers; microcontainers; self-healing; cancer; antibacterial; PCM; antifouling; corrosion

Citation: Kordas, G. All-Purpose Nano- and Microcontainers: A Review of the New Engineering Possibilities. *Eng* **2022**, *3*, 554–572. https://doi.org/10.3390/eng3040039

Academic Editor: Antonio Gil Bravo

Received: 28 October 2022
Accepted: 25 November 2022
Published: 30 November 2022

Publisher's Note: MDPI stays neutral with regard to jurisdictional claims in published maps and institutional affiliations.

Copyright: © 2022 by the author. Licensee MDPI, Basel, Switzerland. This article is an open access article distributed under the terms and conditions of the Creative Commons Attribution (CC BY) license (https://creativecommons.org/licenses/by/4.0/).

1. Introduction

The encapsulation of substances in a protective shell has recently become necessary because of the enormous technological possibilities in medicine [1], materials [2], energy [3], antifouling [4], antimicrobials [5], implants [6], the environment [7], etc. As a result, significant progress has also been made in manufacturing containers [8]. The primary method of their four-step production is firstly, the creation of the core; secondly, the coating of the core with the active shell; thirdly, the removal of the core; and finally, encapsulation with the active material. Reference is made here to production because containers can be produced differently, which is not mentioned in this article, e.g., LDH [9,10]. The present review covers the latest development in this technology via four-step production and gives examples of its industrial application. In this paper, we talk about organic and inorganic nanospheres in which each species is better suited to specific applications. CeMo (MBT or 8-HQ) nanospheres are inhibitors for applications in corrosion because they act simultaneously as a cathodic and anodic corrosion inhibitor. This property cannot be derived from the organic nanospheres that are best suited for cancer-fighting applications where we need artificially intelligent nanocontainers to diagnose and fight cancer. Intelligence cannot be obtained by inorganic nanospheres that are better suited to nontherapeutic applications, except for FeO nanospheres, to which hyperthermia can be applied to improve cancer treatment. Here, we can use the EPR effect to enter cancer via hyperthermia to cause destruction. The choice is made according to the problem we want to solve and the expected results. In addition, there are more applications of the nanocontainers not included in this publication, such as storage of hydrogen, food storage, cosmetic storage, etc., which will be the subject of another extensive review.

2. Materials and Methods

2.1. Inorganic Containers

In producing inorganic nanocontainers, we first produced a polystyrene core with well-known conditions in the bibliography. The polystyrene core's size determines the nanocontainers' final size. An earlier publication investigated the parameters affecting the polystyrene core's size [11]. Terminating the polystyrene core at a negative charge is essential to deposit the metal oxides' salts on it [8]. Then, the sol–gel method deposited the metal oxide coatings, e.g., Ce(acac)$_3$. The third stage involved the removal of polystyrene through combustion at 600 °C. Finally, we obtained a shell consisting of CeO_2 [8], CeMo [12], TiO_2 [13,14], $CeOTiO_2$ [11], Fe_2O_3 [7], SiO_2CaO [15], SiO_2Na_2O [6] and $SiO_2P_2O_5Na_2O$ [6], depending on the alkoxides we used. In the final phase, the nanocontainers entered a vacuum chamber, where we received the maximum vacuum value. Then, we broke it with the materials dissolved in alcohols entering the chamber from a funnel, corrosion inhibitors, filling up the nanocontainers with the desired substance. In one case, paraffin entered the SiO_2 to create phase-change materials [3]. There are cases where the core of SiO_2 [16] nanocontainers consists of super absorbent polymers (SAP) suited for cement "self-healing" [17,18].

2.2. Organic Containers

One uses organic nanocontainers to treat cancer and other diseases [19,20]. The core is composed of PMMA, on which three walls are constructed: one is sensitive to temperature, the second is sensitive to pH, and the third is sensitive to redox. The polymeric nanocontainers are loaded with commercial drugs, such as doxorubicin. The containers are grafted with targeting groups to get bonded to cancer. Furthermore, the nanocontainers are grafted with gadolinium for MRI probes, iron oxide for hyperthermia, Fitch for locating them by fluorescent spectroscopy, etc. The literature has named this system quadrupole stimuli-responsive targeted nanocontainer or Nano4XX (XX = Dox, Daun, Cis, etc.) platforms. The synthesis of such platforms was the subject of several publications in recent literature [1,19]. Figure 1 shows the Nano4XX (XX = Dox, Daun, Cis, etc.) platform and the molecules one uses to produce them [21].

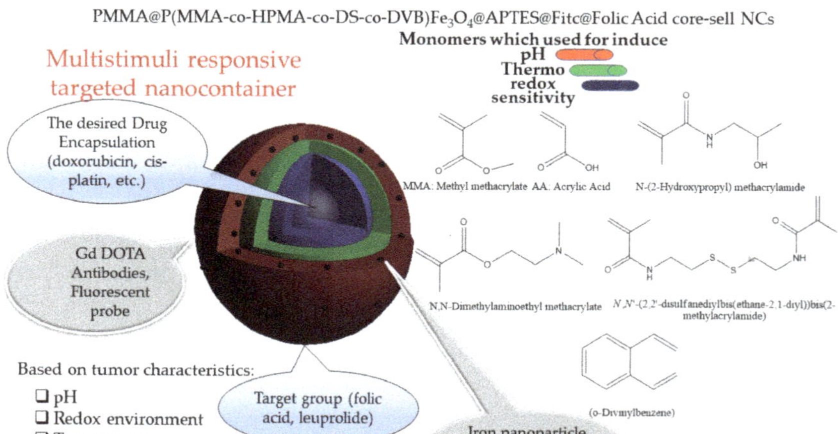

Figure 1. Quadrupole stimuli-responsive targeted nanocontainers loaded with cancer therapeutic drugs to treat cancer: the Nano4XX (Dox, Cis, etc.) platform.

3. Discussion

3.1. Corrosion

Protection against the corrosion of metals is performed by chemical methods designed to replace chromium salts. Several ways have evolved into different metals that offer entirely satisfactory protection. These coatings are, however, passive. They become active when introduced into these nanocontainers filled with inhibitors. With the stimulation caused by corrosion, the inhibitors are released, and thus interrupt the corrosion. This phenomenon is called "self-healing" [22] and is observed in all metals and nanocontainers. [11,13,23–28] Cerium molybdate hollow nanocontainers filled with

2-mercaptobenzothiazole incorporated into epoxy coating deposited onto galvanized steel samples show outstanding inhibition potential after a prolonged corrosion activity. The anodic and cathodic currents determined by SVET showed values close to the noise levels after 20 h of the exhibition to the salt solution until the end of the experiment in 50 h [22]. We attributed the corrosion activity to the organic inhibitor and inorganic inhibitor release of cerium ions from the nanocontainers. Figure 2 shows the SVET measurements of the sample to demonstrate the case.

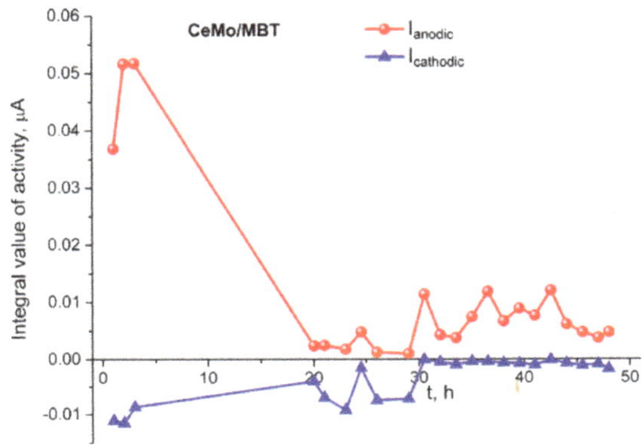

Figure 2. Evolution of the total anodic and cathodic current of the sample containing CeMo (MBT).

Today, the technology of nanocontainers is evolving "intelligence" onto them, containing valves sensitive to the pH change of the environment. With this innovation, we hope for a better and more controlled performance of nanocontainers in stimuli due to corrosion. In a recent paper, mesoporous silicon nanocontainers were prepared and filled with benzotriazole (BTA) corrosion inhibitors. Nanocontainers contain nanovalves consisting of cucurbit [6] uril (CB [6]) rings attached to the surface of the nanocontainers. They do not undergo a corrosion inhibitor release when the pH is neutral. However, the release rate increases with increasing pH values in an alkaline solution. These nanocontainers respond to the pH, but how much better do they work compared to simple nanocontainers? Especially in commercial paints need to be studied better, but such a study is very innovative. Figure 3 shows their function.

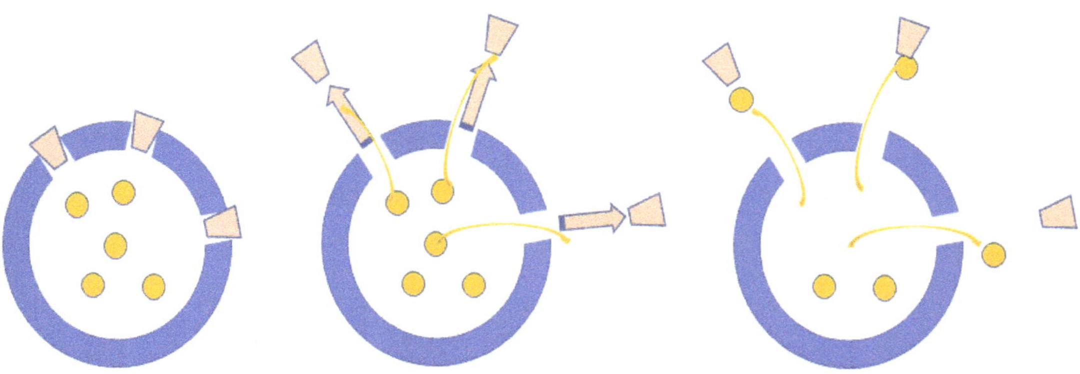

Figure 3. Nanovalves sensitive to pH incorporated onto the surface of silica nanocontainers.

A recent review explains in great detail the state of the art of this technology [2]. We refer the reader to an update from this publication if they want to be informed more about the recent development.

3.2. *Antifouling*

When a metal surface encounters the marine environment, it is then quickly covered by a biofilm of microorganisms and further evolves from invertebrate animals that eventually corrode [29]. The result is that ships develop turbulences on their surface, increasing cruising friction, resulting in decreased vessel efficiency, which at the exact same time increases fuel requirements, and ultimately, increases air pollutants. One realizes that this has enormous economic consequences [30]. To avoid this problem, one can use antifouling coatings containing biocides and copper oxide. In this way, one stops the adhesion of organisms to the surfaces for some time. Most biocides effectively target the microorganisms created in the beginning: bacteria, algae, and barnacles. Biocides can be developed by looking for natural compounds that act as antifouling agents [31–39]. The sea has organisms that defend against biological pollution [40,41]. There are a large number of metabolites at sea that have the potential to grow as antifouling agents.

One such compound is bromosphaerol, isolated in algae cultivated in the marine area of Palaiokastritsas in Corfu. These algae were grown in greenhouse conditions to give large quantities, of which bromosphaerol was chemically isolated. Bromospaerol was encapsulated in copper oxide and zinc oxide nanocontainers to examine the biological aspects of behavior within commercial antifouling paints [4,29]. For this purpose, we used the antifouling bases of the paints of the Wilkens and Re-Turn companies, where we incorporated a small number of CuO and ZnO nanocontainers filled with bromosphaerol. We also used the bases of the two companies and the anticorrosion paints without impurities. We added a small amount of CeMo (8-HQ) to the anticorrosion paint. Figure 4 shows the paint configuration we obtained using basic commercial paints.

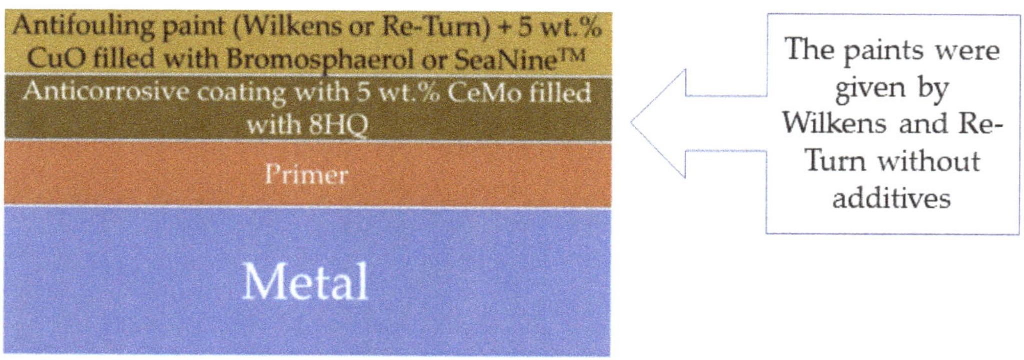

Figure 4. Paint configuration for lab testing and used for real-ship paint testing.

Figure 5a,b show the FRA of the two paints, one (a) consisting of the primer, anticorrosion with CeMo (8HQ) layer, and antifouling paint with CuO (Bromosphaerol); and the other (b) the Wilkens commercial paint. Both samples were exposed to the sea for three months. The FRA curves for the samples were the same before exposure to the seawater. On the contrary, the commercial paint's FRA dropped drastically, while our nanocontainer paint's FRA performed much better.

Figure 6 shows the FRA of another two paints, one (a) doped with CuO (S.N.) + ZnO (S.N.) and the other (b) with CuO (S.N.). R_p is about 10^{10} Ω before exposure of the samples to seawater. However, R_p improves for the samples immersed for three months in seawater. This behavior is known as the "self-healing" phenomenon.

The technology is useful when confirmed in practice. However, it was not easy to convince a paint-trading company to paint commercial ships before, firstly because it is dangerous for new technology not to meet the five-year guarantee given by commercial paints, and secondly, because the financial risk is significant if the new technology does not succeed. However, this was made possible by two companies, one Wilkens and the other Re-Turn, which intervened, and they painted a section of two ships, one traveling to the Adriatic sea and the other to different oceans, with speeds of 14 knots, for one year. Figure 7 shows the sections of the ships painted. It was a pleasant surprise in both cases to see the same results for the parts we painted with the technology of nanocontainers filled with bromophenol. The ships traveled in different marine conditions for a year, and the paints from our laboratory performed better than the commercial paints of the two companies [4,29,42].

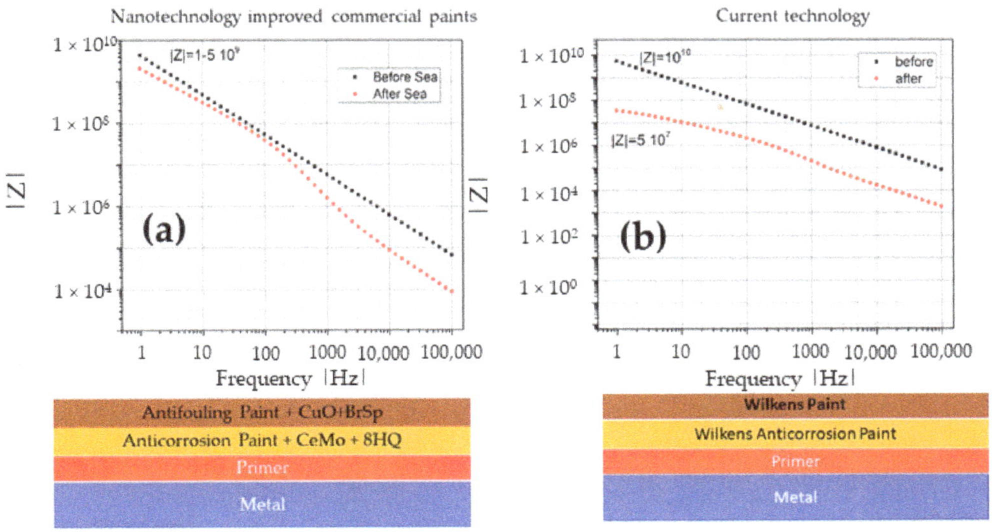

Figure 5. FRA of paints ((**a**) nanocontainer technology and (**b**) commercial paint) before and after immersing the samples in seawater for three paints.

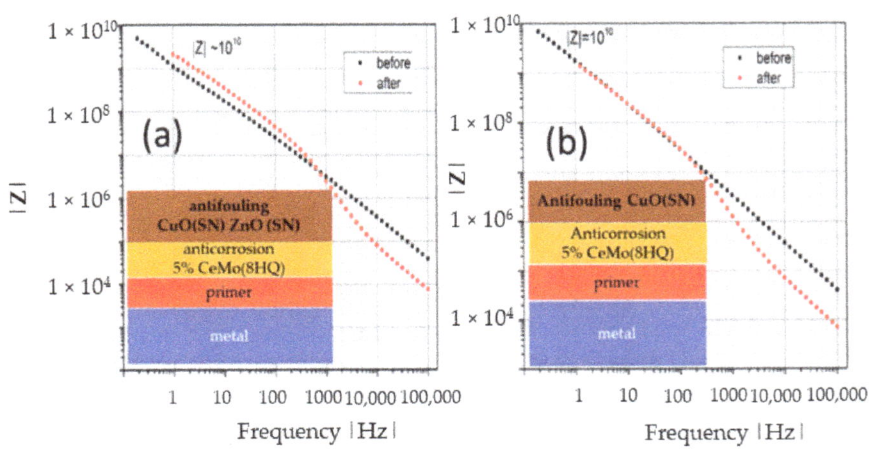

Figure 6. FRA of the paints: (**a**) top layer consisting of CuO (S.N.) and the ZnO (S.N.) nanocontainers, and (**b**) top layer consisting of CuO (S.N.) nanocontainers.

In a recent paper, Al_2O_3 and CuO nanoparticles were incorporated as a pigment using linseed alkyd resin as a binder. The samples were immersed in seawater for 120 days, and the properties were studied with modern spectroscopic techniques where a semantic improvement was observed in the antifouling of steel plates using Al_2O_3 and CuO compared to bare paint. The contact angle increased dramatically, suggesting that the paint becomes more hydrophobic [43]. These new results confirm the impact of nanocontainers on Wilkens and Re-Turn commercial paints [2,4,24,26,29].

Figure 7. The segments of the two ships painted by our nanotechnology.

A relatively recent publication produced a fluorine-free superhydrophobic coating based on the TiO_2 rosin-inoculated nanoparticles. The results were excellent regarding water repellency, which was attributed to the synergistic amplification between natural adhesives and hydrophobic TiO_2 nanoparticles. In addition, the results have shown that such coatings will have great potential to cope with some of the antifouling paints [44].

3.3. *Antibacterial*

Organic and inorganic spheres are the subject of intense scientific activity due to their applications in biology, medicine, photocatalysis, etc. Heterogeneous polymerization methods prepare organic spheres [45–47]. Empty containers are interesting for coatings due to their lower density and optical properties. These can be coated with inorganic shells to modify their properties [48–53]. Photocatalysis uses empty titanium spheres to reduce Cr(VI) to Cr(III), working as an electron acceptor that finally precipitates as solid waste [53]. It is known that TiO_2 appears in nature as brucite (orthorhombic), anatase (quadratic), and rutile (quadratic). Of these three phases, anatase is the most active in photocatalysis. Illumination of TiO_2 by light with an energy higher than 3.2 eV and 3.0 eV for anatase and rutile induces electrons to jump from the valence zone to the conductivity zone, respectively. This transition causes pairs of electrons (e^-) and electrical holes (h^+) via photocatalysis. When an organic compound falls on the surface of the photocatalyst, it will react with the produced $O_2{}^-$ and OH, transforming into carbon dioxide and water. Thus, the photocatalyst decomposes organic matter in the air, including odor molecules, bacteria, and viruses. The Escherichia coli (*E. coli*) bacterium has been used many times for experimental purposes [54].

Hollow-nanosphere titania were used in one study, and their antibacterial activity was evaluated in *E. coli* [55]. Figure 8 shows the survival curve of *E. coli* cells for various conditions. First, the *E. coli* cell concentration was measured in the presence of TiO_2. One observed about a 20% reduction in *E. coli* cells after 79 min of incubation. Furthermore,

E. coli cells were exposed to illumination for 70 min and reduced close to 20%. The *E. coli* cell concentration went down to 0% quickly in the cases illuminated in TiO_2. When the light went out after 30 min, followed by an additional 40 min incubation in the dark, one received the same number of viable cells at 70 min as the sample exposed to light for 70 min continuously.

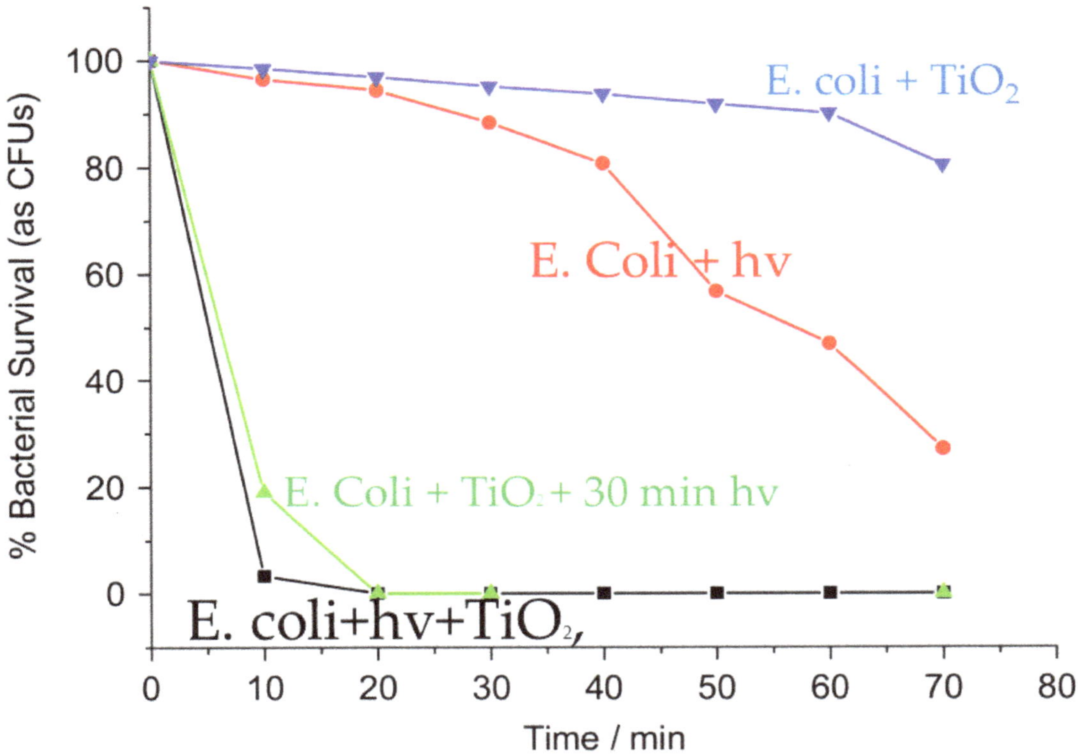

Figure 8. *E. Coli* cell survival in the presence of TiO_2 nanocontainers for *E. coli* + TiO_2, *E. coli* + hv, *E. coli* + TiO_2 + 30 min hv and *E. coli* + hv + TiO_2.

The same experiments were conducted to investigate the antibacterial activity of hollow nanocontainers of cerium molybdenum (CeMo). Again, the hollow nanocontainers were exposed to *E. coli* culture. The study established parameters such as irradiation and time on the antibacterial activity of hollow nanospheres. Figure 9 shows the results of these studies [46]. One can perceive from this work that the E. coli cells in the presence of CeMo nanocontainers diminish after 10 min with or without exposure to the light. The CeMo is a "zero-light" antibacterial compound in the form of a nanocontainer, offering applications in the transport industry, etc.

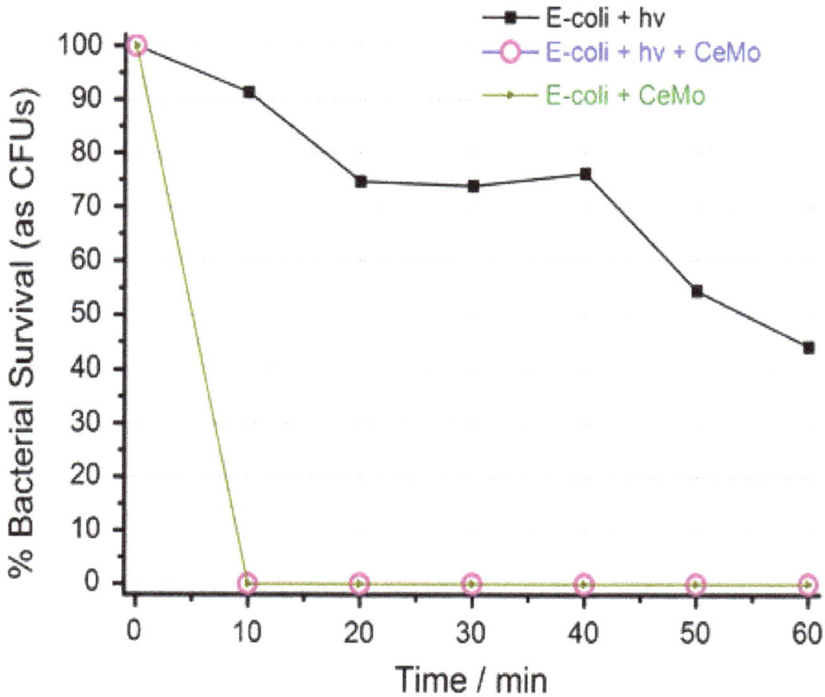

Figure 9. Bacterial survival in CeMo hollow nanospheres' presence with or without light illumination.

3.4. Energy

Thermal energy storage can be carried out in two ways: with latent heat systems (LHS) or thermal energy storage (TES) systems. For LHS, storage is achieved by heat dissipation or by release through a change in the phase of the material. The high energy density and the narrow range of temperature make these materials effective in various applications. Organic phase-change materials include paraffin waxes, fatty acids, and polyethylene glycol. However, they cannot be used freely on devices because of the leakage they sustain that causes severe damage to the device. One solves this problem when LHS materials such as paraffin are trapped in nano- or microcontainers to store the working substance and do not escape into their incorporated material [56].

Nanocontainers Encapsulating PCMs

Implementing the material phase change (PCMs) in thermal energy storage gained significant attention due to the increase in energy consumption and the rescue of the environment from pollution. PCMs absorb, store, and release large amounts of latent heat at specified temperature ranges while phase changes improve device energy efficiency. Depending on the application, the size of the PCMs is selected. Typically, PCMs are classified into nanoPCMs, microPCMs, and macroPCMs, depending on the diameter. The size of the microPCMs usually varies from 1 mm to 1 mm, while capsules less than 100 nm are classified as nanoPCMs and capsules greater than 1 mm as macroPCMs. Encapsulated phase-change materials (EPCM) consist of PCMs with polymer cores and inorganic shells. Microcapsules and nanocapsules containing N-Octadecane in the melamine-formaldehyde shell are manufactured from spot polymerization.

The effects of stirring, the emulsifier's content, the cyclohexane's diameters, morphology, phase-change properties, and thermal stability of PCMs are studied using FT-IR, SEM, DSC, and TGA. For mass production, one can use the spray-drying technique. One can also use the sol–gel method for their production [3]. In this study, the group observed for the latent heat a value of 156 J/g for paraffin and 80% encapsulation into the SiO_2 containers. In a recent survey, n-octadecane paraffin wax as PCM was studied theoretically and experimentally in nanocontainers in terms of size and conditions of measurements. They observed a thin layer of melted PCM between the hot container wall and solid PCM. The concrete PCM sank and the liquid rose to the sphere's top half. Then, the natural convection became dominant at the top half of the sphere, where the melting rate was lower than the bottom half, causing a reduction in the heat transfer and melting rate in general. Encapsulation seaed the nanoparticles to prevent paraffin from being eliminated, and the process was repeated for many cycles [57]. This improvement in nanofluids' heat transfer coefficient impacts the size of the absorbent surface, water-heating time in a water heater, etc. An innovative method was described to encapsulate high-temperature PCM (salts and eutectics, $NaNO_3$, KNO_3, $NaNO_3$-KNO_3, $NaNO_3$-KNO_3-$LiNO_3$) melt in the 120–350 °C temperature range [58]. The study was started to manufacture encapsulated PCMs that can endure the highly corrosive environment of molten alkali metal nitrate-based salts and their eutectics. The established technique does not need a sacrificial layer to lodge the volumetric expansion of the PCMs on melting and reduces the chance of metal corrosion inside the capsule. The encapsulation consists of coating a nonreactive polymer over the PCM pellet, followed by the deposition of a metal layer by a novel nonvacuum (more practical and economically feasible) metal deposition technique (for large-scale fabrication of capsules utilizing commercially available electroless and electroplating chemistry). The fabricated capsules survived more than 2200 thermal cycles (5133 h, equivalent to about 7 years of power plant service) [59]. The thermal cycling test showed no significant degradation in the thermophysical properties of the capsules and PCM on cycling at any testing stage [59].

3.5. Biomaterials

Today, there is a significant need for implants due to the large percentage of diseases, the treatment of which is mainly carried out by a surgical procedure. As far as the surgical procedure is concerned, there is excellent evolution due to the advanced antibiotics, new anesthetics, and stable implants for treating bone defects and motor problems. We call biomaterials the implants made by humans. Their use requires biocompatibility with the body, i.e., not causing thrombosis and toxic or allergic inflammations when used as implants in vital tissue. Furthermore, biomaterials must be stable on the surface of contact with the tissues to avoid breakage. Unfortunately, they do not heal themselves like tissues, which determines the time of their life and proper functioning.

Another category of biomaterials, which we call bioactive, react with their surface during contact with their normal body fluids, through which they develop a bond with the bone and tissues, with the result that the organism assimilates them. L. Hench prepared the first bioactive material in 1971 [60–62]. These materials are regenerative because they can suck and regenerate from the bones without leaving residues and are based on silica, calcium, phosphorus, and sodium elements. These materials should be cell-growth drivers facilitated by having a porous size of 100 µm [15,63]. These materials produce links with the tissues and are histogenic. The material produces only extracellular occlusion on its surface, and its surface is flooded by embryonic cells. A great premise is that these materials are prepared easily, repetitively, and economically. In a relatively recent paper, the synthesis of nanocontainers of the SiO_2-CaO-P_2O_5 (SiCaP) system was performed with a relatively high concentration in Ca and P. The outer diameter was 330 nm and the thickness of the shell was 40 nm, leaving a cavity of about 250 nm. These properties, with their composition, make them candidates for bone tissue-regeneration applications. In another work, nanocontainers of systems: SiO_2–CaO, SiO_2–Na_2OSiO_2–P_2O_5–CaO, and SiO_2–P_2O_5–Na_2O were produced, and their osteogenic properties were examined [6]. Treatments in body fluid revealed their osteogenic properties due to the development of a surface-induced hydroxyapatite layer that resembled in structure the naturally occurring apatite component of bone, enhancing bone development. These systems can be candidates for osteogenic applications tackling bone pathologies such as metabolic bone disease, trauma, and bone cancer ablation.

3.6. Cement

Reinforced concrete is a composite material that results from concrete reinforcement with other materials of greater strength. For example, steel in the form of rods is usually used as a reinforcement, and more rarely, fibers of glass, polymeric materials, and others. The aim is to combine the properties of the above materials into a new one that will meet the needs of the construction. The main disadvantage of concrete is its insufficient tensile strength. Therefore, the reinforcing material must have a high tensile strength to cover the concrete's weakness. In addition, the reinforcing material must have a similar coefficient of thermal expansion. Steel has both of these properties (Figure 10A). On the other hand, a disadvantage of steel is its susceptibility to corrosion (rust) and fire (Figure 10B).

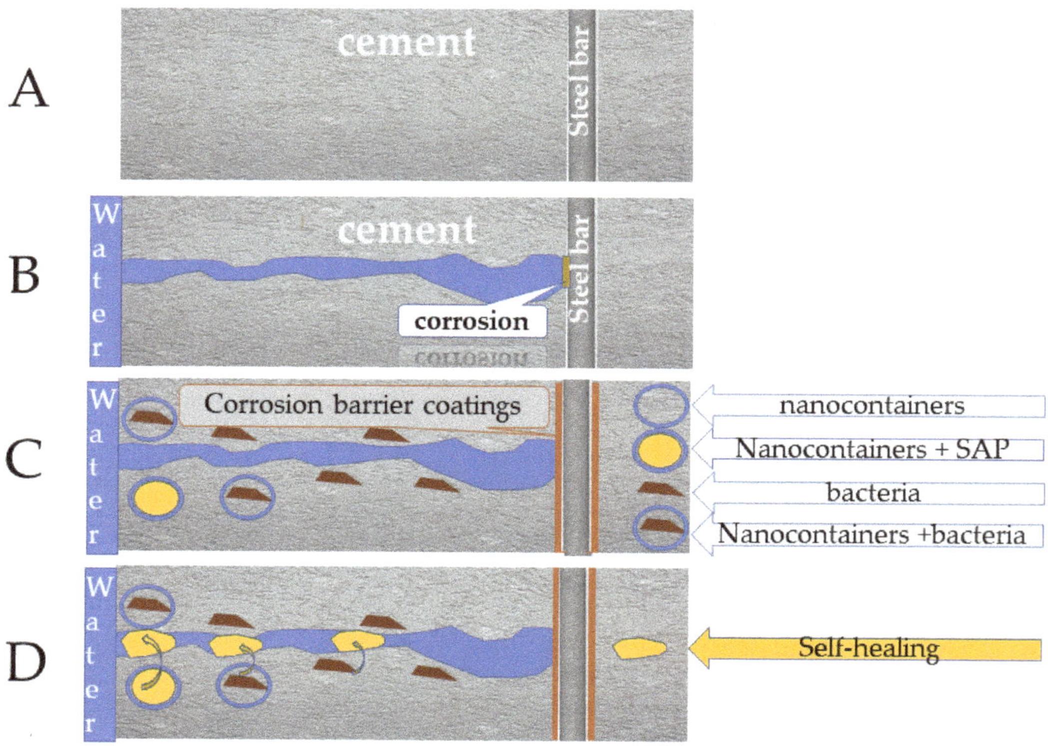

Figure 10. Self-healing mechanism in concrete via bacteria and SAP in nanocontainers. (**A**) Concrete; (**B**) Concrete with crack, water induces corrosion on steel; (**C**) Concrete with bacteria, SAP, Nanocontainers filled with SAP or Bacteria; (**D**) Self-healing.

The microstructure of concrete is porous, which can be isolated or interconnected. Interconnected pores allow water and chemicals to penetrate the concrete. One can understand that permeability plays a vital role in the wear mechanism of concrete. Interconnected pores allow water and chemical compounds to penetrate the concrete matrix. Moreover, CO_2 penetrates the pores to form the cement's alkaline components, e.g., $Ca(OH)_2$. This makes it clear that the number of pores must be reduced to limit the movement of harmful substances into the uterus, resulting in iron corrosion. The bibliography has recently developed iron protection technology with ORMOSIL coatings reinforced with CeO_2 (5-ATDT) nanocontainers. These coatings significantly increase metal protection from corrosion and the appearance of the self-healing effect. Recently, the biological restoration technique has reduced the occlusion of newly formed cracks by introducing bacteria into the concrete. Figure 10C schematically presents this technology. This technology is based on incorporating a bacterium that metabolizes urea and immerses $CaCO_3$ in the crack environment. Microbial immersion of $CaCO_3$ is certified by several factors, such as the concentration of dissolved inorganic carbonate ions and the concentration of Ca^{2+} ions. Bacteria are protected from cement by encapsulation in microcontainers that do not show toxicity.

The spherical poly(methacrylic acids) microspheres of ~700 µm diameter were prepared by distillation–precipitation polymerization. The conversion of carboxylic groups followed this into their sodium salts by treatment with an aqueous sodium hydroxide solution. Figure 11 shows that these water-trap spheres can absorb water 70 times their weight. The absorption and drying cycles are repeated countless times.

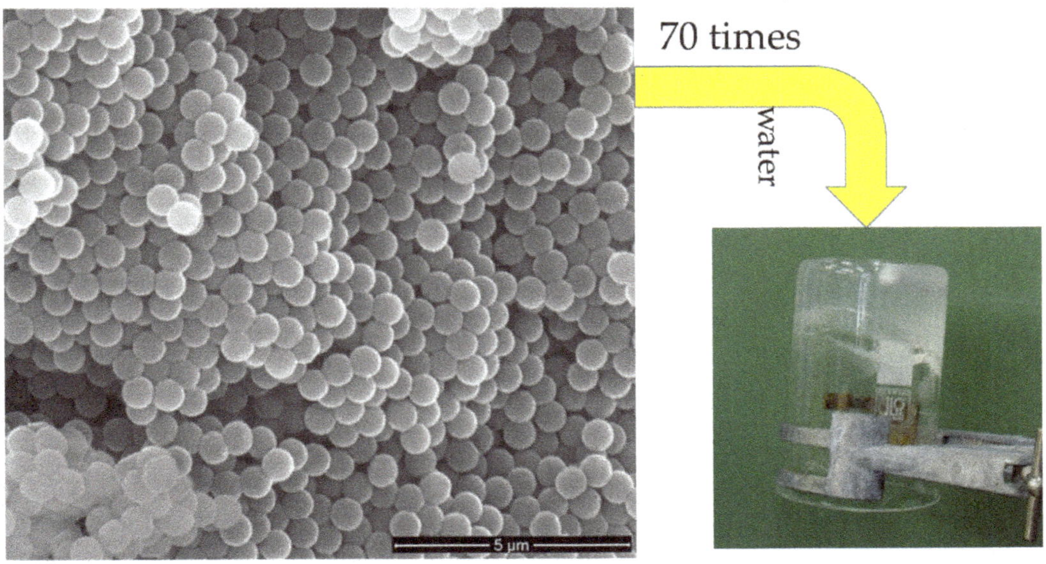

Figure 11. SEM images of water traps capable of absorbing water 70 times their weight.

A recent study expanded previous work and produced P(MAA-co-EGDMA)@SiO_2 by copolymerizing methacrylic acid (MAA) with ethylene glycol dimethacrylate (EGDMA) embedded in the cement slurry, which was found to maintain its structure by exhibiting chemical compatibility with it [16]. The production of P(MAANa-co-EGDMA) @CaO-SiO_2 was an extension of the initial study. Flexural strength and compressive strength of cement-based composites were measured with concentrations bwoc: 0% SAPs, 0.5% SAPs, και 2% SAPs where it was 1.05 MPa, 1.51 MPa, and 1.83 MPa and 63.68 MPa, 59.67 MPa, and 56.27 MPa, respectively. Cracks of cement composites with 2% SAPs healed after 28 days [64].

3.7. Nanomedicine

When a person is diagnosed with cancer, doctors suggest three treatments: surgery, chemotherapy, and radiotherapy. All solutions are painful, with visible and invisible results. The visual effect of chemotherapy is hair loss, heart dysfunction, and many other unfortunate consequences for the patient. Chemotherapy causes a problem to the organs because a small part of the drugs ends up in cancer sites, and a significant fraction of the organs cause severe damage to their functionality. The question is: how can nanomedicine help alleviate the chemotherapy problem? To begin the discussion, let us answer the question: Is cancer the same as other healthy cells? The answer is: no! Cancer has different temperature, pH, and redox values than healthy cells [19,20,65–68]. Can nanocontainers recognize that environment and deliver the chemotherapy drug locally? Another question is: Can nanocontainers target only cancer and provide the drug locally? The answer is yes if we use the nanocontainers of Figure 1. The shell of the Nano4XX platform consists of three polymers sensitive to temperature, pH, and redox. This platform contains magnetic nanoparticles for hyperthermia and targeting groups (folic acid for breast cancer, leuprolide for prostate cancer). The targeting groups can attach cancer-terminating groups, and via endocytosis can help the Nano4XX platform to enter cancer cells. The Nano4XX platform exhibits the same T, pH, and redox as cancer, so they can expand inside cancer and deliver the drug locally.

The realization of this technology involved several individual steps [1,7,19,69–72]. Extensive toxicological studies were conducted on animals [73]. We proved the targeting of cancer cells via positron emission tomography (PET) studies [19]. Figure 12 shows Nano4XX (Dox) functionalized with folic acid (F.A.) to target the cancer cells (HeLa) overexpressed at the surface of the explicit hormone. The nanocontainers enter the cancer cells, illuminating the cells red via Dox. On the contrary, the nanocontainers are not targeted with F.A. on the surface, coloring their site green due to Fitch. DNA replication is canceled by intercalation mode. [74,75] The Nano4XX (Dox) platform enters the cancer cells within 15 min of treatment, contrary to the nonfunctionalized Nano4XX (Dox) platform that agglomerates outside cancer cells.

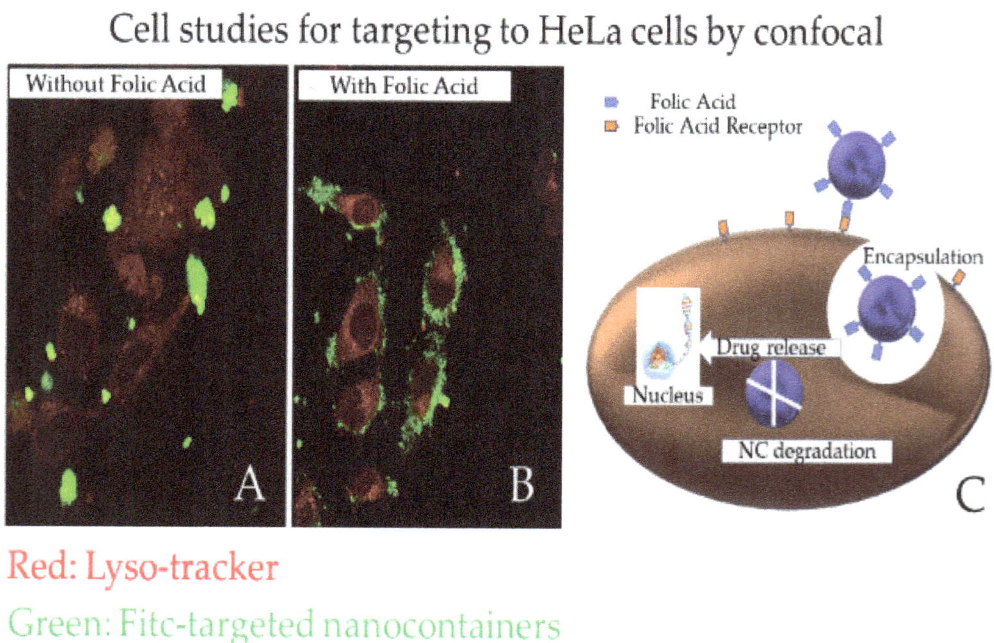

Figure 12. Cell studies for targeting HeLa cells by confocal microscopy [74,76]. (**A**) Agglomeration because the Nano4XX platform is not FA grafted (green color Nano4XX); (**B**) FA grafted Nano4XX color the cancer cell surface green, inside the cancer cells are colored red due to Doxorubicin; (**C**) FA-grafted Nano4Dox enter the cancer cell where they release Doxorubicin inside resulting in a destruction of cancer.

Now that we know that the Nano4XX platform is entering cancer cells, we devised an experiment where the cytotoxicity of the Nano4XX (empty), Nano4Dox platform, and free doxorubicin (0.01, 0.1, 1, 5, 10, and 30 μM) was studied in the cell lines MCF-7 (breast carcinoma) and HeLa. (Cervical carcinoma) [77]. The F.A. receptor recognizes the HeLa cells located on their surface. [24–27] Measurements were made after incubating cells in the presence of Nano4XX with or without F.A. for 72 h. Figure 13 shows that Nano4XX (empty) is not toxic to MCF-7 cells for concentrations from 0.01 to 30 μM. However, once Nano4Dox and doxorubicin are encapsulated in cells, cytotoxicity is practiced in both cases. The same results were obtained in HeLa cells, respectively [76].

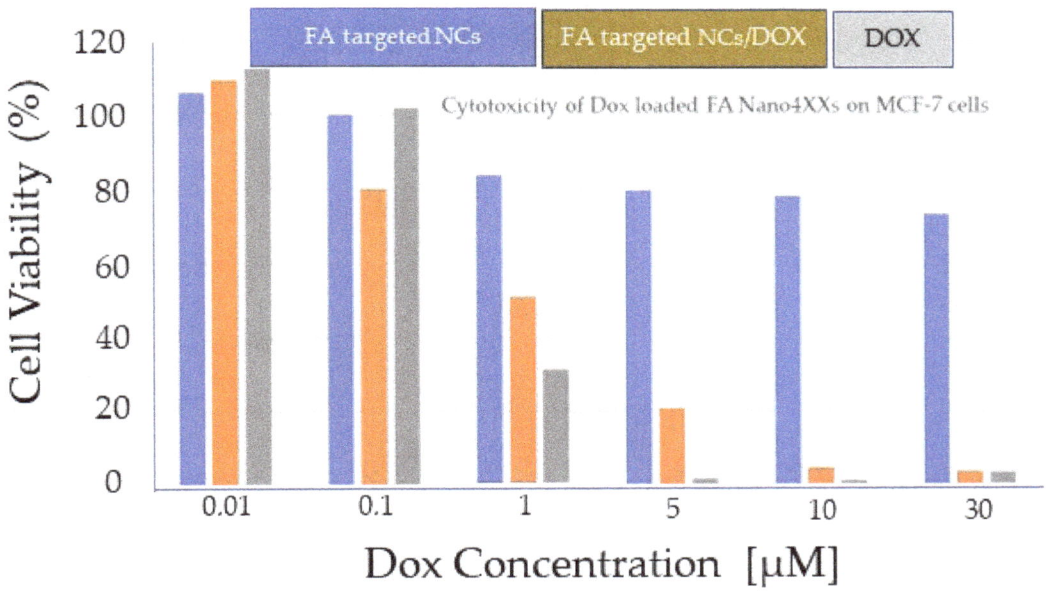

Figure 13. Cytotoxicity of F.A.-Nano4XX, FA-Nano4(Dox), and free DOX in MCF-7 cells repeated three times [5,18].

The PET measured Nano4 (Dox) biodistribution with and without F. A. in Hela mice bearing tumors. Figure 14 shows the distribution of Nano4 (Dox) with or without PET F.A. target groups across various organs and tumors. The measurement was made after a one-hour accumulation where we see the concentration of Nano4 (Dox) in cancer. The concentration in the volume Nano4 (Dox) with F. A. rises to 3.5% after one hour of accumulation [5,18]. Conversely, the attention in the volume without folic acid is zero [21].

Now that we know the Nano4 (Dox) platform with F. A. enters the cancer cells and acts on them; the question is whether they have a therapeutic effect. For this purpose, SCID mice bearing HeLa cervical tumors were studied and used to monitor cancer volume as a function of time in different groups. The experiments were carried out on two types of drugs: doxorubicin and cisplatin. The results are summarized in Figure 15. When the Nano4 (Dox) platform is not equipped with F. A., then there is an increase in cancer volume with time (Figure 15A). The same happens in administering cisplatin to animals with increased volume over time.

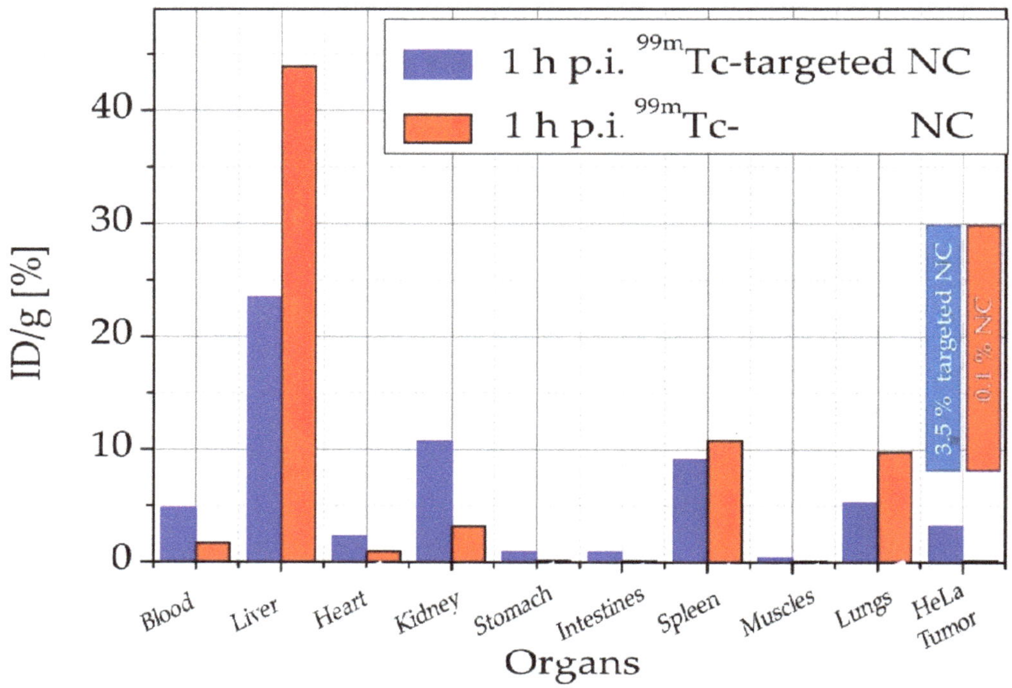

Figure 14. In vivo intake in 1 h in different organs and cancer for Nano4 (Dox) with and without F. A.

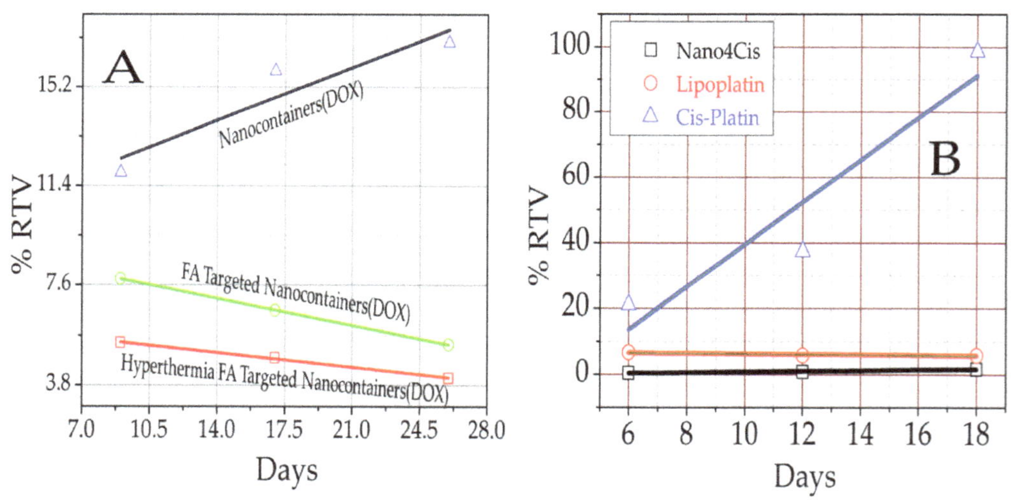

Figure 15. (**A**) The behavior of Nano4Dox. (Black line) increased cancer volume when nanocontainers were delivered with DOX; (green line) decrease in cancer volume when delivering FA-targeted nanocontainers loaded with DOX; (red line) decrease in cancer volume of FA-targeted nanocontainers loaded with DOX with an application of hyperthermia. (**B**) (Blue line) Increase in cancer volume when delivering cisplatin; (red line) decrease in cancer volume when producing cisplatin; (black line) further reduction in cancer volume with disposal of Nano4Cis platforms.

In contrast, the volume decreases with time for the Nano4 (Dox) platform when it incorporates the target molecule. In this case, a decrease in cancer volume by 20% is observed in 25 days (Figure 15A). The result is better when hyperthermia is induced in the treatment. The same effect is obtained in the case of the Nano4Cis platform, which shows better results than even lipoplatin. PET measurements have shown that 3.5% attaches to cancer when the Nano4XX platform (Dox, Cis) contains F.A. Toxicological studies and confocal microscopy measurements have found that the Nano4XX platform (Dox, Cis) enters cancer cells and works therapeutically. All these results suggest that the new system is effective in treating cancer. The Nano4XX platform has the intelligence to recognize cancer and act as a system with "artificial intelligence" because it distinguishes healthy cells from cancer cells. These experiments proved that Nano4XX (Dox, Cis) is significantly safer and more effective in vivo than the current gold standard, Doxil© (doxorubicin liposomal), an absolute nanomedical blockbuster in oncology [20,67,78,79]. This Nano4XX (Dox, Cis, etc.) technology has been patented with European and USA patents (see Patents).

4. Conclusions and Perspectives

Every nanocontainer technology has reached a specific technology readiness level (TRL = 1–9). For example, organic nanocontainers present artificial intelligence and have been tested in terms of their therapeutic efficacy with various anticancer drugs, a worldwide patent has been written, and a business plan has been drawn up. However, this technology has a TRL 7 where there must be human studies and GMP production of nanocontainers from now on. Such a Phase I and IIa clinical study costs EUR 10 million and lasts one year. With this, finding a pharmaceutical company to continue the development in the following phases will be straightforward.

The antifouling paint technology also has a significant technology readiness level, TRL 7, because the technology was tested on two commercial ships, produced on an industrial scale of the nanocontainers, and supported by a patent, and used ecological antifoulants. After a year of sailing, the vessel partially painted with nanotechnology showed much better results than commercial paints.

The technology of anticorrosion painting metals with CeMo (MBT, 8HQ) nanocontainers was made with the funding of two European projects, MULTIPROTECT and MUST, involving DAIMLER, FIAT, EADS, Chemetal, Mankiewicz, and Sika. Prototypes were made in representative metal parts, where nanotechnology paint technology was demonstrated using parts of automobiles and airplanes with a small mix of nanocontainers. In terms of TRL, this paint technology is very advanced because large companies were involved. It is up to the manufacturers to adopt and promote these technologies.

Other technologies, such as nanocontainers in biomaterials, cement self-healing, energy storage, and antimicrobial technology, are in the run-up but are promising technologies. In addition, discussions with industrial partners and funding agencies are underway to develop these technologies further. However, many questions have not been answered regarding the lifetime of these technologies. For example, a building requires the incorporation of self-healing nanocontainers at a time that has not been convincingly verified until today. Furthermore, as far as SiO_2 (paraffin) PCMs are concerned, there are problems with exploiting paraffin by leakage of the nanocontainers that limit their lifetime to applications.

All these achievements, with a research effort of many researchers who have understood the opportunities offered by the nanocontainers, will soon be flooding the commercial world with nanocontainer-based products benefiting human beings. I hope this review will incentivize researchers to engage in this field with many innovations.

5. Patents

W.O. 2015/074762 A1, US2016263221, "MULTI-RESPONSIVE TARGETING DRUG DELIVERY SYSTEMS FOR CONTROLLED-RELEASE PHARMACEUTICAL FORMULATION".

Funding: Support by the grant Self-Healing Construction Materials (contract No. 075-15-2021-590 dated 4 June 2021) is greatly appreciated. The Nano4XX platforms were developed under the two IDEAS ERC Grants with the project acronyms Nanotherapy and grant numbers 232959 (AdG) and 620238 (PoC).

Institutional Review Board Statement: Not applicable.

Informed Consent Statement: Not applicable.

Data Availability Statement: Not applicable.

Acknowledgments: The author is thankful for support from the grant Self-Healing Construction Materials (Contract Nos. 075-15-2021-590 dated 4 June 2021).

Conflicts of Interest: There are no conflict of interest.

References

1. Tapeinos, C.; Efthimiadou, E.K.; Boukos, N.; Kordas, G. Sustained release profile of quatro stimuli nanocontainers as a multi sensitive vehicle exploiting cancer characteristics. *Colloids Surf. B Biointerfaces* **2016**, *148*, 95–103. [CrossRef] [PubMed]
2. Kordas, G. Corrosion Barrier Coatings: Progress and Perspectives of the Chemical Route. *Corros. Mater. Degrad.* **2022**, *3*, 376–413. [CrossRef]
3. Belessiotis, G.V.; Papadokostaki, K.G.; Favvas, E.P.; Efthimiadou, E.K.; Karellas, S. Preparation and investigation of distinct and shape stable paraffin/SiO_2 composite PCM nanospheres. *Energy Convers. Manag.* **2018**, *168*, 382–394. [CrossRef]
4. Kordas, G. Nanocontainers Against Biofouling and Corrosion Degradation of Materials: A Short Review With Prospects. *Front. Nanotechnol.* **2022**, *4*, 1–13. [CrossRef]
5. Kordas, G.; Efthimiadou, E.K. Self-Healing Coatings for Corrosion Protection of Metals. *Sol-Gel Handb.* **2015**, *3*, 1371–1384. [CrossRef]
6. Angelopoulou, A.; Efthimiadou, E.K.; Kordas, G. A new approach to fabricate bioactive silica binary and ternary hybrid microspheres. *Mater. Sci. Eng. C* **2015**, *53*, 76–82. [CrossRef]
7. Tapeinos, C.; Kartsonakis, I.; Liatsi, P.; Daniilidis, I.; Kordas, G. Synthesis and characterization of magnetic nanocontainers. *J. Am. Ceram. Soc.* **2008**, *91*, 1052–1056. [CrossRef]
8. Kordas, G. Protection of HDG Steel Using ORMOSIL Coatings Enhanced with CeO (5-ATDT)-Ceramic Nanocontainers. *Appl. Sci. Eng. Prog.* **2022**, *16*, 6329. [CrossRef]
9. Li, D.; Wang, F.; Yu, X.; Wang, J.; Liu, Q.; Yang, P.; He, Y.; Wang, Y.; Zhang, M. Anticorrosion organic coating with layered double hydroxide loaded with corrosion inhibitor of tungstate. *Prog. Org. Coat.* **2011**, *71*, 302–309. [CrossRef]
10. Shchukina, E.; Shchukin, D.G. Nanocontainer-Based Active Systems: From Self-Healing Coatings to Thermal Energy Storage. *Langmuir* **2019**, *35*, 8603–8611. [CrossRef]
11. Mekeridis, E.D.; Kartsonakis, I.A.; Pappas, G.S.; Kordas, G.C. Release studies of corrosion inhibitors from cerium titanium oxide nanocontainers. *J. Nanoparticle Res.* **2011**, *13*, 541–554. [CrossRef]
12. Kordas, G. ORMOSIL Coatings Enriched with CeO_2 (5-ATDT)-Ceramic Nanocontainers for Enhanced Protection of HDG Steel Used in Concrete. *Materials* **2022**, *15*, 3913. [CrossRef] [PubMed]
13. Mekeridis, E.D.; Kartsonakis, I.A.; Kordas, G.C. Multilayer organic-inorganic coating incorporating TiO 2 nanocontainers loaded with inhibitors for corrosion protection of AA2024-T3. *Prog. Org. Coat.* **2012**, *73*, 142–148. [CrossRef]
14. Kordas, G.C.; Balaskas, A.C.; Kartsonakis, I.A.; Efthimiadou, E.K. A Raman study of 8-Hydroxyquinoline release from loaded TiO_2 nanocontainer. *Int. J. Struct. Integr.* **2013**, *4*, 121–126. [CrossRef]
15. Pappas, G.S.; Liatsi, P.; Kartsonakis, I.A.; Danilidis, I.; Kordas, G. Synthesis and characterization of new SiO_2-CaO hollow nanospheres by sol-gel method: Bioactivity of the new system. *J. Non-Cryst. Solids* **2008**, *354*, 755–760. [CrossRef]
16. Kanellopoulou, I.; Karaxi, E.K.; Karatza, A.; Kartsonakis, I.A.; Charitidis, C. Hybrid superabsorbent polymer networks (SAPs) encapsulated with SiO_2 for structural applications. *MATEC Web Conf.* **2018**, *188*, 01025. [CrossRef]
17. Karatzas, A.; Bilalis, P.; Kartsonakis, I.A.; Kordas, G.C. Reversible spherical organic water microtraps. *J. Non-Cryst. Solids* **2012**, *358*, 443–445. [CrossRef]
18. Krzak, M.; Tabor, Z.; Nowak, P.; Warszyński, P.; Karatzas, A.; Kartsonakis, I.A.; Kordas, G.C.; Warszy, P.; Karatzas, A.; Kartsonakis, I.A.; et al. Water diffusion in polymer coatings containing water-trapping particles. Part 2. Experimental verification of the mathematical model. *Prog. Org. Coat.* **2012**, *75*, 207–214. [CrossRef]
19. Kordas, G. Quadrupole Stimuli-Responsive Targeted Polymeric Nanocontainers for Cancer Therapy: Artificial Intelligence in Drug Delivery Systems. *Nanoeng. Biomater.* **2022**, *1*, 505–522. [CrossRef]
20. Kordas, G. *Nanotechnology in Cancer Treatment as a Trojan Horse: From the Bench to Preclinical Studies*; Sarat Kumar Swain, M.J., Ed.; Elsevier Inc.: London, UK, 2019; ISBN 9780128167717.
21. Rollett, A.; Reiter, T.; Nogueira, P.; Cardinale, M.; Loureiro, A.; Gomes, A.; Cavaco-Paulo, A.; Moreira, A.; Carmo, A.M.; Guebitz, G.M. Folic acid-functionalized human serum albumin nanocapsules for targeted drug delivery to chronically activated macrophages. *Int. J. Pharm.* **2012**, *427*, 460–466. [CrossRef]

22. Montemor, M.F.; Snihirova, D.V.; Taryba, M.G.; Lamaka, S.V.; Kartsonakis, I.A.; Balaskas, A.C.; Kordas, G.C.; Tedim, J.; Kuznetsova, A.; Zheludkevich, M.L.; et al. Evaluation of self-healing ability in protective coatings modified with combinations of layered double hydroxides and cerium molibdate nanocontainers filled with corrosion inhibitors. *Electrochim. Acta* **2012**, *60*, 31–40. [CrossRef]
23. Poornima Vijayan, P.; Al-Maadeed, M.A.S.A. TiO_2 nanotubes and mesoporous silica as containers in self-healing epoxy coatings. *Sci. Rep.* **2016**, *6*, 38812. [CrossRef] [PubMed]
24. Kordas, G. Nanocontainers-Based Anti-Biofouling Coatings—A Pilot Study. *Supramol. Chem. Corros. Biofouling Prot.* **2021**, 383–392. [CrossRef]
25. Kordas, G. Nanocontainers (CeO2): Synthesis, Characterization, Properties, and Anti-corrosive Application. In *Sustainable Corrosion Inhibitors II: Synthesis, Design, and Practical Applications*; ACS Symposium Series; Hussain, C.M., Verma, C., Aslam, J., Eds.; American Chemical Society: Washington, DC, USA, 2021; Chapter 8; pp. 177–185. [CrossRef]
26. Kordas, G. Nanotechnology to improve the biofouling and corrosion performance of marine paints: From lab experiments to real tests in sea. *Int. J. Phys. Res. Appl.* **2019**, *2*, 033–037. [CrossRef]
27. Kartsonakis, I.; Daniilidis, I.; Kordas, G. Encapsulation of the corrosion inhibitor 8-hydroxyquinoline into ceria nanocontainers. *J. Sol-Gel Sci. Technol.* **2008**, *48*, 24–31. [CrossRef]
28. Kartsonakis, I.A.; Athanasopoulou, E.; Snihirova, D.; Martins, B.; Koklioti, M.A.; Montemor, M.F.; Kordas, G.; Charitidis, C.A. Multifunctional epoxy coatings combining a mixture of traps and inhibitor loaded nanocontainers for corrosion protection of AA2024-T3. *Corros. Sci.* **2014**, *85*, 147–159. [CrossRef]
29. Kordas, G. CuO (Bromosphaerol) and CeMo (8 Hydroxyquinoline) microcontainers incorporated into commercial marine paints. *J. Am. Ceram. Soc.* **2020**, *103*, 2340–2350. [CrossRef]
30. Aldred, N.; Clare, A.S. The adhesive strategies of cyprids and development of barnacle-resistant marine coatings. *Biofouling* **2008**, *24*, 351–363. [CrossRef]
31. Guezennec, J.; Herry, J.M.; Kouzayha, A.; Bachere, E.; Mittelman, M.W.; Bellon Fontaine, M.N. Exopolysaccharides from unusual marine environments inhibit early stages of biofouling. *Int. Biodeterior. Biodegrad.* **2012**, *66*, 1–7. [CrossRef]
32. Qian, P.; Lau, S.C.K.; Dahms, H.; Dobretsov, S.; Harder, T. Invited Review Marine Biofilms as Mediators of Colonization by Marine Macroorganisms: Implications for Antifouling and Aquaculture. *Mar. Biotechnol.* **2007**, *9*, 399–410. [CrossRef]
33. Magin, C.M.; Cooper, S.P.; Brennan, A.B. The term fouling generally refers to an undesirable process in which relate to the initial attachment of fouling organisms. *Mater. Today* **2010**, *13*, 36–44. [CrossRef]
34. Bidwell, J.R.; Cherry, D.S.; Farris, J.L.; Petrille, J.C.; Lyons, L.A. Effects of intermittent halogenation on settlement, survival and growth of the zebra mussel, Dreissena polymorpha. *Hydrobiologia* **1999**, *394*, 53–62. [CrossRef]
35. Burgess, J.G.; Boyd, K.G.; Armstrong, E.; Jiang, Z.; Yan, L.; Berggren, M.; May, U.; Pisacane, T.; Granmo, Å.; Adams, D.R. The Development of a Marine Natural Product-based Antifouling Paint. *Biofouling* **2003**, *19*, 197–205. [CrossRef] [PubMed]
36. Rittschof, D.A.N. Natural product antifoulants: One perspective on the challenges related to coatings development Natural Product Antifoulants: One Perspective on the Challenges Related to Coatings Development. *Biofouling* **2009**, *15*, 37–41.
37. Hay, M.E. Marine chemical ecology: What 's known and what 's next? *J. Exp. Mar. Biol. Ecol.* **2009**, *15*, 119–127. [CrossRef]
38. Chapman, J.; Hellio, C.; Sullivan, T.; Brown, R.; Russell, S.; Kiterringham, E.; Le Nor, L.; Regan, F. Bioinspired synthetic macroalgae: Examples from nature for antifouling applications. *Int. Biodeterior. Biodegrad.* **2014**, *86*, 6–13. [CrossRef]
39. Pawlik, J.R. The Development of a Marine Natural Product-based Antifouling Paint. *Oceanogr. Mar. Biol. Rev.* **1992**, *30*, 273–335.
40. Grandgirard, J.; Poinsot, D.; Krespi, L.; Nénon, J.P.; Cortesero, A.M. Chemical ecology of marine microbial defence. *J. Chem. Ecol.* **2002**, *103*, 1971–1985. [CrossRef]
41. Hellio, C.; Berge, J.P.; Beaupoil, C.; Le Gal, Y.; Bourgougnon, N. Screening of marine algal extracts for anti-settlement activities against microalgae and macroalgae. *Biofouling* **2002**, *18*, 205–215. [CrossRef]
42. Kordas, G. Novel Antifouling and Self-Healing Eco-Friendly Coatings for Marine Applications Enhancing the Performance of Commercial Marine Paints. In *Engineering Failure Analysis*; Thanapalan, K., Ed.; IntechOpen: London, UK, 2020; pp. 1–9.
43. Krishna Mohan, M.V.; Bhanuprakash, T.V.K.; Mukherjee, A. Al_2O_3 and CuO nano particulate-based paints for marine applications. *Eng. Res. Express* **2022**, *4*, 035056. [CrossRef]
44. Qing, Y.; Long, C.; An, K.; Liu, C. Natural rosin-grafted nanoparticles for extremely-robust and eco-friendly antifouling coating with controllable liquid transport. *Compos. Part B Eng.* **2022**, *236*, 109797. [CrossRef]
45. Chen, X.; Cui, Z.; Chen, Z.; Zhang, K.; Lu, G.; Zhang, G.; Yang, B. The synthesis and characterizations of monodisperse cross-linked polymer microspheres with carboxyl on the surface. *Polymer* **2002**, *43*, 4147–4152. [CrossRef]
46. Kartsonakis, I.A.; Kontogiani, P.; Pappas, G.S.; Kordas, G. Photocatalytic action of cerium molybdate and iron-titanium oxide hollow nanospheres on Escherichia coli. *J. Nanoparticle Res.* **2013**, *15*, 1–10. [CrossRef]
47. Kartsonakis, I.A.; Liatsi, P.; Danilidis, I.; Bouzarelou, D.; Kordas, G. Synthesis, characterization and antibacterial action of hollow titania spheres. *J. Phys. Chem. Solids* **2008**, *69*, 214–221. [CrossRef]
48. Eiden, S.; Maret, G. Preparation and characterization of hollow spheres of rutile. *J. Colloid Interface Sci.* **2002**, *250*, 281–284. [CrossRef]
49. Wang, D.; Song, C.; Lin, Y.; Hu, Z. Preparation and characterization of TiO_2 hollow spheres. *Mater. Lett.* **2006**, *60*, 77–80. [CrossRef]
50. Song, C.; Wang, D.; Gu, G.; Lin, Y.; Yang, J.; Chen, L.; Fu, X.; Hu, Z. Preparation and characterization of silver/TiO_2 composite hollow spheres. *J. Colloid Interface Sci.* **2004**, *272*, 340–344. [CrossRef]

51. Shiho, H.; Kawahashi, N. Iron compounds as coatings on polystyrene latex and as hollow spheres. *J. Colloid Interface Sci.* **2000**, *226*, 91–97. [CrossRef]
52. Zhao, Y.; Liu, J.; Liu, Q.; Sun, Y.; Song, D.; Yang, W.; Wang, J.; Liu, L. One-step synthesis of SnO_2 hollow microspheres and its gas sensing properties. *Mater. Lett.* **2014**, *136*, 286–288. [CrossRef]
53. Cai, J.; Wu, X.; Zheng, F.; Li, S.; Wu, Y.; Lin, Y.; Lin, L.; Liu, B.; Chen, Q.; Lin, L. Influence of TiO2 hollow sphere size on its photo-reduction activity for toxic Cr(VI) removal. *J. Colloid Interface Sci.* **2017**, *490*, 37–45. [CrossRef]
54. Jackson, G.J.; Merker, R.I.; Bandler, R. *Bacteriological Analytical Manual*; U.S. Food & Drug Administration Center for Food Safety & Applied Nutrition Bacteriological: Silver Spring, MD, USA, 2001.
55. Kartsonakis, I.A.; Liatsi, P.; Daniilidis, I.; Kordas, G. Synthesis, characterization, and antibacterial action of hollow ceria nanospheres with/without a conductive polymer coating. *J. Am. Ceram. Soc.* **2008**, *91*, 372–378. [CrossRef]
56. Umair, M.M.; Zhang, Y.; Iqbal, K.; Zhang, S.; Tang, B. Novel strategies and supporting materials applied to shape-stabilize organic phase change materials for thermal energy storage—A review. *Appl. Energy* **2019**, *235*, 846–873. [CrossRef]
57. Li, R.; Zhou, Y.; Duan, X. Nanoparticle enhanced paraffin and tailing ceramic composite phase change material for thermal energy storage. *Sustain. Energy Fuels* **2020**, *4*, 4547–4557. [CrossRef]
58. Roget, F.; Favotto, C.; Rogez, J. Study of the KNO_3-$LiNO_3$ and KNO_3-$NaNO_3$-$LiNO_3$ eutectics as phase change materials for thermal storage in a low-temperature solar power plant. *Sol. Energy* **2013**, *95*, 155–169. [CrossRef]
59. Wickramaratne, C.; Dhau, J.S.; Kamal, R.; Myers, P.; Goswami, D.Y.; Stefanakos, E. Macro-encapsulation and characterization of chloride based inorganic Phase change materials for high temperature thermal energy storage systems. *Appl. Energy* **2018**, *221*, 587–596. [CrossRef]
60. Hench, L.L.; Best, S.M. *Chapter I. 2.4 Ceramics, Glasses, and Glass-Ceramics: Basic Principles Types of Bioceramics: Tissue*, 3rd ed.; Elsevier: Amsterdam, The Netherlands, 2004.
61. Hench, L.L. Sol-gel materials for bioceramic. *Curr. Opin. Solid State Mater. Sci.* **1997**, *2*, 604–610. [CrossRef]
62. Hench, L.L. The story of Bioglass®. *J. Mater. Sci. Mater. Med.* **2006**, *17*, 967–978. [CrossRef] [PubMed]
63. Pappas, G.S.; Bilalis, P.; Kordas, G.C. Synthesis and characterization of SiO_2-CaO-P_2O_5 hollow nanospheres for biomedical applications. *Mater. Lett.* **2012**, *67*, 273–276. [CrossRef]
64. Kanellopoulou, I.; Karaxi, E.K.; Karatza, A.; Kartsonakis, I.A.; Charitidis, C.A. Effect of submicron admixtures on mechanical and self-healing properties of cement-based composites. *Fatigue Fract. Eng. Mater. Struct.* **2019**, *42*, 1494–1509. [CrossRef]
65. Cheng, R.; Meng, F.; Deng, C.; Klok, H.A.; Zhong, Z. Dual and multi-stimuli responsive polymeric nanoparticles for programmed site-specific drug delivery. *Biomaterials* **2013**, *34*, 3647–3657. [CrossRef]
66. Bilalis, P.; Chatzipavlidis, A.; Tziveleka, L.A.; Boukos, N.; Kordas, G. Nanodesigned magnetic polymer containers for dual stimuli actuated drug controlled release and magnetic hyperthermia mediation. *J. Mater. Chem.* **2012**, *22*, 13451–13454. [CrossRef]
67. Kordas, G.; Efthimiadou, E. Comparison of therapeutic efficacy of quadrupole stimuli-targeted nanocontainers loaded with Doxorubicin (Nano4Dox platform) and cisplatin (Nano4Cis platform) to Doxil and Lipoplatin, respectively. *Ann. Clin. Pharmacol. Toxicol.* **2018**, *1*, 1–5.
68. Sahoo, B.; Devi, K.S.P.; Banerjee, R.; Maiti, T.K.; Pramanik, P.; Dhara, D. Thermal and pH responsive polymer-tethered multifunctional magnetic nanoparticles for targeted delivery of anticancer drug. *ACS Appl. Mater. Interfaces* **2013**, *5*, 3884–3893. [CrossRef] [PubMed]
69. Efthimiadou, E.K.; Fragogeorgi, E.; Palamaris, L.; Karampelas, T.; Lelovas, P.; Loudos, G.; Tamvakopoulos, C.; Kostomitsopoulos, N.; Kordas, G. Versatile quarto stimuli nanostructure based on Trojan Horse approach for cancer therapy: Synthesis, characterization, in vitro and in vivo studies. *Mater. Sci. Eng. C* **2017**, *79*, 605–612. [CrossRef]
70. Kartsonakis, I.A.; Charitidis, C.A.; Kordas, G.C. Synthesis and characterization of ceramic hollow nanocomposites and nanotraps. *Nanocomposites Mater. Manuf. Eng.* **2013**, *2*, 1–31. [CrossRef]
71. Efthimiadou, E.K.; Tapeinos, C.; Tziveleka, L.A.; Boukos, N.; Kordas, G. PH- and thermo-responsive microcontainers as potential drug delivery systems: Morphological characteristic, release and cytotoxicity studies. *Mater. Sci. Eng. C* **2014**, *37*, 271–277. [CrossRef]
72. Efthimiadou, E.K.; Tapeinos, C.; Chatzipavlidis, A.; Boukos, N.; Fragogeorgi, E.; Palamaris, L.; Loudos, G.; Kordas, G. Dynamic in vivo imaging of dual-triggered microspheres for sustained release applications: Synthesis, characterization and cytotoxicity study. *Int. J. Pharm.* **2014**, *461*, 54–63. [CrossRef]
73. Lelovas, P.; Efthimiadou, E.K.; Mantziaras, G.; Siskos, N.; Kordas, G.; Kostomitsopoulos, N. In vivo toxicity study of quatro stimuli nanocontainers in pregnant rats: Gestation, parturition and offspring evaluation. *Regul. Toxicol. Pharmacol.* **2018**, *98*, 161–167. [CrossRef]
74. Yang, H.; Lou, C.; Xu, M.; Wu, C.; Miyoshi, H.; Liu, Y. Investigation of folate-conjugated fluorescent silica nanoparticles for targeting delivery to folate receptor-positive tumors and their internalization mechanism. *Int. J. Nanomed.* **2011**, *6*, 2023–2032. [CrossRef]
75. Roger, E.; Kalscheuer, S.; Kirtane, A.; Guru, B.R.; Grill, A.E.; Whittum-Hudson, J.; Panyam, J. Folic acid functionalized nanoparticles for enhanced oral drug delivery. *Mol. Pharm.* **2012**, *9*, 2103–2110. [CrossRef]

76. Efthimiadou, E.K.; Lelovas, P.; Fragogeorgi, E.; Boukos, N.; Balafas, V.; Loudos, G.; Kostomitsopoulos, N.; Theodosiou, M.; Tziveleka, A.L.; Kordas, G. Folic acid mediated endocytosis enhanced by modified multi stimuli nanocontainers for cancer targeting and treatment: Synthesis, characterization, in-vitro and in-vivo evaluation of therapeutic efficacy. *J. Drug Deliv. Sci. Technol.* **2020**, *55*, 101481. [CrossRef]
77. Mosmann, T. Rapid Colorimetric Assay for Cellular Growth and Survival: Application to Proliferation and Cytotoxicity Assays. *J. OfImmunologicalmethods* **1983**, *65*, 55–63. [CrossRef] [PubMed]
78. Efthimiadou, E.; Tziveleka, L.-A.; Bilalis, P.; Kordas, G. Novel PLA modification of organic microcontainers based on ring opening polymerization: Synthesis, characterization, biocompatibility and drug loading/release properties. *Int. J. Pharm.* **2012**, *428*, 134–142. [CrossRef] [PubMed]
79. Kordas, G. Adjustable Quarto Stimuli (T, pH, Redox, Hyperthermia) Targeted Nanocontainers (Nano4Dox and Nano4Cis) for Cancer Therapy Based on Trojan Horse Approach. *Arch. Pharm. Pharmacol. Res.* **2018**, *1*, 1–7. [CrossRef]

eng

Review

Acoustic-Based Machine Condition Monitoring—Methods and Challenges

Gbanaibolou Jombo [1,*] and Yu Zhang [2]

[1] Centre for Engineering Research, School of Physics, Engineering and Computer Science, University of Hertfordshire, Hatfield AL10 9AB, UK
[2] Department of Aeronautical and Automotive Engineering, Loughborough University, Loughborough LE11 3TU, UK
* Correspondence: g.jombo@herts.ac.uk

Abstract: The traditional means of monitoring the health of industrial systems involves the use of vibration and performance monitoring techniques amongst others. In these approaches, contact-type sensors, such as accelerometer, proximity probe, pressure transducer and temperature transducer, are installed on the machine to monitor its operational health parameters. However, these methods fall short when additional sensors cannot be installed on the machine due to cost, space constraint or sensor reliability concerns. On the other hand, the use of acoustic-based monitoring technique provides an improved alternative, as acoustic sensors (e.g., microphones) can be implemented quickly and cheaply in various scenarios and do not require physical contact with the machine. The collected acoustic signals contain relevant operating health information about the machine; yet they can be sensitive to background noise and changes in machine operating condition. These challenges are being addressed from the industrial applicability perspective for acoustic-based machine condition monitoring. This paper presents the development in methodology for acoustic-based fault diagnostic techniques and highlights the challenges encountered when analyzing sound for machine condition monitoring.

Keywords: machine condition monitoring; anomalous sound detection; industrial sound analysis; detection and classification of acoustic scenes and events

Citation: Jombo, G.; Zhang, Y. Acoustic-Based Machine Condition Monitoring—Methods and Challenges. *Eng* **2023**, *4*, 47–79. https://doi.org/10.3390/eng4010004

Academic Editor: Antonio Gil Bravo

Received: 26 October 2022
Revised: 26 December 2022
Accepted: 28 December 2022
Published: 1 January 2023

Copyright: © 2023 by the authors. Licensee MDPI, Basel, Switzerland. This article is an open access article distributed under the terms and conditions of the Creative Commons Attribution (CC BY) license (https://creativecommons.org/licenses/by/4.0/).

1. Introduction

Unplanned interruption of industrial processes can result in serious financial losses; as such, it becomes of significant relevance to prevent unplanned shutdowns of machinery. The monitoring and diagnosis of the current health state of the machine is crucial in achieving this.

The conventional approach of machine health monitoring involves the use of vibration and other performance monitoring techniques. In these circumstances, sensors such as accelerometer, proximity probe, pressure transducer and temperature transducer are installed on the machine to monitor its health state. However, these methods are of an intrusive nature, requiring physical modification of the machine for their installation. Alternatively, the use of acoustic-based monitoring provides an improved approach which is non-intrusive to the machine operation. Sound signals from a machine contains substantial relevant health information; however, acoustic signals in an industrial environment can be affected by background noise from neighbouring operating machineries; thus, posing a challenge during industrial condition monitoring.

The analysis of sound has been successful in speech and music recognition, especially for creating smart and interactive technologies. Within this context, there exist several large-scale acoustic datasets such as Audio Set [1] and widely available pre-trained deep learning models for audio event detection and classification such as: OpenL3 [2,3], PANNs [4] and VGGish [5]. However, within the context of machine condition monitoring and fault diagnostics, these is a nascent problem for the detection and classification of acoustic scenes and events [6–8].

This paper presents the development in methodology for acoustic-based diagnostic techniques and explores the challenges encountered when analysing sound for machine condition monitoring.

2. Methods—Acoustic-Based Machine Condition Monitoring

2.1. Detection of Anomalous Sound

The goal of anomalous sound detection is to determine if the sound produced by a machine during operation typifies a normal or an abnormal operating state. The ability to detect such automatically is fundamental to machine fault diagnostics using data driven techniques. However, the challenge with this task is that sound produced from anomalous state operation of the machine is rare and varies in nature, hence presenting difficulty in collecting training dataset of such observed abnormal machine operating state. Furthermore, in actual industrial applications, it would be costly and damaging to consider running machines with implanted faults for the sake of data collection. Therefore, the traditional approaches which may be initially apparent such as framing the problem as a two-class classification problem becomes impractical.

In addressing the anomalous sound detection problem, consideration must be given to the fact that only training dataset of the machine running in its normal state would be available. As such, this forms the context within which the problem should be considered. Any such technique would have to learn the normal behaviour of the machine based on this available training dataset.

In furtherance of actualizing anomalous machine sound detection for industrial environment, saw the birth of the Detection and Classification of Acoustic Scenes and Events (DCASE) challenge task "Unsupervised Detection of Anomalous Sounds for Machine Condition Monitoring" in 2020. With the provision of a comprehensive acoustic training dataset combining ToyADMOS [9] dataset and MIMII dataset [10], six categories of machines (i.e., toy and real) of toy car, toy conveyor, valve, pump, fan, and slide rail, operating both in normal and abnormal conditions were considered; researchers were expected to develop and benchmark techniques for detection of anomalous machine sounds. Since the inclusion of this task as part of the DCASE Challenge, over the subsequent years, the task has evolved to account for challenges such as: domain shifted conditions (i.e., accounting for changes in machine operating speed, load, and background noise) [11] and domain generalisation (i.e., invariant to changes in machine operating speed, load, and background noise) [12].

The challenge of machine anomaly detection is to find a boundary between normal and anomalous operating sound. In achieving this, the following methods have emerged.

2.1.1. Autoencoder-Based Anomaly Detection

An autoencoder is a neural network, trained to learn the output as an accurate reconstructed representation of the original input. As an unsupervised learning technique, it has been used by several studies for the detection of anomalous machine operating sound [7–10,13–15].

Autoencoder acts as a multi-layer neural network as shown in Figure 1, consisting of the following segments: encoder network, which accepts a high-dimensional input and transforms to a low-dimensional representation, decoder network, which accepts a latent low-dimensional input to reconstruct the original input, and at least a bottleneck stage within the network architecture. The presence of the bottleneck stage acts to compress the knowledge representation of the original input in order to learn the latent space representation. When the autoencoder is used for anomaly detection the goal during training is to minimize the reconstruction error between the input and the output using the normal machine operating sounds. Herein, the reconstruction error is used as the anomaly score. Anomalies are detected by thresholding the magnitude of the reconstruction error. Based on the application, this threshold could be set. Once an anomalous machine operating sound is provided to the system, it would yield a higher-than-normal reconstruction error, thereby flagging as a fault mode. Table 1 provides baseline autoencoder architecture parameters as applied for anomaly detection. Purohit et al. [10] implemented AE for anomaly detection based on acoustic dataset of malfunctioning industrial machines consisting of faulty valve, pump, fan, and slide rail. Although the dataset used MIMII [10] has been made publicly available, a key part of their work is the adopted architecture of their AE model. Purohit et al. [10] based the input layer on the log-Mel spectrogram. The Mel spectrogram is a spectrogram where frequencies have been transformed to the Mel scale. The Mel spectrogram provides a good correlation with human perception of sound, due to the Mel scale representing scale of pitches that humans would perceive to be equidistant from each other. As such, it not uncommon to find log-Mel spectrogram as performant input feature representation for acoustic event classification amongst others [16]. In [10], the log Mel spectrogram was determined for a frame size of 1024 acoustic time series data points, with a hop size of 512 and 64 Mel filter banks. This results in a log Mel spectrogram of size equal 64. This process was repeated for five consecutive frame sizes. The final input layer feature is formed by concatenating the log Mel spectrogram of five consecutive frames, resulting in an input feature vector size of $5 \times 64 = 320$. This is feed into an auto-encoder network with fully connected layers (FC) such as: encoder section—FC (input, 64, ReLU), FC (64, 64, ReLU), and FC (64, 8, ReLU) and decoder section—FC (8, 64, ReLU), FC (64, 64, ReLU) and FC (64, Output, none). Here, FC (x, y, z) translates fully connected layer with x input neurons, b output neuron, and z activation function such as rectified linear units (ReLU). The implemented AE model is trained for 50 epochs using Adam optimization approach. Similar approach can be adopted using the baseline AE topologies in Table 1.

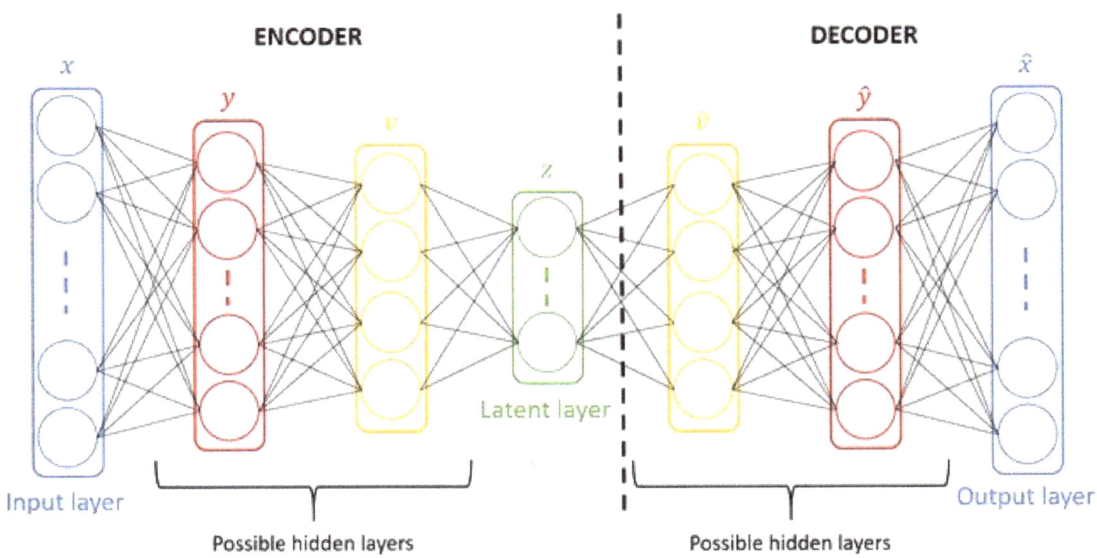

Figure 1. Schematic of an autoencoder [17].

Table 1. Baseline auto encoder system architecture for anomaly detection.

Input	Autoencoder Topology	Ref.
Frequency domain signal analysis: Log Mel spectrogram	Input layer • STFT * frame size 64 ms (50% hop size) • Log Mel-band energies (F = 128 bands) • 5 consecutive frames are concatenated (P = 2, 2P + 1 = 5). • Input dimension (D): 640 (D = F × (2P + 1)) Hidden layers Dense layer (layers 1–4) • Dense layer (units: 128) • Batch Normalization • Activation (ReLU *) Bottleneck/latent layer • Dense layer (units: 8) • Batch Normalization • Activation (ReLU *) Dense layer (layers 5–8) • Dense layer (units: 128) • Batch Normalization • Activation (ReLU *) Output layer • Dense layer (units: 640) • Learning (epochs: 100, batch size: 512, data shuffling between epochs) • Optimizer: Adam (learning rate: 0.001)	[13]
Frequency domain signal analysis: Log Mel spectrogram, MFCC, Spectrogram, Harmonic Percussive Source Separation (HPSS-h [harmonic], HPSS-p [percussive])	Input layer • STFT * (Hanning window size: 1021 samples, hop length: 512 samples) • Log Mel-band energies (128 bands) • Input dimension (D): log-Mel, log-linear, and MFCC* = 640; hpss-h, and hpss-p = 513 Hidden layers Dense layer (layers 1–4) • Dense layer (units: 128) Bottleneck/latent layer • Dense layer (units: 5) Dense layer (layers 5–8) • Dense layer (units: 128) Output layer • Dense layer (units: input dimension = 640 or 513)	[18]
Frequency domain signal analysis: Log Mel spectrogram	Autoencoder architecture as [13]	[9,10]

* STFT: Short-Time Fourier Transform; ReLU: Rectified Linear Unit; MFCC: Mel-Frequency Cepstral Coefficients.

2.1.2. Gaussian Mixture Model-Based Anomaly Detection

Gaussian Mixture Model (GMM) is an unsupervised probabilistic clustering model that assumes each data point belongs to a Gaussian distribution with unknown parameters. As an unsupervised learning technique, it has been used by several studies for the detection of anomalous machine operating sound [19–21].

GMM approach finds a mixture of multi-dimensional Gaussian probability distributions that most likely model the dataset. To achieve this, expectation-maximisation algorithm is used to estimate the parameters of the Gaussian distributions: mean, covariance matrix and mixing coefficients. Expectation-maximisation method is a two-step iterative process which aims to find the maximum likelihood estimates of the Gaussian mixture parameters. It alternates between the expectation step and the maximisation step. Within the expectation step, the responsibilities (which data point belongs to which cluster) are determined using the current estimate of the model parameters, while the maximisation step estimates the model parameters for maximizing the expected log-likelihood function. GMM for anomaly detection uses trained GMM model based on acoustic features as shown in Table 2 to predict the probability of each datapoint being part of one of the k Gaussian distribution clusters. An anomaly is detected by a data point having a probability lower than a threshold which could be either a percentage or a value threshold.

Table 2. Baseline GMM acoustic features.

SN	Signal Analysis Domain	Acoustic Features	Ref.
1	Time Domain Frequency Domain	Zero Crossing Rate, Mean, Max, Min, Covariance of the waveform Short-time Energy, Entropy of Energy, Spectral Centroid, Spectral Spread, Spectral Entropy, Spectral Flux, Spectral Roll-off, MFCC, Chroma Vector, Chroma Deviation	[19]
2	Frequency Domain	Fisher Vectors	[20]
3	Frequency Domain	Log Mel Spectrogram	[21]

2.1.3. Outlier Exposure-Based Anomaly Detection

Outlier Exposure (OE) is an approach for improved anomaly detection in deep learning models [22]. Key in this method is the use of an out-of-distribution dataset, to fine tune a classifier model that enables it to learn heuristics that discriminate in-distribution data points from anomalies. The learned heuristics then has the capability to generalize to new distributions. The OE methodology, first proposed by [22], is achieved by adding a secondary loss to the regular loss for in-distribution training data, which is usually a cross-entropy loss or an error loss term. For classification models, the secondary loss is also a cross-entropy loss computed between the outlier logits and a uniform distribution.

The OE approach has already been applied in the domain of detecting anomalous machine operating sound using classifier models such as MobileNetV2 [11,12]. Herewith, MobileNetV2 [23] is trained to identify from which data segment within both in-distribution and out-of-distribution datasets the observed signal was generated (machine anomaly identification). The trained classifier then outputs the SoftMax value that is the predicted probability for each data segment. The anomaly score becomes the averaged negative logit of the predicted probabilities of the correct data segment. Table 3 shows baseline parameters for an OE approach using MobileNetV2 classifier model.

Table 3. Baseline OE architecture based on MobileNetV2.

Input	OE Topology	Ref.
Frequency domain signal analysis: Log Mel spectrogram	Input layer • STFT frame size 64 ms (50% hop size) • Log Mel-band energies (F = 128 bands) • 64 consecutive frames are concatenated (P) • Input image size (64 × 128) • Hop frames (strides): 8 Triplication layer • Triplicate input image to each color channel MobileNetV2 • Input: 64 × 128 × 3 image • Output: Softmax for sections • Learning (epochs: 20, batch size: 32, data shuffling between epochs) • Optimizer: Adam (learning rate: 0.00001)	[11]

2.1.4. Signal Processing Methods

Acoustic signal processing methods are an adaptation from existing vibration-based approaches reliant on time, frequency, and time-frequency domain analysis of the signal.

Time domain analysis is performed on the acoustic signal time series representation through statistical analysis for calculating feature parameters such as mean, standard deviation, skewness, kurtosis, decibel, crest factor, beta distribution parameters, root mean square, maximum value, etc. These calculated statistical feature parameters from the acoustic signal are used to provide an overall indication of the current health condition of the machine. This approach, although simplistic, has been explored by various investigations for acoustic-based machine fault detection: e.g., Heng and Nor [24] evaluated the applicability of the statistical parameters such as crest factor, kurtosis, skewness, and beta distribution as fault indicators from acoustic signals for monitoring rolling element bearing defect.

For a machine operation under steady state conditions, frequency domain analysis techniques are commonly applied to examine the acoustic signals. Fast Fourier Transform (FFT), a computationally cheap technique to transform time-domain signals to the frequency domain, has been applied in acoustic-based condition monitoring of electric induction motors [25,26], engine intake air leak [27], among others. To capture nonlinear and nonstationary processes in machine operations, Ensemble Empirical Mode Decomposition (EEMD) method has been used [28]. EEMD simulates an adaptive filter, extracting underlying modes in the signal to decompose into a series of intrinsic mode functions (IMF) from high to low frequency content. Spectrum of IMFs has been adopted as a fault indicator for detecting incipient faults in wind turbine blades from acoustic signals [29].

Furthermore, time-frequency domain analysis, such as, short time Fourier transform and wavelet transform, are also powerful approaches for capturing nonstationary processes within machinery acoustic signals. Grebenik et al. [30] used consumer grade microphones and applied EMD and wavelet transform as diagnostic criteria for the acoustic fault diagnostics of transient current instability fault in DC electric motor. Spectral autocorrelation map of acoustic signals has been applied for detection of fault in belt conveyor idler [31]. EMD and wavelet analysis has been applied to extract features from acoustic signals produced by a diesel internal combustion engine for monitoring its combustion dynamics [32,33]. Anami and Pagi [34] used the chaincode of the pseudospectrum to analyse acoustic fault signals from a motorcycle for fault detection.

2.2. *Classification of Anomalous Sound*

The goal of classification of anomalous sound is to categorise a machine sound recording into one of the predefined fault classes that characterises the machine fault state.

Two main approaches have emerged for machine fault diagnostics based on acoustic signal. The first based on feature-based machine learning techniques and the second based on 2D acoustic representation deep learning approaches.

2.2.1. Feature-Based Machine Learning Methods

Feature-based machine learning methods can be broken into three stages. The first stage involves, extracting features from the machine condition acoustic signals. Features are important as fault descriptors are determined using statistical methods, fast Fourier transform, EEMD, or wavelet transform, etc. Extracted features are used to train a machine learning classifier such as Support Vector Machine (SVM), k-Nearest Neighbor (kNN), Random Forest (RF), logistic regression, naïve Bayes, Deep Neural Network (DNN), etc. The trained ML model is then used as a predictor for machine health state based on unknown machine condition acoustic signals.

This approach for machine fault detection based on acoustic inputs is presented in Figure 2. Although the system consists of several steps, the focus here would be in addressing the challenges in engineering feature extraction and for selecting appropriate classifier learning algorithm.

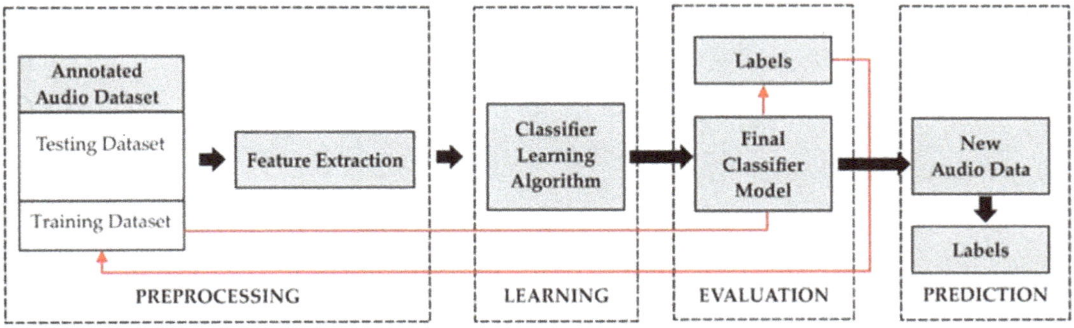

Figure 2. Schematic of feature extraction-based technique for machine fault detection based on acoustic inputs.

(1) Feature Extraction

An approach for acoustic signal representation is required, which is capable to differentiate normal and abnormal operating sound from machinery, utilising low-level features derived from the time domain, frequency domain and time-frequency domain of the acoustic signal. This is achieved as follows and summarized in Table 4:

(a) Time domain-based feature extraction

Time domain features find their basis from descriptive statistical parameters derived from the acoustic signal time-series for representation of both healthy and faulty machine states and training various machine learning models. This approach has been adopted by several investigators [35] and relevant time-domain parameters summarized in Table 4.

(b) Frequency domain-based feature extraction

Frequency domain features take their basis from the Fourier transform spectral transformation of the acoustic signal. Pasha et al. [36] used a band-power ratio as discriminant feature from acoustic signals to monitor air leaks in a sintering plant associated with pallet fault. Here, band-power ratio refers to the ratio of the spectral power within the fault frequency band to the spectral power of the entire signal spectrum. In [36], the feature extraction from a sound recording consisted of the band-power ratio performed repeatedly at fixed sampling window length (i.e., 1024 samples) within the fixed time duration/recording. Other potential parameters can be extracted from the frequency spectrum as demonstrated by [37] and listed in Table 4.

(c) Time-frequency domain-based feature extraction

Time-frequency signal analysis refer to approaches that enable the simultaneous study of signals in both time and frequency domain. The time-frequency representations, such as STFT, wavelet transform, Hilbert-Huang transform, amongst others, provide useful parameters to characterise acoustic signals. Based on the work of [37], relevant time-frequency parameters are provided in Table 4.

Table 4. Feature extraction parameters [37].

SN	Signal Analysis Domain	Features	Summary
1	Time Domain	Zero Crossing Rate	The rate of sign-changes along a signal within a frame length.
2	Frequency Domain	Short-time Energy	The sum of squares of the signal values normalised by frame length.
3	Frequency Domain	Entropy of Energy	Shannon entropy of the normalised energies within a frame length.
4	Frequency Domain	Spectral Centroid	The centre of mass of the spectrum of a frame. Determined by the weighted mean of the frequencies present within the spectrum of a frame length.

Table 4. *Cont.*

SN	Signal Analysis Domain	Features	Summary
5	Frequency Domain	Spectral Spread	The second central moment of the spectrum of a frame length
6	Frequency Domain	Spectral Entropy	Shannon entropy of the normalised spectral energies within the spectrum of a frame length.
7	Frequency Domain	Spectral Flux	The squared difference between the normalised magnitudes of the spectra of the two successive frame length.
8	Frequency Domain	Spectral Roll-off	This is the frequency below which 90% of the spectral distribution for the frame is concentrated.
9	Frequency Domain	MFCC	Mel-Frequency Cepstrum Coefficient (MFCC) provide an effective representation of sound which closely mimics the sound perception of the human ear. MFCC are determined by taking the linear Discrete Cosine Transform (DCT) of the log power spectrum on the nonlinear Mel scale.
10	Frequency Domain	Chroma Vector	A representation of the spectrum projected onto 12 bins representing the 12 distinct semitones (or chroma) of the musical octave.
11	Frequency Domain	Chroma Deviation	Standard deviation of the chroma vector.
12	Frequency Domain	Band-power ratio	Normalised spectral peaks within fault frequency band

(2) Classifier Learning Algorithms

Classifier learning algorithms provide an automated intelligent approach for the detection and classification of machine faults. The generally adopted approach for the development of these machine fault inference systems are based on machine learning classifiers. The machine learning classifier is a supervised learning model that can learn a function that maps an input to a categorical output based on the example input-output pairs [38]. The input for the machine learning classifier model includes the extracted features from the acoustic signal, while the output is the class labels which represent different operational or health state of the machine. To further estimate the optimal classifier model, a cross validation technique can be applied to tune the hyper-parameters of each model.

There are several types of supervised machine learning classifier models, such as: logistic regression, naïve Bayes, decision trees, RF, k-nearest neighbor (kNN), SVM, discriminant analysis, DNN, etc. [39,40]. Each machine learning classifier model has its strengths and weaknesses; for an application, choosing the most appropriate is mostly based on comparing the accuracy and other performance metrics, such as recall rate, F-score, true positive rate, false positive rate, etc. Table 5 highlights exemplar applications of machine learning classifiers for the classification of machine operating sounds.

(a) K-Nearest Neighbors (KNN): KNN is a non-parametric and instance-based machine learning algorithm which can be used for both classification and regression [39,41]. It is classed as a non-parametric method because it makes no explicit assumption about the underlying distribution of the training data and an instance-based method because it does not learn a discriminative function from the training data but memorises it instead [39,41]. When KNN is used for classification, its input consists of the K closest training instances to the unknown instance in the feature space based on a similarity distance metric, e.g., Euclidean distance, hamming distance, Chebyshev distance, Minkowski distance, etc. The output class membership of the unknown instance is determined by a majority vote of its K nearest neighbors. Although KNN is a simplistic classifier model, it is very versatile (i.e., used in many applications), robust (i.e., tight error bounds) and often used as a benchmark for comparison with more complex classifiers [42,43].

(b) Linear Support Vector Machine (SVM): SVM can be viewed as a discriminative classifier model defined by a separating hyperplane [39]. In a nutshell, when an SVM is given labeled training data, the algorithm outputs an optimal hyperplane which classifies new unseen data. The optimal hyperplane is determined by maximising the margin or distance between the nearest points (support vectors) to the hyperplane. Sometimes, the data are not linearly separable, SVM circumvents this by adopting either a soft margin parameter in the optimisation loss or using kernel tricks to transform the feature set into a higher dimensional space.

(c) Random Forest: Random Forest is an ensemble method of learning based on contribution from multiple decision trees [39]. A decision tree is a simple model to classify a dataset, where the data is continuously split based on parameters such as information gain, Gini index, etc. When random forest is used as a classifier, each decision tree in the ensemble, makes a class prediction, and the class with the most vote is the model prediction. A key aspect of the random forest classifier model is that the decision trees are uncorrelated. To achieve uncorrelated decision trees, several techniques such as bagging and feature randomness during tree split are used. Bagging ensures that each individual tree, randomly sample from the dataset with replacement, thus producing different trees in the ensemble.

(d) Decision Tree: Decision tree is used for solving classification problems by crafting a tree-structure where internal nodes represent data attributes, branches represent decision rules and end leaf nodes represent outcomes. It applies a hierarchical structure in determining patterns within data with the intent of creating decision-making rules and predicting regression relationships between dependent and independent variables [39,40]. Optimising the decision tree model, relevant hyperparameters are minimum leaf size, maximum number of split and split criteria, e.g., Gini index, information gain, etc.

(e) Naive Bayes: Naive Bayes classifier rely on Bayes theorem for solving classification problems [39]. Bayes theorem provides a means to formalise the relationship of conditional probabilities or likelihoods of statistical variables. In Naive Bayes classifier, the interest lies in determining the posterior probability of a class label (Y) given some observed features, i.e., $P(Y|features)$. Using Bayes theorem, this posterior probability is expressed as:

$$P(Y|features) = (P(features|Y) \times P(Y))/P(features) \quad (1)$$

where $P(features|Y)$ represent probabilities or likelihood of the features given the class label determined from a naïve assumption of a generative model underlying the dataset such as Gaussian distribution, multinomial distribution, or Bernoulli distribution; $P(Y)$ is the prior probability or initial guess for the occurrence of the class label based on the underlying dataset.

(f) Artificial Neural Network (ANN)/Multi-Layer Perceptron (MLP): ANN or MLP is inspired by the brain biological neural system. It uses the means of simulating the electrical activity of the brain and nervous system interaction to learn a data-driven model. The structure of an ANN comprises of an input layer, one or more hidden layers and an output layer as shown in Figure 3 [39]. Each layer is made up of nodes or neurons and is fully connected to every node in the subsequent layers through weights (w), biases (b), and threshold/activation function. Information in the ANN move in two directions: feed forward propagation (i.e., operating normally) and backward propagation (i.e., during training). In the feedforward propagation, information arrives at the input layer neurons to trigger the connected hidden neurons in subsequent layer. All the neurons in the subsequent layer do not fire at the same time. The node would receive the input from previous node, this is multiplied by the weight of the connection between the neurons; all such inputs from connected previous neurons are summed at each neuron in the next layer. If these values at each neuron is above a threshold value based on chosen activation function, e.g., sigmoid function, hyperbolic tangent (tanh), rectified linear unit (ReLU), etc. the node would fire and pass on the output, or if less than the threshold value, it would not fire. This process is continued for all the layers and nodes in the ANN operating in the feedforward mode from the input layer to the output layer. The backward propagation is used to train the ANN network. Starting from the output layer, this process compares the predicted output with actual output per layer and updates the weights of each neuron connection in the layer by minimize the error using a technique such as gradient descent amongst others as shown in Figure 3. This way, the ANN model learns the relationship between the input and output.

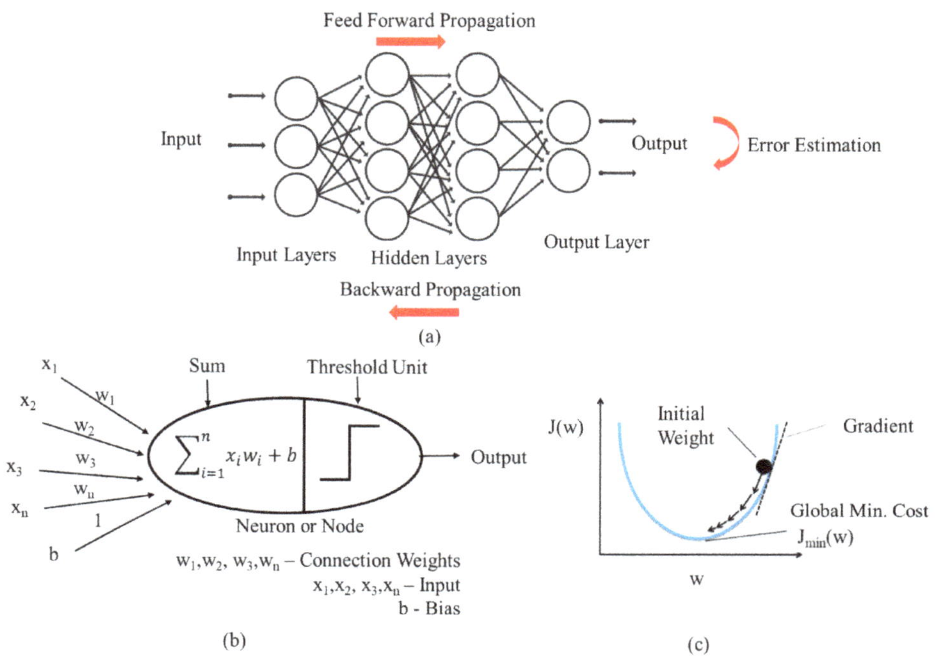

Figure 3. Structure of an artificial neural network (ANN) (**a**) ANN (**b**) single neuron or node (**c**) optimizing weights using gradient descent.

Table 5. Exemplar classifier learning algorithm for classification of machine operating sounds.

SN	Classifier Learning Algorithms	Features	Application	Ref.
1	SVM	Frequency domain signal analysis: Band-power ratio	Detection of air leaks between grate bars lined sinter strand pallets in a sintering plant	[36]
2	Decision Tree (J48/C4.5 Algorithm)	Frequency domain signal analysis: Band-power ratio	Detection of air leaks between grate bars lined sinter strand pallets in a sintering plant	[36]
3	Deep Neural Network (DNN)	Frequency domain signal analysis: Short-Term Fourier Transform (STFT)	Detecting changes in electric motor operational states such as supply voltage and load	[14]
4	Decision tree, Naive Bayes, kNN, SVM, Discriminant Analysis, Ensemble classifier, with Bayesian Optimisation	Frequency domain signal analysis: Wavelet packet transform, with Principal Component Analysis (PCA)	Detecting of internal combustion engine fault	[40]
5	kNN, SVM, and Multi-layer Perceptron (MLP)	Frequency domain signal analysis: Wavelet packet transform with various mother wavelets	Detecting of internal combustion engine fault	[44]
6	Artificial Neural Network (ANN)	Frequency domain signal analysis: Spectral peaks from the fast Fourier Transform of acoustic signal (0–2996.25 Hz)	Detecting loose stator coils in induction electric motors	[6]

2.2.2. Acoustic Image-Based Deep Learning Methods

This approach leverages techniques from the field of machine hearing [45]. Machine hearing involves sound processing considering inherent sound sensing system structures as humans and sound mixtures in realistic context [45].

In emulating human hearing, machine hearing adopts a four-layer architecture within which each layer represents a distinct area of research. The first layer, auditory periphery layer (cochlea model), mimics the representation of the nonlinear sound wave propagation mechanism in the cochlea as cascading filter systems; the second layer, auditory image computation, provides a projection of one or more forms of auditory images to the auditory cortex mimicking the auditory brain stem operation; the third layer abstracts the operation within the auditory cortex via extraction of application-dependent features from the auditory images; the final and forth layer addresses the application specific problem using appropriate machine learning system [46].

For application in classifying anomalous machine operating sound, variations are made in the auditory image computation representation; as such, best referred to as acoustic image representation. From the literature, there have been several possibilities for the 2D acoustic image representation such as: spectrogram (from STFT), Mel-spectrogram, cochleagram, amongst others [47,48]. Table 6 provides a summary of acoustic image representation in combination with deep learning models for classifying anomalous machine operating sounds and Figure 4 shows examples of acoustic image representations.

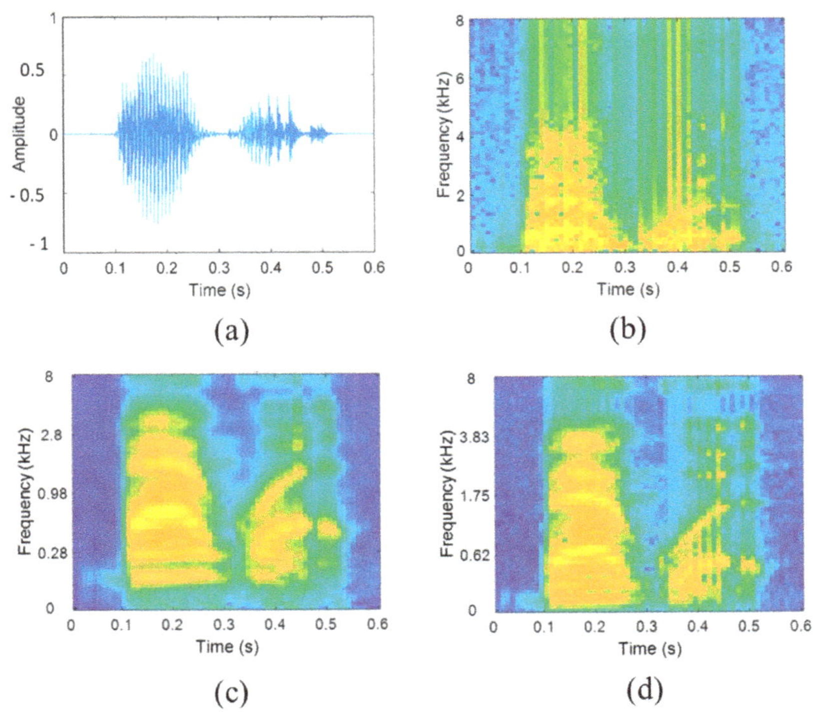

Figure 4. Acoustic image representation (**a**) acoustic input (**b**) spectrogram of acoustic input (**c**) cochleagram of acoustic input (**d**) Mel spectrogram of acoustic input [16].

(1) Acoustic Image Representation

(a) Spectrogram: This is a two-dimensional representation of the frequency characteristics of a time-domain signal as it changes over time as shown in Figure 4. Spectrogram is generated using Fourier transform of the time-domain signal; the time-domain signal is first divided into smaller segments of equal length with some overlap; then, fast Fourier transform (FFT) is applied to each segment to determine its frequency spectrum; the resulting spectrogram becomes a side-by-side overlay of the frequency spectrum of each segment over time. FFT represents an algorithm to compute the discrete Fourier transform (DFT) of the windowed time-domain signal, represented as [16]:

$$F_n = \sum_{k=0}^{N-1} x_n w_n e^{-2\pi ink/N},\ n = 0, \cdots, N-1 \quad (2)$$

where F_n is discrete Fourier transform, N is number of sample points within the window, f_k is the discrete time-domain signal, and w_n is the window function. The spectrogram is obtained as the logarithm of the DFT, as such [16]:

$$S_n = log|F_n|^2 \quad (3)$$

where S_n is spectrogram, and F_n is discrete Fourier transform.

(b) Mel Spectrogram: This is a spectrogram where frequencies have been transformed to the Mel scale as shown in Figure 5. The Mel scale is a linear scale model of the human auditory system, represented as [49,50]:

$$f_{mel} = 2595 \times log_{10}(1 + f/700) \quad (4)$$

where f_{mel} is frequency on the Mel scale, and f is frequency from the spectrum.

As shown in Figure 5, Mel spectrogram is computed by passing the result of windowed times-series signal FFT for each smaller segment of the divided signal through a set of half-overlapped triangular band-pass filter bank equally spaced on the Mel scale. The spectral values outputted from the Mel band-pass filter bank are summed and concatenated into a vector of size dependent on the number of Mel filters, e.g., 128, 512, etc. The resulting Mel spectrogram becomes a side-by-side overlay of the resulting vector representation from each consecutive time-series signal segment over time.

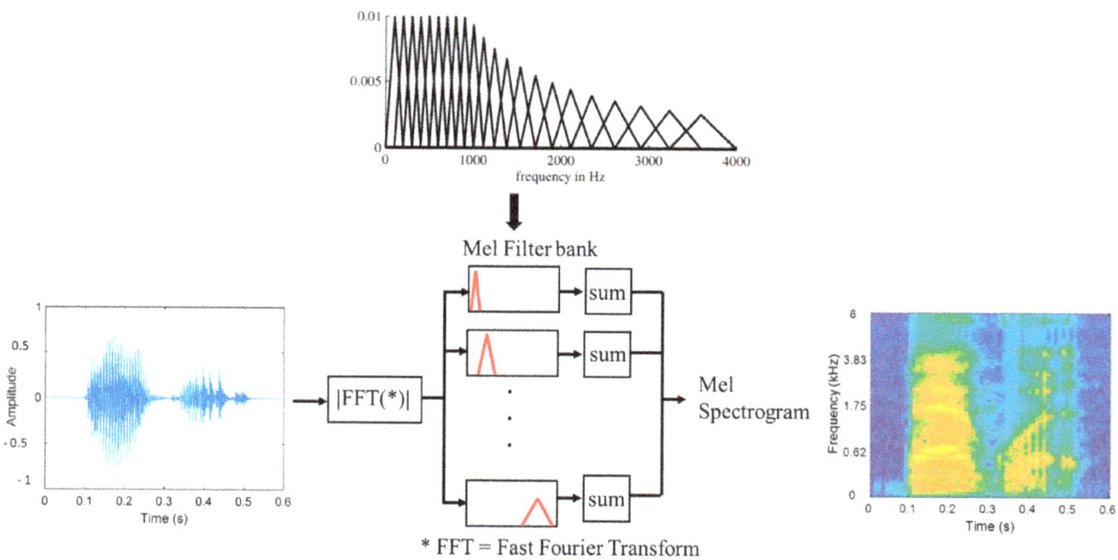

Figure 5. Mel spectrogram operation.

(c) Cochleagram: A cochleagram is a time-frequency representation of the frequency filtering response of the cochlea (in the inner ear) as simulated by a bank of Gammatone filters [48]. The Gammatone filter represents a pure sinusoidal tone that is modulated by a Gamma distribution function; the impulse response of the Gammatone filter is expressed as [16]:

$$h(t) = At^{n-1}e^{-2\pi bt}\cos(2\pi f_{cm}t + \phi) \quad (5)$$

where A is amplitude, n is filter order, b is filter bandwidth, f_{cm} is filter centre frequency, ϕ is phase shift between filters, and t is time.

As shown in Figure 6, cochleagram is computed by passing the result of windowed times-series signal FFT for each smaller segment of the divided signal through a series of overlapping band-pass Gammatone filter bank. The spectral values outputted from the Gammatone filter bank are further transformed by logarithmic and discrete cosine transform operations before been summed and concatenated into a vector of size dependent on the number of Gammatone filters, e.g., 128, etc. The resulting cochleagram becomes a side-by-side overlay of the resulting vector representation from each consecutive time-series signal segment over time.

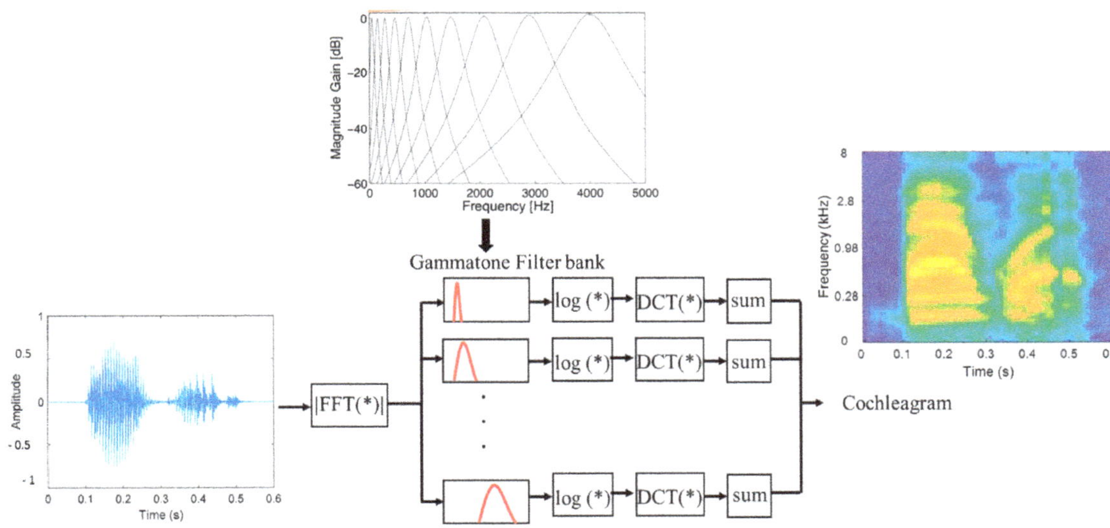

* FFT = Fast Fourier Transform, DCT = Discrete Cosine Transform, log = Logarithm

Figure 6. Cochleagram operation.

(2) Deep Learning Methods

(a) Convolution Neural Network (CNN): CNN is inspired from the operation of the mammalian visual cortex. As shown in Figure 7, CNN is a multi-stage neural network made up of key stages: filter stage (i.e., convolution layer, pooling layer, normalisation layer and activation layer) and classification stage (i.e., fully connected layer of multilayer perceptron) [51]. The convolution layer functions to extract feature set from acoustic image representation into a feature map, pooling layer reduces the dimensionality of the feature map, and the classification stage performs the classification task using the multilayer perceptron. [47] has applied CNN with a combination of log-spectrogram, short-time Fourier transform and log-Mel spectrogram features to classify rolling-element bearing cage fault based on acoustics signals. Implemented CNN model consisted of three stage feature extraction layers: fully connected layer (shape = 16 × 16, rectified linear unit (ReLU) activation function, max. pooling = 2 × 2), fully connected layer (shape

= 32 × 32, ReLU, max. pooling = 2 × 2), and fully connected layer (shape = 64 × 64, ReLU, max. pooling = 2 × 2) and a final classification stage based on multi-layer perception with 512 hidden nodes, ReLU and sigmoid activation function. Dataset was very sparse, and model was not optimized; therefore, impacting model performance on training accuracy. Table 6 highlights other applications of acoustic image-based classifiers of anomalous machine operating sounds.

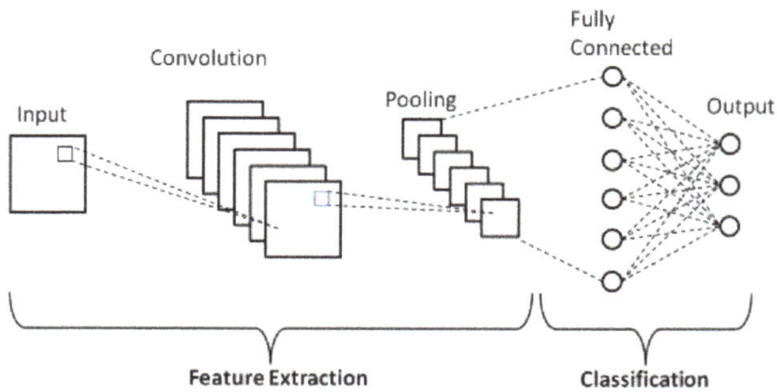

Figure 7. CNN basic architecture [51].

(b) Recurrent Neural Network (RNN): RNN is a type of neural network which uses sequential data or time series data to learn. Unlike CNN, RNN have internal memory state (i.e., can be trained to hold knowledge about the past); this is possible as inputs and outputs are not independent of each other, prior inputs influence the current input and output; simply put, output from previous layer state are feed back to the input of the next layer state. As shown in Figure 8, x is input layer, h is middle layer (i.e., consist of multiple hidden layers) and y is output layer. W, V and U are the parameters of the network such as weights and biases. At any given time (t), the current input is constituted from the input x(t) and previous x(t − 1); as such the output from x(t − 1) is feedback into the input x(t) to improve the network output. This way, information cycles through a loop within the hidden layers in the middle layer. RNN uses the same network parameters for every hidden layer, such as: activation function, weights, and biases (W, V, U). Despite the flexibility of the basic RNN model to learning sequential data, they suffer from the vanishing gradient problem (i.e., difficulty training the model when the weights get too small, and the model stops learning) and exploding gradient problem (i.e., difficulty training the model due to very high weight assignment). To overcome these challenges, the long short-term memory (LSTM) network variant of RNN is normally used. LSTM has the capability to learn long-term dependencies between time steps of sequential data. LSTM can read, write and delete information from its memory. It achieves this via a gating process made up of three stages: forget gate, update/input gate and output gate which interacts with is long-term memory and short-term memory pathways used to feedback its memory states amongst hidden layers. As shown in Figure 9, "c" represents the cell state and long-term memory, "h" represents the hidden state and short-term memory, and "x" represent the sequential data input. The forget gate determines how much of the cell state "c" is thrown away or forgotten. The update gate determines how much of new information is going to be stored in the cell state, and output gate determines what is going to be outputted. [52] has applied LSTM RNN with cochleagram features to classify varying rolling-element bearing faults based on 60 s acoustics signals. Implemented model consisted of an input feature set based on 128 gammatone filter bank cochleagram; Considering a 1 s. duration as a frame, the 60 s dataset generated 60-time frames. Each frame is represented as a cochleagram. 67% of the dataset was used to train the LSTM RNN model and 33% for testing. Model accuracy on fault classification task was 94.7%. Table 6 highlights other applications of acoustic image-based classifiers of anomalous machine operating sounds.

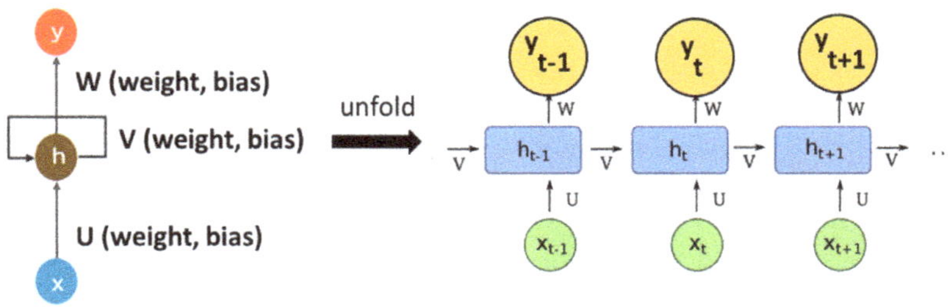

Figure 8. RNN basic architecture.

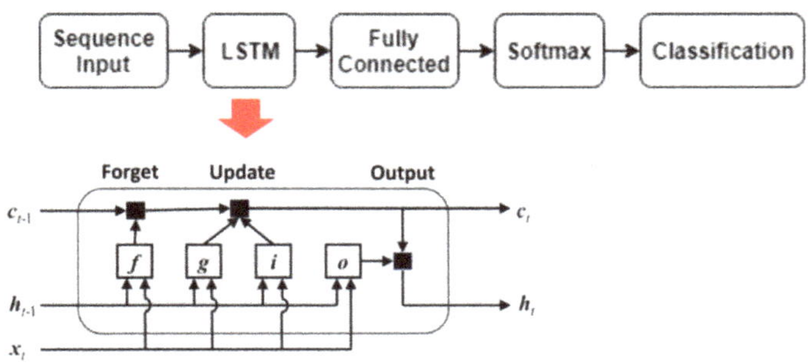

Figure 9. LSTM RNN architecture [52].

(c) Spiking Neural Network (SNN): SNN is a brain-inspired neural network where information is represented as binary events (spikes). It shares similarity with concepts such as event potentials in the brain. SNN incorporates time into its propagation model for information; SNN only transmit information when neuronal potential exceeds a threshold value. Working only with discrete timed events, SNS accepts as input spike train and outputs spike train. As such, information is required to be encoded into the spikes which is achieved via different encoding means: binary coding (i.e., all-or-nothing encoding with neurons active or inactive per time, rate coding, fully temporal codes (i.e., precise timing of spikes), latency coding, amongst others [53]. As shown in Figure 10, SNN is trained with the margin maximization technique, described in [54]. During first epoch, SNN hidden layer is developed based on neuron addition scheme. In subsequent epochs, the weights and biases of the hidden layer neurons are updated further using the margin maximization technique. Here, weights of the winner neuron are strengthened, while those of the others are inhibited; this reflects the Hebbian learning rule of the natural neurons; as a result, neurons are only connected to their local neurons, so they process the relevant input patterns together. This approach maximizes the margin among the classes which lends itself to training the spike patterns. Ref. [48] has applied SNN with cochleagram features to classify varying rolling-element bearing faults based on 10 s acoustics signals. Implemented model consisted of an input feature set based on 128 gammatone filter bank cochleagram; later reduced to 50 using principal component analysis (PCA). Considering a 10 ms duration as a frame, the 10 s dataset generated 1000-time frames. Each frame was encoded into a spike train using the population coding method. 90% of the dataset was used to train the SNN model and 10% for testing. Model accuracy was above 85%. Table 6 highlights other applications of acoustic image-based classifiers of anomalous machine operating sounds.

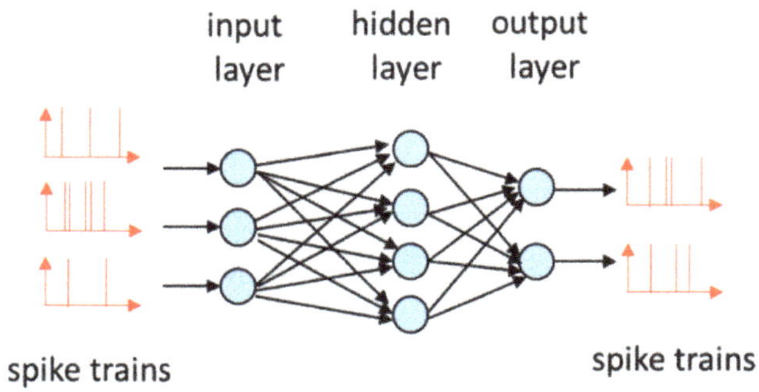

Figure 10. SNN architecture [48].

Table 6. Exemplar acoustic image representation and classifier models.

SN	Acoustic Image Representation	Deep Learning Methods	Application	Ref.
1	Spectrogram/Log-Spectrogram	CNN *	Detection of rolling-element bearing fault such as cage defect	[47]
		RNN *	Detection of air leaks between grate bars lined sinter strand pallets in a sintering plant	[36]
2	Cochleagram	RNN *	Detection of rolling-element bearing fault such as inner race defect, outer race defect, rolling-element defect, combined defect, and heavily worn bearing	[52]
3	Cochleagram	SNN *	Detection of rolling-element bearing fault such as inner race defect, outer race defect, rolling-element defect, combined defect, and heavily worn bearing	[48]
4	Spectrogram (from STFT)	CNN *	Detection of rolling-element bearing fault such as cage defect	[47]
5	Log-Mel Spectrogram	CNN *	Detection of rolling-element bearing cage fault	[47,55]

* CNN: Convolutional Neural Network, RNN: Recurrent Neural Network, SNN: Spiking Neural Network.

3. Datasets for Detection and Classification of Anomalous Machine Sound (DCAMS)

Openly available datasets are vital for progress in the data-driven machine condition monitoring approaches. In recent time, there have been significant progress in the corollary area of acoustic scene classification mainly due to opensource dataset such as: AudioSet dataset [1], which provides a collection over 2 million manually labelled 10 s sound segments from YouTube within 632 audio event classes. However, nothing of such large scale is available for Detection and Classification of Anomalous Machine Sounds (DCAMS). Within limited scale, several research projects are beginning to lay the foundation as they were bridging the dataset gap for DCAMS.

3.1. ToyADMOS Dataset

This dataset provided by [9], is a collection of anomalous machine sounds produced by miniaturised machines (i.e., toy car, toy conveyor, and toy train) as shown in Figure 11. It is designed to provide scenarios such as: inspecting machine condition (toy car), fault diagnostics for a static machine (toy conveyor) and fault diagnostics for a dynamic machine (toy train). The data acquisition setup for each scenario is performed using four microphones sampled at 48 kHz and measurement locations are shown in Figure 12. To provide anomalous operating conditions for the miniaturised machines, systematic fault modes as shown in Table 7 are imbedded in the various toy machines.

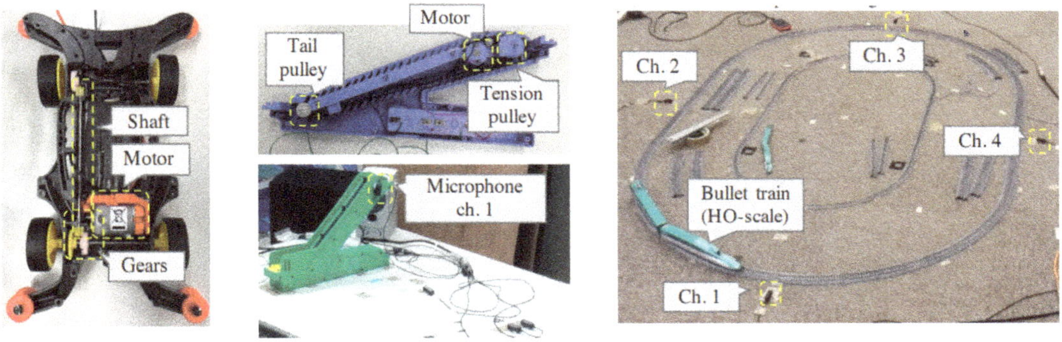

(a) (b) (c)

Figure 11. Schematic of ToyADMOS miniaturised machines (**a**) toy car (**b**) toy conveyor (**c**) toy train [9].

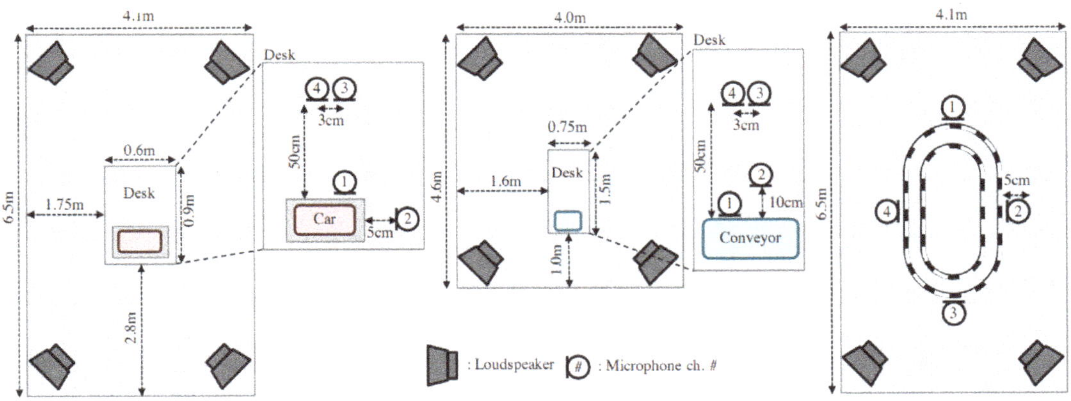

(a) (b) (c)

Figure 12. Schematic of microphone installation setup for ToyADMOS miniaturised machines (**a**) toy car (**b**) toy conveyor (**c**) toy train [9].

Table 7. Imbedded faults in ToyADMOS miniaturized machines [9].

Toy Car		Toy Conveyor		Toy Train	
Parts	**Condition**	**Parts**	**Condition**	**Parts**	**Condition**
Shaft	Bent	Tension pulley	Excessive tension	First carriage	Chipped wheel axle
Gears	Deformed Melted	Tail pulley	Excessive tension Removed	Last carriage	Chipped wheel axle
Tires	Coiled (plastic ribbon) Coiled (steel ribbon)	Belt	Attached metallic object 1 Attached metallic object 2 Attached metallic object 3	Straight railway track	Broken Obstructing stone Disjointed
Voltage	Over voltage Under voltage	Voltage	Over voltage Under voltage	Curved railway track	Broken Obstructing stone Disjointed

3.2. MIMII Dataset

The MIMII (Malfunctioning Industrial Machine Investigation and Inspection) dataset comprises normal and anomalous machine operating sounds of four types of real machines such as valves, pumps, fans, and slide rails [10]. The dataset was captured using an 8-microphone circular array with machine configuration in Figure 13 and sampled at 16 kHz. Each recording consists of 10 s. segments recordings of the machines with various faults as shown in Table 8.

Table 8. Imbedded faults in MIMII real machines [10].

Machine Type	Operations	Examples of Anomalous Conditions
Valve	Open/close repeat with different timing	More than two kinds of contamination
Pump	Suction from discharge to a water pool	Leakage, contamination, clogging, etc.
Fan	Normal work	Unbalanced, voltage change, clogging, etc.
Slide rail	Slide repeat at different speeds	Rail damage, loose belt, no grease, etc.

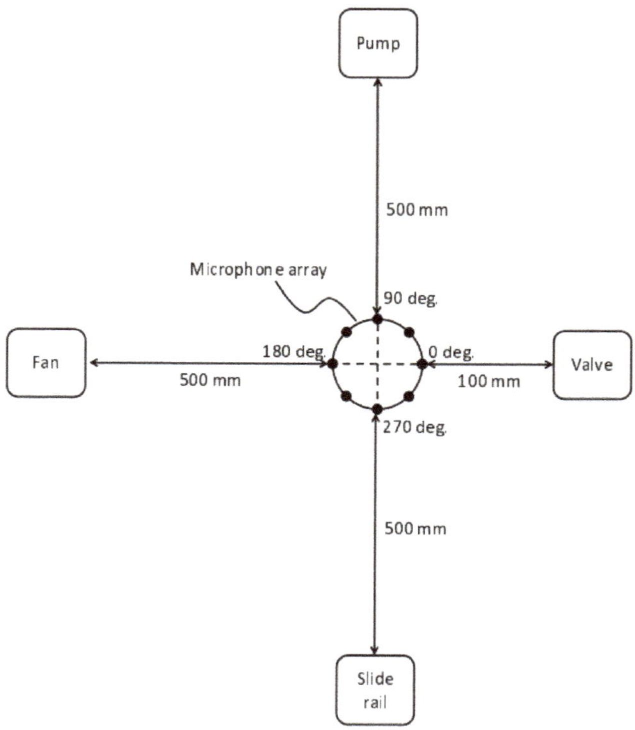

Figure 13. Schematic of microphone installation setup for MIMII [10].

3.3. DCASE Dataset

The DCASE dataset [13] is a merge of subset of ToyADMOS and MIMII dataset comprising both normal and anomalous machine operating sounds. To harmonise both datasets, each audio file includes a single channel and 10 s in duration. All the audio files are resampled at 16 kHz. The dataset relates to the following machine operating sounds: toy car (ToyADMOS), toy conveyor (ToyADMOS), valve (MIMII), pump (MIMII), fan (MIMII) and slide rail (MIMII).

3.4. IDMT-ISA-ELECTRIC-ENGINE Dataset

The IDMT-ISA-ELECTRIC-ENGINE dataset [14] consists of anomalous operating sounds of three brushless electric motors. Different operational states such as good, heavy load and broken are simulated within the electric motors by changing the supply voltage and loads. The dataset provides mono audio for each sound file sampled at 44.1 kHz. For each of the operational states, IDMT-ISA-ELECTRIC-ENGINE dataset provides 774 sound files for "good" state, 789 for "broken" state and 815 for "heavy load". Figure 14 shows the setup for acoustic data acquisition in the electric motor machines.

Figure 14. Three electric motor setups for IDMT-ISA-ELECTRIC-ENGINE dataset [14].

3.5. MIMII DUE Dataset

The MIMII DUE (Malfunctioning Industrial Machine Investigation and Inspection with domain shifts due to changes in operational and environmental conditions) provides a sound dataset for training and testing anomalous sound detection techniques and their invariance to domain shifts [56]. This builds on the authors' previous released MIMII dataset [10] which had the limitation of not representing industrial scenarios with changes in machine operational speed and background noises.

MIMII DUE provides normal and anomalous sounds for five industrial machines: fan, gearbox, pump, slide rail and valve. For each of the machines, six sub-division is provided referred to as sections. Each section refers to a unique instance of machine product; this provides for manufacturing variability within machine type. Furthermore, each section has its dataset is split into source domain and target domain. The source domain contains machine operating sound running at design point while target domain contains machine operating sound running at off-design point.

3.6. ToyADMOS2 Dataset

ToyADMOS2 dataset also provides for training and testing anomalous machine sound detection techniques for their performance in domain shifted conditions [57]. As opposed to ToyADMOS its predecessor, it only carters for two types of miniature machines: toy car and toy trains. The recording and system setup is same for ToyADMOS [9]; however, a key difference, ToyADMOS2 has the normal and anomalous machine operating sounds recorded with machines operating under different speeds. This provides for a source domain consisting of machines with specified operating conditions and the target domain with machines having different operating conditions. Suitable for training and testing with the different domains.

3.7. MIMII DG Dataset

MIMII DG dataset provides normal and anomalous machine operating sounds for benchmark Domain Generalisation techniques [58]. It comprises five groups of machines including valve, gearbox, fan, slide rail and bearing. The audio recording for each machine consists of three sections representing different types of domain shift conditions, which for each machine could be operating condition change and environmental background noise change.

4. Challenges

4.1. Sound Mixtures with Background Noise

The presence of background noise interfering with machine fault signature during acquisition of acoustic data poses a challenge in terms of accuracy and repeatability of machine fault diagnostics. Background noise in this context refers to sound from other operating machines that are different from the target machine. Additionally, it includes the sounds from other activities in the industrial environment.

Approaches are therefore required to eliminate background noise from the collected acoustic data. The challenge lies in the fact that the background noise sources are uncorrelated, as such, filtering techniques are not applicable. Techniques, such as Blind Signal Separation (BSS) and Independent Component Analysis (ICA), have the potential to address this challenge by recovering the signal of interest out of the observed sound mixtures. BSS has been applied in [59] for extracting the unobserved fault acoustic signal during metal stamping with a mechanical press. Wang et al. [60] also applied BSS using sparse component analysis for separating sound mixtures of power transformer origin. In [48], ICA was applied together with variational mode decomposition, to separate the independent components hidden in the observation low signal-to-noise ratio signals, for an intelligent diagnosis application.

In practice, the mixture of acoustic signals is formed by the random mixing of multiple sound sources resulting in non-linear mixture models, which is an area requiring further attention for acoustic-based machine condition monitoring.

4.2. Domain Shift with Changes in Machine Operation and Background Noise

Domain shift represents the change in machine operating and environmental conditions. This is common in industrial settings as machines would not always operate in their design point conditions. There is always a need for the machine to run at an off-design point, indicating changes to both speed and loading as well as changes in the background noise from auxiliaries during operation. Tackling the domain shift problem is important for effective anomaly detectors applicable to machine operating sound.

The concept of domain adaptation is gaining prominence as an approach for anomaly detection in domain shifted conditions [11,61]. Domain adaptation addresses the problem as: when provided with a set of normal data from a source domain and a limited set of normal data from a target domain, how do you develop a performant anomaly detector in the target domain. From the literature, the following approaches for domain adaptation have emerged: learning the transformation from the source domain to the target domain [62,63], learning invariant representations between the source and the target domains [64–67] and few-shot domain adaptation [68,69]. With the option of domain adaptation, it opens opportunities for application to acoustic-based machine condition monitoring and fault diagnostics.

4.3. Domain Generalisation Invariant to Changes in Machine Operation and Background Noise

Domain generalisation is an attempt to provide an alternative to the domain adaptation techniques when dealing with domain shift due to the computational cost of the domain adaptation techniques. Domain generalisation poses the problem of learning commonalities across various domains (i.e., source and target) to enable the model to generalize across the domains. Such generalisation would need to account for domain shift caused by differences in environmental conditions, machine physical conditions, changes due to maintenance, and differences in recording devices for instance.

Fundamentally, domain generalisation attempts the out-of-distribution generalisation by using only the source domain data. In the literature, several techniques have emerged such as [70]: domain alignment, meta-learning, ensemble learning, data augmentation, self-supervised learning, learning disentangled representations, regularisation strategies, and reinforcement learning. With the development and application of domain generalisation techniques for machine fault diagnostics problem, it would open compelling opportunities for the applicability of the acoustic-based approaches.

4.4. Effect of Measurement Distance, Measurement Device and Sampling Parameters

4.4.1. Measurement Distance (Microphones Positions)

Sound propagates through air as a longitudinal wave; as it moves through the air medium, from the source to the listener or observer, sound as characterised by sound intensity, experiences attenuation, i.e., loss in energy. For a point source (i.e., uniformly radiating sound in all directions), this attenuation follows the inverse square law as shown in Figure 15, which is dependent on the measurement distance. In practice, for every doubling of measurement distance, the sound intensity reduces by a factor of 4; alternatively, the sound pressure level reduces by 6 dB. From sound propagation theory, it is evident that, the measurement distance of anomalous machine operating sound is important [71]. However, very little consideration has been given to this effect during experimental setup for anomalous machine sound data acquisition as corroborated by the benchmarking open-source datasets such as ToyADMOS, MIMII, IDMT-ISA-ELECTRIC ENGINE, MIMII DUE, ToyADMOS2, and MIMII DG. One can argue, the measurement distance effect can be accounted for within domain adaptation or domain generalisation challenges. Yet, the various datasets do not provide a systematic grouping of the dataset based on the measurement distance for this to be considered. The parameters often considered are changes in machine operating parameters (i.e., rotating speed and load) and environmental/background noise.

An important question is then raised; how far should the microphones be from the sound source considering the measurement distance effect?

In acoustics, two physical regions exist that shed light to the above question: the acoustics near field and acoustics far field as shown in Figure 16. The transition from near field to far field occur in at least 1 wavelength of the sound source [72]. It is important, to note, as wavelength is a function of frequency, this transition distance would change as the frequency content of the sound source changes. The near field exist very close to the sound source with no fixed relationship between sound intensity and distance. Within the far field, the inverse square law of sound propagation holds true. In practice, this is the region where the measuring microphone should ideally be located. As a minimum, a single microphone can suffice for accurate and repeatable measurement of sound. Although fundamental acoustics theory would place the far field at least 1 wavelength of the sound source [72]; ISO 3745, provides several guidelines or criteria for microphone placement within the far field for sound power measurement [73]:

$$\text{(a) } r \geq 2d_o \tag{6}$$

$$\text{(b) } r \geq \lambda/4 \tag{7}$$

$$\text{(c) } r \geq 1 \textit{ metre} \tag{8}$$

where r is measurement distance, d_o is characteristic dimension or largest dimension of the sound source, and λ is the lowest wavelength of the sound source.

For small, low-noise sound sources with measurement over a limited frequency range, the measurement distance can be less than 1 m, but not less than 0.5 m, provided consideration for criteria (a) and (b) above are adhered to [73].

Within the near field, measurement is feasible; but would require multi-microphone array. For the measurement of anomalous machine operating sound, guidelines are lacking in the literature and further research is required.

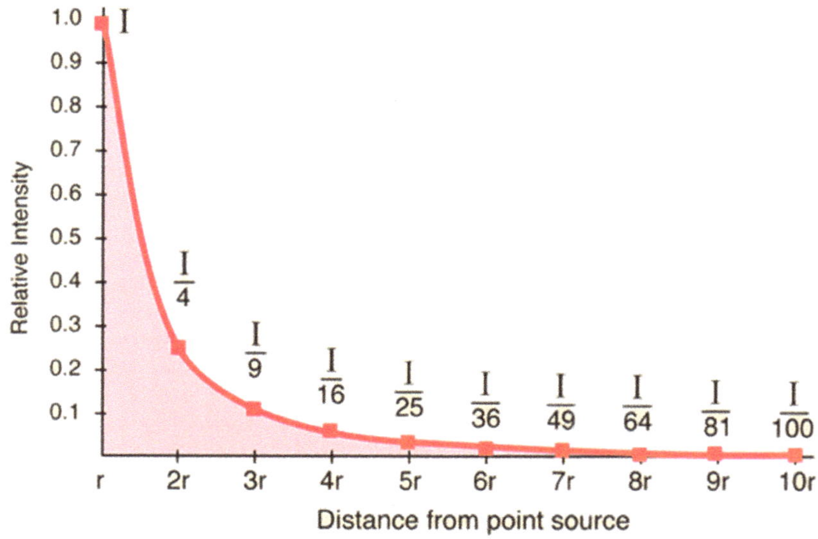

Figure 15. Distance effect on sound intensity propagation and attenuation [74].

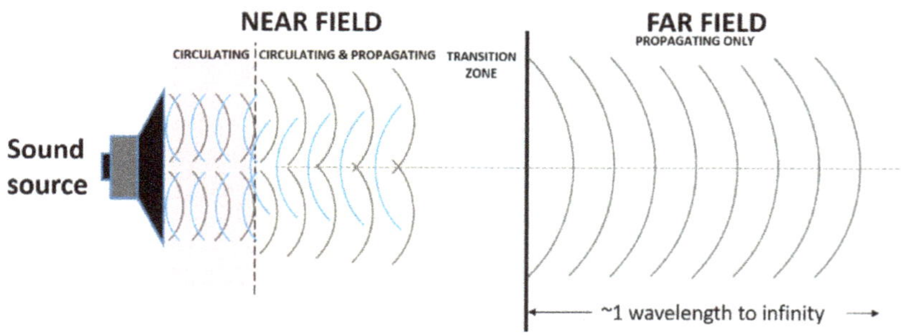

Figure 16. Acoustic sound field consideration [72].

4.4.2. Single Microphone Measurement Device and Sampling Parameters

Acoustic measuring device mismatch between development data acquisition and testing can occur in practice. As every microphone have its unique transfer function which dictates its frequency response and perception of sound, measuring device mismatch needs to be considered. Very little has been done in considering this challenge in the detection and classification of anomalous machine operating sound. However, such consideration is already attracting attention in the corollary field of acoustic scene classification [75]. Key to this consideration in acoustic scene classification field, is the realization of the TUT Urban Acoustic Scenes dataset which consists of ten different acoustic scenes, recorded in six large European cities with four different microphone devices: highlighting the importance of considering the acoustic measuring device for robust pattern learning algorithm [75].

As very little work has been explored on the effect of recording device mismatch in anomalous machine operating sound detection and classification to inform device choice; still, some learning can be gleaned from the choice of microphones, sampling frequency and sample duration as shown in Table 9 from the opensource dataset projects on DCAMS.

Table 9. Exemplar acoustic measurement devices and sampling parameters.

Datasets	Microphone Brand	Microphone Type	Sampling Frequency	Sample Duration	Ref.
ToyADMOS	Shure SM11-CN	Omni-directional Microphone	16 kHz (Downsampled)	10 s, and 10 min	[9]
MIMII	TAMAGO-03 (Circular microphone array with 8 distinct microphones)	-	16 kHz	10 s.	[10]
IDMT-ISA-ELECTRIC-ENGINE	-	-	44.1 kHz	3 s.	[14]
MIMII DUE	TAMAGO-03 (Circular microphone array with 8 distinct microphones)	-	16 kHz	10 s.	[56]
ToyADMOS2	Shure SM11-CN TOMOCA EM-700	Omni-directional Microphone Condenser Microphone	48 kHz	12 s.	[57]
MIMII DG	TAMAGO-03 (Circular microphone array with 8 distinct microphones)	-	16 kHz	10 s.	[58]

4.4.3. Microphone Array Measurement (Acoustic Camera)

Acoustic camera measurement provides the capability for sound source localisation, quantification and visualization using multi-dimensional acoustic signals processed from a microphone array unit and overlaid on either image or video of the sound source as shown in Figure 17 [76]. An acoustic camera, is a collection of several microphones, acting as a microphone array unit, where the microphones within the array can be arranged either as uniform circular configuration, uniform linear configuration, uniform square configuration or customized array configuration for specific application. Acoustic camera can provide acoustic scene measurement both in the near and far acoustic fields.

For localizing anomalous machine operating sound in application, acoustic camera has been used to map the variation in machine emitted sound for fault detection as follows: localizing sources of aircraft fly by noise [77], characterising emitted sound from internal combustion engine running idle in a vehicle [78], fault detection in a gearbox unit [79], fault localisation in rolling-element bearing [80], etc.

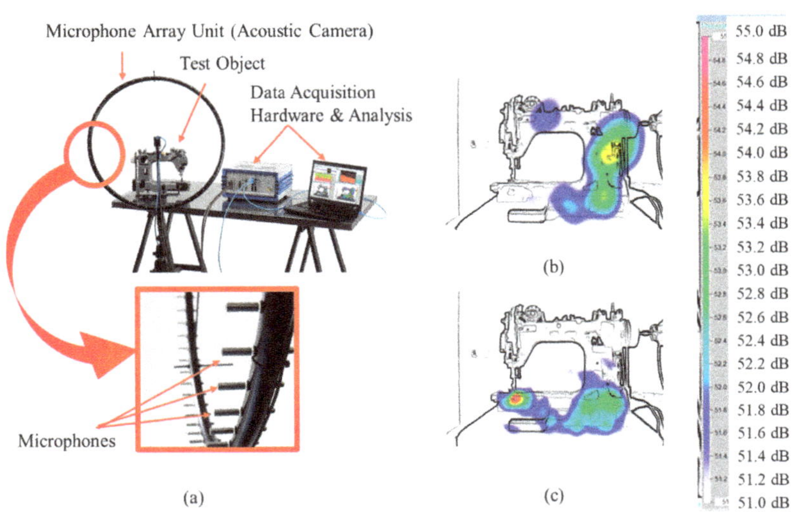

Figure 17. Acoustic camera for fault detection based on variation in emitted sound (**a**) Acoustic camera setup (**b**) test object without a fault (**c**) test object with a fault [81].

Central to the analysis and interpretation of the multi-dimensional acoustic signals is acoustic beamforming technique [76,82]. Ref. [82] provides an extensive review on acoustic beamforming theory including consideration for acoustic beamforming test design criteria.

Acoustic beamforming is a spatial filtering technique used in far field acoustic domain, for localisation and quantification of the sound source; where it amplifies the acoustic signal of interest while suppressing interfering sound sources (e.g., background noise) [82]. In principle, the beamforming algorithm works by summing individual acoustic signals based on their arrival times from the sound source to the microphone array. This summation process suppresses the interfering signals while enhancing the acoustic signal of interest. The technique can be performed both in the time-domain and frequency domain [82].

(1) Delay and Sum Beamforming in the Time-Domain: This is demonstrated in Figure 18 as follows, considering only two sound sources as an example (i.e., source 1 and source 2). For each sound source, the travel path of emitted sound to the microphone array would be different; as such, captured signals by the microphone array would show different delays and phases for the measured signals from both sources. As both parameters, delay, and phase, are proportional to the travelled distance between microphone array and source; with the knowledge of the speed of sound in the medium (e.g., air), the runtime delay is estimated for the signal of interest (source 1) reaching all the microphone locations. The measured signal for every microphone in the array is then shifted by the calculated runtime delay for that channel, creating an alignment in phase in the time-domain for the signal of interest (source 1). The resulting signals from every microphone channel are summed and normalised by the number of microphones in the array; As shown in Figure 18, the signal of interest (source 1) is amplified due to constructive interference while source 2 is minimized due to destructive interference. To create the final acoustic scene representation, for each microphone channel, the root mean square (RMS) amplitude value or the maximum amplitude value of the time-domain acoustic signal can be evaluated for visualization as an acoustic map.

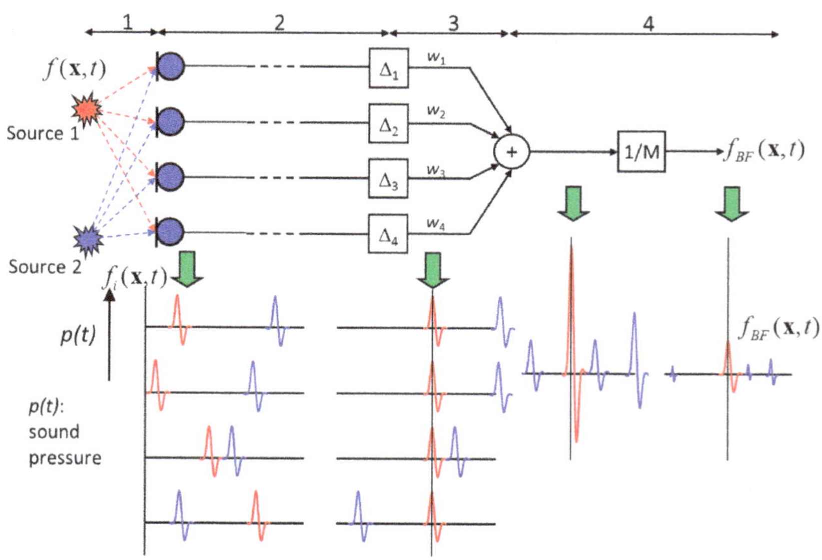

Figure 18. Schematic of delay and sum beamforming in the time domain for acoustic sources [83].

(2) Delay and Sum Beamforming in the Frequency Domain: This is demonstrated in Figure 19 as follows, considering only two sound sources as an example (i.e., source 1 and source 2). For each sound source, the travel path of emitted sound to the microphone array would be different; as such, captured signals by the microphone array would show different delays and phases for the measured signals from both sources. The delay for the signal of interest can be determined using information such as, distance between source and microphone and the speed of sound in the medium. Fourier transform is performed at all microphone channel resulting in a complex spectrum for amplitude and phase. To eliminate the delay in phase for the signal of interest at all microphone location, the complex spectra is multiplied by a complex phase term as shown in Figure 19, bringing the interested acoustic source in phase without impacting the amplitude of the spectra. The resulting complex spectra from all the microphone channels are summed and normalised by the number of microphone channels. The interest sound source signal (source 1) is enhanced due to constructive interference, while source 2 is diminished due to destructive interference.

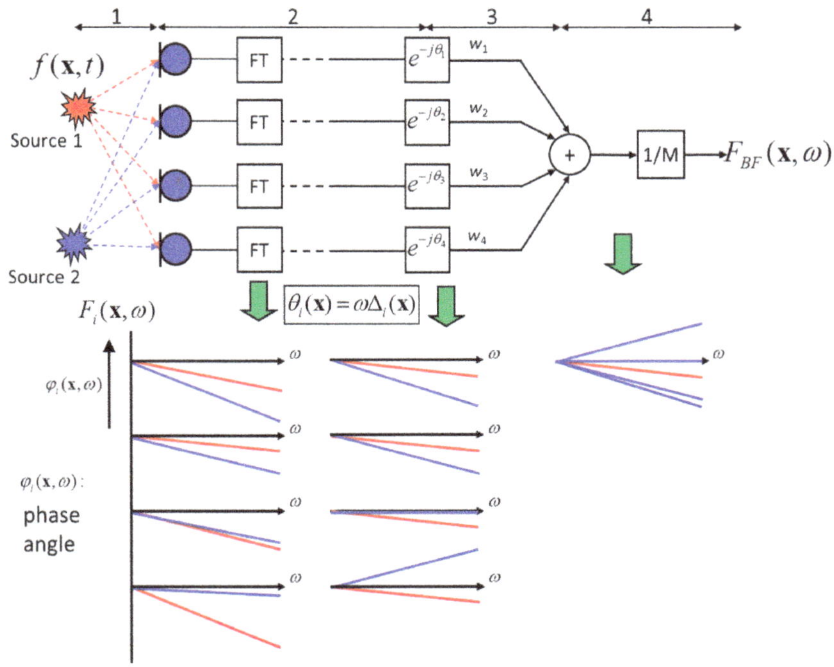

Figure 19. Schematic of delay and sum beamforming in the frequency domain for acoustic sources [84].

Application of acoustic camera to machine diagnostics have been attracting increasing interest [77–80,85,86]. Of note, is the approach proposed by [85,86] to localise faults in rotating machinery using acoustic beamforming and spectral kurtosis (i.e., spectral kurtosis is an effective indicator of machine fault [87,88]). As shown in Figure 20, spectral kurtosis is used as a post-processor of the multi-dimensional acoustic time-domain signals from the microphone array to identify and localise fault-related frequency bands (i.e., frequency bands that are impulsive); the resulting kurtogram having a spatial dimension provides the capability to localise the high kurtosis region providing indication of machine fault.

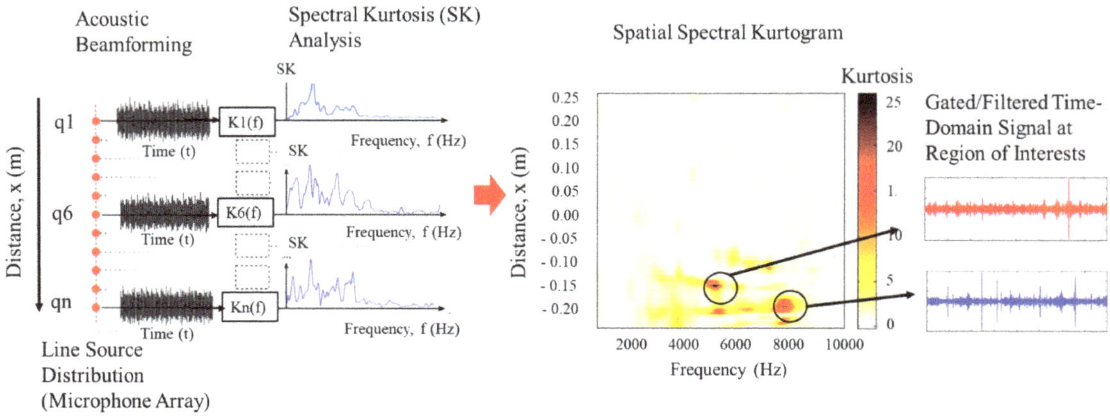

Figure 20. Application of spectral kurtosis to acoustic beamforming for machine fault diagnosis [85,86].

5. Outlook

Anomalous machine operating sound provides a rich set of information about a machine's current health state upon which to automate the detection and classification of machinery faults. Despite advances in data-driven machine learning and deep learning approaches as currently applied for acoustic-based machine condition monitoring, there still exist areas for further research for this technique to be industrially applicable.

5.1. Addressing Pitfalls in Acoustic Data Collection

The performance of data-driven models and their ability to generalize during training and testing depends on the available datasets being a representative of the actual fault scenario. However, generating machine fault dataset for actual machines is a costly endeavor. If the training dataset is too small, the model learns sampling noise. As a work around, most of the opensource dataset for the detection and classification of anomalous machine operating sounds have focused on either toy machines or scaled down machine models. This approach has provided initial seeding to be able to benchmark currently developed techniques. Generally, available datasets account for steady-state changes in machine operational parameters such as speed and load, consideration of varying degree of background noise during acoustic signal measurement, and different models of similar machine class. These datasets are lacking in the following areas: consideration of the distance effect during grouping of the dataset (i.e., it would be relevant to have measurements at different distances from the source to test the robustness of developed approaches working in the field where it would be difficult to maintain repeatable measurement distance), consideration of transient operation regime of machines during dataset grouping (i.e., steady-state dataset alone is a non-representative training data; developed approach need to be able to differentiate transient operation from anomalous operation), and consideration of device mismatch during data acquisition (i.e., recording for same machine fault with different types of microphones, such as omni-directional microphone, pressure-free field microphone, condenser microphone, etc.; Furthermore, it would be relevant to specify a standard reference microphone such as the omni-directional microphone, in other for spectrum correction coefficients for various microphones to be provided with respect to this [89]; using spectrum correction coefficients opens up the possibility of data transformation to account for device mismatch).

5.2. Addressing Measurement Artifacts (i.e., Background Noise, and Distance Effect)

In the industrial environment, acoustic-based machine condition monitoring is often plagued with the problem of having multiple signals mixing such as acoustic signal of interest indicative of anomalous machine operation and the background noise, i.e., neighboring machinery, factory noise, etc. It is required for the sound mixture to be separable, i.e., separating the acoustic signal of interest from the background noise. Conventional approaches such as spectral subtraction methods which rely on the background noise having a constant magnitude spectrum and acoustic signal of interest been short-time stationary would not be applicable as there is the possibility of removing fault frequencies from the spectrum of the acoustic signal of interest [90]. Blind signal separation can be useful as it offers sound mixture separation without prior knowledge of either of the signals or the way in which they are mixed [91]. Application and optimisation of blind signal separation for acoustic-based machine condition monitoring provides an area for further research.

The effect of distance between the acoustic source and microphone leads to attenuation of the measured sound intensity. Furthermore, it places a burden of repeatability between laboratory conditions and industrial conditions, impacting data-driven model accuracy for application. Eliminating or minimizing the distance effect on the acquired acoustic signal is an area requiring further research. [71] proposed a normalisation scheme (i.e., d-normalization) in the frequency domain using the spectrum representation of the acoustic signal which minimized the distance effect as shown in Figure 21 and expressed as:

$$I(f) = \overline{I}(f)/\mu_I \quad (9)$$

where $I(f)$ is the normalised spectrum of the measured sound intensity, $\overline{I}(f)$ is the unnormalised spectrum of the measured sound intensity (i.e., determined from fast Fourier transform of the time-domain acoustic signal), and μ_I is the mean of the rectified time-domain acoustic signal intensity, given as:

$$\mu_I = (1/N) \times \sum_{i=1}^{N} |X_i| \quad (10)$$

where N is number of sample points in the acoustic time-domain signal, $|X_i|$ is the absolute amplitude value of the acoustic time-domain signal.

Although the result is promising, it is applicable to the spectral representation of the acoustic signal. Alternative normalisation scheme be required for other acoustic image representation such as cochleagram, Mel-spectrogram, amongst others? Furthermore, what would be the impact on the data-driven model accuracy due to normalisation of the input acoustic representation? These are open questions for further research.

5.3. Improving Data-Driven Model Accuracy for Application: Domain Adaptation versus Domain Generalisation

Domain shift (i.e., changes in machinery operating speed and load) is inevitable in industrial processes due to machines operating in off-design conditions and harsh environment. As such, training data-driven models for the DCAMS problem to account for this system dynamics is a must have. However, learning robust model representation by using data from multiple domains to identify invariant relationships between the various domains is still a challenging problem. Two schools of thought have emerged to address the domain shift problem in acoustic-based machine condition monitoring: domain adaptation [92,93] and domain generalisation [94]. Both approaches tackle the same problem based on the available dataset. Domain adaptation assumes you have dataset from the source domain (i.e., machine operating at design point) and some set of data in the target domain (i.e., machine operating at off-design point), it attempts to learn the mapping between the source and target domain based on these criteria. Alternatively, domain generalisation assumes you have dataset from two different source domains, it attempts to learn the mapping to an unseen domain. Although several domain adaptation and generalization techniques have been proposed in the literature, the model performance for both approaches is yet to reach satisfactory level in applications as evident from DCASE2021 and DCASE2022 Task 2 challenges [11,12].

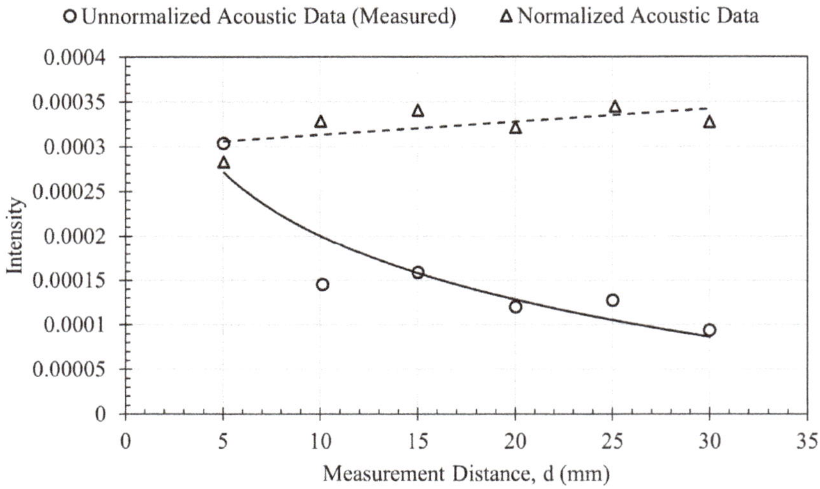

Figure 21. Minimizing distance effect on measured acoustic signal using d-normalisation [71].

5.4. Addressing Multi-Fault Diagnosis

In industrial environment, machinery may need to operate in both off-design conditions and harsh conditions continuously for extended periods of time. As such, machine components are liable to the occurrence of multiple faults at the same time. When these multi-faults occur, their impact to machine performance and lifespan is more severe as compared to the presence of a single fault due to fault interactions [95]. Fault diagnosis approaches needs to be able to accommodate both single fault and multi-faults detection scenarios. From the literature, within acoustic-based condition monitoring methodology, the focus has been on addressing the single-fault diagnosis problem; multi-fault diagnosis of machinery is still lacking. This area of research needs consideration for viable industrial applications, e.g., fault diagnosis in gearbox, electric motor, compressor, pump, amongst others.

5.5. Improving Acoustic Camera Spatial Detection of Machine Faults

Acoustic camera for machine fault diagnosis provides spatial information not possible with conventional condition monitoring approaches such vibration analysis. However, interpreting the visualization of the emitted sound field from the machine from acoustic beamforming is very limited; It is important to note that regions of high sound pressure level does not necessarily correlate with the presence of a fault. Further research is required to analyse the multi-dimensional acoustic time-domain signals as a function of space from the acoustic beamforming analysis using either signal processing methods or data-driven machine learning/deep learning approaches. Pioneering in this regard, [85,86] have proposed spectral kurtosis as means to filter the multi-dimensional acoustic time-domain signals from acoustic beamforming to localise impulsive-related machine faults, e.g., gearbox faults, rolling-element bearing faults, etc., as well as extract the time-domain acoustic signals from the region of high spectral kurtosis. This area of research is still limited in correlating regions of high spectral kurtosis to a fault. The extract time-domain signal provides an opportunity to be explored for evaluation using data-driven approaches. Furthermore, beyond spectral kurtosis, what other signal processing approaches are relevant with improved sensitivity to localizing machine faults from the multi-domain acoustic signals provided by the acoustic camera?

6. Conclusions

Acoustic-based machine condition monitoring has been attracting increasing attention, especially with the annual DCASE challenge task on unsupervised anomalous sound detection for identifying machine conditions. Given the industrial relevance and significance of this research area, it becomes important in this paper to address the following questions: (i) are there commonalities or differences amongst the developed methodologies for detecting and classifying anomalous machine operating sounds, (ii) what open datasets are available for benchmarking the developed techniques, and (iii) what challenges are still there for the applicability of acoustic-based machine condition monitoring. Hopefully, this review of the state-of-the-arts can inspire more advancement in the acoustic-based machine condition monitoring research area.

Author Contributions: Conceptualization, G.J. and Y.Z.; writing—original draft preparation, G.J.; writing—review and editing, Y.Z. All authors have read and agreed to the published version of the manuscript.

Funding: This research received no external funding.

Data Availability Statement: Data sharing not applicable.

Conflicts of Interest: The authors declare no conflict of interest.

References

1. Gemmeke, J.F.; Ellis, D.P.W.; Freedman, D.; Jansen, A.; Lawrence, W.; Moore, R.C.; Plakal, M.; Ritter, M. Audio Set: An Ontology and Human-Labeled Dataset for Audio Events. In Proceedings of the 2017 IEEE International Conference on Acoustics, Speech and Signal Processing (ICASSP), New Orleans, LA, USA, 5–9 March 2017; pp. 776–780.
2. Cramer, J.; Wu, H.-H.; Salamon, J.; Bello, J.P. Look, Listen, and Learn More: Design Choices for Deep Audio Embeddings. In Proceedings of the IEEE International Conference on Acoustics, Speech and Signal Processing (ICASSP), Brighton, UK, 12–17 May 2019; pp. 3852–3856.
3. Arandjelović, R.; Zisserman, A. Look, Listen and Learn. In Proceedings of the IEEE International Conference on Computer Vision (ICCV), Venice, Italy, 22–29 October 2017.
4. Kong, Q.; Cao, Y.; Iqbal, T.; Wang, Y.; Wang, W.; Plumbley, M. PANNs: Large-Scale Pretrained Audio Neural Networks for Audio Pattern Recognition (Pretrained Models). *IEEE/ACM Trans. Audio Speech Lang. Process.* **2020**, *28*, 2880–2894. [CrossRef]
5. Hershey, S.; Chaudhuri, S.; Ellis, D.P.W.; Gemmeke, J.F.; Jansen, A.; Moore, R.C.; Plakal, M.; Platt, D.; Saurous, R.A.; Seybold, B.; et al. CNN architectures for large-scale audio classification. In Proceedings of the 2017 IEEE International Conference on Acoustics, Speech and Signal Processing (ICASSP), New Orleans, LA, USA, 5–9 March 2017; pp. 131–135. [CrossRef]
6. Gaylard, A.; Meyer, A.; Landy, C. Acoustic Evaluation of Faults in Electrical Machines. In Proceedings of the 1995 Seventh International Conference on Electrical Machines and Drives (Conf. Publ. No. 412), Durham, UK, 11–13 September 1995; pp. 147–150.
7. Kawaguchi, Y.; Endo, T. How Can We Detect Anomalies from Subsampled Audio Signals? In Proceedings of the 2017 IEEE 27th International Workshop on Machine Learning for Signal Processing (MLSP), Tokyo, Japan, 25–28 September 2017; pp. 1–6.
8. Koizumi, Y.; Saito, S.; Uematsu, H.; Harada, N. Optimizing Acoustic Feature Extractor for Anomalous Sound Detection Based on Neyman-Pearson Lemma. In Proceedings of the 2017 25th European Signal Processing Conference (EUSIPCO), Kos, Greece, 28 August–2 September 2017; pp. 698–702.
9. Koizumi, Y.; Saito, S.; Uematsu, H.; Harada, N.; Imoto, K. ToyADMOS: A Dataset of Minia-ture-Machine Operating Sounds for Anomalous Sound Detection. In Proceedings of the 2019 IEEE Workshop on Applications of Signal Processing to Audio and Acoustics (WASPAA), New Paltz, NY, USA, 20–23 October 2019; pp. 313–317.
10. Purohit, H.; Tanabe, R.; Ichige, T.; Endo, T.; Nikaido, Y.; Suefusa, K.; Kawaguchi, Y. MIMII Dataset: Sound Dataset for Malfunctioning Industrial Machine Investigation and Inspection. In Proceedings of the Detection and Classification of Acoustic Scenes and Events 2019 Workshop (DCASE2019), New York, NY, USA, 25–26 October 2019; pp. 209–213.
11. Kawaguchi, Y.; Imoto, K.; Koizumi, Y.; Harada, N.; Niizumi, D.; Dohi, K.; Tanabe, R.; Purohit, H.; En-do, T. Description and Discussion on DCASE 2021 Challenge Task 2: Unsupervised Anomalous Sound Detection for Machine Condition Monitoring under Domain Shifted Conditions. *arXiv* **2021**, arXiv:2106.04492.
12. Dohi, K.; Imoto, K.; Harada, N.; Niizumi, D.; Koizumi, Y.; Nishida, T.; Purohit, H.; Endo, T.; Yamamoto, M.; Kawaguchi, Y. Description and Discussion on DCASE 2022 Challenge Task 2: Unsupervised Anomalous Sound Detection for Machine Condition Monitoring Applying Domain Generalization Tech-niques. *arXiv* **2022**, arXiv:2206.05876.
13. Koizumi, Y.; Kawaguchi, Y.; Imoto, K.; Nakamura, T.; Nikaido, Y.; Tanabe, R.; Purohit, H.; Suefusa, K.; Endo, T.; Yasuda, M.; et al. Description and Discussion on DCASE2020 Challenge Task2: Unsupervised Anomalous Sound Detection for Machine Condition Monitorin. In Proceedings of the Detection and Classification of Acoustic Scenes and Events 2020 Workshop (DCASE2020), Tokyo, Japan, 2–4 November 2020; pp. 81–85.

14. Grollmisch, S.; Abeßer, J.; Liebetrau, J.; Lukashevich, H. Sounding Industry: Challenges and Datasets for Industrial Sound Analysis. In Proceedings of the 2019 27th European Signal Processing Conference (EUSIPCO), A Coruña, Spain, 2–6 September 2019; pp. 1–5.
15. Koizumi, Y.; Saito, S.; Uematsu, H.; Kawachi, Y.; Harada, N. Unsupervised Detection of Anomalous Sound Based on Deep Learning and the Neyman–Pearson Lemma. *IEEE/ACM Trans Audio Speech Lang Process* **2019**, *27*, 212–224. [CrossRef]
16. Sharan, R.; Xiong, H.; Berkovsky, S. Benchmarking Audio Signal Representation Techniques for Classification with Convolutional Neural Networks. *Sensors* **2021**, *21*, 3434. [CrossRef]
17. Roche, F.; Hueber, T.; Limier, S.; Girin, L. Autoencoders for Music Sound Modeling: A Com-parison of Linear, Shallow, Deep, Recurrent and Variational Models. In Proceedings of the 16th Sound & Music Computing Conference (SMC 2019), Malaga, Spain, 28–31 May 2019.
18. Bai, J.; Chen, C.; Chen, J. Feature Based Fusion System for Anomalous Sounds Monitoring. In Proceedings of the 5th Workshop on Detection and Classification of Acoustic Scenes and Events (DCASE), Tokyo, Japan, 2–4 November 2020.
19. Ahmed, F.; Nguyen, P.; Courville, A. An Ensemble Approach for Detecting Machine Failure from Sound. In Proceedings of the 5th Workshop on Detection and Classification of Acoustic Scenes and Events (DCASE), Tokyo, Japan, 2–4 November 2020.
20. Alam, J.; Boulianne, G.; Gupta, V.; Fathan, A. An Ensemble Approach to Unsupervised Anomalous Sound Detection. In Proceedings of the 5th Workshop on Detection and Classification of Acoustic Scenes and Events (DCASE), Tokyo, Japan, 2–4 November 2020.
21. Morita, K.; Yano, T.; Tran, K.Q. Anomalous Sound Detection by Using Local Outlier Factor and Gaussian Mixture Model. In Proceedings of the 5th Workshop on Detection and Classification of Acoustic Scenes and Events (DCASE), Tokyo, Japan, 2–4 November 2020.
22. Hendrycks, D.; Mazeika, M.; Dietterich, T. Deep Anomaly Detection with Outlier Exposure. In Proceedings of the Seventh International Conference on Learning Representations, Vancouver, BC, Canada, 30 April–3 May 2018.
23. Sandler, M.; Howard, A.G.; Zhu, M.; Zhmoginov, A.; Chen, L.-C. Inverted Residuals and Lin-ear Bottlenecks: Mobile Networks for Classification, Detection and Segmentation. In Proceedings of the IEEE conference on computer vision and pattern recognition, Salt Lake City, UT, USA, 18–22 June 2018.
24. Heng, R.B.W.; Nor, M.J.M. Statistical Analysis of Sound and Vibration Signals for Monitor-ing Rolling Element Bearing Condition. *Appl. Acoust.* **1998**, *53*, 211–226. [CrossRef]
25. Van Riesen, D.; Schlensok, C.; Henrotte, F.; Hameyer, K. Acoustic Measurement for Detecting Manufacturing Faults in Electrical Machines. In Proceedings of the 17th International Conference on Electrical Machines (ICEM), Chania, Greece, 2–5 September 2006.
26. Benko, U.; Petrovcic, J.; Juričić, Ð.; Tavčar, J.; Rejec, J. An Approach to Fault Diagnosis of Vac-uum Cleaner Motors Based on Sound Analysis. *Mech. Syst. Signal Process.* **2005**, *19*, 427–445. [CrossRef]
27. Mishra, R.; Gu, F.; Fazenda, B.; Stubbs, C.; Ball, A. Measurement and Characterisation of Faults in the Intake System of a Turbocharged Engine Using a Directional Acoustic Probe. In Proceedings of the COMADEM 2009, San Sebastian, Spain, 9–11 June 2009.
28. Wu, Z.; Huang, N.E. Ensemble Emprical Mode Decomposition: A Noise-Assisted Data Analysis Method. *Adv. Adapt. Data Anal.* **2009**, *1*, 1–41. [CrossRef]
29. Fazenda, B.M. Acoustic Based Condition Monitoring of Turbine Blades. In Proceedings of the 18th International Congress on Sound and Vibration, Rio de Janeiro, Brazil, 10–14 July 2011.
30. Grebenik, J.; Bingham, C.; Srivastava, S. Acoustic Diagnostics of Electrical Origin Fault Modes with Readily Available Consumer-Grade Sensors. *IET Electr. Power Appl.* **2019**, *13*, 1946–1953. [CrossRef]
31. Shiri, H.; Wodecki, J.; Ziętek, B.; Zimroz, R. Inspection Robotic UGV Platform and the Proce-dure for an Acoustic Signal-Based Fault Detection in Belt Conveyor Idler. *Energies* **2021**, *14*, 7646. [CrossRef]
32. Fang, S.; Li, S.-C.; Zhen, D.; Shi, Z.; Gu, F.; Ball, A.D. Acoustic Feature Extraction for Monitor-ing the Combustion Process of Diesel Engine Based on EMD and Wavelet Analysis. *Int. J. COMADEM* **2017**, *20*, 25–30.
33. Zhen, D.; Wang, T.; Gu, F.; Tesfa, B.; Ball, A. Acoustic Measurements for the Combustion Di-agnosis of Diesel Engines Fuelled with Biodiesels. *Meas. Sci. Technol.* **2013**, *24*, 055005. [CrossRef]
34. Anami, B.S.; Pagi, V.B. Acoustic Signal-Based Approach for Fault Detection in Motorcycles Using Chaincode of the Pseudospectrum and Dynamic Time Warping Classifier. *IET Intell. Transp. Syst.* **2014**, *8*, 21–27. [CrossRef]
35. Amarnath, M.; Sugumaran, V.; Kumar, H. Exploiting Sound Signals for Fault Diagnosis of Bearings Using Decision Tree. *Measurement* **2013**, *46*, 1250–1256. [CrossRef]
36. Pasha, S.; Ritz, C.; Stirling, D.; Zulli, P.; Pinson, D.; Chew, S. A Deep Learning Approach to the Acoustic Condition Monitoring of a Sintering Plant. In Proceedings of the APSIPA Annual Summit and Conference 2018, Hawaii, HI, USA, 12–15 November 2018.
37. Giannakopoulos, T. PyAudioAnalysis: An Open-Source Python Library for Audio Signal Analysis. *PLoS ONE* **2015**, *10*, e0144610. [CrossRef]
38. Russell, S.J.; Norvig, P. *Artificial Intelligence: A Modern Approach*, 3rd ed.; Pearson: Boston, MA, USA, 2010.
39. Raschka, S. *Python Machine Learning*; Packt Publishing: Birmingham, UK, 2015.
40. Mathew, S.K.; Zhang, Y. Acoustic-Based Engine Fault Diagnosis Using WPT, PCA and Bayesian Optimization. *Appl. Sci.* **2020**, *10*, 6890. [CrossRef]

41. Altman, N.S. An Introduction to Kernel and Nearest-Neighbor Nonparametric Regression. *Am. Stat.* **1992**, *46*, 175–185.
42. Yao, Z.; Ruzzo, W.L. A Regression-Based K Nearest Neighbor Algorithm for Gene Function Prediction from Heterogeneous Data. *BMC Bioinform.* **2006**, *7*, S11. [CrossRef]
43. Kleyko, D.; Osipov, E.; Papakonstantinou, N.; Vyatkin, V.; Mousavi, A. Fault Detection in the Hy-perspace: Towards Intelligent Automation Systems. In Proceedings of the 2015 IEEE 13th International Conference on Industrial Informatics (INDIN), Cambridge, UK, 22–24 July 2015.
44. Ghaderi, H.; Kabiri, P. Automobile Engine Condition Monitoring Using Sound Emission. *Turk. J. Electr. Eng. Comput. Sci.* **2017**, *25*, 1807–1826. [CrossRef]
45. Lyon, R.F. Machine Hearing: An Emerging Field [Exploratory DSP]. *IEEE Signal Process. Mag.* **2010**, *27*, 131–139. [CrossRef]
46. Lyon, R.F. Machine Hearing: Audio Analysis by Emulation of Human Hearing. In Proceedings of the 2011 IEEE Workshop on Applications of Signal Processing to Audio and Acoustics (WASPAA), New Paltz, NY, USA, 13 May 2011; p. viii.
47. Jombo, G.; Shriram, A. Evaluation of 2D Acoustic Signal Representations for Acoustic-Based Machine Condition Monitoring. In Proceedings of the PECS 2022 Physics, Engineering and Computer Science Research Conference, Kavala, Greece, 21–23 June 2022.
48. Zhang, Y.; Dora, S.; Martinez-Garcia, M.; Bhattacharyaand, S. Machine Hearing for Industrial Acoustic Monitoring Using Cochleagram and Spiking Neural Network. In Proceedings of the 2022 IEEE/ASME International Conference on Advanced Intelligent Mechatronics (AIM), Sapporo, Japan, 11–15 July 2022; pp. 1047–1051.
49. Tanveer, M.H.; Zhu, H.; Ahmed, W.; Thomas, A.; Imran, B.M.; Salman, M. Mel-Spectrogram and Deep CNN Based Representation Learning from Bio-Sonar Implementation on UAVs. In Proceedings of the 2021 Interna-tional Conference on Computer, Control and Robotics (ICCCR), Singapore, 23–26 April 2021; pp. 220–224.
50. Li, J.; Zhang, X.; Huang, L.; Li, F.; Duan, S.; Sun, Y. Speech Emotion Recognition Using a Du-al-Channel Complementary Spectrogram and the CNN-SSAE Neutral Network. *Appl. Sci.* **2022**, *12*, 9518. [CrossRef]
51. Ogundokun, R.O.; Maskeliunas, R.; Misra, S.; Damaševičius, R. Improved CNN Based on Batch Normalization and Adam Optimizer. In *International Conference on Computational Science and Its Applications*; Gervasi, O., Murgante, B., Misra, S., Rocha, A.M.A.C., Garau, C., Eds.; Springer International Publishing: Cham, Switzerland, 2022; pp. 593–604.
52. Zhang, Y.; Martínez-García, M. Machine Hearing for Industrial Fault Diagnosis. In Proceedings of the 2020 IEEE 16th International Conference on Automation Science and Engineering (CASE), Hong Kong, China, 20–21 August 2020; pp. 849–854.
53. Guo, W.; Fouda, M.E.; Eltawil, A.M.; Salama, K.N. Neural Coding in Spiking Neural Net-works: A Comparative Study for Robust Neuromorphic Systems. *Front. Neurosci.* **2021**, *15*, 638474. [CrossRef] [PubMed]
54. Dora, S.; Kasabov, N. Spiking Neural Networks for Computational Intelligence: An Overview. *Big Data Cogn. Comput.* **2021**, *5*, 67. [CrossRef]
55. Mittel, D.; Pröll, S.; Kerber, F.; Schöler, T. Mel Spectrogram Analysis for Punching Machine Operating State Classification with CNNs. In Proceedings of the 2021 26th IEEE International Conference on Emerging Technologies and Factory Automation (ETFA), Vasteras, Sweden, 7–10 September 2021; pp. 1–4.
56. Tanabe, R.; Purohit, H.; Dohi, K.; Endo, T.; Nikaido, Y.; Nakamura, T.; Kawaguchi, Y. MIMII DUE: Sound Dataset for Malfunctioning Industrial Machine Investigation and Inspection with Domain Shifts Due to Changes in Operational and Environmental Conditions. *arXiv* **2021**, arXiv:2006.05822.
57. Harada, N.; Niizumi, D.; Takeuchi, D.; Ohishi, Y.; Yasuda, M.; Saito, S. ToyADMOS2: Another Dataset of Miniature-Machine Operating Sounds for Anomalous Sound Detection under Domain Shift Conditions. *arXiv* **2021**, arXiv:2106.02369.
58. Dohi, K.; Nishida, T.; Purohit, H.; Tanabe, R.; Endo, T.; Yamamoto, M.; Nikaido, Y.; Kawaguchi, Y. MIMII DG: Sound Dataset for Malfunctioning Industrial Machine Investigation and Inspection for Domain Generalization Task. *arXiv* **2022**, arXiv:2205.13879.
59. Ubhayaratne, I.; Xiang, Y.; Pereira, M.; Rolfe, B. An Audio Signal Based Model for Condition Monitoring of Sheet Metal Stamping Process. In Proceedings of the 2015 IEEE 10th Conference on Industrial Electronics and Ap-plications (ICIEA), Auckland, New Zealand, 15–17 June 2015; pp. 1267–1272.
60. Wang, G.; Wang, Y.; Min, Y.; Lei, W. Blind Source Separation of Transformer Acoustic Signal Based on Sparse Component Analysis. *Energies* **2022**, *15*, 6017. [CrossRef]
61. Michau, G.; Fink, O. Domain Adaptation for One-Class Classification: Monitoring the Health of Critical Systems Under Limited Information. *arXiv* **2019**, arXiv:1907.09204.
62. Wang, W.; Wang, H.; Ran, Z.-Y.; He, R. Learning Robust Feature Transformation for Domain Adaptation. *Pattern Recognit.* **2021**, *114*, 107870. [CrossRef]
63. Schneider, J. Domain Transformer: Predicting Samples of Unseen, Future Domains. *arXiv* **2021**, arXiv:2106.06057.
64. Yang, Z.; Bozchalooi, I.S.; Darve, E. Anomaly Detection with Domain Adaptation. *arXiv* **2020**, arXiv:2006.03689.
65. Kumagai, A.; Iwata, T.; Fujiwara, Y. Transfer Anomaly Detection by Inferring Latent Domain Representations. In Proceedings of the 33rd Conference on Neural Information Processing Systems (NeurIPS 2019), Vancouver, BC, Canada, 8–14 December 2019; pp. 2471–2481.
66. Wang, Q.; Michau, G.; Fink, O. Domain Adaptive Transfer Learning for Fault Diagnosis. In Proceedings of the 2019 Prognostics and System Health Management Conference (PHM-Paris), Paris, France, 2–5 May 2019; pp. 279–285.
67. Yamaguchi, M.; Koizumi, Y.; Harada, N. AdaFlow: Domain-Adaptive Density Estimator with Application to Anomaly Detection and Unpaired Cross-Domain Translation. In Proceedings of the ICASSP 2019–2019 IEEE International Conference on Acoustics, Speech and Signal Processing (ICASSP), Brighton, UK, 12–17 May 2019; pp. 3647–3651.

68. Motiian, S.; Jones, Q.; Iranmanesh, S.M.; Doretto, G. Few-Shot Adversarial Domain Adaptation. In Proceedings of the 31st Conference on Neural Information Processing Systems (NIPS 2017), Long Beach, CA, USA, 4–9 December 2017; pp. 6673–6683.
69. Zhang, W.; Shen, L.; Zhang, W.; Foo, C.-S. Few-Shot Adaptation of Pre-Trained Networks for Domain Shift. *arXiv* **2022**, arXiv:2205.15234.
70. Zhou, K.; Liu, Z.; Qiao, Y.; Xiang, T.; Loy, C.C. Domain Generalization: A Survey. *IEEE Trans. Pattern Anal. Mach. Intell.* **2022**. [CrossRef]
71. Li, W.; Tsai, Y.P.; Chiu, C.L. The Experimental Study of the Expert System for Diagnosing Unbalances by ANN and Acoustic Signals. *J. Sound Vib.* **2004**, *272*, 69–83. [CrossRef]
72. Siemens. Sound Fields: Free versus Diffuse Field, Near versus Far Field. 2020. Available online: https://community.sw.siemens.com/s/article/sound-fields-free-versus-diffuse-field-near-versus-far-field (accessed on 10 December 2022).
73. *ISO 3745:2012*; Acoustics—Determination of Sound Power Levels and Sound Energy Levels of Noise Sources Using Sound Pressure—Precision Methods for Anechoic Rooms and Hemi-Anechoic Rooms. International Standards Organization: Geneva, Switzerland, 2012.
74. Nave, C.R. Inverse Square Law, Sound. HyperPhysics. 2002. Available online: http://hyperphysics.phy-astr.gsu.edu/hbase/Acoustic/invsqs.html (accessed on 9 December 2022).
75. Mesaros, A.; Heittola, T.; Virtanen, T. A Multi-Device Dataset for Urban Acoustic Scene Clas-sification. In Proceedings of the Detection and Classification of Acoustic Scenes and Events 2018 Workshop (DCASE2018), Surrey, UK, 19–20 November 2018; pp. 9–13.
76. Erić, M.M. Some Research Challenges of Acoustic Camera. In Proceedings of the 2011 19thTelecommunications Forum (TELFOR) Proceedings of Papers, Belgrade, Serbia, 22–24 November 2011; pp. 1036–1039.
77. Cariou, C.; Delverdier, O. Localizing Aircraft Noise Sources with Large Scale Acoustic An-tenna. In Proceedings of the 27th International Congress of the Aeronautical Sciences, Nice, France, 19–24 September 2010.
78. Guidati, S. Advanced Beamforming Techniques in Vehicle Acoustics. In Proceedings of the 3rd Berlin Beamforming Conference, Berlin, Germany, 21–24 February 2010.
79. Belsak, A.; Prezelj, J. Analysis of Vibrations and Noise to Determine the Condition of Gear Units. In *Advances in Vibration Analysis Research*; IntechOpen: London, UK, 2011; Available online: https://www.intechopen.com/chapters/14665 (accessed on 9 December 2022).
80. Coutable, P.; Thomas, J.-H.; Pascal, J.-C.; Eveilleau, F. Bearing Fault Detection Based on Near-Field Acoustic Holography. In Proceedings of the International Conference Surveillance 6, Compiegne, France, 25–26 October 2011.
81. Gfai Tech. The Acoustic Camera as an Innovative Tool for Fault Detection and Quality Control. 2022. Available online: https://www.gfaitech.com/applications/quality-control (accessed on 22 December 2022).
82. Chiariotti, P.; Martarelli, M.; Castellini, P. Acoustic Beamforming for Noise Source Localization—Reviews, Methodology and Applications. *Mech. Syst. Signal. Process.* **2019**, *120*, 422–448. [CrossRef]
83. Gfai Tech. How Does Delay-and-Sum Beamforming in the Time Domain Work? 2022. Available online: https://www.gfaitech.com/knowledge/faq/delay-and-sum-beamforming-in-the-time-domain (accessed on 22 December 2022).
84. Gfai Tech. How Does Delay-and-Sum Beamforming in the Frequency Domain Work? 2022. Available online: https://www.gfaitech.com/knowledge/faq/delay-and-sum-beamforming-in-the-frequency-domain (accessed on 22 December 2022).
85. Cabada, E.C.; Hamzaoui, N.; Leclere, Q.; Antoni, J. Acoustic Imaging Applied to Fault Detection in Rotating Machine. In Proceedings of the International Conference Surveillance 8, Roanne, France, 21–22 October 2015.
86. Cardenas Cabada, E.; Leclere, Q.; Antoni, J.; Hamzaoui, N. Fault Detection in Rotating Ma-chines with Beamforming: Spatial Visualization of Diagnosis Features. *Mech. Syst. Signal. Process.* **2017**, *97*, 33–43. [CrossRef]
87. Antoni, J.; Randall, R.B. The Spectral Kurtosis: Application to the Vibratory Surveillance and Diagnostics of Rotating Machines. *Mech. Syst. Signal. Process.* **2006**, *20*, 308–331. [CrossRef]
88. Antoni, J. The Spectral Kurtosis: A Useful Tool for Characterising Non-Stationary Signals. *Mech. Syst. Signal. Process.* **2006**, *20*, 282–307. [CrossRef]
89. Kosmider, M. Spectrum Correction: Acoustic Scene Classification with Mismatched Recording Devices. In Proceedings of the Interspeech 2020, Shanghai, China, 25–29 October 2020.
90. Upadhyay, N.; Karmakar, A. Speech Enhancement Using Spectral Subtraction-Type Algorithms: A Comparison and Simulation Study. *Procedia Comput. Sci.* **2015**, *54*, 574–584. [CrossRef]
91. Wildeboer, R.R.; Sammali, F.; Van Sloun, R.J.G.; Huang, Y.; Chen, P.; Bruce, M.; Rabotti, C.; Shulepov, S.; Salomon, G.; Schoot, B.C.; et al. Blind Source Separation for Clutter and Noise Suppression in Ultrasound Imaging: Review for Different Applications. *IEEE Trans. Ultrason. Ferroelectr. Freq. Control* **2020**, *67*, 1497–1512. [CrossRef] [PubMed]
92. Farahani, A.; Voghoei, S.; Rasheed, K.; Arabnia, H.R. A Brief Review of Domain Adaptation. In *Advances in Data Science and Information Engineering. Transactions on Computational Science and Computational Intelligence*; Stahlbock, R., Weiss, G.M., Abou-Nasr, M., Yang, C.Y., Arabnia, H.R., Deligiannidis, L., Eds.; Springer: Cham, Switzerland, 2021. [CrossRef]
93. Liu, X.; Yoo, C.; Xing, F.; Oh, H.; el Fakhri, G.; Kang, J.-W.; Woo, J. Deep Unsupervised Domain Adaptation: A Review of Recent Advances and Perspectives. *APSIPA Trans. Signal. Inf. Process* **2022**, *11*, e25. [CrossRef]

94. Wang, J.; Lan, C.; Liu, C.; Ouyang, Y.; Qin, T. Generalizing to Unseen Domains: A Survey on Domain Generalization. In Proceedings of the 30th International Joint Conference on Artificial Intelligence (IJCAI-21), Montreal, QC, Canada, 19–26 August 2021.
95. Li, Z.; Lv, Y.; Yuan, R.; Zhang, Q. Multi-Fault Diagnosis of Rotating Machinery via Iterative Multivariate Variational Mode Decomposition. *Meas. Sci. Technol.* **2022**, *33*, 125104. [CrossRef]

Disclaimer/Publisher's Note: The statements, opinions and data contained in all publications are solely those of the individual author(s) and contributor(s) and not of MDPI and/or the editor(s). MDPI and/or the editor(s) disclaim responsibility for any injury to people or property resulting from any ideas, methods, instructions or products referred to in the content.

eng

Article

Analysis of Grid Disturbances Caused by Massive Integration of Utility Level Solar Power Systems

Esteban A. Soto [1], Lisa B. Bosman [1,*], Ebisa Wollega [2] and Walter D. Leon-Salas [1]

[1] Purdue Polytechnic Institute, Purdue University, West Lafayette, IN 47907, USA; soto34@purdue.edu (E.A.S.); wleonsal@purdue.edu (W.D.L.-S.)
[2] Department of Engineering, Colorado State University–Pueblo, Pueblo, CO 81001, USA; ebisa.wollega@csupueblo.edu
* Correspondence: lbosman@purdue.edu

Abstract: Solar generation has increased rapidly worldwide in recent years and it is projected to continue to grow exponentially. A problem exists in that the increase in solar energy generation will increase the probability of grid disturbances. This study focuses on analyzing the grid disturbances caused by the massive integration to the transmission line of utility-scale solar energy loaded to the balancing authority high-voltage transmission lines in four regions of the United States electrical system: (1) California, (2) Southwest, (3) New England, and (4) New York. Statistical analysis of equality of means was carried out to detect changes in the energy balance and peak power. Results show that when comparing the difference between hourly net generation and demand, energy imbalance occurs in the regions with the highest solar generation: California and Southwest. No significant difference was found in any of the four regions in relation to the energy peaks. The results imply that regions with greater utility-level solar energy adoption must conduct greater energy exchanges with other regions to reduce potential disturbances to the grid. It is essential to bear in mind that as the installed solar generation capacity increases, the potential energy imbalances created in the grid increase.

Keywords: photovoltaic systems; grid disturbances; energy market; renewable energy systems

Citation: Soto, E.A.; Bosman, L.B.; Wollega, E.; Leon-Salas, W.D. Analysis of Grid Disturbances Caused by Massive Integration of Utility Level Solar Power Systems. *Eng* **2022**, *3*, 236–253. https://doi.org/10.3390/eng3020018

Academic Editor: Antonio Gil Bravo

Received: 7 March 2022
Accepted: 18 April 2022
Published: 29 April 2022

Publisher's Note: MDPI stays neutral with regard to jurisdictional claims in published maps and institutional affiliations.

Copyright: © 2022 by the authors. Licensee MDPI, Basel, Switzerland. This article is an open access article distributed under the terms and conditions of the Creative Commons Attribution (CC BY) license (https://creativecommons.org/licenses/by/4.0/).

1. Introduction

1.1. Proposed Solution

In the last decade, solar energy generation has grown enormously around the world. At the end of 2019, the installed capacity in the world of photovoltaic systems was more than 635 GW [1]. By 2050, it is predicted that solar energy will become the second-largest renewable generation source in the world after wind. In 2050, it is also predicted that the installed capacity in the world will exceed 8000 GW [2]. The increase in renewable generation, particularly solar energy, increases the probability of grid disturbances. Due to the above issues, it is necessary to quantify the potential impact of grid disturbances produced by the integration to the transmission line utility-scale solar energy loaded to the balancing authority high-voltage transmission lines (not utility-scale solar powering low voltage local distribution). In this way, electric power companies can size the problem and justify implementing solutions. This study proposes an analysis of the impact on the grid considering integrating solar energy plants to the grid in four regions of the United States electrical system: (1) California (high solar generation), (2) Southwest (moderate solar generation), (3) New England (low solar generation), and (4) New York (null solar generation). These four regions were selected because there is variation between them, ranging from the region with the most solar generation, California, to one with no utility-level solar generation (according to the Energy Information Administration [3]), New York. The impact analysis of the grid was completed using hourly increments, considering net generation changes, net generation error with demand, and energy power peaks. The findings contribute to solving the problem by quantifying the impact on the energy balance and the power peaks caused by the massive integration of solar power plants. This study seeks to answer the following research question.

Research Question: How does the penetration of solar energy utility level affect energy imbalances and the peak of power in the grid?

1.2. Current Approaches to the Problem and Gaps in Current Approaches

Among the current approaches, several technologies support the integration of solar energy into the grid to reduce potential disturbances, including technological advancements of inverters, solar grid protection plants, better forecasts of solar energy generation, net metering policies, and peer-to-peer energy trading [4].

The function of the inverters is to convert the direct current (DC) produced by the solar panels into alternating current (AC) and control its output voltage [5]. These inverter features are validated at the manufacturing stage, where the devices are subjected to loads that simulate their operation and interaction with the network [4]. However, a gap exists in that implementing inverters that allow voltage control to maintain a more stable grid can only be applicable to small-scale and small-sized photovoltaic installations, generally used in solar plants of less than 30 MW. In larger plants, grid and plant protection is required.

Other technologies include grid and plant protections, which are devices that monitor all the critical parameters of the grid and disconnect the plant from the network in the event of a disturbance [4]. Grid plant protection is a solution for today's grid. However, a gap exists in moving to a smart grid and increasing solar energy penetration, and this technology will need to be adapted for future solutions [6].

Solar forecasting is another technology used to reduce disturbances on the grid; it consists of predicting the behavior of solar generation to react quickly to any problem on the grid. Generally, solar forecasting uses historical data of generation and weather conditions [7]. However, a gap exists in that a few countries have established standards on performing solar forecasting, whereby the methodologies vary from one electrical system to another. Additionally, there are unique local factors in each region that can impact the prediction of solar generation. Also, there is still a gap in the analysis of the integration of solar energy considering the hourly operation of the electricity market at the transmission level.

Another approach is net metering policies which allow users to load excess energy production into the local grid. Some studies have shown that net metering can improve the quality of the power, which would help reduce disturbances in the grid [8,9]. However, a gap exists in that the penetration of solar energy at the residential level is still shallow in the world and most of the states of the United States. As a result of an imminent massive increase in solar generation at the utility level, net metering will not be enough to reduce disturbances in the grid. In addition, it has been reported that net metering is being phased-out [10].

Finally, Peer-to-Peer (P2P) energy trading is another approach that refers to the fact that energy prosumers can sell their electricity surplus to other users in the same grid. P2P has some benefits for the grid, reducing peak demand and improving the grid's reliability [11,12]. However, a gap exists in that P2P still does not have the technical validation, security levels, and regulations necessary to be implemented on a large scale [11]. Furthermore, there are only a few pilot projects worldwide that have not been fully validated [13,14].

The rest of this paper is organized as follows: the next section (Section 2, Background) presents the concepts relevant to the study: grid disturbances, solar energy integration, and power grid in the US, followed by Section 3, Methods, in which the data collection and data analysis are presented. Section 4, Results, presents the main findings of the study. Subsequently, in Section 5, the results are discussed and compared to other studies. Finally, Section 6, Conclusions, summarizes the article.

2. Background

2.1. Problem Indentification

It is essential to resolve connectivity issues in the grid for a smooth transition to renewable energy [15]. It is also vital to analyze new methods to correctly integrate renewable energies into the grid [16]. Here, a grid disturbance means tripping one or more elements of the grid energy system such as a generator, transmission line, or transformer, ultimately shutting down electricity access from the grid. As an example, in early 2021, Europe suffered a massive disruption on the grid, which caused concern in the energy-intensive industry in Germany [17,18]. The event occurred after a sudden drop in frequency (from 50 Hz to 0.25 Hz), which caused the European interconnected system to split in two. In some regions, sensitive machines automatically stopped working. In addition, the network operators in Italy and France had to disconnect some power plants in an effort to maintain grid stability [17]. While this event has not been linked to an increase in renewable energy, as generation from wind and solar units increases, incidents like this will become more frequent [18,19]. Experts who delivered the report on the incident mentioned that in terms of the transition to renewable energy, a more robust electrical system is required to guarantee a stable supply of power to citizens [18]. Events like this are not limited to Europe. In Australia, there have been problems with integrating renewable energies into the power grid. For example, in 2016, there was a massive blackout because of a wind energy disruption [19]. On the day of the event, Australia experienced an abnormally violent storm, which caused a decrease in the outage of a number of wind farms and the disconnection of several wind towers due to the high wind speed [19]. The increase in the generation of solar energy and its participation in the generation of electricity in the world and the United States is inevitable. With this massive increase in solar generation, it is expected that large amounts of intermittent electricity produced by renewable energies will create huge oscillations in the grid supply. Because of this, when planning the distribution of energy in the energy market, the changes in the supply and demand for energy and the different sources of energy used to meet the users' needs must be considered. However, little is known about the rate (i.e., quantity of solar energy adoption) or tipping point where the greatest potential impact could occur and its implications for the grid, particularly the potential disruptions created by increased solar power generation at the utility level. The following sections will focus on three relevant areas: First, an explanation of the United States power grid. Second a review of grid disturbances. Finally, the last subsection is dedicated to integrating solar energy into the grid.

2.2. The United States Power Grid

The United States electrical system includes power plants, transmission and distribution lines, sub-stations, and end-users. The system uses a wide variety of energy sources to produce electricity, including coal, natural gas, nuclear power, and renewable energy sources. These components form a complex electrical power grid [20]. The US electricity grid is one of the most complex and technologically demanding systems due to its interconnectivity that requires long-distance power transmission. This long-distance energy transmission has the potential for associated disturbances in the network [21]. In the lower 48 states, the US power system comprises three primary interconnected systems, operating largely independently of each other with a limited interchange of energy between them [22]. The Eastern Interconnection ranges from the east coast to the Rocky Mountains. The Western Interconnection ranges from the Rocky Mountains to the west coast. The third interconnection covers most of the state of Texas, the largest state in the United States [23].

The power grid in the United States has several challenges that are anticipated to arise in the future: first, increase in user demand; second, infrastructure renewal; third, greater risk of a cyberattack; and fourth, greater frequency of grid interruptions [24]. In recent years, the electrical grid has become more fragile and vulnerable to interruptions [25]. Although the US has developed a capacity to protect electrical infrastructure from cyber-attacks, it has been impossible to eliminate risks due to its complexity. Distributed energy resources and more micro-resources are essential to decentralize the electricity grid and thus increase supply security and supply capacity during cyber-attacks [26], yet they can also be problematic when considering grid disturbances. Also, increased solar power generation increases the likelihood of grid disturbances at balancing authority levels in the US electrical systems [27]. Finally, the massive increase in renewable energy generation will also cause interruptions due to intermittent power generation. One of the biggest problems of the massive incorporation of photovoltaic energy is the disturbances that can be created in the grid.

2.3. Grid Disturbances

The variation in the quality of power in the grid due to the presence of disturbances in the voltage wave of the network is an issue that has increased its intensity due to the energy transition. In technical terms, to maintain a stable grid, the voltage waves must be pure sine waves with a constant frequency [28]. However, the grid is generally unstable since the voltage wave exhibits disturbances, such as noise in the differential, electrical impulses, fast or slow voltage variations, flickering, harmonic distortion, and frequency variations [29]. When a massive number of distributed energy sources are connected to the grid, the grid is subjected to various electrical loads, altering the voltage. This phenomenon increases with the intermittent and often unpredictable generation produced by renewable energy, such as solar or wind [30]. The increase in renewable energy can cause severe problems to the grid, such as power fluctuations; imbalances in the grid that can increase overcurrent, thereby affecting energy efficiency; and efficiency decreases in photovoltaic systems [31]. Although solar generation is currently the third most significant renewable energy source, only after hydro and wind, few countries have implemented technical standards or contingency plans to prevent and reduce disturbances in the grid. Yet, grid disturbances are increasingly becoming a problem due to the growth in solar plants around the world [32]. One of the biggest problems resulting from the addition of photovoltaic-generated electricity into electrical systems is the disturbances caused by voltage variations [33]. Mahela et al. analyzed the behavior in common coupling points of the voltage, the current, and the power and the relationship of these variables with the disturbances in the grid [34]. The authors found that the resistive–inductive load disconnection affects the current, the voltage, and the voltage in photovoltaic systems [34]. Purnamaputra and colleagues analyzed the total distortion of solar systems connected to the grid, considering the frequency of disturbances as the primary variable [35]. They found that the disturbances in the voltage are constant at frequencies of 10 kHz and 30 kHz [35]. Most studies have focused on technical aspects and solar plant disturbances on the local grid. Yet, there is a lack of literature that analyzes the integration of photovoltaic installations considering electrical systems or subsystems as a whole.

2.4. Solar Energy Integration

In recent years, the decarbonization of the electrical system in the United States has been promoted to lead a transition to a cleaner energy matrix and reduce polluting emissions [36]. In conjunction with decarbonization, it is necessary to expand renewable resources to meet the increased demand for electricity [37]. Also, renewable energy will significantly reduce carbon emissions and greenhouse gases [38]. There is a broad technical consensus that renewable energy resources need the support of multiple critical actors (generation, transmission, and distribution) in the electrical system to be effectively integrated into the grid [39]. A series of profound changes are necessary for the electrical network architecture, including energy distribution and storage [40]. In addition, it is considered that renewable energy technologies for the production of electricity, such as solar energy, wind energy, geothermal energy, and hydroelectric energy, among others, have great potential to satisfy the demand for electric energy when implemented on a large scale [41]. The integration of wind and solar energy has a negative marginal impact on the reliability of an electrical system at low levels of electricity generation [42]. As the penetration of renewables into the grid increases, the integration challenges will increase. Following an analysis, a mismatch between supply and demand was predicted due to the overproduction of energy at certain times of the day [43]. Solar energy would be a fundamental source when integrating renewable energies into the grid. The supreme competitiveness of solar energy is reflected in the long-term forecast high-penetration levels of solar energy above that of wind and hydro [44]. The increase in solar energy generation is not without its problems. By having greater penetration of solar energy in the grid, higher peaks of generation of gas plants will occur at sunset, which is when solar generation decreases [45]. In this way, the massive integration of solar energy will cause potential disturbances in the electrical system of the United States. Thus, it is necessary to study the potential impacts of the massive integration of solar generation in the US electrical grid.

3. Methods

3.1. Study Design

This study includes a comparison of four regions of the United States electrical system (California, Southwest, New England, and New York) before and after the massive incorporation to the transmission line of utility-scale solar energy loaded to the balancing authority high voltage transmission lines (not utility-scale solar powering low voltage local distribution). Data from the Energy Information Administration (EIA) and the National Renewable Energy Laboratory (NREL) were used for the comparative analysis. The EIA data include hourly data from energy generation by source and energy demand; the NREL data include hourly generation from hypothetical solar plants. Statistical analysis was performed to compare the mean at different levels of solar energy penetration. Table 1 shows the list of the 13 regions in which the EIA data were divided and their respective codes; additionally, the percentage of solar generation of each region is shown. Four representative regions were selected for this study, the two regions with the highest utility level solar generation (California and Southwest) and the two regions with low solar generation utility levels (New England and New York). Statistical analysis compared net generation, the difference between net generation and demand, and the power peaks before and after incorporating hypothetical solar plants in the four analyzed regions.

Table 1. List of regions in the US electric power system.

N	Code	Name	% of Solar Generation
1	CAL	California	16.20%
2	SW	Southwest	3.20%
3	CAR	Carolinas	2.80%
4	NW	Northwest	1.90%
5	FLA	Florida	1.50%
6	SE	Southeast	1.10%
7	TEX	Texas	1.00%
8	MIDA	Mid-Atlantic	0.30%
9	TEN	Tennessee	0.30%
10	CENT	Central	0.20%
11	NE	New England	0.20%
12	MIDW	Midwest	0.10%
13	NY	New York	0.00%

3.2. Data Collection

The EIA is the Department of Energy's statistical and analytical agency in the United States. The EIA provides centralized and complete hourly information on the high voltage electrical power grid in 48 of the contiguous United States (Hawaii and Alaska are excluded). The data (EIA-930) are compiled by the electricity balance authorities and include forecast demand, actual demand, net generation, net interchange, and net generation from the following: coal, natural gas, nuclear energy, all petroleum derivatives, hydroelectric, solar, wind, and other energy sources [3].

For this study, hourly data from actual demand, net generation, and net generation from the following were used between 1 January 2019, and 31 December 2019 [3,46]: coal, natural gas, nuclear, hydropower, solar, wind, and other energy sources. For this study, four regions of the United States electrical system were used: (1) California, (2) New England, (3) New York, and (4) Southwest. These regions were considered to have a broad spectrum of solar generation percentages. According to the EIA data [3], California is the region that generates the most solar energy (16.2%), and New York does not have solar utility generation. Southwest and New England have 3.2% and 0.2% of solar generation between the two extremes.

NREL has decades of leadership focused on clean energy research, development, and implementation. The expertise of NREL is essential for the transition to clean energy [47]. NREL has a hypothetical photovoltaic solar plant database for renewable energy integration studies [48]. The database consists of 1 year (2006) of solar energy generation every 5 min and daily hourly forecasts of about 6000 hypothetical PV plants. For the purpose of this study, the data were aligned by hour. Solar power plant locations were determined based on the capacity expansion plan for renewable energy. The database has three data types: real power output, day-ahead forecast, and 4 h-ahead forecast. For this study, the real data power output and the day-ahead forecast were considered. The number of hypothetical solar plants considered for each analyzed region in the study is shown in Table 2.

Table 2. Number of hypothetical utility level solar plants per region included in the study [48].

Region	Number of Solar Plants
California	167
Southwest	149
New England	68
New York	62

3.3. Data Analysis

The EIA-930 data were used as input of a new hourly energy balance (see Equation (1)) [3,46]. For this new energy balance, the new solar energy plants had preference over the existing plants that use fossil fuels to cover the real total net generation. According to the EIA, each fossil fuel produces a different amount of carbon emissions. In decreasing order, the fuels that produce the most carbon dioxide are coal, diesel, gasoline, propane, and natural gas [49]. The selection criteria to replace the fossil fuels with the new solar generation were based on the amount of carbon emissions generated by each fuel. Coal plants are the first to be replaced, followed by petroleum products and natural gas plants. After fossil fuels, nuclear energy was considered along with hydropower and other sources. Solar energy was not selected to replace wind energy. Additionally, different levels of presentation of solar energy, 100%, 75%, 50%, and 25%, were considered when performing the analysis. Data analysis in this study was carried out using the statistical software RStudio Desktop version 1.3.1093 (open-source edition, RStudio, Boston, MA, USA).

Equation (1) represents the energy balance according to the EIA-930 data [3,46].

$$NG = COL + NGA + NUC + PET + WAT + SUN + WND + OTH \tag{1}$$

where

NG = *Net generation*
COL = *Net generation from Coal in* MWh
NGA = *Net generation from Natural Gas in* MWh
NUC = *Net generation from Nuclear Energy in* MWh
PET = *Net generation from Petroleum products in* MWh
WAT = *Net generation from Hydro in* MWh
SUN = *Net generation from Solar Energy in* MWh
WND = *Net generation from Wind in* MWh

OTH = Net generation from others energy sources in MWh.

Equation (2) describes net generation, including the forecast solar generation of the hypothetical solar plants minus the difference (delta) in generation from the other sources. The delta in coal generation (ΔCOL_{FH}), natural gas generation (ΔNGA_{FH}), petroleum generation (ΔPET_{FH}), nuclear generation (ΔNUC_{FH}), other energy sources (ΔOTH_{FH}), and hydro (ΔWAT_{FH}) are functions of the forecast generation of the hypothetical solar power plants (note that the FH subscript represents forecast hypothetical). The percentage decrease in coal and natural gas, petroleum, nuclear, and other energy sources is offset by the same percentage increase in solar energy.

$$NG(SUN_{FH}) = \Delta COL_{FH} + \Delta NGA_{FH} + \Delta NUC_{FH} + \Delta PET_{FH} + \Delta WAT_{FH} + SUN + WND + \Delta OTH_{FH} + SUN_{FH} \quad (2)$$

where $SUN_{FH} = Forecast\ net\ generation\ from\ hypothetical\ solar\ plants\ in$ MWh.

The following net-generation balance Equation (3) is a function of the forecast generation of the hypothetical solar plants and the actual generation. The forecast generation of hypothetical solar plants (SUN_{FH}) is replaced by the actual generation of the hypothetical solar plants (SUN_H).

$$NG(SUN_{FH},\ SUN_H\ ) = \Delta COL_{FH} + \Delta NGA_{FH} + \Delta NUC + \Delta PET + \Delta WAT + SUN + WND + \Delta OTH + SUN_H \quad (3)$$

After establishing the new energy balances Equations (1)–(3), statistical *t*-tests of two samples means, assuming equal variances, were carried out. The two-sample *t*-test is a method used to test whether the unknown population means of two groups are equal or not.

The hypotheses tested were the following:

Hypothesis 1.

$$H_0 : \mu_{NG} - \mu_{FH,H} = 0 H_1 : \mu_{FH} - \mu_{FH,H} \neq 0 \quad (4)$$

where $\mu_{NG} = mean\ of\ NG;\ \mu_{FH,H} = mean\ of\ NG(SUN_{FH},\ SUN_H)$.

The null hypothesis is that the mean of the net generation according to the EIA-930 data (denoted by μ_{NG}) is the same as the mean of the net forecast generation of the hypothetical solar plant and the current generation (represented by $\mu_{FH,H}$), and the alternative is that they are not equal.

Hypothesis 2.

$$H_0 : \mu_{NG-D} - \mu_{(FH,H)-D} = 0 H_1 : \mu_{FH-D} - \mu_{(FH,H)-D} \neq 0 \quad (5)$$

where $\mu_{NG-D} = mean\ of\ (NG - Demand);\ \mu_{(FH,H)} -D = mean\ of\ (NG(SUN_{FH},\ SUN_H\ ) - Demand)$.

Hypothesis 3.

$$H_0 : \mu_{|NG-D|} - \mu_{|(FH,H)-D|} = 0 H_1 : \mu_{|NG-D|} - \mu_{|(FH,H)-D|} > 0 \quad (6)$$

where $\mu_{|NG-D|} = mean\ of\ (NG - Demand)\ in\ absolute\ value;\ \mu_{|(FH,H)-D|} = mean\ of\ (NG(SUN_{FH},\ SUN_H\ ) - Demand)\ in\ absolute\ value$.

Hypotheses 1–3 were tested using one year-long hourly data.

Hypothesis 4.

$$H_0 : \mu_{Peak,\ NG} - \mu_{Peak,FH,H} = 0 H_1 : \mu_{Peak,\ NG} - \mu_{Peak,FH,H} > 0 \quad (7)$$

where $\mu_{Peak,NG} = mean\ of\ daily\ peak\ of\ energy\ from\ NG;\ \mu_{Peak,FH,H} = mean\ of\ daily\ peak\ of\ energy\ from\ NG(SUN_{FH},\ SUN_H\ )$.

Hypothesis 5.

$$H_0 : \mu_{Peak,\ NG-D} - \mu_{Peak,(FH,H)-D} = 0 H_1 : \mu_{Peak,\ NG-D} - \mu_{Peak,(FH,H)-D} > 0 \quad (8)$$

where $\mu_{Peak,NG-D} = mean\ of\ daily\ peak\ of\ energy\ from(NG - Demand)$; $\mu_{Peak,(FH,H-D)} = mean\ of\ daily\ peak\ of\ energy\ from\ (NG(SUN_{FH},\ SUN_H\ ) - Demand)$.

Hypotheses 4 and 5 were tested using a one-year horizon with daily data.

4. Results

4.1. Results at Different Levels of Solar Energy Penetration

Table 3 shows the before and after of adding hypothetical solar plants, considering a solar energy penetration of 100%. In Table 3, it is observed that the California region generates the highest percentage of solar energy with 16.2%, followed by the Southwest region with 3.4%. On the other hand, New England only generates 0.2% of solar energy, and the New York region does not have solar generation. This reflects the different levels of solar generation considered in the study. Also, Table 3 shows that the primary energy source in California, New England, and New York is natural gas, with 42.4%, 49.8%, and 34.7%, respectively. In the Southwest region, the main energy-generation resources are nuclear (39.3%) and natural gas (37.5%). The generation from coal and petroleum products, the largest carbon emitters, are less than 4% in California, less than 1% in New England, and less than 3% in New York. However, in the Southwest, the generation from coal reaches almost 15%, being the third most used source, and the generation from petroleum products is zero. This demonstrates that each region has a different energy matrix, with various levels of fossil fuel use.

Table 3. 100% Solar Penetration—Generation by region and sources before and after adding the hypothetical solar plants.

		COL	*NGA*	*NUC*	*PET*	*WAT*	*SUN*	*WND*	*OTH*
California	Before	3.9%	42.4%	8.3%	0.3%	17.1%	16.2%	8.5%	3.2%
	After	2.2%	32.0%	6.9%	0.2%	16.4%	30.8%	8.6%	2.9%
New England	Before	0.5%	49.8%	31.2%	0.2%	8.4%	0.2%	3.6%	6.2%
	After	0.3%	46.2%	31.2%	0.1%	8.4%	4.0%	3.6%	6.2%
New York	Before	0.0%	34.7%	33.6%	2.9%	21.9%	0.0%	3.3%	3.5%
	After	0.0%	32.2%	33.6%	1.7%	22.0%	3.7%	3.3%	3.5%
Southwest	Before	14.7%	37.5%	39.3%	0.0%	3.2%	3.2%	1.9%	0.2%
	After	8.1%	28.2%	37.3%	0.0%	3.2%	21.0%	1.9%	0.2%

Table 3 also shows the results with 100% hypothetical penetration of solar energy. Solar generation almost doubled from 16.2% to more than 30% in the California region. With this increase, solar generation becomes the second source of energy. In the case of New England, where solar generation was only 0.2%, it increased to 4%. This implies that the increase in solar generation helped reduce the consumption of fossil fuels in the New England area. The New York region had no solar generation. However, after the incorporation of 100% of the hypothetical solar plants, it reached a solar generation of 3.7%. The second area with the highest solar generation, the Southwest region, increased from 3.2% to 21%. This implies that Southwest has tremendous potential for solar generation (Arizona, New Mexico, and Southern Nevada).

Table 4 shows the solar generation in each region, and by source, with a penetration of 75%. Even with a 75% penetration of solar energy in the California area, solar energy is the second most used source. Solar energy remains the third most widely used source in the Southwest region. It is essential to mention that even with a 75% penetration, it is still possible to reduce use of fossil fuels significantly. For example, the use of coal in the

Southwest decreased from 14.7% to 8.3%. In New England, coal and oil products were cut by almost half. Solar generation becomes the third renewable energy source in the New York region, below hydropower and wind. This implies that even with a 75% penetration of solar energy, a significant reduction in the use of fossil fuels can be achieved in the analyzed regions.

Table 4. 75% solar penetration—Generation by region and sources before and after adding the hypothetical solar plants.

		COL	*NGA*	*NUC*	*PET*	*WAT*	*SUN*	*WND*	*OTH*
California	Before	3.9%	42.4%	8.3%	0.3%	17.1%	16.2%	8.5%	3.2%
	After	2.2%	34.2%	7.5%	0.2%	17.1%	27.2%	8.6%	3.0%
New England	Before	0.5%	49.8%	31.2%	0.2%	8.4%	0.2%	3.6%	6.2%
	After	0.3%	47.1%	31.2%	0.1%	8.4%	3.1%	3.6%	6.2%
New York	Before	0.0%	34.7%	33.6%	2.9%	21.9%	0.0%	3.3%	3.5%
	After	0.0%	33.0%	33.6%	1.8%	22.0%	2.8%	3.3%	3.5%
Southwest	Before	14.7%	37.5%	39.3%	0.0%	3.2%	3.2%	1.9%	0.2%
	After	8.3%	30.9%	39.0%	0.0%	3.2%	16.6%	1.9%	0.2%

Table 5 shows the results in the energy balances before and after the incorporation of hypothetical solar plants, with a 50% penetration of solar energy. While in California and the Southwest solar power generation remains the leading renewable source, in the New England and New York regions, it is the third-largest renewable source behind hydro and wind power. With 50% solar energy penetration, the New England and New York regions only have 2.1% and 1.8% solar generation, respectively. In the Southwest region, there is still a significant decrease in the generation of coal, from 14.7% to 8.7%. This implies that by reducing the penetration of solar generation to 50%, the impacts on the grid are less significant, particularly in New York and New England.

Table 5. 50% solar penetration—Generation by region and sources before and after adding the hypothetical solar plants.

		COL	*NGA*	*NUC*	*PET*	*WAT*	*SUN*	*WND*	*OTH*
California	Before	3.9%	42.4%	8.3%	0.3%	17.1%	16.2%	8.5%	3.2%
	After	2.3%	37.0%	8.1%	0.2%	17.2%	23.5%	8.6%	3.2%
New England	Before	0.5%	49.8%	31.2%	0.2%	8.4%	0.2%	3.6%	6.2%
	After	0.3%	48.0%	31.2%	0.1%	8.4%	2.1%	3.6%	6.2%
New York	Before	0.0%	34.7%	33.6%	2.9%	21.9%	0.0%	3.3%	3.5%
	After	0.0%	33.7%	33.6%	2.0%	22.0%	1.8%	3.3%	3.5%
Southwest	Before	14.7%	37.5%	39.3%	0.0%	3.2%	3.2%	1.9%	0.2%
	After	8.7%	34.5%	39.5%	0.0%	3.2%	12.1%	1.9%	0.2%

Table 6 shows the generation by region and source considering a 25% penetration of integration of photovoltaic generation plants. Solar generation in the California area reaches 19.9%, more than 10% less when compared with 100% penetration of solar energy. After hypothetical solar plant integration, solar generation in the New England and New York regions is about 1% higher than baseline. In the Southwest region, solar generation remains the main source of renewable energy, being 7.7% higher than hydropower (3.2%), and wind (1.9%). This implies that the reduction in fossil fuel use is noticeably less than the other scenarios (higher percentage of solar power generation), particularly in New York and New England, which have the smallest percentage increases in solar power generation.

Table 6. 25% solar penetration—Generation by region and sources before and after adding the hypothetical solar plants.

		COL	*NGA*	*NUC*	*PET*	*WAT*	*SUN*	*WND*	*OTH*
California	Before	3.9%	42.4%	8.3%	0.3%	17.1%	16.2%	8.5%	3.2%
	After	2.4%	40.4%	8.3%	0.2%	17.2%	19.9%	8.6%	3.2%
New England	Before	0.5%	49.8%	31.2%	0.2%	8.4%	0.2%	3.6%	6.2%
	After	0.4%	48.9%	31.2%	0.1%	8.4%	1.2%	3.6%	6.2%
New York	Before	0.0%	34.7%	33.6%	2.9%	21.9%	0.0%	3.3%	3.5%
	After	0.0%	34.2%	33.6%	2.4%	22.0%	0.9%	3.3%	3.5%
Southwest	Before	14.7%	37.5%	39.3%	0.0%	3.2%	3.2%	1.9%	0.2%
	After	10.4%	37.3%	39.4%	0.0%	3.2%	7.7%	1.9%	0.2%

4.2. Results of t-Test for Each Hypothesis

Table 7 shows the results of the equal means t-tests when comparing net generation before and after incorporating the solar plants. Table 7 details the results of Hypothesis 1 by region and solar energy penetration level. Only one p-value in the Southwest region with 100% solar energy penetration is significant ($p < 0.05$). The rest of the p-values are not significant (using $p < 0.05$), which means that the null hypothesis is not rejected and that there is no evidence to establish that the means are different. The Southwest region had the most significant increase in solar power generation, from 3.2% to 21.0%. The substantial increase in solar generation resulted in a difference in means when considering 100% penetration of solar energy. However, when reducing the percentage of solar energy penetration, there is not enough evidence to reject the null hypothesis that the means are equal. On the other hand, it is observed that in the four regions, California, New England, New York, and Southwest, as the penetration of solar energy decreases, the p-value of the statistical test increases. This implies that the equality of the means fails to be rejected as the percentage of the solar penetration decreases. Another important insight from the table is that of the regions analyzed, California produces the most energy on average, followed by Southwest, New York, and New England.

Table 7. Hypothesis 1—t-test results by region and level of solar penetration.

Region	Solar Penetration	Mean 1 [MWh]	Mean 2 [MWh]	p-Value
California	100%	22,350	22,185	0.0575 *
	75%	22,350	22,226	0.1532
	50%	22,350	22,267	0.3399
	25%	22,350	22,309	0.6326
New England	100%	10,906	10,902	0.9020
	75%	10,906	10,903	0.9264
	50%	10,906	10,904	0.9509
	25%	10,906	10,905	0.9755
New York	100%	14,955	14,953	0.9602
	75%	14,955	14,954	0.9701
	50%	14,955	14,954	0.9801
	25%	14,955	14,955	0.9900
Southwest	100%	17,769	17,652	0.0327 **
	75%	17,769	17,682	0.1069
	50%	17,769	17,711	0.2797
	25%	17,769	17,740	0.5871

* <0.1 and ** <0.05.

Table 8 shows the results of Hypothesis 2 by region and level of solar penetration. This part of the analysis shows that when comparing the difference between net generation and demand before and after incorporating the hypothetical solar plants, there is evidence to reject the null hypothesis in the California and Southwest regions. Particularly in the California area, when considering 100% and 75% solar energy penetration, the *p*-value is less than 0.05, so there is evidence to establish a significant difference between the means (net generation—demand) when comparing before and after incorporating the hypothetical solar plants. In the Southwest area, when solar penetration levels of 100%, 75%, and 50% are considered, there is evidence to reject the null hypothesis ($p < 0.05$). This implies that the means of the difference between net generation and demand before and after incorporating solar plants are different. On the other hand, in the New England and New York regions, at all levels of solar energy penetration, none of the *p*-values is significant ($p < 0.05$), which means that the null hypothesis is not rejected and that there is no evidence to establish that there is a difference between the means (difference between net generation and demand).

Table 8. Hypothesis 2—*t*-test results by region and level of solar penetration.

Region	Solar Penetration	Mean 1 [MWh]	Mean 2 [MWh]	*p*-Value
California	100%	−7828	−7993	0.0014 **
	75%	−7828	−7952	0.0166 **
	50%	−7828	−7911	0.1108
	25%	−7828	−7869	0.4264
New England	100%	−2596	−2601	0.6780
	75%	−2596	−2600	0.7540
	50%	−2596	−2598	0.8338
	25%	−2596	−2597	0.9163
New York	100%	−2832	−2834	0.8952
	75%	−2832	−2834	0.9210
	50%	−2832	−2833	0.9471
	25%	−2832	−2833	0.9735
Southwest	100%	5842	5725	<0.05 **
	75%	5842	5754	<0.05 **
	50%	5842	5783	0.0062 **
	25%	5842	5813	0.1664

** <0.05.

Table 9 shows the results when comparing the absolute error of the difference between net generation and demand before and after incorporating the solar plants into the system. The *p*-values of the analyses carried out to test Hypothesis 3 of the study indicate a significant difference (using $p < 0.05$) between the means in the California and Southwest regions. In the California region, for penetration levels of 100% and 75%, a *p*-value of less than 0.05 was found. This means that the absolute value of the difference between net generation and demand is different when comparing before and after the massive integration of solar plants. In the case of the Southwest area, a significant difference was found in the means ($p < 0.05$) for penetration levels of 100%, 75%, and 50% of solar energy. In the New England and New York regions, the *p*-values are greater than 0.05, so the null hypothesis cannot be rejected. This implies that based on the given data, there is no strong evidence to suggest that the absolute value of the difference between the net generation and demand is not different.

Additionally, it is observed that the *p*-values in each of the regions increase with increasing levels of solar energy penetration. This means that the massive integration of solar energy impacts the absolute value of the difference between net generation and demand. In other words, solar energy affects the energy interexchange between balancing authorities in absolute value.

Table 9. Hypothesis 3—*t*-test results by region and level of solar penetration.

Region	Solar Penetration	Mean 1 [MWh]	Mean 2 [MWh]	*p*-Value
California	100%	7896	8061	0.0004 **
	75%	7896	8017	0.0072 **
	50%	7896	7975	0.0553 *
	25%	7896	7934	0.2187
New England	100%	2596	2601	0.3438
	75%	2596	2600	0.3823
	50%	2596	2598	0.4212
	25%	2596	2597	0.4604
New York	100%	2832	2835	0.4441
	75%	2832	2834	0.4579
	50%	2832	2833	0.4718
	25%	2832	2833	0.4859
Southwest	100%	5842	5727	<0.05 **
	75%	5842	5754	<0.05 **
	50%	5842	5783	0.0031 **
	25%	5842	5813	0.0832 *

* <0.1 and ** <0.05.

Table 10 shows the results associated with Hypothesis 4 by region and level of solar energy penetration. Hypothesis 4 analyzes the peak energy, considering net generation as a variable before and after integrating solar plants into the system. As a result, it is recognized that the daily peak of energy does not have a statistically significant difference (at $p < 0.05$) when analyzing the net generation. However, it can be observed in the table that as the penetration of solar energy decreases, the *p*-value increases. This implies that the greater the penetration of solar energy, the greater the probability of increasing the power peaks in the balancing authorities.

Table 10. Hypothesis 4—*t*-test results by region and level of solar penetration.

Region	Solar Penetration	Mean 1 [MWh]	Mean 2 [MWh]	*p*-Value
California	100%	27,730	27,876	0.3863
	75%	27,730	27,798	0.4461
	50%	27,730	27,746	0.4876
	25%	27,730	27,719	0.4912
New England	100%	13,317	13,308	0.4791
	75%	13,317	13,306	0.4743
	50%	13,317	13,308	0.4782
	25%	13,317	13,311	0.4869
New York	100%	17,464	17,482	0.4673
	75%	17,464	17,469	0.4900
	50%	17,464	17,462	0.4966
	25%	17,464	17,461	0.4936
Southwest	100%	20,579	20,850	0.1803
	75%	20,579	20,740	0.2919
	50%	20,579	20,659	0.3915
	25%	20,579	20,602	0.4685

Finally, Table 11 shows the results when the energy peaks of the difference between net generation and demand are analyzed before and after the integration of solar plants. As a result, the null hypothesis is not rejected in the four analyzed regions, California, New England, New York, and Southwest. This means that the difference between net generation and demand before and after the massive integration of solar plants is not different. These results are explained due to the high mean value in the baseline of the difference between net generation and demand (the exchange of energy between balancing authorities). From Table 8, it can be seen that, for example, in California, there are exchanges of almost 8000 MW, in New England and New York of more than 2500 MW, and in Southwest of nearly 6000 MW on average. This implies, as the difference between net generation and demand is significantly high, the impact of solar energy generation is much smaller.

Table 11. Hypothesis 5—*t*-test results by region and level of solar penetration.

Region	Solar Penetration	Mean 1 [MWh]	Mean 2 [MWh]	*p*-Value
California	100%	−4245	−4393	0.2814
	75%	−4245	−4384	0.2907
	50%	−4245	−4360	0.3228
	25%	−4245	−4316	0.3872
New England	100%	−2084	−2007	0.0591 *
	75%	−2084	−2040	0.1809
	50%	−2084	−2065	0.3447
	25%	−2084	−2081	0.4709
New York	100%	−1933	−1854	0.1004
	75%	−1933	−1885	0.2152
	50%	−1933	−1907	0.3380
	25%	−1933	−1924	0.4380
Southwest	100%	6842	6968	0.1060
	75%	6842	6887	0.3260
	50%	6842	6837	0.4829
	25%	6842	6823	0.4245

* <0.1.

5. Discussion

This study aimed to analyze how the penetration of solar energy utility levels affects energy imbalances and the peak of power in the grid. Subsequently, in response to the research objective, the study shows the following main results. The difference between before and after (the massive integration of solar power plants) is not statistically significant ($p < 0.05$) in most of the regions analyzed when including net generation as a variable (see Table 7). The exception is the Southwest, as it had a considerable increase in solar power generation (3.2% to 21.0%). In the case of the New England and New York regions, it may be that the percentage of solar power generation is too low, 4.0% and 3.7%, respectively, to generate a significant impact on the imbalance of the systems. On the other hand, the California area, in the baseline, already had a high percentage of solar energy generation (16.2%), which increased almost the double (30.8%). For this reason, the impact when analyzing the net generation before and after the massive integration of solar plants was not significant in the California region (see Table 7). For massive integration of renewable energies in the grid, it is necessary to maintain a balance [50].

Unlike this study, a study conducted in Texas found significant differences in energy balances and peak power after the massive integration of solar plants [51]. Although this study did not find a significant impact on the energy balance in most regions, one of the solutions suggested to improve the balance in the grid was the implementation of storage systems [50]. The difference between net generation and demand is analyzed as a variable, the total amount of energy that each region must exchange with other balancing authorities. The results show that the difference when analyzing the hourly data is significant between before and after the integration of the hypothetical solar plants in the system in the regions with the highest generation of solar energy, California and Southwest. This implies an imbalance in energy exchanges with other balance authorities due to incorporating more solar generation into the systems. Having to carry out more significant energy exchanges with other external systems could cause disturbances due to voltage and frequency differences. Like this study, NREL studies have shown that cooperation between balancing areas is essential when significantly increasing renewable energy generation [52,53]. Studies have shown that by having greater solar energy penetration, more significant power fluctuations in the network are produced due to the changes in solar irradiance, which impacts the energy balance and energy peaks [54].

Additionally, when analyzing different penetration levels, it is observed that the probability of generating disturbances in the system increases when solar generation increases. However, this depends on the percentage of solar generation of the system. For instance, in the case of New England and New York, it is not significant (the portion in both regions is 4% or less). Also, when analyzing the absolute error of the difference between net generation and demand, the results support an increase in the absolute exchanges of energy with other interconnected systems in the California and Southwest regions. This difference decreases as the level of penetration of solar energy decreases. It has been found that power peaks are generated on the grid, which are produced by large solar systems [55]. However, studies have shown that peaks of power can be minimized by incorporating a large number of small solar plants instead of a few large solar plants [55]. In contrast, regarding the daily peak of energy in the net generation, no significant difference was found (at $p < 0.05$, see Table 10) in any of the four regions analyzed. Particularly, in the New England and New York regions, the lack of difference in power peaks is due to the low percentage of solar generation (4% or less). In the cases of California and Southwest, although solar generation is much higher (30.8% and 21.0%, respectively), it is still not enough to generate an impact on energy peaks. The highest energy peaks can be produced at midday, which is when a greater amount of electrical energy is generated from solar systems [33]. For this reason, future studies would benefit from an hourly peak-energy analysis. In addition to being physically adjacent systems connected to the same substation, energy imbalances increase [33]. An analysis at the substation level could generate different results on the peak of energy in the grid.

In previous studies, it was found that substituting traditional electricity generation by photovoltaic systems impacts the stability of the electrical network about energy peaks [56]. In this study, fossil fuels plants were substituted by solar generation. However, no significant impact was found in the energy peaks when the peaks of energy in the difference between net generation and demand were analyzed, that is, the exchange of energy with other areas. It was found that power peaks did not increase with the massive incorporation of solar plants into the systems. The four regions analyzed have a high dependence on energy interchanges with the other areas of the United States electrical system. As interchanges occur hourly, which in some cases may exceed 30% of the energy demanded or produced in the area, it would be necessary for the generation of solar energy to be one of the main sources of generation to create an impact on the energy peaks. Finally, like all studies, this study has limitations. Only four regions were analyzed, and not all the interconnected systems in the United States. Hypothetical data from solar plants were used, which may differ from reality. Furthermore, only four levels of solar energy penetration were considered.

6. Conclusions

6.1. Summary

This study provides findings from a comparative analysis considering the massive integration of solar power plants in four regions of the United States electrical system: (1) California, (2) Southwest, (3) New England, and (4) New York. The analysis in the network was carried out per hour, considering the changes in net generation, the net generation error with demand, and the energy power peaks. Figure 1 summarizes the percentages of generation by energy resource before and after the massive incorporation of photovoltaic plants (100% penetration). These plots show that having 100% penetration of solar generation impacts the reduction of the use of fossil fuels. The greatest changes between before and after were found in the region of California and Southwest, as seen in Figure 1, are those that generate the greatest amount of solar energy.

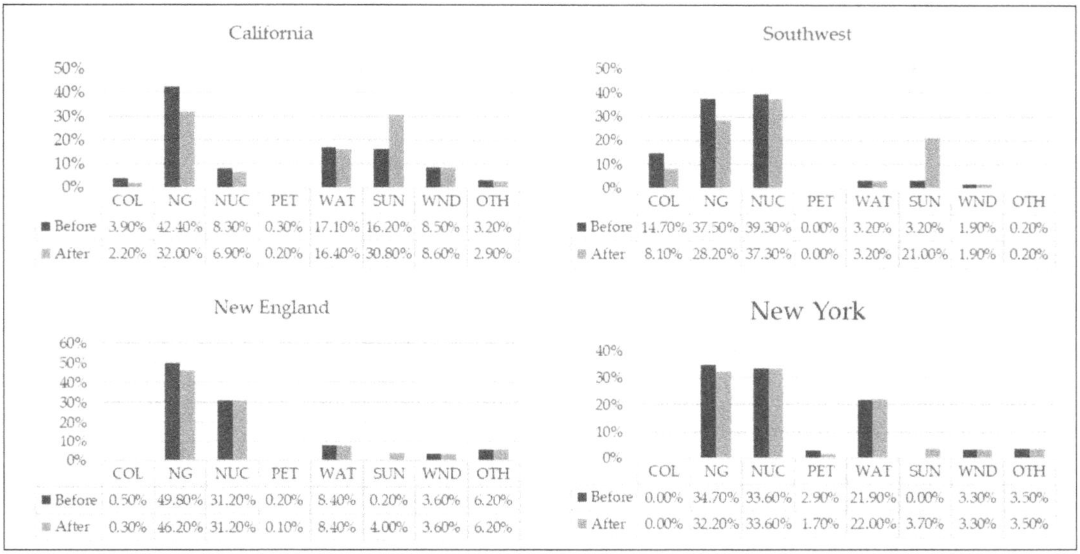

Figure 1. 100% Solar Penetration by region before and after.

The study sought to verify that there are more significant energy imbalances and higher peak power in the grid at a higher level of solar energy penetration. The findings show that when comparing the difference between net generation and demand before and after the massive integration of solar plants, there is a significant difference in the energy balance in the regions with the highest solar generation, namely California and the Southwest. Additionally, by increasing the penetration levels of solar energy, the results are intensified. On the other hand, no significant difference was found in any of the analyzed regions in relation to the energy peaks. However, the *p*-value of the statistical analysis decreases with increasing penetration levels of solar energy in each area. This indicates that when considering a higher penetration of solar energy, there is a greater probability that the energy peak will increase.

6.2. Practical Implications

The practical implications of this study are of vital importance for increasing solar energy adoption in different regions of the United States electrical system. It is essential to bear in mind that as the installed solar generation capacity increases, the potential energy imbalances that can be created in the electrical network also increase. By having a greater penetration of solar energy, the probability of generating disturbances in the grid will increase. There are several solutions to this problem. First, problems could be overcome by improving solar grid protection plants. By improving protection systems in solar plants and including them as a requirement for future solar plants installations, disturbances in the grid could be reduced. Second, energy storage systems can significantly help grid disturbances. With the development of new storage technologies, costs should decrease and make their deployment feasible on a large scale in the electrical system. A third solution is to improve and standardize the solar forecasting technologies in each of the balancing authorities in the US electrical system. By increasing the precision of solar generation and demand forecasting, the probability of events expected in the electrical grid will decrease. Fourth, the problem could be overcome with the future implementation of peer-to-peer energy trading at the utility level. However, there are still no regulations and security levels necessary for the massive implementation of P2P models.

6.3. Future Research

Future studies should be carried out in different regions and with different statistical approaches. First, different time horizons may be included. The current study analyzes data hourly; nevertheless, future studies should be extrapolated to an analysis every 10 or 30 minutes. In this way, the energy fluctuations that could cause more significant energy peaks can be detected. Second, future research can include more levels of solar energy penetration and, in this way, achieve a more detailed sensitivity analysis. In addition, studies can be extended to other regions of the United States and other countries to compare and contrast the reality under different conditions of generation and consumption of energy. Fourth, there are several ways to reduce the impacts caused by the massive integration of solar energy. For example, it would be beneficial to incorporate technologies that reduce the adverse impacts on the electrical grid due to solar energy integration. Integrating energy storage in conjunction with solar plants would be an interesting scenario to assess the real impact of energy storage systems in reducing grid disturbances.

Author Contributions: Conceptualization, E.A.S.; methodology, E.A.S.; software, E.A.S.; validation, E.A.S.; formal analysis, E.A.S.; investigation, E.A.S.; resources, E.A.S.; data curation, E.A.S.; writing—original draft preparation, E.A.S.; writing—review and editing, L.B.B., E.W., W.D.L.-S.; visualization, E.A.S.; supervision, L.B.B.; project administration, L.B.B.; funding acquisition, E.A.S. All authors have read and agreed to the published version of the manuscript.

Funding: This work was funded by the National Agency for Research and Development (ANID)/ Scholarship Program/DOCTORADO FULBRIGHT BIO/2015-*56150019*.

Conflicts of Interest: The authors declare no conflict of interest.

Abbreviations

CAL	California
SW	Southwest
NE	New England
NY	New York
NG	Net generation
FH	Forecast hypothetical
COL	Net generation from coal
NGA	Net generation from natural gas
NUC	Net generation from nuclear
PET	Net generation from petroleum products
WAT	Net generation from hydro
SUN	Net generation from solar energy
WND	Net generation from wind
OTH	Net generation from other energy sources

References

1. Jäger-Waldau, A. Snapshot of photovoltaics—February 2020. *Energies* **2020**, *13*, 930. [CrossRef]
2. IRENA. *Future of Solar Photovoltaic—Deployment, Investment, Technology, Grid Integration and Socio-Economic Aspects (A Global Energy Transformation Paper)*; International Renewable Energy Agency: Abu Dhabi, United Arab Emirates, 2019.
3. EIA. About the EIA-930 Data. Available online: https://www.eia.gov/beta/electricity/gridmonitor/about (accessed on 20 January 2020).
4. Nwaigwe, K.; Mutabilwa, P.; Dintwa, E. An overview of solar power (PV systems) integration into electricity grids. *Mater. Sci. Energy Technol.* **2019**, *2*, 629–633. [CrossRef]
5. Wenham, S.R.; Green, M.A.; Watt, M.E.; Corkish, R.; Sproul, A. *Applied Photovoltaics*; Routledge: London, UK, 2013.
6. Jafari, M.; Olowu, T.O.; Sarwat, A.I.; Rahman, M.A. Study of smart grid protection challenges with high photovoltaic penetration. In Proceedings of the 2019 North American Power Symposium (NAPS), Wichita, KS, USA, 13–15 October 2019; pp. 1–6.

7. Li, B.; Zhang, J. A review on the integration of probabilistic solar forecasting in power systems. *Sol. Energy* **2020**, *210*, 68–86. [CrossRef]
8. Thakur, J.; Chakraborty, B. Smart net metering models for smart grid in India. In Proceedings of the 2015 International Conference on Renewable Energy Research and Applications (ICRERA), Palermo, Italy, 22–25 November 2015; pp. 333–338.
9. Bedi, H.S.; Singh, N.; Singh, M. A technical review on solar-net metering. In Proceedings of the 2016 7th India International Conference on Power Electronics (IICPE), Patiala, India, 17–19 November 2016; pp. 1–5.
10. REN21. *Renewables 2020 Global Status Report*; REN21 Secretariat: Paris, France, 2020.
11. Soto, E.A.; Bosman, L.B.; Wollega, E.; Leon-Salas, W.D. Peer-to-peer energy trading: A review of the literature. *Appl. Energy* **2020**, *283*, 116268. [CrossRef]
12. Zhang, C.; Wu, J.; Zhou, Y.; Cheng, M.; Long, C. Peer-to-Peer energy trading in a Microgrid. *Appl. Energy* **2018**, *220*, 1–12. [CrossRef]
13. Zhang, C.; Wu, J.; Long, C.; Cheng, M. Review of existing peer-to-peer energy trading projects. *Energy Procedia* **2017**, *105*, 2563–2568. [CrossRef]
14. Park, C.; Yong, T. Comparative review and discussion on P2P electricity trading. *Energy Procedia* **2017**, *128*, 3–9. [CrossRef]
15. Ali, M.A.S. LMI-Based State Feedback Control Structure for Resolving Grid Connectivity Issues in DFIG-Based WT Systems. *Eng* **2021**, *2*, 562–591. [CrossRef]
16. Nikolaidis, P.; Poullikkas, A. Evolutionary Priority-Based Dynamic Programming for the Adaptive Integration of Intermittent Distributed Energy Resources in Low-Inertia Power Systems. *Eng* **2021**, *2*, 643–660. [CrossRef]
17. Appunn, K. European Power Grid Disturbance Has German Energy Intensive Industry Worried. Available online: https://www.cleanenergywire.org/news/european-power-grid-disturbance-has-german-energy-industry-worried (accessed on 20 September 2021).
18. ICS Investigation Expert Panel. *Continental Europe Synchronous Area Separation on 08 January 2021—Final Report*; The Expert Panel on the Separation of the Continental Europe Synchronous Area of 08 January 2021, 2021.
19. Starn, J.; Parkin, B.; Vilcu, I. The Day Europe's Power Grid Came Close to a Massive Blackout. Available online: https://www.bloomberg.com/news/articles/2021-01-27/green-shift-brings-blackout-risk-to-world-s-biggest-power-grid (accessed on 3 January 2022).
20. EPA. About the U.S. Electricity System and Its Impact on the Environment. Available online: https://www.epa.gov/energy/about-us-electricity-system-and-its-impact-environment#about (accessed on 12 December 2021).
21. Kinney, R.; Crucitti, P.; Albert, R.; Latora, V. Modeling cascading failures in the North American power grid. *Eur. Phys. J. B-Condens. Matter Complex Syst.* **2005**, *46*, 101–107. [CrossRef]
22. EIA. U.S. Electric System Is Made Up of Interconnections and Balancing Authorities. Available online: https://www.eia.gov/todayinenergy/detail.php?id=27152 (accessed on 12 December 2021).
23. EIA. Electricity Explained—How Electricity Is Delivered to Consumers. Available online: https://www.eia.gov/energyexplained/electricity/delivery-to-consumers.php (accessed on 13 December 2021).
24. Barrett, J.M. *Challenges and Requirements for Tomorrow's Electrical Power Grid*; Lexington Institute: Arlington, VA, USA, 2016.
25. D'Agostino, D.M. *Defense Critical Infrastructure: Actions Needed to Improve the Consistency, Reliability, and Usefulness of DoD's Tier 1 Task Critical Asset List*; United States Government Accountability Office: Washington, DC, USA, 2010.
26. Sullivan, J.E.; Kamensky, D. How cyber-attacks in Ukraine show the vulnerability of the US power grid. *Electr. J.* **2017**, *30*, 30–35.
27. Soto, E.A.; Bosman, L.B.; Wollega, E. Quantification of Solar Energy Grid Disturbances in the United States. In Proceedings of the 2021 IEEE Green Technologies Conference (GreenTech), Denver, CO, USA, 7–9 April 2021; pp. 13–18.
28. Liu, L.; Chai, J.; Qi, H.; Liu, Y. Power grid disturbance analysis using frequency information at the distribution level. In Proceedings of the 2014 IEEE International Conference on Smart Grid Communications (SmartGridComm), Venice, Italy, 3–6 November 2014; pp. 523–528.
29. Mariani, E.; Murthy, S.S. *Control of Modern Integrated Power Systems*; Springer Science & Business Media: Berlin/Heidelberg, Germany, 2012.
30. Santofimia, M.J.; del Toro, X.; Roncero-Sánchez, P.; Moya, F.; Martinez, M.A.; Lopez, J.C. A qualitative agent-based approach to power quality monitoring and diagnosis. *Integr. Comput. -Aided Eng.* **2010**, *17*, 305–319. [CrossRef]
31. Al-Shetwi, A.Q.; Hannan, M.; Jern, K.P.; Mansur, M.; Mahlia, T. Grid-connected renewable energy sources: Review of the recent integration requirements and control methods. *J. Clean. Prod.* **2020**, *253*, 119831. [CrossRef]
32. Moreno-Munoz, A. *Large Scale Grid Integration of Renewable Energy Sources*; Institution of Engineering & Technology: London, UK, 2017.
33. Steffel, S.; Caroselli, P.; Dinkel, A.; Liu, J.; Sackey, R.; Vadhar, N. Integrating solar generation on the electric distribution grid. *IEEE Trans. Smart Grid* **2012**, *3*, 878–886. [CrossRef]
34. Mahela, O.P.; Ola, S.R. Impact of grid disturbances on the output of grid connected solar photovoltaic system. In Proceedings of the 2016 IEEE Students' Conference on Electrical, Electronics and Computer Science (SCEECS), Bhopal, India, 5–6 March 2016; pp. 1–6.
35. Purnamaputra, R.; Sudiarto, B. Disturbance Frequency 9–150 kHz Characteristics towards Total Demand Distortion on On-Grid Solar Panel System in the Electrical System of Kuningan Gas Station. In Proceedings of the IOP Conference Series: Earth and Environmental Science, Pangkal Pinang, Indonesia, 3–4 September 2019; p. 012011.

36. Thompson, D.J.; Schoonenberg, W.C.; Farid, A.M. A Hetero-Functional Graph Resilience Analysis of the Future American Electric Power System. *IEEE Access* **2021**, *9*, 68837–68848. [CrossRef]
37. Van der Wardt, T.J.; Farid, A.M. A hybrid dynamic system assessment methodology for multi-modal transportation-electrification. *Energies* **2017**, *10*, 653. [CrossRef]
38. Doblinger, C.; Soppe, B. Change-actors in the US electric energy system: The role of environmental groups in utility adoption and diffusion of wind power. *Energy Policy* **2013**, *61*, 274–284. [CrossRef]
39. Bloom, A.; Townsend, A.; Palchak, D.; Novacheck, J.; King, J.; Barrows, C.; Ibanez, E.; O'Connell, M.; Jordan, G.; Roberts, B. *Eastern Renewable Generation Integration Study*; National Renewable Energy Laboratory (NREL): Golden, CO, USA, 2016.
40. Lannoye, E.; Flynn, D.; O'Malley, M. Evaluation of power system flexibility. *IEEE Trans. Power Syst.* **2012**, *27*, 922–931. [CrossRef]
41. Neuhoff, K. Large-scale deployment of renewables for electricity generation. *Oxf. Rev. Econ. Policy* **2005**, *21*, 88–110. [CrossRef]
42. Steele, A.J.H.; Burnett, J.W.; Bergstrom, J.C. The impact of variable renewable energy resources on power system reliability. *Energy Policy* **2021**, *151*, 111947. [CrossRef]
43. Schaber, K.; Steinke, F.; Mühlich, P.; Hamacher, T. Parametric study of variable renewable energy integration in Europe: Advantages and costs of transmission grid extensions. *Energy Policy* **2012**, *42*, 498–508. [CrossRef]
44. Ozoegwu, C.; Mgbemene, C.; Ozor, P.A. The status of solar energy integration and policy in Nigeria. *Renew. Sustain. Energy Rev.* **2017**, *70*, 457–471. [CrossRef]
45. Badakhshan, S.; Hajibandeh, N.; Shafie-khah, M.; Catalão, J.P. Impact of solar energy on the integrated operation of electricity-gas grids. *Energy* **2019**, *183*, 844–853. [CrossRef]
46. EIA. *EIA-930 Data Users Guide and Known Issues*; U.S. Energy Information Administration: Washington, DC, USA, 2018.
47. Flores-Espino, F.; Tian, T.; Chernyakhovskiy, I.; Chernyakhovskiy, I.; Miller, M. *Competitive Electricity Market Regulation in the United States: A Primer*; National Renewable Energy Laboratory (NREL): Golden, CO, USA, 2016.
48. NREL. Solar Power Data for Integration Studies. Available online: https://www.nrel.gov/grid/solar-power-data.html (accessed on 30 March 2021).
49. EIA. Carbon Dioxide Emissions Coefficients. Available online: https://www.eia.gov/environment/emissions/co2_vol_mass.php (accessed on 16 September 2021).
50. Etxeberria, A.; Vechiu, I.; Camblong, H.; Vinassa, J.-M. Hybrid energy storage systems for renewable energy sources integration in microgrids: A review. In Proceedings of the 2010 Conference Proceedings IPEC, Singapore, 27–29 October 2010; pp. 532–537.
51. Soto, E.A.; Bosman, L.B. Grid disturbances caused by massive integration of solar systems at utility level. In Proceedings of the 2021 International Conference on Electrical, Computer and Energy Technologies (ICECET), Cape Town, South Africa, 9–10 December 2021; pp. 1–6.
52. Lew, D.; Piwko, D.; Miller, N.; Jordan, G.; Clark, K.; Freeman, L. *How Do High Levels of Wind and Solar Impact the Grid? The Western Wind and Solar Integration Study*; National Renewable Energy Laboratory (NREL): Golden, CO, USA, 2010.
53. Energy, G. *Western Wind and Solar Integration Study*; NREL/SR-550-47434; National Renewable Energy Laboratory (NREL): Golden, CO, USA, 2010.
54. Tan, Y.T.; Kirschen, D.S. Impact on the power system of a large penetration of photovoltaic generation. In Proceedings of the 2007 IEEE Power Engineering Society General Meeting, Tampa, FL, USA, 24–28 June 2007; pp. 1–8.
55. Anees, A.S. Grid integration of renewable energy sources: Challenges, issues and possible solutions. In Proceedings of the 2012 IEEE 5th India International Conference on Power Electronics (IICPE), Delhi, India, 6–8 December 2012; pp. 1–6.
56. Hoballah, A. Power system dynamic behavior with large scale solar energy integration. In Proceedings of the 2015 4th International Conference on Electric Power and Energy Conversion Systems (EPECS), Sharjah, United Arab Emirates, 24–26 November 2015; pp. 1–6.

eng

Article

Efficient Identification of Jiles–Atherton Model Parameters Using Space-Filling Designs and Genetic Algorithms

Varun Khemani *, Michael H. Azarian and Michael G. Pecht

Center for Advanced Life Cycle Engineering (CALCE), University of Maryland, College Park, MD 20740, USA
* Correspondence: vkheman@terpmail.umd.edu

Abstract: The Jiles–Atherton model is widespread in the hysteresis description of ferromagnetic, ferroelectric, magneto strictive, and piezoelectric materials. However, the determination of model parameters is not straightforward because the model involves numerical integration and the solving of ordinary differential equations, both of which are error prone. As a result, stochastic optimization techniques have been used to explore the vast ranges of these parameters in an effort to identify the parameter values that minimize the error differential between experimental and modelled hysteresis curves. Because of the time-consuming nature of these optimization techniques, this paper explores the design space of the parameters using a space-filling design. This design provides a narrower range of parameters to look at with optimization algorithms, thereby reducing the time required to identify the optimal Jiles–Atherton model parameters. This procedure can also be carried out without using expensive hysteresis measurement devices, provided the desired transformer's secondary voltage is known.

Keywords: genetic algorithm; Jiles–Atherton model; space-filling design

Citation: Khemani, V.; Azarian, M.H.; Pecht, M.G. Efficient Identification of Jiles–Atherton Model Parameters Using Space-Filling Designs and Genetic Algorithms. *Eng* **2022**, 3, 364–372. https://doi.org/10.3390/eng3030026

Academic Editor: Huanyu Cheng

Received: 25 June 2022
Accepted: 16 August 2022
Published: 18 August 2022

Publisher's Note: MDPI stays neutral with regard to jurisdictional claims in published maps and institutional affiliations.

Copyright: © 2022 by the authors. Licensee MDPI, Basel, Switzerland. This article is an open access article distributed under the terms and conditions of the Creative Commons Attribution (CC BY) license (https://creativecommons.org/licenses/by/4.0/).

1. Introduction

Hysteresis phenomena are prevalent in various technological domains, resulting in a growing interest in models of the magnetization processes. Jiles and Atherton [1] proposed a model for describing the magnetization of soft magnetic materials. From an engineering point of view, this model is attractive because of the physical interpretation of the parameters that define it and the fact that it is based on the physical insight of hysteresis. However, as noted in [2], the iterative procedure of estimating model parameters poses convergence problems. The model is also extremely sensitive to initial parameter values and hence requires physical experimentation on the material to identify the starting point. Therefore, researchers have tried a host of different techniques to reduce the sum of squared errors (SSE) between experimental and modelled hysteresis curves. Some of these attempts include implementations of global optimization techniques, for example, simulated annealing methods [3], metaheuristic techniques such as genetic algorithms [4,5], machine learning techniques such as neural networks [6], or an exhaustive search in the solution space [7].

Section 2 shows that Jiles–Atherton model parameters are clearly connected with the physical properties of magnetic materials. However, there is no definitive method to calculate the value of each Jiles–Atherton model parameter. All of the methods use optimization algorithms to minimize the objective function, which is defined as the sum of squares of differences between the experimental hysteresis curve and the hysteresis curve as a result of modeling. Unfortunately, this sum exhibits many local minima, and hence gradient optimization techniques strongly depend on the starting point.

The traditional method of estimating model parameters [1] assumed knowledge of measured slopes dH/dM on several characteristic points on the hysteresis curve. This information facilitated the development of a set of nonlinear equations, which were solved iteratively to obtain the values of model parameters using numerical Runge–Kutta-algorithm-based methods. The anhysteric magnetization equation for anisotropic materials has been solved with the Gauss–Konrod method. Optimization methods have been explored by fitting Jiles–Atherton hysteresis curves to measurement data by using techniques including nonlinear least-squares, simulated annealing [3], genetic algorithms (binary and real-coded) [4,5], levy whale optimization [8], particle swarm optimization [9], cuckoo search [10,11], the covariance matrix adaptation evolution strategy, and other differential evolution algorithms [12]. Trapanese [6] trained a neural network with the hysteresis data and corresponding Jiles–Atherton model parameters of several materials. The trained network was used to predict the unknown Jiles–Atherton model parameters for a test material.

As optimization algorithms are time-consuming, a trial-and-error approach to Jiles–Atherton model parameter determination has also been suggested. The original paper [1] provided plots of different Jiles–Atherton model parameters held constant while one of them was varied for isotropic materials, whereas Prigozy [13] provided plots for anisotropic materials.

This paper aims to reduce the solution space of the Jiles–Atherton model parameters to a smaller and more manageable set using space-filling designs. In essence, we are reducing the area of the solution space that these algorithms explore, thereby cutting down drastically on the solution space and the time required to reach an optimal solution. The solution space exploration is done efficiently with a space-filling design that is described in the following sections. Once a reduced solution space is obtained, any of the aforementioned algorithms can be used to exploit and further explore the reduced solution space in order to arrive at the optimal solution. However, we focus on the genetic algorithm because it is the most robust among the algorithms [14].

The remainder of the paper is organized as follows. Section 2 describes the Jiles–Atherton model in detail. Section 3 describes space-filling designs and its application to the Jiles–Atherton Model, and Section 4 describes the genetic algorithm for identifying the parameters of the Jiles Atherton Model. The conclusions follow in Section 5.

2. Jiles–Atherton Model

The Jiles–Atherton [1] accounts for all important features of the hysteresis curve (Figure 1)—initial magnetization curve, saturation of magnetization, coercivity, remanence/retentivity. The Jiles–Atheron model was developed for anhysteric magnetization (M_{an}) using the mean field approach [1]. The effects of magnetic domain wall pinning on defect sites are then considered to account for hysteresis. Examples of defect sites include grain inhomogeneity, for example, angles of dislocation, inhomogeneous strain regions, etc.

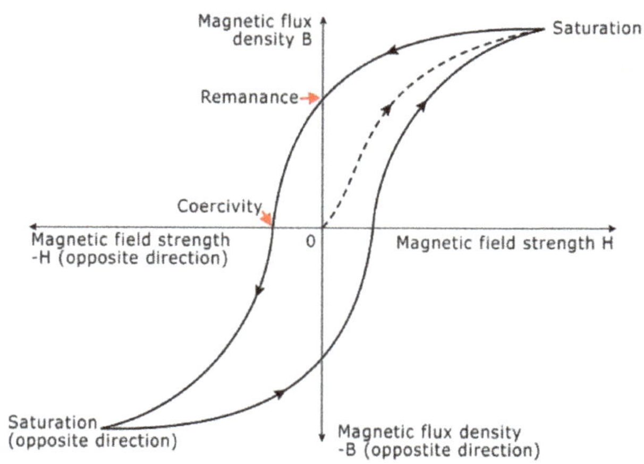

Figure 1. Typical hysteresis loop.

The following expression represents the effective magnetic field,

$$H_e = H + \alpha M_{an} \tag{1}$$

where α is a parameter representing the experimentally determined inter-domain coupling, and H is the magnetizing field. For an isotropic material, the magnetization response to this effective field is expressed as

$$M = MS * f(H_e) \tag{2}$$

where MS is the saturation magnetization of the ferromagnetic material with unit A/m, and f is a function to be defined. This magnetization expression accounts for the magnetic field response and the mean magnetic interaction with the rest of the material using the term αM, and hence is only a statistical domain distribution. Hence, it does not account for pinning and is considered the anhysteric magnetization. The modified Langevin equation is considered for f, which leads to the following expression for anhysteric magnetization,

$$M_{an} = MS\left[\coth\left[\frac{H_e}{A}\right] - \left[\frac{A}{H_e}\right]\right] \tag{3}$$

Here, A is the anhysteric behavior parameter which characterizes the shape of the anhysteric magnetization. When the work done by the field equals the magnetization energy of the sample, the domain wall displacement stops. On removal of the field, the domain wall returns to its original location. The domain wall translation causes an energy loss called the irreversible magnetization component M_{irr}, and is given by

$$\frac{dM_{\text{irr}}}{dH} = \frac{M_{\text{an}} - M}{\frac{\delta K}{\mu} - \alpha(M_{\text{an}} - M)} \tag{4}$$

Here, K is the average energy to break the pinning location, μ is the initial permeability of the material, and

$$\delta = \begin{cases} +1 \text{ for } \frac{\text{dH}}{\text{dt}} > 0 \\ -1 \text{ for } \frac{\text{dH}}{\text{dt}} < 0 \end{cases} \tag{5}$$

The total magnetization is given by

$$\frac{dM}{dH} = \frac{1}{1+C}\frac{M_{\text{an}} - M}{\frac{\delta k}{\mu} - \alpha(M_{\text{an}} - M)} + \frac{C}{1+C} \tag{6}$$

Here, C is the magnetization reversibility proportion.

Various modifications/additions were made to the Jiles–Atherton model by multiple researchers, for example:

(a) The original Jiles–Atherton model only considered isotropic materials. The anhysteric magnetization for anistropic materials is given by

$$M_{\text{ah}}^{\text{aniso}} = MS\left[\frac{\int_0^{\pi} e^{\frac{E(1)+E(2)}{2}} \sin\theta\cos\theta d\theta}{\int_0^{\pi} e^{\frac{E(1)+E(2)}{2}} \sin\theta d\theta}\right] \tag{7}$$

where E (i) is given by

$$E(i) = \frac{H_{\text{eff}}}{a}\cos\theta - \frac{K_{\text{an}}}{M_s\mu_0 a}\sin^2\phi_i \tag{8}$$

where ϕ is the angle between the applied field and easy magnetization axis, θ is the angle between the atomic magnetic moment and magnetizing field direction, and K_{an} is the magnetic anisotropy energy density with units J/m^3. In some materials such as constructional steels, isotropic and anisotropic phases can be mixed. In these cases, the total anhysteric magnetization is calculated as per Equation (9), where t is between 0 and 1.

$$M_{\text{an}} = tM_{\text{an}}^{\text{aniso}} + (1-t)M_{\text{an}}^{\text{iso}} \tag{9}$$

(b) Rotating electrical machines experience rotational fluxes, which are more complicated as compared to pulsating fluxes. For the same flux amplitude, magnetic losses due to rotating fluxes are almost 3 to 5 times that of pulsating fluxes. This necessitates a vector hysteresis like the vector Jiles–Atherton model, which retains the simplicity of the original model, which is scalar, but requires the original number of parameters in each spatial direction considered [9].

(c) In some scenarios, the induction is known before the field is applied. A classic example is a finite element model, where the vector Jiles–Atherton model is employed. For these simulations, an inverse Jiles–Atherton model presenting the magnetic induction as an independent variable [10] is used, with the main equation of this model as

$$\frac{dM}{dB} = \frac{(1-C)\frac{dM_{\text{irr}}}{dB} + \frac{C}{\mu_0}\frac{dM_{\text{an}}}{dH_e}}{1 + \mu_0(1-C)(1-\alpha)\frac{dM_{\text{irr}}}{dB} + C(1-\alpha)\frac{dM_{\text{an}}}{dH_e}} \tag{10}$$

3. Space-Filling Design

In deterministic modeling problems such as circuit SPICE simulations, the variability is negligible, so the traditional design of experiment features such as replication, randomization, and blocking to reduce experimental variability are unnecessary. Computer deterministic models are complex and can involve hundreds of variables with interactions. Space-filling designs are used to find a simpler model form of the complex computer model called a surrogate model. Space-filling designs find accurate representations of complex computer models by spreading out the design points as far apart from each other as possible while staying within the model parameter boundaries. As opposed to traditional designed experiments that have fixed levels for each factor for each simulation, space-filling designs explore the design space between two levels more thoroughly by having different levels in every simulation. This leads to a higher coverage of the parametric space, which is extremely important in the case of the Jiles–Atherton model, which has multiple local minima.

Latin hypercube designs [15] spread out the points in the design space more evenly across all possible values as compared to sphere packing designs. The parametric space is partitioned into intervals, and a sample is selected from each interval. Uniform design [16] minimizes the discrepancy between the design points (which have an empirical distribution that approximates uniformity) and a theoretical uniform distribution.

The Latin hypercube was used to set up a space-filling design that explores the design space of the four Jiles–Atherton parameters MS, A, C, and K because of its computational efficiency compared to the other types of space-filling designs. By default, the number of simulations that need to be run is 10 times the number of factors, which, in this case, means 40 simulations need to be run. First, a linear model (11) is fit where the response SCORE represents the sum of squared errors (SSE) between the actual transformer's secondary voltage and the transformer's secondary voltage simulated on PSpice. The analysis is carried out using SAS JMP software.

$$\text{SCORE} = \beta_0 + \beta_1 MS + \beta_2 A + \beta_3 C + \beta_4 K \tag{11}$$

The estimates of the regression coefficients of the linear model (11) are given in the 'Estimate' column in Table 1, whereas the column 'Std Error' gives the standard deviation of each of the parameters. The 't Ratio' column gives the t ratio metric, which tests whether the true value of the parameter is zero. It is a ratio of the estimate to its standard error, and under the null hypothesis (true value of parameter is zero), has a Student's t distribution. The 'Prob > |t|' column lists the p-value for the test where the true parameter value is zero. A p value of less than 0.05 implies that the parameter is statistically significant at the 95% confidence level. The goodness of fit of the linear model (11), as measured by the metric RSquare Adjusted, which is the coefficient of determination adjusted to account for overfitting, is 0.71 or 71%. As can be seen from Table 2, as expected, all Jiles–Atherton model parameters are statistically significant to the SCORE at the 95% confidence level. JMP has the option of a prediction profiler (Figure 2) that can be used to vary the parameter values simultaneously to bring the SCORE (SSE) to zero (Figure 3).

Table 1. Parameter estimates of linear model effects.

Term	Estimate	Std Error	t Ratio	Prob > |t|
Intercept	11,283.9	134.26	84	<0.0001
MS	5534.8	224.5	24.6	<0.0001
A	−3641.7	226	−16.1	<0.0001
C	2034	223.4	9.1	<0.0001
K	−1812.9	222.6	−8.14	<0.0001

Table 2. Parameter estimates of response surface model effects.

Term	Estimate	Std Error	t Ratio	p Value
Intercept	11,283.9	134.26	84	<0.0001
MS	5534.8	224.5	24.6	<0.0001
A	−3641.7	226	−16.1	<0.0001
C	2034	223.4	9.1	<0.0001
K	−1812.9	222.6	−8.14	<0.0001
MS * MS	2200	220	10	<0.0001
MS * A	240.813	297.3	0.81	0.6365
A * A	2698.8	224.9	12	<0.0001
MS * C	3207.198	240.6	13.33	0.0152
A * C	288.86	262.6	1.1	0.2562
C * C	12,621.42	222.6	56.7	<0.0001
MS * K	559.86	266.6	2.1	0.3768
A * K	16,936.888	221.6	76.43	<0.0001
C * K	10,124.95	225.5	44.9	<0.0001
K * K	7925.79	227.1	34.9	<0.0001

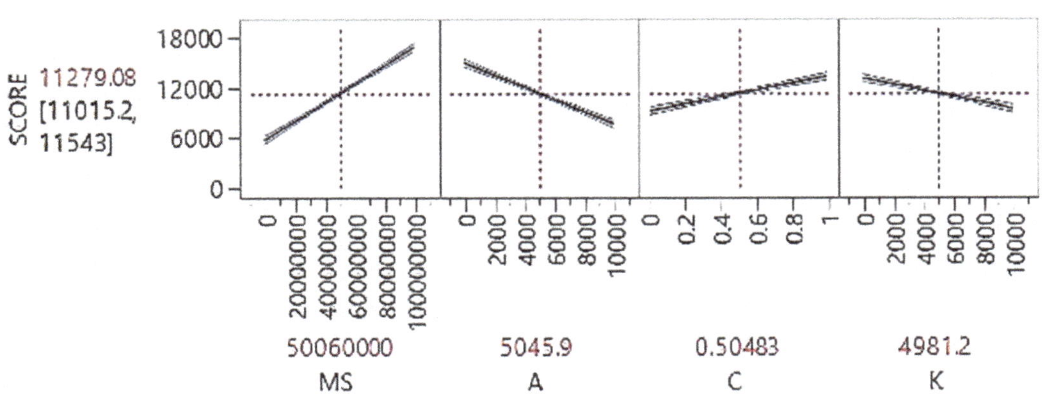

Figure 2. Prediction profiler for the linear model.

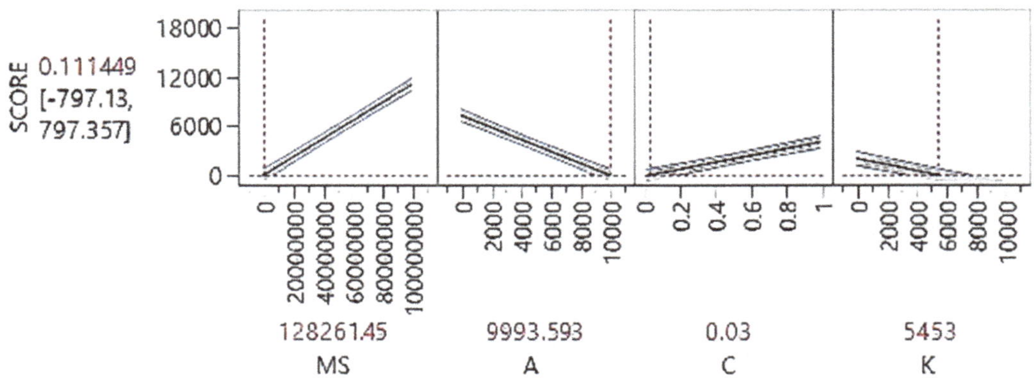

Figure 3. Prediction profiler for the optimized linear model.

Figures 2 and 3 show the individual response surfaces of the different Jiles–Atherton model parameters. The slopes of the response surfaces are the values in the 'Estimate' column of Table 1, which in turn are the regression coefficients of the linear model (11). The larger the absolute value in the 'Estimate column' of Table 1 of the Jiles Atherton model parameter, the larger its statistical significance and larger the slope of its response surface. The cross-hairs on the prediction profiler can be moved to reduce the response towards zero as much as possible. For example, reducing *MS* and increasing *A* would cause the response to move towards zero as shown in Figure 3. This results in a SCORE (SSE) value of 0.111449, but with a wide confidence interval from −797.13 to 797.357.

To improve the goodness of fit, we fit a response surface model (12) to check if there are any significant interactions or quadratic effects among the Jiles–Atherton model parameters.

$$\begin{aligned}\text{SCORE} = {} & \beta_0 + \beta_1 MS + \beta_2 A + \beta_3 C + \beta_4 K + \beta_{11} MS * MS + \beta_{12} MS * A + \beta_{13} MS * C + \beta_{14} MS * K \\ & + \beta_{22} A * A + \beta_{23} A * C + \beta_{23} A * K + \beta_{33} C * C + \beta_{34} C * K + \beta_{44} K * K\end{aligned} \tag{12}$$

As can be seen from Table 2, as expected, all the Jiles–Atherton model parameters are significant to the SSE. However, the interaction effects between *MS* and *C*, *A* and *K*, and *C* and *K* are significant, too. Additionally, the quadratic effects of all the Jiles–Atherton model parameters are significant, too. The same is evident from the quadrature of the individual response surfaces in the prediction profiler, as can be seen in Figures 4 and 5.

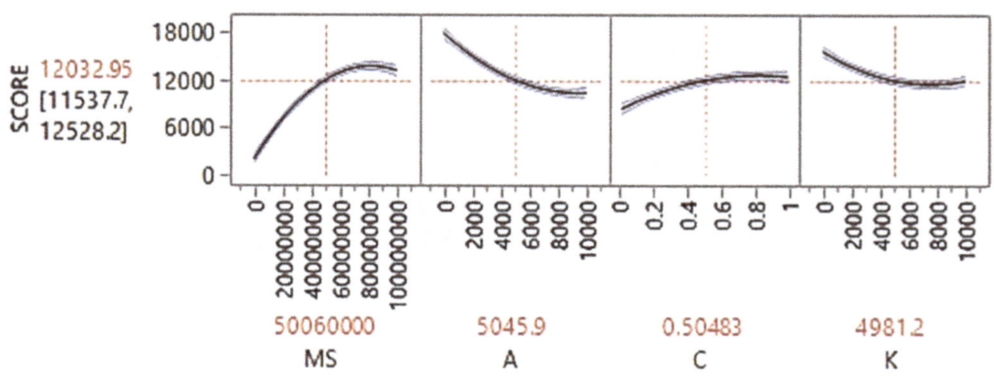

Figure 4. Prediction profiler for the response surface model.

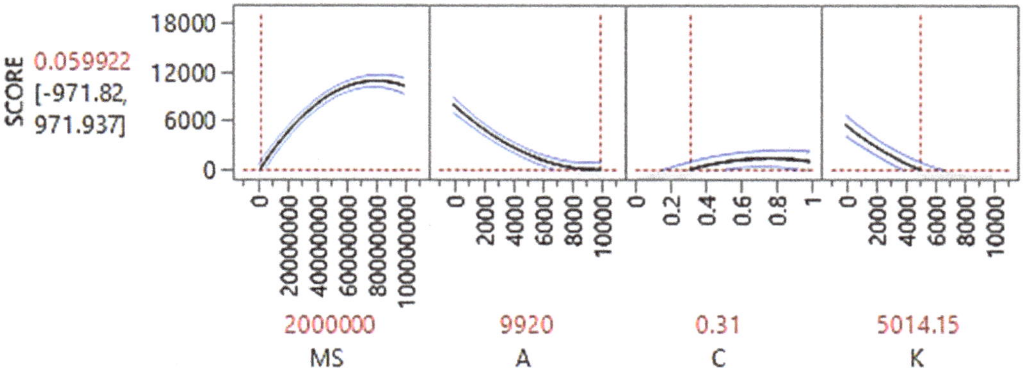

Figure 5. Prediction profiler for the optimized response surface model.

The goodness of fit of the response surface model, as measured by the metric RSquare Adjusted, is 0.84 or 84%. Since it is not a perfect fit i.e., 100%, we further explored the region near the values that give a zero response in the prediction profiler using stochastic optimization.

4. Parameter Identification of The Jiles–Atherton Model

Genetic algorithms [17] are based on the theory of evolution. A population consists of individuals with genetic material called genes, which reproduce to create the next generation. Genes from two parent individuals are combined using various crossover procedures to create offspring individuals for the next generation and so on and so forth. The selection of individuals in the parent generation to reproduce is dependent on their fitness, i.e., their evaluation of the objective function of the optimization problem. Usually, the individuals with the best fitness move on to the next generation without reproduction in order to propagate the best solution through a process called elitism. Individuals that are not elite reproduce through crossover. To introduce variety in the genes, random changes are introduced into the genes of a fraction of the individuals in the offspring generation. This is analogous to mutation in evolution and helps in avoiding local minima in the optimization of the objective function. This evolution process continues until there is no improvement in the fitness in consecutive generations or until the predefined number of generations is reached.

The genetic algorithm was implemented with 50 individuals in each generation and 50 maximum generations. A crossover probability of 90% and a mutation probability of 5% was used. The full ranges in SPICE and the reduced ranges for each variable after the space-filling design are shown in Table 3. The fitness function to be minimized corresponds to the total SSE between the actual and simulated transformer's secondary voltage. The optimal values in those ranges as found by the genetic algorithm are also shown in the table. SAS JMP Pro 15 was used for the space-filling design, and MATLAB was used to implement the genetic algorithm and communicate with the SPICE simulator (OrCAD PSpice). The code required for conducting the approach is available in the supplementary material. Due to the significant reduction (by about 85%) of the solution space of the Jiles–Atherton model parameters that the stochastic optimization algorithms have to explore, the computational time using this approach is significantly smaller than without the approach. The exact time required for the approach depends on the simulation time for the circuit of which the transformer is a part of.

Table 3. Allowable ranges and optimized values for Jiles–Atherton model parameters.

Parameter	SPICE Range	Reduced Range	Optimized Value
MS	0–1,000,000	150,000–270,000	252,037
A	0–10,000	9000–10,000	9985
C	0–1	0.25–0.5	0.31
K	0–10,000	4000–6000	5010

This modeling technique was developed to be able to simulate a large analog circuit with five transformers. The Jiles–Atherton model parameters learnt by the proposed technique were used to implement the transformers in the SPICE circuit model. The circuit output as a result of the usage of these Jiles–Atherton model parameters was verified with the circuit output of the actual circuit. This procedure confirmed the validity of the developed method. Additionally, this procedure also speaks to the advantage of this technique —estimation of the Jiles–Atherton model parameters without resorting to the need for hysteresis parameter measurement.

5. Conclusions

This paper demonstrated a novel way to identify parameters for the Jiles–Atherton model. A space-filling design was used to search the solution space optimally, which is advantageous because the Jiles–Atherton model has multiple local minima. The overall method can ascertain Jiles–Atherton model parameters without having to use expensive hysteresis measurement devices. The only prerequisite for the application of this method is that the transformer's secondary voltage (or current) waveforms must be known. This information can be measured by relatively inexpensive measurement devices.

Another advantage of this method is that it significantly reduces (by about 85%) the solution space of the Jiles–Atherton model parameters that the stochastic optimization algorithms have to explore. Additionally, by using space-filling designs, we have been able to discover previously unknown relations between Jiles–Atherton model parameters. For example, in addition to linear effects, the quadratic effects of the Jiles–Atherton model parameters are statistically significant at the 95% confidence level. We have observed that some of these interactions are also significant. A detailed simulation study is required to confirm these observations and possibly modify the Jiles–Atherton model to account for the new observations. This will be the focus of our future work.

Supplementary Materials: The following are available online at https://www.mdpi.com/article/10.3390/eng3030026/s1, Code.

Author Contributions: Conceptualization, methodology, investigation, software, writing—original draft, V.K.; writing—review and editing, M.H.A.; writing—review and editing, supervision, M.G.P. All authors have read and agreed to the published version of the manuscript.

Funding: This research received no external funding.

Institutional Review Board Statement: Not applicable.

Informed Consent Statement: Not applicable.

Conflicts of Interest: The authors declare no conflict of interest.

References

1. Jiles, D.C.; Atherton, D.L. Theory of ferromagnetic hysteresis. *J. Magn. Magn. Mater.* **1986**, *61*, 48–60. [CrossRef]
2. Szewczyk, R. Computational Problems Connected with Jiles-Atherton Model of Magnetic Hysteresis. In *Recent Advances in Automation, Robotics and Measuring Techniques*; Springer: Cham, Switzerland, 2014; pp. 275–283.
3. Bai, B.; Wang, J.; Zhu, K. Identification of the Jiles-Atherton model parameters using simulated annealing method. In Proceedings of the IEEE Electrical Machines and Systems (ICEMS), Beijing, China, 20–23 August 2011.
4. Chwastek, K.; Szczygłowski, J. Identification of a hysteresis model parameters with genetic algorithms. *Math. Comput. Simul.* **2006**, *71*, 206–211. [CrossRef]

5. Wilson, P.; Ross, J.; Brown, A. Optimizing the Jiles-Atherton model of hysteresis by a genetic algorithm. *IEEE Trans. Magn.* **2001**, *37*, 989–993. [CrossRef]
6. Chen, Z.; Yu, Y.; Wang, Y. Parameter Identification of Jile-Atherton Model based on Levy Whale Optimization Algorithm. *IEEE Access* **2022**, *10*, 66711–66721. [CrossRef]
7. Chen, L.; Feng, Y.; Li, R.; Chen, X.; Jiang, H. Jiles-Atherton Based Hysteresis Identification of Shape Memory Alloy-Actuating Compliant Mechanism via Modified Particle Swarm Optimization Algorithm. *Complexity* **2019**, *2019*, 7465461. [CrossRef]
8. Trapanese, M. Identification of parameters of the Jiles–Atherton model by neural networks. *J. Appl. Phys.* **2011**, *109*, 07D355. [CrossRef]
9. Chwastek, K.; Szczyglowski, J.; Najgebauer, M. A direct search algorithm for estimation of Jiles–Atherton hysteresis model parameters. *Mater. Sci. Eng. B* **2006**, *131*, 22–26. [CrossRef]
10. Chwastek, K.; Szczyglowski, J. Estimation methods for the Jiles-Atherton model parameters–A review. *Prz. Elektrotechniczny* **2009**, *84*, 145–147.
11. Coelho, L.S.; Guerra, F.; Batistela, N.J.; Leite, J.V. Multiobjective cuckoo search algorithm based on Duffing's oscillator applied to Jiles-Atherton vector hysteresis parameters estimation. *IEEE Trans. Magn.* **2013**, *49*, 1745–1748. [CrossRef]
12. Sadowski, N.; Batistela, N.J.; Bastos, J.P.A.; Lajoie-Mazenc, M. An inverse Jiles-Atherton model to take into account hysteresis in time-stepping finite-element calculations. *IEEE Trans. Magn.* **2002**, *38*, 797–800. [CrossRef]
13. Prigozy, S. PSPICE computer modeling of hysteresis effects. *IEEE Trans. Educ.* **1993**, *36*, 2–5. [CrossRef]
14. McKay, M.D.; Beckman, R.J.; Conover, W.J. A Comparison of Three Methods for Selecting Values of Input Variables in the Analysis of Output from a Computer Code. *Technometrics* **1979**, *21*, 239–245.
15. Fang, K.-T.; Lin, D.K.J.; Winker, P.; Zhang, Y. Uniform Design: Theory and Application. *Technometrics* **2000**, *42*, 237. [CrossRef]
16. *JMP®*; Version 12; SAS Institute Inc.: Cary, NC, USA, 2019.
17. Holland, J.H. *Adaption in Natural and Artificial Systems*; University Michigan Press: Ann Arbor, MI, USA, 1975.

eng

Article

Preliminary Siting, Operations, and Transportation Considerations for Licensing Fission Batteries in the United States

DaeHo Lee and Mihai A. Diaconeasa *

Department of Nuclear Engineering, North Carolina State University, Raleigh, NC 27695, USA
* Correspondence: madiacon@ncsu.edu

Abstract: Nuclear energy is currently in the spotlight as a future energy source all over the world amid the global warming crisis. In the current state of miniaturization, through the development of advanced reactors, such as small modular reactors (SMRs) and micro-reactors, a fission battery is inspired by the idea that nuclear energy can be used by ordinary people using the "plug-and-play" concept, such as chemical batteries. As for design requirements, fission batteries must be economical, standardized, installed, unattended, and reliable. Meanwhile, the commercialization of reactors is regulated by national bodies, such as the United States (U.S.) Nuclear Regulatory Commission (NRC). At an international level, the International Atomic Energy Agency (IAEA) oversees the safe and peaceful use of nuclear power. However, regulations currently face a significant gap in terms of their applicability to advanced non-light water reactors (non-LWRs). Therefore, this study investigates the regulatory gaps in the licensing of fission batteries concerning safety in terms of siting, autonomous operation, and transportation, and suggests response strategies to supplement them. To figure out the applicability of the current licensing framework to fission batteries, we reviewed the U.S. NRC Title 10, Code of Federal Regulations (CFR), and IAEA INSAG-12. To address siting issues, we explored the non-power reactor (NPR) approach for site restrictions and the permit-by-rule (PBR) approach for excessive time burdens. In addition, we discussed how the development of an advanced human-system interface augmented with artificial intelligence and monitored by personnel for fission batteries may enable successful exemptions from the current regulatory operation staffing requirements. Finally, we discovered that no transportation regulatory challenge exists.

Keywords: fission battery; regulation; licensing; siting; transportation; autonomous operation

Citation: Lee, D.; Diaconeasa, M.A. Preliminary Siting, Operations, and Transportation Considerations for Licensing Fission Batteries in the United States. *Eng* **2022**, 3, 373–386. https://doi.org/10.3390/eng3030027

Academic Editor: Antonio Gil Bravo

Received: 17 June 2022
Accepted: 30 August 2022
Published: 4 September 2022

Publisher's Note: MDPI stays neutral with regard to jurisdictional claims in published maps and institutional affiliations.

Copyright: © 2022 by the authors. Licensee MDPI, Basel, Switzerland. This article is an open access article distributed under the terms and conditions of the Creative Commons Attribution (CC BY) license (https://creativecommons.org/licenses/by/4.0/).

1. Introduction

1.1. Background

Nuclear energy is one of the eco-friendly and low-carbon energy sources for our world currently struggling with pollution, severe climate change, and the resulting natural disasters. Historically, the power of nuclear energy was recognized and started to be used in the 1940s, and through continuous development, it has become a major energy source, accounting for 10% of the global electricity production and 20% of the United States' (U.S.) electricity production [1].

However, historical accidents from the previous generation, large-scale nuclear power plants (NPPs), have taken away trust in nuclear energy and instilled fear. As a result, the United Kingdom, France, South Korea, and Japan declared a gradual reduction in NPPs, although some have reconsidered their position due to recent global events and climate goals. In the U.S., cost considerations are forcing the early retirement of NPPs and weakening the national nuclear supply chain [2].

In this trend, nuclear experts are conducting research on the miniaturization of NPPs to reduce huge damage in the event of an accident and the economic burden from the large capital cost per plant of the current NPPs. Accordingly, advanced reactors, such as SMRs and micro-reactors, are under development, where SMRs are expected to be commercialized in 2029 [1]. Going one step further, Idaho National Laboratory (INL) took the idea from batteries and established the fission battery initiative to make nuclear energy accessible to the public in any location with the vision of "plug and play", just like in chemical batteries, without the need for licensed operators.

Meanwhile, the commercialization of reactors is regulated by the U.S. Nuclear Regulatory Commission (NRC), and the safe and peaceful use of nuclear energy in terms of safety, security, and safeguards are supervised by the International Atomic Energy Agency (IAEA). However, current regulations focusing on current NPPs are facing significant regulatory gaps of applicability to advanced reactors. Therefore, this research investigates the regulatory challenges of the licensing of fission batteries concerning safety in terms of siting, autonomous operation, and transportation, and suggests potential response strategies to supplement it.

1.2. Fission Batteries

1.2.1. Fission Battery Attributes

Five attributes, economical, standardized, installed, unattended, and reliable, support the vision and suggest the direction for development [3]. The fission battery attributes are defined as follows:

- Economical: Fission batteries will have cost competitiveness, compared to energy sources that operate only on a specific platform, through a wide range of use and multiple deployments.
- Standardized: Fission batteries will be developed in standardized sizes, power outputs, and manufacturing processes for extensive use, and will be fully assembled in the factory to ensure low-cost and quality assurance.
- Installed: Fission batteries will be ready for deployment to implement "plug-and-play".
- Unattended: Fission batteries will be operated without the need for on-site operators based on a resilient and autonomous system.
- Reliable: Fission batteries will have high reliability during their lifetime based on a robust, resilient, fault-tolerant, and durable system to achieve fail-safe operation.

1.2.2. Fission Battery Design Features

The fission battery design is expected to follow the micro-reactor design features, mainly gas-cooled reactors with tri-structural isotropic (TRISO) fuel or heat-pipe reactors with metal, oxide, or TRISO fuels. Fission batteries will be designed to be used for less than 1 year with an output of less than 25 MWth and cheaper than 0.1 billion USD to meet midsize customer energy demands [4], such as isolated grids, military bases, and electricity supply to electric vehicles [5]. A design example of an autonomous micro-reactor currently is the eVinci design, currently under development by Westinghouse [6,7].

The most notable feature of this design is that it aims for a dramatically enhanced safety performance compared to the current large light water reactors. This is achieved by active and passive safety features for reactivity control, heat removal, and containment for redundancy and diversity [7]. For reactivity control, three strategies were designed: control drum subsystem, emergency shutdown subsystem, and passive release of hydrogen from the moderator. The design includes two strategies for decay heat removal, one using heat channels through the power conversion system and the other through the reliance on the conduction of heat through the core block to the canister with natural air convection heat removal from the outside surface of the canister to an air duct system that channels air to the surrounding environment. For the containment function, the eVinci design includes three barriers to prevent the release of radioactive material: a monolith encapsulation of fuel, a solid core block, and a canister containment subsystem.

Fuel with high-assay low-enriched uranium (HALEU), enriched from 5% to 20%, is the most typically considered fuel type for advanced reactors, such as the eVinci micro-reactor design described above [8]. TRISO fuel is one of the representative fuels using

HALEU with the uranium form of uranium oxycarbide (UCO) or uranium dioxide (UO_2). The TRISO particles are encapsulated with three layers of carbon and ceramic-based materials that prevent the release of most radioactive fission products and withstand very high temperatures without melting, ensuring good fission product retention even under extremely severe conditions, including temperatures of 1600 °C for hundreds of hours [9].

Moreover, since the thermal power level is hundreds of times smaller compared to light water reactors, the number of fission products that are produced and could potentially be released to the environment is also much smaller. This can be seen from the radiological consequence evaluations performed for the eVinci micro-reactor design configurations having thermal power output levels of 1 MWth and 14 MWth [10]. When considering three barriers, the maximum total effective dose equivalent (TEDE) to a dose receptor within 1 m was between 6.33×10^{-12} rem and 6.84×10^{-8} rem for a 1 MWth reactor, and for a 14 MWth reactor, it was between 6.33×10^{-12} rem and 9.58×10^{-7} rem in all accident scenarios [10]. This shows that the released doses are essentially zero for all practical purposes when compared to the average U.S. resident's annual background radiation from natural and anthropologic sources of approximately 6.2×10^{-1} rem [11]. Even when assuming only one barrier, the maximum total effective dose equivalent (TEDE) to a dose receptor within 1 m was between 5.11×10^{-4} rem and 5.59 rem for a 1 MWth reactor, and for a 14 MWth reactor, it was between 5.11×10^{-4} rem and 78.3 rem in all accident scenarios [10]. To put it into perspective, this is about the same order of magnitude of doses below which we have no data to establish a firm link between radiation exposure and cancer [11] and thousands smaller than the high doses 134 workers received while on the site during the early morning of 26 April 1986 after the Chernobyl accident, of which 28 were confirmed dead in the first three months due to radiation exposure [12].

In addition, the eVinci micro-reactor design supports the fission battery unattended attribute by including only one operator action of tripping the control drum system [10]. However, during emergency conditions, the reactivity control function is achieved without operator actions through the automatic emergency shutdown subsystem and the passive release of hydrogen from the moderator.

2. Regulatory Review Methodology

The regulatory review methodology used in this study is shown in Figure 1.

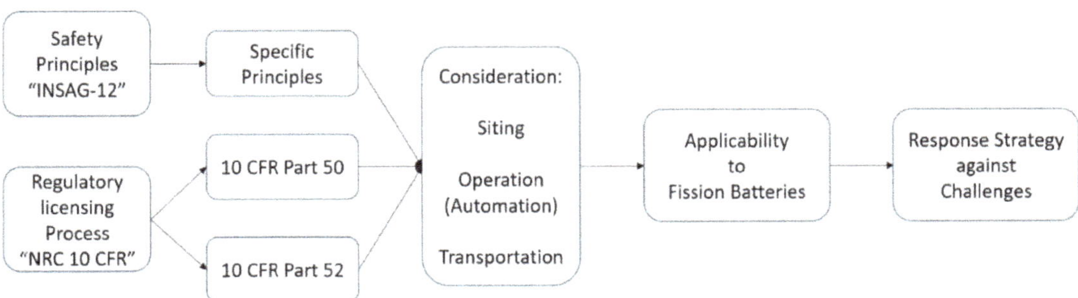

Figure 1. Regulatory review process in this study.

INSAG-12 "Basic Safety Principles for NPPs" [13] and the U.S. NRC Title 10, Code of Federal Regulations (CFR) [14] were reviewed to evaluate the specific safety principles essential for the licensing of NPPs and obtain regulatory information on the licensing process of 10 CFR Part 50 and 52 (Table 1).

The next step was to apply the current regulatory licensing framework to fission batteries in terms of siting, operation, and transportation, and figure out regulatory challenges considering the characteristics of fission batteries.

Table 1. List of U.S. NRC 10 CFR Part related to licensing commercial nuclear reactors.

Part	Title
PART 50	Domestic licensing of production and utilization facilities
PART 51	Environmental protection regulations for domestic licensing and related regulatory functions
PART 52	Licenses, certifications, and approvals for NPPs
PART 53 (Reserved)	Licensing and regulations of advanced nuclear reactors

Finally, response strategies were presented to support the licensing of fission batteries against the challenges. In this step, the non-power reactor approach was cited from the "Regulatory review of microreactors-Initial considerations" [15] and the permit-by-rule approach was referenced from "Key regulatory issues in nuclear microreactor transport and siting" [16]. In addition, an advanced human-system interface (HSI) for autonomous operation approach was derived from "Human-system interface to automatic systems: Review guidance and technical basis" [17] and "A method to select human-system interface for NPPs" [18].

3. The Current Licensing Framework for NPPs

3.1. Basic Safety Principles for NPPs (INSAG-12)

Internationally, the IAEA oversees the safe and peaceful use of nuclear energy, ensuring the protection of people and the environment from the harmful effects of radiation. INSAG-12 [13], written by the International Nuclear Safety Advisory Group (INSAG), provides three safety objectives, three fundamental safety principles, and eight specific principles. The specific principles present eight types of safety principles applied during the main design phases, from the early conceptual design phase to the decommissioning phase (Figure 2). Above all, siting and operation are recognized as top research priorities for licensing issues considering the features of fission batteries that would be used anywhere without on-site licensed operators.

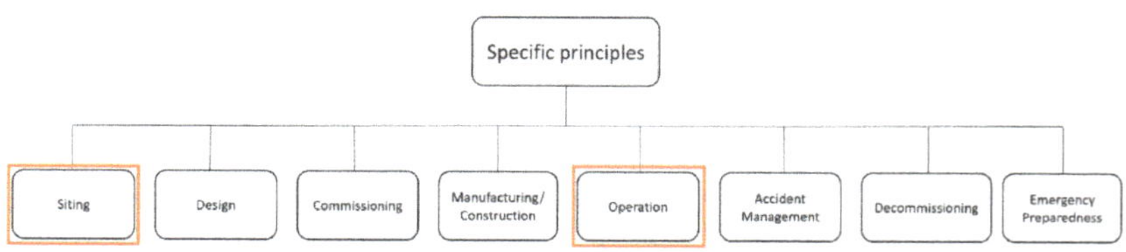

Figure 2. Structure of specific principles.

3.2. The Regulatory Licensing Framework of the U.S. NRC

Title 10 of the CFR, established by the U.S. NRC, contains the requirements that need to be met by organizations using nuclear materials or operating nuclear facilities in the U.S. Currently, there are two ways to achieve a commercial license regulated by the U.S. NRC; 10 CFR Part 50 dividing construction permit (CP) and operating license (OL) or 10 CFR Part 52 supporting a combined process of construction and operating license (COL) [19].

According to Figures 3 and 4 describing the process of Part 50, initial public hearings, an environmental report, and an NRC review of the preliminary design for a CP are required, which is a pre-requisite for obtaining an OL issued with final mandatory public hearings and final safety and environmental requirements [20]. Through this process, obtaining a license normally takes more than 10 years [21].

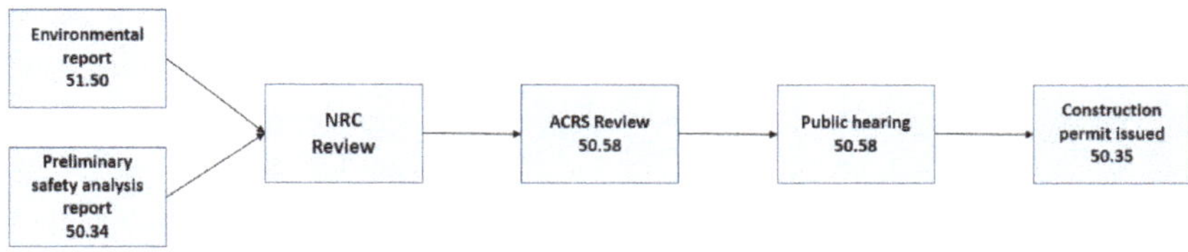

Figure 3. CP process of 10 CFR Part 50.

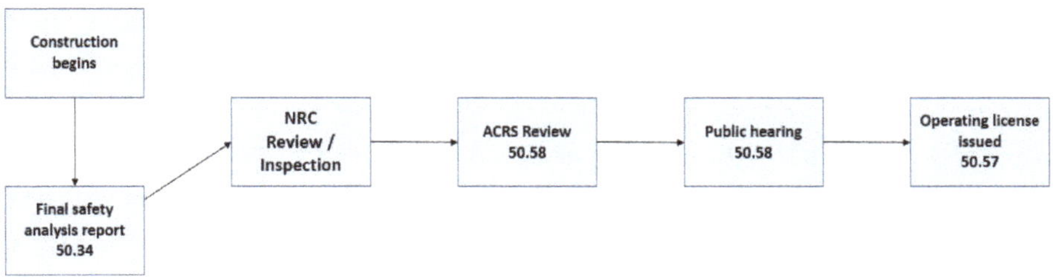

Figure 4. OL process of 10 CFR Part 50.

In order to reduce the various economic and regulatory risks that may arise during the long period of the Part 50 process, Part 52 was introduced [21], combining CP and OL steps. As seen in Figure 5, the Part 52 process is conducted with an early site permit (ESP) and a design certification (DC) together prior to issuing a COL [20]. Through this streamlined process, it shortens the period to within 10 years. Figure 6 graphically shows, for power reactors, how the two kinds of licenses can be obtained with the process of Part 50 and Part 52, that is, prototype license and license through analysis and test [22].

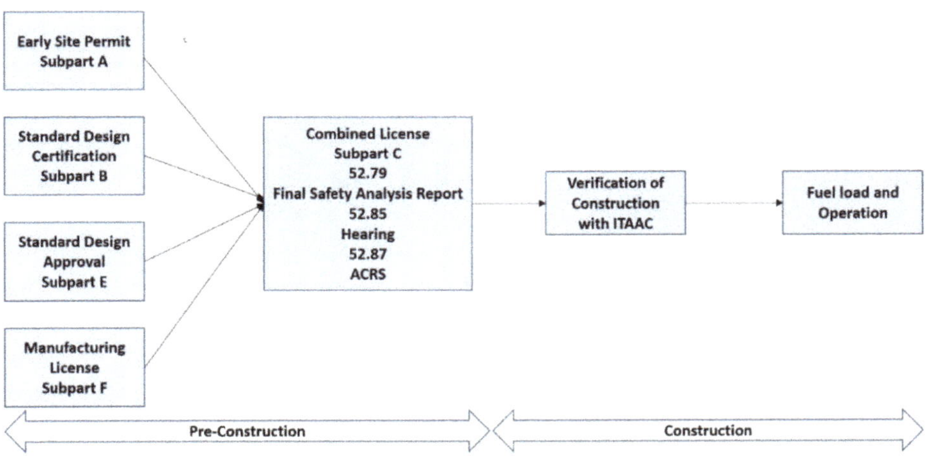

Figure 5. COL process of 10 CFR Part 52.

However, the current licensing framework of Part 50 and Part 52 does not fully consider the features of advanced reactors, and so the U.S. NRC is taking the process for 10 CFR Part 53 "Licensing and Regulations of advanced nuclear reactors" mandated by the Nuclear Energy Innovation and Modernization Act (NEIMA). Currently, the preliminary rule language consists of 10 subparts based on a risk-informed and performance-based regulatory approach [23].

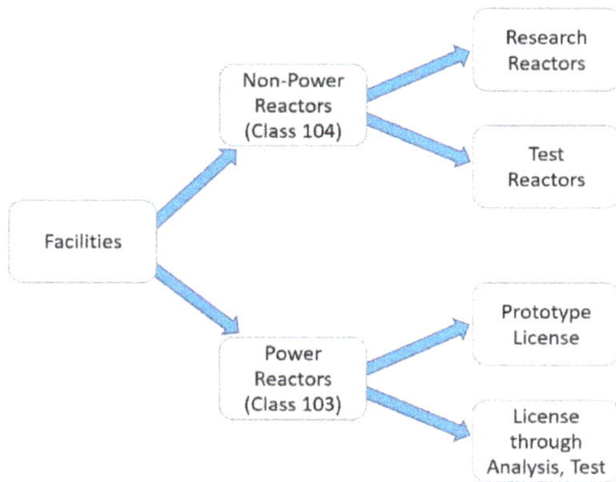

Figure 6. The U.S. NRC licensing structure.

4. Applicability of the Current U.S. Licensing Framework to Fission Batteries

4.1. Siting Regulations

Siting is used to select an appropriate location for a safe operation, including the process of analyzing natural and anthropogenic hazards, such as the radiological impact on the public and the environment [13]. Accordingly, the NRC requires an environmental report (ER) during the licensing process for CP, which takes several years for the site investigation and requires detailed site-specific information, including impacts on area populations and surrounding environmental conditions. To minimize the impacts on the site, regulations and guidance are strictly stipulated by the U.S. NRC and IAEA. Table 2 shows in detail the current regulations and regulatory guides related to siting.

Table 2. Regulations and regulatory guides related to siting.

Source	Contents
10 CFR Part 50.34 and 52.79	Radiation dose to an individual located on the boundary of the exclusion area for any 2-h period would not exceed a total effective dose of 25 rem.
10 CFR Part 100.3	Residence within the exclusion area surrounding the reactor shall normally be prohibited. A low population zone surrounding the exclusion area requires an appropriate protective measure in a serious accident.
10 CFR Part 53.53	Every site must have an exclusion area and low population zone. Reactor sites should be located away from the public.
10 CFR Part 100.21	A reactor should be located more than 1 mile away from any commercial rail line.
RG 4.7	A reactor should be located within 20 square miles and not exceeding 500 people per square mile.
RG. 1.7	A reactor should be located away from population centers of more than 25,000 people.
NUREG-0800	A reactor should be located more than 10 miles away from an airport and 5 miles from a hazardous site.

4.1.1. Applicability of Siting Regulations to Fission Batteries

Considering the expansive use of target electricity markets by military bases, isolated grids, and electricity supply to electric vehicles, fission batteries are expected to be developed to enable multi-site deployment with the concept of "plug-and-play". However, the current regulations and guidance presented above, directly contradict the vision for fission batteries, designed to be used anywhere, due to numerous prescriptive rules on the site selection.

In response to performance rules, such as doses at the exclusion areas under normal operation and emergency conditions, technology suppliers are designing enhanced passive safety systems for advanced reactors. Fission batteries are expected to have additional attributes such that any abnormal events will result in a significantly reduced source term and limit any radioactive materials release to within the site boundary or be limited to within a short distance of the exclusion area boundary [24]. Therefore, site restrictions may not be suitable for the universal use of fission batteries equipped with enhanced passive safety systems.

Additionally, the long duration, on the order of multiple years, for site evaluation in the current licensing process would interfere with the multi-site deployment and expedient site transfer required by user needs.

As a result, we conclude that the current site regulations and licensing process do not apply to the characteristics of fission batteries in terms of site restrictions and excessive time burdens for on-site evaluations.

4.1.2. Response Strategy to Site Restrictions: Non-Power Reactor Approach

The IAEA suggests an emergency planning zone (EPZ) where preparations are made to promptly shelter in place to perform environmental monitoring and to implement urgent protective actions. Table 3 shows the represented EPZ size [25]. Therefore, it can be inferred that the EPZ size of fission batteries whose output is [4] less than 25 MWth would be within 1.5 km.

Table 3. Suggested EPZ size for NPPs.

Authorized Power Level	Acceptable EPZ Size
2 MWth < Output ≤ 10 MWth	500 m
10 MWth < Output ≤ 100 MWth	0.5~5 km
100 MWth < Output ≤ 1000 MWth	5~30 km
Output > 1000 MWth	5~30 km

However, it seems to be reasonable to re-analyze the EPZ size of fission batteries that are expected to be equipped with enhanced passive safety systems, so those doses could be under the regulatory limit of 1 rem for non-power reactors [15]. Accordingly, applying the rules to non-power reactors is appropriate, and Table 4 shows the EPZ size of non-power reactors [26]. If it is applied to fission batteries, the EPZ size of the fission batteries would be reduced to approximately 400 m for power levels up to 20 MWth and even the operation boundary for power levels below 2 MWth.

Table 4. EPZ size for NPRs.

Authorized Power Level	Acceptable EPZ Size
Output ≤ 2 MWth	Operations boundary
2 MWth < Output ≤ 10 MWth	100 m
10 MWth < Output ≤ 20 MWth	400 m
20 MWth < Output ≤ 50 MWth	800 m
Output > 50 MWth	Case-by-Case

Meanwhile, SMR developers insist that the EPZ size for SMRs with an output of 300 MWth should be within 300 m or less to replace fossil fuel power plants located near big cities [27]. In Tables 3 and 4, we can see the relationships between EPZ size and power output. In Table 3, it shows that when power output increases 10 times from 10 MWth to

100 MWth, EPZ size also becomes 10 times larger, from 0.5 km to 5 km. Therefore, if we assume that the power level is proportional to the fission products that are produced and potentially released and that the relationship above could apply to advanced reactors, such as SMRs and fission batteries, we could expect that the EPZ size of fission batteries whose power output may be less than 25 MWth would be 25 m. These assumptions need to be confirmed by analysis; however, since the sudden request for a big regulatory change can be burdensome to the regulatory authorities, starting with the officially proven non-power reactor approach, it is desirable to request gradual deregulation, as the design of fission batteries is materialized, and its safety systems are verified and validated. Finally, the zero-EPZ concept should be applicable in the future [28], such that fission batteries can be widely used in highly populated areas.

4.1.3. Response Strategy to Excessive Time Burdens: Permit-By-Rule Approach

A permit-by-rule is a pre-construction permit issued by a reviewing authority that may be applied to a number of similar emissions units or sources within a designated category [29]. It is widely used for safety-guaranteed facilities, for example, on-site power generation. Sites for fuel-burning equipment are applied to permit-by-rule in Georgia State [30]. The key to applying permit-by-rule is to prove high levels of safety and reliability at the design stage. Therefore, considering the enhanced safety features of fission batteries, permit-by-rule would be a fast and reliable regulatory approach for achieving multi-site deployment and expedient site transfer by reducing the time for siting to a few days or weeks instead of several years within the current regulation [16].

The permit-by-rule approach would be conducted with the analysis of the plant parameter envelope and site parameter envelope. Major steps for it may include [16]:

1. Defining a safety design plant parameter envelope of mandatory requirements for construction and operation under permit-by-rule.
2. Defining a site parameter envelope related to plant design safety parameters.
3. Safety assessment with a plant parameter envelope and a site parameter envelope.
4. Plant parameter envelope site acceptance criteria would be created based on the above steps.

Therefore, defining a well-developed plant parameter envelope and a hypothetical site parameter envelope are essential for the permit-by-rule approach. A plant parameter envelope may be analyzed in the design process, and a site parameter envelope may be analyzed by modeling, applying simulation tools, and applying unsupervised machine learning technology for expected areas.

As a result, when fission battery design is mature enough and a high-quality enhanced safety system is verified and validated, permit-by-rule could be a reasonable approach that would sufficiently replace or complement the current years-long siting evaluation process for fission batteries that may require hundreds or thousands to be deployed simultaneously.

4.2. Operations Staffing Regulations

Operation is the key phase in the lifecycle of NPPs. Once NPPs begin operation, their safety performance depends on the reliability and capability of the facility equipment and human personnel, especially during abnormal conditions. As shown in Table 5, therefore, the composition of the control room and related staffing regulations are specified in 10 CFR Part 50 and Part 55. What stands out is that regulations prescriptively set the minimum required number of licensed operators on site during normal operation and emergencies. Even the preliminary language of 10 CFR Part 53 for advanced reactors still requires licensed human operators. As such, the operation and response to emergencies for NPPs are highly dependent on licensed human personnel.

Table 5. Regulations and regulatory guides related to operations staffing.

<table>
<tr><th>Source</th><th colspan="3">Contents</th></tr>
<tr><td>10 CFR Part 55.4</td><td colspan="3">An operator is any individual licensed to manipulate a control of a facility.</td></tr>
<tr><td>10 CFR Part 50.54 (k)</td><td colspan="3">An operator or senior operator shall be present at the controls at all times during operation.</td></tr>
<tr><td>10 CFR Part 50.54 (m)(1)</td><td colspan="3">A senior operator shall be present at the facility during initial start-up and approach to power, recovery from an unplanned or un-scheduled shut-down or significant reduction in power, and re-fueling.</td></tr>
<tr><td rowspan="5">10 CFR Part 50.54 (m)(2)(i)</td><td>Number of operating units</td><td>Position</td><td>One unit</td></tr>
<tr><td rowspan="2">None</td><td>Senior operator</td><td>1</td></tr>
<tr><td>Operator</td><td>1</td></tr>
<tr><td rowspan="2">One</td><td>Senior operator</td><td>2</td></tr>
<tr><td>Operator</td><td>2</td></tr>
<tr><td>10 CFR Part 53.80</td><td colspan="3">Each licensee must establish and implement a facility safety program (FSP) that routinely and systematically evaluates potential hazards, operating experience relate to human actions and programmatic controls affecting the safety functions.</td></tr>
<tr><td>NUREG-0654</td><td colspan="3">Addresses the minimum staffing requirements for emergencies, 10 staff on-site, and 11 additional staff within 30 min, and 15 additional staff within 60 min.</td></tr>
</table>

4.2.1. Applicability of Operations Staffing Regulations to Fission Batteries

According to the un-attended fission battery attribute, fission batteries are expected to be developed for un-attended operation through resilience and automation. Investigations in the aftermath of the Three Mile Island and Chernobyl accidents showed that human errors resulted from equipment design and human factor deficiencies [31]. Therefore, the development of automation should be attained with the advanced design of passive safety systems, simplicity of operation, and limited important human actions based on innovative un-supervised machine learning technology [3].

However, the current operations staffing regulations covering licensed operators seem to be highly dependent on personnel and do not fully capture current automation capabilities. The exemption process for control room staffing requirements shows some benefits. For example, NuScale Power successfully obtained an exemption to reduce the number of staff for its SMR light-water design, but it was not a complete elimination [22].

Nevertheless, the designer community is still expected to develop fission batteries with high automation and remote monitoring and with no operator control or at least partial control [32]. This is because the un-attended operation is the most important attribute of fission batteries, that is, aiming to enable their use by ordinary people, such as chemical batteries. Therefore, current regulations related to operators cannot be applied to un-attended operations of any advanced reactor, including fission batteries, for which a change in regulations is necessary.

4.2.2. Response Strategy to Operations Staffing Regulations: Advanced Human-System Interface

Human-system interface technology is defined as the part of the nuclear reactor through which personnel interact to perform their functions and tasks with the systems. The primary purpose of the human-system interface is to provide the operator with a means to monitor and control the nuclear reactors and to restore them to a safe state when adverse conditions occur [18], and so it is widely used at present.

In advanced human-system interface systems with improved telecommunication technologies, an off-site space equipped with a set of computer displays and input devices may replace the control rooms and make it feasible for remote monitoring and control. Moreover, the enhanced safety systems and simplified design may allow one controller to manage multiple reactors. In the current state, human-system interface technology still requires minimum human functions. However, for fission batteries equipped with un-supervised machine learning technology, un-supervised machine learning could replace human functions. Therefore, an advanced human-system interface managed by un-supervised machine learning would be the core technology required for autonomous operation of fission batteries.

Meanwhile, the advancements in automation systems and the development of computer performance have had a tremendous impact on the deployment of automation technologies and systems in many industries over the past years, such as a nearly autonomous management and control (NAMAC) system [33], where the digital twin (DT) and advanced machine learning algorithms play key roles in replacing human personnel.

Therefore, optimistic expectations on the progress of a complete remote-control system with advanced telecommunication technologies and human-system interface operated by un-supervised machine learning could make it possible for fission batteries to be exempted from current regulations related to operations staffing.

4.3. Transportation Regulations

Transportation in the nuclear industry means moving radioactive materials to the desired places. Related regulations are co-managed by 10 CFR 71 of the U.S. NRC and 49 CFR 173 of the U.S. Department of Transportation (DOT). Table 6 shows the currently regulated packaging types for the transportation of radioactive materials.

Table 6. Classification of packaging type for transportation of radioactive materials.

Packaging Type		
Industrial	Type A	Type B
Little hazardous materials (e.g., contaminated clothing)	Small quantities of radioactive materials (e.g., medical use isotopes)	Large quantities and the highest levels of radioactivity materials (e.g., used fuel)

The packaging type is determined by the quantities of radioactive materials, and each package is required to resist certain conditions. In order to verify the safety performance of each package, the U.S. NRC requires specific tests on the normal conditions of transport (NCT) and hypothetical accident conditions (HAC). Especially, tests on the HAC for Type B packaging assuming possible severe transportation accidents are specified in 10 CFR 71.73 [34]. The need for the safety performance test is to prevent the leakage of radioactive material or to control it below a prescribed regulatory dose limit as described in Table 7 under all conditions.

Table 7. Regulated radiation dose limits for transportation of radioactive materials.

Source	Contents
10 CFR Part 71.47	2 mSv/h on the external surface package 0.02 mSv/h at normally occupied space 0.1 mSv/h at any point 2 m from the outer lateral surface of vehicle
10 CFR Part 71.51 [1]	10 mSv/h at 1 m from the external surface of the package on the HAC

[1] Additional requirements for Type B packaging.

Therefore, even in the most severe transportation accidents, packages should be able to maintain their containment function and prevent the release of radioactive materials under regulatory the dose limit.

Applicability of Transportation Regulations to Fission Batteries

According to the installed fission battery attribute, fission batteries are expected to be installed at the factory, delivered to multiple users, and decommissioned with the fuel loaded. Considering this design goal, three transportation phases could be analyzed:

- Transporting fresh fuel to the manufacturing factory.
- Deploying new fission batteries to users with fresh fuel.
- Transferring the fission battery's location after an operation with a used fuel either to a different user or for decommissioning.

These new transportation phases for mobile reactors equipped with fuel pose technical and regulatory challenges. The third phase is especially critical for fission batteries since the used fuel will contain highly radioactive fission products.

When it comes to technical challenges, the key for fission battery transportation is to achieve complete safety reliability for radiation shielding, decay heat removal, and maintaining subcriticality, and capability for preventing the release of radioactive materials. However, when fission batteries are fully developed and commercialized, the technical challenges are expected to be addressed. Therefore, this study assumed that fission batteries would have adequate safety systems to meet the technical challenges associated with fission battery transportation.

For regulatory considerations, the current regulations for transporting radioactive materials stipulate the packaging type and the dose limit for packages. Currently, one of the most hazardous radioactive materials is used or spent nuclear fuel, which requires the use of Type B packaging as seen in Table 6.

Moreover, since fission battery transportation will include used fuel during the third transportation phase, it seems reasonable to assume the designers may try to meet the Type B packaging requirements, which is the safest and most robust cask in use nowadays. Accordingly, if fission battery designs meet the Type B packaging requirements at a reasonable cost and, implicitly, meet the regulatory dose limits, safe fission battery transportation is possible under current regulations without any foreseeable burdens due to the performance-based nature of transportation regulations [35].

5. Discussion on the Applicability of the Current U.S. Licensing Framework to Fission Batteries

This research indicates that the current regulatory framework is facing considerable challenges in terms of its applicability to fission batteries for siting and operations staffing; however, under certain design constraints for the fission batteries, it is feasible to meet the current transportation rules.

For siting regulations, strict site restrictions and excessive analysis and review-time burdens were presented as a limitation for the deployment of fission batteries. Thus, suggesting that the non-power reactors approach to resolve siting regulatory limitations. However, before applying the results of the non-power reactor approach, the fundamental difference should be considered, that is, a fission battery is a power reactor, unlike non-power reactors. Nevertheless, since the safety features of fission batteries would be more adequate than non-power reactors, the non-power reactors approach may be reasonable considering that site inspection is focused on the safety aspects. Next, we proposed the permit-by-rule approach as a countermeasure to excessive analysis and review-time burdens. Similar to the non-power reactors approach, the permit-by-rule approach requires a reliable safety performance. Therefore, if regulatory authorities accept a permit-by-rule approach for fission batteries, multi-site deployment could be achievable.

In the case of operations staffing regulations, autonomous operation is an essential feature for fission batteries, thus, fission battery developers are working on applying un-supervised machine learning technologies to fission batteries to replace licensed operators. Therefore, for fission batteries controlled by un-supervised machine learning, human personnel will have the role of only monitoring the un-supervised machine learning control systems as necessary. As a result, an advanced human-system interface was suggested to support the role of an off-site un-supervised machine learning specialist for the successful exemption from regulations on control room operations staffing. If an advanced human-system interface is applied to fission batteries controlled by un-supervised machine learning, the regulatory authorities will need to change their rules to accommodate the new role of human personnel from licensed operators controlling the fission batteries directly to the certified personnel monitoring un-supervised machine learning control systems.

Finally, for transportation regulations, we have identified no regulatory gaps for the licensing of fission batteries, but it is necessary to consider whether to use Type B packaging requirements for fission batteries, which have their own safety features unlike used nuclear fuel, which may be too conservative. Therefore, using Type B packaging may be un-economical for nuclear vendors. As a result, it is necessary to carry out dedicated studies on the development of specific packaging for fission batteries and figure out which packaging is more reasonable to use in terms of safety and economic production of fission batteries.

6. Conclusions

In the development of innovative technologies, numerous regulatory barriers exist in all industries. Now that the transformation of nuclear reactors is taking place, fission batteries are at the peak of innovation, and accordingly, many challenges are expected. For this reason, this research was aimed at identifying possible regulatory challenges for the licensing of fission batteries and suggesting countermeasures to support their successful development and licensing. Among the many licensing topics, siting, operations staffing, and transportation were intensively studied, considering the five attributes of fission batteries.

For siting challenges, strict site restrictions to prevent impact on the public and several years of site inspections were presented. Considering the expansive use of fission batteries equipped with enhanced safety systems, the non-power reactor approach to relax site limitations and the permit-by-rule approach to shorten review time periods were proposed for site inspections to support simultaneous multi-site deployment.

Regarding operations staffing issues, un-attended operation is the core attribute of fission batteries. Currently, nuclear reactors are highly dependent on control room operators. However, fission batteries are envisioned to be operated by un-supervised machine learning control systems without the need for on-site staff. Therefore, the development of an advanced human-system interface supporting remote monitoring for fission batteries controlled by un-supervised machine learning may enable successful exemptions from the current regulatory requirements.

In terms of transportation regulations, fission batteries have the characteristic of needing to be transportable without removing the used fuel after an operation. If the fission battery designs meet the regulatory dose limits by adopting the certified Type B packaging requirements used for the transportation of used fuel from current reactors, fission battery transportation is achievable within the current regulatory framework.

Overall, the development of fission batteries in the U.S. is facing other regulatory challenges than the three discussed above. However, the status of present regulations should not hinder the development of innovative technologies in the future. Therefore, the necessary regulatory changes and the development of fission batteries should evolve in parallel through an open regulatory engagement process for the safe and practical deployment of advanced nuclear energy.

Author Contributions: Conceptualization, M.A.D.; Formal analysis, D.L.; Methodology, D.L. and M.A.D.; Supervision, M.A.D.; Validation, M.A.D.; Writing—original draft, D.L. and M.A.D. All authors have read and agreed to the published version of the manuscript.

Funding: This research was performed as part of DaeHo Lee's Master of Science degree at North Carolina State University in the Department of Nuclear Engineering supported by the Republic of Korea Army.

Data Availability Statement: Not applicable.

Conflicts of Interest: The authors declare no conflict of interest.

References

1. Baranwal, R. *Office of Nuclear Energy: Strategic Vision*; DOE, Office of Nuclear Energy: Washington, DC, USA, 2021. Available online: https://www.energy.gov/ne/downloads/office-nuclear-energy-strategic-vision (accessed on 15 January 2022).
2. Birol, F. *World Energy Outlook 2018*; IEA: Paris, France, 2018; p. 661.
3. Agarwal, V.; Ballout, Y.A.; Gehin, J.C. *Fission Battery Initiative*; Idaho National Laboratory: Idaho Falls, ID, USA, 2021; p. 24.
4. Forsberg, C. Co-siting Fission Battery Refurbishment, Nuclear Hydrogen and Fuel-Cycle Facilities with Waste Disposal Sites. In Proceedings of the 2021 ANS Winter Meeting, Washington, DC, USA, 30 November–3 December 2021.
5. Christensen, J.; Avramova, M.; Wang, D.; Palmtag, S.; Diaconeasa, M.A.; Hou, J. *Safety & Licensing Workshop*; Idaho National Laboratory: Idaho Falls, ID, USA, 2021; p. 147.
6. Westinghouse Nuclear > Energy Systems > eVinci^{TM} Micro-Reactor. Available online: https://www.westinghousenuclear.com/energy-systems/evinci-micro-reactor (accessed on 3 August 2022).
7. Westinghouse Global Technology Office, Westinghouse eVinci Micro Reactor Factsheet. Westinghouse Electric Company, Oct. 2017. Available online: https://www.westinghousenuclear.com/Portals/0/new%20plants/evincitm/GTO-0001%20eVinci%20flysheet.pdf (accessed on 15 January 2022).
8. Arafat, Y. Westinghouse eVinci^{TM} Micro-Reactor Program. Idaho National Laboratory, Jun. 2019. Available online: https://gain.inl.gov/SiteAssets/Micro-ReactorWorkshopPresentations/Presentations/13-Arafat-GAINMicro-reactorWorkshop_June2019_Westinghouse_RSB.pdf (accessed on 15 January 2022).
9. Demkowicz, P.A.; Liu, B.; Hunn, J.D. Coated particle fuel: Historical perspectives and current progress. *J. Nucl. Mater.* **2018**, *515*, 434–450. [CrossRef]
10. Maioli, A.; Detar, H. *Westinghouse eVinci Micro-Reactor Licensing Modernization Project Demonstration*; Southern Company: Atlanta, GA, USA, 2019.
11. Backgrounder on Biological Effects of Radiation. NRC Web. Available online: https://www.nrc.gov/reading-rm/doc-collections/fact-sheets/bio-effects-radiation.html (accessed on 17 June 2022).
12. The Chernobyl Accident. United Nations: Scientific Committee on the Effects of Atomic Radiation. Available online: https://www.unscear.org/unscear/en/areas-of-work/chernobyl.html (accessed on 17 June 2022).
13. International Nuclear Safety Advisory Group; International Atomic Energy Agency (Eds.) *Basic Safety Principles for Nuclear Power Plants: 75-INSAG-3 Rev. 1*; International Atomic Energy Agency: Vienna, Austria, 1999.
14. NRC Regulations Title 10, Code of Federal Regulations. U.S. NRC. Available online: https://www.nrc.gov/reading-rm/doc-collections/cfr/index.html (accessed on 15 January 2022).
15. Samanta, P.; Diamond, D.; O'Hara, J. *Regulatory Review of Micro-Reactors—Initial Considerations*; Brookhaven National Laboratory: Upton, NY, USA, 2020; p. 45.
16. Moe, W. *Key Regulatory Issues in Nuclear Micro-Reactor Transport and Siting*; INL/EXT–19-55257-Rev.000; Idaho National Laboratory: Idaho Falls, ID, USA, 2019; p. 1616515. [CrossRef]
17. Hara, J.M.O.; Higgins, J.C. *Human-System Interfaces to Automatic Systems: Review Guidance and Technical Basis*; BNL–91017-2010; Brookhaven National Laboratory: Upton, NY, USA, 2010; p. 1013461. [CrossRef]
18. Hugo, J.V.; Gertman, D.I. A Method to Select Human–System Interfaces for Nuclear Power Plants. *Nucl. Eng. Technol.* **2016**, *48*, 87–97. [CrossRef]
19. Frantz, S.; Tegeler, B.; Hughes, J. *Micro Reactor Regulatory Issues*; Nuclear Energy Institute: Washington, DC, USA, 2019.
20. Williams, J. *Existing NRC Regulations, Policies, and Guidance for Licensing*; ML15245A744; U.S. Nuclear Regulatory Commission: Rockville, MD, USA, 2015; p. 16.
21. Owusu, D.; Holbrook, M.; Sabharwall, P. *Regulatory and Licensing Strategy for Microreactor Technology*; INL/EXT–18-51111-Rev000; Idaho National Laboratory: Idaho Falls, ID, USA, 2018; p. 1565916. [CrossRef]
22. Belles, R.; Muhlheim, M.D. *Licensing Challenges Associated with Autonomous Control*; ORNL/SPR–2018/1071; Idaho National Laboratory: Idaho Falls, ID, USA, 2018; p. 1492160. [CrossRef]
23. *10 CFR Part 53, "Licensing and Regulation of Advanced Nuclear Reactors"*; Preliminary Proposed Rule Language; U.S. Nuclear Regulatory Commission: Rockville, MD, USA, 2021.
24. Belles, R.; Flanagan, G.; Hale, R.; Holcomb, D.; Huning, A.; Poore, W., III. *Advanced Reactor Siting Policy Considerations*; ORNL/TM–2019/1197; Idaho National Laboratory: Idaho Falls, ID, USA, 2019; p. 1542213. [CrossRef]
25. *IAEA Safety Standards GS-G-2.1*; Arrangements for Preparedness for a Nuclear or Radiological Emergency. IAEA: Vienna, Austria, 2007.
26. *NUREG-1537*; Guidelines for Preparing and Reviewing Applications for the Licensing of Non-Power Reactors. U.S. NRC: Rockville, MD, USA, 1996; p. 521.
27. Small Nuclear Power Reactors. WNA, Aug. 2021. Available online: https://www.world-nuclear.org/information-library/nuclear-fuel-cycle/nuclear-power-reactors/small-nuclear-power-reactors.aspx (accessed on 15 January 2022).
28. Park, G. *Nuclear Future Prospects and Countermeasures*; Korean Nuclear Society: Seoul, Korea, 2020; p. 93.
29. 40CFR 49.156-General Permits and Permits by Rule. Available online: https://www.law.cornell.edu/cfr/text/40/49.156 (accessed on 15 January 2022).
30. Permit-by-Rule. Environmental Protection Division. Available online: https://epd.georgia.gov/permit-rule (accessed on 15 January 2022).

31. Hara, J.O. *NUREG-0711 Rev 3 'Human Factors Engineering Program Review Model'*; Human Factors; U.S. Nuclear Regulatory Commission: Rockville, MD, USA, 2012; p. 147.
32. Arafat, Y. Technology Innovation for Fission Batteries: Autonomous Controls and Operation. 20 January 2021. Available online: https://nucl.inl.gov/SiteAssets/Fission%20Battery%20Initiative/Presentations/01-20-21%20Technology%20Innovation%20for%20Fission%20Batteries.pdf (accessed on 2 August 2022).
33. Lin, L.; Avramova, M.; Dinh, N. *Development and Assessment of a Nearly Autonomous Management and Control System for Advanced Reactors*; Elsevier: Amsterdam, The Netherlands, 2020; p. 22.
34. 10 CFR Part 71.73 Hypothetical Accident Conditions. U.S. NRC. Available online: https://www.nrc.gov/reading-rm/doc-collections/cfr/part071/part071-0073.html (accessed on 15 January 2022).
35. Transportation and Siting for Fission Batteries. [Online Video]. Available online: https://nucl.inl.gov/SiteAssets/Forms/AllItems.aspx?RootFolder=%2FSiteAssets%2FFission%20Battery%20Initiative%2FWorkshop%20Recordings&FolderCTID=0x0120002155053CDC369346A5967CE94F91126D&View=%7BDE629CBE%2D978D%2D4967%2D8ECB%2DCC5CB741D6B5%7D (accessed on 2 August 2022).

eng

MDPI

Article

Investigating The Impact of Roadway Characteristics on Intersection Crash Severity

Mostafa Sharafeldin [1,*], Ahmed Farid [2] and Khaled Ksaibati [1]

1 Wyoming Technology Transfer Center (WYT2/LTAP), Department of Civil and Architectural Engineering, University of Wyoming, Laramie, WY 82071, USA
2 Department of Civil and Environmental Engineering, California Polytechnic State University, San Luis Obispo, CA 93407, USA
* Correspondence: msharafe@uwyo.edu

Abstract: Intersections are commonly recognized as crash hot spots on roadway networks. Therefore, intersection safety is a major concern for transportation professionals. Identifying and quantifying the impact of crash contributing factors is crucial to planning and implementing the appropriate countermeasures. This study covered the analysis of nine years of intersection crash records in the State of Wyoming to identify the contributing factors to crash injury severity at intersections. The study involved the investigation of the influence of roadway (intersection) and environmental characteristics on crash injury severity. The results demonstrated that several parameters related to intersection attributes (pavement friction; urban location; roadway functional classification; guardrails; right shoulder width) and two environmental conditions (road surface condition and lighting) influence the injury severity of intersection crashes. This study identified the significant roadway characteristics influencing crash severity and explored the key role of pavement friction, which is a commonly omitted variable.

Keywords: crash injury severity; intersection safety; pavement friction; roadway characteristics; roadway geometric characteristics; intersection crash analysis

Citation: Sharafeldin, M.; Farid, A.; Ksaibati, K. Investigating The Impact of Roadway Characteristics on Intersection Crash Severity. *Eng* **2022**, *3*, 412–423. https://doi.org/10.3390/eng3040030

Academic Editor: Antonio Gil Bravo

Received: 20 September 2022
Accepted: 7 October 2022
Published: 8 October 2022

Publisher's Note: MDPI stays neutral with regard to jurisdictional claims in published maps and institutional affiliations.

Copyright: © 2022 by the authors. Licensee MDPI, Basel, Switzerland. This article is an open access article distributed under the terms and conditions of the Creative Commons Attribution (CC BY) license (https://creativecommons.org/licenses/by/4.0/).

1. Introduction

Intersection-related crashes are responsible for more than 20% of road traffic fatalities, and more than 40% of total crash injuries in the United States. Complex traffic movements and the interaction between different transportation modes establish the intersections as hazardous locations for all road users [1–4]. In addition, intersection safety is facing new challenges with rapidly increasing traffic volumes and developing technologies. Planning for traffic control and safety at intersections can be even more challenging with multimodal operations. Therefore, the Federal Highway Administration (FHWA) and state Departments of Transportations (DOTs) are continuously striving to mitigate crash injury severity and reduce traffic-related fatalities [5–9]. The injury severity of traffic crashes can be related to different categories of contributing factors including crash, driver, environmental, and roadway attributes. While roadway characteristics are usually considered in traffic safety studies, pavement surface friction is a commonly omitted variable in crash analysis.

Pavement friction is the force resisting the relative motion between the vehicle tires and the pavement surface. The loss of skid resistance prevents drivers from safely maneuvering or stopping their vehicles, which leads to increased crash frequency and severity [10–12]. Tire-pavement interaction leads to aggregate polishing in the pavement surface layer, which reduces pavement friction supply over time. Thus, transportation agencies ought to consistently monitor the pavement surface friction levels. A recent survey of the roadway surface friction management practices revealed that only eleven of the surveyed thirty-two DOTs collect pavement friction data on specific road locations (such as ramps, curves, and intersections) to investigate safety concerns. The increase in wet pavement crashes is a grave safety concern, urging the need for friction-related data collection at locations experiencing such crashes [13].

Wet pavement crashes are typically linked to poor skid resistance, since wet pavement surfaces can substantially reduce the frictional force. Insufficient friction levels may lead to a similar issue on dry pavement surfaces [14,15]. The State of Wyoming has one of the highest snowfall rates in the United States. Therefore, non-dry pavement conditions are common in the state. A third of traffic crashes in Wyoming occur on non-dry pavement surface conditions, including snowy, icy, wet, or slushy road surfaces [16–18]. Pavement surface treatments play a critical role in supplying sufficient friction levels across the roadway network. Surface treatments are commonly applied to specific locations with high friction demand, such as intersections. These surface treatments include hot-mix asphalt (HMA) overlays, chip seals, open-graded friction courses (OGFC), micro-milling, and high friction surface treatments (HFST) [14,19–22].

The objective of this study was to investigate the roadway risk factors, including pavement friction, influencing the injury severity of intersection crashes. A set of roadway-related characteristics and critical environmental conditions were considered in the analysis. This paper is organized as follows. A review of the relevant studies in the traffic safety literature is discussed. Afterward, the methodology, data description, results of the empirical analysis, conclusion, and recommendations are all discussed.

2. Literature Review

This section provides a review of multiple studies related to the influence of roadway characteristics and pavement friction on crash injury severity. The limitations of the reviewed studies are identified, and the contribution of this research is discussed. The following studies did not consider pavement friction as a risk factor in the analysis.

Abdel-Aty and Keller [23] addressed different contributing factors to crash injury severity at signalized intersections in Florida using ordinal probit models. The results demonstrated that a combination of crash-related attributes and intersection characteristics influence crash injury severity. It was found that an increase in the number of lanes, the presence of medians, and right-turn channelization reduces the risk of sustaining severe injury. Even though the authors examined a wide set of roadway attributes, the pavement surface friction and road surface condition variables were not incorporated in the study.

Haleem and Abdel-Aty [24] selected multiple approaches, including two ordinal probit models, to examine crash injury severity at unsignalized intersections in Florida. The authors incorporated the driver's characteristics, intersection attributes, pavement surface type (concrete, asphalt, etc.), and road surface condition (dry, wet, etc.), among other factors. This study included the number of thorough lanes as a surrogate measure for traffic volume on the minor road. The authors identified several significant factors influencing the crash severity including intersection and driver characteristics. The left shoulder width, right shoulder width, and number of turning lanes were found to be among the influential factors. Yet, the authors did not account for pavement friction as a potential risk factor.

Anowar et al. [25] investigated the contributing factors to intersection crash severity in Bangladesh by utilizing a generalized ordinal logit model. The authors examined the impact of various crash, environmental, and roadway attributes. As per the results, undivided roads, dry pavement surfaces, and rural areas were found to raise the risk of incurring severe injury. Even though the authors considered the road surface condition, pavement friction was not considered in the analysis.

Oh [26] examined the contributing factors to crash injury severity at four-leg signalized intersections in rural areas. The author applied ordinal probit models to investigate the crash, weather, and roadway risk factors. The results indicated that tighter horizontal curves and higher speed limits contribute to severe crashes, while wider medians and the presence of protected left-turn phases are associated with less severe crashes. Yet, the author did not incorporate pavement surface friction as a potential contributing factor to crash injury severity.

Lee et al. [27] developed Bayesian ordinal logistic regression models to explore the impact of pavement surface conditions on the crash injury severity. The findings indicated that poor pavement surfaces increase the severity of multiple-vehicle crashes regardless of the posted speed limit. The findings also demonstrated that deteriorated pavement surfaces decrease the severity of single-vehicle crashes on low-speed roads (having posted speed limits of 35 mi/h or below) and increase such severity on high-speed roads (having speed limits of 50 mi/h or above). It should be noted that this study incorporated the pavement condition variable instead of the pavement friction in the analyses.

Zhao et al. [28] employed a multivariate Poisson log-normal model to analyze traffic crashes on the approaches of urban signalized intersections in the State of Nebraska. The study was focused on traffic and roadway geometric risk factors. The study's results demonstrated that intersection approaches on urban arterial roads have more frequent and higher severity crashes compared to collector roads. The results also indicated that the number of right-turn, left-turn, and through lanes influences crash frequency. The study did not consider any factors related to pavement condition or pavement friction.

The following study considered pavement friction as a risk factor, but they had other limitations, as follows.

Hussien et al. [29] investigated the effects of pavement resurfacing on intersection safety by conducting a before-after study on signalized intersections that were subjected to resurfacing in Melbourne, Australia. The authors incorporated multiple pavement condition data variables including roughness, skid resistance, and rutting. The authors also considered roadway characteristics and environmental conditions. The results demonstrated that pavement maintenance and improving skid resistance reduce the frequency and severity of crashes at signalized intersections. The results also identified other significant factors, including lighting, road surface condition, and interaction parameters, such as approach width interactions with the presence of a median, bus stop, or shared lane. Even though the authors incorporated pavement condition information including skid resistance and various roadway characteristics, the study's scope did not encompass rural intersections and the authors omitted several roadway characteristics. They include the roadway grade, horizontal curvature, roadway functional classification, and right shoulder attributes.

Sharafeldin et al. [30] developed a Bayesian ordinal probit model to investigate the impact of pavement friction, among other risk factors, on injury severity of the intersection crashes. The study concluded that insufficient pavement friction supply is one of the main contributors to severe crashes at intersections. Even though the study considered pavement friction as a potential risk factor, the study analyzed a limited data set and did not include other roadway attributes. Other related studies to this research topic include those of Chen et al. [31], Sharafeldin et al. [32], Karlaftis and Golias [33], Roy et al. [12], Chen et al. [34], Papadimitriou et al. [35], and Zhao et al. [36].

Generally, there is a growing interest in research about the relationship between pavement friction and traffic safety. However, to the best of the authors' knowledge, the investigation of the pavement friction's effect, among the other roadway attributes, on intersection crash severity is insufficient. In this research, the risk of observing severe injury crashes at intersections is modeled as a function of environmental and roadway factors, especially pavement surface friction.

3. Research Methodology

Ordered response modeling techniques have been widely adopted in crash injury severity studies to account for the ordinal nature of the injury severity levels. Ordinal probit and logit models were extensively utilized to study the risk factors of crash injury severity [37,38]. The ordinal probit model structure estimates the latent propensity, y_i^*, for each crash, i, as follows [39]:

$$y_i^* = \beta_0 + \beta_1 X_{1i} + \beta_2 X_{2i} + \ldots + \beta_p X_{pi} + \epsilon_i \tag{1}$$

The predictors are described by the X's, while their regression coefficients are described by the β's, which are estimated using the maximum likelihood estimation (MLE) method. The random error term is defined by $\epsilon_{i,}$ and it is assumed to be normally distributed. The response formulation is stated as follows [39], where ψ is a threshold that is estimated via the MLE technique.

$$y_i = \begin{cases} O, & y_i^* < 0 \\ BC, & 0 < y_i^* < \psi \\ KA, & y_i^* > \psi \end{cases} \tag{2}$$

The outcome probabilities, P(.)'s, are calculated by the following equations where F(.) is the cumulative standard normal distribution function [40]:

$$P(y_i = O) = F\left(-\left(\beta_0 + \sum_{p=1}^{P} \beta_p X_{pi}\right)\right) \tag{3}$$

$$P(y_i = BC) = F\left(\psi - \left(\beta_0 + \sum_{p=1}^{P} \beta_p X_{pi}\right)\right) - F\left(-\left(\beta_0 + \sum_{p=1}^{P} \beta_p X_{pi}\right)\right) \tag{4}$$

$$P(y_i = KA) = 1 - F\left(\psi - \left(\beta_0 + \sum_{p=1}^{P} \beta_p X_{pi}\right)\right) \tag{5}$$

Confidence intervals of 90th were utilized to identify the statistically significant variables instead of the 95th intervals. This was to retain the valuable information usually lost by utilizing narrower confidence intervals. Marginal effects are estimated to identify the influences of contributing factors on crash injury severity. The marginal effect is the average change in the probability of incurring an injury of severity j, ΔP (y = j), as a result of the variable's influence, provided that all other variables are controlled [40].

4. Data Collection

This study involved the examination of crash data obtained from the Critical Analysis Reporting Environment (CARE) package of the Wyoming Department of Transportation (WYDOT). The crash records were collected by WYDOT from police crash reports and inputted into the package. The data were prepared such that each data point represented a unique intersection crash record, including the pavement friction number measured at the intersection in the crash year. The data included records of 9108 unique crashes at 359 intersections from January 2007 through December 2017, except for the years 2010 and 2011 due to friction data availability. Crashes specified as intersection crashes are those located within 250 feet (76.2 m) from the center of the intersection, as per the American Association of State Highway and Transportation Officials [41]. The crash records included information on the roadway and other characteristics as well.

WYDOT personnel collected the pavement friction data across the state using the locked-wheel tester. The locked-wheel tester is a trailer with two wheels having standard tires. The device tests the longitudinal friction by using either one or two wheels. The testing tires can be either smooth or ribbed. The smooth tire is sensitive to macrotexture while the ribbed tire is more sensitive to microtexture [10]. The locked-wheel device measures pavement friction by fully locking the testing wheel(s) and recording the average sliding force at which the fully locked state is achieved. Accordingly, the locked-wheel device can only measure friction at specific time intervals due to the full-lock requirement [42]. The friction number are usually reported as (FN40R), which is measured by using a locked wheel tester, fitted by a standard ribbed tire at 40 mile per hour. The Federal Highway

Administration (FHWA) promotes utilizing the continuous pavement friction measurement (CPFM) technique to collect friction data continually along road networks, including special locations, such as curves, ramps, and intersections [43].

The field data were calibrated by WYDOT at the regional calibration center. Friction data were integrated with the obtained intersection crash data by matching the mile post of the intersection with the friction measurement's locations identified by mile posts. When the friction measure was not gauged exactly at the intersection location, the nearest two measurements (before and after the intersection) along the major route were averaged to calculate the friction at the intersection. Moreover, the friction numbers were estimated at the years with no friction data collection by averaging the measurements of the previous and the subsequent years at the study location. This approach was only applied at locations where friction numbers were declining. This indicated that no maintenance work was performed and the difference in friction numbers (FN40R) between the previous and the subsequent years was 10 or less, to ensure the validity of the averaged measurements. This method assumes that the pavement friction was deteriorating at a steady rate over the three years. The friction measurements were matched to the crash records that occurred in the same year of the friction measurement. Each row of the dataset represented a unique crash record with the friction number at the intersection, measured in the crash year.

5. Data Description

In this study, the crash injury severity is classified into three categories: O for property damage only (PDO) or no injury crashes, BC for possible or minor injury crashes, and KA for the highest severity level, which is disabling or fatal injury crashes. PDO crashes represented 75.9% of the total crash records. Possible and minor injury crashes accounted for 22.3%, while disabling and fatal injury crashes comprised 1.8% of the data. The investigated roadway attributes were pavement surface friction, intersection type, intersection location attributes, number of lanes, grade (uphill, downhill, and level), horizontal curvature, roadway functional classification, roadway surface type, guardrail presence, presence of rumble strips, median type, median width, right shoulder type, and right shoulder width. The pavement friction values (FN40R) ranged from 19 to 71 with an average of 40. Figure 1 illustrates the friction number distribution for the crash records.

Table 1 presents summary statistics of this study's data. As for the other roadway characteristics, the majority of the examined crashes occurred at signalized intersections, intersections with four or more legs, and intersections in urban areas. The intersections were identified as urban or rural according to the US Census Bureau's definition [44]. Limited proportions of crashes were at uphill and downhill intersections in contrast to crashes on level intersections. The data included the functional classification of the major roadways of the intersections. Most recorded crashes occurred at intersections of principal arterial roads, while smaller proportions occurred on interstate, minor arterial, and collector roads. Local roads were considered as the reference category for this variable in the modeling. It should be noted that the intersections with a functional classification of "Interstate" refer to intersections with interstate interchanges (on/off ramps).

Low proportions of crashes occurred at intersections having horizontal curves. The number of through lanes and road surface type were also considered. Low proportions of crashes occurred at intersections with guardrails or rumble strips. A total of 45% of the collected crash records occurred at intersections near schools, while a low proportion of them occurred near liquor stores.

As for the median type, almost half of the crashes involved a raised median, while a limited proportion involved a depressed median. The absence of a median was considered as the reference category in the modeling. Medians wider than 100 feet (30.5 m) were estimated as 120 feet (36.6 m) wide. As for the right shoulder type, more than a quarter of the crash records occurred at sites having asphalt shoulders, while a third of them occurred at sites having concrete shoulders. On the contrary, a small percentage occurred at sites with unpaved right shoulders. The absence of a right shoulder was considered as the reference category for this variable in the modeling. Moreover, right shoulders that were wider than 8.5 feet (2.6 m) were estimated as 10 feet (3 m) wide.

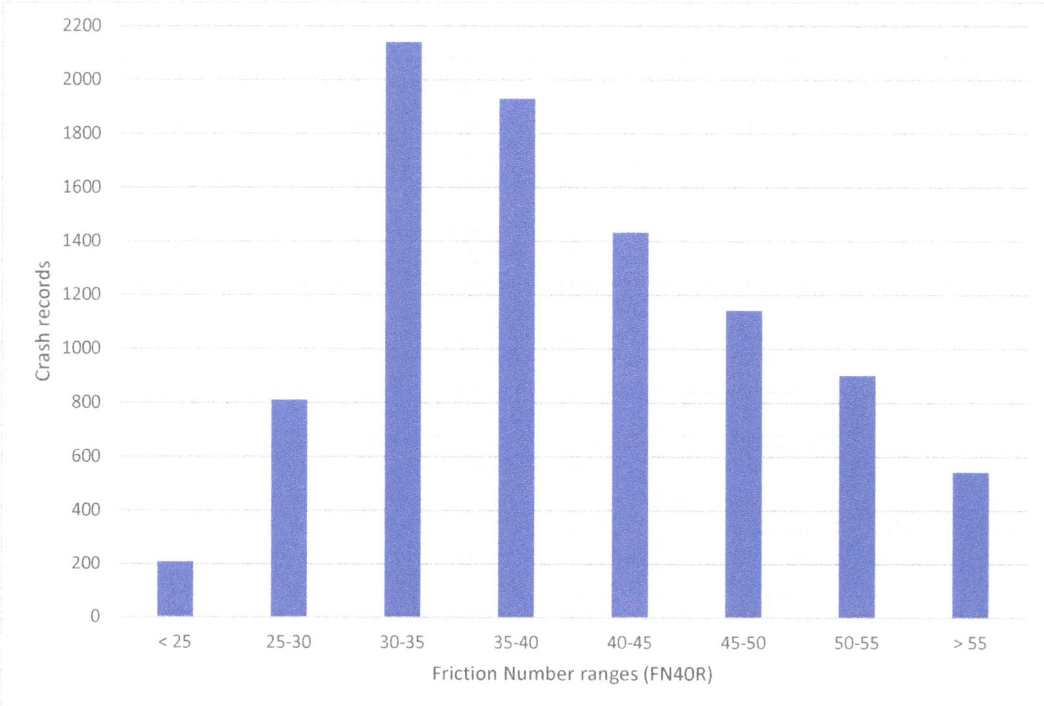

Figure 1. Friction number distribution by crash frequency.

Table 1. Data's descriptive statistics.

Binary Variables		
Response	**Count**	**Percent**
No injury (property damage only, PDO, or O)	6915	75.9
Possible or suspected minor injury (BC)	2032	22.3
Fatal or suspected serious injury (KA)	161	1.8
Roadway Characteristics		
Type: four or more legs (1 if yes or 0 otherwise)	7833	86.0
Location: urban (1 if yes or 0 otherwise)	8530	93.7
Traffic control: signalized (1 if yes or 0 otherwise)	8556	93.9
Grade: uphill (1 if yes or 0 otherwise)	193	2.1
Grade: downhill (1 if yes or 0 otherwise)	627	6.9
Functional classification: interstate (1 if yes or 0 otherwise)	213	2.3
Functional classification: principal arterial (1 if yes or 0 otherwise)	7700	84.5
Functional classification: minor arterial (1 if yes or 0 otherwise)	532	5.8
Functional classification: collector (1 if yes or 0 otherwise)	42	0.5

Table 1. *Cont.*

Binary Variables		
Response	**Count**	**Percent**
Road surface: concrete (1 if yes or 0 otherwise)	4623	50.8
Guardrail Present (1 if yes or 0 otherwise)	87	1.0
Median: depressed (1 if yes or 0 otherwise)	525	5.8
Median: raised (1 if yes or 0 otherwise)	4292	47.1
Right shoulder type: asphalt (1 if yes or 0 otherwise)	2522	27.7
Right shoulder type: concrete (1 if yes or 0 otherwise)	3101	34.0
Right shoulder type: unpaved (1 if yes or 0 otherwise)	337	3.7
Slight curve with a radius between 1500 and 5000 ft (1 if yes or 0 otherwise)	843	9.3
Moderately tight curve with a radius between 750 and 1500 ft (1 if yes or 0 otherwise)	327	3.6
Tight curve with a radius between 300 and 750 ft (1 if yes or 0 otherwise)	371	4.1
Severely tight curve with a radius shorter than 300 ft (1 if yes or 0 otherwise)	70	0.8
Rumble strips (1 if yes or 0 otherwise)	143	1.6
Near school (1 if yes or 0 otherwise)	4088	44.9
Near liquor store (1 if yes or 0 otherwise)	424	4.7
Environmental Characteristics		
Road condition: non-dry surface (1 if yes or 0 otherwise)	2523	27.7
Adverse weather (1 if yes or 0 otherwise)	1746	19.2
Lighting: non-daylight (1 if yes or 0 otherwise)	2054	22.6
Continuous Variables		
	Mean	Minimum	Maximum
Pavement friction	40	19	71
Number of lanes	3.7	2	4
Median width	9.7	0	120
Right shoulder width	3.6	0	10

When it comes to environmental conditions, almost a quarter of the crashes occurred under non-daylight conditions, such as nighttime, dawn, or dusk conditions. Moreover, a considerable proportion of crashes occurred during adverse weather conditions. The adverse weather categories included rain, snow, blizzard, hail, fog, and any other inclement weather conditions. Concerning the road surface condition, over a quarter of the crashes occurred on non-dry road surfaces such as wet, snowy, icy, slushy, and any other adverse conditions.

6. Empirical Analysis

An ordinal probit model was developed to analyze the intersection crash data. The aforementioned explanatory variables were all considered in the model. The 90th percentile confidence level was selected for ascertaining statistically significant variables and the log-likelihood ratio test was conducted to test for the model's significance. The results of the model are presented in Table 2. Note that statistically insignificant variables are not shown in the table.

Table 2. Ordinal probit model results.

Coefficients	Estimate	Standard Error	*p*-Value
Constant	−0.463	0.101	<0.001
Pavement friction	−0.003	0.002	0.083
Location: urban	−0.233	0.058	<0.001
Functional classification: principal arterial	0.083	0.050	0.096
Functional classification: minor arterial	0.146	0.074	0.050
Guardrail present	0.242	0.139	0.082
Right shoulder width	0.016	0.004	<0.001
Road condition: non-dry surface	−0.253	0.033	<0.001
Lighting: non-daylight	0.073	0.034	0.031
ψ	1.416	0.032	<0.001
Log-likelihood		−5551	
Log-Likelihood of constant-only model		−5603	
Log-Likelihood ratio χ^2		104	
Degrees of freedom		8	
P-Value		<0.001	

The modeling results indicated that several roadway attributes, including pavement surface friction, and two environmental conditions, are significantly impacting the crash injury severity at intersections. The marginal effects of the significant risk factors are presented in Table 3. In Table 3, the $\Delta P(.)$'s represent the changes in the risks of observing crash severity j, whether KA, BC, or O are as a result of the explanatory variable's effect. Each variable's effect on the injury severity was estimated assuming all other variables were controlled, and the continuous variables (pavement friction and right shoulder width) were at their average values.

The findings demonstrated that several intersection attributes had a strong impact on crash severity risk. As shown in Table 3, pavement surface friction substantially influences the severity of intersection crashes. It was estimated that, on average, increasing the pavement friction numbers (FN40R) at intersections from 25 to 45 reduces the risk of observing BC and KA injuries by 1.65% and 0.36%, respectively. Sharafeldin et al. [30] and Hussien et al. [29] reported relevant findings indicating that insufficient friction levels increase the risks of crash frequency and severity. This finding emphasizes the significance of maintaining sufficient pavement friction levels on roadway networks, especially at high-risk crash locations with a larger friction demand, such as intersections, ramps, and curves, to alleviate severe injury concerns.

Crashes at urban intersections were found to be associated with lower injury severity risk compared to crashes at rural intersections. It was estimated that, on average, an urban intersection crash would have a 6.44% and a 1.14% lower chance of resulting in BC and KA injuries, respectively, relative to rural intersection crashes. The higher severity of rural crashes is plausibly related to higher speed limits, higher chances of driver distraction, non-compliance with safety measures, and driver fatigue due to longer travel distances. In addition, medical assistance has better access to crash victims in urban areas compared to rural locations. These findings align with those of Anowar et al. [25] and Oh [26]. This finding shed light on the premise that crashes at rural intersections have higher injury severities. This is critical to the State of Wyoming, since it has a higher number of rural and semirural intersections.

Table 3. Marginal effects of the intersection crash severity factors.

Variable	Marginal Effects (%)		
	P (y = O)	P (y = BC)	P (y = KA)
Pavement friction	2.01	−1.65	−0.36
Location: urban	7.58	−6.44	−1.14
Functional classification: principal arterial	−2.93	2.39	0.55
Functional classification: minor arterial	−5.25	4.23	1.02
Guardrail present	−8.88	7.02	1.86
Right shoulder width	3.36	2.73	0.64
Road condition: non-dry surface	8.18	−6.97	−1.21
Lighting: non-daylight	−2.59	2.11	0.48

Notes: ΔP (y = O) = change in the likelihood of observing no injury, ΔP (y = BC) = change in the likelihood of observing possible or suspected minor injury, ΔP (y = KA) = change in the likelihood of observing fatal or suspected serious injury.

The roadway functional classification was found to be a significant contributing factor to crash severity. Intersection crashes on principal and minor arterial roads were found to be severe compared to those that occurred on local roads. Crashes on principal arterials were found to have higher severity levels with marginal effects of 2.39% and 0.55% for BC and KA injuries, respectively. Crashes on minor arterials were found to have higher severity levels with marginal effects of 4.23% and 1.02% for BC and KA injuries, respectively. These findings are possibly attributed to the higher percentage of trucks and more complex traffic mixes on arterial roads compared to those on local roads. Zhao et al. [28] reported similar findings. The higher severity of crashes on minor arterials compared to that of principal arterials is plausibly related to the higher speed differentials among vehicles on minor arterials. The presence of guardrails was found to be associated with higher injury severity levels. Crashes on intersections with guardrails would have 7.02% and 1.86% higher chances of resulting in BC and KA injuries, respectively. Plausibly, this is because of the correlation between higher speed facilities and guardrail installation.

The right shoulder width was found to significantly impact intersection crash injury severity. It was estimated that, on average, widening the right shoulders at intersections from 4 to 10 feet (1.2 to 3 m) raises the risk of observing BC and KA injuries by 2.73% and 0.64%, respectively. This finding may be attributed to the improper use of wide shoulders, which increases the risk of observing sideswipe and rear-end crashes. Such crashes are possibly severe at high-impact speeds. It should be noted that wider shoulders are typically utilized on high-speed roads. Haleem and Abdel-Aty [24] reported similar findings.

As for the environmental factors, two environmental conditions were found to have a significant impact on injury severity risk. Non-dry road surfaces were found to be inversely related to crash injury severity. It was estimated that, on average, crashes on non-dry road surfaces have 6.97% and 1.21% lower chances of resulting in BC and KA injuries, respectively, compared to crashes on dry surfaces. This finding is possibly attributed to the cautious driving behavior and lower speeds observed on non-dry roads. Comparable findings were reported by Anowar et al. [25]. The lighting condition at the time of the crash was found to influence crash injury severity. Crashes that occurred during non-daylight conditions would have 2.11% and 0.48% higher chances of resulting in BC and KA injuries, respectively, compared to crashes that occurred under daylight conditions. Haleem and Abdel-Aty [24] and Oh [26] reported similar findings.

7. Conclusions and Recommendations

In this study, an attempt was made to explore the influencing factors of crash injury severity at intersections. That is, intersection and environmental contributing factors were examined. An ordinal probit model was developed to investigate the crash severity risk factors. The analysis results demonstrated that several parameters significantly impact crash injury severity. Pavement friction was found to be a substantial effect, as increasing friction numbers at intersections was found to mitigate crash injury severity. It was also concluded that fatal and disabling injury crashes are more likely to occur at rural intersections compared to urban intersections. Therefore, rural intersections require more attention when it comes to maintaining adequate pavement friction levels and implementing crash mitigation measures. This finding is particularly valuable to the State of Wyoming, since it is characterized by rural and semi-rural areas. The functional classification of the roadway was also found to influence crash severity, as intersection crashes on arterial roads tend to have higher injury severity likelihoods compared to local roads. The widening of right shoulders and the deployment of guardrails were found to be associated with severe crashes. On the other hand, non-dry road surfaces were found to reduce the likelihood of observing severe crashes. Finally, crashes that occurred during daylight conditions were found to be less severe than those that occurred during other conditions.

It is recommended to raise pavement friction levels at intersections to adequate levels to mitigate crash injury severity and crash probability. It is also recommended to provide proper lighting at intersections, especially rural intersections, to lower the risk of observing severe crashes. Intersections on arterials, high-speed facilities, and rural intersections require more attention for countermeasure planning and implementation, since they are linked to high injury severity crashes. In addition, the findings related to the intersection characteristics can be further investigated to plan for the appropriate treatments. Implementing countermeasures that reduce severe crashes, such as those documented in the Crash Modification Factors (CMF) Clearinghouse [45], may be extensively reviewed.

8. Study Limitations and Future Research

The study had one main limitation, which is not including the traffic volume at intersection approaches due to data availability.

Author Contributions: Conceptualization, M.S., A.F. and K.K.; methodology, M.S.; software, M.S.; validation M.S., A.F. and K.K.; formal analysis, M.S.; investigation, M.S. and A.F.; resources, K.K.; data curation, M.S.; writing—original draft preparation, M.S.; writing—review and editing, M.S., A.F. and K.K.; visualization, M.S.; supervision, K.K.; project administration, K.K.; funding acquisition, K.K. All authors have read and agreed to the published version of the manuscript.

Funding: This research was funded by the Wyoming Department of Transportation (WYDOT), grant number: RS05221.

Institutional Review Board Statement: Not applicable.

Informed Consent Statement: Not applicable.

Data Availability Statement: The data were collected from the Critical Analysis Reporting Environment (CARE) package, supported by the Wyoming Department of Transportation (WYDOT).

Acknowledgments: The authors gratefully acknowledge the effective financial support of WYDOT. All opinions are solely of the authors. The subject matter, all figures, tables, and equations, not previously copyrighted by outside sources, are copyrighted by WYDOT, the State of Wyoming, and the University of Wyoming. All rights reserved copyrighting in 2022.

Conflicts of Interest: The authors declare that they have no conflict of interest with all parties.

References

1. PDO; TPDO. *The National Intersection Safety Problem*; U.S. Department of Transportation, Federal Highway Administration, Office of Safety: Washington, DC, USA, 2004.
2. Arafat, M.; Hadi, M.; Raihan, M.A.; Iqbal, M.S.; Tariq, M.T. Benefits of connected vehicle signalized left-turn assist: Simulation-based study. *Transp. Eng.* **2021**, *4*, 100065. [CrossRef]
3. Stevanovic, A.; Dobrota, N.; Mitrovic, N. *NCHRP 20-07/Task 414: Benefits of Adaptive Traffic Control Deployments—A Review of Evaluation Studies*; NCHRP: Washington, DC, USA, 2019.

4. Reza, I.; Ratrout, N.T.; Rahman, S.M. Calibration protocol for paramics microscopic traffic simulation model: Application of neuro-fuzzy approach. *Can. J. Civ. Eng.* **2016**, *43*, 361–368. [CrossRef]
5. Cvijovic, Z.; Zlatkovic, M.; Stevanovic, A.; Song, Y. Multi-Level Conditional Transit Signal Priority in Connected Vehicle Environments. *J. Road Traffic Eng.* **2021**, *67*, 1–12. [CrossRef]
6. Dobrota, N.; Stevanovic, A.; Mitrovic, N. Development of assessment tool and overview of adaptive traffic control deployments in the US. *Transp. Res. Rec.* **2020**, *2674*, 464–480. [CrossRef]
7. Dobrota, N.; Stevanovic, A.; Mitrovic, N. A novel model to jointly estimate delay and arrival patterns by using high-resolution signal and detection data. *Transp. A Transp. Sci.* **2022**, 1–33. [CrossRef]
8. Cvijovic, Z.; Zlatkovic, M.; Stevanovic, A.; Song, Y. Conditional transit signal priority for connected transit vehicles. *Transp. Res. Rec.* **2022**, *2676*, 490–503. [CrossRef]
9. Reza, I.; Ratrout, N.T.; Rahman, S.M. Artificial Intelligence-Based Protocol for Macroscopic Traffic Simulation Model Development. *Arab. J. Sci. Eng.* **2021**, *46*, 4941–4949. [CrossRef]
10. Hall, J.W.; Smith, K.L.; Titus-Glover, L.; Wambold, J.C.; Yager, T.J.; Rado, Z. *Guide for Pavement Friction*; Final Report for NCHRP Project; NCHRP: Washington, DC, USA, 2009; Volume 1, p. 43.
11. Hafez, M.; Farid, A.; Ksaibati, K.; Director, P.E.; Rickgauer, S.; Carlson, M. *Managing Pavement Friction of Wyoming's Roads Considering Safety*; Wyoming Department of Transportation: Cheyenne, WY, USA, 2020.
12. Roy, U.; Farid, A.; Ksaibati, K. Effects of Pavement Friction and Geometry on Traffic Crash Frequencies: A Case Study in Wyoming. *Int. J. Pavement Res. Technol.* **2022**, 1–14. [CrossRef]
13. Elkhazindar, A.; Hafez, M.; Ksaibati, K. Incorporating Pavement Friction Management into Pavement Asset Management Systems: State Department of Transportation Experience. *CivilEng* **2022**, *3*, 541–561. [CrossRef]
14. FHWA. Evaluation of Pavement Safety Performance. Available online: https://www.fhwa.dot.gov/publications/research/safety/14065/003.cfm (accessed on 31 July 2022).
15. Abaza, O.A.; Chowdhury, T.D.; Arafat, M. Comparative analysis of skid resistance for different roadway surface treatments. *Am. J. Eng. Appl. Sci.* **2017**, *10*, 890–899. [CrossRef]
16. NHTSA. National Highway Traffic Safety Administration (NHTSA) Motor Vehicle Crash Data Querying and Reporting. Available online: https://www.nhtsa.gov/research-data/fatality-analysis-reporting-system-fars (accessed on 19 July 2022).
17. Alrejjal, A.; Farid, A.; Ksaibati, K. Investigating factors influencing rollover crash risk on mountainous interstates. *J. Saf. Res.* **2022**, *80*, 391–398. [CrossRef]
18. Alrejjal, A.; Farid, A.; Ksaibati, K. A correlated random parameters approach to investigate large truck rollover crashes on mountainous interstates. *Accid. Anal. Prev.* **2021**, *159*, 106233. [CrossRef]
19. Abdalla, A.; Faheem, A.F.; Walters, E. Life cycle assessment of eco-friendly asphalt pavement involving multi-recycled materials: A comparative study. *J. Clean. Prod.* **2022**, *362*, 132471. [CrossRef]
20. Abaza, O.A.; Arafat, M.; Uddin, M.S. Physical and economic impacts of studded tyre use on pavement structures in cold climates. *Transp. Saf. Environ.* **2021**, *3*, tdab022. [CrossRef]
21. Rezapour, M.; Hafez, M.; Ksaibati, K. Evaluating the Complex Relationship between Environmental Factors and Pavement Friction Based on Long-Term Pavement Performance. *Computation* **2022**, *10*, 85. [CrossRef]
22. Abdalla, A.; Faheem, A.F.; Hosseini, A.; Titi, H. *Performance Related Asphalt Mixtures Characterization*; No. TRBAM-21-03073; National Academies of Sciences, Engineering, and Medicine: Washington, DC, USA, 2021.
23. Abdel-Aty, M.; Keller, J. Exploring the overall and specific crash severity levels at signalized intersections. *Accid. Anal. Prev.* **2005**, *37*, 417–425. [CrossRef]
24. Haleem, K.; Abdel-Aty, M. Examining traffic crash injury severity at unsignalized intersections. *J. Saf. Res.* **2010**, *41*, 347–357. [CrossRef]
25. Anowar, S.; Yasmin, S.; Tay, R. Factors influencing the severity of intersection crashes in Bangladesh. *Asian Transp. Stud.* **2014**, *3*, 143–154.
26. Oh, J.T. Development of severity models for vehicle accident injuries for signalized intersections in rural areas. *KSCE J. Civ. Eng.* **2006**, *10*, 219–225. [CrossRef]
27. Lee, J.; Nam, B.; Abdel-Aty, M. Effects of pavement surface conditions on traffic crash severity. *J. Transp. Eng.* **2015**, *141*, 04015020. [CrossRef]
28. Zhao, M.; Liu, C.; Li, W.; Sharma, A. Multivariate Poisson-lognormal model for analysis of crashes on urban signalized intersections approach. *J. Transp. Saf. Secur.* **2018**, *10*, 251–265. [CrossRef]
29. Hussein, N.; Hassan, R.; Fahey, M.T. Effect of pavement condition and geometrics at signalised intersections on casualty crashes. *J. Saf. Res.* **2021**, *76*, 276–288. [CrossRef] [PubMed]
30. Sharafeldin, M.; Albatayneh, O.; Farid, A.; Ksaibati, K. A Bayesian Approach to Examine the Impact of Pavement Friction on Intersection Safety. *Sustainability* **2022**, *14*, 12495. [CrossRef]
31. Chen, S.; Saeed, T.U.; Alqadhi, S.D.; Labi, S. Safety impacts of pavement surface roughness at two-lane and multi-lane highways: Accounting for heterogeneity and seemingly unrelated correlation across crash severities. *Transp. A Transp. Sci.* **2019**, *15*, 18–33. [CrossRef]
32. Sharafeldin, M.; Farid, A.; Ksaibati, K. Examining the Risk Factors of Rear-End Crashes at Signalized Intersections. *J. Transp. Technol.* **2022**, *12*, 635–650. [CrossRef]
33. Karlaftis, M.G.; Golias, I. Effects of road geometry and traffic volumes on rural roadway accident rates. *Accid. Anal. Prev.* **2002**, *34*, 357–365. [CrossRef]

34. Chen, H.; Cao, L.; Logan, D.B. Analysis of risk factors affecting the severity of intersection crashes by logistic regression. *Traffic Inj. Prev.* **2012**, *13*, 300–307. [CrossRef]
35. Papadimitriou, E.; Filtness, A.; Theofilatos, A.; Ziakopoulos, A.; Quigley, C.; Yannis, G. Review and ranking of crash risk factors related to the road infrastructure. *Accid. Anal. Prev.* **2019**, *125*, 85–97. [CrossRef]
36. Zhao, G.; Jiang, Y.; Li, S.; Tighe, S. Exploring implicit relationships between pavement surface friction and vehicle crash severity using interpretable extreme gradient boosting method. *Can. J. Civ. Eng.* **2022**, *99*, 1–14. [CrossRef]
37. Farid, A.; Alrejjal, A.; Ksaibati, K. Two-lane highway crash severities: Correlated random parameters modeling versus incorporating interaction effects. *Transp. Res. Rec.* **2021**, *2675*, 565–575. [CrossRef]
38. Farid, A.; Ksaibati, K. Modeling severities of motorcycle crashes using random parameters. *J. Traffic Transp. Eng. (Engl. Ed.)* **2021**, *8*, 225–236. [CrossRef]
39. Eluru, N.; Bhat, C.R.; Hensher, D.A. A mixed generalized ordered response model for examining pedestrian and bicyclist injury severity level in traffic crashes. *Accid. Anal. Prev.* **2008**, *40*, 1033–1054. [CrossRef]
40. Fountas, G.; Anastasopoulos, P.C.; Abdel-Aty, M. Analysis of accident injury-severities using a correlated random parameters ordered probit approach with time variant covariates. *Anal. Methods Accid. Res.* **2018**, *18*, 57–68. [CrossRef]
41. National Research Council. *Highway Safety Manual*, 1st ed.; AASHTO: Washington, DC, USA, 2010.
42. De León Izeppi, E.; Flintsch, G.; Katicha, S.; McCarthy, R.; McGhee, K. *Locked-Wheel and Sideway-Force CFME Friction Testing Equipment Comparison and Evaluation Report*; No. FHWA-RC-19-001; U.S. Federal Highway Administration: Washington, DC, USA, 2019.
43. FHWA. Pavement Friction Management. Federal Highway Administration (FHWA). Available online: https://safety.fhwa.dot.gov/roadway_dept/pavement_friction/cpfm/ (accessed on 31 July 2022).
44. U.S. Census Bureau. 2010 Census Urban Area Reference Maps. Available online: https://www.census.gov/geographies/reference-maps/2010/geo/2010-census-urban-areas.html (accessed on 19 July 2022).
45. FHWA. Crash Modification Factors Clearinghouse. Available online: http://www.cmfclearinghouse.org (accessed on 19 July 2022).

eng

MDPI

Article

A Novel Method for Controlling Crud Deposition in Nuclear Reactors Using Optimization Algorithms and Deep Neural Network Based Surrogate Models †

Brian Andersen [1], Jason Hou [1,*], Andrew Godfrey [2] and Dave Kropaczek [2]

[1] Department of Nuclear Engineering, North Carolina State University, Raleigh, NC 27965, USA
[2] Oak Ridge National Laboratory, 1 Bethel Valley Road, Oak Ridge, TN 37830, USA
* Correspondence: jason.hou@ncsu.edu; Tel.: +1-919-513-6705

† Notice: This manuscript has been authored in part by UT-Battelle LLC, under contract DE-AC05-00OR22725 with the US Department of Energy (DOE). The publisher acknowledges the US government license to provide public access under the DOE Public Access Plan (http://energy.gov/downloads/doe-public-access-plan).

Abstract: This work presents the use of a high-fidelity neural network surrogate model within a Modular Optimization Framework for treatment of crud deposition as a constraint within light-water reactor core loading pattern optimization. The neural network was utilized for the treatment of crud constraints within the context of an advanced genetic algorithm applied to the core design problem. This proof-of-concept study shows that loading pattern optimization aided by a neural network surrogate model can optimize the manner in which crud distributes within a nuclear reactor without impacting operational parameters such as enrichment or cycle length. Several analysis methods were investigated. Analysis found that the surrogate model and genetic algorithm successfully minimized the deviation from a uniform crud distribution against a population of solutions from a reference optimization in which the crud distribution was not optimized. Strong evidence is presented that shows boron deposition in crud can be optimized through the loading pattern. This proof-of-concept study shows that the methods employed provide a powerful tool for mitigating the effects of crud deposition in nuclear reactors.

Keywords: convolutional neural network; genetic algorithm; crud; surrogate model; optimization

Citation: Andersen, B.; Hou, J.; Godfrey, A.; Kropaczek, D. A Novel Method for Controlling Crud Deposition in Nuclear Reactors Using Optimization Algorithms and Deep Neural Network Based Surrogate Models. *Eng* **2022**, 3, 504–522. https://doi.org/10.3390/eng3040036

Academic Editor: Antonio Gil Bravo

Received: 17 October 2022
Accepted: 20 November 2022
Published: 23 November 2022

Publisher's Note: MDPI stays neutral with regard to jurisdictional claims in published maps and institutional affiliations.

Copyright: © 2022 by the authors. Licensee MDPI, Basel, Switzerland. This article is an open access article distributed under the terms and conditions of the Creative Commons Attribution (CC BY) license (https://creativecommons.org/licenses/by/4.0/).

1. Introduction

Crud is a unique form of fouling in light water reactors (LWRs) caused by particulates—such as iron and nickel—depositing on fuel rods in the reactor as a result of system corrosion [1]. Crud imposes operational challenges to the current fleet of operating LWRs [2] and is strongly associated with subcooled boiling and high-power fuel regions, such as within fresh fuel assemblies loaded into the reactor [3]. For pressurized water reactors (PWRs), the primary issues caused by crud deposition are crud-induced localized corrosion (CILC) and crud-induced power shift (CIPS) caused by the uptake of soluble boron within the crud layer. Crud deposition is also associated with a pressure drop in nuclear reactors as well [4]. Methods of managing crud are based on conservatively bounding the risk associated with the occurrence of CIPS and CILC through the reactor core reload design. Some of these techniques, such as flattening the power distribution to reduce the overall steaming rate, increase the fuel cycle cost of the reactor due to an increase in the number of fresh fuel assemblies.

CIPS, also known as the axial offset anomaly (AOA), is depicted in Figure 1. CIPS is an unexpected downward shift in the power distribution, which manifests as a decrease in the axial offset (AO) of the reactor [5] with the potential for rapid AO increase in the event of crud burst. CIPS is caused by boron coming out of solution from the moderator and uptaking into the crud layer. This introduces an extraneous neutron absorber in the upper portion of crud-impacted assemblies [6]. Significant AO deviations from the operating target AO force reactor operators to decrease operating power to bring the reactor to a more stable operating regime. For example, Cycle 9 of the Callaway Plant had to reduce to 70% of rated operating power due to CIPS [7]. For a 1000 MWe PWR, every 1% decrease in operating power corresponds to an approximate loss of $10,000 per day in revenue due to the need for replacement power purchases [8].

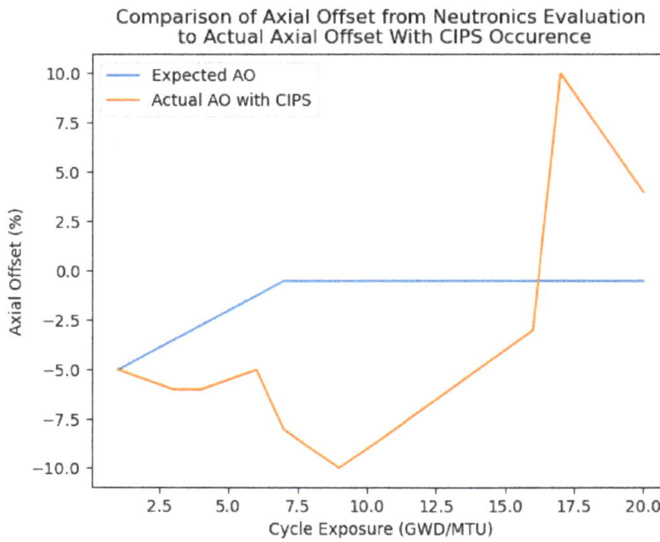

Figure 1. Comparison of the expected AO based on a neutronics analysis to the AO when CIPS occurs [9].

CILC is an increased rate of corrosion in the fuel cladding that arises from the insulating effect of crud on the fuel rod [10] and a higher temperature distribution across the fuel rod [11]. In addition, chemical interactions between the crud and cladding also play a role in increasing the rate of corrosion [12].

It is estimated that nuclear power plants spend $2 million annually in operations and maintenance costs related to crud [13]. Several methods have been proposed for preventing crud deposition. Strict PH control of the moderating water has been proposed as a possible solution, but PH control becomes more difficult as fuel assemblies achieve high burnup levels [2]. Advanced material coatings for fuel rods have also been proposed as a method for preventing crud deposition [8,14]. Ultrasonic cleaning of fuel assemblies has become a standard practice for mitigating crud deposition in the operating PWR fleet. Ultrasonic cleaning effectively reduces the thickness of the crud layer on fuel assemblies reloaded into the reactor, and it reduces the total mass of crud within the reactor [2]. The aforementioned methods have successfully reduced occurrences of CIPS and CILC in the operating reactor fleet with limitations. For example, ultrasonic cleaners do not remove enough particulate to prevent crud-induced effects.

Therefore, advanced simulation tools for predicting crud growth and its impact on reactor performance and safety are highly desired by reactor core designers in order to minimize or completely eradicate the adverse effects of crud. Some efforts have been made in the nuclear industry in this regard. The boron offset anomaly toolbox (BOA) is a tool developed by the Electric Power Research Institute (EPRI) and Westinghouse capable of predicting the mass of boron that deposits within crud on a nodal basis. BOA is often used to check a PWR loading pattern's susceptibility to CIPS [2]. Finally, utilities have developed their own proprietary operating limits for lowering instances of CIPS and CILC based on correlated parameters such as number of fuel rods predicted to undergo subcooled boiling or maximum soluble boron concentration. However, the use of BOA or utility operating limits for preventing CIPS and CILC can lead to overly conservative core loading patterns that require an increase in fresh fuel loading.

This work provides the initial progress of a new methodology for reducing the effects of crud deposition within PWRs by using a genetic algorithm and a neural network surrogate model based on the crud chemistry code MAMBA [9] for designing loading patterns that mitigate the effects of CIPS and CILC without penalizing other loading pattern objectives such as enrichment or cycle length. The crud chemistry code MAMBA was chosen for this work because it performs the crud calculations at the pin level. This level of depth is desired in order to readily understand how the optimization algorithm is changing the loading pattern design in order to account for crud deposition. CrUdNET, the neural network surrogate model developed for this work, is a necessary replacement of MAMBA because MAMBA simulations are computational cost prohibitive for use with optimization algorithms. For example, solutions were evaluated in sixteen parallel processes. Directly running MAMBA in this way would require 896 processors and take approximately 400 h to perform the optimization. In addition, an experimental database suitable for training a neural network based crud chemistry code is not available, which makes the simulation data the only feasible option.

A neural network, a popular family of machine learning (ML) algorithms, is used in this work as a surrogate model for crud evaluation. ML is observing increasing application in the field of nuclear engineering. ML algorithms have been applied to cross section predictions [15], neutron transport acceleration [16], and accident classification [17]. Neural networks were used as surrogate models in loading pattern optimizations. For example, they have been applied as a core simulator for evaluating loading pattern solutions [18,19]. Additionally, ML algorithms have been a preemptive evaluator to reduce the computational burden of the optimization [20].

Genetic algorithms (GAs) were chosen to perform core loading pattern optimization due to their long history of application in the field. They were one of the first optimization algorithms applied to the core loading pattern problem [21]. They have been successfully applied numerous times to PWRs [22,23], and they have also been used for the loading pattern optimization of boiling water reactors [24–27]. Moreover, GAs have served as the yardstick by which new optimization methodologies are measured. For example, GA was one of the benchmarks used to evaluate development of Tabu algorithms for fuel loading pattern optimization [28,29]. Likewise, it has been used to test the development of various particle swarm algorithms [30,31]. GAs were also used in a wide comparison of optimization methodologies for BWR loading pattern optimization [32].

2. Optimization Tools and Methods

This work made use of the neural network surrogate model crUdNET and a GA within the Modular Optimization Framework (MOF) [33,34]. This section provides a brief discussion of these tools, and how they are employed for crud optimization.

2.1. Neural Network for CRUD Modeling

Pin-level crud calculations are desired to understand how changes in the loading pattern affect the crud distribution. This necessitates the use of the crud chemistry code MAMBA for its capability of calculating crud deposition on a pin level basis, as opposed to BOA which provides results on a nodal level.

MAMBA has been integrated into the core simulator VERA [9,35]. Through VERA, MAMBA is coupled to the neutronics solver MPACT [36] and subchannel thermal hydraulics code CTF [37]. MAMBA uses information provided by these two codes and Equation (1) to calculate the surface deposition of crud on every fuel rod across a PWR [9].

$$C_{dens}(t+\delta t) = C_{dens}(t) + \delta t\Big((k^{p}_{s,nonboil} + k^{p}_{s,boil}q''_{s,boil})N_{cool} - \gamma k_{tke}\Big) \quad (1)$$

In Equation (1), $C_{dens}(t+\delta t)$ represents the deposited crud molar density after timestep δt. $k^p_{s,nonboil}$ represents the non-boiling crud deposition rate. $k^p_{s,boil}$ represents the boiling deposition rate, and $q''_{s,boil}$ represents the boiling heat flux obtained through the VERA coupling. $k^p_{s,nonboil}$, $k^p_{s,boil}$, and $q''_{s,boil}$ are multiplied by the term N_{cool}, which represents the concentration of nickel-ferrite particulate in the reactor coolant. Lastly, γ and k_{tke} represent the erosion rate and surface kinetic energy which account for the crud that erodes from the surface of the fuel rod [9].

MAMBA coupled within VERA requires significant numbers of processors and wall-clock time. This makes a fully coupled MAMBA analysis unsuitable for use in an optimization algorithm. For this reason, MAMBA is replaced with a convolutional neural network (CNN) based on the U-NET neural network architecture [38] to assess crud deposition. Reference [33] details why the U-NET neural network architecture was selected for the surrogate model.

Figure 2 shows the architecture of the CNN surrogate model, crUdNET. CrUdNET was designed to predict the change in the crud distribution at a single axial layer of the 3D CTF mesh in a reactor core. In essence, crUdNET can be thought of as replacing Equation (1) with Equation (2) for performing reactor core crud deposition calculations.

$$C_{sur-dens}(t+\delta t) = C_{sur-dens}(t) + F(\Delta P, N_{cool}, B_{cool}, E), \tag{2}$$

In Equation (2), F represents the change in the crud surface density as predicted by crUdNET, and the crud densities are altered from molar densities to surface mass densities at the beginning and end of a time step, $C_{sur-dens}(t)$, $C_{sur-dens}(t+\delta t)$. The density is altered because VERA reports the surface mass density in units g/cm^2, rather than the molar density mol/cm^3. As the primary driver of crud deposition in MAMBA, F is naturally a function of the nickel-ferrite particulate in the coolant, N_{cool}, in parts per billion [9]. By using multiple trained networks, developed uniquely for each axial layer, a reconstruction of the three-dimensional (3D) crud distribution is obtainable. Thus, the use of crUdNET reformulates the crud deposition analysis from an analytical problem to a pattern recognition problem. The soluble boron concentration in the coolant in parts per million (ppm), B_{cool}, and end of time step cycle exposure in Giga-Watt-days/Metric-Ton-Uranium (GWD/MTU), E are also provided to aid in pattern recognition. In broad terms, cycle exposure accounts for the nuclear fuel residence time in the core, while soluble boron concentration reflects the reactivity of the fuel. The higher the soluble boron concentration, the more reactive the fuel is, and so more crud should likely deposit. In other words, less and less crud will deposit in the reactor towards the end of the cycle exposure. Lastly, F is a function of the change in the whole core pin power distribution, ΔP, given by Equation (3).

$$\Delta P = P(t+\delta t) - P(t) \tag{3}$$

The leakyReLU activation function, given in Equation (4), is used as the activation between all layers in crUdNET [39]. The number of nodes used in the layers of crUdNET are provided in Table 1. Convolutional layers used a window size of 3 × 3.

$$f(x) = \begin{cases} -0.1 * x & x < 0 \\ x & x \geq 0 \end{cases} \tag{4}$$

The difference in the pin power distribution, ΔP, is provided as input to the "U" portion of the neural network. Here, the data is first normalized in the batch normalization layer before being transformed by a series of 2D convolutional and averaging nodes. These nodes transform the data, shaping it from a core-wide matrix of data to an assembly wide matrix which identifies which assemblies are most likely to see significant changes in crud deposition. Through a series of more convolutional nodes, upsampling nodes, and concatenation nodes, the network then transforms this data into the core-wide crud distribution, in relative quantities, for a single layer in the 3D CTF mesh. Meanwhile, the inputs N_{cool}, B_{cool}, and E are fed into the linear dense connections of the neural network where they are transformed and normalized to determine the scale of the change in the crud deposition. Lastly the spatial distribution and scaling terms are multiplied together and output from the neural network in order to get the final change in the crud distribution. As Equation (2) shows, the cycle crud deposition can then be calculated by summing the outputs of the network over each timestep [40].

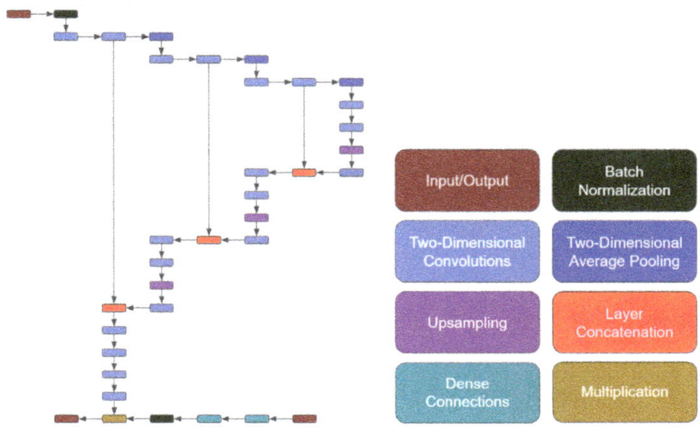

(a) (b)

Figure 2. (**a**) The neural network architecture of the crUdNET surrogate model. (**b**) Key proivding each of the neural network layers used in crUdNET.

Table 1. Number of nodes used in the layers of crUdNET.

Layer Numbers	Number of Nodes	Layer Types
1,2	136 × 136 × 49	Convolution Input, Convolution Normalization
3, 4, 5	136 × 136 × 16	Convolution 2D, Average Pooling 2D
6, 7, 8	68 × 68 × 32	Convolution 2D, Average Pooling 2D
9, 10, 11	34 × 34 × 64	Convolution 2D, Average Pooling 2D
12, 13, 14, 15, 16	17 × 17 × 128	Convolution 2D, Upsampling, Concatenation
17, 18, 19, 20, 21	34 × 34 × 64	Convolution 2D, Upsampling, Concatenation
22, 23, 24, 25, 26	68 × 68 × 32	Convolution 2D, Upsampling, Concatenation
27, 28, 29	128 × 128 × 16	Convolution 2D
30, 35, 36	128 × 128 × 1	Convolution 2D, Multiplication, Output
31	3	Dense Input
32, 33, 34	16	Dense, Dense Normalization

CrUdNET was trained based on a fixed time step, δt, of 0.5 GWd/MTU. Pin-powers, soluble boron concentrations, and cycle exposures are provided by a nodal analysis code. which was used to replace MPACT for nuclear analysis to reduce the computational burden of developing the training library, while this means that the CTF+MAMBA analyses are decoupled from neutronics, this does not impact this work [33].

CrUdNET was trained on a library of 6600 unique sets of input and output data in an 80/20 training/validation split. The performance of crUdNET was then tested against a further 1500 unique samples [33]. The training and testing inputs were developed through repeated core loading pattern optimizations to obtain ΔP, B_{cool}, and E. N_{cool} was obtained through random sampling. Figure 3 provides a comparison of MAMBA and crUdNET for a crud distribution at end of cycle (EOC) for a reactor predicted by crUdNET. Figure 3 shows that crUdNET provides acceptable agreement with MAMBA in predicting crud distributions. Figure 3 also represents a computational power reduction from 540 processors and 1 h of wall clock time to a single processor and 30 s of wall clock time. More detailed explanations on the development and training process for crUdNET used in this work were previously presented in reference [33].

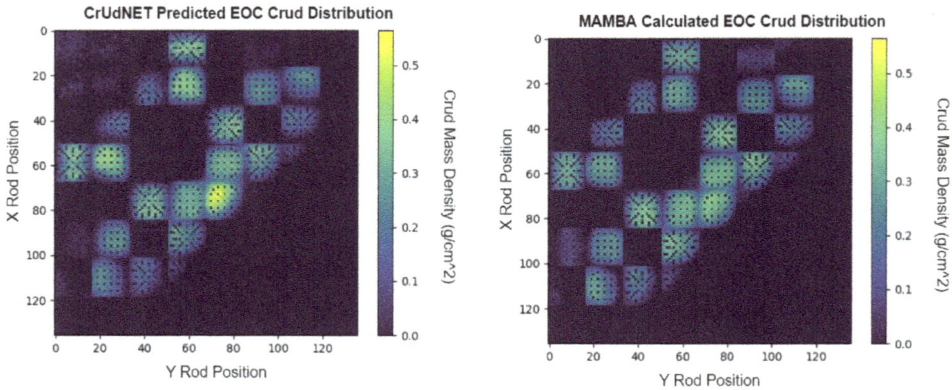

(a) (b)

Figure 3. (**a**) The crud distribution at EOC as predicted by crUdNET. (**b**) The crud distribution at EOC as calculated by MAMBA for a reactor core.

2.2. Genetic Algorithm

The main features of a GA are crossover, mutation, and selections [41]. For this work, the GA was developed using MOF, an object-oriented code for facilitating the rapid development and application of optimization algorithms [34]. This GA utilized is relatively standard.A flowchart of the GA is provided in Figure 4.

The initial population of solutions is generated by randomly selecting assembly types allowed for each core location in all initial solutions. All solutions that pass the selection process become parents to the next generation of solutions. Solutions are selected to undergo either mutation or crossover, based on the mutation rate. For each solution a random number is drawn. If the number is less than the current mutation rate, the solution is selected to undergo mutation. Otherwise the parent solution will create a new solution through crossover. All solutions selected for crossover are designed to mate and undergo crossover with the most genetically similar solution. This mating is performed by selecting the first un-mated solution, and examining the remaining un-mated solutions for the highest number of fuel assembly types in the same position. These solutions are then mated for crossover. A solution can only be mated once for undergoing crossover.

Crossover is performed by exchanging fresh fuel assemblies in the same core location between the two genomes while the positions of reloaded fuel assemblies are shuffled within the core. These restrictions on crossover ensure that the inventory on fresh and burned fuel assemblies is preserved throughout the entire optimization. This is done in place of other techniques, such as throwing out solutions that violate the used fuel inventory and desired number of fresh fuel assemblies.

Mutation is performed in two ways. Fresh fuel assemblies are allowed to be freely replaced with other available fresh fuel assembly designs, or fresh fuel assemblies can swap their position in the core with another fuel assembly. Reloaded fuel assemblies are allowed to exchange positions only within the other fuel assemblies in the solution. The number of solutions that undergo mutation is determined by the mutation rate R, and Equation (5) [34].

$$R_{new} = 1 - \Delta_{mutation}(1 - R_{current}), \tag{5}$$

where R_{new} and $R_{current}$ are the updated and current mutation rate, respectively, and $\Delta_{mutation}$ is defined by

$$\Delta_{mutation} = \frac{\ln\left(\frac{1 - R_{final}}{1 - R_{initial}}\right)}{N}, \tag{6}$$

where $R_{initial}$ and R_{final} are the initial and final mutation rate, respectively, [34].

Selection is performed using the tournament method [42]. The tournament method is completely random, allowing parents to compete against child solutions, child solutions to compete against other child solutions, and parent solutions to compete against other parents. Specific parameters of the GA, such as the mutation rate and population size, are provided with the relevant optimization performed.

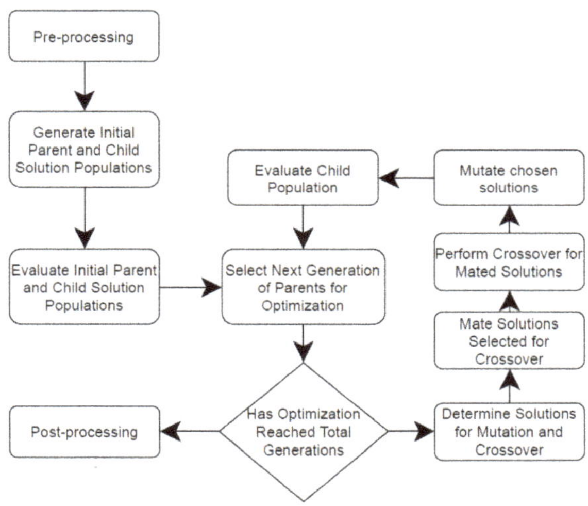

Figure 4. Flowchart of the GA used for loading pattern optimization developed through MOF.

2.3. *Crud Optimization Methodologies*

Solutions generated by MOF are evaluated in a two-step process. In the first step, a neutronic analysis is performed to evaluate the loading pattern designed by the GA. This provides information on the radial rod power peaking, soluble boron concentrations, and cycle length. This also provides soluble boron concentrations, exposures, and the quarter-core power distribution for crUdNET. In the second step, these values are combined with a nickel particulate concentration history to calculate the crud distribution produced by the solution. As previously mentioned, solutions were evaluated in sixteen parallel processes. This was a limit set by a system limit on the number of parallel nuclear simulation evaluations that could be performed in parallel.

In order to evaluate the effectiveness of optimizing crud, an initial optimization is first performed without any optimization objectives related to crud. The optimization is then re-performed. This second optimization includes an objective related to crud, and it begins from an initial random population just as the first optimization. The crud optimization is considered successful if the optimized solution provides similar results to the initial optimization in regard to the non-crud objectives, and must show improved performance in regard to crud over the first optimized solution when evaluated using CTF+MAMBA. For this work, the results of MAMBA calculations are taken as the true crud deposition.

Three optimization objectives unrelated to crud were used in each optimization. The first objective was maximizing the cycle length based on a fixed number of fresh fuel assemblies. This was used in place of meeting requirements on a specified cycle length and minimizing the core-wide enrichment. The second objective was minimizing the cycle peak soluble boron concentration. The third objective was minimizing radial rod power peaking (FΔH). This is calculated using Equation (7).

$$F\Delta H = \frac{\text{Peak Rod Power}}{\text{Core Average Rod Power}} = \frac{Max \frac{1}{L}\int_0^L P(x,y,z)dz}{\frac{1}{V_{Core}}\int\int\int_{V_{Core}} P(x,y,z)dxdydz}. \tag{7}$$

Core loading pattern optimization was performed on the third cycle of a four-loop, 193 assembly, Westinghouse PWR. This model used geometry and operating conditions from the publicly available P9 progression problem published by CASL [43]. Heuristic restrictions were imposed on where certain types of fuel assemblies could be placed when generating initial solutions. This is a requirement and limitation of MOF in order to maintain the desired fuel inventory because MOF can only directly track the number of decision variables in a group, not how the placement of those decision variables affect total assembly count in a full core arrangement. Fuel assemblies were divided into four symmetry groups based on allowed location in the core (i.e., major and minor axis, non-axis), and whether they were a fresh or previously burned fuel assembly. Figure 5 provides the allowed locations of assemblies for the four groups. 1's denote allowed locations and 0's denote prohibited locations. Figure 5 shows that fresh fuel assemblies are grouped based on octant or quarter symmetry depending on whether they are axis or non-axis locations.

0
0 0
0 1 0
0 1 1 0
0 1 1 1 0
0 1 1 1 1 0
0 1 1 0 0 0
0 0 0 0

(**a**) Fresh Octant

0
1 1
1 0 1
1 0 0 1
1 0 0 0 1
1 0 0 0 0 1
1 0 0 0 0 0
0 0 0 0

(**b**) Fresh Quarter

1
1 1
1 0 1
1 0 0 1
1 0 0 0 1
1 0 0 0 0 1
1 0 0 0 0 0
0 0 0 0

(**c**) Burned Quarter

1
1 1
1 1 1
1 1 1 1
1 1 1 1 1
1 1 1 1 1 1
1 1 1 1 1 1
1 1 1 1

(**d**) Burned Octant

Figure 5. Decision variable maps for the four fuel assembly groups used in each optimization case: (**a**) fresh fuel assemblies in octant symmetry, (**b**) fresh fuel assemblies in quarter symmetry, (**c**) burned fuel assemblies previously placed in quarter symmetry, (**d**) burned fuel assemblies previously placed in octant symmetry.

3. Optimization Methodologies and Results

Two optimization methodologies were tested. The first sought to reduce the total mass of crud in the core. The second sought to have a uniform amount of crud distribute on all fuel rods in the core.

3.1. Total Crud Mass Reduction

The logical first objective regarding crud deposition would be minimizing the total mass of crud that deposits within the reactor core. Reducing the total mass of crud that deposits in the core reduces the risk of CIPS and CILC. A methodology for reducing the total crud mass was proposed, however it turned out that this optimization objective could not be optimized using the chosen toolset.

The proposed optimization objective formulation for reducing the total core crud mass was quite simple. Three crUdNET models were trained. Each model predicted the crud distribution at a different axial elevation. Per Equation (8), these three predictions, N_{planes}, are summed across all fuel rods, N_{rods}, to produce a single crud mass value.

$$m_{total}^{crud} = \sum_{j=1}^{N_{planes}} \sum_{i=1}^{N_{rods}} m_{ij}^{crud} \tag{8}$$

The crud mass optimization methodology was explored using six test cases. Cases differed in two ways: (1) whether the limiting value FΔH was 1.55 or 1.60, and (2) whether the case used 84, 88, or 92 fresh fuel assemblies. This exploratory study consisted of generating an optimized loading pattern using MOF based on each of the six cases using the previously described non-crud related optimization objectives. These objectives were maximizing cycle length, measured in effective full power days (EFPD), and meeting the described limits on maximum FΔH and a maximum soluble boron concentration less than 1300 ppm. These cases were then repeated with the inclusion of crUdNET and the crud mass objective described in Equation (8). The twelve optimized cases were then re-analyzed using CTF+MAMBA to determine if the combination of MOF and crUdNET had noticeably reduced the total mass of crud in the core. For this work, the results of MAMBA calculations are taken as the true crud deposition.

The loading pattern parameters for the cases optimized without crud are presented in Table 2. Results for the cases optimized with crud as an optimization objective, are presented in Table 3. Both tables also provide the mass of crud as predicted by crUdNET. This mass is significantly smaller than the total core crud mass because of the use of a subset of axial planes used in the analysis, as described previously. Figure 6 shows the FΔH values over the course of the cycle for the twelve highest fitness solutions, and Figure 7 provides the soluble boron concentration.

Tables 2 and 3 show that in five of the six cases optimized, crUdNET evaluation of the crud objective within the GA lowered the total mass of crud deposited. These tables also show that the highest fitness loading patterns for the 12 optimizations performed are unique. This is further reinforced by Figures 6 and 7, which show unique FΔH and soluble boron concentration histories for each of the 12 cases analyzed.

Table 2. Optimization objective values, including crud mass predicted by crUdNET, for highest fitness solution for optimizations performed without total crud mass optimization objective.

LP Number	Number Fresh Assemblies	Limiting FΔH Value	Maximum Boron Concentration (PPM)	Maximum FΔH	Cycle Length (EFPD)	Predicted Crud Mass (g)
1	84	1.55	1290.9	1.512	447.3	535.95
2	88	1.55	1297.9	1.550	474.4	711.17
3	92	1.55	1295.1	1.530	485.9	638.64
4	84	1.60	1271.7	1.594	461.1	858.50
5	88	1.60	1267.4	1.574	473.6	722.67
6	92	1.60	1289.6	1.596	482.8	925.37

Table 3. Optimization objective values, including crud mass predicted by crUdNET, for highest fitness solution for optimizations performed with total crud mass optimization objective.

LP Number	Number Fresh Assemblies	Limiting FΔH Value	Maximum Boron Concentration (PPM)	Maximum FΔH	Cycle Length (EFPD)	Predicted Crud Mass (g)
7	84	1.55	1238.6	1.543	445.3	464.14
8	88	1.55	1287.2	1.544	470.2	478.97
9	92	1.55	1323.6	1.542	486.0	765.92
10	84	1.60	1298.1	1.572	461.1	846.86
11	88	1.60	1288.6	1.590	471.9	680.65
12	92	1.60	1290.1	1.594	482.6	640.67

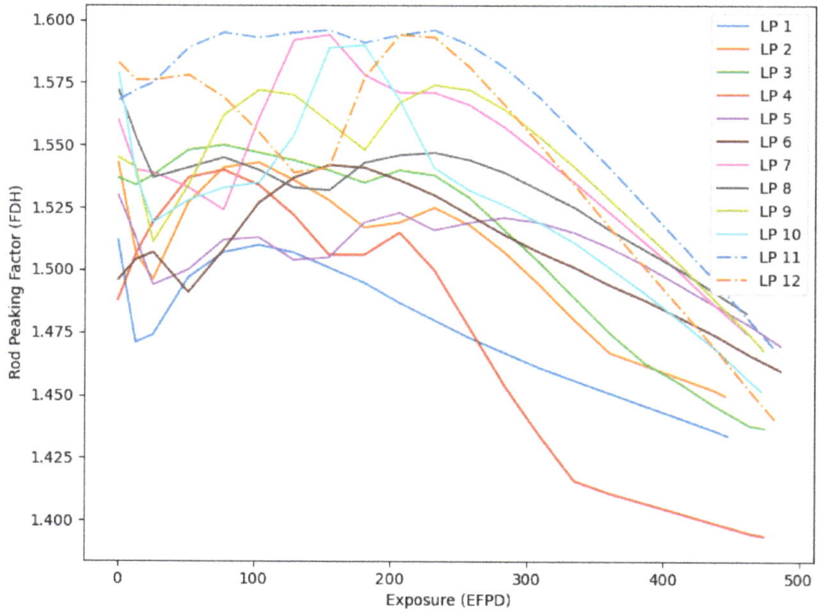

Figure 6. Comparison of FΔH versus exposure for the highest fitness solutions for the six optimization cases with and without crud objectives.

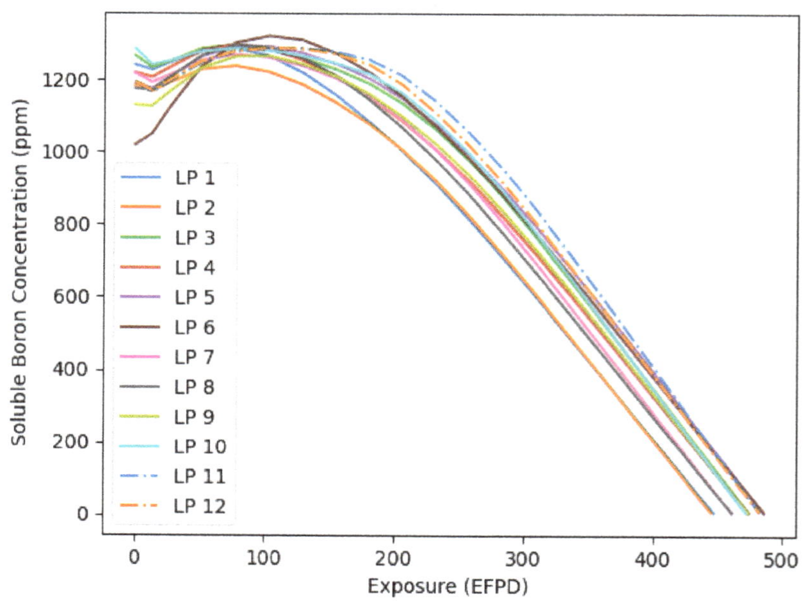

Figure 7. Comparison of the soluble boron concentration for the highest fitness solutions for the six optimization cases with and without crud objectives.

Figure 8 shows the total mass of crud within the core, as calculated by MAMBA, over the length of the cycle for the 12 cases. Table 4 provides the EOC total crud mass for the twelve cases.

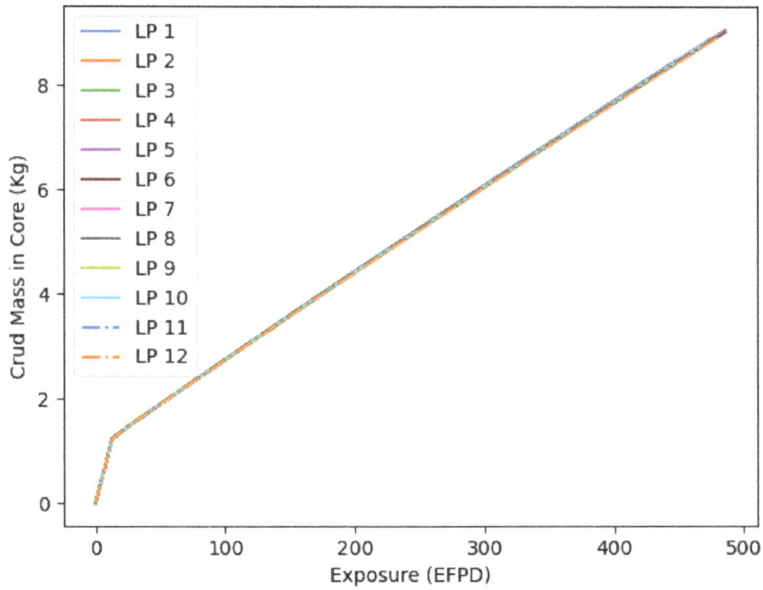

Figure 8. Whole crud mass, as calculated by MAMBA, versus exposure for loading patterns analyzed in crud mass optimization.

Table 4. Comparison of total core crud mass, as calculated by MAMBA, at cycle exposure of 438 EFPD for the 12 highest fitness solutions for the total core crud mass optimization demonstrations.

Fresh Assembly Count	Limiting FΔH Value	Crud Mass without Crud Optimization (Kg)	Crud Mass with Crud Optimization (Kg)
84	1.55	8.328	8.308
88	1.55	8.308	8.315
92	1.55	8.311	8.268
84	1.60	8.290	8.296
88	1.60	8.279	8.296
92	1.60	8.277	8.259

Tables 2–4 and Figure 8 mean several things. The significant difference in mass between the crUdNET predictions and MAMBA calculations indicate that the use of a three-layer modeled by crUdNET is not sufficient to represent the whole-core crud mass. Additionally, MAMBA calculating the same crud mass for all twelve loading pattern designs indicate MAMBA is not mature in regard to the total crud mass deposited. It is unlikely for the nickel particulate concentration in the coolant to be the sole factor in determining the core wide crud mass, and for the power distribution to not significantly impact the core-wide crud mass.

Improvements in MAMBA will improve both itself and crUdNET, and further refinement of crUdNET will improve its predictive capability when it comes to total core crud mass. This work will make it possible to use crUdNET, in conjunction with an optimization algorithm, to design loading patterns that drive down the mass of crud that deposits within the reactor core. In the short term however, this means that a different optimization objective is required to demonstrate that crUdNET can successfully be used to optimize loading patterns in regard to crud deposition.

3.2. Crud Deposition Analysis

To demonstrate that crUdNET in conjunction with optimization algorithms could control crud deposition, an optimization objective to maximize the uniformity of the crud distribution over the entire reactor was adopted. In other words, the objective is to have as many fuel rods with the same amount of crud as possible. This methodology modeled a single axial layer. Deviation from a uniform distribution was measured via Equations (9) and (10):

$$M_{ave} = \frac{\sum_{i=1}^{N_{rods}} M_i}{N_{rods}^2}, \tag{9}$$

$$D = \sum_{i=1}^{N_{rods}^2} |M_i - M_{ave}|. \tag{10}$$

For Equations (9) and (10), M_i is the crud mass density for fuel pin i, M_{ave} is the average crud mass density, and D is the deviation from the average value.

Equation (11) is the fitness equation for the deviation from uniform methodology analysis.

$$Fitness = D - 10 \cdot max(L_{cycle} - T_{cycle}, 0) - 1500 \cdot \max(F\Delta H_m - 1.55, 0) - 2 \cdot \max(C_m^{sb} - 1300, 0) \tag{11}$$

In Equation (11), D is the deviation from a uniform distribution given by Equation (10), L_{cycle} represents the solution cycle length, $F\Delta H_m$ is the maximum rod power peaking, and C_m^{sb} is the maximum soluble boron concentration.

The GA used a population size of 30 and iterated solutions over 200 generations. Initially, 25% of solutions were mutated, but the number of solutions grew to 55% of solutions by the end of the optimization. Assemblies used in the optimization had enrichments of 4.4, 4.7, or 4.9 w/o. IFBA, gadolinium, and pyrex were used as burnable poisons in the fuel assemblies for all three enrichments. Fuel assemblies containing IFBA used 80 or 120 IFBA rods. Fuel assemblies using gadolinium had either 12 or 24 rods containing gadolinium. Gadolinium was utilized at 3%, 5%, and 8% w/o. Finally, assemblies containing pyrex as a burnable poison used a 12, 16, or 24 pyrex rod configuration.

3.2.1. Optimization Results

Figure 9 provides the deviation of crud from a uniform distribution—as calculated by Equations (9) and (10)—in the reactor over the course of the cycle as predicted by crUdNET for the reference and test methodology optimizations. The reference optimization did not include the optimization objective related to crud. The test optimization methodology sought minimize the deviation from a uniform crud distribution. Figure 9 shows that the use of crUdNET significantly improved the uniformity of the crud distribution across the reactor core.

As mentioned, crUdNET and the optimization algorithm are considered to have successfully optimized the crud distribution if the optimization objective for the test methodology is improved over the reference value when solutions for both optimizations are evaluated using MAMBA. Figure 10 is reproduced Figure 9 using MAMBA, rather than crUdNET, to calculate the crud distribution for the population of solutions. It shows that although the difference between the two populations in terms of deviance from a uniform crud distribution is significantly smaller, the population of solutions optimized for crud clearly shows lower values, and thus a more uniform crud distribution than the reference optimization population.

Figure 10 shows that crUdNET and optimization algorithms can be combined to directly optimize crud distributions through the fuel loading pattern. This provides a significant advancement toward reducing the effects of CIPS and CILC within PWRs by providing core design engineers with a direct means of evaluating and manipulating the crud distribution when designing loading patterns. This is an improvement over current methods that seek to reduce crud deposition based on correlated parameters or evaluating crud as part of a post processing analysis.

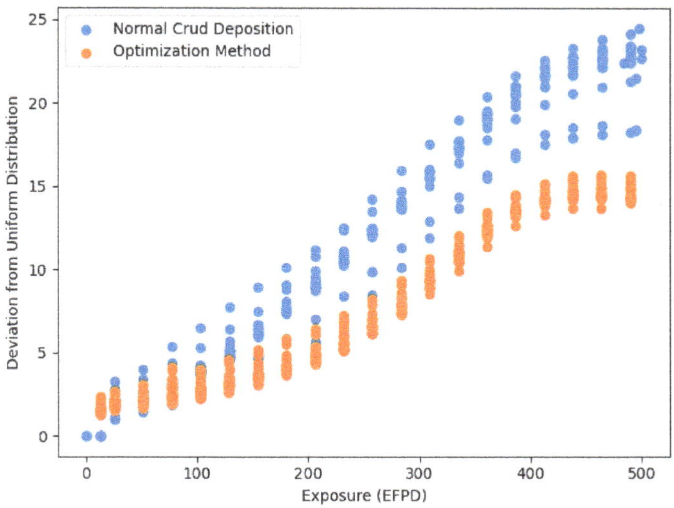

Figure 9. Deviation from a uniform crud distribution using crUdNET with and without considering crud as an optimization objective.

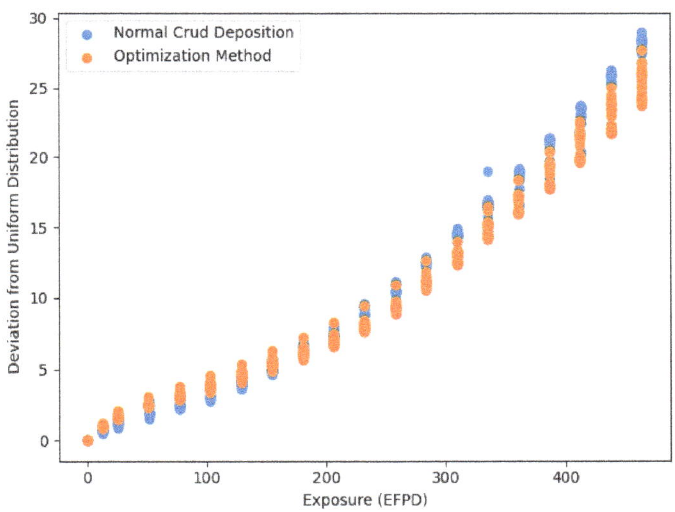

Figure 10. Deviation from a uniform crud distribution using MAMBA with and without considering crud as an optimization objective

It is also important to understand how the genetic algorithm optimized the loading pattern in regard to crud. Figure 11 provides the loading patterns for the highest fitness solutions for the reference optimization and crud optimization test methodology. The comparison shows that the loading pattern for the test methodology has a larger concentration of fresh fuel assemblies toward the outer edge of the reactor core.

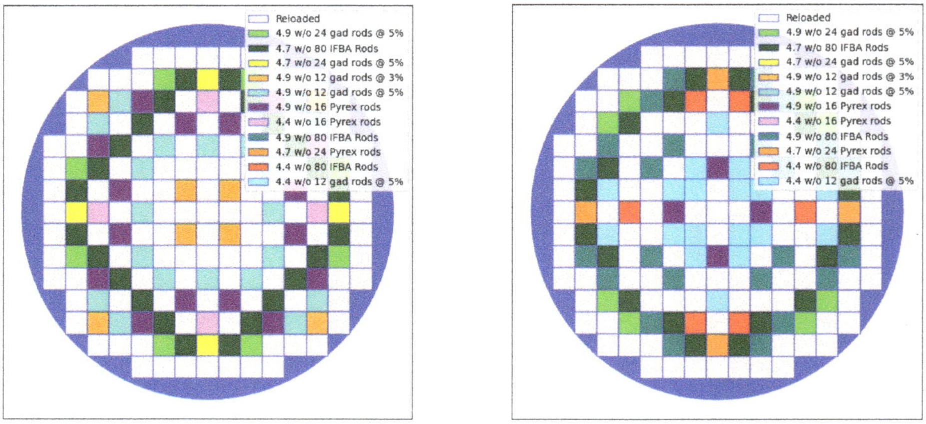

(a) (b)

Figure 11. (**a**) The loading pattern for the highest fitness solution in the reference optimization. (**b**) The loading pattern for the highest fitness solution in the test methodology optimization. The loading pattern optimized in the test methodology has far fewer fresh assemblies towards the center of the core than the reference optimization loading pattern.

The effects are shown in Figures 12 and 13, which compare the power distributions between the loading patterns at beginning of cycle (BOC) and middle of cycle (MOC), respectively. The figures show that the loading pattern optimized for crud maintains higher and denser power concentrations over the reference optimization.

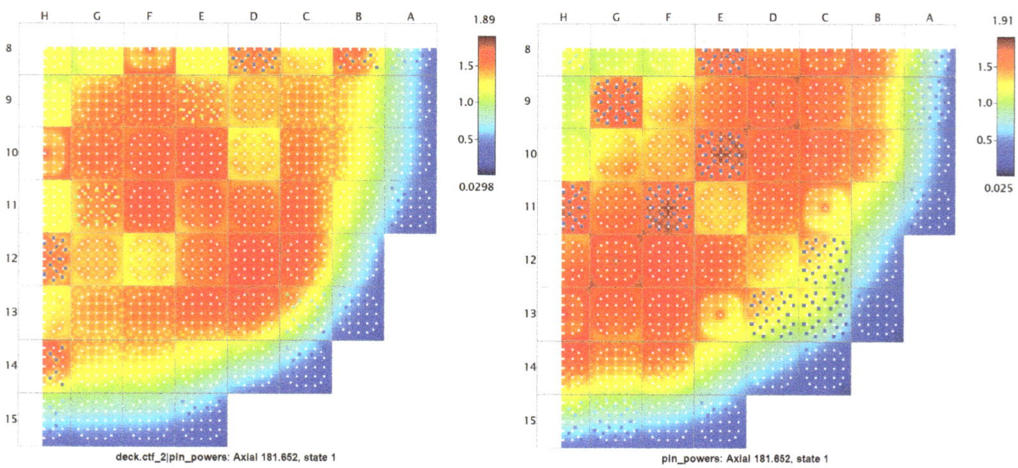

(a) (b)

Figure 12. (**a**) The BOC power distribution for the highest fitness solution of the reference optimization. (**b**) The BOC power distribution for the highest fitness solution of the crud test methodology. The solution optimized for crud shows higher power concentrations over the reference solution.

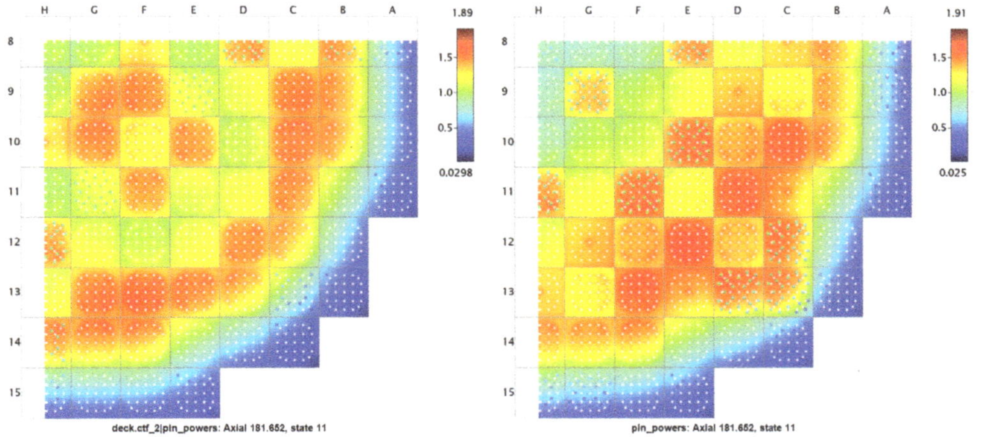

(a) (b)

Figure 13. (**a**) The MOC power distribution for the highest fitness solution in the reference population. (**b**) The MOC power distribution for the highest fitness solution in the crud test methodology population. The crud optimized solution shows continued higher power concentrations than the reference case, which has more distributed power.

Figure 14 illustrates the effect of these higher power concentrations on the crud distribution in the loading patterns. Figure 14 indicates that the GA optimized the loading pattern to have a minimal deviation from uniform crud distribution by concentrating the power distribution so that crud grows very densely on a small number of assemblies. This results in most fuel rods having no crud on them, and so the deviation from a uniform distribution is minimized.

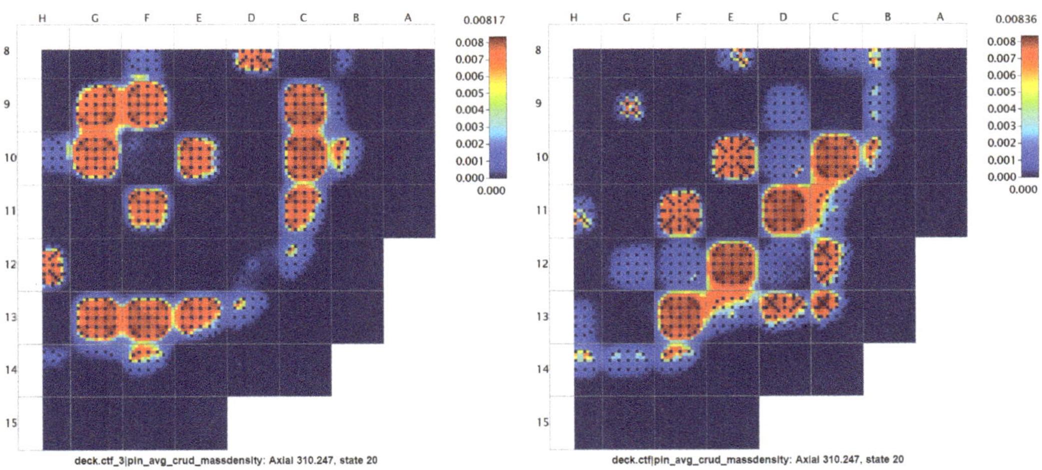

(a) (b)

Figure 14. (**a**) The EOC crud distribution for the highest fitness solution in the reference optimization population. (**b**) The EOC crud distribution for the highest fitness solution in the crud test methodology. The GA optimized the deviation from uniform crud distribution objective by designing the loading pattern to concentrate the power. This causes crud to grow in only a few assemblies.

This shows that there is room for improvement in both MAMBA and the surrogate model crUdNET. However, the needs for improvement do not discount this work. Improvements in both crUdNET and MAMBA will only increase the effectiveness of the methods discussed here. The optimization objective demonstrated that the combination of crUdNET and a GA can be successfully used for a multi-objective optimization that designs the crud distribution in a given way while also meeting other requirements such as rod power peaking and cycle length.. However, in practice, dense distributions of crud, such as the one shown in Figure 14, created through the optimization are undesirable. Refinement of MAMBA, crUdNET, and the crud optimization objective will significantly improve reactor performance with respect to CIPS and CILC.

3.2.2. Note on the Mass of Boron in Crud

Although not the focus, the boron mass deposited in crud was calculated as part of the MAMBA crud deposition analysis. The total boron mass in crud over cycle exposure for cases analyzed in MAMBA are presented in Figure 15. Figure 15 shows that there is nearly a 50 g range mass of boron that uptakes into the crud layer. This implies that loading pattern optimization with a reduced-order model or fast surrogate model trained for boron in crud prediction could effectively reduce the impact or occurrence of CIPS without penalizing other parameters such as enrichment or cycle length.

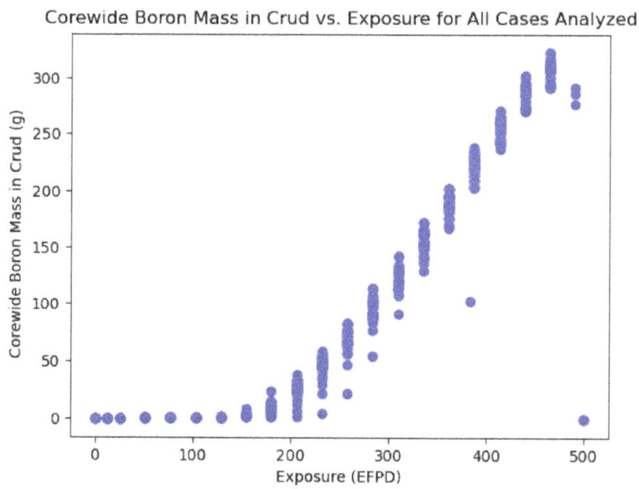

Figure 15. The boron mass in crud values for all cases analyzed using MAMBA show a large amount of variation, indicating that loading pattern optimization can be utilized to reduce instances of CIPS.

4. Conclusions and Future Work

This work presents a proof of concept demonstration that neural network surrogate models combined with optimization algorithms such as the GA can optimize properties related to crud deposition in nuclear reactors via loading pattern optimization. Deficiencies in both MAMBA and the modeling approach taken with crUdNET prevented optimization of the mass of crud that deposits in the core. However, by setting a crud related optimization objective to minimize the deviation from a uniform crud distribution, it was shown that the GA could successfully use crUdNET to develop loading patterns that outperformed a reference optimization regarding this parameter without sacrificing other objectives of the loading pattern optimization including power peaking, cycle length, and maximum soluble boron concentration.

CrUdNET's accuracy requires some improvement. The most immediate way in which the fidelity of crUdNET could be improved is through the introduction of ensemble modeling. Ensemble modeling is a powerful tool for increasing the predictive capability of neural networks, and the expansion of the surrogate modeling used in this work to multiple neural network architectures trained using varying data sets would greatly increase the accuracy of crUdNET. Additionally, it was shown that there is need for improvement in MAMBA, particularly regarding the crud mass calculations. As MAMBA matures the strength of the neural network surrogate models and the efficacy of the methods demonstrated here will only improve.

These improvements will also allow for the further exploration of optimization objectives related to crud. The deviation from a uniform crud distribution was used here based on the current capabilities of both MAMBA and crUdNET. Improvements to MAMBA and crUdNET will allow for the inclusion of other optimization objectives such as the core crud mass or maximum density of crud on fuel assemblies. Lastly, Figure 15 showed that there is a significant amount of variance between loading pattern designs in the total mass of boron that uptakes into the crud distribution. This provides the motivation for developing a surrogate model dedicated to predicting the boron uptake into the crud layer. The development of such a model would allow for the direct analysis and inclusion within the optimization of a loading patterns risk to CIPS using the methods demonstrated in this work.

Author Contributions: Conceptualization, J.H., A.G. and D.K.; Methodology, B.A., J.H., A.G. and D.K.; Software, J.H. and A.G.; Formal analysis, B.A.; Investigation, B.A. and D.K.; Data curation, B.A.; Writing—original draft, B.A.; Writing—review & editing, J.H. and D.K.; Visualization, B.A.; Supervision, D.K.; Funding acquisition, D.K. All authors have read and agreed to the published version of the manuscript.

Funding: This research was supported by the Consortium for Advanced Simulation of Light Water Reactors (www.casl.gov), an Energy Innovation Hub (http://www.energy.gov/hubs) for Modeling and Simulation of Nuclear Reactors under US Department of Energy (DOE) contract no. DE-AC05-00OR22725. This research used resources of the Compute and Data Environment for Science at the Oak Ridge National Laboratory, which is supported by the DOE Office of Science under contract no. DE-AC05-00OR22725. This research used the resources of the High Performance Computing Center at Idaho National Laboratory, which is supported by the DOE Office of Nuclear Energy and the Nuclear Science User Facilities under contract no. DE-AC07-05ID14517.

Data Availability Statement: Not applicable.

Acknowledgments: The authors wish to express their thanks to researchers at Oak Ridge National Laboratory for their work on VERA and MAMBA, as well as for their help on this project.

Conflicts of Interest: The authors declare no conflict of interest.

References

1. Thomas Wellock, U.S. NRC Blog, Crud: Another Acronym Bites the Dust. Available online: https://public-blog.nrc-gateway.gov/2015/03/31/crud-another-acronym-bites-the-dust/ (accessed on 10 September 2022).
2. Frattini, P.L.; Blok, J.; Chauffriat, S.; Sawicki, J.A. Axial offset anomaly: Coupling PWR primary chemistry with core design. *Nucl. Energy* **2001**, *40*, 123–135. [CrossRef]
3. Uchida, S.; Asakura, Y.; Suzuki, H. Deposition of boron on fuel rod surface under subcooled boiling conditions- An approach toward understanding AOA occurrence. *Nucl. Eng. Des.* **2011**, *241*, 2398–2410. [CrossRef]
4. Johnson, N.; Wu, J.; Morrison, J.; Connolly, B.; Banks, A. Mechanisms of Crud Deposition in Pressurised Water Nuclear Plant. In Proceedings of the 17th International Conference on Environmental Degradation of Materials in Nuclear Power Systems—Water Reactors, Ottawa, Canada, 9–13 August 2015.
5. Roe, J. Effects of Crud Buildup and Boron Deposition on Power Distribution and Shutdown Margin. Available online: https://www.nrc.gov/reading-rm/doc-collections/gen-comm/info-notices/1997/in97085.html (accessed on 10 September 2022).
6. Joe, J.H.; Kim, S.J.; Jones, B.G. A study of solute transport of radiolysis products in crud and its effects on crud grown on PWR fuel pin. *Nucl. Eng. Des.* **2016**, *300*, 433–451. [CrossRef]
7. Sawicki, J.A. Evidence of Ni_2FeBO_5 and $m-ZrO_2$ precipitates in fuel rod deposits in AOA-affected high boiling duty PWR core. *J. Nucl. Mater.* **2008**, *374*, 248–269. [CrossRef]
8. Dumnernchanvanit, I.; Zhang, N.Q.; Robertson, S.; Delmore, A.; Carlson, M.B.; Hussey, D.; Short, M.P. Initial Experimental evaluation of crud-resistant materials for light water reactors. *J. Nucl. Mater.* **2018**, *498*, 1–8. [CrossRef]

9. Collins, B.; Galloway, J.; Salko, R., Jr.; Clarno, K.; Wysocki, A.; Okhuysen, B.; Andersson, A.D. Whole Core Crud-Induced Power Shift Simulations Using VERA. In Proceedings of the Physor 2018: Reactor Physics paving the way towards more efficient systems, Cancun, Mexico, 22–26 April 2018.
10. Short, M.P.; Hussey, D.; Kendrick, B.K.; Besmann, T.M.; Stanek, C.R.; Yip, S. Multiphysics modeling of porous CRUD deposits in nuclear reactors. *J. Nucl. Mater.* **2013**, *443*, 579–587. [CrossRef]
11. Jin, M.; Short, M. Multiphysics modeling of two-phase film boiling within porous corrosion deposits. *J. Comput. Phys.* **2016**, *316*, 504–518. [CrossRef]
12. Park, M.-S.; Shim, H.-S.; Baek, S.H.; Kim, J.G.; Hur, D.H. Effects of oxidation states of fuel cladding surface on crud deposition in simulated primary water of PWRs. *Ann. Nucl. Energy* **2017**, *103*, 275–281. [CrossRef]
13. Short, M.P. The particulate nature of the crud source term in light water reactors. *J. Nucl. Mater.* **2018**, *509*, 478–481. [CrossRef]
14. Shim, H.-S.; Park, M.-S.; Baek, S.H.; Hur, D.-H. Effect of aluminum oxide coated on fuel cladding surface on crud deposition in simulated PWR primary water. *Ann. Nucl. Energy* **2018**, *121*, 607–614. [CrossRef]
15. Pawel, A.; Collins, B.; Maldonado, G.I. Machine Learning Algorithms for Nodal Method Cross-Section Functionalization. In Proceedings of the Physor 2018: Reactor Physics paving the way towards more efficient systems, Cancun, Mexico, 22–26 April 2018.
16. Tano, M.E.; Ragusa, J.C. Accelerating Radiation S_n Transport Solves Using Artificial Neural Networks. In Proceedings of the Transactions of the American Nuclear Society, Washington, DC, USA, 17–21 November 2019; Volume 121, pp. 825–827.
17. Mena, P.; Kirby, L. Machine Learning Accident Classification Using Nuclear Reactor Data. In Proceedings of the Transactions of the American Nuclear Society, Washington, DC, USA, 17–21 November 2019; Volume 121, pp. 825–827.
18. Ortiz, J.J.; Requena, I. Using a multi-state recurrent neural network to optimize loading patterns in BWRs. *Ann. Nucl. Energy* **2004**, *31*, 789–803. [CrossRef]
19. Erdogan, A.; Geckinli, M. A PWR reload optimisation code (XCore) using artificial neural networks and genetic algorithms. *Ann. Nucl. Energy* **2003**, *30*, 35–53. [CrossRef]
20. Ortiz-Servin, J.J.; Pelta, D.A.; Cadenas, J.M.; Castillo, A.; Montes-Tadeo, J.L. Methodology for integrated fuel lattice and fuel load optimization using population-based metaheuristics and decision trees. *Prog. Nucl. Energy* **2018**, *104*, 264–270. [CrossRef]
21. Poon, P.W.; Parks, G.T. Optimizing PWR reload core design. In Proceedings of the Parallel Problem Solving from Nature, 2, Brussels, Belgium, 28–30 September 1992.
22. Alim, F.; Ivanov, K.; Levine, S.H. New genetic algorithms to optimize PWR reactors Part I: Loading Pattern and burnable poison placement optimization techniques for PWRs. *Ann. Nucl. Energy* **2008**, *35*, 93–112. [CrossRef]
23. Israeli, E.; Gilad, E. Novel genetic algorithm for loading pattern optimization based on core physics heuristics. *Ann. Nucl. Energy* **2018**, *118*, 35–48. [CrossRef]
24. Martin-del-Campo, C.; Francois, J.-L; Avendano, L.; Gonzalez, M. Development of a BWR loading pattern design system based on modified genertic algorithms and knowledge. *Ann. Nucl. Energy* **2004**, *31*, 1901–1911. [CrossRef]
25. Martin-del-Campo, C.; Palomera-Perez, M.-A.; Francois, J.-L. Advanced and flexible genetic algorithms for BWR fuel loading pattern optimization. *Ann. Nucl. Energy* **2009**, *36*, 1553–1559. [CrossRef]
26. Kobayashi, Y.; Aiyoshi, E. Optimization of a Boiling Water Reactor Loading Pattern Using an Improved Genetic Algorithm. *Nucl. Technol.* **2003**, *143*, 144–151. [CrossRef]
27. Francois, J.L.; Lopez, H.A. SOPRAG: A system for boiling water reactors reload pattern optimization using genetic algorithms. *Ann. Nucl. Energy* **1999**, *26*, 1053–1063. [CrossRef]
28. Mawdsley, M.; Parks, G. In-Core PWR Loading Pattern Optimization Via Tabu Search. In Proceedings of the Physor 2018: Reactor Physics paving the way towards more efficient systems, Cancun, Mexico, 22–26 April 2018.
29. Hill, N.J.; Parks, G.T. Pressurized water reactor in-core nuclear fuel management by tabu search. *Ann. Nucl. Energy* **2015**, *75*, 64–71. [CrossRef]
30. Safarzadeh, O.; Zolfaghari, A.; NOrouzi, A.; Minuchehr, H. Loading pattern optimization of PWR reactors using Artificial Bee Colony. *Ann. Nucl. Energy* **2011**, *38*, 2218–2226. [CrossRef]
31. Khoshahval, F.; Minuchehr, H.; Zolfaghari, A. Performance evaluation of PSO and GA in PWR core loading pattern optimization. *Nucl. Eng. Des.* **2011**, *241*, 799–808. [CrossRef]
32. Francois, J.-L.; Ortiz-Servin, J.J.; Martin-del-Campo, C.; Castillo, A.; Esquivel-Estrada, J. Comparison of metaheuristic optimization techniques for BWR fuel reloads pattern design. *Ann. Nucl. Energy* **2013**, *51*, 189–195. [CrossRef]
33. Andersen, B.; Godfrey, A.; Hou, J.; Kropaczek, D.J. Application of Deep Learning Networks to Surrogate Modeling of Crud Deposition. In Proceedings of the International Conference on Mathematics and Computational Methods Applied to Nuclear Science and Engineering, Raleigh, NC, USA, 3–7 October 2021.
34. Andersen, B.; Delipei, G.; Kropaczek, D.; Hou, J. MOF: A Modular Framework for Rapid Application of Optimization Methodologies to General Engineering Design Problems. *arXiv* **2022**, arXiv:2204.00141.
35. Turner, J.A.; Clarno, K.; Sieger, M.; Bartlett, R.; Collins, B.; Pawlowski, R.; Schmidt, R.; Summers, R. The Virtual Environment for Reactor Applications (VERA): Design and architecture. *J. Comput. Phys.* **2016**, *326*, 544–568. [CrossRef]
36. Kochunas, B.; Collins, B.; Stimpson, S.; Salko, R.; Jabaay, D.; Graham, A.; Liu, Y.; Kim, K.-S.; Wieselquist, W.; Godfrey, A.; et al. VERA Core Simulator Methodology for Pressurized Water Reactor Cycle Depletion. *Nucl. Sci. Eng.* **2017**, *185*, 616–628. [CrossRef]

37. Salko, R.; Lange, T.L.; Kucukboyaci, V.; Sung, Y.; Palmtag, S.; Gehin, J.C.; Avromova, M. Development of COBRA-TF for modeling full-core, reactor operating cycles. In Proceedings of the Advances in Nuclear Fuel Management V (ANFM 2015), Hilton Head Island, SC, USA, 29 March–1 April 2015.
38. Ronneberger, O.; Fischer, P.; Brox, T. U-NET: Convolutional Networks for Biomedical Image Segmentation. *arXiv* **2015**, arXiv:1505.04597.
39. Chollet, F. *Deep Learning with Python*; Manning Publications Co.: Shelter Island, NY, USA, 2018.
40. Andersen, B. A Machine Learning Based Approach to Minimize Crud Induced Effects in Pressurized Water Reactors. Ph.D. Dissertation, North Carolina State University, Raleigh, NC, USA, 2021.
41. Goldberg, D.E. *Genetic Algorithms in Search, Optimization, and Machine Learning*; Pearson India Education, Inc.: Noida, India, 1989.
42. Miller, B.; Goldberg, D. Genetic Algorithms, Tournament Selection, and the Effects of Noise. *Complex Syst.* **1995**, *9*, 193–212.
43. Godfrey, A. *CASL-U-2012-0131-004, VERA Core Physics Benchmark Progression Problem Specifications*; U.S. Department of Energy: Washington, DC, USA, 2014.

eng

MDPI

Article

Dispersive Optical Solitons for Stochastic Fokas-Lenells Equation with Multiplicative White Noise

Elsayed M. E. Zayed [1], Mahmoud El-Horbaty [1], Mohamed E. M. Alngar [2,*] and Mona El-Shater [1]

[1] Department of Mathematics, Faculty of Science, Zagazig University, Zagazig 44519, Egypt
[2] Basic Science Department, Faculty of Computers and Artificial Intelligence, Modern University for Technology & Information, Cairo 11585, Egypt
* Correspondence: mohamed.hassan@cs.mti.edu.eg

Abstract: For the first time, we study the Fokas–Lenells equation in polarization preserving fibers with multiplicative white noise in Itô sense. Four integration algorithms are applied, namely, the method of modified simple equation (MMSE), the method of sine-cosine (MSC), the method of Jacobi elliptic equation (MJEE) and ansatze involving hyperbolic functions. Jacobi-elliptic function solutions, bright, dark, singular, combo dark-bright and combo bright-dark solitons are presented.

Keywords: stochastic F L equation; modified simple equation method; sine-cosine method; Jacobi-elliptic function expansion method; ansatze method

Citation: Zayed, E.M.E.; El-Horbaty, M.; Alngar, M.E.M.; El-Shater, M. Dispersive Optical Solitons for Stochastic Fokas-Lenells Equation with Multiplicative White Noise. *Eng* **2022**, *3*, 523–540. https://doi.org/10.3390/eng3040037

Academic Editor: Antonio Gil Bravo

Received: 28 October 2022
Accepted: 22 November 2022
Published: 28 November 2022

Publisher's Note: MDPI stays neutral with regard to jurisdictional claims in published maps and institutional affiliations.

Copyright: © 2022 by the authors. Licensee MDPI, Basel, Switzerland. This article is an open access article distributed under the terms and conditions of the Creative Commons Attribution (CC BY) license (https://creativecommons.org/licenses/by/4.0/).

1. Introduction

Nonlinear differential equations (NLDEs) play a very important role in scientific fields and engineering such as optical fibers, the heat flow, plasma physics, solid-state physics, chemical kinematics, the proliferation of shallow water waves, fluid mechanics, quantum mechanics, wave proliferation phenomena, etc. One of the fundamental physical problems for these models is to obtain their traveling wave solutions. As a consequence, the search for mathematical methods to create exact solutions of NLDEs is an important and essential activity in nonlinear sciences. In recent years, many articles have studied optical solitons' form in telecommunications industry. These soliton molecules form the information transporter across intercontinental distances around the world. Lastly, the nonlinear Schrödinger's equation (NLSE) has been discussed with the help of many models [1–38]. The aspect of stochasticity is one of the features that is less touched upon and there are hardly any papers that have debated this point [3–9]. The Fokas–Lenells equation (FLE) appears as a model which appoints nonlinear pulse propagation in optical fibers. The FLE is a completely integrable equation which has arisen as an integrable generalization of the NLSE using bi-Hamiltonian methods [10]. On the other hand, the FLE models have the propagation of nonlinear light pulses in monomode optical fibers when certain higher-order nonlinear effects are considered in optics field [11]. The complete integrability of the FLE has been presented by using the inverse scattering transform (IST) method [12]. In the main, a Lax pair and a few conservation laws connected to it have been obtained using the bi-Hamiltonian structure and the multi-soliton solutions have been derived by using the dressing method [13]. One more main characteristic of the FLE is that it is the first negative flow of the integrable hierarchy of the derivative NLSE [14].

In the present article, we will study the FLE with multiplicative white noise in the Itô sense. Our results are presented after a comprehensive analysis obtained in this article.

2. Governing Model

The dimensionless structure of the stochastic perturbed FLE in polarization preserving fiber with multiplicative white noise in the Itô sense is written, for the first time, as:

$$iq_t + a_1 q_{xx} + a_2 q_{xt} + |q|^2(bq + icq_x) + \sigma(q - ia_2 q_x)\frac{dW(t)}{dt} = i\left[\alpha q_x + \lambda\left(|q|^2 q\right)_x + \mu\left(|q|^2\right)_x q\right], \tag{1}$$

where $q(x,t)$ is a complex-valued function that represents the wave profile, while $a_1, a_2, b, c, \sigma, \alpha, \lambda, \mu$ are real-valued constants and $i = \sqrt{-1}$. The first term in Equation (1) is the linear temporal evolution, a_1 is the coefficient of chromatic dispersion (CD), a_2 is the coefficient of spatio-temporal dispersion (STD), b is the coefficient of self-phase modulation (SPM), c is the coefficient of nonlinear dispersion term, σ is the coefficient of the strength of noise, the Wiener process is denoted by $W(t)$, while $dW(t)/dt$ represents the white noise. Also, the term $dW(t)/dt$ is the time derivative of the standard Wiener process $W(t)$ which is called a Brownian motion and has the following properties [7]: (i) $W(t), t \geq 0$, is a continuous function of t, (ii) For $s < t, W(t) - W(s)$ is independent of increments. (iii) $W(t) - W(s)$ has a normal distribution with mean zero and variance $(t - s)$.

Next, α is the coefficient of inter-modulation dispersion (IMD), λ is the coefficient of self-steepening (SS) term, and finally μ is the coefficient of higher-order nonlinear dispersion term. If $\sigma = 0$, Equation (1) reduces to the familiar FLE which is studied in [1,2,37]. The authors [37] studied Equation (1) with variable coefficients and $\sigma = 0$. The motivation of adding the stochastic term $\sigma(q - ia_2 q_x)\frac{dW(t)}{dt}$ to Equation (1) is to formulate the stochastic FLE with noise or fluctuations depending on the time, which has been recognized in many areas via physics, engineering, chemistry and so on. This stochastic term has been constructed with the help of the two terms iq_t and $a_2 q_{xt}$. Therefore, in general, the stochastic model means that the model of differential equations should contain the white noise term $(\sigma \neq 0)$. The physical importance of the stochastic FL Equation (1) is to find its traveling wave stochastic solutions which appoint the nonlinear pulse propagations in optical fibers.

The aim of this article is to use the method of MMSE in Section 3, the method of MSC in Section 4, the method of MJEE in Section 5 and the ansatze involving hyperbolic functions in Section 6 to find the bright, dark, singular soliton solutions, as well as the Jacobi elliptic function solutions of Equation (1). Some numerical simulations are obtained in Section 7. Finally, conclusions are illustrated in Section 8.

3. On Solving Equation (1) by MMSE

In order to solve the stochastic Equation (1), we use a wave transformation involving the noise coefficient σ and the Wiener process $W(t)$ in the form:

$$q(x,t) = \phi(\xi) \exp i\left[-\kappa x + wt + \sigma W(t) - \sigma^2 t\right], \tag{2}$$

where the transformation $\xi = x - vt$ is used. Here, κ, w, v, are real constants, such that κ represents the wave number, w represents the frequency and v represents the soliton velocity. The function $\phi(\xi)$ is real function which represents the amplitude part. When we put Equation (2) into Equation (1), we obtain the ordinary differential equation (ODE):

$$[a_1 - a_2 v]\phi'' + \Upsilon\phi + [b + \kappa(c - \lambda)]\phi^3 = 0, \tag{3}$$

and the soliton velocity,

$$v = \frac{\Upsilon}{(a_2\kappa - 1)}, \quad a_2\kappa \neq 1 \tag{4}$$

as well as the constraint condition,

$$c - 3\lambda - 2\mu = 0, \tag{5}$$

where $Y = \left[(w - \sigma^2)(a_2\kappa - 1) - a_1\kappa^2 - \alpha\kappa\right]$ and $'' = \frac{d^2}{d\xi^2}$. We have the balance number $N = 1$ by balancing ϕ'' with the ϕ^3 in Equation (3). According to the method of MSE [15–20], the solution of Equation (3) is written as:

$$\phi(\xi) = A_0 + A_1\left[\frac{\psi'(\xi)}{\psi(\xi)}\right], \tag{6}$$

where $\psi(\xi)$ is a new function of ξ, and A_0, A_1 are constants to be determined later, provided $A_1 \neq 0$, $\psi(\xi) \neq 0$ and $\psi'(\xi) \neq 0$.

Inserting Equation (6) into Equation (3), and collecting all the coefficients of $\psi^{-i}(\xi)$ $(i = 0, 1, 2, 3)$, we obtain the equations:

$$\psi^0 : A_0 Y + A_0^3[b + \kappa(c - \lambda)] = 0, \tag{7}$$

$$\psi^{-1} : A_1\psi'''[a_1 - a_2 v] + A_1\psi Y + 3A_0^2 A_1\psi'[b + \kappa(c - \lambda)] = 0, \tag{8}$$

$$\psi^{-2} : -3A_1\psi'\psi''[a_1 - a_2 v] + 3A_0 A_1^2\psi'^2[b + \kappa(c - \lambda)] = 0, \tag{9}$$

$$\psi^{-3} : 2A_1\psi'^3[a_1 - a_2 v] + A_1^3\psi'^3[b + \kappa(c - \lambda)] = 0. \tag{10}$$

By solving Equations (7) and (10), we obtain:

$$A_0 = 0,\ \ A_0 = \pm\sqrt{-\frac{Y}{[b + \kappa(c - \lambda)]}},\ \ A_1 = \pm\sqrt{-\frac{2[a_1 - a_2 v]}{[b + \kappa(c - \lambda)]}}, \tag{11}$$

provided $[b + \kappa(c - \lambda)]Y < 0$ and $[b + \kappa(c - \lambda)][a_1 - a_2 v] < 0$.

By solving Equations (8) and (9), we conclude that $A_0 = 0$ is rejected. Therefore, $A_0 \neq 0$. Now, Equation (9) reduces to the ODE :

$$[a_1 - a_2 v]\psi'' - A_0 A_1[b + \kappa(c - \lambda)]\psi' = 0, \tag{12}$$

which has the solution

$$\psi'(\xi) = \xi_0 \exp\left[\frac{A_0 A_1[b + \kappa(c - \lambda)]}{[a_1 - a_2 v]}\xi\right], \tag{13}$$

where $\xi_0 \neq 0$ is a constant. From Equation (11) and Equation (13), we can show that Equation (8) is valid. Hence, we have the results:

$$\psi(\xi) = \frac{\xi_0[a_1 - a_2 v]}{A_0 A_1[b + \kappa(c - \lambda)]}\exp\left[\frac{A_0 A_1[b + \kappa(c - \lambda)]}{[a_1 - a_2 v]}\xi\right] + \xi_1, \tag{14}$$

where ξ_1 is a nonzero constant of integration. Now, the exact solution of Equation (1) has the form:

$$q(x,t) = \left\{A_0 + A_1\frac{\xi_0\exp\left[\frac{A_0 A_1[b+\kappa(c-\lambda)]}{[a_1 - a_2 v]}(x - vt)\right]}{\xi_1 + \frac{\xi_0[a_1 - a_2 v]}{A_0 A_1[b+\kappa(c-\lambda)]}\exp\left[\frac{A_0 A_1[b+\kappa(c-\lambda)]}{[a_1 - a_2 v]}(x - vt)\right]}\right\}\exp i\left[-\kappa x + wt + \sigma W(t) - \sigma^2 t\right]. \tag{15}$$

In particular, if we set,

$$\xi_1 = \frac{\xi_0[a_1 - a_2 v]}{A_0 A_1[b + \kappa(c - \lambda)]}, \tag{16}$$

we have the dark soliton solution:

$$q(x,t) = \pm\sqrt{-\frac{Y}{[b+\kappa(c-\lambda)]}}\tanh\left[\sqrt{\frac{Y}{2[a_1 - a_2 v]}}(x - vt)\right]\exp i\left[-\kappa x + wt + \sigma W(t) - \sigma^2 t\right], \tag{17}$$

while, if we set,

$$\xi_1 = -\frac{\xi_0[a_1 - a_2 v]}{A_0 A_1[b + \kappa(c - \lambda)]}, \tag{18}$$

we have the singular soliton solution:

$$q(x,t) = \pm\sqrt{-\tfrac{Y}{[b+\kappa(c-\lambda)]}}\coth\left[\sqrt{\tfrac{Y}{2[a_1-a_2 v]}}(x - vt)\right]\exp i\left[-\kappa x + wt + \sigma W(t) - \sigma^2 t\right], \tag{19}$$

provided,

$$[b + \kappa(c - \lambda)]Y < 0, [a_1 - a_2 v]Y > 0. \tag{20}$$

On comparing our above results (17) and (19) with the results (19) and (20) obtained in [37], we deduce that they are equivalent when $\sigma = 0$.

4. On Solving Equation (1) by MSC

To apply this method according to [21–25], assume that Equation (3) has the sine-solution form:

$$\phi(\xi) = \begin{cases} \lambda_1 \sin^{\beta_1}(\mu_1 \xi), \text{ if } |\xi| < \frac{\pi}{\mu_1}, \\ 0 \quad , \text{ otherwise.} \end{cases} \tag{21}$$

Substituting Equation (21) into Equation (3), we obtain:

$$\begin{gathered}[a_1 - a_2 v]\left[\lambda_1 \mu_1^2 \beta_1(\beta_1 - 1)\sin^{\beta_1 - 2}(\mu_1 \xi) - \lambda_1 \mu_1^2 \beta_1^2 \sin^{\beta_1}(\mu_1 \xi)\right] \\ +Y\lambda_1 \sin^{\beta_1}(\mu_1 \xi) + [b + \kappa(c - \lambda)]\lambda_1^3 \sin^{3\beta_1}(\mu_1 \xi) = 0.\end{gathered} \tag{22}$$

From (22), we deduce that $\beta_1 - 2 = 3\beta_1$ which leads $\beta_1 = -1$. Consequently, we have the results:

$$\mu_1^2 = \frac{Y}{[a_1 - a_2 v]}, \quad \lambda_1^2 = -\frac{2Y}{[b + \kappa(c - \lambda)]}. \tag{23}$$

Now, the periodic solution of Equation (1) is:

$$q(x,t) = \pm\sqrt{-\tfrac{2Y}{[b+\kappa(c-\lambda)]}}\csc\left[\sqrt{\tfrac{Y}{[a_1-a_2 v]}}(x - vt)\right]\exp i\left[-\kappa x + wt + \sigma W(t) - \sigma^2 t\right], \tag{24}$$

provided $[b + \kappa(c - \lambda)]Y < 0$, $[a_1 - a_2 v]Y > 0$.

Since $\csc(ix) = -i\operatorname{csch}x$, then the singular soliton solution of Equation (1) is written as:

$$q(x,t) = \pm\sqrt{\tfrac{2Y}{[b+\kappa(c-\lambda)]}}\operatorname{csch}\left[\sqrt{-\tfrac{Y}{[a_1-a_2 v]}}(x - vt)\right]\exp i\left[-\kappa x + wt + \sigma W(t) - \sigma^2 t\right], \tag{25}$$

provided $Y[b + \kappa(c - \lambda)] > 0$, $[a_1 - a_2 v]Y < 0$.

In parallel, if we allow that Equation (3) has the cosine-solution:

$$\phi(\xi) = \begin{cases} \lambda_1 \cos^{\beta_1}(\mu_1 \xi), \text{ if } |\xi| < \frac{\pi}{2\mu_1}, \\ 0 \quad , \text{ otherwise.} \end{cases} \tag{26}$$

Putting Equation (26) into Equation (3), we obtain

$$\begin{gathered}[a_1 - a_2 v]\left[-\mu_1^2 \beta_1^2 \lambda_1 \cos^{\beta_1}(\mu_1 \xi) + \lambda_1 \mu_1^2 \beta_1(\beta_1 - 1)\cos^{\beta_1 - 2}(\mu_1 \xi)\right] \\ +Y\lambda_1 \cos^{\beta_1}(\mu_1 \xi) + [b + \kappa(c - \lambda)]\lambda_1^3 \cos^{3\beta_1}(\mu_1 \xi) = 0.\end{gathered} \tag{27}$$

From Equation (27), we deduce that $\beta_1 - 2 = 3\beta_1$, which leads $\beta_1 = -1$. Therefore, we have the solutions:

$$q(x,t) = \pm\sqrt{-\tfrac{2Y}{[b+\kappa(c-\lambda)]}}\sec\Big[\sqrt{\tfrac{Y}{[a_1-a_2v]}}(x-vt)\Big]\exp i\Big[-\kappa x + wt + \sigma W(t) - \sigma^2 t\Big], \quad (28)$$

with conditions $[b+\kappa(c-\lambda)]Y < 0$, $[a_1 - a_2 v]Y > 0$.

Since, $\sec(ix) = \text{sech}x$, we have the bright soliton solution:

$$q(x,t) = \pm\sqrt{-\tfrac{2Y}{[b+\kappa(c-\lambda)]}}\text{sech}\Big[\sqrt{-\tfrac{Y}{[a_1-a_2v]}}(x-vt)\Big]\exp i\Big[-\kappa x + wt + \sigma W(t) - \sigma^2 t\Big], \quad (29)$$

provided$[b+\kappa(c-\lambda)]Y < 0$, $[a_1 - a_2 v]Y < 0$.

5. On Solving Equation (1) by MJEE

If we multiply Equation (3) by $\phi'(\xi)$ and integrate, we have the JEE as:

$$\phi'^2(\xi) = l_0 + l_2\phi^2(\xi) + l_4\phi^4(\xi), \quad (30)$$

where,

$$l_0 = \frac{2c_1}{[a_1 - a_2 v]}, l_2 = -\frac{Y}{[a_1 - a_2 v]}, l_4 = -\frac{[b+\kappa(c-\lambda)]}{2[a_1 - a_2 v]}, \quad (31)$$

and c_1 is the integration constant, $[a_1 - a_2 v] \neq 0$. It is noted [26–30] that Equation (30) has the Jacobi-elliptic solutions in the forms:

(1) If $l_0 = 1, l_2 = -(1+m^2), l_4 = m^2, 0 < m < 1$, then,

$$\phi(\xi) = \text{sn}(\xi) \text{ or } \phi(\xi) = \text{cd}(\xi). \quad (32)$$

Then, Equation (1) has the JEE solution:

$$\begin{aligned} &q(x,t) = \text{sn}(x-vt)\exp i[-\kappa x + wt + \sigma W(t) - \sigma^2 t], \\ &\text{or} \\ &q(x,t) = \text{cd}(x-vt)\exp i[-\kappa x + wt + \sigma W(t) - \sigma^2 t], \end{aligned} \quad (33)$$

where,

$$\begin{aligned} &c_1 = \tfrac{1}{2}(a_1 - a_2 v), \\ &Y = (1+m^2)(a_1 - a_2 v), \\ &b + \kappa(c-\lambda) = -2m^2(a_1 - a_2 v), \end{aligned} \quad (34)$$

and consequently, we obtain

$$Y = -\frac{(1+m^2)}{2m^2}[b+\kappa(c-\lambda)].$$

Particularly, if $m \to 1$, we get,

$$q(x,t) = \tanh(x-vt)\exp i\Big[-\kappa x + wt + \sigma W(t) - \sigma^2 t\Big]. \quad (35)$$

Note that the solution Equation (35) is equivalent to the solution Equation (17) under the conditions of Equation (34).

(2) If $l_0 = m^2, l_2 = -(1+m^2), l_4 = 1, 0 < m < 1$, then,

$$\phi(\xi) = \text{ns}(\xi) \text{ or } \phi(\xi) = \text{dc}(\xi). \quad (36)$$

Then, we obtain the JEE solution for Equation (1),

$$\begin{aligned} &q(x,t) = \text{ns}(x-vt)\exp i[-\kappa x + wt + \sigma W(t) - \sigma^2 t], \\ &\text{or} \\ &q(x,t) = \text{dc}(x-vt)\exp i[-\kappa x + wt + \sigma W(t) - \sigma^2 t], \end{aligned} \quad (37)$$

where,

$$\begin{aligned} c_1 &= \tfrac{1}{2}m^2(a_1 - a_2 v), \\ Y &= (1+m^2)(a_1 - a_2 v), \\ b + \kappa(c - \lambda) &= -2(a_1 - a_2 v), \end{aligned} \tag{38}$$

and consequently, we have,

$$Y = -\frac{(1+m^2)}{2}[b + \kappa(c-\lambda)].$$

Particularly, if $m \to 1$, we obtain,

$$q(x,t) = \coth(x - vt)\exp i\Big[-\kappa x + wt + \sigma W(t) - \sigma^2 t\Big]. \tag{39}$$

Note that the solution in Equation (39) is equivalent to the solution in Equation (19) under the conditions in Equation (38).

(3) If $l_0 = 1 - m^2, l_2 = 2m^2 - 1, l_4 = -m^2, 0 < m < 1$, then,

$$\phi(\xi) = \text{cn}(\xi). \tag{40}$$

Now, we have the JEE solution for Equation (1),

$$q(x,t) = \text{cn}(x - vt)\exp i\Big[-\kappa x + wt + \sigma W(t) - \sigma^2 t\Big], \tag{41}$$

where,

$$\begin{aligned} c_1 &= \tfrac{1}{2}(1-m^2)(a_1 - a_2 v), \\ Y &= -(2m^2-1)(a_1 - a_2 v), \\ b + \kappa(c - \lambda) &= 2m^2(a_1 - a_2 v), \end{aligned} \tag{42}$$

and consequently, we have,

$$Y = -\frac{(2m^2-1)}{2m^2}[b + \kappa(c-\lambda)].$$

Particularly, if $m \to 1$, we obtain,

$$q(x,t) = \text{sech}(x - vt)\exp i\Big[-\kappa x + wt + \sigma W(t) - \sigma^2 t\Big] \tag{43}$$

Note that the solution of Equation (43) is equivalent to the solution of Equation (29) under the conditions of Equation (42).

(4) If $l_0 = -m^2(1-m^2), l_2 = 2m^2 - 1, l_4 = 1, 0 < m < 1$, then,

$$\phi(\xi) = \text{ds}(\xi). \tag{44}$$

Consequently, we have the JEE solution for Equation (1),

$$q(x,t) = \text{ds}(x - vt)\exp i\Big[-\kappa x + wt + \sigma W(t) - \sigma^2 t\Big], \tag{45}$$

where,

$$\begin{aligned} c_1 &= -\tfrac{m^2}{2}(1-m^2)(a_1 - a_2 v), \\ Y &= -(2m^2-1)(a_1 - a_2 v), \\ b + \kappa(c - \lambda) &= -2(a_1 - a_2 v), \end{aligned} \tag{46}$$

and we have,

$$Y = \frac{1}{2}(2m^2-1)[b + \kappa(c-\lambda)].$$

Particularly, if $m \to 1$, we obtain

$$q(x,t) = \text{csch}(x - vt) \exp i\left[-\kappa x + wt + \sigma W(t) - \sigma^2 t\right], \tag{47}$$

Note that the solution of Equation (47) is equivalent to the solution of Equation (25) under the conditions of Equation (46).

(5) If $l_0 = \frac{1}{4}, l_2 = \frac{1}{2}(1 - 2m^2), l_4 = \frac{1}{4}, 0 < m < 1$, then,

$$\phi(\xi) = \frac{\text{sn}(\xi)}{1 \pm \text{cn}(\xi)}. \tag{48}$$

Now, we have the JEE solution for the Equation (1),

$$q(x,t) = \frac{\text{sn}(x - vt)}{1 \pm \text{cn}(x - vt)} \exp i\left[-\kappa x + wt + \sigma W(t) - \sigma^2 t\right], \tag{49}$$

where,

$$\begin{aligned} c_1 &= \tfrac{1}{8}(a_1 - a_2 v), \\ Y &= -\tfrac{1}{2}(1 - 2m^2)(a_1 - a_2 v), \\ b + \kappa(c - \lambda) &= -\tfrac{1}{2}(a_1 - a_2 v), \end{aligned} \tag{50}$$

and consequently, we have,

$$Y = (1 - 2m^2)[b + \kappa(c - \lambda)].$$

Particularly, if $m \to 1$, we obtain the combo dark-bright soliton solutions:

$$q(x,t) = \frac{\tanh(x - vt)}{1 \pm \text{sech}(x - vt)} \exp i\left[-\kappa x + wt + \sigma W(t) - \sigma^2 t\right]. \tag{51}$$

(6) If $l_0 = \frac{1-m^2}{4}, l_2 = \frac{1+m^2}{2}, l_4 = \frac{1-m^2}{4}, 0 < m < 1$, then,

$$\phi(\xi) = \frac{\text{cn}(\xi)}{1 \pm \text{sn}(\xi)}. \tag{52}$$

Then, we have the JEE solution for Equation (1),

$$q(x,t) = \frac{\text{cn}(x - vt)}{1 \pm \text{sn}(x - vt)} \exp i\left[-\kappa x + wt + \sigma W(t) - \sigma^2 t\right], \tag{53}$$

where

$$\begin{aligned} c_1 &= \tfrac{1}{8}(1 - m^2)(a_1 - a_2 v), \\ Y &= -\tfrac{1}{2}(1 + m^2)(a_1 - a_2 v), \\ b + \kappa(c - \lambda) &= -\tfrac{1}{2}(1 - m^2)(a_1 - a_2 v). \end{aligned} \tag{54}$$

Particularly, if $m \to 1$, we obtain the combo bright-dark soliton solutions:

$$q(x,t) = \frac{\text{sech}(x - vt)}{1 \pm \tanh(x - vt)} \exp i\left[-\kappa x + wt + \sigma W(t) - \sigma^2 t\right]. \tag{55}$$

Finally, there are many other Jacobi elliptic solutions which are omitted here for simplicity.

6. Ansatze Involving Hyperbolic Functions

To this aim, we first write Equation (3) in the simple form,

$$A\phi'' + Y\phi + C\phi^3 = 0, \tag{56}$$

where,

$$\begin{aligned} A &= a_1 - a_2 v, \\ C &= b + \kappa(c - \lambda). \end{aligned} \quad (57)$$

Along these lines, the main steps of the proposed ansatze have been presented according to the ansatze involving the hyperbolic functions method [31].

6.1. Combo Bright-Dark Solitons

We assume the ansatz,

$$\phi(\xi) = \frac{\alpha_1 \text{sech}(\mu_1 \xi)}{1 + \lambda_1 \tanh(\mu_1 \xi)}. \quad (58)$$

where $\alpha_1, \lambda_1, \mu_1$ are parameters to be determined. Now, we obtain

$$\phi''(\xi) = \frac{\alpha_1 \mu_1^2 (2\lambda_1^2 - 1)\text{sech}(\mu_1\xi) + 2\alpha_1 \lambda_1 \mu_1^2 \text{sech}(\mu_1\xi)\tanh(\mu_1\xi) + \alpha_1 \mu_1^2 (2 - \lambda_1^2)\text{sech}(\mu_1\xi)\tanh^2(\mu_1\xi)}{(1 + \lambda_1 \tanh(\mu_1\xi))^3}. \quad (59)$$

Substituting Equations (58) and (59) into Equation (56), combining all the coefficients of $\text{sech}^p(\xi)\tanh^q(\xi)$ $(p = 1, q = 0, 1, 2)$, we obtain the set of equations:

$$\begin{aligned} A\alpha_1 \mu_1^2 (2\lambda_1^2 - 1) + Y\alpha_1 + C\alpha_1^3 &= 0, \\ 2A\alpha_1 \lambda_1 \mu_1^2 + 2Y\alpha_1 \lambda_1 &= 0, \\ A\alpha_1 \mu_1^2 (2 - \lambda_1^2) + Y\alpha_1 \lambda_1^2 - C\alpha_1^3 &= 0. \end{aligned} \quad (60)$$

By resolving the Equation (60), we have the results:

$$\mu_1^2 = \frac{-Y}{A}, AY < 0, \ \lambda_1^2 = \frac{2Y + C\alpha_1^2}{2Y} > 0, \alpha_1 \neq 0.$$

Now, we obtain

$$q(x,t) = \left\{ \frac{\alpha_1 \text{sech}\left[\sqrt{\frac{-Y}{A}}(x - vt)\right]}{1 \pm \sqrt{\frac{2Y + C\alpha_1^2}{2Y}} \tanh\left[\sqrt{\frac{-Y}{A}}(x - vt)\right]} \right\} \exp i\left[-\kappa x + wt + \sigma W(t) - \sigma^2 t\right]. \quad (61)$$

which represent the combo-bright-dark soliton solutions and are equivalent to the solutions Equation (55) of Section 5, if $A = -Y, C = 0$ and $\alpha_1 = 1$.

6.2. Combo Dark-Bright Solitons

We assume the ansatz

$$\phi(\xi) = \frac{\alpha_1 \tanh(\mu_1 \xi)}{1 + \lambda_1 \text{sech}(\mu_1 \xi)}, \quad (62)$$

where $\alpha_1, \lambda_1, \mu_1$ are parameters to be determined. Now, we obtain

$$\phi''(\xi) = \frac{\alpha_1 \mu_1^2 (\lambda_1^2 - 2)\text{sech}^2(\mu_1\xi)\tanh(\mu_1\xi) - \alpha_1 \lambda_1 \mu_1^2 \text{sech}(\mu_1\xi)\tanh(\mu_1\xi)}{(1 + \lambda_1 \text{sech}(\mu_1\xi))^3}. \quad (63)$$

Substituting Equations (62) and (63) into Equation (56), combining all the coefficients of $\tanh^p(\xi)\text{sech}^q(\xi)$ $(p = 1, q = 0, 1, 2)$, we obtain the algebraic equations:

$$\begin{aligned} Y\alpha_1 + C\alpha_1^3 &= 0, \\ -A\alpha_1 \lambda_1 \mu_1^2 + 2Y\alpha_1 \lambda_1 &= 0, \\ A\alpha_1 \mu_1^2 (\lambda_1^2 - 2) + Y\alpha_1 \lambda_1^2 - C\alpha_1^3 &= 0. \end{aligned} \quad (64)$$

Solving the algebraic Equation (64), we obtain the results:

$$\mu_1^2 = \frac{2Y}{A}, AY > 0, \ \alpha_1^2 = \frac{-Y}{C}, YC < 0, \lambda_1^2 = 1.$$

Now, Equation (1) has the combo dark-bright soliton solutions:

$$q(x,t) = \pm\sqrt{\frac{-Y}{C}}\left\{\frac{\tanh\left[\sqrt{\frac{2Y}{A}}(x-vt)\right]}{1\pm\operatorname{sech}\left[\sqrt{\frac{2Y}{A}}(x-vt)\right]}\right\}\exp i\left[-\kappa x + wt + \sigma W(t) - \sigma^2 t\right]. \quad (65)$$

which are equivalent to the solutions Equation (51) of Section 5, if $A = 2Y$ and $Y = -C$.

7. Numerical Simulations

In this section, we present the graphs of some solutions for Equation (1). Let us now examine Figures 1–15, as it illustrates some of our solutions obtained in this paper. To this aim, we select some special values of the obtained parameters.

Figure 1: The numerical simulations of the solutions (17) 3D and 2D (with $t = \frac{1}{2}$) with the parameter values

$a_1 = 1, a_2 = 1, b = 1, \sigma = 0, \alpha = 1, \kappa = 2, w = 2, \lambda = 1, \mu = 1, c = 5, v = 3,$ $-5 \leq x, t \leq 5$.

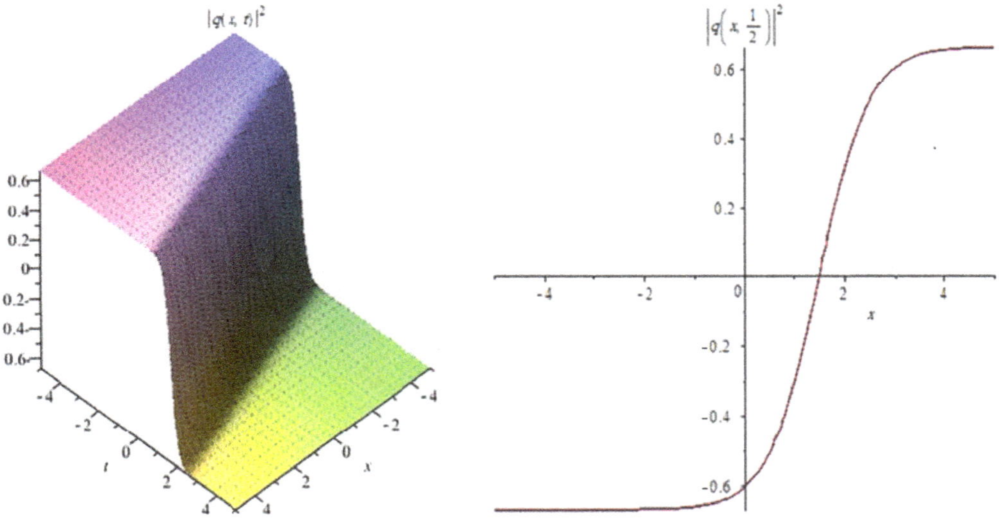

Figure 1. The profile of the dark soliton solutions (17).

Figure 2: The numerical simulations of the solutions (17) 3D and 2D (with $t = \frac{1}{2}$) with the parameter values $a_1 = 1$, $a_2 = 1$, $b = 1$, $\sigma = 1$, $\alpha = 1$, $\kappa = 2, w = 2, \lambda = 1, \mu = 1$, $c = 5, v = 4$, $-5 \leq x, t \leq 5$.

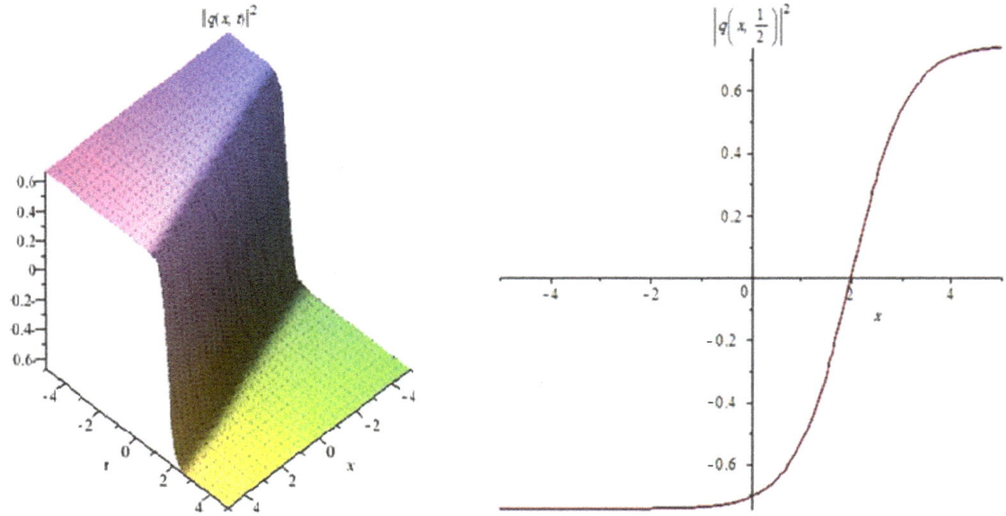

Figure 2. The profile of the dark soliton solutions (17).

Figure 3: The numerical simulations of the solutions (17) 3D and 2D (with $t = \frac{1}{2}$) with the parameter values $a_1 = 1$, $a_2 = 1$, $b = 1$, $\sigma = 2$, $\alpha = 1$, $\kappa = 2, w = 2, \lambda = 1, \mu = 1$, $c = 5, v = 8$, $-5 \leq x, t \leq 5$.

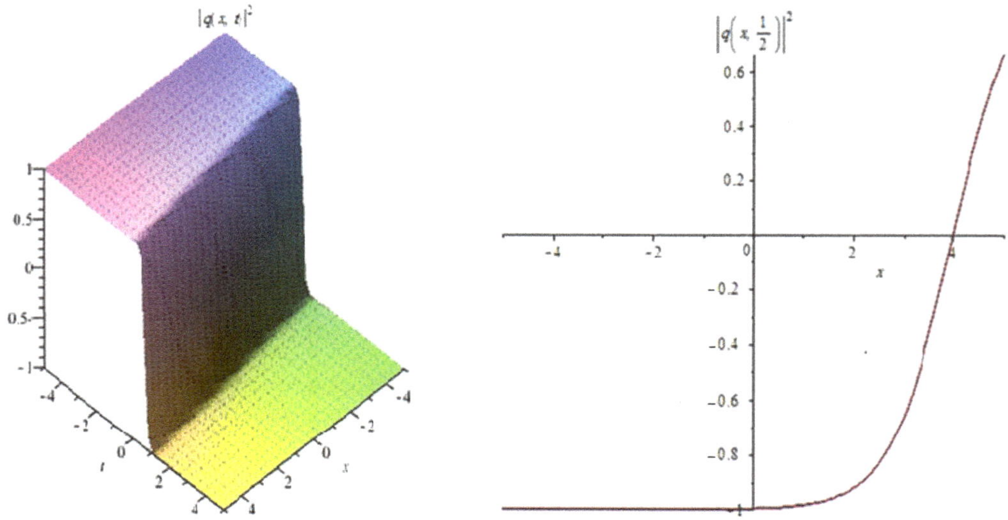

Figure 3. Shows the profile of the dark soliton solutions (17).

Figure 4: The numerical simulations of the solutions (19) 3D and 2D (with $t = \frac{1}{2}$) with the parameter values $a_1 = 1$, $a_2 = 1$, $b = 1$, $\sigma = 0$, $\alpha = 1$, $\kappa = 2$, $w = 2$, $\lambda = 1, \mu = 1$, $c = 5, v = 3$, $-5 \leq x, t \leq 5$.

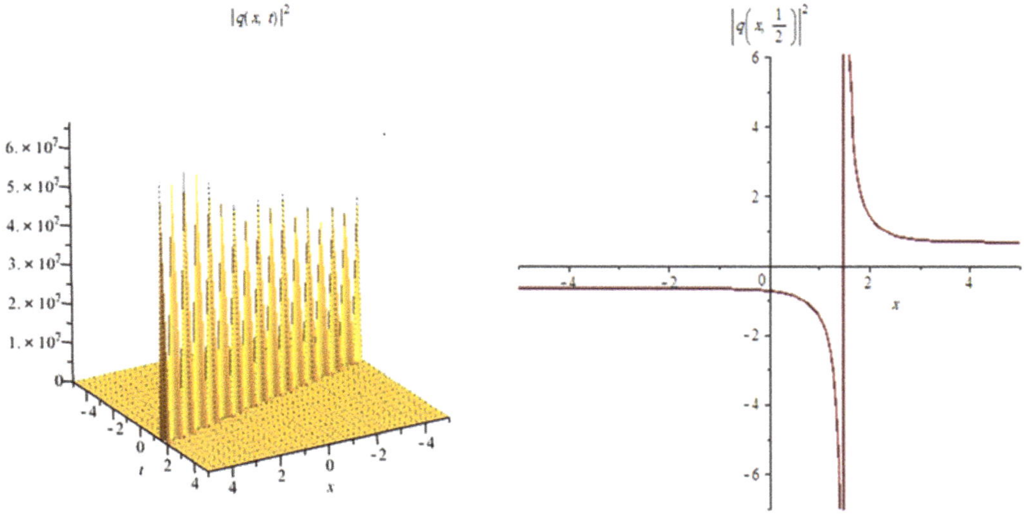

Figure 4. Shows the profile of the singular soliton solutions (19).

Figure 5: The numerical simulations of the solutions (19) 3D and 2D (with $t = \frac{1}{2}$) with the parameter values $a_1 = 1, a_2 = 1, b = 1, \sigma = 1, \alpha = 1, \kappa = 2, w = 2, \lambda = 1, \mu = 1,$ $c = 5, v = 4, \ -5 \leq x, t \leq 5$.

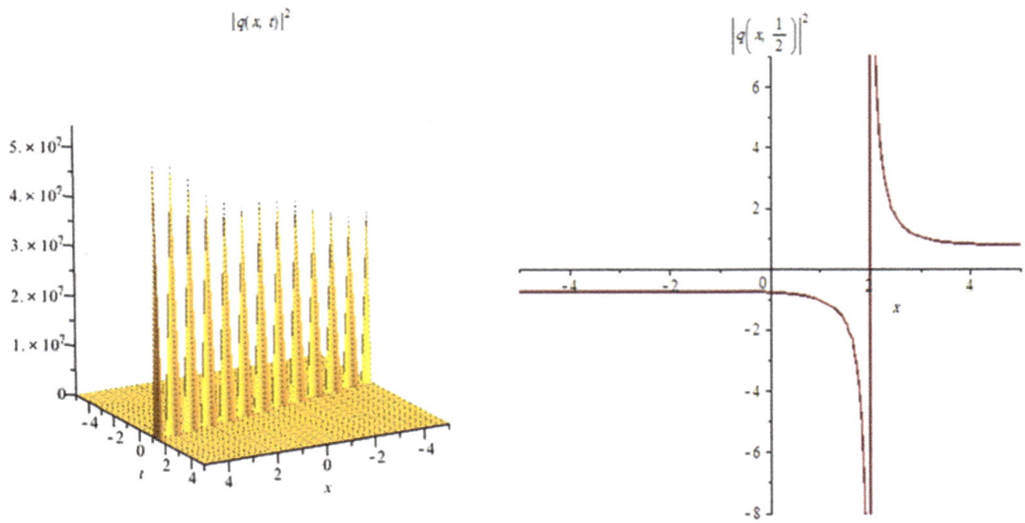

Figure 5. Shows the profile of the singular soliton solutions (19).

Figure 6: The numerical simulations of the solutions (19) 3D and 2D (with $t = \frac{1}{2}$) with the parameter values $a_1 = 1, a_2 = 1, b = 1, \sigma = 2, \alpha = 1, \kappa = 2, w = 2, \lambda = 1, \mu = 1,$ $c = 5, v = 8, \ -5 \leq x, t \leq 5$.

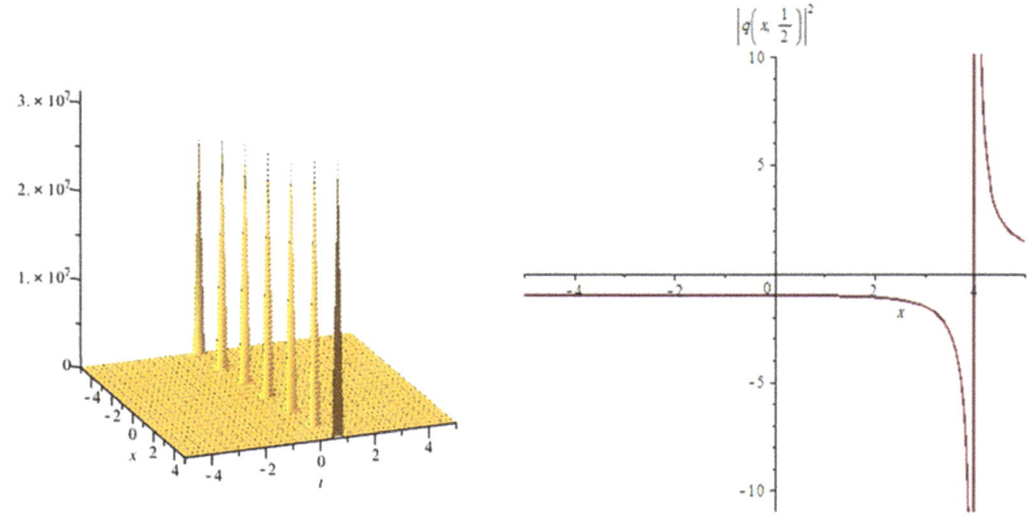

Figure 6. Shows the profile of the singular soliton solutions (19).

Figure 7: The numerical simulations of the solutions (29) 3D and 2D (with $t = \frac{1}{2}$) with the parameter values $a_1 = 1, a_2 = 1, b = 1, \sigma = 0, \alpha = 1, \kappa = \frac{1}{2}, w = 2, \lambda = 1, \mu = 1,$ $c = 5, v = -2,\ -5 \leq x, t \leq 5$.

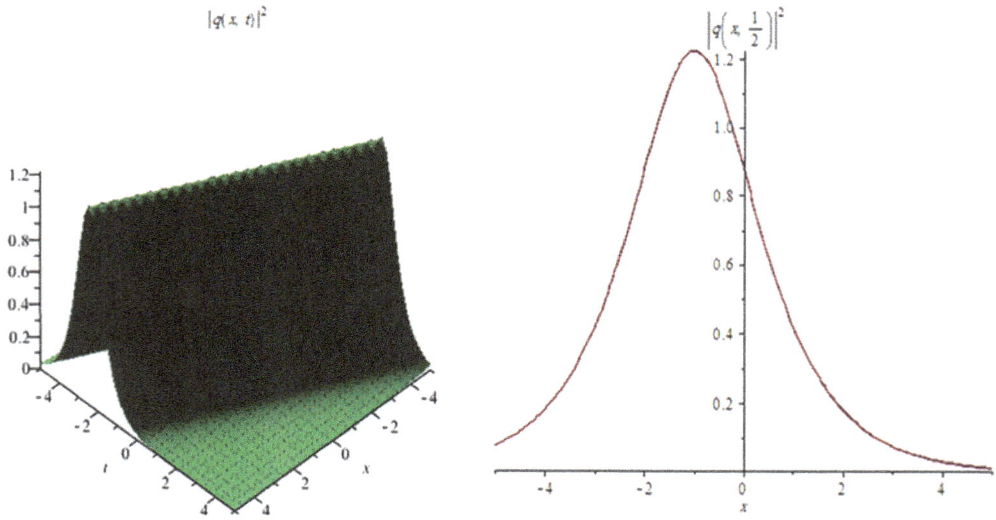

Figure 7. Shows the profile of the bright soliton solutions (29).

Figure 8: The numerical simulations of the solutions (29) 3D and 2D (with $t = \frac{1}{2}$) with the parameter values $a_1 = 1, a_2 = 1, b = 1, \sigma = 4, \alpha = 4, \kappa = \frac{1}{4}, w = 16, \lambda = 2, \mu = 2,$ $c = 10, v = -6,\ -5 \leq x, t \leq 5$.

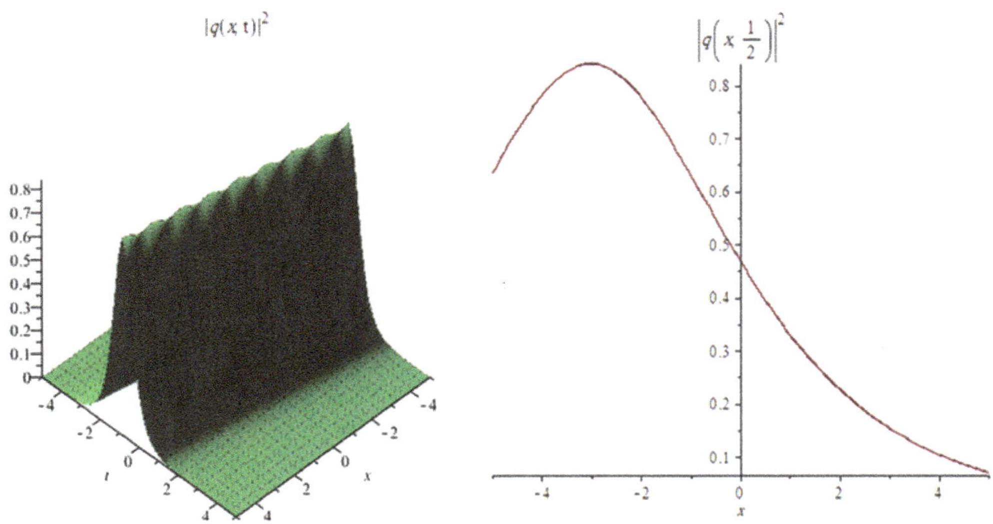

Figure 8. Shows the profile of the bright soliton solutions (29).

Figure 9: The numerical simulations of the solutions (29) 3D and 2D (with $t = \frac{1}{2}$) with the parameter values $a_1 = 1, a_2 = 1, b = 1, \sigma = 2, \alpha = 2, \kappa = \frac{1}{2}, w = 2, \lambda = 1, \mu = 1,$ $c = 5, v = -10,\ -5 \leq x, t \leq 5$.

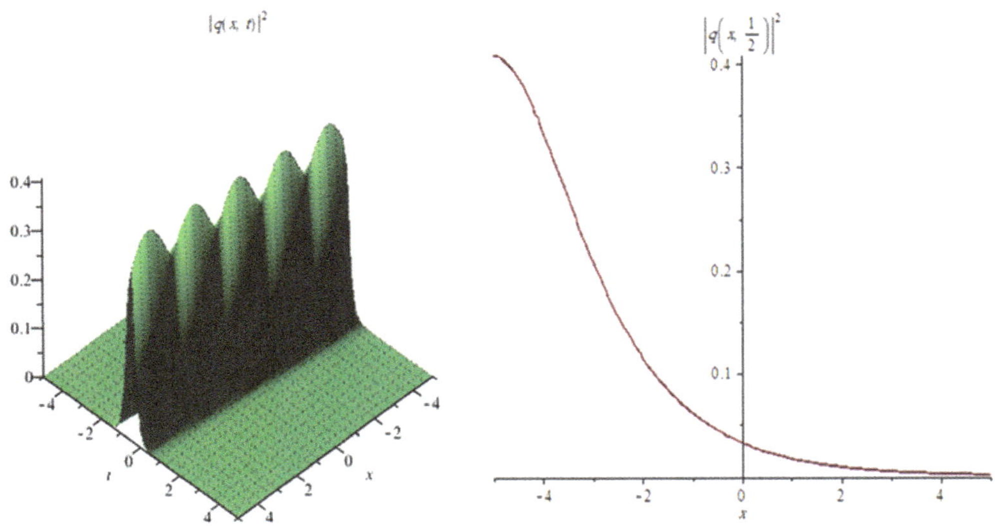

Figure 9. Shows the profile of the bright soliton solutions (29).

Figure 10: The numerical simulations of the solutions (51) 3D and 2D (with $t = \frac{1}{2}$) with the parameter values $a_1 = 4, a_2 = 2, b = -16, \sigma = 0, \alpha = 1, \kappa = 2, w = 10, \lambda = 2,$ $\mu = 2, c = 10, v = 2,\ -5 \leq x, t \leq 5$.

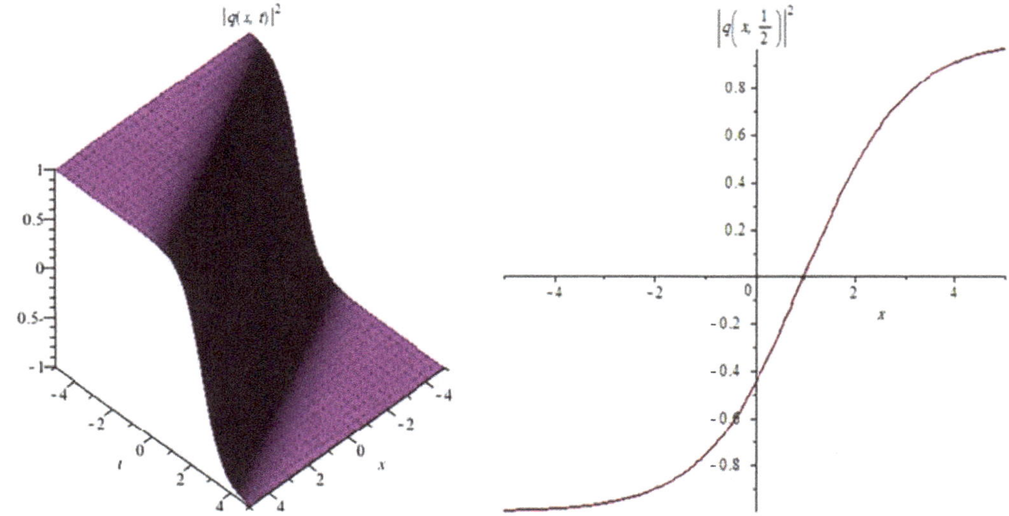

Figure 10. The profile of the combination of dark-bright soliton solutions (51).

Figure 11: The numerical simulations of the solutions (51) 3D and 2D (with $t = \frac{1}{2}$) with the parameter values $a_1 = 4, a_2 = 1, b = -21, \sigma = 1, \alpha = 1, \kappa = 2, w = 24, \lambda = 2$, $\mu = 2, c = 10, v = -6$, $-5 \leq x, t \leq 5$.

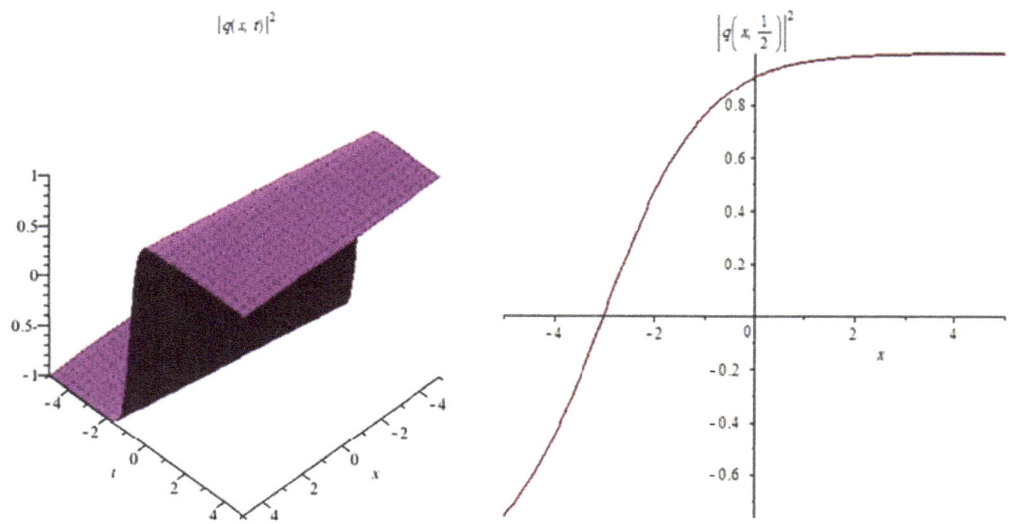

Figure 11. The profile of the combination of dark-bright soliton solutions (51).

Figure 12: The numerical simulations of the solutions (51) 3D and 2D (with $t = \frac{1}{2}$) with the parameter values $a_1 = 6, a_2 = 1, b = -16, \sigma = 2, \alpha = 1, \kappa = 2, w = 30, \lambda = 2$, $\mu = 2, c = 10, v = 6$, $-5 \leq x, t \leq 5$.

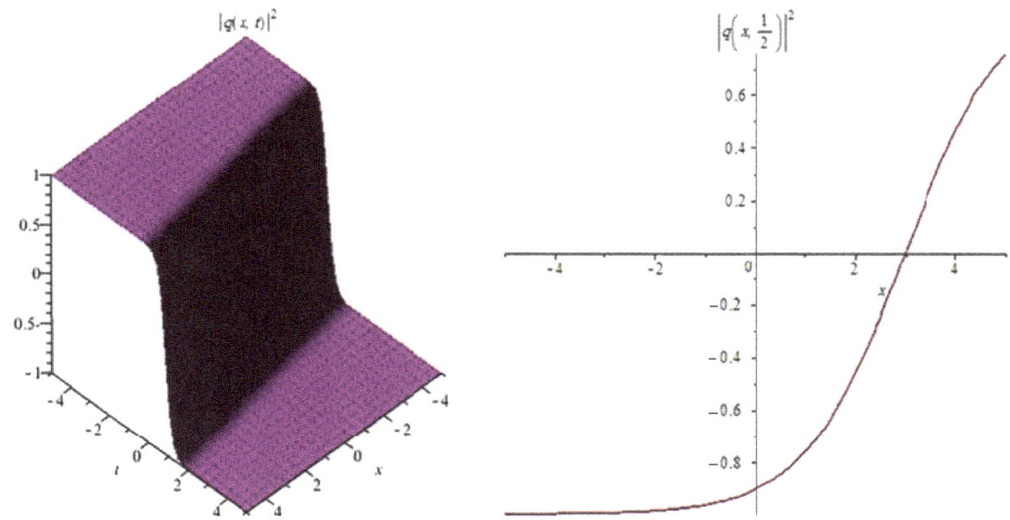

Figure 12. The profile of the combination of dark-bright soliton solutions (51).

Figure 13: The numerical simulations of the solutions (55) 3D and 2D (with $t = \frac{1}{2}$) with the parameter values $a_1 = 1, a_2 = 1, b = -16, \sigma = 0, \alpha = 2, \kappa = 2, w = 2, \lambda = 2, \mu = 2, c = 10, v = -7, \ -5 \leq x, t \leq 5$.

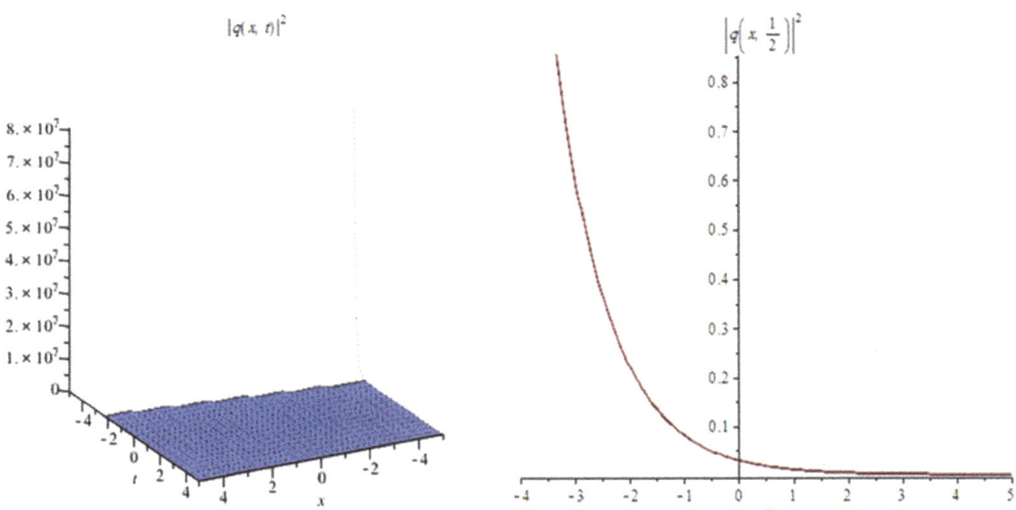

Figure 13. The profile of the combination of bright-dark soliton solutions (55).

Figure 14: The numerical simulations of the solutions (55) 3D and 2D (with $t = \frac{1}{2}$) with the parameter values $a_1 = 1, a_2 = 1, b = -16, \sigma = 1, \alpha = \frac{-5}{3}, \kappa = 2, w = 2, \lambda = 2, \mu = 2, c = 10, v = \frac{4}{3}, \ -5 \leq x, t \leq 5$.

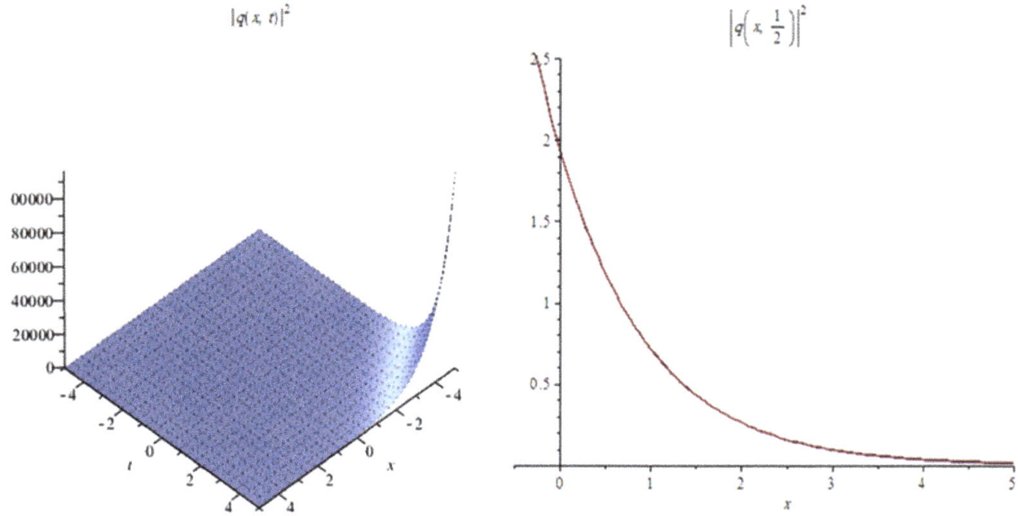

Figure 14. The profile of the combination of bright-dark soliton solutions (55).

Figure 15: The numerical simulations of the solutions (55) 3D and 2D (with $t = \frac{1}{2}$) with the parameter values $a_1 = 1, a_2 = 1, b = -16, \sigma = 2, \alpha = 2, \kappa = 2, w = 2, \lambda = 2, \mu = 2, c = 10, v = -9, c_1 = 0, \ -5 \leq x, t \leq 5$.

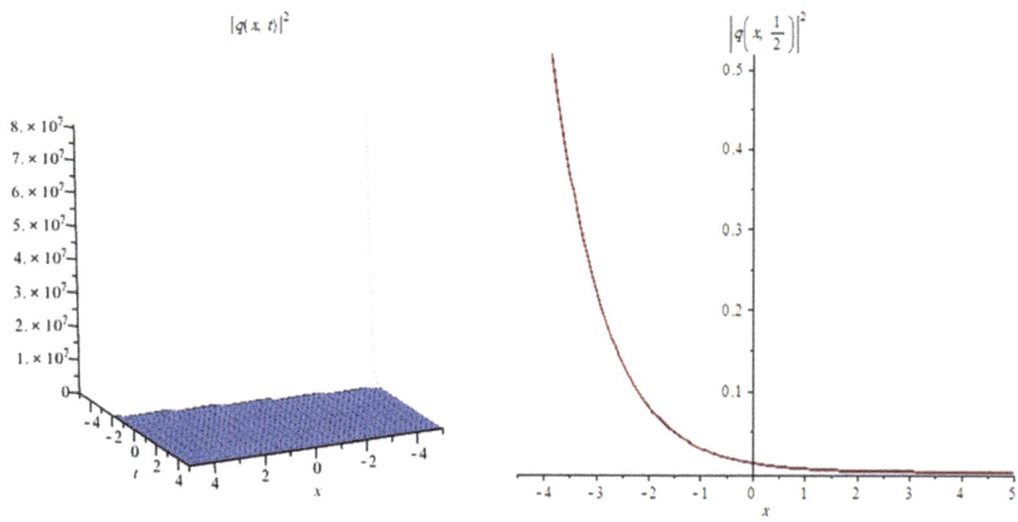

Figure 15. The profile of the combination of bright-dark soliton solutions (55).

Let us now explain the effect of multiplicative white noise in the obtained solutions as follows:

In Figures 1, 4, 7, 10 and 13 when the noise $\sigma = 0$, we note that the surface is less planer. But in Figures 2, 3, 5, 6, 8, 9, 11 and 12 when the noise σ increases ($\sigma = 1, 2, 4$), we note that the surface becomes more planer after small transit behaviors. This means the multiplicative noise effects on the solutions and it makes the solutions stable.

8. Conclusions

In this article, we have obtained the solutions of the stochastic FLE in the presence of multiplicative white noise in the Itô sense. The modified simple equation method, the sine-cosine method, the Jacobi-elliptic function expansion method and the ansatze method are applied. Dark solitons, bright solitons, singular solitons, combo dark-bright solitons, combo bright-dark solitons, as well as Jacobi-elliptic solutions are given. Without noise ($\sigma = 0$) the authors [1,2,37] studied a number of methods to get the exact solutions of FL equation while the stochastic FL Equation (1) is not yet studied. So, on comparing our stochastic solutions ($\sigma \neq 0$) obtained in our present article with the non- stochastic solutions ($\sigma = 0$) obtained in [1,2,37] we deduce that the stochastic solutions are more general than the non-stochastic solutions. Finally, in future, this work will be extended in birefringent fibers, in fiber Bragg gratings and in magneto-optic waveguides. Also, we will study the stochastic FL Equation (1) with variable coefficients [37] when $\sigma \neq 0$, to get stochastic solutions.

Author Contributions: Conceptualization, E.M.E.Z. and M.E.-S.; methodology, M.E.-S. and M.E.-H.; software, M.E.-S.; writing—original draft preparation, M.E.-S. and E.M.E.Z.; writing—review and editing, M.E.M.A. and M.E.-H. All authors have read and agreed to the published version of the manuscript.

Funding: This research received no external funding.

Institutional Review Board Statement: Not applicable.

Informed Consent Statement: Not applicable.

Data Availability Statement: All data generated or analyzed during this study are included in this manuscript.

Acknowledgments: The authors thank the anonymous referees whose comments helped to improve the paper.

Conflicts of Interest: The authors declare no conflict of interest.

References

1. Biswas, A.; Yakup, Y.; Yaşar, E.; Zhou, Q.; Moshokoa, S.P.; Belic, M. Optical soliton solutions to Fokas-lenells equation using some different methods. *Optik* **2018**, *173*, 21–31. [CrossRef]
2. Biswas, A. Chirp-free bright optical soliton perturbation with Fokas–Lenells equation by traveling wave hypothesis and semi-inverse variational principle. *Optik* **2018**, *170*, 431–435. [CrossRef]
3. Albosaily, S.; Mohammed, W.W.; Aiyashi, M.A.; Abdelrahman, A.A.E. Excat solutions of the (2+1)-dimensional stochastic chiral nonlinear Schrödinger's Equation. *Symmetry* **2020**, *12*, 1874–1886. [CrossRef]
4. Mohammed, W.W.; El-Morshedy, M. The influence of multiplicative noise on the stochastic exact solutions of the Nizhnik-Novikov-Veselov system. *Math. Comput. Simul.* **2021**, *190*, 192–202. [CrossRef]
5. Mohammed, W.W.; Albosaily, S.; Iqbal, N.; El-Morshedy, M. The effect of multiplicative noise on the exact solutions of the stochastic Burger equation. *Wave Random Complex Media* **2021** , 1–13. [CrossRef]
6. Abdelrahman, A.A.E.; Mohammed, W.W.; Alesemi, M.; Albosaily, S. The effect of multiplicative noise on the exact solutions of Nonlinear Schr ödinger's Equation. *AIMS Math.* **2021**, *6*, 2970–2980. [CrossRef]
7. Mohammed, W.W.; Ahmad, H.; Boulares, H.; Khelifi, F.; El-Morshedy, M. Exact solution of Hirota-Maccari system forced by multiplicative noise in the Itô sense. *J. Low FrEquation Noise Vib Act. Control.* **2021**, *41*, 74–84. [CrossRef]
8. Zayed, E.M.E.; Alngar, M.E.M.; Shohib, R.M.A.; Biswas, A.; Yildirim, Y.; Alshomrani, A.S.; Alshehri, H.M. Optical solitons having Kudryashov's self-phase modulation with multiplicative white noise via It ô Calculus using new mapping approach. *Optik* **2022**, *264*, 169369. [CrossRef]
9. Zayed, E.M.E.; Shohib, R.M.A.; Alngar, M.E.M.; Biswas, A.; Moraru, L.; Khan, S.; Yildirim, Y.; Alshehri, H.M.; Belic, M.R. Dispersive optical solitons with Schrödinger-Hirota model having multiplicative white noise via Itô calculus. *Phys. Lett. A* **2022**, *445*, 128268. [CrossRef]
10. Fokas, A.S. On a class of physically important integrable equations. *Phys. D Nonlinear Phenom.* **1995**, *87*, 145–150. [CrossRef]
11. Jonatan, L. Exactly Solvable Model for Nonlinear Pulse Propagation in Optical Fibers. *Stud. Appl. Math.* **2009**, *123*, 215–232.
12. Jonatan, L.; Fokas, A.S. On a Novel Integrable Generalization of the Nonlinear Schrödinger Equation. *Nonlinearity* **2009**, *22*, 11–27.
13. Jonatan, L. Dressing for a Novel Integrable Generalization of the Nonlinear Schrödinger Equation. *J. Nonlinear Sci.* **2010**, *20*, 709–722.

14. Kundu, A. Two-fold Integrable Hierarchy of Nonholonomic Deformation of the Derivative Nonlinear Schrödinger and the Lenells-Fokas Equation. *J. Math. Phys.* **2010**, *51*, 1–17. [CrossRef]
15. Zayed, E.M.E. A note on the modified simple equation method applied to Sharma-Tasso-Olver equation. *Appl. Math. Comput.* **2011**, *218*, 3962–3964. [CrossRef]
16. Jawad, A.J.M.; Petkovic, M.D.; Biswas, A. Modified simple equation method for nonlinear evolution equations. *Appl. Math. Comput.* **2010**, *217*, 869–877.
17. Zayed, E.M.E.; Al-Nowehy, A.G. The modified simple equation method, the exp-function method and the method of soliton ansatz for solving the long-short wave resonance equations. *Z. Naturforsch.* **2016**, *71*, 103–112. [CrossRef]
18. El-Borai, M.; El-Owaidy, H.; Arnous, A.H.; Moshokoa, S.; Biswas, A.; Belic, M. Dark and singular optical solitons with spatio-temporal dispersion using modified simple equation method. *Optik* **2017**, *130*, 324–331. [CrossRef]
19. Arnous, A.H.; Ullah, M.Z.; Moshokoa, S.P.; Zhou, Q.; Triki, H.; Mirzazadeh, M.; Biswas, A. Optical solitons in birefringent fibers with modified simple equation method. *Optik* **2017**, *130*, 996–1003. [CrossRef]
20. Arnous, A.H.; Ullah, M.Z.; Asma, M.; Moshokoa, S.P.; Zhou, Q.; Mirzazadeh, M.; Biswas, A.; Belic, M. Dark and singular dispersive optical solitons of Schrödinger-Hirota equation by modified simple equation method. *Optik* **2017**, *136*, 445–450. [CrossRef]
21. Biswas, A. The tanh and the sine-cosine methods for the complex modified KdV and the generalized KdV equations. *Comput. Math. Appl.* **2005**, *49*, 1101–1112.
22. Biswas, A. A sine–cosine method for handing nonlinear wave equations. *Math. Comput. Model.* **2004**, *40*, 499–509.
23. Zayed, E.M.E.; Al-Nowehy, A.G. Solitons and other solutions for the generalized KdV-mKdV equation with higher-order nonlinear terms. *J. Part. Diff. Equ.* **2016**, *29*, 218–245.
24. Yusufuglu, E.; Bekir, A.; Alp, M. Periodic and Solitary wave solutions of Kawahara and modifed Kawahara equations by using sine-cosine method. *Chaos Solitons Fract.* **2008**, *37*, 1193–1197. [CrossRef]
25. Tascan, F.; Bekir, A. Analytical solutions of the (2 + 1)-dimensional nonlinear equations using the sine–cosine method. *Appl. Math. Comput.* **2009**, *215*, 3134–3139.
26. Zayed, E.M.E.; Amer, Y.A.; Shohib, R.M.A. The Jacobi elliptic function expansion method and its applications for solving the higher order dispersive nonlinear Schrödinger's equation. *Sci. J. Math. Res.* **2014**, *4*, 53–72.
27. Liu, S.; Fu, Z.; Liu, S.; Zhao, Q. Jacobi elliptic function expansion method and periodic wave solutions of nonlinear wave equations. *Phys. Lett. A* **2001**, *289*, 69–74. [CrossRef]
28. Xiang, C. Jacobi elliptic function solutions for (2+1)-dimensional Boussinesq and Kadomtsev-Petviashvilli equation. *Appl. Math.* **2011**, *2*, 1313–1316. [CrossRef]
29. Lu, D.; Shi, Q. New Jacobi elliptic functions solutions for the combined KdV-mKdV equation. *Int. J. Nonlinear Sci.* **2010**, *10*, 320–325.
30. Zheng, B.; Feng, Q. The Jacobi elliptic equation method for solving fractional partial differential equations. *Abs. Appl. Anal.* **2014**, *228*, 249071–249080. [CrossRef]
31. Wazwaz, A.-M. New solitons and kink solutions for the Gardner equation. *Commun. Nonlinear Sci. Numer.* **2007**, *12*, 1395–1404. [CrossRef]
32. Palencia, J.L.D. Travelling waves approach in a parabolic coupled system for modeling the behaviour of substances in a fuel tank. *App. Sci.* **2021**, *11*, 5846. [CrossRef]
33. Jiao, Y.; Yang, J.; Zhang, H. Traveling wave solutions to a cubic predator-prey diffusion model with stage structure for the prey. *AIMS Math.* **2022**, *7*, 16261–16277. [CrossRef]
34. Hu, Y.; Liu, Q. On traveling wave solutions of a class of KdV-Burgers-Kuramoto type equations. *AIMS Math.* **2019**, *4*, 1450–1465. [CrossRef]
35. Bracken, P. The quantum Hamilton–Jacobi formalism in complex space. *Quantum Stud. Math. Found.* **2022**, *7*, 389–403. [CrossRef]
36. Palencia, J. L. D.; ur Rahman, S.; Redond, A.N. Regularity and reduction to a Hamilton-Jacobi equation for a MHD Eyring-Powell fluid. *Alex. Eng.* **2022**, *61*, 12283–12291. [CrossRef]
37. Gomez, C.A.; Roshid, S.H.O.; Inc, M.; Akinyemi, L.; Rezazadeh, H. On soliton solutions for perturbed Fokas–Lenells equation. *Opt. Quantum Electron.* **2022**, *54*, 370. [CrossRef]
38. Nandi, D. C.; Safi Ullah, M.; Roshid, H.O.; Ali, M. Z. Application of the unified method to solve the ion sound and Langmuir waves model. *Heliyon* **2022**, *8*, e10924. [CrossRef]

eng

MDPI

Article

Evaluation of Color Anomaly Detection in Multispectral Images for Synthetic Aperture Sensing

Francis Seits, Indrajit Kurmi and Oliver Bimber *

Institute of Computer Graphics, Johannes Kepler University Linz, 4040 Linz, Austria
* Correspondence: oliver.bimber@jku.at; Tel.: +43-732-2468-6631

Abstract: In this article, we evaluate unsupervised anomaly detection methods in multispectral images obtained with a wavelength-independent synthetic aperture sensing technique called Airborne Optical Sectioning (AOS). With a focus on search and rescue missions that apply drones to locate missing or injured persons in dense forest and require real-time operation, we evaluate the runtime vs. quality of these methods. Furthermore, we show that color anomaly detection methods that normally operate in the visual range always benefit from an additional far infrared (thermal) channel. We also show that, even without additional thermal bands, the choice of color space in the visual range already has an impact on the detection results. Color spaces such as HSV and HLS have the potential to outperform the widely used RGB color space, especially when color anomaly detection is used for forest-like environments.

Keywords: multispectral; image processing; anomaly detection; search and rescue; unmanned aerial vehicles; airborne optical sectioning

Citation: Seits, F.; Kurmi, I.; Bimber, O. Evaluation of Color Anomaly Detection in Multispectral Images for Synthetic Aperture Sensing. *Eng* **2022**, 3, 541–553. https://doi.org/10.3390/eng3040038

Academic Editor: Antonio Gil Bravo

Received: 3 November 2022
Accepted: 25 November 2022
Published: 29 November 2022

Publisher's Note: MDPI stays neutral with regard to jurisdictional claims in published maps and institutional affiliations.

Copyright: © 2022 by the authors. Licensee MDPI, Basel, Switzerland. This article is an open access article distributed under the terms and conditions of the Creative Commons Attribution (CC BY) license (https://creativecommons.org/licenses/by/4.0/).

1. Introduction

Color anomaly detection methods identify pixel regions in multispectral images that have a low probability of occurring in the background landscape and are therefore considered to be outliers. Such techniques are used in remote sensing applications for agriculture, wildlife observation, surveillance, or search and rescue. Occlusion caused by vegetation, however, remains a major challenge.

Airborne Optical Sectioning (AOS) [1–13] is a synthetic aperture sensing technique that computationally removes occlusion in real-time by registering and integrating multiple images captured within a large synthetic aperture area above the forest (cf. Figure 1). With the resulting shallow-depth-of-field integral images, it becomes possible to locate targets (e.g., people, animals, vehicles, wildfires, etc.) that are otherwise hidden under the forest canopy. Image pixels that correspond to the same target on the synthetic focal plane (i.e., the forest ground) are computationally aligned and enhanced, while occluders above the focal plane (i.e., trees) are suppressed in strong defocus. AOS is real-time and wavelength-independent (i.e., it can be applied to images in all spectral bands), which is beneficial for many areas of application. Thus far, AOS has been applied to the visible [1,11] and the far-infrared (thermal) spectrum [4] for various applications, such as archeology [1,2], wildlife observation [5], and search and rescue [8,9]. By employing a randomly distributed statistical model [3,10,12], the limits of AOS and its efficacy with respect to its optimal sampling parameters can be explained. Common image processing tasks, such as classification with deep neural networks [8,9] or color anomaly detection, [11] are proven to perform significantly better when applied to AOS integral images compared with conventional aerial images. We also demonstrated the real-time capability of AOS by deploying it on a fully autonomous and classification-driven adaptive search and rescue drone [9]. In [11,13], we presented the first solutions to tracking moving people through densely occluding foliage.

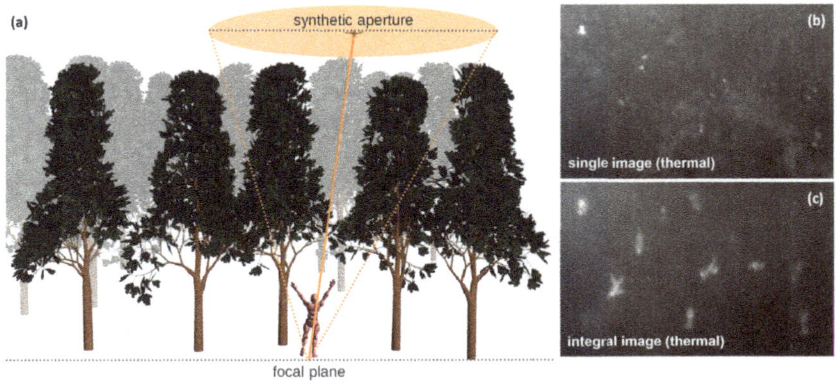

Figure 1. Airborne optical sectioning (AOS) is a synthetic aperture sensing technique that computationally combines multiple aerial images captured within a synthetic aperture area (**a**) to an integral image, which enhances targets on the synthetic focal plane while suppressing occluders above it. Right: People covered by forest canopy. Single aerial image (thermal channel) that suffers from strong occlusion (**b**), and corresponding integral image of the same environment with occlusion removed (**c**).

Anomaly detection methods for wilderness search and rescue have been evaluated earlier [14], and bimodal systems using a composition of visible and thermal information were already used to improve detection rates of machine learning algorithms [15,16]. However, none of the previous work considered occlusion.

With AOS, we are able to combine multispectral recordings into a single integral image. Our previous work has shown that image processing tasks, such as person classification with deep neural networks [8–10], perform significantly better on integral images when compared to single images. These classifiers are based on supervised architectures, which have the disadvantage that training data must be collected and labeled in a time-consuming manner and that the trained neural networks do not generalize well into other domains. It was also shown in [11] that the image integration process of AOS decreases variance and covariance, which allows better separation of target and background pixels when applying the Reed–Xiaoli (RX) unsupervised anomaly detection [17].

In this article, we evaluate several common unsupervised anomaly detection methods being applied to multispectral integral images that are captured from a drone when flying over open and occluded (forest) landscapes. We show that their performance can significantly be improved by the right combination of spectral bands and choice of color space input format. Especially for forest-like environments, detection rates of occluded people can be consistently increased if visible and thermal bands are combined and if HSV or HLS color spaces are used for the visible bands instead of common RGB. Furthermore, we also evaluate the runtime behavior of these methods when considered for time-critical applications, such as search and rescue.

2. Materials and Methods

For our evaluation, we applied the dataset from [8], which was used to prove that integral images improve people classification under occluded conditions. It consists of RGB and thermal images (pairwise simultaneously) captured with a drone prototype over multiple forest types (broadleaf, conifer, mixed) and open landscapes, as shown in Figure 2. In all images, targets (persons laying on the ground) are manually labeled. Additional telemetry data (GPS and IMU sensor values) of the drone during capturing are also provided for each image.

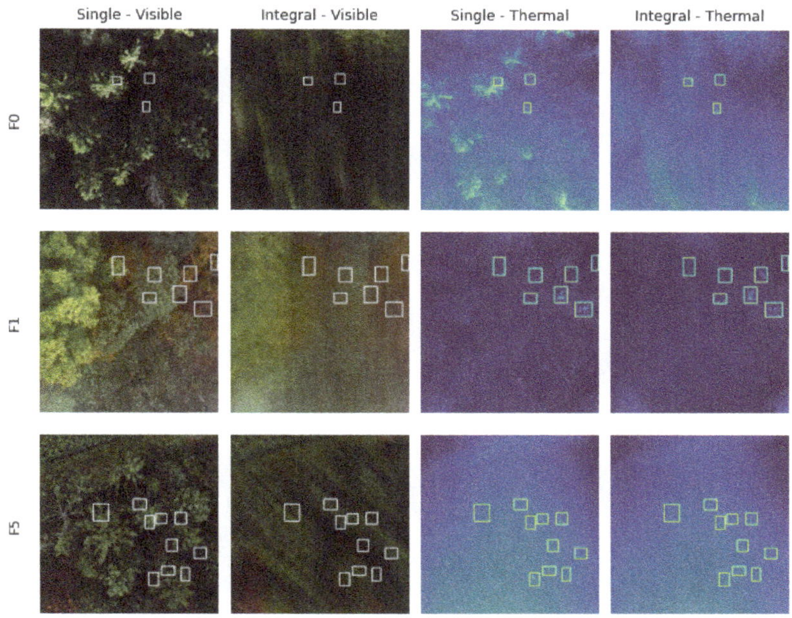

(**a**) Forest Landscapes

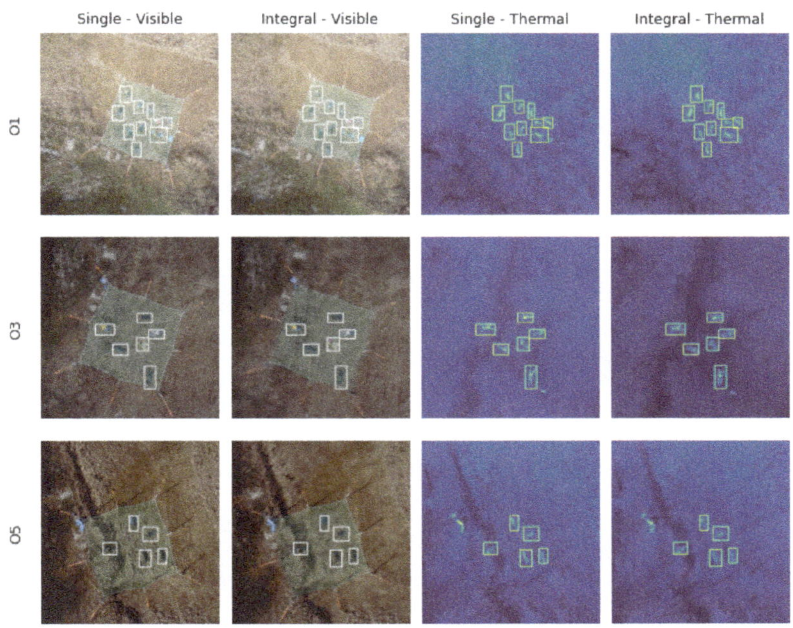

(**b**) Open Landscapes

Figure 2. Our evaluation dataset consists of several forest (**a**) and open (**b**) landscape images captured with a drone from an altitude of about 35m AGL. Each scenery (F0, F1, F5, O1, O3, O5) contains about 20 consecutive single images taken in the visible (RGB) and thermal spectrum, which are combined into integral images. Rectangles indicate manually labeled persons lying on the ground.

While the visible bands were converted from RGB to other color spaces (HLS, HSV, LAB, LUV, XYZ, and YUV), the thermal data were optionally added as a fourth (alpha) channel, resulting in additional input options (RGB-T, HLS-T, HSV-T, LAB-T, LUV-T, XYZ-T, and YUV-T).

All images had a resolution of 512x512 pixels, so the input dimensions where either (512, 512, 3) or (512, 512, 4). Methods that do not require spatial information used flattened images with (262144, 3) or (262144, 4) dimensions.

The publicly available C/C++ implementation of AOS Source Code: https://github.com/JKU-ICG/AOS (accessed on 28 November 2022) was used to compute integral images from single images.

2.1. Color Anomaly Detectors

Unsupervised color anomaly detectors have been widely used in the past [17–22], with the Reed–Xiaoli (RX) detector [17] being commonly considered as a benchmark. Several variations of RX exist, where the standard implementation calculates global background statistics (over the entire image) and then compares individual pixels based on the Mahalanobis distance. In the further course of this article, we will refer to this particular RX detector as Reed–Xiaoli Global (RXG).

The following briefly summarizes the considered color anomaly detectors, while details can be found through the provided references:

The Reed–Xiaoli Global (RXG) detector [17] computes a $K_{n\times n}$ covariance matrix of the image, where n is the number of input channels (e.g., for RGB, $n = 3$ and for RGB-T, $n = 4$). The pixel under test is the n-dimensional vector r, and the mean is given by the n-dimensional vector μ:

$$\alpha_{RXG}(r) = (r-\mu)^T K_{n\times n}^{-1}(r-\mu).$$

The Reed–Xiaoli Modified (RXM) detector [18] is a variation of RXG, where an additional constant $\kappa = ||r-\mu||^{-1}$ is used for normalization:

$$\alpha_{RXM}(r) = \kappa \cdot \alpha_{RXG}(r) = \left(\frac{r-\mu}{||r-\mu||}\right)^T K_{n\times n}^{-1}(r-\mu).$$

The Reed–Xiaoli Local (RXL) detector computes covariance and mean over smaller local areas and, therefore, does not use global background statistics. The areas are defined by an inner window (*guard_win*) and an outer window (*bg_win*). The mean μ and covariance K are calculated based on the outer window but excludes the inner window. Window sizes were chosen to be $guard\_win = 33$ and $bg\_win = 55$, based on the projected pixel sizes of the targets in the forest landscape.

The principal component analysis (PCA) [19] uses singular value decomposition for a linear dimensionality reduction. The covariance matrix of the image is decomposed into eigenvectors and their corresponding eigenvalues. A low-dimensional hyperplane is constructed by selected (*n_components*) eigenvectors. Outlier scores for each sample are then obtained by their euclidean distance to the constructed hyperplane. The number of eigenvectors to use was chosen to be $n\_components = n$, where n is the number of input channels (e.g., for RGB, $n = 3$ and for RGB-T, $n = 4$).

The Gaussian mixture model (GMM) [20] is a clustering approach, where multiple Gaussian distributions are used to characterize the data. The data are fit to each of the single Gaussians (*n_components*), which are considered as a representation of clusters. For each sample, the algorithm calculates the probability of belonging to each cluster, where low probabilities are an indication of being an anomaly. The number of Gaussians to use was chosen to be $n\_components = 2$.

The cluster based anomaly detection (CBAD) [21] estimates background statistics over clusters instead of sliding windows. The image background is partitioned (using any clustering algorithm) into clusters (*n_cluster*), where each cluster can be modeled as a Gaussian distribution. Similar to GMM, anomalies have values that deviate significantly from the cluster distributions. Samples are each assigned to the nearest background cluster, becoming an anomaly if their value deviates farther from the mean than background pixel values in that cluster. The number of clusters to use was chosen to be $n\_cluster = 2$.

The local outlier factor (LOF) [22] uses a distance metric (e.g., Minkowski distance) to determine the distances between neighboring ($n\_neighbors$) data points. Based on the inverse of those average distances, the local density is calculated. This is then compared to the local densities of their surrounding neighborhood. Samples that have significantly lower densities than their neighbors are considered isolated and, therefore, become outliers. The number of neighbors to use was chosen to be $n\_neighbors = 200$.

2.2. Evaluation

The evaluation of the methods summarized above was carried out on a consumer PC (Intel Core i9-11900H @ 2.50GHz) for the landscapes shown in Figure 2.

Precision (Equation (1)) vs. recall (Equation (2)) was used as metrics for performance comparisons.

The task can be formulated as a binary classification problem, where positive predictions are considered anomalous pixels. The data we want to classify (image pixels) is highly unbalanced, as most of the pixels are considered as background (majority class) and only some of the pixels are considered anomalies (minority class).

The true positive (TP) pixels are determined by checking whether they lie within one of the labeled rectangles, as shown in Figure 2. Pixels detected outside these rectangles are considered false positives (FP) and pixels inside the rectangle but not classified anomalously are considered false negatives (FN). Since the dataset only provides rectangles for labels and not perfect masks around the persons, the recall results are biased (in general, not as good as expected). As we are mainly interested in the performance difference between individual methods and the errors introduced are always constant (rectangle area—real person mask), the conclusions drawn from the results should be the same, even if perfect masks were used instead.

The precision (Equation (1)) quantifies the number of correct positive predictions made, and recall (Equation (2)) quantifies the number of correct positive predictions made out of all positive predictions that could have been made.:

$$Precision = \frac{TP}{TP + FP}, \tag{1}$$

$$Recall = \frac{TP}{FN + TP}. \tag{2}$$

Precision and recall both focus on the minority class (anomalous pixels) and are therefore less concerned with the majority class (background pixels), which is important for our unbalanced dataset.

Since the anomaly detection methods provide probabilistic scores on the likelihood of a pixel being considered anomalous, a threshold value must be chosen to obtain a final binary result.

The precision–recall curve (PRC) in Figure 3 shows the relationship between precision and recall for every possible threshold value that could be chosen. Thus, a method performing well would have high precision and high recall over different threshold values. We use the area under the precision–recall curve (AUPRC), which is simply the integral of the PRC, as the final evaluation metric.

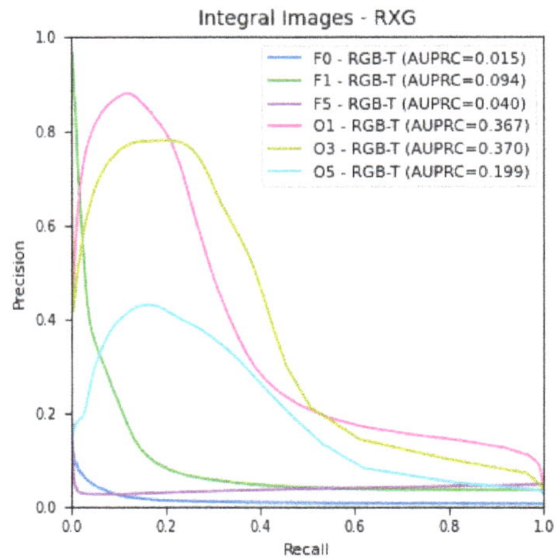

Figure 3. The area under the precision–recall curve (AUPRC) is used as a metric for comparing the performance of the evaluated anomaly detection methods. We consider true positives (TP), false positives (FP), and false negatives (FN) pixels for each image and calculate precision and recall. The example above illustrates precision–recall curves of all landscapes for RXG and with RGB-T input.

The AUPRC metric provides comparable results on the overall performance of a method but is not well suited when it comes to finding the best threshold for a single image. To obtain the best threshold value for a single image, we use the F_β-score (Equation (3)), which is also calculated from precision and recall:

$$F_\beta = \left(1 + \beta^2\right) \cdot \frac{Precision \cdot Recall}{(\beta^2 \cdot Precision) + Recall}, \tag{3}$$

where β is used as a weighting factor that can be chosen such that recall is considered β-times more important than precision.

The balanced F_1-score is the harmonic mean of precision and recall and is widely used. However, as we care more about minimizing false positives than minimizing false negatives, we would select a $\beta < 1$. A grid search of β values has given the best results for $\beta = \frac{1}{2}$. With this setting, precision is weighted more heavily than recall. The F_β metric is only used to threshold the image scores for comparison purposes, as shown in Figure 6.

3. Results

Figure 4 shows the AUPRC values across different color spaces and methods. The methods are evaluated on each color space, once with three channels (visible spectrum only) and once with four channels (visible and thermal spectrum). The results of the forest landscape are average values over F0, F1 and F5, and the results of the open landscape are average values over O1, O3 and O5.

As expected (and as we have also seen in Figure 3), the overall AUPRC of the open landscapes is much higher than the AUPRC values of the more challenging forest landscapes. The reason is an occlusion in the presence of forests.

The AUPRC values of the four-channel (color + thermal) and three-channel (color only) inputs are overlayed in the same bar. The slightly lighter colored four-channel results are always higher than the three-channel results—regardless of the method or the color space used. However, the difference is more pronounced for the forest landscapes than for the open landscapes. This shows that regardless of the scenery and regardless of the method and the color space used, the additional thermal information always improves the performance of anomaly detection.

With a look at the AUPRC values in the forest landscapes, we can observe that RXL gives the overall best results and outperforms all other methods. Utilizing the additional thermal information gives, in this case, even a 2× gain. This can also be observed visually in the anomaly detection scores shown in Figure 6, where FP's detections highly decrease and TP's detections highly increase if the thermal channel is added (e.g., F1, in the visible spectral band, many background pixels are considered anomalous, with the additional thermal information those misclassified pixels are eliminated).

Looking at the AUPRC values in the open landscapes, we can observe that the difference between the methods is not as pronounced as in the forest landscapes. An obvious outlier, however, seems to be LOF, which nevertheless performs very well (second best) in the forest landscapes. This can be explained by the fact that hyper-parameters of the methods were specifically chosen for the forest landscape. In the case of LOF, the *n_neighbors* parameter was set to be 200, which seems suboptimal for the open landscapes. The same holds for RXL (window sizes), CBAD (number of clusters), GMM (number of components) and PCA (number of components). All other methods do not require hyper-parametrization.

Another observation that can be made is that some color spaces consistently give better results than others. In the forest landscapes, HSV(-T) usually gives the best results, regardless of the methods being used. In the open landscapes, it is not as clear which color space performs best, but HSV(-T) still gives overall good results. In general, and especially for RXM, the improvements achieved by choosing HSV(-T) over other color spaces are clearly noticeable.

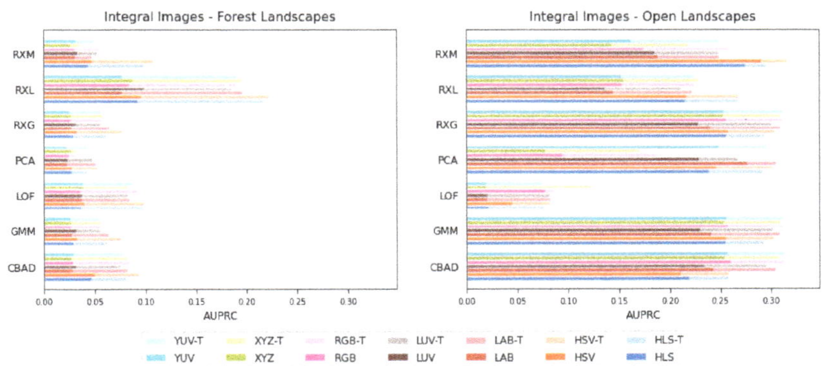

Figure 4. Results of area under the precision–recall curve (AUPRC) values for multiple color spaces and color anomaly detection methods. The results of the forest landscape are average values over F0, F1, and F5, and the results of the open landscape are average values over O1, O3, and O5. The stacked bar charts highlight the improvement gains caused by the additional thermal channel.

The individual results plotted in Figure 4 are also shown in Tables 1 and 2, where the mean values over all color spaces (last row) may give a useful estimate of the method's overall performance. The highest AUPRC value for the forest and open scenery is highlighted in bold.

Since anomaly detection for time-critical applications should deliver reliable results in real-time, we have also measured their runtimes, as shown in Table 3. The best-performing methods on the forest landscapes in terms of AUPRC values are RXL and LOF. In terms of runtime, both are found to be very slow, as they consume 20 to 35 s for computations, where all other algorithms provide anomaly scores in under a second (cf. Figure 5).

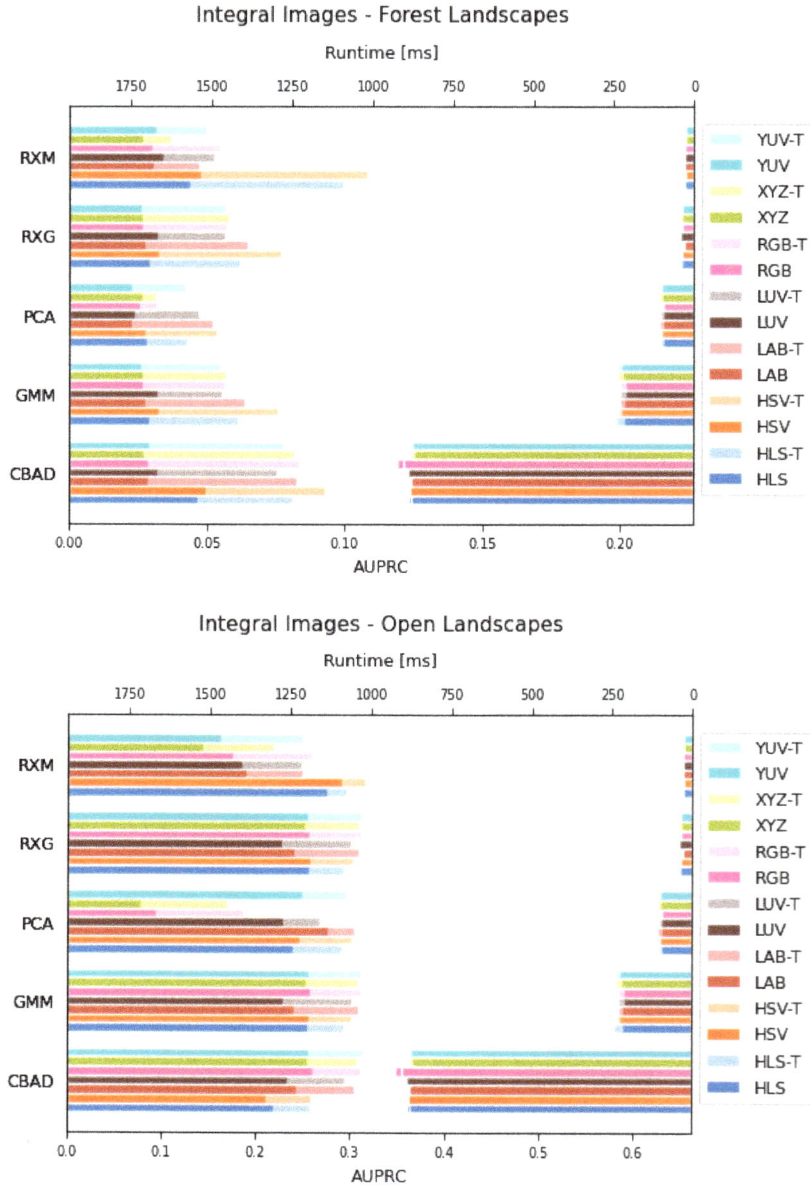

Figure 5. Performance in AUPRC (left bars) vs. runtime in *ms* (right bars): Color anomaly detection methods that produce results in less than a second. Reed–Xiaoli local (RXL) and local outlier factor (LOF) performed well but needed more than 20 seconds and are, therefore, not practicable for applications with real-time demands.

Table 1. Area under the precision–recall curve (AUPRC) values for each color space and color anomaly detection method. The scores are obtained from integral images and are averaged over forest landscapes. The last row is the mean AUPRC value over all color spaces. The best performance is highlighted in bold.

	CBAD	GMM	LOF	PCA	RXG	RXL	RXM
HLS	0.047	0.029	0.037	0.028	0.029	0.093	0.044
HSV	0.050	0.033	0.040	0.028	0.033	0.096	0.048
LAB	0.029	0.028	0.038	0.023	0.028	0.077	0.031
LUV	0.032	0.032	0.038	0.024	0.032	0.098	0.034
RGB	0.029	0.027	0.036	0.026	0.027	0.084	0.030
XYZ	0.027	0.027	0.039	0.027	0.027	0.088	0.027
YUV	0.029	0.026	0.038	0.023	0.026	0.077	0.032
AUPRC	0.035	0.029	0.038	0.025	0.029	0.088	0.035
	CBAD	**GMM**	**LOF**	**PCA**	**RXG**	**RXL**	**RXM**
HLS-T	0.081	0.061	0.098	0.043	0.062	0.218	0.099
HSV-T	0.093	0.076	0.099	0.053	0.077	**0.223**	0.108
LAB-T	0.083	0.064	0.084	0.052	0.065	0.196	0.047
LUV-T	0.076	0.056	0.083	0.047	0.056	0.186	0.052
RGB-T	0.084	0.057	0.092	0.032	0.057	0.191	0.055
XYZ-T	0.082	0.057	0.083	0.032	0.058	0.196	0.037
YUV-T	0.078	0.055	0.088	0.042	0.056	0.190	0.050
AUPRC	0.082	0.061	0.090	0.043	0.062	0.200	0.064

Table 2. Area under the precision–recall curve (AUPRC) values for each color space and anomaly detection method. The scores are obtained from integral images and are averaged over open landscapes. The last row is the mean AUPRC value over all color spaces. Best performance is highlighted in bold.

	CBAD	GMM	LOF	PCA	RXG	RXL	RXM
HLS	0.219	0.255	0.022	0.239	0.256	0.215	0.275
HSV	0.211	0.256	0.044	0.246	0.258	0.217	0.291
LAB	0.243	0.242	0.021	0.276	0.241	0.144	0.189
LUV	0.234	0.230	0.020	0.229	0.228	0.136	0.185
RGB	0.261	0.257	0.078	0.094	0.256	0.153	0.175
XYZ	0.254	0.254	0.019	0.078	0.253	0.155	0.144
YUV	0.257	0.256	0.019	0.249	0.255	0.153	0.162
AUPRC	0.240	0.250	0.032	0.202	0.249	0.168	0.203
	CBAD	**GMM**	**LOF**	**PCA**	**RXG**	**RXL**	**RXM**
HLS-T	0.258	0.293	0.077	0.292	0.294	0.268	0.296
HSV-T	0.259	0.303	0.082	0.302	0.304	0.268	0.316
LAB-T	0.305	0.309	0.082	0.306	0.310	0.222	0.250
LUV-T	0.295	0.302	0.082	0.267	0.300	0.212	0.249
RGB-T	0.312	0.312	0.086	0.187	0.312	0.225	0.260
XYZ-T	0.308	0.310	0.123	0.169	0.310	0.229	0.219
YUV-T	**0.314**	0.312	0.075	0.297	0.312	0.225	0.250
AUPRC	0.293	0.306	0.087	0.260	0.306	0.236	0.263

Table 3. Runtime for each input format and method in milliseconds. The input format (color spaces) does not have an influence on the runtime, but addition channels (thermal) may increase the runtime for some algorithms. The last row is the mean runtime of an algorithm. Best performance is highlighted in bold.

	CBAD	GMM	LOF	PCA	RXG	RXL	RXM
HLS	887	219	18,440	98	39	36,647	28
HSV	883	225	18,094	102	40	36,346	**27**
LAB	877	219	19,833	99	36	36,495	29
LUV	888	212	19,975	96	41	36,166	29
RGB	925	216	19,367	96	40	35,732	29
XYZ	872	224	19,107	100	40	36,367	**27**
YUV	874	228	20,485	100	42	36,409	**27**
Runtime	887	221	19,329	99	40	36,309	28
	CBAD	**GMM**	**LOF**	**PCA**	**RXG**	**RXL**	**RXM**
HLS-T	878	243	28,054	104	43	36,002	30
HSV-T	882	237	28,235	107	40	35,801	30
LAB-T	879	230	28,005	108	32	35,876	31
LUV-T	893	229	27,506	105	45	36,059	32
RGB-T	904	230	27,323	98	37	35,618	32
XYZ-T	880	241	24,395	108	43	36,047	29
YUV-T	877	238	27,146	107	40	36,180	29
Runtime	885	236	27,238	105	40	35,940	31

4. Discussion

The AUPRC results in Figure 4 show that all color anomaly detection methods benefit from additional thermal information, but especially in combination with the forest landscapes.

In challenging environments, where the distribution of colors has a much higher variance (e.g., F1 in Figure 6, due to bright sunlight), the additional thermal information improves results significantly. If the temperature difference between targets and the surrounding is large enough, the thermal spectral band may add spatial information (e.g., distinct clusters of persons), which is beneficial for methods that calculate results based on locality properties (e.g., RXL, LOF).

In forest-like environments, the RXL anomaly detector performs best regardless of the input color space. This could be explained by the specific characteristics of an integral image. In the case of occlusion, the integration process produces highly blurred images caused by defocused occluders (forest canopy) above the ground, which results in a much more uniformly distributed background. Since target pixels on the ground stay in focus, anomaly detection methods such as RXL, which calculate background statistics on a smaller window around the target, benefit from the uniform distributed (local) background. The same is true for LOF, where the local density in the blurred background regions is much higher than the local density in the focused target region, resulting in overall better outlier detection rates. Since most objects in open landscapes are located near the focal plane (i.e., at nearly the same altitude above the ground), there is no out-of-focus effect caused by the integration process. Thus, these methods do not produce similarly good results for open landscapes.

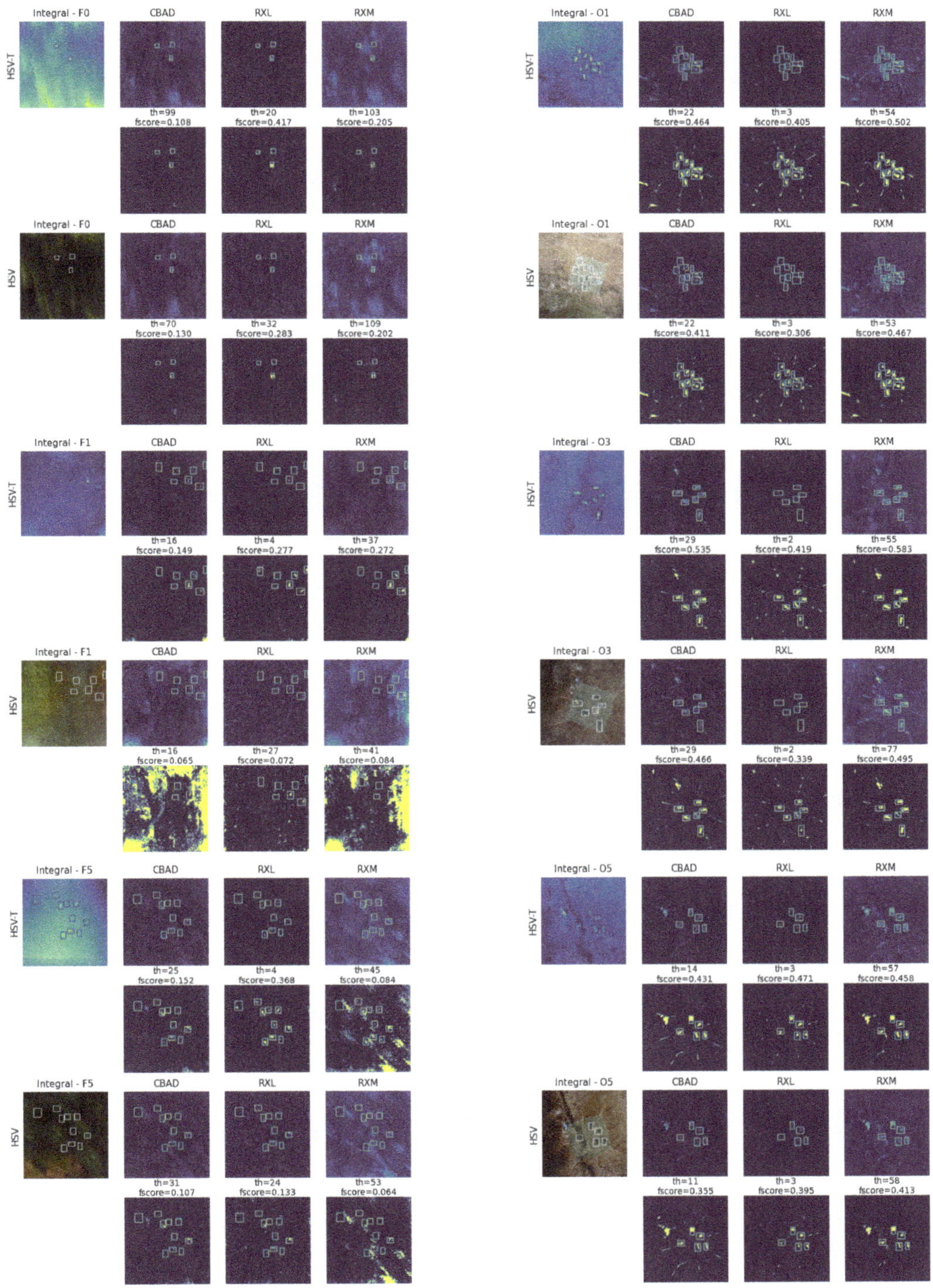

(a) Forest Landscapes

(b) Open Landscapes

Figure 6. Color anomaly detection scores for forest (**a**) and open (**b**) landscapes, comparing the overall good performing HSV(-T) inputs. The first rows per scenery show anomaly scores of the best

(RXL) algorithm without considering runtime and the best (RXM) and second-best (CBAD) method when considering runtime. The second rows per scenery are the anomaly scores after thresholding.

For the forest landscapes, the HSV(-T) and HSL(-T) color spaces consistently give better results than others. The color spaces HSV (hue, saturation, value) and HSL (hue, saturation, lightness) are both based on cylindrical color space geometries and differ mainly in their last dimension (brightness/lightness). The first two dimensions (hue, saturation) can be considered more important when distinguishing colors, as the last dimension only describes the value (brightness) or lightness of a color. We assume that the more uniform background resulting from the integration process also has a positive effect on the distance metric calculations when those two color spaces are used, especially if the background mainly consists of a very similar color tone. This is again more pronounced for the forest landscapes than for the open landscapes.

Although the AUPRC results obtained from RXL and LOF are best for forest landscapes, the high runtime indicates that these methods are impractical for real-time applications. A trade-off must be made between good anomaly detection results and fast runtime; therefore, we consider the top-performing methods that provide reliable results within milliseconds further.

Based on the AUPRC and runtime results shown in Figure 5, one could suggest that the RXM method may be used. The AUPRC results combined with HSV-T are the best among methods that run under one second, regardless of the landscape. Since this method does not require a-priory settings to be chosen (only the final thresholding value) and the runtime is one of the fastest, it would be well suited for usage in forests and open landscapes. The second-best algorithm based on the AUPRC values would be CBAD, with the disadvantage that it requires a hyper-parameter setting and does not generalize well for open landscapes.

5. Conclusions

In this article, we have shown that the performance of unsupervised color anomaly detection methods applied to multispectral integral images can be further improved by an additional thermal channel. Each of the evaluated methods performs significantly better when thermal information is utilized in addition, regardless of the landscape (forest or open). Another finding is that even without the additional thermal band, the choice of input color space (for the visible channels) already has an influence on the results. Color spaces such as HSV and HLS can outperform the widely used RGB color space, especially in forest-like landscapes.

These findings might guard decisions on the choice of color anomaly detection method, input format, and applied spectral band, depending on individual use cases. Occlusion cause by vegetation, such as forests, remains challenging for many of them. In the future, we will investigate anomalies caused by motion in the context of synthetic aperture sensing. In combination with color and thermal anomaly detection, motion anomaly detection has the potential to further improve detection results for moving targets, such as people, animals, or vehicles.

Author Contributions: Conceptualization, O.B. and F.S.; methodology, F.S.; software, F.S. and I.K.; validation, F.S., I.K. and O.B.; formal analysis, F.S.; investigation, F.S.; resources, I.K.; data curation, I.K.; writing—original draft preparation, F.S. and O.B.; writing—review and editing, F.S. and O.B.; visualization, F.S.; supervision, O.B.; project administration, O.B.; funding acquisition, O.B. All authors have read and agreed to the published version of the manuscript.

Funding: This research was funded by the Austrian Science Fund (FWF) under grant number P32185-NBL and by the State of Upper Austria and the Austrian Federal Ministry of Education, Science and Research via the LIT–Linz Institute of Technology under grant number LIT-2019-8-SEE114.

Data Availability Statement: The data and source code used in the experiments can be downloaded from https://doi.org/10.5281/zenodo.3894773 and https://github.com/JKU-ICG/AOS (accessed on 28 November 2022).

Conflicts of Interest: The authors declare no conflict of interest.

References

1. Kurmi, I.; Schedl, D.; Bimber, O. Airborne optical sectioning. *J. Imaging* **2018**, *4*, 102. [CrossRef]
2. Bimber, O.; Kurmi, I.; Schedl, D. Synthetic aperture imaging with drones. *IEEE Comput. Graph. Appl* **2019**, *39*, 8–15. [CrossRef] [PubMed]
3. Kurmi, I.; Schedl, D.; Bimber, O. A statistical view on synthetic aperture imaging for occlusion removal. *IEEE Sens. J* **2019**, *19*, 9374–9383. [CrossRef]
4. Kurmi, I.; Schedl, D.; Bimber, O. Thermal airborne optical sectioning. *Remote Sens.* **2019**, *11*, 1668. [CrossRef]
5. Schedl, D.; Kurmi, I.; Bimber, O. Airborne optical sectioning for nesting observation. *Sci. Rep.* **2020**, *10*, 7254. [CrossRef]
6. Kurmi, I.; Schedl, D.; Bimber, O. Fast Automatic Visibility Optimization for Thermal Synthetic Aperture Visualization. *IEEE Geosci. Remote Sens. Lett.* **2021**, *18*, 836–840. [CrossRef]
7. Kurmi, I.; Schedl, D.; Bimber, O. Pose Error Reduction for Focus Enhancement in Thermal Synthetic Aperture Visualization. *IEEE Geosci. Remote. Sens. Lett.* 2021, *to be published.* [CrossRef]
8. Schedl, D.; Kurmi, I.; Bimber, O. Search and rescue with airborne optical sectioning. *Nat. Mach. Intell.* **2020**, *2*, 783–790. [CrossRef]
9. Schedl, D.; Kurmi, I.; Bimber, O. An autonomous drone for search and rescue in forests using airborne optical sectioning. *Sci. Robot* **2021**, *6*, 1188. [CrossRef]
10. Kurmi, I.; Schedl, D.; Bimber, O. Combined person classification with airborne optical sectioning. *Sci. Rep.* **2022**, *12*, 3804. [CrossRef]
11. Nathan, R.; Kurmi, I.; Schedl, D.; Bimber, O. Through-Foliage Tracking with Airborne Optical Sectioning. *J. Remote Sens.* **2022**, *2022*, 9812765. [CrossRef]
12. Seits, F.; Kurmi, I.; Nathan, R.; Ortner, R.; Bimber, O. On the Role of Field of View for Occlusion Removal with Airborne Optical Sectioning. *arXiv* **2022**, arXiv:2204.13371. https://doi.org/10.48550/ARXIV.2204.13371.
13. Nathan, R.; Kurmi, I.; Bimber, O. Inverse Airborne Optical Sectioning. *Drones* **2022**, *6*, 231. [CrossRef]
14. Morse, B.; Thornton, D.; Goodrich, M. Color anomaly detection and suggestion for wilderness search and rescue. In Proceedings of the Seventh Annual ACM/IEEE International Conference on Human-Robot Interaction, Boston, MA, USA, 5–8 March 2012; Association for Computing Machinery: New York, NY, USA, 2012; pp. 455–462. [CrossRef]
15. Rudol, P.; Doherty, P. Human Body Detection and Geolocalization for UAV Search and Rescue Missions Using Color and Thermal Imagery. In Proceedings of the 2008 IEEE Aerospace Conference, Big Sky, MT, USA, 1–8 March 2008; pp. 1–8. [CrossRef]
16. Hinzmann, T.; Stegemann, T.; Cadena, C.; Siegwart, R. Deep Learning-based Human Detection for UAVs with Optical and Infrared Cameras: System and Experiments. *arXiv* **2020**, arXiv:2008.04197. https://doi.org/10.48550/ARXIV.2008.04197.
17. Reed, I.; Yu, X. Adaptive Multiple-Band CFAR Detection of an Optical Pattern with Unknown Spectral Distribution. *IEEE Trans. Acoust. Speech Signal Process.* **1990**, *38*, 1760–1770. [CrossRef]
18. Chang, C.; Chiang, S. Anomaly detection and classification for hyperspectral imagery. *Geosci. Remote Sens. IEEE Trans.* **2002**, *40*, 1314–1325. [CrossRef]
19. Shyu, M.; Chen, S.; Sarinnapakorn, K.; Chang, L. A Novel Anomaly Detection Scheme Based on Principal Component Classifier. In Proceedings of the IEEE Foundations and New Directions of Data Mining Workshop, in conjunction with the Third IEEE International Conference on Data Mining (ICDM'03), Melbourne, FL, USA, 19–22 November 2003;
20. Bishop, C.M.; Nasrabadi, N.M. Pattern Recognition and Machine Learning. *J. Electron. Imaging* **2007**, *16*, 049901.
21. Carlotto, M. A cluster-based approach for detecting man-made objects and changes in imagery. *IEEE Trans. Geosci. Remote Sens.* **2005**, *43*, 374–387. [CrossRef]
22. Breunig, M.; Kriegel, H.; Ng, R.; Sander, J. LOF: Identifying Density-Based Local Outliers. *ACM SIGMOD Rec.* **2000**, *29*, 93–104. [CrossRef]

eng

Article

Infrared Spectroscopy for the Quality Control of a Granular Tebuthiuron Formulation

Joel B. Johnson [1,*], Hugh Farquhar [2], Mansel Ismay [3,†] and Mani Naiker [1]

1 School of Health, Medical & Applied Sciences, Central Queensland University, Bruce Highway, Rockhampton, QLD 4701, Australia
2 Cirrus Ag, 171 Alexandra St, Kawana, Rockhampton, QLD 4701, Australia
3 Independent Researcher, Gladstone, QLD 4680, Australia
* Correspondence: joel.johnson@cqumail.com
† Previous affiliation: Cirrus Ag, 171 Alexandra St, Kawana, Rockhampton, QLD 4701, Australia.

Abstract: Tebuthiuron is a selective herbicide for woody species and is commonly manufactured and sold as a granular formulation. This project investigated the use of infrared spectroscopy for the quality analysis of tebuthiuron granules, specifically the prediction of moisture content and tebuthiuron content. A comparison of different methods showed that near-infrared spectroscopy showed better results than mid-infrared spectroscopy, while a handheld NIR instrument (MicroNIR) showed slightly improved results over a benchtop NIR instrument (Antaris II FT-NIR Analyzer). The best-performing models gave an R^2_{CV} of 0.92 and RMSECV of 0.83% w/w for moisture content, and R^2_{CV} of 0.50 and RMSECV of 7.5 mg/g for tebuthiuron content. This analytical technique could be used to optimise the manufacturing process and reduce the costs of post-manufacturing quality assurance.

Keywords: process analytical technology; quality assurance; non-destructive assessment; NIRS

Citation: Johnson, J.B.; Farquhar, H.; Ismay, M.; Naiker, M. Infrared Spectroscopy for the Quality Control of a Granular Tebuthiuron Formulation. *Eng* **2022**, *3*, 596–619. https://doi.org/10.3390/eng3040041

Academic Editor: Antonio Gil Bravo

Received: 14 November 2022
Accepted: 30 November 2022
Published: 2 December 2022

Publisher's Note: MDPI stays neutral with regard to jurisdictional claims in published maps and institutional affiliations.

Copyright: © 2022 by the authors. Licensee MDPI, Basel, Switzerland. This article is an open access article distributed under the terms and conditions of the Creative Commons Attribution (CC BY) license (https://creativecommons.org/licenses/by/4.0/).

1. Introduction

Tebuthiuron is a thiadiazolyl urea herbicide (Figure 1) primarily used for the control of woody plants. Application is typically via pellet-type (granular) formulations containing tebuthiuron (200–400 mg/g), which may be applied from the ground (either by hand or mechanically), or aerially dispersed if a large area is to be treated. As tebuthiuron is highly water-soluble [1], it leaches from the granules into the soil [2], where it is subsequently absorbed by the roots and translocated to the leaf tissue [3]. Its mode of action is through inhibition of Photosystem II, thus preventing photosynthesis in the affected plant [4]. The degradation and persistence of tebuthiuron is still an area under investigation, with studies reporting half-lives between 20 days [5] and 16–22 days [6], to as high as one year [7], 12.9 months [8], 'considerably greater' than 15 months [9] and even 2–7 years [10]. du Toit and Sekwadi [11] reported that tebuthiuron residue remained active in soil for 8 years after application.

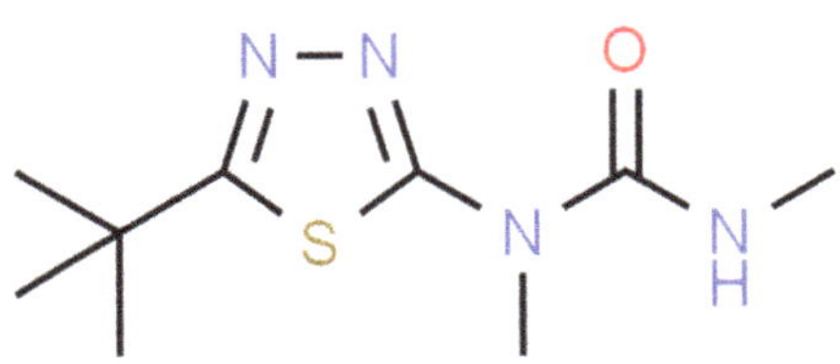

Figure 1. The chemical structure of tebuthuiron. Retrieved from http://www.chemspider.com/ (accessed on 28 September 2022) under Creative Commons 4.0 license.

The structure of tebuthiuron includes two amide bonds (sharing the one carbonyl group) and a unique aromatic thiadiazolyl group (Figure 1), which should make it well-suited to detection by non-destructive analytical techniques such as infrared spectroscopy. Both of these chemical moieties would be expected to show distinct spectral characteristics in the mid-infrared (MIR) and near-infrared (NIR) regions. Additionally, the high concentration of tebuthiuron in most commercial tebuthiuron products (200–400 mg/g), should make it relatively simple to detect using MIR/NIR spectroscopy. The benefits of infrared spectroscopy over traditional analytical techniques include its speed (real-time), no ongoing costs related to traditional analysis costs, non-destructiveness and versatility (many units are portable; some are even handheld).

The previous literature has documented the use of NIR spectroscopy for the prediction of pesticide concentrations in liquid solutions [12], although this particular study did not analyse pesticides present in granular/powder form. Other authors have used MIR spectroscopy for the production quality control of numerous pesticides in liquid and solid forms [13]. Consequently, there is increasing interest in the use of infrared spectroscopy as a process analytical technology (PAT) tool [14]. However, no previous studies were found using infrared spectroscopy for the analysis of tebuthiuron content in any sample matrices.

Consequently, the aim of this work was to investigate the prospect of using infrared spectroscopy for a rapid, non-invasive method for the assessment of tebuthiuron and moisture, another important analyte in the manufacturing quality assurance process. This included a comparison of the performance of different infrared spectrophotometers for this purpose.

2. Materials and Methods

2.1. Sample Description

Sixty-eight (68) granular tebuthiuron samples (Regain™ brand) were manufactured by Cirrus Ag (North Rockhampton, Australia) over a period of approximately 10 months (April 2021–February 2022). These were each collected at specific time points as the final product came off the manufacturing line. The samples were stored in an air-conditioned room (approx. 20 °C) after receipt at the laboratory. Additionally, five samples of tebuthiuron powder (>95% purity) were included for comparative purposes, although they were not included in the quantitative modelling.

Two types of Regain formulation were produced by Cirrus Ag: one containing 200 mg/g of the active ingredient (i.e., tebuthiuron) and the other containing 400 mg/g. Throughout the manuscript, these are abbreviated at Regain200 and Regain400, respectively. The exact composition of the Regain granules is a trade secret, hence cannot be disclosed in this paper.

2.2. Sample Preparation

To ensure representative sampling of the granules upon receipt at the laboratory, each sample bag was thoroughly shaken and pellets were subsampled from at least 6 different locations throughout the bag. Approximately half of the sample (20–30 g) was subsampled, with the pellets then ground to a fine powder (Breville Coffee & Spice Grinder; Botany, NSW, Australia). When weighing out the required mass of powder for each extraction, care was taken to ensure that powder from at least 3 different locations within the sample subset was included.

2.3. Analysis of Moisture Content

The moisture content was measured on the intact (unground) Regain samples. Approximately 3 g of each sample was weighed into a pre-weighed aluminium foil tray, before being dried in a laboratory oven (Memmert 400; Buechenbach, Germany) at 110 °C overnight (16 h) until reaching a constant mass. After cooling to room temperature, the samples were reweighed, with the moisture content was determined as the loss in mass upon drying. Only one replicate was performed for each sample; however, triplicate analyses were performed on one sample to assess the reproducibility of the method.

To provide an extra data point for the NIR prediction of moisture content, the dried granules from all 68 samples were combined and mixed thoroughly. This aggregate granule sample was taken to have a moisture content of 0%, as it had been already dried at 110 °C for 16 hrs and did not lose any mass upon further drying.

2.4. Tebuthiuron Extraction Protocol

The extraction method was adapted from Lydon et al. [15] and validated by our laboratory. Approximately, 30 mg of the finely ground Regain powder was weighed into a 50 mL centrifuge tube. The total mass of the tube + sample was then recorded, before 20 mL of 90% *v*/*v* methanol was added using a calibrated bottle-top dispenser. The tube + sample + methanol was then re-weighed, with the determined mass of methanol converted to volume using the density of 90% methanol (0.823 ± 0.001 g/mL; n = 6 independent measurements). This allowed for the added volume of methanol to be determined with a much higher level of accuracy than that which could be obtained using a calibrated pipette.

The powder was extracted using an end-over-end shaker (Ratek RM-4; Boronia, VIC, Australia) operating at 50 rpm for 30 min. The extract was then centrifuged (1000 rcf for 5 min) and the supernatant collected for direct HPLC analysis (no dilution required). All samples were extracted and analysed in triplicate.

2.5. Tebuthiuron Analysis by HPLC

The tebuthiuron content of the methanol extracts was determined by high-performance liquid chromatography (HPLC). The HPLC analysis method used was adapted from Ferracini et al. [16]. Tebuthiuron was quantified on an Agilent 1100 HPLC system, comprising a G1313A autosampler, G1322A vacuum degasser, G1311A quaternary pump and G1315B diode array detector. A reversed-phase C_{18} column was used (Agilent Eclipse XDB-C18; 250 × 4.6 mm; 5 μm pore size; Agilent Technologies, Santa Clara, CA, USA) along with a C_{18} guard column (Agilent Eclipse XDB-C18; 12.5 × 4.6 mm; 5 μm pore size). The injection volume was 5 μL and the detection wavelength was 254 nm. The elution method was an isocratic mixture of 50% methanol/50% water, at a flow rate of 1 mL min^{-1}. The total run time was 15 min.

The tebuthiuron concentration of the samples was determined using an external calibration of analytical-grade tebuthiuron standard (Sigma-Aldrich Australia; North Ryde. NSW, Australia), ranging between 100–1000 mg L^{-1}. Results were expressed as mg/g, on an as-is basis.

Figure 2 shows a typical chromatogram obtained from the extract of a Regain400 sample, demonstrating the absence of any interfering compounds at the selected wavelength.

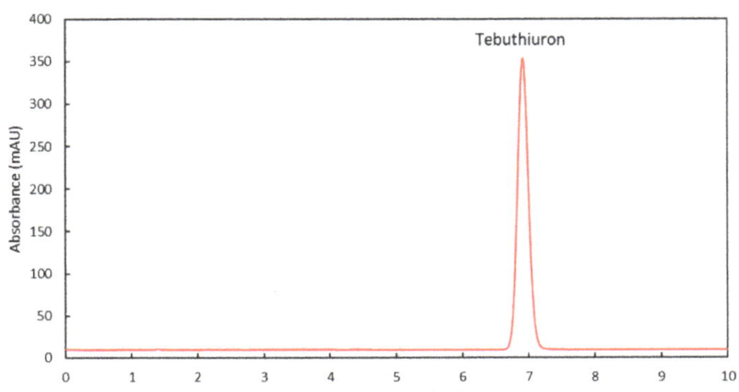

Figure 2. A typical HPLC chromatogram of a Regain extract, with the location of the tebuthiuron peak indicated.

There was no clear consistency in the literature for the λmax or detection wavelength for tebuthiuron. Weber [17] reported λmax values of 252 nm in neutral solution and 261 nm in acidic conditions for tebuthiuron. Lourencetti et al. [18] found a λmax of 255 nm, while more recently, Ferreira et al. [19] reported the λmax of tebuthiuron to be 253 nm. Similarly, Lydon et al. [15] used a wavelength of 254 nm for the quantification of tebuthiuron via HPLC, while other authors have used 245 nm [20] and 247 nm [16,18].

Consequently, preliminary investigations were conducted to determine the optimum detection wavelength to use for HPLC analysis, through HPLC-DAD analysis and scanning tebuthiuron solutions using a Thermo Scientific Genesys 10S UV-Vis spectrophotometer (Sydney, Australia).

2.6. Validation of the HPLC Method

To assess the linearity of the HPLC method, tebuthiuron standards were prepared between 0.1–1000 mg L^{-1} and analysed using HPLC.

To assess the intra-day precision of the method, six replicate injections of 100 mg L^{-1} tebuthiuron standard were analysed on the same day. Similarly, the inter-day precision was assessed by injecting a sample of 100 mg L^{-1} tebuthiuron standard over six different days and comparing the peak areas.

To assess the stability of the methanolic tebuthiuron extracts, one Regain extract was analysed immediately following extraction and re-analysed after it had been stored for 8 days at room temperature.

The reproducibility of the finalised extraction method was determined by extracting and analysing 20 Regain samples each in triplicate, with the %CV calculated for each sample. In addition, the reproducibility of the extraction and HPLC method was assessed by performing 7 replicate extractions on one sample of homogenised Regain400 powder.

2.7. Assessment of Sample Variation

Based on preliminary results and observations during the manufacturing process, it was thought that there could be a high level of variation in tebuthiuron content between different portions of each Regain batch. To test this hypothesis, ten samples were randomly selected from different parts of one Regain400 sample (see Figure 3). These were then ground, extracted and analysed separately (each in triplicate).

Figure 3. Sampling of the ten spatial replicates of the Regain400 sample. The red circles show the sampled areas.

2.8. Collection of FTIR Spectra

Mid-infrared (MIR) spectra were collected from the powdered samples using a Bruker Alpha FTIR (Fourier transform infrared) spectrophotometer (Bruker Optics Gmbh, Ettlingen, Germany) fitted with a platinum diamond attenuated total reflectance (ATR) single reflection module. FTIR spectroscopy requires the samples to be in a powdered form, as it necessitates firm contact between the sample and the ATR platform used to collect the spectra. Consequently, the FTIR spectra could only be collected from the powdered Regain samples, not the whole granular product.

The reflection module was covered with powder (approximately 100–200 mg) and pressure was applied to achieve uniform contact between the ATR interface and powder. Air was used as a reference background; the background measurement was performed every 15 min. Cross-contamination of samples was minimised by cleaning and drying the platform with isopropyl alcohol and laboratory Kimwipes® between samples. Using the OPUS software version 7.5 (Bruker Optics Gmbh, Ettlingen, Germany), the FTIR spectra were recorded between 4000 and 400 cm^{-1} as the average of 24 scans at a resolution of 4 cm^{-1}. Three spectra were collected from each sample, repacking the instrument with fresh powder each time.

2.9. Collection of NIR Spectra—Benchtop Instrument

Near-infrared (NIR) spectra were collected from both the granular and powdered Regain samples using Antaris II FT-NIR Analyzer (Thermo Scientific; Madison, WI, USA). This instrument provides a high level of accuracy and reproducibility, making it highly suitable for method development purposes. Throughout the report, this is referred to as the "benchtop" NIR method.

The instrument was operated in reflectance mode, using the integrating sphere with a rotating sample cup (30 mm diameter). Spectra were collected between 1000–2500 nm (10,000–4000 cm^{-1}), as the mean of 32 scans (resolution of 8 cm^{-1}). The optimised gain was found to be 2×, with an empty attenuator screen. Background (dark) reference measurements were collected every hour. Spectra were collected in triplicate, repacking the sample cup with fresh granules or powder each time. The spectra were exported in *.csv format, with the mean of the triplicate spectra for each sample used in subsequent data analysis.

2.10. Collection of NIR Spectra—Handheld Instrument

NIR spectra were also collected from the granular Regain samples using a handheld NIR instrument, in order to determine the typical accuracy that could be obtained using portable NIR instrumentation. Handheld instrumentation would be greatly beneficial in an industrial setting due to its portability and lower cost (less than 1/3 of the benchtop Antaris instrument). Furthermore, instruments such as the MicroNIR may be suitable for installation as in-line sensors in an industrial setting.

A MicroNIR OnSite handheld spectrometer (Viavi; Santa Rosa, CA, USA) was used for this work. Spectra were collected across the full wavelength range of this instrument (908–1676 nm); the integration time was set to 100 ms. Reference dark and light spectra were collected every 10 min. Again, spectra were collected in triplicate and exported in *.csv format. The mean of the triplicate spectra for each sample was used in subsequent data analysis.

2.11. Independent Test Set—Using Handheld NIR

The best-performing infrared spectrophotometer was applied to an independent test set (i.e., samples not used in the model calibration). For this, thirteen additional samples (12 of Regain400 and 1 of Regain200) were sourced from Cirrus Ag (Rockhampton, Australia). NIR spectra were collected from the granules using the MicroNIR instrument and predictions made using the optimum model for each analyte.

2.12. Data Analysis

Chemometric analysis of the infrared spectra was conducted in the Unscrambler X 10.5 software (Camo ASA, Oslo, Norway). A variety of pre-processing methods were trialled, including the use of the standard normal variate (SNV) algorithm and the 1st and 2nd derivatives were calculated using a Savitzky–Golay algorithm with varying numbers of smoothing points. These are abbreviated as number–letter combinations showing the derivative number and number of smoothing points, e.g., 1d5 indicates 1st derivative with 5 smoothing points.

Partial least squares regression (PLSR) was used as the regression method. The maximum number of components considered in each model was set to 7, to reduce the possibility of overfitting. Full cross-validation of the PLSR models was conducted using the leave-one-out method.

To avoid creating a 'two-point' model, only Regain400 samples were included in the models for tebuthiuron content. The spectra and loadings were plotted using R Studio running R 4.2.2 (R Foundation for Statistical Computing, Vienna, Austria) [21].

3. Results and Discussion

3.1. Validation of the HPLC Method

Analysis of the tebuthiuron peak using HPLC-DAD showed that the maximum absorbance was located at 254 nm (Figure 4). This was confirmed by subsequent UV spectral scans of pure tebuthiuron standard in 100% methanol, which revealed a λ_{max} of 253.5 nm. Hence, a detection wavelength of 254 nm was chosen for this work, agreeing with Lydon et al. [15].

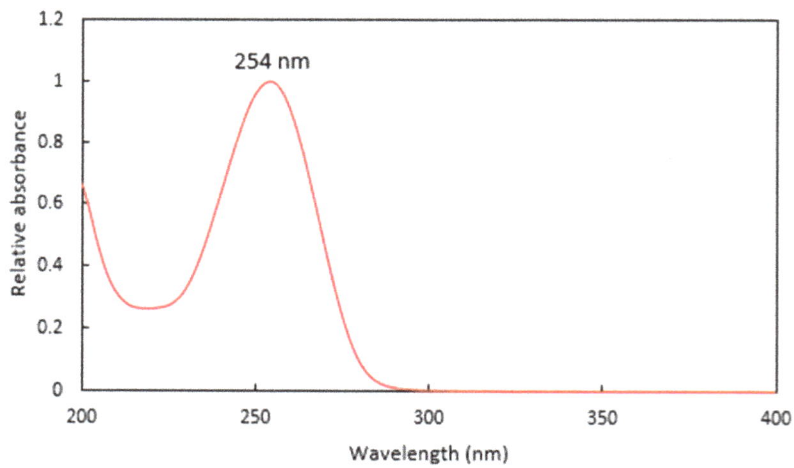

Figure 4. The UV spectra of a Regain extract, showing the λ_{max} at 254 nm.

For the analysis of linearity, the tebuthiuron standards were found to be linear over the range of 0.1–1000 mg L^{-1}, with an R^2 value of 0.9999 (Figure 5).

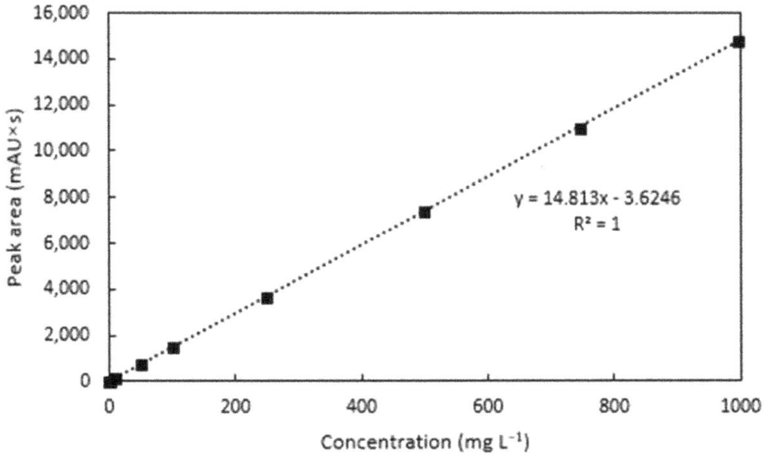

Figure 5. Linearity of the tebuthiuron standards.

As can be seen in Table 1, the intra-day precision of the HPLC method was quite high, with a mean relative error (coefficient of variation) of 0.35%. The inter-day precision was slightly poorer than the intra-day precision (Table 1), with a mean coefficient of variation of 0.95%.

Table 1. Intra-day and inter-day precision of replicate injections of 100 mg L^{-1} tebuthiuron standards analyzed using the HPLC method.

Injection	Peak Area	Day	Peak Area
1	1451.1	1	1463.2
2	1460.8	2	1466.3
3	1464.0	3	1460.2
4	1466.3	4	1453.8
5	1465.8	5	1428.2
6	1468.3	6	1450.5
Mean ± SD	1463.2 ± 5.1	Mean ± SD	1453.7 ± 13.8
%CV	0.35%	%CV	0.95%

For the assessment of extract stability at room temperature, there was virtually no change in the tebuthiuron concentration (CV = 0.07%), indicating high stability of the tebuthiuron content over the 8 day storage period (Table 2).

Table 2. Stability of the Regain methanolic extract after 8 days of storage at room temperature.

Sample	Peak Area	Tebuthiuron (mg/g)
Initial extract	4019.6	403.5
After 8 days at room temperature	4023.4	403.9

Reproducibility of the Extraction and HPLC Method

In order to create accurate prediction models using infrared spectroscopy, it is essential to have accurate analytical protocols for quantifying the analyte in question. In other words, the "reference" values used to calibrate the IR models must be accurate; otherwise, the IR models will be of no use.

For the reproducibility of the finalised extraction method, the mean %CV of the 20 samples extracted and analysed in triplicate was 2.1 ± 1.8% (n = 20), indicating an acceptable level of reproducibility.

Additionally, for the seven extractions performed on a single homogenised, powdered sample of Regain400, most of the replicate measurements showed good agreement with one another. One result (Replicate 2) was identified as an outlier and removed from subsequent calculations (Table 3). The mean content of the remaining samples was 401.3 ± 7.0 mg/g tebuthiuron, corresponding to a coefficient of variation of 1.74%.

Table 3. Replicate tebuthiuron content measurements performed on replicate extracts from one powdered, homogenised Regain400 sample.

Replicate	Tebuthiuron Content (mg/g)
1	396.5
2	422.9 #
3	399.1
4	392.1
5	406.6
6	411.5
7	402.1
Mean	401.3
SD	7.0
%CV	1.74%

Flagged as an outlier and excluded from subsequent analysis

Furthermore, the mean coefficient of variation across 104 samples—each analysed in triplicate—was 1.46%, corresponding to an error ±5.8 mg/g for a sample with a nominal tebuthiuron content of 400 mg/g. This supported the reproducibility of the HPLC method for measuring tebuthiuron content in the Regain samples. In addition, the routine use of triplicate analyses for the assessment of tebuthiuron content allows for the detection and removal of any outlier results.

3.2. *Observations on Moisture Content*

The standard deviation (i.e., laboratory error) of the triplicate moisture analysis was found to be 0.42% *w*/*w* (Table 1), with a coefficient of variation of 3.9%, indicating acceptable analytical error.

In the samples of Regain400, a moderate negative correlation between the tebuthiuron and moisture content was found (Figure 6); however, this relationship was very weak (R^2 = 0.05).

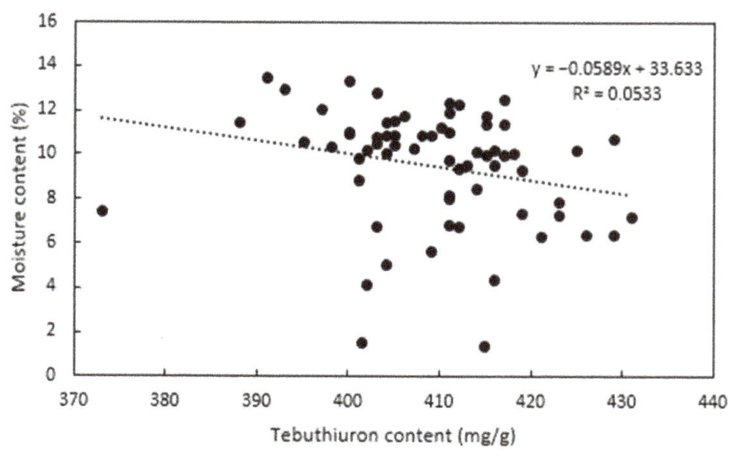

Figure 6. Relationship between the tebuthiuron content and the moisture content of the Regain400 samples.

3.3. *Assessment of Spatial Variation in the Sample*

The results from the sample variation experiment indicated a moderate amount of spatial variation in the tebuthiuron content, with the mean content of each sampling replicate ranging from 412.3–421.0 mg/g (Table 4). The mean content was 416.1 ± 3.3 mg/g, giving a coefficient of variation of 0.79%. The results are also shown visually in Figure 7.

Table 4. Assessment of the variation in tebuthiuron content between different portions of one Regain400 sample.

		Tebuthiuron Content (mg/g)		
Location	**Replicate**	**Intra-Replicate Analyses**	**Mean Content**	**Within-Sample %CV**
Location 1	Replicate A	405.9		
	Replicate B	416.8		
	Replicate C	421.4	414.7	1.91%
Location 2	Replicate A	411.8		
	Replicate B	414.0		
	Replicate C	461.3 [#]	412.9	0.39%
Location 3	Replicate A	419.7		
	Replicate B	413.8		
	Replicate C	404.3	412.6	1.89%
Location 4	Replicate A	424.0		
	Replicate B	404.5		
	Replicate C	408.5	412.3	2.49%
Location 5	Replicate A	420.1		
	Replicate B	416.8		
	Replicate C	409.9	415.6	1.26%
Location 6	Replicate A	431.1		
	Replicate B	417.3		
	Replicate C	414.5	421.0	2.12%
Location 7	Replicate A	417.2		
	Replicate B	420.1		
	Replicate C	417.6	418.3	0.38%
Location 8	Replicate A	417.1		
	Replicate B	412.2		
	Replicate C	412.6	414.0	0.65%
Location 9	Replicate A	415.9		
	Replicate B	419.4		
	Replicate C	422.6	419.3	0.79%
Location 10	Replicate A	417.4		
	Replicate B	428.7		
	Replicate C	413.5	419.9	1.87%
Mean		416.2 (n = 29)	416.1 (n = 10)	1.40% (n = 10)
SD		6.3 (n = 29)	3.3 (n = 10)	-
%CV		1.52% (n = 29)	0.79% (n = 10)	-

[#] Flagged as an outlier and excluded from subsequent analysis

Figure 7. The mean tebuthiuron contents for the ten spatial replicates analysed. The numbers over each sample location provide the mean tebuthiuron content in mg/g.

3.4. *Analysis of FTIR Spectra*

Figure 8 shows the FTIR spectra of one of the Regain samples, overlaid with the FTIR spectra of a sample of pure tebuthiuron powder. The tebuthiuron spectrum appeared similar to literature records on SpectraBase (https://spectrabase.com/spectrum/Gwi6TOkmgNy; accessed on 22 June 2022), with a complex pattern of peaks. The major peaks and their aetiological bonds are summarised in Table 5. Not all of the peaks between 1500–900 cm^{-1} were able to be confidently assigned identities, due to the complexity in this region. The Regain400 sample showed additional peaks at 1026, 1002, 910 and <600 cm^{-1}, due to absorptions from the other matrix constituents. In contrast, the pure tebuthiuron powder samples showed minimal absorbance in this region (1026–910 cm^{-1}).

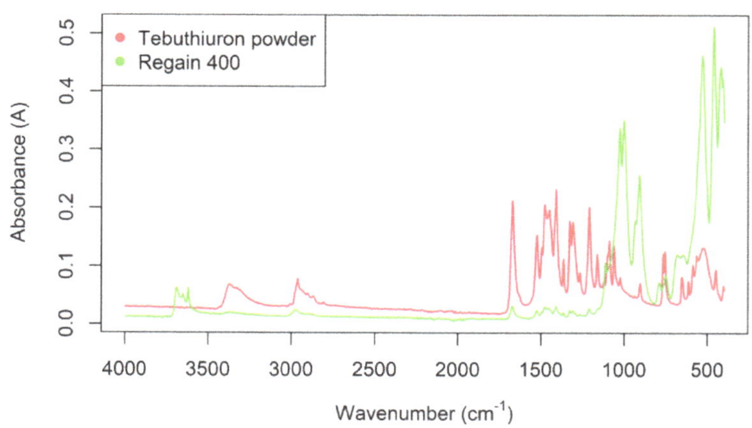

Figure 8. FTIR spectra of a typical powdered 400 mg/g Regain sample and a sample of pure tebuthiuron powder (99.6% purity).

Table 5. The major peaks observed in the FTIR spectrum of pure tebuthiuron powder and their assigned bonds.

Peak (cm^{-1})	Appearance	Bond	Reference
3366	Medium	N-H stretch of secondary amine	[22]
2961	Medium	C-H stretch of methyl group (shoulder may be due to sulfide bond)	[22]
1672	Strong	C=O stretch of secondary and tertiary amide (Amide I band) C=N stretch of imine may also contribute	[22]
1526	Strong	N-H bend and C-N stretch (Amide II band) N-H bend may also be contributed from secondary amine	[23]
1478	Strong	Unknown (thiadiazyl group likely contributing)	
1451	Strong	C-H bend of methyl group	[22]
1410	Strong	Unknown (thiadiazyl group contributing)	
1328	Strong	-	
1238	Strong	Tentative: Amide III band	[24]
1210	Strong	-	
1161	Medium	Tentative: C-N stretch of amine	[22]
1090	Medium	C=N stretch of imine	[25]
1062	Medium	Unknown (thiadiazyl group contributing)	
905	Weak	=C-N bond	
767 & 756	Strong	C-S stretch May be some contribution from N-H wag of secondary amine	[26]
654	Medium	C-S stretch	[27]

Partial least squares regression performed on the FTIR spectra indicated that this technique could be used to predict the moisture content of the samples with reasonable accuracy (R^2_{cv} = 0.85; RMSECV = 1.12%; Table 6). The RPD (ratio of performance to deviation; equal to the standard deviation of the entire calibration set divided by the RMSECV) was 2.55, indicating good predictive ability for this analyte [28].

Table 6. Optimum pre-processing methods for the prediction of moisture and tebuthiuron content in powdered Regain samples using FTIR spectroscopy. The best-performing model for each analyte is highlighted in bold.

	Moisture			Tebuthiuron		
Pre-Processing	**Factors**	**R^2_{cv}**	**RMSECV (%)**	**Factors**	**R^2_{cv}**	**RMSECV (mg/g)**
None	7	0.82	1.22	6	0.38	8.3
SNV	**7**	**0.85**	**1.12**	**7**	**0.43**	**8.0**
1d5	4	0.77	1.39	5	0.37	8.4
1d11	7	0.82	1.23	5	0.39	8.3
1d15	7	0.82	1.22	6	0.40	8.2
1d21	7	0.82	1.21	6	0.41	8.2
1d31	7	0.83	1.19	7	0.41	8.1
1d41	7	0.84	1.15	5	0.39	8.3
2d11	4	0.75	1.42	4	0.42	8.1
2d15	7	0.80	1.28	5	0.39	8.3
2d21	7	0.81	1.25	5	0.40	8.2
2d31	7	0.81	1.23	6	0.42	8.1
2d41	7	0.82	1.23	6	0.41	8.2

However, FTIR spectroscopy was unable to predict the tebuthiuron content of the Regain400 samples with a high level of accuracy (R^2_{cv} = 0.43; RMSECV = 8.0 mg/g; RPD = 1.32). The best-performing results for both moisture and tebuthiuron content were found using standard normal variate (SNV) pre-processing of the FTIR spectra (Figures 9 and 10).

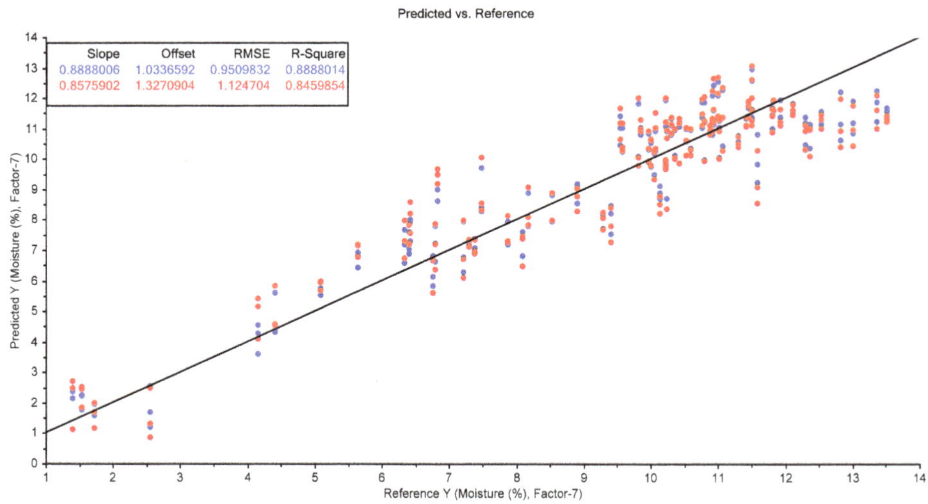

Figure 9. The results of the best-performing PLSR model for the prediction of moisture content using the FTIR spectra of the Regain samples (using SNV pre-processing). The blue points show the model calibration, while the red points show the model cross-validation.

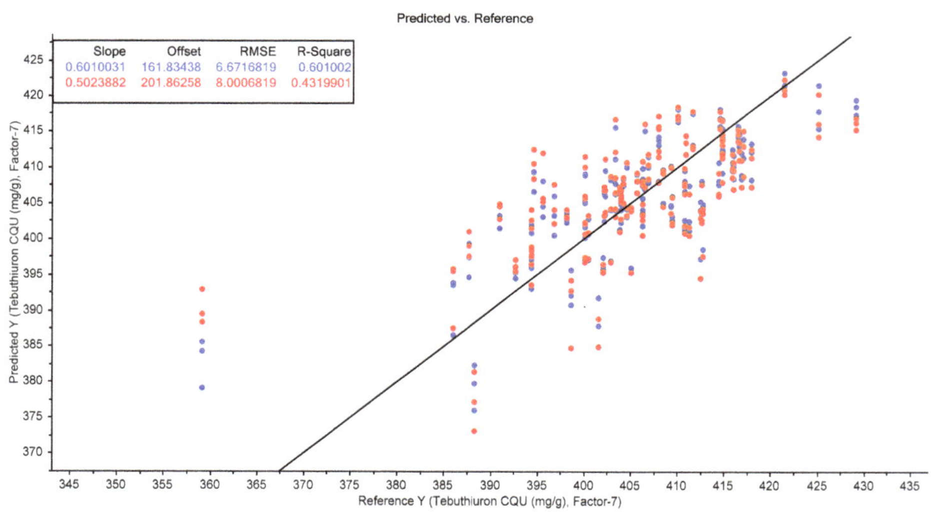

Figure 10. The results of the best-performing PLSR model for the prediction of tebuthiuron content using the FTIR spectra of the powdered Regain400 samples (using SNV pre-processing). The blue points show the model calibration, while the red points show the model cross-validation.

3.5. *Analysis of Benchtop NIR Spectra—Granules*

The NIR spectra collected from the whole Regain granules is shown in Figure 11. Four of the oldest samples (shown in blue) had much lower absorbances across the NIR spectrum; possibly due to a different matrix composition for the non-active ingredients.

These samples also contained the lowest moisture contents. However, the NIR spectra of the remaining samples were more consistent.

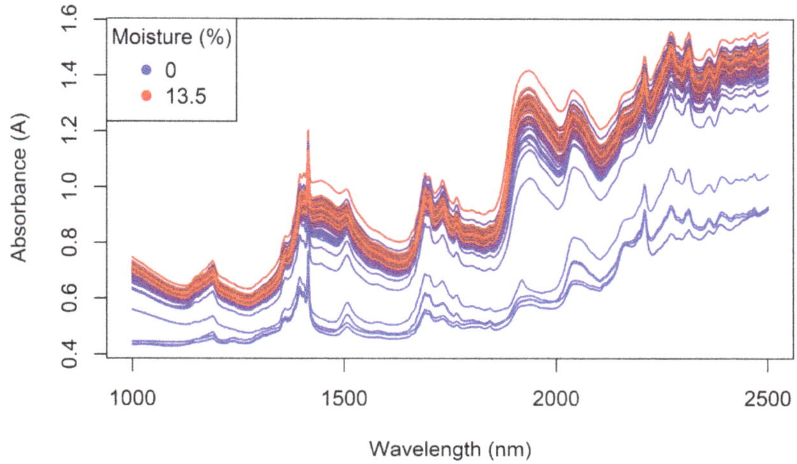

Figure 11. The benchtop NIR spectra of the Regain granules, coloured by moisture content.

The main peaks observed in a pure (99.6% assay) sample of tebuthiuron powder (Figure 12) were at 1190, 1380, 1506, 1733 (shoulder), 2042 and 2269 nm. These were attributed to CH_3 (second overtone), CH_3 (second overtone), amide bond (second overtone), S-H (first overtone), C-S-C stretch and amide bond (amide I and III region), respectively [29].

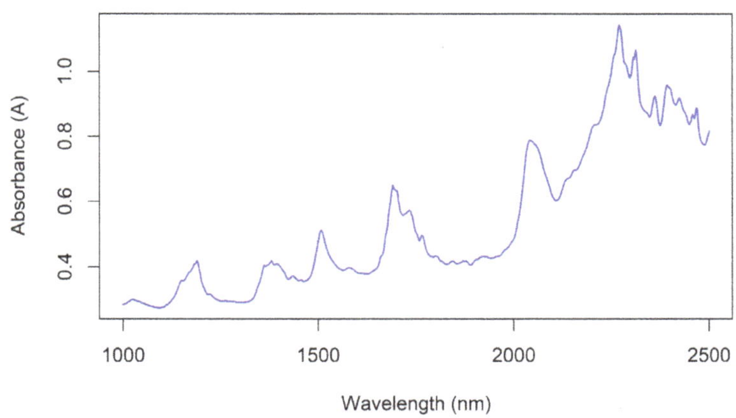

Figure 12. The NIR spectra of a sample of pure tebuthiuron powder (99.6% assay), measured using the benchtop Antaris instrument.

The Regain400 samples showed additional peaks at 1396 (weak), 1415 (strong), and 2210 nm (strong) (cf. Figure 11). These are most likely attributable to CH_3 (second overtone), H_2O (second overtone) and CHO or CH_3 combination bands, arising from non-active ingredients of the formulation (e.g., from organic matter in clay components or from polymer-like ingredients) [30].

The best-performing PLSR models for the prediction of moisture and tebuthiuron content using the benchtop Antaris NIR instrument are shown in Table 7. The highest accuracy for the prediction of moisture content was found using no pre-processing (Figure 13); however, the best model for tebuthiuron content used 2d11 pre-processing (Figure 14).

The model accuracy was quite high for moisture content (R^2_{cv} of 0.93 and RMSECV of 0.85%); however, the prediction results for tebuthiuron content were somewhat poorer (R^2_{cv} of 0.46 and RMSECV of 7.9 mg/g). However, the RMSECV of the tebuthiuron calibration was only slightly lower than the standard deviation of the analytical (HPLC) reference method (7.0 mg/g). Consequently, this indicates that the accuracy of the NIR method was approaching the maximum accuracy expected using this reference method.

Table 7. Optimum pre-processing methods for the prediction of moisture and tebuthiuron content in Regain granules using the benchtop Antaris NIR instrument. The best-performing model for each analyte is highlighted in bold.

	Moisture			Tebuthiuron		
Pre-Processing	**Factors**	**R^2_{cv}**	**RMSECV (%)**	**Factors**	**R^2_{cv}**	**RMSECV (mg/g)**
None	**3**	**0.93**	**0.85**	5	0.24	9.3
SNV	4	0.91	0.92	5	0.42	8.2
1d5	2	0.92	0.86	5	0.40	8.3
1d11	2	0.92	0.86	5	0.35	8.7
1d15	2	0.92	0.87	5	0.33	8.8
1d21	2	0.92	0.89	4	0.31	8.9
1d31	2	0.91	0.90	4	0.30	8.9
1d41	2	0.91	0.91	4	0.30	9.0
2d11	3	0.91	0.93	**5**	**0.46**	**7.9**
2d15	3	0.91	0.94	5	0.45	8.0
2d21	3	0.91	0.91	5	0.41	8.3
2d31	2	0.92	0.87	5	0.38	8.5
2d41	2	0.92	0.89	5	0.37	8.5

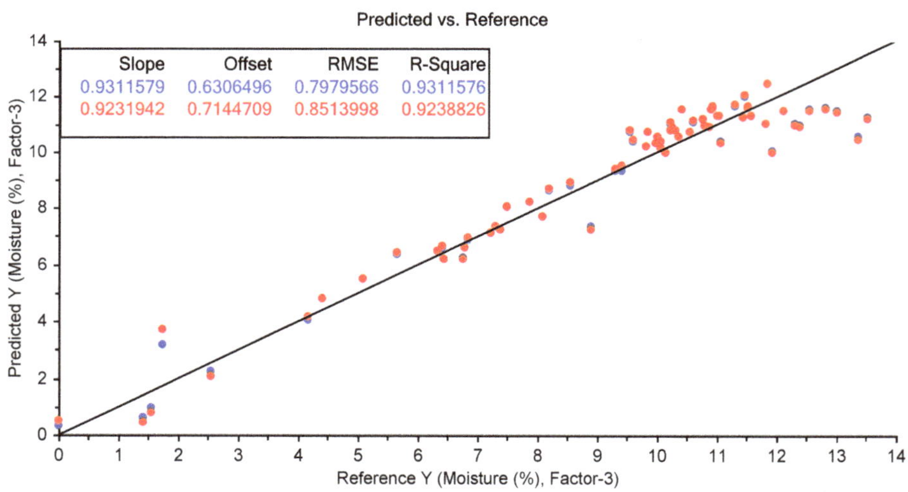

Figure 13. The results of the best-performing PLSR model for the prediction of moisture content using the benchtop NIR spectra collected from the Regain granules (using no pre-processing). The blue points show the model calibration, while the red points show the model cross-validation.

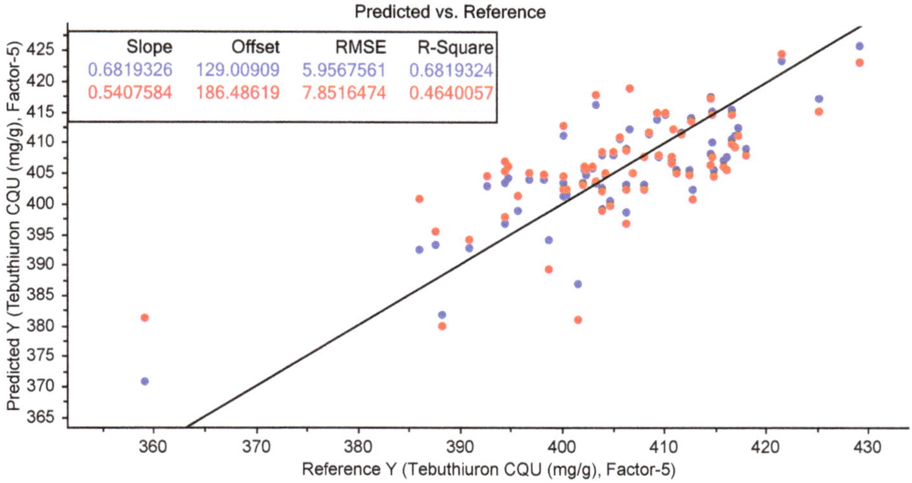

Figure 14. The results of the best-performing PLSR model for the prediction of tebuthiuron content using the benchtop NIR spectra collected from the Regain400 samples (using 2d11 pre-processing). The blue points show the model calibration, while the red points show the model cross-validation.

The model loadings for the moisture content PLSR model showed positive contributions from the peaks at 1936, 1450 and 1190 nm (Figure 15). These regions correspond to the H_2O 1st overtone, amide second overtone and C-H second overtone, respectively. The greatest contribution was observed from the H_2O 1st overtone, confirming that the model was principally measuring moisture in the samples. The smaller contribution from the amide region should be considered in future investigations, as it appears that the tebuthiuron content may be moderately influencing the model.

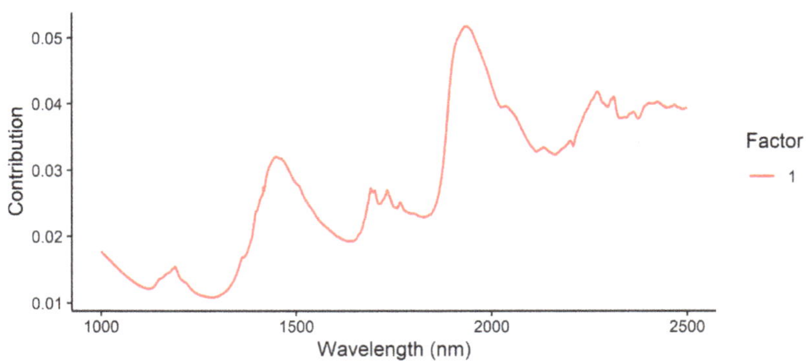

Figure 15. Loadings plot for the prediction of moisture content in Regain granules using the benchtop Antaris NIR instrument.

The loading plot for the tebuthiuron content model (Figure 16) showed major contributions in the 2200–2400 nm region, corresponding to the CH_3 combination band and amide I and III regions. Contributions were also observed around 1420 nm (H_2O second overtone) and 1700 nm (S-H first overtone). Again, this confirmed that the model was principally looking at tebuthiuron in the sample, with a possible minor influence of moisture content.

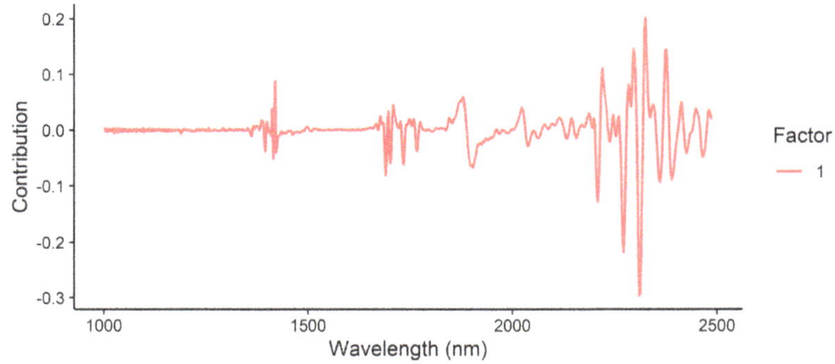

Figure 16. Loadings plot for the prediction of tebuthiuron content in Regain granules, using the benchtop Antaris NIR instrument.

3.6. *Analysis of Benchtop NIR Spectra—Powder*

Generally, the accuracy of results obtained by NIR spectroscopy is improved by having a more homogenous sample matrix. Large particle sizes can lead to light scattering effects, which in turn biases the resultant NIR spectra obtained [31,32]. Using finely powdered matrices reduces this scattering effect and may consequently provide improved insight into the true sample composition. Consequently, the granule samples were ground to a fine powder and the NIR spectra collected from the powdered samples using the benchtop Antaris instrument. Figure 17 shows the NIR spectra of the powdered Regain samples, which were quite similar to the spectra collected from the granules (cf. Figure 11).

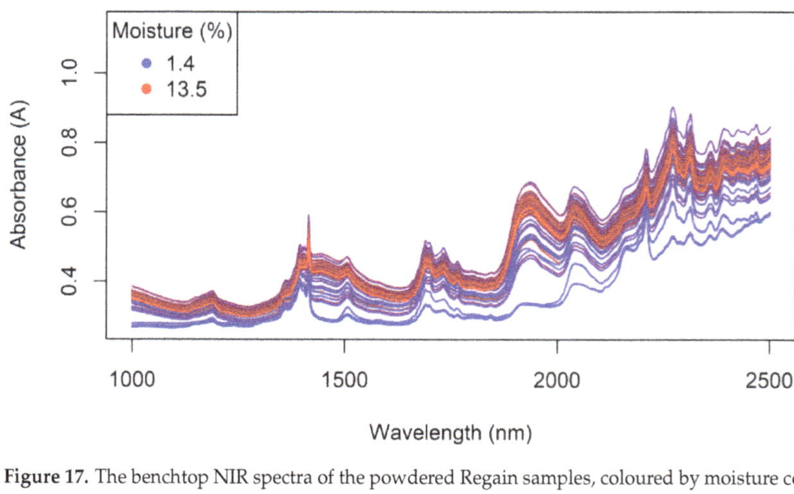

Figure 17. The benchtop NIR spectra of the powdered Regain samples, coloured by moisture content.

The results for the prediction of moisture and tebuthiuron content from the powdered samples are shown in Table 8. Interestingly, the moisture prediction was less accurate when using the powdered samples compared to the whole granules, with an RMSECV of 1.02% for the best-performing model, compared to an RMSECV of 0.85% for the granules. This is an important observation from a rapid quality control viewpoint, as it indicates that NIR spectra obtained from the intact/whole Regain granules will perform just as well—if not better—than spectra obtained from the powdered samples, thus removing the need for the time-consuming grinding process.

In contrast, the prediction accuracy for tebuthiuron content was slightly increased compared to NIR spectra collected from the whole granules (RMSECV of 7.7 mg/g, compared to 7.9 mg/g). However, this slight increase in accuracy would not be significant enough to justify the additional sample preparation time in most practical applications.

Table 8. Optimum pre-processing methods for the prediction of moisture and tebuthiuron content in powdered Regain samples using the benchtop Antaris NIR instrument. The best-performing model for each analyte is highlighted in bold.

	Moisture			Tebuthiuron		
Pre-Processing	**Factors**	R^2_{cv}	**RMSECV (%)**	**Factors**	R^2_{cv}	**RMSECV (mg/g)**
None	7	0.86	1.09	6	0.49	7.8
SNV	**6**	**0.88**	**1.02**	4	0.40	8.3
1d5	6	0.84	1.15	4	0.45	7.9
1d11	6	0.85	1.12	5	0.45	8.0
1d15	6	0.85	1.12	5	0.45	8.0
1d21	6	0.85	1.11	5	0.44	8.0
1d31	7	0.86	1.07	6	0.44	8.0
1d41	6	0.85	1.12	6	0.45	8.0
2d11	6	0.81	1.28	**4**	**0.48**	**7.7**
2d15	6	0.83	1.21	4	0.46	7.9
2d21	6	0.83	1.18	4	0.45	8.0
2d31	6	0.84	1.18	4	0.44	8.0
2d41	6	0.84	1.15	4	0.43	8.1

3.7. Analysis of Handheld NIR Spectra—Granules

Given the promising results observed for the benchtop NIR spectra collected from the granules, the MicroNIR spectra were only collected from the granular Regain samples, not the powdered Regain samples. Additionally, the analysis of powder samples using this handheld instrument would require the instrument to be cleaned after every sample to prevent cross-contamination, adding to the time taken to collect the NIR spectra from each sample.

The NIR spectra of the Regain granules collected with the MicroNIR instrument are shown in Figure 18. The wavelength range of this instrument (908–1676 nm) is narrower than that of the benchtop Antaris instrument (1000–2500 nm); however, it still contains important information in the regions of 1185 nm (CH_3 second overtone), 1360 nm (shoulder; CH_3 second overtone), 1408 nm (H_2O second overtone) and 1509 nm (amide bon d second overtone). Consequently, it was expected that this wavelength range could still be used for the prediction of moisture and tebuthiuron content.

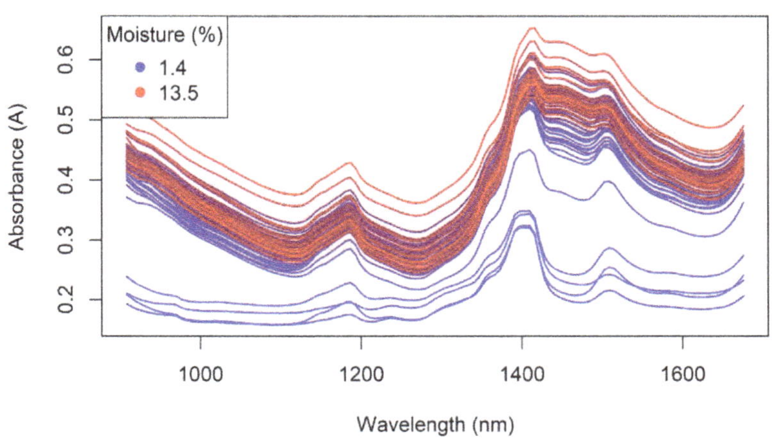

Figure 18. The handheld MicroNIR spectra of the Regain granules, coloured by moisture content.

The performance statistics for the models developed for the prediction of moisture and tebuthiuron content using the MicroNIR spectra are shown in Table 9. Notably, the best models for moisture and tebuthiuron content were both slightly better (lower mean error) than those previously found using the benchtop NIR instrument. This supports the proposition of using handheld NIR instrumentation for real-time quality assessment of Regain products.

Table 9. Optimum pre-processing methods for the prediction of moisture and tebuthiuron content in Regain granules using the handheld MicroNIR instrument. The best-performing model for each analyte is highlighted in bold.

	Moisture			Tebuthiuron		
Pre-Processing	**Factors**	**R^2_{cv}**	**RMSECV (%)**	**Factors**	**R^2_{cv}**	**RMSECV (mg/g)**
None	4	0.91	0.86	6	0.24	9.3
SNV	4	0.91	0.88	7	0.35	8.6
1d3	2	0.92	0.83	5	0.44	8.0
1d5	2	0.92	0.83	5	0.45	8.0
1d7	2	0.92	0.83	5	0.44	8.0
1d9	2	0.92	0.83	5	0.44	8.0
1d11	**2**	**0.92**	**0.83**	5	0.44	8.0
2d5	2	0.91	0.89	5	0.45	7.9
2d9	1	0.90	0.91	5	0.46	7.9
2d11	1	0.91	0.89	5	0.46	7.8
2d15	2	0.91	0.87	5	0.48	7.7
2d21	2	0.91	0.85	**5**	**0.50**	**7.5**

As seen from the calibration graph in Figure 19, the MicroNIR instrument was able to predict the moisture content of the Regain samples with reasonably high accuracy across most of the range tested. Towards the higher range of moisture contents (>12%), the accuracy flattened off, with the model under-predicting the moisture content of all of these samples.

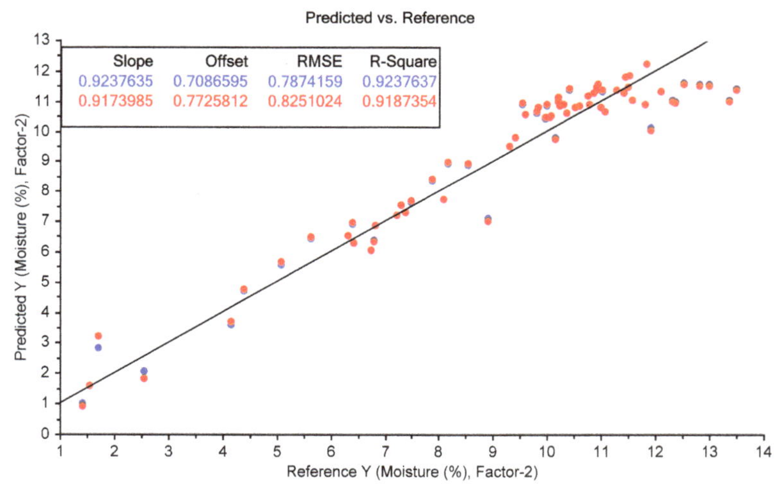

Figure 19. MicroNIR prediction of moisture content. The blue points show the model calibration, while the red points show the model cross-validation.

The loading plot of the PLSR model for moisture content (Figure 20) shows the greatest contribution at 1410 nm, corresponding to the second overtone of H_2O. A minor positive contribution was also observed at 1156 nm (CH_3 second overtone) as well as negative contributions around 1497 nm (likely corresponding to the second overtone of the amide region).

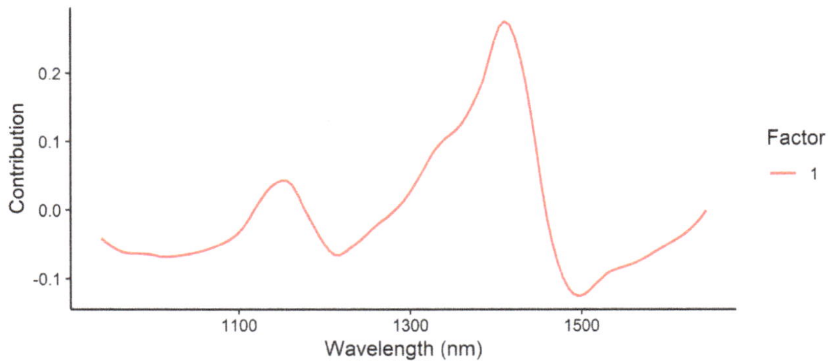

Figure 20. Loadings plot for the prediction of moisture content using the MicroNIR instrument.

The calibration graph for tebuthiuron content showed increased variability compared to the moisture content prediction (Figure 21), with a few samples which could potentially be outliers. The R^2 for cross-validation was not particularly high (0.53), but the RMSECV was quite good (7.5 mg/g), particularly compared to the mean laboratory error of 7.0 mg/g for tebuthiuron analysis using HPLC. Inclusion of a larger number of samples, particularly covering a wider calibration range, could potentially improve the prediction accuracy.

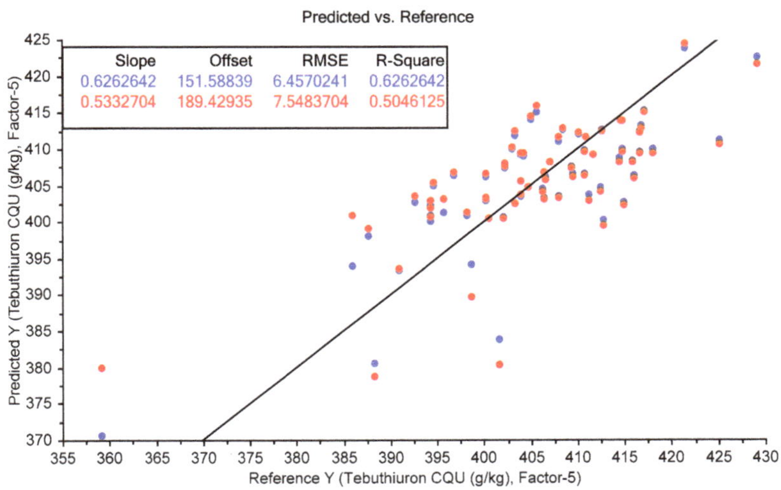

Figure 21. MicroNIR prediction of tebuthiuron content. The blue points show the model calibration, while the red points show the model cross-validation.

The loadings plot for tebuthiuron prediction (Figure 22) showed the largest contributions at 1379 and 1453 nm. These likely correspond to the CH_3 second overtone and amide band overtone. Both of these bonds are found in the structure of tebuthiuron (Figure 1).

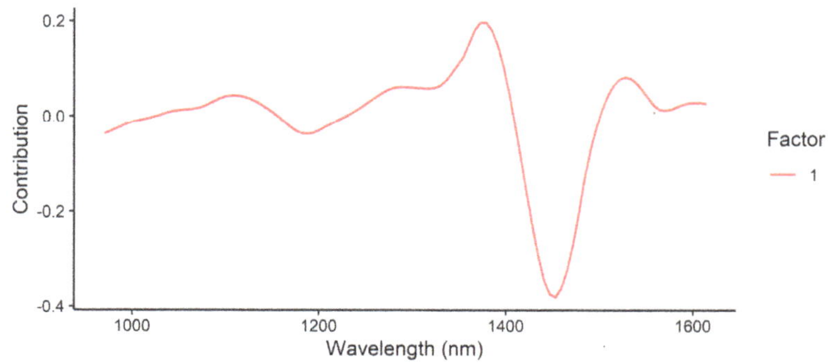

Figure 22. Loadings plot for the prediction of tebuthiuron content using the MicroNIR instrument.

3.8. Comparison of Different Instruments and Sample Matrices

Table 10 provides a succinct summary of the best-performing models from all the different methods of infrared spectroscopy trialled, as well as the different matrix types (powder vs. granules).

Table 10. Optimum pre-processing methods for the prediction of moisture and tebuthiuron content in powdered and granular Regain samples, using three different IR instruments. The best-performing model for each analyte is highlighted in bold.

Instrument	Matrix	Moisture			Tebuthiuron		
		Factors	R^2_{cv}	RMSECV (%)	Factors	R^2_{cv}	RMSECV (mg/g)
Bruker FTIR (benchtop)	Powder	7	0.85	1.12	7	0.43	8.0
Antaris (benchtop)	Powder	6	0.88	1.02	4	0.48	7.7
Antaris (benchtop)	Granules	3	0.93	0.85	5	0.46	7.9
MicroNIR (handheld)	Powder	-	-	-	-	-	-
MicroNIR (handheld)	Granules	**2**	**0.92**	**0.83**	**5**	**0.50**	**7.5**

The best-performing models were found using the handheld MicroNIR instrument, applied to the whole Regain granule samples. This was true for both moisture and tebuthiuron content. Consequently, this demonstrates that the handheld MicroNIR instrument should be highly suitable for the rapid, in-situ quality assessment of Regain granules throughout the manufacturing process.

3.9. Independent Test Set—Handheld NIR

The final stage of this work was to test the performance of the handheld NIR models using an independent set of samples (i.e., ones not used in the model calibrations). NIR spectra were collected from the granules using the MicroNIR instrument and predictions made using the optimum model for each analyte.

All of the PLS scores of the test set lay within the bounds of the scores of the calibration set (Figure 23), indicating that the spectra of the test set samples were similar enough to the spectra of the calibration set to allow accurate predictions to be made.

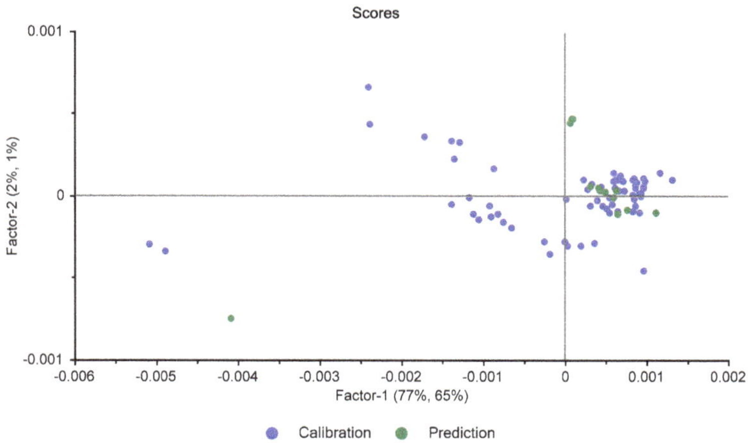

Figure 23. Scores plot of the calibration and test set spectra.

The prediction results are shown in Table 11, alongside the reference measurements. Most of the moisture predictions were relatively close, with a mean prediction error of +0.36% *w*/*w*. The RMSEP (root mean square error of prediction) was 0.93% *w*/*w*, with a R^2_{pred} of 0.52.

Table 11. Prediction and reference results for the independent test set.

Moisture by NIR (%)	Moisture by Drying (%)	Prediction Error (%)	Tebuthiuron by NIR (mg/g)	Tebuthiuron by HPLC (mg/g)	Prediction Error (mg/g)
9.74 ± 1.34	11	−1.26	414 ± 12	409	+5
9.91 ± 1.39	ND		418 ± 16	413	+5
9.84 ± 1.39	ND		416 ± 20	414	+2
3.51 ± 2.86	ND		275 ± 85	214	+61 #
10.86 ± 0.92	10	0.86	422 ± 34	425	−3
10.9 ± 0.85	10.7	0.2	418 ± 37	419	−1
10.63 ± 0.79	10	0.63	414 ± 35	416	−2
10.74 ± 0.81	10	0.74	411 ± 37	410	+1
11.21 ± 0.58	10.8	0.41	410 ± 15	405	+5
10.81 ± 0.45	11.5	−0.69	407 ± 10	414	−7
11.74 ± 0.57	11.6	0.14	408 ± 8	411	−3
10.47 ± 0.27	8.38	2.09	406 ± 9	404	+2
10.28 ± 0.43	9.79	0.49	402 ± 12	402	0
	Mean error	**0.36%**		**Mean error**	**+0.3**

outlier value excluded; ND = no data available.

Similarly, the prediction of tebuthiuron was quite accurate, with a mean prediction error of just +0.3 mg/g. The range of prediction errors varied between −7 and +5 mg/g. As shown in Figure 24, the prediction results were relatively linear (R^2_{pred} = 0.63), with an RMSEP of 3.8 mg/g. Again, this was well within the expected range of error associated with the HPLC reference method (7 mg/g).

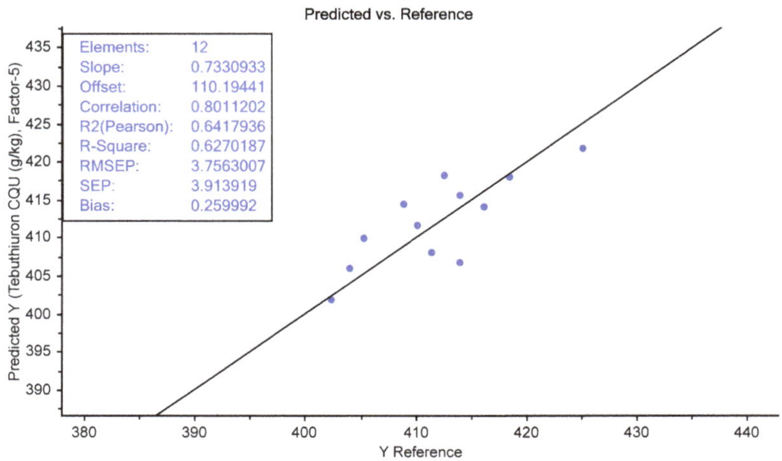

Figure 24. The predicted vs. measured tebuthiuron content of the Regain 400 samples in the independent test set (n = 12).

It should be noted that the single Regain200 sample was excluded from the tebuthiuron prediction results, as the model had only previously been trained on the Regain400 samples; therefore, we could not predict the tebuthiuron content of the Regain200 sample with acceptable accuracy.

4. Conclusions

The results from this study suggest that NIRS is quite accurate for the rapid prediction of moisture content and moderately accurate for the prediction of tebuthiuron content. Handheld and even benchtop NIR devices could not only allow for rapid quality control, but also for improvement of the manufacturing process. This form of rapid, on-site testing with sufficient accuracy could allow for isolation of sources of process variation and guide targeted efforts to minimize their effects. This is particularly important as unwanted variations in the processes can be costly, cause manufacturing downtime, or be an indication of phenomena that reduce the plant reliability and performance. Overall, the results found here support the use of the handheld MicroNIR instrument for future studies and potential real-time implementation. Furthermore, the use of a larger calibration set is likely to moderately improve the prediction accuracy of the tebuthiuron model.

Author Contributions: Conceptualization, J.B.J., M.I. and M.N.; methodology, J.B.J.; software, J.B.J.; validation, J.B.J.; formal analysis, J.B.J.; investigation, J.B.J.; resources, J.B.J., M.N. and H.F.; data curation, J.B.J.; writing—original draft preparation, J.B.J.; writing—review and editing, J.B.J., H.F., M.I. and M.N.; visualization, J.B.J.; supervision, M.N.; project administration, J.B.J.; funding acquisition, M.N. All authors have read and agreed to the published version of the manuscript.

Funding: This research received no external funding.

Institutional Review Board Statement: Not applicable.

Informed Consent Statement: Not applicable.

Data Availability Statement: The data presented in this study are available on request from the corresponding author.

Acknowledgments: Thanks to Kerry Walsh for providing access to the NIR instrumentation.

Conflicts of Interest: Hugh Farquhar is a current employee and Mansel Ismay is a past employee of Cirrus Ag, the manufacturer of Regain™. Aside from supplying the samples, Cirrus Ag had no role in the collection, analysis or interpretation of data.

References

1. Bovey, R.W.; Burnett, E.; Meyer, R.E.; Richardson, C.; Loh, A. Persistence of Tebuthiuron in Surface Runoff Water, Soil, and Vegetation in the Texas Blacklands Prairie. *J. Environ. Qual.* **1978**, *7*, 233–236. [CrossRef]
2. Faria, A.T.; Souza, M.F.; Rocha de Jesus Passos, A.B.; da Silva, A.A.; Silva, D.V.; Zanuncio, J.C.; Rocha, P.R.R. Tebuthiuron leaching in three Brazilian soils as affected by soil pH. *Environ. Earth Sci.* **2018**, *77*, 214. [CrossRef]
3. McNeil, W.K.; Stritzke, J.F.; Basler, E. Absorption, Translocation, and Degradation of Tebuthiuron and Hexazinone in Woody Species. *Weed Sci.* **1984**, *32*, 739–743. [CrossRef]
4. Hatzios, K.K.; Penner, D.; Bell, D. Inhibition of Photosynthetic Electron Transport in Isolated Spinach Chloroplasts by Two 1,3,4-Thiadiazolyl Derivatives 1. *Plant Physiol.* **1980**, *65*, 319–321. [CrossRef]
5. Cerdeira, A.L.; Desouza, M.D.; Queiroz, S.C.N.; Ferracini, V.L.; Bolonhezi, D.; Gomes, M.A.F.; Rosa, M.A.; Balderrama, O.; Rampazzo, P.; Queiroz, R.H.C.; et al. Leaching and half-life of the herbicide tebuthiuron on a recharge area of Guarany aquifer in sugarcane fields in Brazil. *J. Environ. Sci. Health Part B* **2007**, *42*, 635–639. [CrossRef]
6. Qian, Y.; Matsumoto, H.; Liu, X.; Li, S.; Liang, X.; Liu, Y.; Zhu, G.; Wang, M. Dissipation, occurrence and risk assessment of a phenylurea herbicide tebuthiuron in sugarcane and aquatic ecosystems in South China. *Environ. Pollut.* **2017**, *227*, 389–396. [CrossRef]
7. Helling, C.S. The science of soil residual herbicides. In *Soil Residual Herbicides: Science and Management. Topics in Canadian Weed Science*; van Acker, R., Ed.; Canadian Weed Science Society: Saint-Anne-de-Bellevue, QC, Canada, 2005; Volume 3, pp. 3–22.
8. Elanco. *Environmental Fate of Graslan in Rangeland Ecosystems and Safety to Non-Target Organisms*; Elanco: Greenfield, FL, USA, 1988.
9. Chang, S.S.; Stritzke, J.F. Sorption, Movement, and Dissipation of Tebuthiuron in Soils. *Weed Sci.* **1977**, *25*, 184–187. [CrossRef]
10. Johnsen, T.N., Jr.; Morton, H.L. Tebuthiuron Persistence and Distribution in Some Semiarid Soils. *J. Environ. Qual.* **1989**, *18*, 433–438. [CrossRef]
11. du Toit, J.C.O.; Sekwadi, K.P. Tebuthiuron residues remain active in soil for at least eight years in a semi-arid grassland, South Africa. *Afr. J. Range Forage Sci.* **2012**, *29*, 85–90. [CrossRef]
12. Metrohm. *Quantification of Five Effective Components in Pesticides by Visible Near-Infrared Spectroscopy*; Version 1, NIR Application Note NIR-056; Metrohm: Herisau, Switzerland, 2017.
13. Armenta, S.; Quintás, G.; Garrigues, S.; de la Guardia, M. Mid-infrared and Raman spectrometry for quality control of pesticide formulations. *TrAC Trends Anal. Chem.* **2005**, *24*, 772–781. [CrossRef]
14. Wu, Z.; Peng, Y.; Chen, W.; Xu, B.; Ma, Q.; Shi, X.; Qiao, Y. NIR spectroscopy as a process analytical technology (PAT) tool for monitoring and understanding of a hydrolysis process. *Bioresour. Technol.* **2013**, *137*, 394–399. [CrossRef] [PubMed]
15. Lydon, J.; Engelke, B.F.; Helling, C.S. Simplified high-performance liquid chromatography method for the simultaneous analysis of tebuthiuron and hexazinone. *J. Chromatogr. A* **1991**, *536*, 223–228. [CrossRef]
16. Ferracini, V.L.; Queiroz, S.C.; Gomes, M.A.; Santos, G.L. Método para a determinação de hexazinone e tebutiuron em água (Method for determination of hexazinone and tebuthiuron in water). *Química Nova* **2005**, *28*, 380–382. [CrossRef]
17. Weber, J.B. Ionization of Buthidazole, VEL 3510, Tebuthiuron, Fluridone, Metribuzin, and Prometryn. *Weed Sci.* **1980**, *28*, 467–474. [CrossRef]
18. Lourencetti, C.; de Marchi, M.R.R.; Ribeiro, M.L. Determination of sugar cane herbicides in soil and soil treated with sugar cane vinasse by solid-phase extraction and HPLC-UV. *Talanta* **2008**, *77*, 701–709. [CrossRef]
19. Figueiredo Ferreira, A.V.D.T.P.; Barbosa, L.V.; de Souza, S.D.; Ciuffi, K.J.; Vicente, M.A.; Trujillano, R.; Korili, S.A.; Gil, A.; de Faria, E.H. Titania-triethanolamine-kaolinite nanocomposites as adsorbents and photocatalysts of herbicides. *J. Photochem. Photobiol. Chem.* **2021**, *419*, 113483. [CrossRef]
20. Su, M.; Jia, L.; Wu, X.; Sun, H. Residue investigation of some phenylureas and tebuthiuron herbicides in vegetables by ultra-performance liquid chromatography coupled with integrated selective accelerated solvent extraction–clean up in situ. *J. Sci. Food Agric.* **2018**, *98*, 4845–4853. [CrossRef]
21. Team, R.C. *A Language and Environment for Statistical Computing*; Version 4.0.2; R Foundation for Statistical Computing: Vienna, Austria, 2020.
22. Sigma Aldrich. IR Spectrum Table & Chart. Available online: https://www.sigmaaldrich.com/technical-documents/articles/biology/ir-spectrum-table.html (accessed on 31 March 2022).
23. Jabs, A. Determination of Secondary Structure in Proteins by Fourier Transform Infrared Spectroscopy (FTIR). Available online: http://jenalib.leibniz-fli.de/ImgLibDoc/ftir/IMAGE_FTIR.html (accessed on 28 March 2022).
24. Riaz, T.; Zeeshan, R.; Zarif, F.; Ilyas, K.; Muhammad, N.; Safi, S.Z.; Rahim, A.; Rizvi, S.A.A.; Rehman, I.U. FTIR analysis of natural and synthetic collagen. *Appl. Spectrosc. Rev.* **2018**, *53*, 703–746. [CrossRef]
25. Long, F.; Chen, Z.; Han, K.; Zhang, L.; Zhuang, W. Differentiation between Enamines and Tautomerizable Imines Oxidation Reaction Mechanism using Electron-Vibration-Vibration Two Dimensional Infrared Spectroscopy. *Molecules* **2019**, *24*, 869. [CrossRef]
26. Norimasa, N.; Hiromu, S.; Tatsuo, M. Vibrational Spectra and Molecular Structure of Ethyl Methyl Sulfide. *Bull. Chem. Soc. Jpn.* **1975**, *48*, 3573–3575. [CrossRef]
27. Rao, C.N.R.; Venkataraghavan, R.; Kasturi, T.R. Contribution to the infrared spectra of organosulphur compounds. *Can. J. Chem.* **1964**, *42*, 36–42. [CrossRef]

28. Nicolaï, B.M.; Beullens, K.; Bobelyn, E.; Peirs, A.; Saeys, W.; Theron, K.I.; Lammertyn, J. Nondestructive measurement of fruit and vegetable quality by means of NIR spectroscopy: A review. *Postharvest Biol. Technol.* **2007**, *46*, 99–118. [CrossRef]
29. Ishigaki, M.; Ito, A.; Hara, R.; Miyazaki, S.-i.; Murayama, K.; Yoshikiyo, K.; Yamamoto, T.; Ozaki, Y. Method of Monitoring the Number of Amide Bonds in Peptides Using Near-Infrared Spectroscopy. *Anal. Chem.* **2021**, *93*, 2758–2766. [CrossRef] [PubMed]
30. Tominack, R.L.; Tominack, R. Herbicide Formulations. *J. Toxicol. Clin. Toxicol.* **2000**, *38*, 129–135. [CrossRef]
31. Maleki, M.R.; Mouazen, A.M.; Ramon, H.; De Baerdemaeker, J. Multiplicative Scatter Correction during On-line Measurement with Near Infrared Spectroscopy. *Biosyst. Eng.* **2007**, *96*, 427–433. [CrossRef]
32. Pilorget, C.; Fernando, J.; Ehlmann, B.L.; Schmidt, F.; Hiroi, T. Wavelength dependence of scattering properties in the VIS–NIR and links with grain-scale physical and compositional properties. *Icarus* **2016**, *267*, 296–314. [CrossRef]

eng

MDPI

Article

On the "Thixotropic" Behavior of Fresh Cement Pastes

Youssef El Bitouri * and Nathalie Azéma

Laboratoire de Mécanique et Génie Civil, LMGC, IMT Mines Ales, University of Montpellier, CNRS, 34000 Montpellier, France

* Correspondence: youssef.elbitouri@mines-ales.fr; Tel.: +33-4-66-78-53-67

Abstract: Thixotropic behavior describes a time-dependent rheological behavior characterized by reversible changes. Fresh cementitious materials often require thixotropic behavior to ensure sufficient workability and proper casting without vibration. Non-thixotropic behavior induces a workability loss. Cementitious materials cannot be considered as an ideal thixotropic material due to cement hydration, which leads to irreversible changes. However, in some cases, cement paste may demonstrate thixotropic behavior during the dormant period of cement hydration. The aim of this work is to propose an approach able to quantify the contribution of cement hydration during the dormant period and to examine the conditions under which the cement paste may display thixotropic behavior. The proposed approach consists of a succession of stress growth procedures that allow the static yield stress to be measured. For an inert material, such as a calcite suspension, the structural build-up is due to the flocculation induced by attractive Van der Waals forces. This structural build-up is reversible. For cement paste, there is a significant increase in the static yield stress due to cement hydration. The addition of superplasticizer allows the thixotropic behavior to be maintained during the first hours due to its retarding effect. However, an increase in the superplasticizer dosage leads to a decrease in the magnitude of the Van der Waals forces, which can erase the thixotropic behavior.

Keywords: thixotropy; yield stress; cement paste; hydration; superplasticizer

Citation: El Bitouri, Y.; Azéma, N. On the "Thixotropic" Behavior of Fresh Cement Pastes. *Eng* **2022**, *3*, 677–692. https://doi.org/10.3390/eng3040046

Academic Editors: Antonio Gil Bravo and F. Pacheco Torgal

Received: 16 November 2022
Accepted: 14 December 2022
Published: 14 December 2022

Publisher's Note: MDPI stays neutral with regard to jurisdictional claims in published maps and institutional affiliations.

Copyright: © 2022 by the authors. Licensee MDPI, Basel, Switzerland. This article is an open access article distributed under the terms and conditions of the Creative Commons Attribution (CC BY) license (https://creativecommons.org/licenses/by/4.0/).

1. Introduction

In rheology, thixotropy characterizes a time-dependent behavior [1–3]. This phenomenon, which is generally characteristic of flocculated suspensions, reflects the progressive breakdown (under a constant shear rate) of the structure formed at rest. The rheograms (shear stress as a function of shear rate) of thixotropic materials generally display a hysteresis loop. This evolution of the rheological behavior is reversible since the structural build-up occurs if the material is left at rest.

For cementitious materials, thixotropy was used to ensure proper casting and workability, especially for self-compacting concretes or printable concretes [4,5]. In addition, it allows the maintenance of workability and fluidity to be evaluated [6,7], which is very important from a practical point of view.

During the dormant or low activity period of cement hydration, the rheological behavior of cement pastes is often considered to be reversible. However, it appears that the initial structure can never be fully restored, even during this dormant period [6,8–10]. This is why cement pastes cannot be considered as typically thixotropic materials. In fact, due to the chemical evolution induced by the initial hydration reactions, the structural build-up (or breakdown) is not reversible. Roussel et al. [1] found that the structural build-up of cement pastes may be due to two origins: colloidal interactions between cement particles, which are reversible (thixotropy), and early hydrates, which form preferentially at the contact points between cement grains (irreversible). It can be noted that the irreversible changes in fresh cement paste structures can affect workability in time. This permanent change is thus defined as workability loss.

Furthermore, the addition of a superplasticizer decreases the contribution of hydration on the structural build-up when the cement paste is left at rest [11] and thus contributes to the decrease in the workability loss. The effect of the superplasticizer can be explained by the retarding effect.

The assessment of the contributions of the reversible flocculation (thixotropy) and the irreversible chemical evolution to the structural build-up is a very interesting challenge. Different methods based on rheological measurements have been developed to assess these contributions. One of these approaches consists of the determination of the evolution of the shear stress as a function of an ascendant and descendant shear rate. The hysteresis loop, i.e., the area between the up and down curves, is an indicator of thixotropy [12–14]. Another relevant approach to assess the structural build-up is to use oscillatory measurements, such as small amplitude oscillating shear (SAOS), which allow measurements of the viscoelastic properties of suspensions (storage modulus G'; loss modulus G") within the linear viscoelastic region [1,15–18].

Another method consists of determining the evolution of the static yield stress (the minimum stress that induces flow) by a stress growth procedure [8,10,19–21]. The slope (A_{thix}) of the static yield versus the resting time curve is a relevant indicator of the structural build-up. In literature, this slope represents the flocculation rate and describes the reversible part of the structural build-up (thixotropy) due to particle flocculation. The order of magnitude is between 0.1 and 1.7 Pa/s [1,15].

Furthermore, the contribution of the chemical evolution during the dormant period is generally neglected, and the application of a strong shearing or remixing is considered sufficient for erasing the structural build-up. However, it appears that the structural build-up during the dormant period is not fully reversible. Recently, by using oscillatory measurements, Zhang et al. [15] found that the irreversible part of the structural build-up cannot be neglected and suggested that the structural build-up can be quantified by A_{struct}, which is the sum of the thixotropic part and the chemical part:

$$A_{struct} = A_{thix} + A_{chem} \tag{1}$$

However, Zhang et al. did not provide quantification of A_{struct}. Based on their results, A_{struct} is about 0.07 Pa/s, A_{thix} is about 0.06, and A_{chem} is 0.01 Pa/s.

The aim of this study is to propose another approach based on the static yield stress measurement able to quantify the contribution of the chemical evolution to the structural build-up during the dormant period of cement hydration. This method is tested on ordinary Portland cement and calcite. The effect of a superplasticizer on the structural build-up is examined.

2. Materials

In this study, an ordinary Portland cement (CEM I 52.5 R CE CP2 NF) provided by Lafarge Holcim is used. This cement is composed of clinker (95%) and gypsum (5%). Its specific surface (Blaine) measures at 4420 cm^2/g, and its density is about 3.14 g/cm^3. In addition to cement, an inert carbonate of calcium (calcite) provided by Omya BL is used. Its density is about 2.75 g/cm^3, and its BET-specific surface is 2.25 m^2/g. Calcite is commonly used as a model material to mimic the behavior of complex cementitious materials during the dormant period [16,22–24].

The particle size distributions of the cement and calcite are determined in water using a laser granulometer (LS 13320) from Beckman Coulter Company with an adapted optical model (Figure 1). The physical properties are summarized in Table 1.

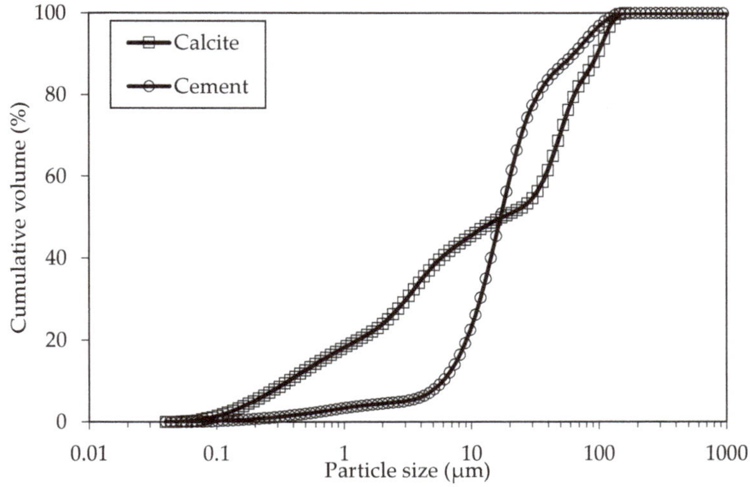

Figure 1. Particle size distributions of cement and calcite.

Table 1. Physical properties of cement and calcite.

Parameter	Cement	Calcite
Mean diameter (μm)	21.2	37.9
d_{10} (μm)	3.9	0.4
d_{50} (μm)	15.3	19.2
d_{90} (μm)	45.6	108.4
Density (g/cm^3)	3.14	2.75
Blaine-specific surface (cm^2/g)	4420	-
BET-specific surface (m^2/g)	-	2.25

A commercial polycarboxylate-based superplasticizer (PCE) from Masters Builders with an equivalent dry extract content of 19.5 wt% is used.

The cement and calcite pastes were mixed with deionized water with a water-to-solid ratio (E/C) of 0.4 in a planetary agitator according to the following sequence: 5 min mixing at 500 rpm, 30 s scraping the mixer walls, and 1 min mixing at 1000 rpm. Two dosages of superplasticizer are used: 0.05 and 0.1 wt% of dry substance. A delayed addition of the superplasticizer is performed after 5 min of mixing.

The samples' preparations are performed at ambient temperature (20 °C ± 2).

3. Methods

3.1. Rheological Measurements

The rheological measurements were carried out using a rotational rheometer AR2000Ex from TA Instruments equipped with a four-blade vane geometry. The internal diameter of this geometry is 28 mm, and the outer cup diameter is 30 mm. The resulting gap is 1 mm. The geometry constants were calibrated using the Couette analogy suggested by Aït-Kadi et al. [25].

The proposed testing method consists of a succession of stress growth measurements [19,26,27] with different resting times, as shown in Tables 2 and 3. The testing procedures begin with a strong pre-shear (100 s^{-1}) to homogenize the paste in the rheometer cup and are followed by a resting time (10 min, 20 min, and 40 min). Then, stress growth (1, 2, 3, and 4) is applied to the paste. Procedure 1 allows for measurement of the rate of increase in the static yield stress due to the total structural build-up (reversible thixotropy and irreversible chemical evolution), while Procedure 2 (Table 3) allows for erasure of the reversible part of the structural build-up via application of a strong pre-shear before the stress growth measurements.

Table 2. The proposed testing method for the total structural build-up (Procedure 1).

Time (min)	Hydration Time (min)	Procedure	Shear Rate (s^{-1})	Duration (s)
0.0	7	Pre-shear	100	30
0.5	7.5	Resting time	0	120
2.5	9.5	Stress growth 1	0.01	180
6.0	13.0	Resting time	0	630
16.0	23.0	Stress growth 2	0.01	180
19.5	26.5	Resting time	0	1230
39.5	46.5	Stress growth 3	0.01	180
43.0	50.0	Resting time	0	2430
83.0	90.0	Stress growth 4	0.01	180

Table 3. The proposed testing method for the chemical structural build-up (Procedure 2).

Time (min)	Hydration Time (min)	Procedure	Shear Rate (s^{-1})	Duration (s)
0.0	7	Pre-shear	100	30
0.5	7.5	Resting time	0	120
2.5	9.5	Stress growth 1	0.01	180
5.5	12.5	Pre-shear	100	30
6.0	13.0	Resting time	0	600
16.0	23.0	Stress growth 2	0.01	180
19.0	26.0	Pre-shear	100	30
19.5	26.5	Resting time	0	1200
39.5	46.5	Stress growth 3	0.01	180
42.5	49.5	Pre-shear	100	30
43.0	50.0	Resting time	0	2400
83.0	90.0	Stress growth 4	0.01	180

The stress growth experiment consists of measuring the shear stress evolution under a very low constant low shear rate (0.01 s^{-1}). The typical stress growth curve (Figures 2, A1 and A2 and Appendix A) displays two domains. The first domain, in which the shear stress increases almost linearly with the strain until it reaches a peak, is followed by a second domain (plateau) representing the steady-state flow. The peak defines the static yield stress, which is the minimum stress required to induce the first evidence of flow. The static yield stress originates from interparticle forces and direct contacts [28] and constitutes a relevant parameter for examining the workability of cementitious materials.

The experiments are carried out in triplicate, and the average values with their standard deviation are represented.

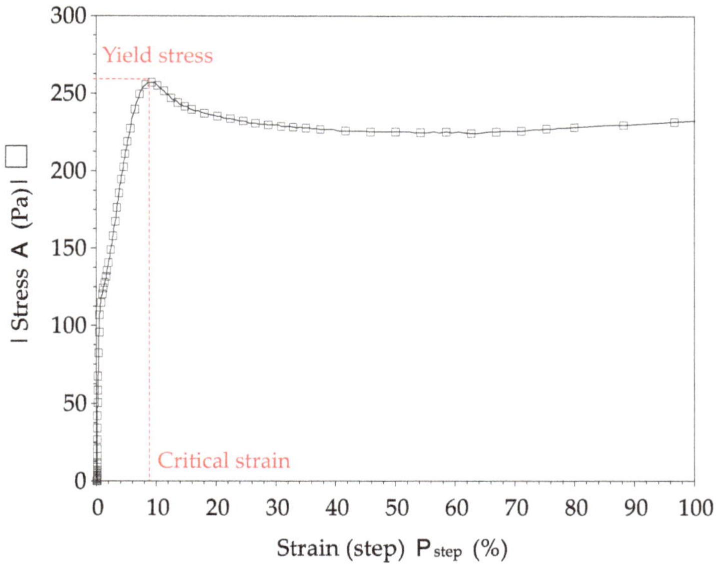

Figure 2. Typical evolution of shear stress under a low constant shear rate.

3.2. Isothermal Calorimetry

The addition of the superplasticizer leads to a retarding effect. To assess this effect, the hydration heat flow of the cement pastes is explored with an isothermal calorimeter TAM Air from TA Instruments. Pastes are prepared by external mixing at w/c = 0.4 and then introduced into the device. The calorimeter measures the difference in the heat flow between 5 g of cement paste and a reference (deionized water) at 25 °C.

4. Results and Discussion

4.1. Thixotropy vs. Non-Reversible Structural Build-Up

Thixotropy describes a time-dependent rheological behavior with reversible changes [2]. In fact, when a thixotropic material is left at rest for a long time, its yield stress (or viscosity) gradually increases. Shearing or mixing then makes it possible to recover the initial yield stress (or viscosity). At rest, there is a reversible structural build-up (flocculation) leading to an increase in the yield stress (or viscosity), whereas, under shearing, the structure formed at rest is broken (deflocculation).

For chemically inert colloidal suspensions, the structural build-up is almost reversible and is due to physicochemical interparticle interactions that lead to reversible agglomeration/dispersion phenomena. For cementitious materials, due to the chemical changes induced by hydration reactions, the structural build-up is not completely reversible; this is why they cannot be considered thixotropic materials. A part of the structural build-up induces permanent changes in fresh cement paste.

In order to assess the contribution of hydration reactions to the structural build-up, the proposed approach based on a succession of stress growth procedures is performed (Tables 2 and 3). The behavior of an inert carbonate calcium (calcite) suspension is compared to that of cement paste. The evolution of the static yield stress as a function of time is shown in Figure 3 for calcite and Figure 4 for cement paste.

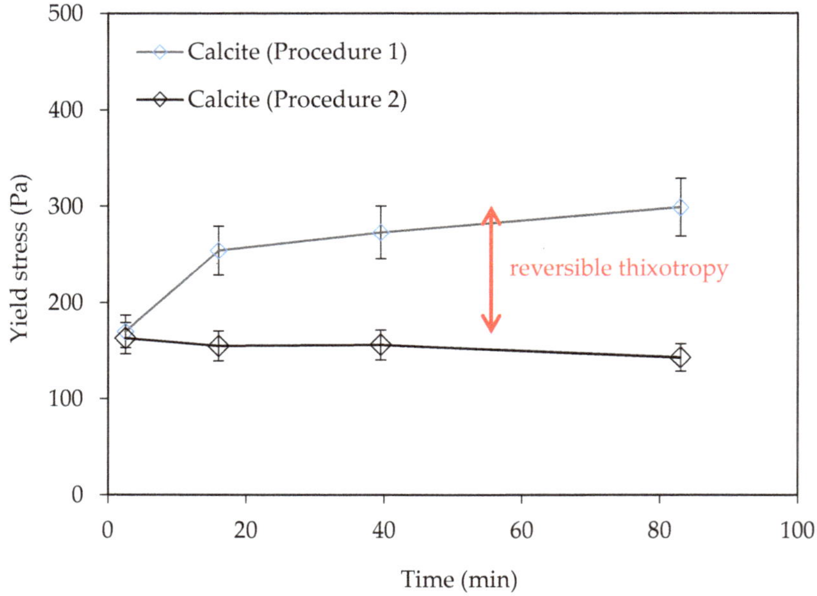

Figure 3. Evolution of the yield stress of the calcite suspension.

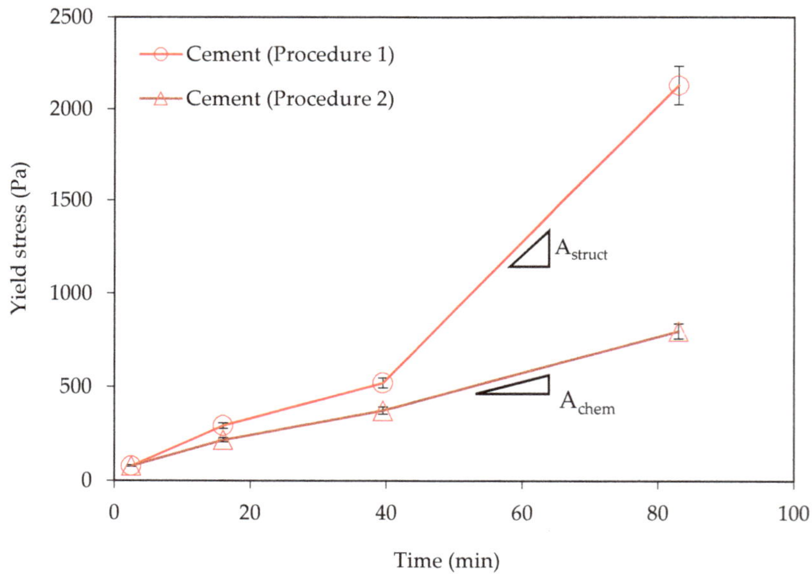

Figure 4. Evolution of the yield stress of the cement paste.

For the calcite suspension, the application of a strong pre-shear before the stress growth measurements allows for the erasure of the structural build-up, as shown in Figure 3. In fact, without pre-shearing (procedure 1), the yield stress increases with time, which is characteristic of a structural build-up at rest. Moreover, the static yield stress remains almost constant with the application of shearing (procedure 2). In fact, after the first pre-shear (Table 3), the suspension is left at rest for 2 min, and then an initial static yield stress of an order of 160 Pa is measured. A second pre-shear is applied to the suspension in order to break down the structure formed at rest, and then the suspension is left at rest again for 10 min. The second static yield stress is about 155 Pa. After resting times of 20 min and 40 min, the calcite suspension successively exhibits static yield stresses of about 156 and 143 Pa. It thus appears that a thixotropic material, such as the calcite suspension, displays a constant static yield stress that is not time dependent since the structure formed at rest is broken by the application of a strong pre-shear. The calcite suspension, therefore, represents a reference for the reversible part of the structural build-up, since no chemical reactions occur.

Furthermore, the static yield stress of the cement paste increases even with the application of a strong pre-shear able to erase the structural build-up. In fact, after 2 min of resting, the cement paste displays a static yield stress of 79 Pa. This static yield stress increases almost linearly with time to reach 661 Pa after 40 min of resting despite the strong pre-shear applied. This shows the irreversible nature of the structural build-up during the dormant period of cement hydration. In fact, as observed for the calcite suspension, if the cement paste is behaving as a thixotropic material, the static yield stress should remain constant with the application of the pre-shear. Figure 4 shows a significant increase in the static yield stress, which demonstrates that the irreversible structural build-up during the dormant period cannot be considered negligible.

As suggested by Zhang et al. [15], it thus appears that the structural build-up in the cement paste is the sum of a reversible part (thixotropy) and a chemical part (early age hydration). Procedure 1 provides the total structural build-up (A_{struct}), while Procedure 2 allows for the evaluation of the contribution of the chemical evolution (A_{chem}). For a thixotropic material such as calcite, the A_{struc} is equal to the A_{thix} and ranges from 0.14 to 0.22 Pas/s.

4.2. Effect of the Superplasticizer on the Structural Build-Up

Superplasticizers are usually used to improve the workability of cementitious materials [29,30]. They adsorb onto cement particles and act by electro-steric repulsion to enhance their dispersion [31–34]. This leads to the release of water trapped between agglomerated particles. Cement paste without a superplasticizer commonly exhibits a shear-thinning behavior (i.e., a non-linear behavior with a viscosity that decreases with the shear rate), while, with the addition of a superplasticizer, the rheological behavior becomes Newtonian. In addition, the yield stress decreases with the superplasticizer dosage.

In addition to their dispersive action, superplasticizers are known to retard cement hydration [35]. The retarding effect increases with the superplasticizer dosage [36,37]. Thus, the irreversible structural build-up during the dormant period is expected to be lower than that of the cement paste without a superplasticizer.

The approach presented in Tables 2 and 3 is applied to examine the effect of the superplasticizer on the structural build-up during the first 2 h of cement hydration. The results are presented in Figure 5.

First, it can be noted that the addition of the superplasticizer leads to a decrease in the static yield stress. This effect can be explained by the dispersive action. In fact, the superplasticizer allows for the deflocculation of the cement particles via the decrease in attractive Van der Waals forces, which leads to a decrease in the yield stress.

Then, contrary to the cement paste without a superplasticizer, it can be observed that the static yield stress remains almost constant during the first hour of cement hydration for the cement paste with a superplasticizer when a strong pre-shear is applied (Procedure 2). This indicates that the contribution of the chemical part to the structural build-up is negligible during this period. This effect may be due to the retarding effect induced by the presence of the superplasticizer, as shown in Figure 6. The static yield stress then starts increasing. Thus, the cement paste with the superplasticizer can be considered a thixotropic material during the first hour (or more, depending on the superplasticizer dosage). After this period, the contribution of cement hydration to the structural build-up cannot be neglected.

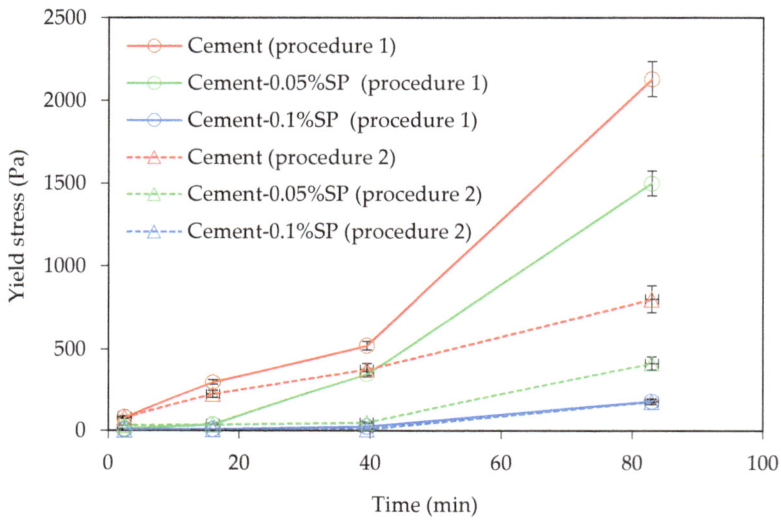

Figure 5. Effect of the superplasticizer dosage on the evolution of the static yield stress.

Thus, it appears that the addition of the superplasticizer affects the structural build-up during the dormant period since the superplasticizer induces a retarding effect. In this case, the contribution of cement hydration during the dormant period can be neglected. The cement paste thus behaves similarly to a thixotropic material with reversible changes. The use of the superplasticizer allows for a reduction in the workability loss during the dormant period.

The rheological procedures proposed in this work allow for quantification of the contribution of the structural build-up during the dormant period of cement hydration through the slope of the static yield stress–time curve (a derivative of the curve). As shown in Figure 7, the contribution of the irreversible chemical part (A_{chem}) is almost constant for the cement paste without the superplasticizer. The contribution of the reversible part (A_{thix}) increases with the resting time. In fact, when the cement paste is left at rest, attractive Van der Waals forces lead to a reversible structuration. For the cement paste with 0.05% SP, both the A_{thix} and A_{chem} increase with time. However, due to the retarding effect (Figure 6), during the first 40 min, this paste behaves similarly to a thixotropic material since the contribution of the chemical part is negligible. Furthermore, for the cement paste with 0.1% SP, the contribution of the chemical part increases with time, while the thixotropic part remains negligible. In fact, an increase in the SP dosage leads to a decrease in the magnitude of the attractive Van der Waals forces [38,39], which are no longer able to contribute to the structuration of the cement paste at rest.

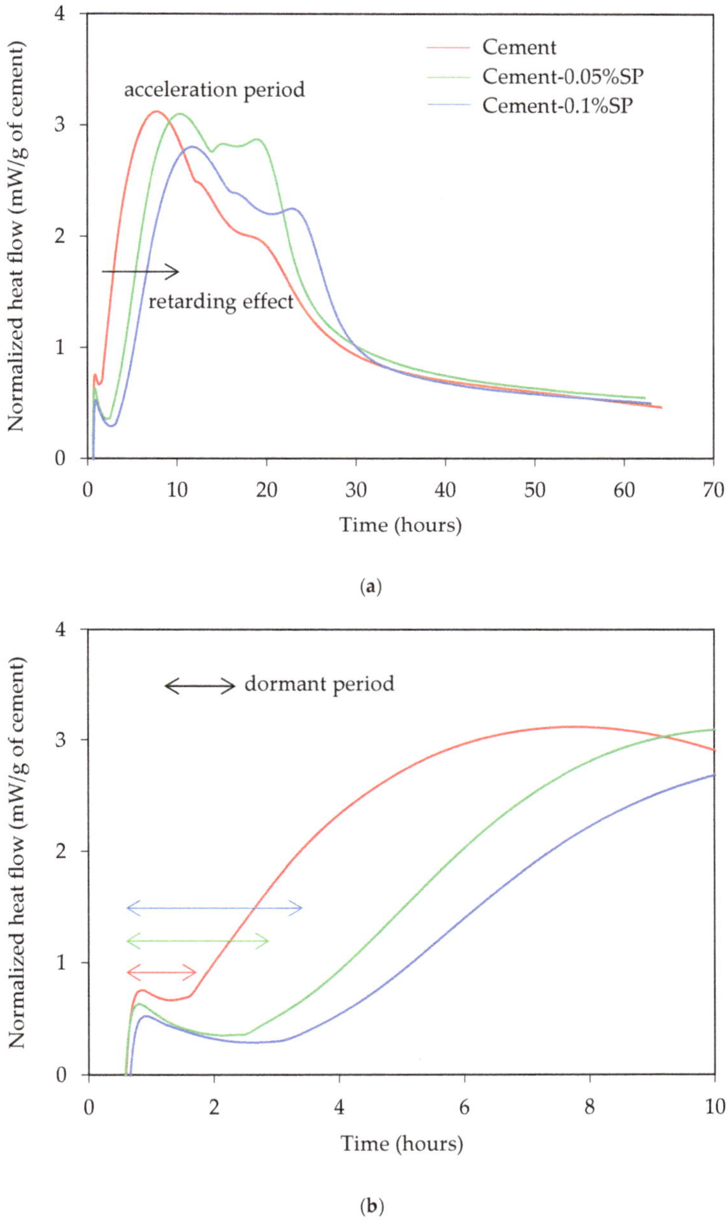

Figure 6. Evolution of the heat flow due to cement hydration. (**a**) retarding effect on the main peak; (**b**) retarding effect on the dormant period.

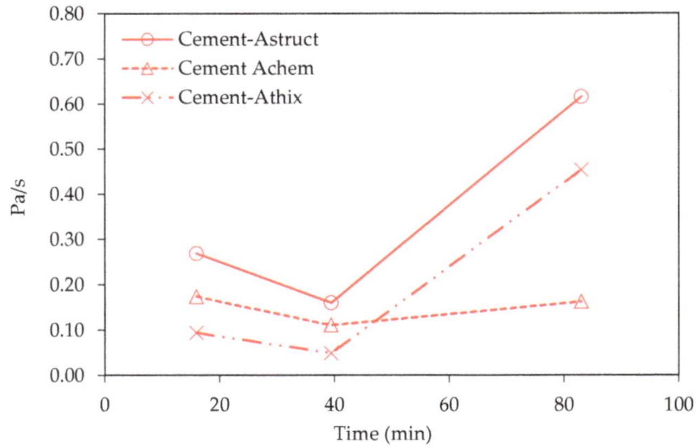

(**a**)

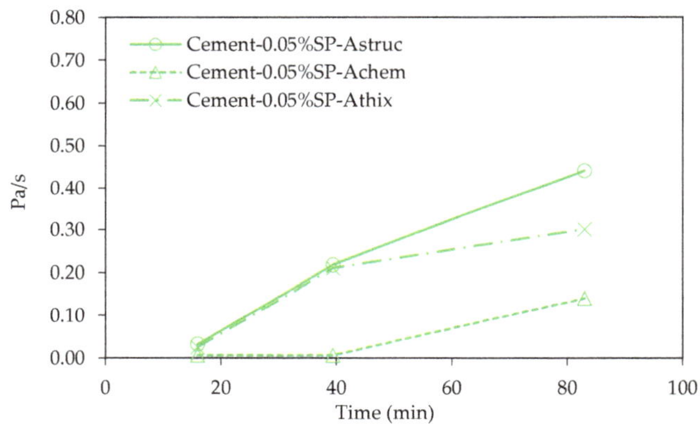

(**b**)

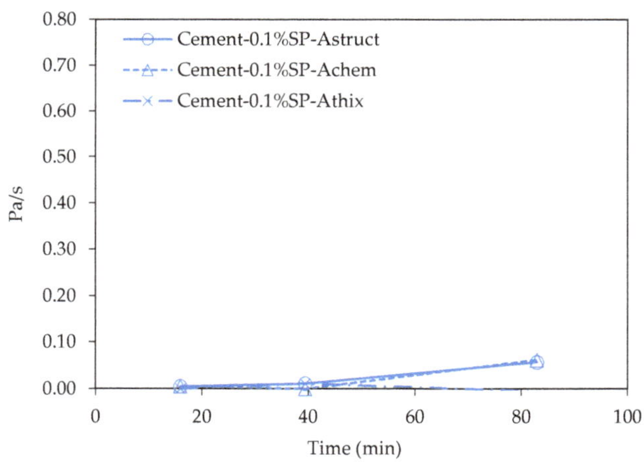

(**c**)

Figure 7. Evolution of the contribution of the structural build-up: (**a**) cement paste; (**b**) cement paste with 0.05% SP; (**c**) cement paste with 0.1% SP.

The approach proposed in this study is thus able to quantify the contribution of the thixotropic part and the chemical changes on the structural build-up of cementitious materials at rest. In addition, it allows the effect of the superplasticizer during the dormant period to be examined and quantified. Very few techniques make it possible to quantify the effect of the chemical changes during the dormant period. In fact, the isothermal calorimetry allows us to follow cement hydration via an indirect method (heat flow, Figure 6a), but it does not detect significant differences during the dormant period, except for its extension (Figure 6b). Combining the calorimetric data with in situ XRD patterns may detect changes during the early stage of cement hydration [40]. This proposed procedure can complete the chemical data by quantifying the rheological effect of the chemical evolution during the first hours of cement hydration.

5. Conclusions

In this work, an approach has been proposed to quantify the contribution of the structural build-up during the dormant period of cement hydration.

This approach has been validated on an inert thixotropic calcite suspension in which the structural build-up is almost reversible. Then, the thixotropic behavior of a cement paste without a superplasticizer has been examined. It appears that the contribution of cement hydration during the so-called "dormant period" cannot be considered negligible. In fact, there is a significant increase in the static yield stress during this period despite the strong pre-shear performed. This increase in the static yield stress describes a loss of workability that can be detrimental from a practical point of view.

Furthermore, the proposed approach allowed the effect of the superplasticizer to be investigated. It seems that the cement paste with the superplasticizer behaved similarly to a thixotropic material during the first hours of cement hydration. The structural build-up during this period can be considered reversible. Beyond this period, there are permanent changes characterized by a significant increase in the static yield stress. In addition, an increase in the superplasticizer dosage leads to a decrease in the magnitude of the attractive Van der Waals forces, which can erase the thixotropic structural build-up.

The proposed approach can thus be applied to complete the chemical data provided by other techniques, such as in situ XRD patterns and calorimetric data, to examine the chemical changes during the dormant period.

Author Contributions: Y.E.B.: conceptualization, methodology, validation, investigation, writing—original draft preparation. N.A.: validation, conceptualization. All authors have read and agreed to the published version of the manuscript.

Funding: This research received no external funding.

Conflicts of Interest: The authors declare no conflict of interest.

Appendix A

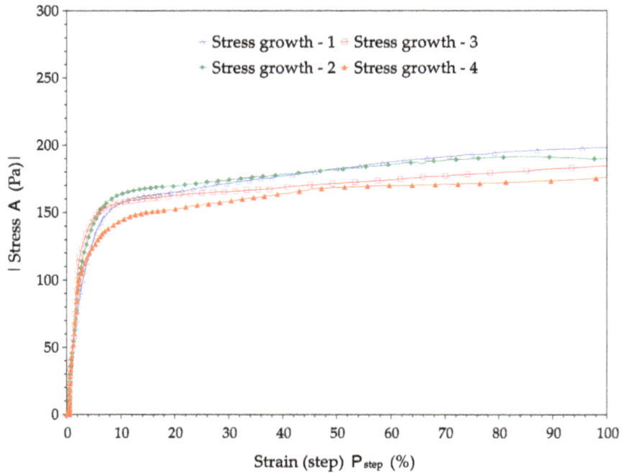

(**a**) Calcite

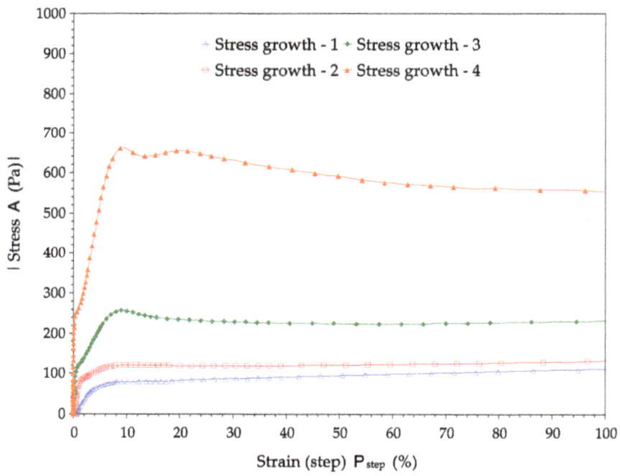

(**b**) Cement

Figure A1. *Cont.*

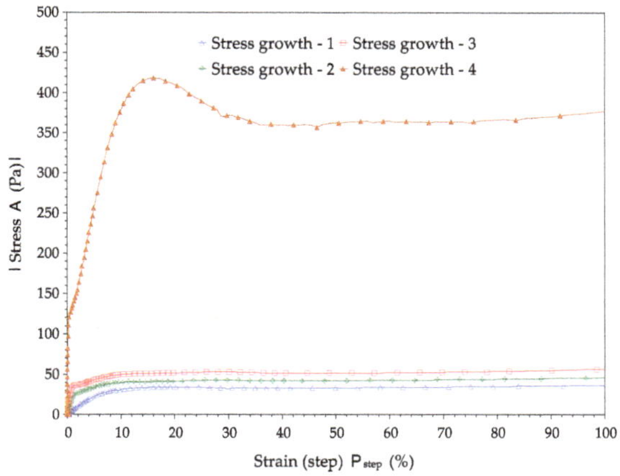

(**c**) cement paste with 0.05% SP

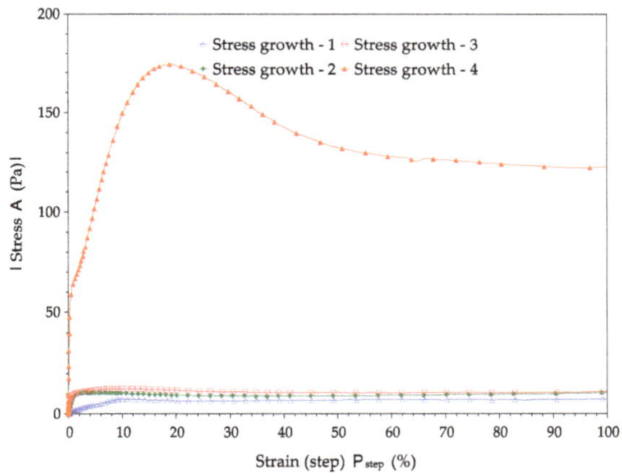

(**d**) cement paste with 0.1% SP

Figure A1. Example of the yield stress measurements of calcite (**a**), cement paste (**b**), cement paste with 0.05% SP (**c**), and cement paste with 0.1% SP (**d**) during Procedure 2.

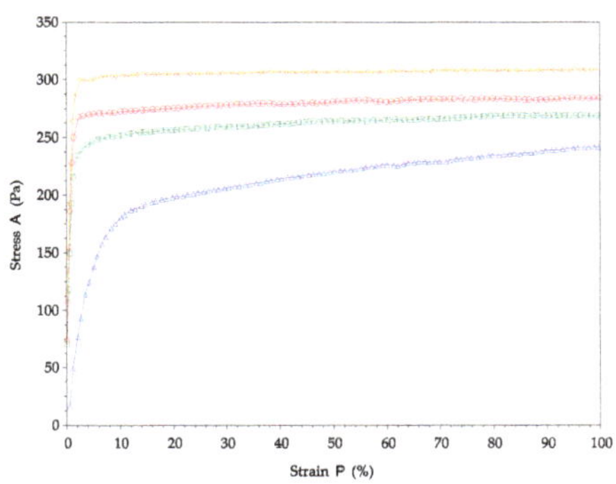

(**a**) calcite

Figure A2. *Cont.*

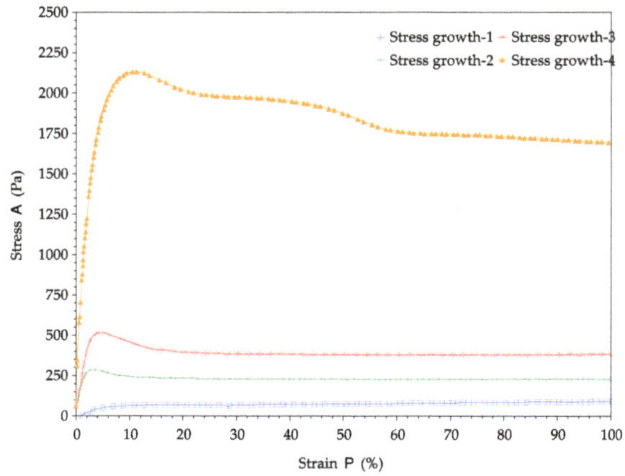

(**b**) cement

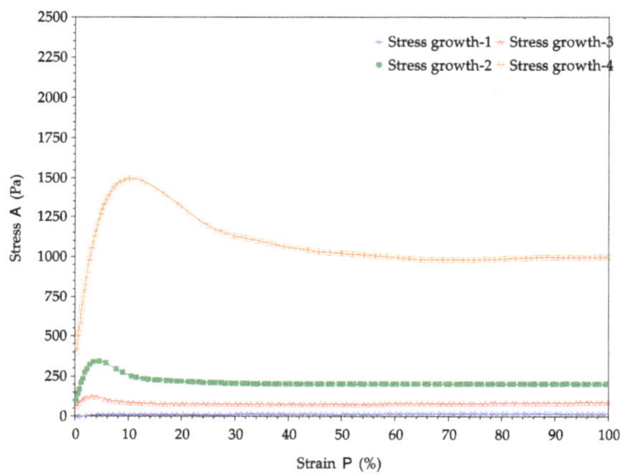

(**c**) cement paste with 0.05% SP

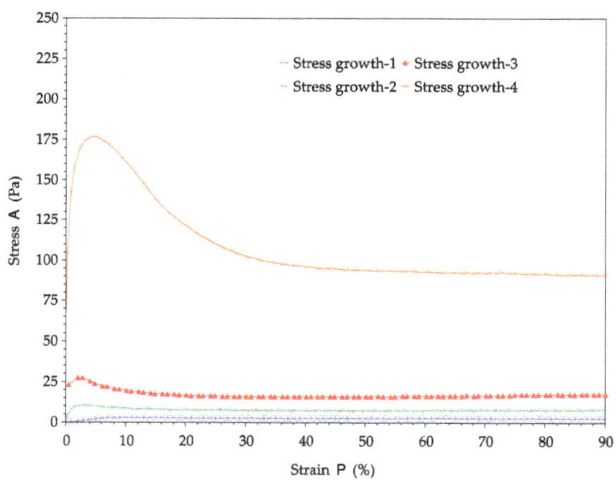

(**d**) cement paste with 0.1% SP

Figure A2. Example of the yield stress measurements of calcite (**a**), cement paste (**b**), cement paste with 0.05% SP (**c**), and cement paste with 0.1% SP (**d**) during Procedure 1.

References

1. Roussel, N.; Ovarlez, G.; Garrault, S.; Brumaud, C. The origins of thixotropy of fresh cement pastes. *Cem. Concr. Res.* **2012**, *42*, 148–157. [CrossRef]
2. Barnes, H.A. Thixotropy—A review. *J. Nonnewton. Fluid Mech.* **1997**, *70*, 1–33. [CrossRef]
3. Wallevik, J.E. Rheological properties of csement paste: Thixotropic behavior and structural breakdown. *Cem. Concr. Res.* **2009**, *39*, 14–29. [CrossRef]
4. Roussel, N. Rheological requirements for printable concretes. *Cem. Concr. Res.* **2018**, *112*, 76–85. [CrossRef]
5. Biricik, Ö.; Mardani, A. Parameters affecting thixotropic behavior of self compacting concrete and 3D printable concrete; a state-of-the-art review. *Constr. Build. Mater.* **2022**, *339*, 127688. [CrossRef]
6. Kovler, K.; Roussel, N. Properties of fresh and hardened concrete. *Cem. Concr. Res.* **2011**, *41*, 775–792. [CrossRef]
7. Bani Ardalan, R.; Joshaghani, A.; Hooton, R.D. Workability retention and compressive strength of self-compacting concrete incorporating pumice powder and silica fume. *Constr. Build. Mater.* **2017**, *134*, 116–122. [CrossRef]
8. Lapasin, R.; Longo, V.; Rajgelj, S. Thixotropic behaviour of cement pastes. *Cem. Concr. Res.* **1979**, *9*, 309–318. [CrossRef]
9. Yahia, A.; Mantellato, S.; Flatt, R.J. Concrete rheology: A basis for understanding chemical admixtures. In *Science and Technology of Concrete Admixtures*; Woodhead Publishing: Sawston, UK, 2016; pp. 97–127. ISBN 9780081006962.
10. Roussel, N. Steady and transient flow behaviour of fresh cement pastes. *Cem. Concr. Res.* **2005**, *35*, 1656–1664. [CrossRef]
11. Khayat, K.H.; Assaad, J.J. Effect of w/cm and High-Range Water-Reducing Admixture on Formwork Pressure and Thixotropy of Self-Consolidating Concrete. *ACI Mater. J.* **2006**, *103*, 186. [CrossRef]
12. Roussel, N. Rheology of fresh concrete: From measurements to predictions of casting processes. *Mater. Struct. Constr.* **2007**, *40*, 1001–1012. [CrossRef]
13. Banfill, P.F.G. A viscometric study of cement pastes containing superplasticizers with a note on experimental techniques. *Mag. Concr. Res.* **1981**, *33*, 37–47. [CrossRef]
14. Petkova, V.; Samichkov, V. Some influences on the thixotropy of composite slag Portland cement suspensions with secondary industrial waste. *Constr. Build. Mater.* **2007**, *21*, 1520–1527. [CrossRef]
15. Zhang, K.; Mezhov, A.; Schmidt, W. Chemical and thixotropic contribution to the structural build-up of cementitious materials. *Constr. Build. Mater.* **2022**, *345*, 128307. [CrossRef]
16. Mostafa, A.M.; Yahia, A. New approach to assess build-up of cement-based suspensions. *Cem. Concr. Res.* **2016**, *85*, 174–182. [CrossRef]
17. Kawashima, S.; Chaouche, M.; Corr, D.J.; Shah, S.P. Rate of thixotropic rebuilding of cement pastes modified with highly purified attapulgite clays. *Cem. Concr. Res.* **2013**, *53*, 112–118. [CrossRef]
18. Schultz, M.A.; Struble, L.J. Use of oscillatory shear to study flow behavior of fresh cement paste. *Cem. Concr. Res.* **1993**, *23*, 273–282. [CrossRef]
19. Mahaut, F.; Mokéddem, S.; Chateau, X.; Roussel, N.; Ovarlez, G. Effect of coarse particle volume fraction on the yield stress and thixotropy of cementitious materials. *Cem. Concr. Res.* **2008**, *38*, 1276–1285. [CrossRef]
20. Otsubo, Y.; Miyai, S.; Umeya, K. Time-dependent flow of cement pastes. *Cem. Concr. Res.* **1980**, *10*, 631–638. [CrossRef]
21. Roussel, N. A thixotropy model for fresh fluid concretes: Theory, validation and applications. *Cem. Concr. Res.* **2006**, *36*, 1797–1806. [CrossRef]
22. Pourchet, S.; Pochard, I.; Brunel, F.; Perrey, D. Chemistry of the calcite/water interface: Influence of sulfate ions and consequences in terms of cohesion forces. *Cem. Concr. Res.* **2013**, *52*, 22–30. [CrossRef]
23. Tramaux, A.; Azéma, N.; El Bitouri, Y.; David, G.; Negrell, C.; Poulesquen, A.; Haas, J.; Remond, S. Synthesis of phosphonated comb-like copolymers and evaluation of their dispersion efficiency on $CaCO_3$ suspensions part II: Effect of macromolecular structure and ionic strength. *Powder Technol.* **2018**, *334*, 163–172. [CrossRef]
24. Mikanovic, N.; Khayat, K.; Pagé, M.; Jolicoeur, C. Aqueous $CaCO_3$ dispersions as reference systems for early-age cementitious materials. *Colloids Surfaces A Physicochem. Eng. Asp.* **2006**, *291*, 202–211. [CrossRef]
25. Aït-Kadi, A.; Marchal, P.; Choplin, L.; Chrissemant, A.S.; Bousmina, M. Quantitative analysis of mixer-type rheometers using the couette analogy. *Can. J. Chem. Eng.* **2002**, *80*, 1166–1174. [CrossRef]
26. Dzuy, N.Q.; Boger, D.V. Direct Yield Stress Measurement with the Vane Method. *J. Rheol.* **1985**, *29*, 335–347. [CrossRef]
27. El Bitouri, Y.; Azéma, N. Potential Correlation between Yield Stress and Bleeding. In *Proceedings of the SP-349: 11th ACI/RILEM International Conference on Cementitious Materials and Alternative Binders for Sustainable Concrete, Online, 7–10 June 2021*; American Concrete Institute: Farmington Hills, MI, USA, 2022; Volume 349, pp. 479–494.
28. Roussel, N.; Lemaître, A.; Flatt, R.J.; Coussot, P. Steady state flow of cement suspensions: A micromechanical state of the art. *Cem. Concr. Res.* **2010**, *40*, 77–84. [CrossRef]
29. Flatt, R.; Schober, I. *Superplasticizers and the Rheology of Concrete*; Woodhead Publishing Limited: Sawston, UK, 2011; ISBN 9780857090287.
30. Houst, Y.F.; Flatt, R.J.; Bowen, P.; Hofmann, H.; Mader, U.; Widmer, J.; Sulser, U.; Bürge, T.A. Influence of Superplasticizer Adsorption on the Rheology of Cement Paste. In Proceedings of the International RILEM Conference on "The Role of Admixtures in High Performance Concrete", Monterrey, Mexico, 21–26 March 1999; pp. 387–402, ISBN 2-912143-05-5.
31. Yoshioka, K.; Sakai, E.; Daimon, M.; Kitahara, A. Role of steric hindrance in the performance of superplasticizers for concrete. *J. Am. Ceram. Soc.* **1997**, *80*, 2667–2671. [CrossRef]

32. Hanehara, S.; Yamada, K. Interaction between cement and chemical admixture from the point of cement hydration, absorption behaviour of admixture, and paste rheology. *Cem. Concr. Res.* **1999**, *29*, 1159–1165. [CrossRef]
33. Plank, J.; Sakai, E.; Miao, C.W.; Yu, C.; Hong, J.X. Chemical admixtures—Chemistry, applications and their impact on concrete microstructure and durability. *Cem. Concr. Res.* **2015**, *78*, 81–99. [CrossRef]
34. Nkinamubanzi, P.C.; Mantellato, S.; Flatt, R.J. Superplasticizers in practice. In *Science and Technology of Concrete Admixtures*; Woodhead Publishing: Sawston, UK, 2016; pp. 353–377. ISBN 9780081006962.
35. Marchon, D.; Flatt, R.J. Impact of chemical admixtures on cement hydration. In *Science and Technology of Concrete Admixtures*; Woodhead Publishing: Sawston, UK, 2016; pp. 279–304. ISBN 9780081006962.
36. Zhu, W.; Feng, Q.; Luo, Q.; Bai, X.; Lin, X.; Zhang, Z. Effects of pce on the dispersion of cement particles and initial hydration. *Materials* **2021**, *14*, 3195. [CrossRef] [PubMed]
37. Singh, N.B.; Sarvahi, R.; Singh, N.P. Effect of superplasticizers on the hydration of cement. *Cem. Concr. Res.* **1992**, *22*, 725–735. [CrossRef]
38. Flatt, R.J. Dispersion forces in cement suspensions. *Cem. Concr. Res.* **2004**, *34*, 399–408. [CrossRef]
39. Flatt, R.J.; Bowen, P. Yodel: A yield stress model for suspensions. *J. Am. Ceram. Soc.* **2006**, *89*, 1244–1256. [CrossRef]
40. Hesse, C.; Goetz-Neunhoeffer, F.; Neubauer, J. A new approach in quantitative in-situ XRD of cement pastes: Correlation of heat flow curves with early hydration reactions. *Cem. Concr. Res.* **2011**, *41*, 123–128. [CrossRef]

eng

MDPI

Article

Real-Time Human Authentication System Based on Iris Recognition

Huma Hafeez [1], Muhammad Naeem Zafar [2], Ch Asad Abbas [1,3], Hassan Elahi [4,5,*] and Muhammad Osama Ali [5]

1 Key Laboratory of High-Efficiency and Clean Mechanical Manufacturing, Ministry of Education, School of Mechanical Engineering, Shandong University, 17923 Jingshi Road, Jinan 250061, China
2 Department of Mechatronics Engineering, University of Engineering and Technology Taxila, Rawalpindi 47050, Pakistan
3 Department of Mechatronics Engineering, University of Chakwal, Chakwal 48800, Pakistan
4 Department of Mechanical and Aerospace Engineering, Sapienza University of Rome, 00184 Rome, Italy
5 Department of Mechatronics Engineering, College of Electrical & Mechanical Engineering, National University of Science and Technology, Islamabad 44000, Pakistan
* Correspondence: hassan.elahi@uniroma1.it

Abstract: Biometrics deals with the recognition of humans based on their unique physical characteristics. It can be based on face identification, iris, fingerprint and DNA. In this paper, we have considered the iris as a source of biometric verification as it is the unique part of eye which can never be altered, and it remains the same throughout the life of an individual. We have proposed the improved iris recognition system including image registration as a main step as well as the edge detection method for feature extraction. The PCA-based method is also proposed as an independent iris recognition method based on a similarity score. Experiments conducted using our own developed database demonstrate that the first proposed system reduced the computation time to 6.56 sec, and it improved the accuracy to 99.73, while the PCA-based method has less accuracy than this system does.

Keywords: biometrics; iris recognition; security system; image processing; pattern recognition; iris image acquisition; image registration; PCA

Citation: Hafeez, H.; Zafar, M.N.; Abbas, C.A.; Elahi, H.; Ali, M.O. Real-Time Human Authentication System Based on Iris Recognition. *Eng* **2022**, *3*, 693–708. https://doi.org/10.3390/eng3040047

Academic Editor: Antonio Gil Bravo

Received: 4 October 2022
Accepted: 24 November 2022
Published: 15 December 2022

Publisher's Note: MDPI stays neutral with regard to jurisdictional claims in published maps and institutional affiliations.

Copyright: © 2022 by the authors. Licensee MDPI, Basel, Switzerland. This article is an open access article distributed under the terms and conditions of the Creative Commons Attribution (CC BY) license (https://creativecommons.org/licenses/by/4.0/).

1. Introduction

The classical human identification system was based on physical keys, passwords or ID cards, etc., which can be lost or be forgotten easily, while the modern identification system is based on distinct and unique traits, i.e., physical or behavioral characteristics. The human eye has a very sensitive part named the iris, which has unique pattern in every individual. The iris has a thin structure, and it has a sphincter muscle lying in between the sclera and the pupil of the human eye. It is just like a person who has a living password which can never be altered. Although, fingerprints, face and voice recognition have been also widely used as proofs of identity [1], the iris pattern is more reliable, non-invasive and has higher recognition accuracy rate [2–4]. The iris pattern does not change significantly throughout the human's life, and even the left and right eyes have different iris patterns [5]. Every eye has its own iris features with a very high degree of freedom [6]. These are some benefits of iris recognition, which make it better than the other recognition systems [7]. It began in 1936 when Dr. Frank Burch proposed the innovative idea of using iris patterns as a method to recognize an individual. In 1995, the first commercial product was made available [8]. Nowadays, iris recognition is extensively applied in many corporations for identification such as in security systems, immigration systems, border control systems, attendance systems, and many more [9].

The iris recognition framework is divided into four sections: iris segmentation, iris normalization, iris feature extraction and matching [10]. Daugman proposed the method of capturing the image at a very close range using camera and a point light source [2,11].

After an iris image has been captured, a series of integro-differential operators can be used for its segmentation [12]. In [13], the author proposed that the active contour method is better than fixed shape modelling is for describing the inner and outer boundaries of iris [14]. Wildes proposed the localization of the iris boundaries using Hough transform, and they represented the iris pattern via Laplacian pyramid. The author used a normalized correlation to find the goodness of matching between two iris patterns [15]. Boles et al. proposed a WT zero-crossing representation for a finer approximation of the iris features at different resolution levels, and an average dissimilarity at each resolution level was calculated to determine the overall dissimilarity between two irises [16].

Zhu et al. used a multi-scale texture analysis for global feature extraction, 2D texture analysis for feature extraction and a weighted Euclidean distance classifier for iris matching [17]. Daouk et al. proposed the Hough transform and canny edge detector to detect the inner and outer boundaries of an iris [18]. Tan et al. proposed the iterative pulling and pushing method for the localization of the iris boundaries, and they used key local variations and ordinal measure encoding to represent the iris pattern [18–21]. Patil et al. proposed the lifting wavelet scheme for the iris features [22]. Sundaram et al. used the circular Hough transform for the iris localization, 2D Haar wavelet and the Grey Level Co-occurrence Matrix (GLCM) to describe the iris features and the Probabilistic Neural Network (PNN) for matching the computed iris features [23].

Shin et al. proposed pre-classification based on the left or right eye and the color information of the iris, and then, they authenticated the iris by comparing the texture information in terms of binary code [24]. The authors proposed the Contrast Limited Adaptive Histogram Equalization (CLAHE) to remove the noise and occlusions from the image, and they used SURF (Speeded Up Robust Features)-based descriptors for the feature extraction [25,26]. Jamaludin et al. proposed the improved Chan–Vese active contour method for iris localization and the 1D log-Gabor filter for feature extraction for non-ideal iris recognition [9]. Kamble et al. proposed their own developed database and used Fourier descriptors to make an image quality assessment [27]. Dua et al. proposed an integro-differential operator and Hough transform for segmentation, a 1D log-wavelet filter for feature encoding and a Radial basis function neural network (RBFNN) for classification and matching [28]. This algorithm presented very high precision value, but it involved massive calculations which increased the computation time, and as well as this, the method was not tested practically on humans or animals. So, in order to improve the recognition efficiency and reduce the computation time, we sum up the state-of-the-art technology in the domain of iris recognition. We propose an improved state-of-the-art Iris recognition system based on image registration along with feature extraction which employs the physiological characteristics of the human eye. We propose two different methods, i.e., the classical image processing-based method and the PCA-based method to determine which will handle the noisy conditions, the illumination problems, as well as camera-to-face distance problems better in the real-time implementation of the systems.

In this paper, Section 2 describes the steps of proposed iris recognition method in which we have added image registration (to align the image) as a compulsory step to reduce FAR (False Acceptance Rate) and FRR (False Rejection Rate). Section 3 describes the principle component analysis, which is our own proposed method to describe the iris texture in terms of its Eigen vector, Eigen value and similarity score. Section 4 describes the evaluation part, and last section concludes the paper.

2. Proposed System

The proposed system based on iris recognition consists of six main steps: data acquisition, pre-processing, image registration, segmentation, feature extraction and matching. The principle component analysis method is the second proposed method for iris recognition. These methods are proposed in context of them having less computation time and a high level of accuracy.

2.1. Data Acquisition

The data in our case are based on real-time images, while generally, pre-captured images are used. It consists of two steps, either using the images that are already available for the testing system, i.e., the CASIA database, or using images that were taken from camera directly. The iris pattern is only visible in the presence of infrared rays so an ordinary camera cannot be used for this purpose. The real-time implementation of the system proves the system's effectiveness, so we have used our own database based on real-time images that were taken using an IR-based iris scanner, which were taken instantly. The core characteristics of the iris scanner are: auto-iris-detection, auto-capturing and auto-storage to a directory that has been assigned. The specifications of the iris scanner include a monocular IR camera, a capture distance of 4.7 cm to 5.3 cm using an image sensor, a capturing speed of 1 s, an image dimension of 640 $\times$ 480 (Grayscale) and the compression format was BMP.

Iris database created using this scanner consists of 454 images of 43 persons. For 12 persons, we have captured ten images of the right eye and five images of the left eye. For 15 persons, we have captured five images of the right eye and five images of the left eye, and for 16 persons, we have captured two images of the left eye and two images of the right eye. All of the images were captured at different angles and different iris locations.

2.2. Pre-Processing

After loading the images, the next step was pre-processing. This mainly involved RGB-to-grayscale conversion, contrast adjustment, brightness adjustment and deblurring. Figure 1 shows the four right eye images of same person taken using the iris scanner. As our images are already in grayscale, there was no need to conduct the grayscale conversion. The iris scanner took images after focusing, so there was very little chance that image would be blurred, but still, we have used a weighted average filter to perform deblurring.

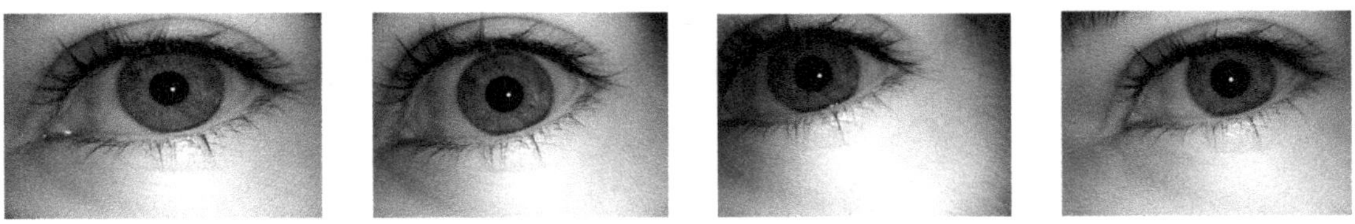

Figure 1. Four right eye images of one person.

In weighted average filters, there will be higher intensities of pixels in the center of the image. The center values (pixels) of the mask are multiplied with highest values, thus, the intensities of the pixels at center are the highest. The weighted average filter was implemented for filtering an image in the order of $M \times N$ with filter of size $m \times n$, which is given by Equation (1).

$$g(x,y) = \frac{\sum_{s=-a}^{a} \sum_{t=-b}^{b} w(s,t) f(x+s, y+t)}{\sum_{s=-a}^{a} \sum_{t=-b}^{b} w(s,t)} \tag{1}$$

where w (s, t) is the weight, f $(x + s, y + t)$ is the input for which $x = 0, 1, 2 \ldots ., m - 1$ and $y = 0, 1, 2 \ldots ., n - 1$ and g (x, y) are the output image. This will result in the image having better intensities in the iris portion than those that are shown in Figure 2. During the iris recognition process, the pre-processing steps can also be used again and again if they are required.

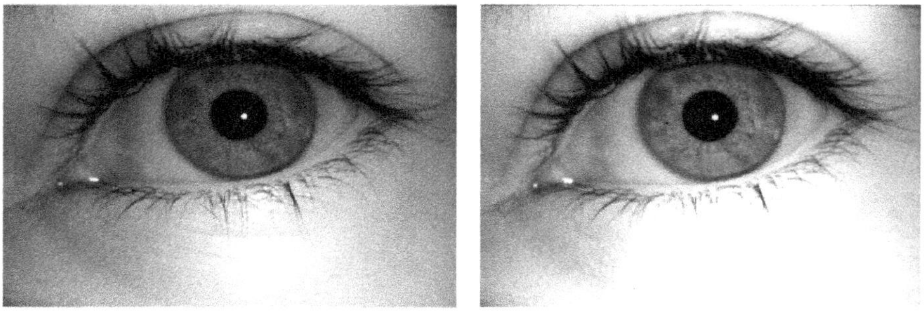

(a) (b)

Figure 2. (**a**) Original Image. (**b**) Output of weighted average filter image.

2.3. Image Registration

Image registration is a process during image processing that overlaps two or more images from various imaging equipment or sensors which are taken at different angles to geometrically align the images for an analysis and to reduce the problems of misalignment, rotation and scale, etc. If the angle of the iris is changed (i.e., the person kept their eye near the scanner at a different angle or a different position), then the iris pattern at the same position as in other image will be changed, and it can cause mismatch. As it can be seen in Figure 1, the images were taken at different positions as well as angles, so we needed to perform image registration. If the new image I_n (x, y) is rotated or tilted at any angle, it will be compared with the sample image I_s (x, y), and it will automatically be rotated to the ideal position. There are different processes in image registration based on point mapping, multimodal configurations and feature matching. These points will be detected in both of the images to find whether they are at same angle and position or not, and if they are not, then the images will be aligned by adjusting these points, respectively. When we were choosing a mapping function $(u\ (x, y), v\ (x, y))$ for the ordinal coordinates transformation, the intensity values of the new image were made to be close to the corresponding points in the sample image. The mapping function must simplify to Equation (2).

$$\int_x \int_y (I_s(x,y) - I_n(x-u, y-v))^2 dxdy \tag{2}$$

It is constrained to capture the similarity transformation of the image coordinates from (x, y) to (x', y') as shown in Equation (3).

$$\begin{pmatrix} x' \\ y' \end{pmatrix} = \begin{pmatrix} x \\ y \end{pmatrix} - sR() \begin{pmatrix} x \\ y \end{pmatrix} \tag{3}$$

where s is a scaling factor and $R()$ is the rotation matrix, which is represented by . Practically, when a pair of iris images I_n and I_d are given, the warping parameters s and are recovered via an iterative minimization procedure. The output of image registration is shown in Figure 3, and image registration data are represented in Algorithm 1.

Algorithm 1: Image Registration.

Step 1: Read sample iris image and new (i.e., tilted or rotated) grayscale eye image.
Step 2: Detect surface features of both images.
Step 3: Extract features from both images.
Step 4: Find the matching features using Equation (2).
Step 5: Retrieve location of corresponding points for both images using Equation (3).
Step 6: Find a transformation corresponding to the matching point pairs using M-estimator Sample Consensus (MSAC) algorithm.
Step 7: Use geometric transform to recover the scale and angle of new image corresponding to the sample image. Let sc = scale ∗ cos (theta) and ss = scale ∗ sin (theta), then: Tinv = [sc-ss 0; ss sc 0; tx ty 1]
where tx and ty are x and y translations of new image relative to the sample image, respectively.
Step 8: Make the size of new image same as that of sample and display in same frame.

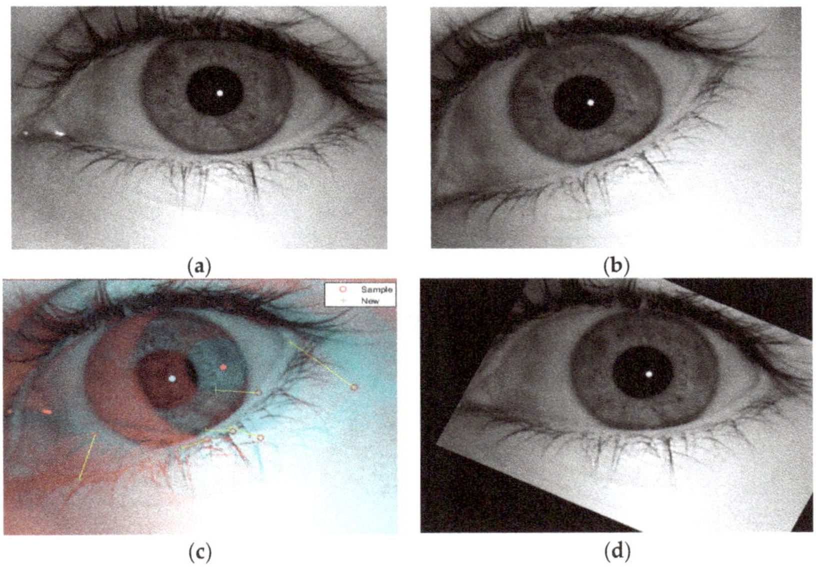

Figure 3. Image Registration. (**a**) Sample Image. (**b**) New Image. (**c**) Matching Points Between both images. (**d**) Rotated/Registered Image.

2.4. Segmentation

Segmentation mainly involves the separation of the iris portion from the eye. The iris region consists of two circles, one of them is the outer iris–sclera boundary, and another one is the interior iris–pupil boundary. The eyelids and eyelashes sometimes hide the upper and lower parts of the iris region. It is considered to be a very crucial stage to achieve the correct detection the of outer and inner boundaries of the iris. There are different methods which are commonly used for this section including the integro-differential integrator [29], moving agent [30], Hough transform [31], circular Hough transform [32], iterative algorithm [33], Chan–Vese active contour method [34] and Fourier spectral density ones, [35] etc. We have used the circular Hough transform (CHT) one for the detection of the iris boundaries due to its robust performance even in noise, occlusion and varying illumination. It depends upon the equation of circle which is described by Equation (4):

$$(x-a)^2 + (y-b)^2 = r^2 \tag{4}$$

where a, b is the center, r is the radius and x, y represents the coordinates of the circle. Equations (5) and (6) shows the parametric representation of this circle:

$$x = a + r * \cos\theta \tag{5}$$

$$y = b + r * \sin\theta \tag{6}$$

The CHT use a 3D array with the first two dimensions, representing the coordinates of the circle, and the last third of it specifies the radii. When a circle of a desired radii is drawn at every edge point, the values in the accumulator (the array which will find the intersection point) will increase. The accumulator, which keeps count of the circles passing through the coordinates of each edge point, will vote for the highest count as shown in Figure 4.

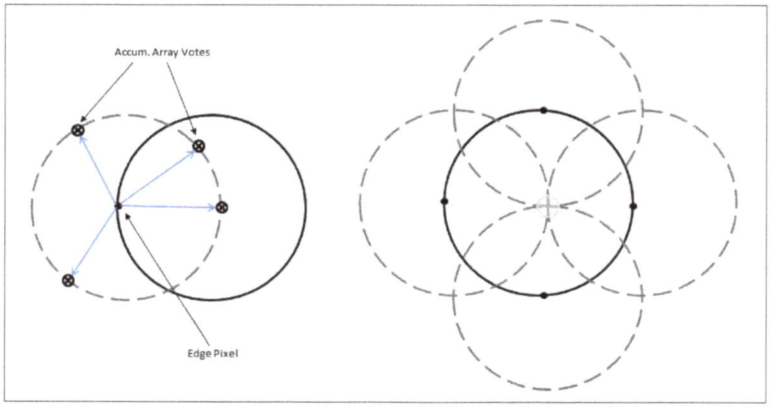

Figure 4. Circular Hough transform voting pattern.

The coordinates of the center of circles in the images will be the coordinates with the highest count. For efficient recognition, the circular Hough transform was performed on the iris–sclera boundary first, and then, on the iris–pupil boundary. The segmentation using the circular Hough transform method is shown in Figure 5. After the circular portion was detected, the next step was to separate this circular portion from the eye. We have applied the mask of zeros to extract the iris from the eye as shown in Figure 6, and the circle detection is represented in Algorithm 2.

Algorithm 2: Circle Detection Using Circular Hough Transform.

Step 1: Define iris radius range [50, 155] and pupil radius range [20, 55].
Step 2: Define object polarity bright as dark.
Step 3: Define sensitivity of 0.98.
Step 4: Define edge threshold value of 0.05.
Step 5: Apply circular Hough transform for boundary detection.
Step 6: Find centers and radii and display only required circles.
Step 7: Display the detected iris portion.
Step 8: Apply mask to separate the iris from eye.

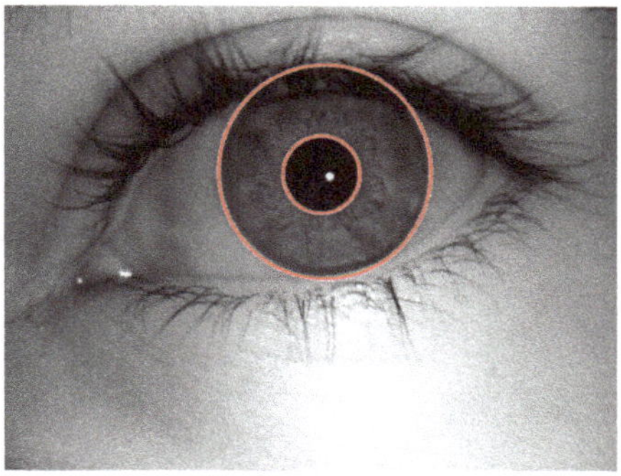

Figure 5. Output of circle detection.

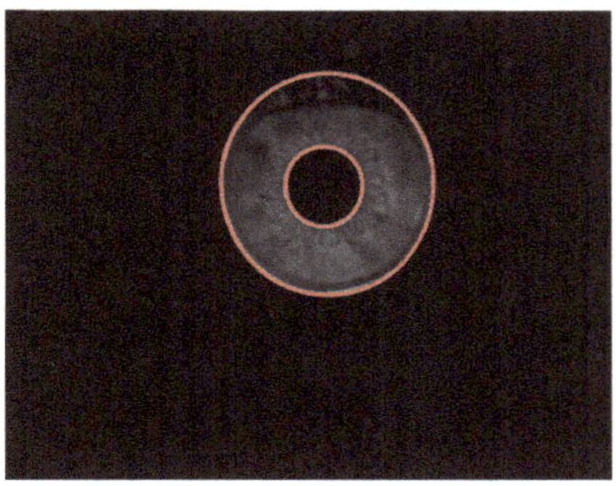

Figure 6. Output of segmentation.

2.5. Feature Extraction

Feature extraction is one of the most important steps involved in process. We have used two different methods for the feature extraction. One is the two-dimensional Discrete Wavelet Transform (DWT) and other is edge detection.

2.5.1. Two-Dimensional Discrete Wavelet Transform (2-D DWT)

The 2D wavelet and scaling functions were obtained by taking the vector product of the 1D wavelet and the scaling functions. This leads to the decomposition of the approximate coefficients at level j in four components, i.e., the approximation at level $j + 1$, and the details in three orientations (the horizontal, vertical, and diagonal ones). Two-dimensional wavelet transform is generally obtained by separable products of scaling functions Ø and wavelet functions Ψ as in Equation (7):

$$C_{j+1}[k,l] = \sum_{m,n} h[m-2k]h[n-2]C_j[m,n] \tag{7}$$

The detail coefficient images, which are obtained from three wavelets, are given by Equations (8)–(10), and they are shown in Figure 7.

$$\text{Vertical Wavelet}: \ \Psi^1(t1,t2) = (t1)\Psi(\text{t}2) \tag{8}$$

$$\text{Horizontal Wavelet}: \ \Psi^2(t1,t2) = (t2)\Psi(\text{t}1) \tag{9}$$

$$\text{Diagonal Wavelet}: \ \Psi^3(t1,t2) = \Psi(\text{t}1)\Psi(\text{t}1) \tag{10}$$

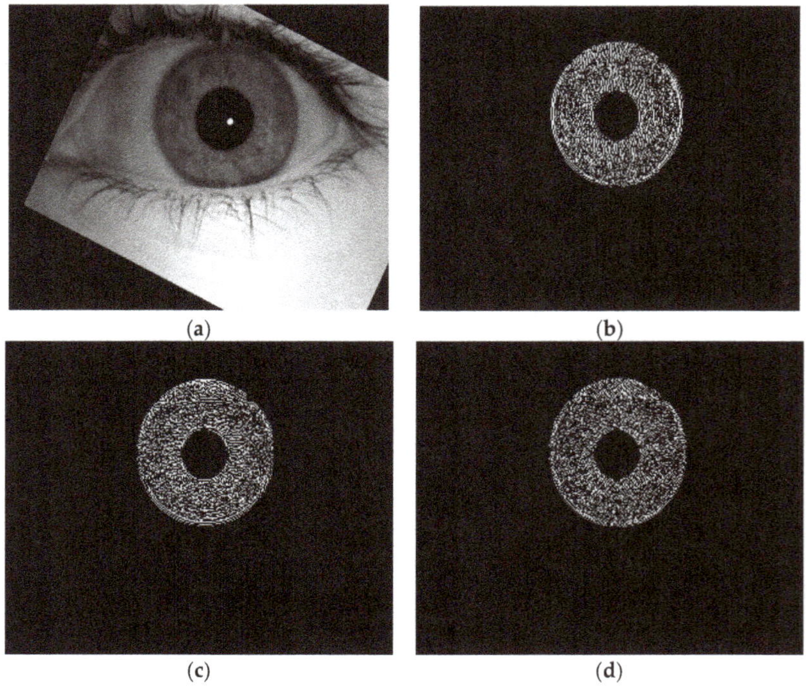

Figure 7. Feature extraction using 2D DWT. (**a**) Original eye image. (**b**) Vertical wavelet. (**c**) Horizontal wavelet. (**d**) Diagonal wavelet.

The energy is computed to approximate the three detailed coefficients (the horizontal, vertical and diagonal ones) by Equation (11):

$$\text{Energy} = \sum_{m=0}^{M-1} \sum_{n=0}^{N-1} |X(m,n)| \tag{11}$$

where $X(m, n)$ is a discrete function whose energy is to be computed.

2.5.2. Edge Detection:

Edge detection consists of a variety of mathematical methods that can be used to identify the points in an image where the image brightness changes sharply, which are generally organized into set of curved line segments, or they have discontinuities. It can also be used for finding out those points where the intensities change rapidly. There are a lot of different edge detection techniques, but we have used a zero-crossing-based second-order derivative edge detector named the canny edge detector. This technique uses two thresholds: a high threshold for low edge sensitivity and a low threshold for high edge sensitivity to detect strong as well as weak edges which enables the edge detector to not be affected by noise, and thus, it is more likely to detect the true weak edges. Bing Wang and Shaosheng Fan developed a filter which evaluated the discontinuity between the grayscale values of each pixel [36]. For higher discontinuity, a lower weight value was set to smooth the filter at the corresponding point, and for lower a discontinuity between the grayscale values, the higher weight value was set to the filter. The resultant image after applying the edge detection is shown in Figure 8, and feature extraction is represented in Algorithm 3.

Algorithm 3: Feature Extraction using Edge Detection.

Step 1: Convolve the Gaussian filter with image to smooth the image using:
$H_{ij} = \frac{1}{2\pi\sigma^2} exp\left(-\frac{(i-(k+1))^2+(j-(k+1))^2}{2\sigma^2}\right); 1 \le i,j \le (2k+1)$
where σ is standard deviation and kernel size is $(2k + 1) \times (2k + 1)$.
Step 2: Compute the local gradient $\left[g_x^2 + g_y^2\right]^{\frac{1}{2}}$ at each point.
Step 3: Find edge direction $tan^{-1}(\frac{g_x}{g_y})$ at each point.
Step 4: Apply an edge thinning technique to get more accurate representation of real edges.
Step 5: Apply hysteresis thresholding based on two thresholds, T_1 and T_2 with $T_1 < T_2$, to determine potential edges in image.
Step 6: Perform edge linking by incorporating the weak pixels connected to the strong pixels.

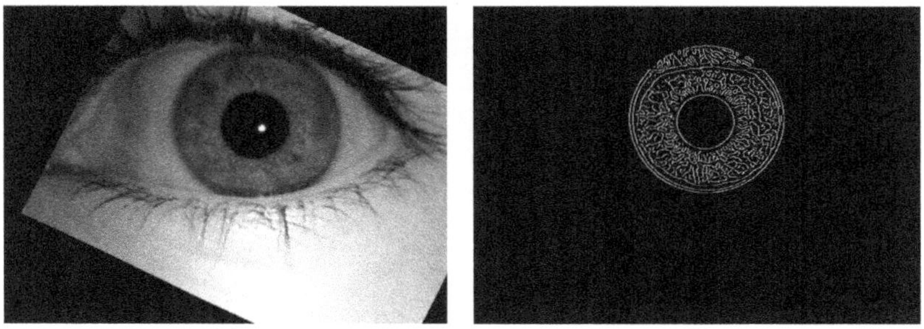

(**a**) (**b**)

Figure 8. Feature extraction using canny edge detector. (**a**) Original eye image. (**b**) Output image of feature extraction using edge detection.

2.6. Feature Matching

For matching, we have used two different methods, i.e., the Hamming distance and Absolute differencing method.

2.6.1. Hamming Distance

Hamming distance makes use of only those parts in both of the iris patterns which corresponds to "0". It is calculated by using the formula in Equation (12). Its value will be zero when both of the iris patterns match exactly, but unfortunately this never happens because of light variation while we are capturing the image, noise which will remain undetected during normalization and environmental effects on the sensor, etc. So, a value up to a 0.5 distance which was chosen in the hit and trial method is usually considered to be accurate. If the hamming distance is below 0.5, it means that the both iris patterns are the same, but if the distance is greater than 0.5, it means that the iris patterns may be matched or not matched. If the distance has value of 1, it clearly means that the iris patterns are not matched.

$$Hamming\ Distance = \frac{1}{N}\sum_{i-1}^{1}|X_i - Y_i| \quad (12)$$

where N is the number of parts to be compared and $|X_i - Y_i|$ is the difference between the two iris patterns.

2.6.2. Absolute Differencing

The absolute differencing method will find the absolute difference between each element in one iris pattern from the corresponding element in other iris pattern, and it returns the absolute difference in the corresponding element of the output. If one pattern is similar to the other, then the absolute difference will be zero.

We have mixed both of these techniques in a way that first, it finds the hamming distance, and then, find out the absolute difference between the two images. If the distance between the two images is zero, the display feature are matched, but if it is not zero, then, we must calculate the absolute difference between the two patterns. However, if the absolute difference is zero, it will display the non-matched features.

3. Principal Component Analysis (PCA)

Principal component analysis is a method used to extract strong patterns from a given dataset. The data become easy to visualize using this technique and it converts set of correlated variables into a linearly correlated variable. This process gives the differences and similarities in the dataset. The dataset which has highest variance becomes the first axis, which is called the first principal component. The dataset which has the second highest variance becomes the second axis, which is called the second principal component and so on. PCA reduces the dimensions of the dataset, but it retains the features and characteristics of the dataset. We have used PCA to reduce the steps and obtain the desired results as by using the traditional image processing steps. It does not detect the inner features which is very important step for our system. It makes a decision based on the Eigenvectors, Eigenvalues and matching score between the two images. The results obtained using PCA are shown in Figure 9, and the PCA is represented in Algorithm 4.

Algorithm 4: Principal Component Analysis (PCA).

Step 1: Create MAT file of the database and load database.
Step 2: Find the mean of images using $\frac{1}{2}\sum_{i=1}^{n} X_{ij}$.
Step 3: Find the mean shifted input image.
Step 4: Calculate the Eigen vector and Eigen values using $A\upsilon = \lambda\upsilon$, where matrix λ is the Eigen value of non-zero square matrix (A) corresponding to υ.
Step 5: Find the cumulative energy content for each Eigen vector by $g_j = \sum_{k=1}^{j} D_{kk}$, j = 1, 2, 3, ..., p. It will retain the top principal components only.
Step 6: Create the feature vector by taking the product of cumulative energy content of Eigen vector and mean shifted input image.
Step 7: Separate out feature vector (iris section) from input image.
Step 8: Find the similarity score with images in database.
Step 9: Display the image having highest similarity score with input image.

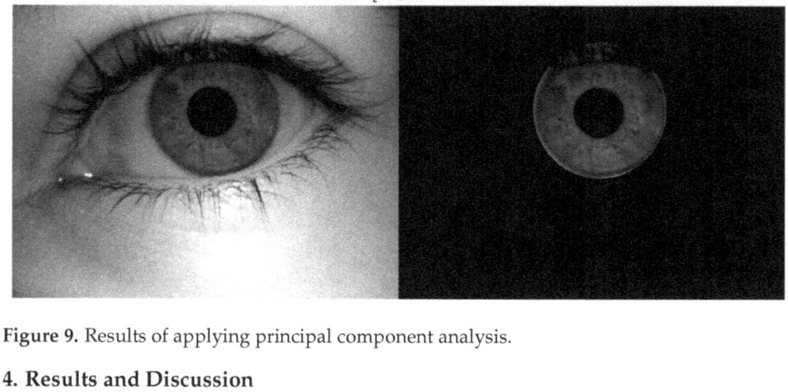

Figure 9. Results of applying principal component analysis.

4. Results and Discussion

Our proposed work was implemented using same laptop with the specifications of an Intel Core 5Y10C, 4 GB RAM, Windows 8, IriCore software and MATLAB R2017a. About 454 images of 43 persons taken using a camera, MK-2120U, using IriCore software were used to test the performance of the proposed system. Multiple samples of individual eyes were recorded in our database. Each image was captured at 640 × 480, 8-bit grayscale image, and they were saved in the BMP format. Figure 10 shows the left and right eye images of different persons.

1. Image registration aligns the input image with the reference image. In this method, we have taken an eye image as shown in Figure 11a, and the image registration was performed to align the image, as shown in Figure 11b. It is one of the major step that is performed to align the images for the analysis, and it reduce the problems of misalignment, rotation and scale.
2. Segmentation involves the circular portion detection and extraction from an eye image. Iris segmentation combines the technique of edge detection and Hough transform to detect the circular edges in the image. The segmented iris is shown in Figure 11c. It also involves the extraction of the iris region from an eye image, as shown in Figure 11d, which was evaluated by the combination of the circular Hough transform and masking methods, and this resulted in the circular iris portion extraction.
3. Feature extraction was realized by using 2-Dimensional Discrete Wavelet transformation and canny edge detection. The 2D DWT results in the horizontal, vertical and diagonal components of the iris feature contain a lot of information which are difficult to handle, while on other hand, the edge detection technique provides all of the features in a single matrix. If we compared both of the techniques, 2D DWT takes more time to execute and it provides the desired results in three matrices, which will further take more time in matching, while the edge detection technique takes less time to provide the desired result, and it will be easy to find matched features, as it can be seen in Figure 11e. So, the results of the edge detection technique were employed for further processing.
4. Matching comprises of two different methods to avoid FAR and FRR as much as possible. These methods include the Hamming distance and absolute differencing method. First, the Hamming distance method was implemented to find the distance between the iris features, and if this distance is zero, then access is granted. This is difficult to achieve as occlusions such as eyelids, eyelashes, change under different lightening conditions and noise effect the features of an iris. So, threshold of 0.5 was adjusted to pass the barrier (gate), but still, the iris pattern was subjected to the absolute differencing method. This method checks the difference between the features of the two iris patterns, and it grant access or denies access to the barrier. If the Hamming distance is greater than 0.5 and there also exists absolute difference, the access to the system will be denied.
5. Principal component analysis comprises all of the steps from reading an image to matching it. It consists of iris extraction from an eye image and calculating the mean Eigen values, Eigen vectors and similarity score of an image to compare them with those of the images in the database. The decision is based on the similarity score between two images as shown in Figure 9. The similarity score is 1 with ID 13_L1, which means its features are similar to a person having this ID. This method takes the surface features mainly, while the iris pattern has detailed information hidden which needs to be involved during processing.

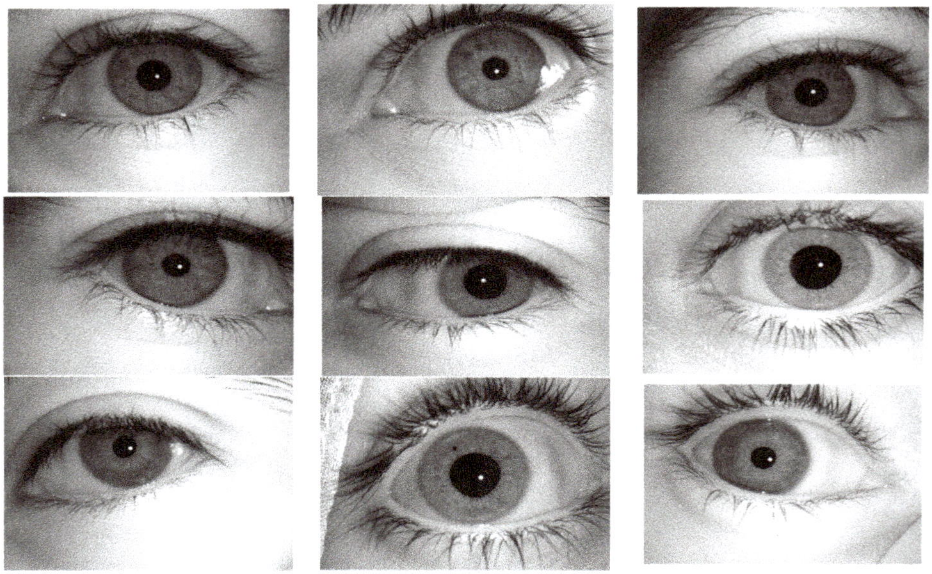

Figure 10. Iris images of different persons in our database.

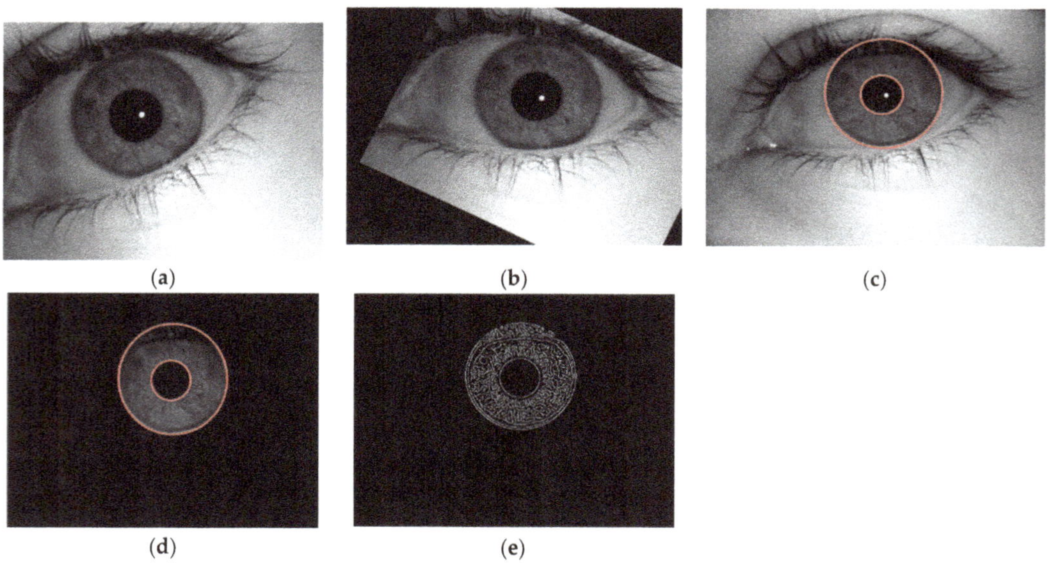

Figure 11. (**a**) Original eye image. (**b**) Image after registration. (**c**) Outer and inner boundaries of image. (**d**) Iris portion separation from an eye image. (**e**) Features of an Iris.

To check the accuracy of the proposed system, it was evaluated using the false acceptance rate, the false rejection rate and the equal error rate. The equal error rate (EER) is achieved at a point where the FAR and FRR overlaps; the lower the EER is, the better the performance accuracy of the system will be. The matching algorithm uses a threshold value which will determine the closeness of the input iris to the database iris. The lower the threshold value is, the lower the FRR will be, while the FAR will be higher, and a higher threshold value will lead to a lower FAR and higher FRR, as shown in Figures 12 and 13. EER is the point at which the FRR equals the FAR, and it is considered to be the most important measure of the biometric system's accuracy. The proposed iris recognition system gives an EER = 0.134 as shown in Figure 12, while using a PCA-based method for iris recognition gives an EER = 0.384 as shown in Figure 13.

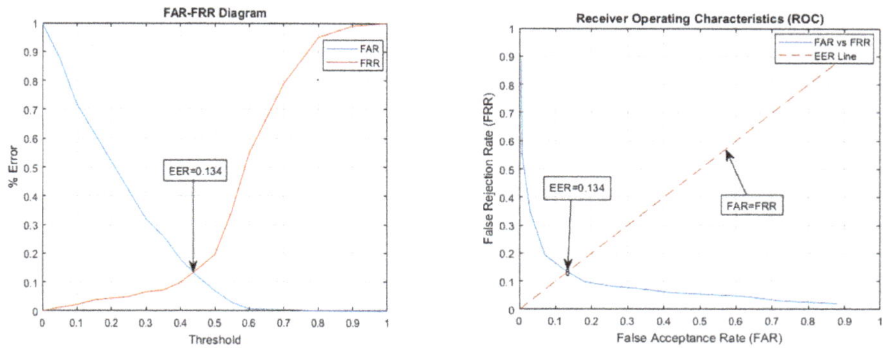

(a) (b)

Figure 12. Performance graph for proposed method. (**a**) % Error for threshold distance (**b**) ROC curve.

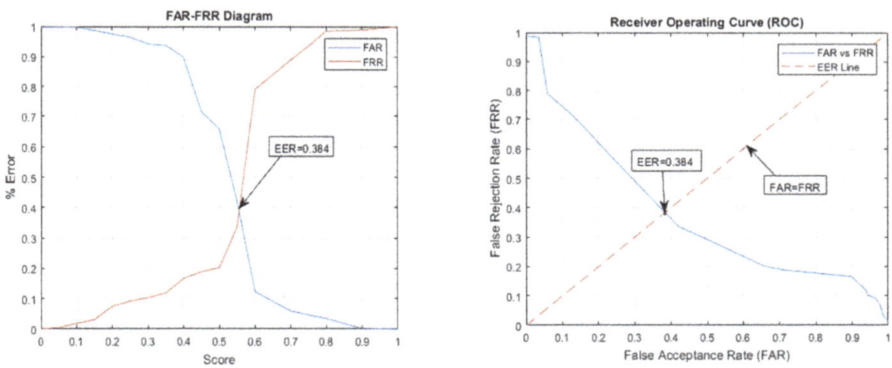

(a) (b)

Figure 13. Performance graph for PCA-based method. (**a**) % Error for threshold distance. (**b**) ROC curve.

Table 1 describes the performance of different methodologies for iris recognition. It shows the false acceptance rates (FAR), the false rejection rates (FRR) and the recognition accuracy taken using different methodologies. It can be observed from the table that our proposed system has outperformed the already existing techniques in terms of the processing time. Our proposed system has recognition rate of 99.73%, and the execution time is 6.56 s, while the other system based on PCA has a recognition rate of 88.99%, and the execution time is 21.52 s. Therefore, the proposed system without PCA is more proficient for identification than the proposed system with PCA because it has less accuracy.

Table 1. Performance comparison of different methodologies.

Methodology	FAR(%)	FRR(%)	Average Accuracy(%)
Ma et al. [20]	0.020	010980	98
Sanchez et al. [3]	0.030	2.080	97.89
Tisse et al. [4]	1.840	8.790	89.37
Proposed System	0.12	0.1452	99.73
Proposed system based on PCA	8.980	1.030	89.99

It can be seen from the above table that our proposed system's accuracy is better than the other methods are as we have involved image registration which provide much better results, while the PCA-based system is less efficient than the other proposed system is, but it can be made more efficient by using a camera with a higher resolution. So, the proposed system with image registration is proficient for the identification and verification of the iris.

5. Conclusions

Iris recognition is an emerging field in biometrics as the iris has a data-rich unique structure, which makes it one of the best ways to identify an individual. The designed project is an innovation in the current modes of security systems that are being used today. Due to the unique nature of the iris, it can be used as a password for life. As the iris is the only part of human that can never be altered, there are no chances of trespassing when one is using an iris detection system, by any means. In this paper, an efficient approach for an iris recognition system using image registration and PCA is presented using a database that was built using images taken using an iris scanner. The iris characteristics enhance its suitability in the automatic identification which includes the ease of image registration, the natural protection from external environment and surgical impossibilities without the loss of vision.

The application of the iris recognition system has been seen in various areas of life such as in crime detection, airport, business application, banks and industries. Image registration adjust the angle and alignment of the input image to the reference image. The iris segmentation uses the circular Hough transform method for the iris portion detection, and then, the mask is applied to extract the iris segment from the eye. Feature extraction is achieved by using the two-dimensional discrete wavelet transform (2-D DWT) method and by an edge detection technique so that the most supreme areas of the iris pattern can be extracted, and hence, a high recognition rate and less computation time is achieved. The Hamming distance and absolute differencing methods were applied on the extracted features which give us the accuracy of 99.73%.

PCA was also applied on the same database, and the decision is purely based on the matching score. This system gives us the recognition rate of 89.99% as it does not analyses the deep features. Then, the doors were automated using serial communication between MATLAB and Arduino. The door automation using iris recognition has reduced the labor work for opening and closing the barrier. The system proved to be efficient, and it saved time as it had a processing time of 6.56 s which can be reduced further if an advance programming software had been used. In the future, an improved system can be accessible by investigating with the proposed iris recognition system under different constraints and environments.

Author Contributions: H.H.: Proposed topic, basic study Design, Data Collection, methodology, statistical analysis and interpretation of results; M.N.Z.: Manuscript Writing, Review and Editing; C.A.A.: Literature review, Writing and Referencing; H.E.: Review and Editing; M.O.A.: Review and Editing. All authors have read and agreed to the published version of the manuscript.

Funding: This research received no external funding.

Institutional Review Board Statement: Not applicable.

Informed Consent Statement: Not applicable.

Data Availability Statement: Not applicable.

Acknowledgments: The author acknowledges the National Development Complex (NDC, NESCOM) Islamabad, Pakistan, for giving sponsorship for the project and authorized faculty for technical and non-technical help to accomplish the goal.

Conflicts of Interest: The authors declare no conflict of interest.

References

1. Zhang, D. Biometrics technologies and applications. In Proceedings of the International Conference on Image and Graphics, Vancouver, BC, Canada, 10–13 September 2000.
2. Daugman, J.G. High confidence visual recognition of persons by a test of statistical independence. *IEEE Trans. Pattern Anal. Mach. Intell.* **1993**, *15*, 1148–1161. [CrossRef]
3. Sanchez-Avila, C.; Sanchez-Reillo, R. Two different approaches for iris recognition using Gabor filters and multiscale zero-crossing representation. *Pattern Recognit.* **2005**, *38*, 231–240. [CrossRef]
4. Tisse, C.-L.; Martin, L.; Torres, L.; Robert, M. Person identification technique using human iris recognition. In *Proceeding Vision Interface*; Citeseer: Princeton, NJ, USA, 2002.
5. Daugman, J.; Downing, C. Epigenetic Randomness, Complexity and Singularity of Human Iris patterns. *Proc. R. Soc. Lond. Ser. B Biol. Sci.* **2001**, *268*, 1737–1740. [CrossRef] [PubMed]
6. Jeong, D.S.; Hwang, J.W.; Kang, B.J.; Park, K.R.; Won, C.S.; Park, D.-K.; Kim, J. A new iris segmentation method for non-ideal iris images. *Image Vis. Comput.* **2010**, *28*, 254–260. [CrossRef]
7. Szewczyk, R.; Grabowski, K.; Napieralska, M.; Sankowski, W.; Zubert, M.; Napieralski, A. A reliable iris recognition algorithm based on reverse biorthogonal wavelet transform. *Pattern Recognit. Lett.* **2012**, *33*, 1019–1026. [CrossRef]
8. Sreekala, P.; Jose, V.; Joseph, J.; Joseph, S. The human iris structure and its application in security system of car. In Proceedings of the 2012 IEEE International Conference on Engineering Education: Innovative Practices and Future Trends (AICERA), Kottayam, India, 19–21 July 2012; IEEE: Piscataway, NJ, USA, 2012.
9. Jamaludin, S.; Zainal, N.; Zaki, W.M.D.W. Sub-iris Technique for Non-ideal Iris Recognition. *Arab. J. Sci. Eng.* **2018**, *43*, 7219–7228. [CrossRef]
10. Daugman, J. Demodulation by complex-valued wavelets for stochastic pattern recognition. *Int. J. Wavelets Multiresolution Inf. Process.* **2003**, *1*, 1–17. [CrossRef]
11. Daugman, J. Statistical richness of visual phase information: Update on recognizing persons by iris patterns. *Int. J. Comput. Vis.* **2001**, *45*, 25–38. [CrossRef]
12. Daugman, J. How iris recognition works. In *The Essential Guide to Image Processing*; Elsevier: Amsterdam, The Netherlands, 2009; pp. 715–739.
13. Daugman, J. New methods in iris recognition. *IEEE Trans. Syst. Man Cybern. Part B* **2007**, *37*, 1167–1175. [CrossRef]
14. Daugman, J. Information theory and the iriscode. *IEEE Trans. Inf. Forensics Secur.* **2015**, *11*, 400–409. [CrossRef]
15. Wildes, R.P. Iris recognition: An emerging biometric technology. *Proc. IEEE* **1997**, *85*, 1348–1363. [CrossRef]
16. Boles, W.W.; Boashash, B. A human identification technique using images of the iris and wavelet transform. *IEEE Trans. Signal Process.* **1998**, *46*, 1185–1188. [CrossRef]
17. Zhu, Y.; Tan, T.; Wang, Y. Biometric personal identification based on iris patterns. In Proceedings of the 15th International Conference on Pattern Recognition. ICPR-2000, Barcelona, Spain, 3–7 September 2000; IEEE: Piscataway, NJ, USA, 2000.
18. Ma, L.; Wang, Y.; Tan, T. Iris recognition using circular symmetric filters. In *Object Recognition Supported by User Interaction for Service Robots*; IEEE: Piscataway, NJ, USA, 2002.
19. Ma, L.; Tan, T.; Wang, Y.; Zhang, D. Efficient iris recognition by characterizing key local variations. *IEEE Trans. Image Process.* **2004**, *13*, 739–750. [CrossRef]
20. He, Z.; Tan, T.; Sun, Z.; Qiu, X. Toward accurate and fast iris segmentation for iris biometrics. *IEEE Trans. Pattern Anal. Mach. Intell.* **2008**, *31*, 1670–1684.
21. Sun, Z.; Tan, T. Ordinal measures for iris recognition. *IEEE Trans. Pattern Anal. Mach. Intell.* **2008**, *31*, 2211–2226.
22. Patil, C.M.; Patilkulkarani, S. Iris feature extraction for personal identification using lifting wavelet transform. In Proceedings of the 2009 International Conference on Advances in Computing, Control, and Telecommunication Technologies, Bangalore, India, 28–29 December 2009; IEEE: Piscataway, NJ, USA, 2009.
23. Sundaram, R.M.; Dhara, B.C. Neural network based Iris recognition system using Haralick features. In Proceedings of the 2011 3rd International Conference on Electronics Computer Technology, Kanyakumari, India, 8–10 April 2011; IEEE: Piscataway, NJ, USA, 2011.
24. Shin, K.Y.; Nam, G.P.; Jeong, D.S.; Cho, D.H.; Kang, B.J.; Park, K.R.; Kim, J. New iris recognition method for noisy iris images. *Pattern Recognit. Lett.* **2012**, *33*, 991–999. [CrossRef]

25. Ismail, A.I.; Hali, S.; Farag, F.A. Efficient enhancement and matching for iris recognition using SURF. In Proceedings of the 2015 5th national symposium on information technology: Towards new smart world (NSITNSW), Riyadh, Saudi Arabia, 17–19 February 2015; IEEE: Piscataway, NJ, USA, 2015.
26. Ali, H.S.; Ismail, A.I.; Farag, F.A.; El-Samie, F.E.A. Speeded up robust features for efficient iris recognition. *Signal Image Video Process.* **2016**, *10*, 1385–1391. [CrossRef]
27. Kamble, U.R.; Waghmare, L. Person Identification Using Iris Recognition: CVPR_IRIS Database. In Proceedings of the International Conference on ISMAC in Computational Vision and Bio-Engineering, Palladam, India, 16–17 May 2018; Springer: Berlin/Heidelberg, Germany, 2018.
28. Dua, M.; Gupta, R.; Khari, M.; Crespo, R.G. Biometric iris recognition using radial basis function neural network. *Soft Comput.* **2019**, *23*, 11801–11815. [CrossRef]
29. Labati, R.D.; Genovese, A.; Piuri, V.; Scotti, F. Iris segmentation: State of the art and innovative methods. In *Cross Disciplinary Biometric Systems*; Springer: Berlin/Heidelberg, Germany, 2012; pp. 151–182.
30. Rankin, D.M.; Scotney, B.W.; Morrow, P.J.; McDowell, D.R.; Pierscionek, B.K. Dynamic iris biometry: A technique for enhanced identification. *BMC Res. Notes* **2010**, *3*, 182. [CrossRef]
31. Proença, H.; Alexandre, L.A. Iris segmentation methodology for non-cooperative recognition. *IEE Proc.-Vis. Image Signal Process.* **2006**, *153*, 199–205. [CrossRef]
32. de Martin-Roche, D.; Sanchez-Avila, C.; Sanchez-Reillo, R. Iris recognition for biometric identification using dyadic wavelet transform zero-crossing. In Proceedings of the IEEE 35th Annual 2001 International Carnahan Conference on Security Technology (Cat. No. 01CH37186), London, UK, 16–19 October 2001; IEEE: Piscataway, NJ, USA, 2001.
33. Sreecholpech, C.; Thainimit, S. A robust model-based iris segmentation. In Proceedings of the 2009 International Symposium on Intelligent Signal Processing and Communication Systems (ISPACS), Kanazawa, Japan, 7–9 December 2009; IEEE: Piscataway, NJ, USA, 2009.
34. Puhan, N.B.; Sudha, N.; Kaushalram, A.S. Efficient segmentation technique for noisy frontal view iris images using Fourier spectral density. *Signal Image Video Process.* **2011**, *5*, 105–119. [CrossRef]
35. Teo, C.C.; Ewe, H.T. An efficient one-dimensional fractal analysis for iris recognition. In Proceedings of the 13th WSCG International Conference in Central Europe on Computer Graphics, Visualization and Computer Vision, Plzen-Bory, Czech Republic, 31 January–4 February 2005.
36. Ad, S.; Sasikala, T.; Kumar, C.U. Edge Detection Algorithm and its application in the Geo-Spatial Technology. In Proceedings of the 2017 IEEE International Conference on Computational Intelligence and Computing Research (ICCIC), Coimbatore, India, 14–16 December 2017; IEEE: Piscataway, NJ, USA, 2017.

eng

Article

Strategic Participation of Active Citizen Energy Communities in Spot Electricity Markets Using Hybrid Forecast Methodologies

Hugo Algarvio

LNEG—National Laboratory of Energy and Geology, Est. Paço Lumiar 22, 1649-038 Lisbon, Portugal; hugo.algarvio@lneg.pt

Abstract: The increasing penetrations of distributed renewable generation lead to the need for Citizen Energy Communities. Citizen Energy Communities may be able to be active market players and solve local imbalances. The liberalization of the electricity sector brought wholesale and retail competition as a natural evolution of electricity markets. In retail competition, retailers and communities compete to sign bilateral contracts with consumers. In wholesale competition, producers, retailers and communities can submit bids to spot markets, where the prices are volatile or sign bilateral contracts, to hedge against spot price volatility. To participate in those markets, communities have to rely on risky consumption forecasts, hours ahead of real-time operation. So, as Balance Responsible Parties they may pay penalties for their real-time imbalances. This paper proposes and tests a new strategic bidding process in spot markets for communities of consumers. The strategic bidding process is composed of a forced forecast methodology for day-ahead and short-run trends for intraday forecasts of consumption. This paper also presents a case study where energy communities submit bids to spot markets to satisfy their members using the strategic bidding process. The results show that bidding at short-term markets leads to lower forecast errors than to long and medium-term markets. Better forecast accuracy leads to higher fulfillment of the community programmed dispatch, resulting in lower imbalances and control reserve needs for the power system balance. Furthermore, by being active market players, energy communities may save around 35% in their electrical energy costs when comparing with retail tariffs.

Keywords: Balance Responsible Parties; Citizen Energy Communities; electricity markets; forecast methodologies; imbalance penalties; strategic bidding

Citation: Algarvio, H. Strategic Participation of Active Citizen Energy Communities in Spot Electricity Markets Using Hybrid Forecast Methodologies. *Eng* **2023**, *4*, 1–14. https://doi.org/10.3390/eng4010001

Academic Editor: Antonio Gil Bravo

Received: 23 November 2022
Revised: 15 December 2022
Accepted: 16 December 2022
Published: 21 December 2022

Copyright: © 2022 by the authors. Licensee MDPI, Basel, Switzerland. This article is an open access article distributed under the terms and conditions of the Creative Commons Attribution (CC BY) license (https://creativecommons.org/licenses/by/4.0/).

1. Introduction

The liberalization process brought full competition to the electricity supply industry in both wholesale and retail markets [1]. As a consequence, the market agents have the option to trade electricity in different markets [2]: spots, continuous, derivatives, non-organized, and ancillary services markets.

In spot markets, agents can submit bids to electricity pools based on day-ahead and intraday or real-time marginal auctions. In continuous intraday markets, players can negotiate energy based on the pay-as-bid scheme, i.e., an automatic match of opposite bids [3]. These markets were designed for dispatchable players, i.e., players that can comply with a programmed dispatch, which means that players like consumers and variable generation without storage capacity will have real-time deviations [4,5]. Real-time deviations from the schedules of Balance Responsible Parties (BRPs) have to be balanced at balancing markets. Balancing markets are part of the ancillary services of the system, managed by transmission system operators (TSOs) to guarantee the secure operation of power systems. BRPs with deviations from their schedules may need to pay penalties concerning spot markets. They will receive the down/up balancing prices according to the direction of their deviations [6]. Those penalties are computed considering each country's imbalance settlement (IS) mechanism [7]. In derivatives markets, agents can sign standard financial and physical contracts [8]. For non-standard contracts, agents can negotiate and set the terms and conditions of the private bilateral agreements [9].

Normally, in retail competition, retailers sign private bilateral contracts with clients [10]. Citizen Energy Communities (CECs) are a new market player that competes with retailers for signing private bilateral contracts with end-use consumers [11]. The main problem of retailers is that they usually follow a business-as-usual strategy, proposing high tariffs, equal in each consumer segment [12]. So, being part of a CEC is more economically attractive than signing retail tariffs, but also more demanding by considering the active participation of their members. CECs may be composed of local consumers, prosumers, distributed generation, and storage assets. Considering the global goal of a carbon-neutral society and the increasing penetration of distributed generation, CECs aim to achieve energy sustainability by managing local resources [13]. Against this background, new European legislation supports the active participation of consumers through CECs by providing significant discounts on their grid usage and access costs [14–16]. To satisfy the needs of their members, CECs can enter into the wholesale competition, submitting bids to spot markets, signing private bilateral contracts with producers, and standard contracts on the exchanges or OTC [17,18]. Algarvio [13] presented a review of CECs, as power system alliances that need resource management and coordination.

To avoid future losses, forecasting market prices is one of the aspects that CECs have to consider when participating in wholesale markets. Furthermore, forecasting their energy needs is one of the biggest issues that CECs have to face. The consumption dynamic of members is very dependent on the meteorological conditions, the type of days, and the segment type of consumers [19]. So, minimizing the consumption volatility of members can be a good solution to avoid high forecast errors, which can result in unbalances, and, consequently, in the payment of penalties by CECs. Thus, CECs should have an appropriate trading strategy to mitigate those errors. An adequate short-run strategic bidding on spot markets is crucial to mitigate potential consumption unbalances, since bilateral transactions are usually made in the long run (months prior to real-time consumption). Accordingly, monitoring the real local dispatch with smart meters is a critical aspect of communities with members composed of consumers, prosumers, and distributed generation [20,21]. Furthermore, it enables them to control their net load by using demand response programs, i.e., controlling the local energy production or consumption in case of shortages or excesses of energy [22,23].

Ayón et al. [24] indicated that large and diversified quantities of end-use clients might reduce load forecast errors. Furthermore, they concluded that aggregations of flexible loads are typically beneficial to reduce their forecast errors. Therefore, load aggregations may benefit market players concerning individual loads. Wei et al. [25] presented a complete review of 128 forecast models of energy load. They considered that highly accurate forecasts have a maximum mean absolute percentage error (MAPE) of 10%. Naturally, they concluded that forecasting small-scale loads have larger errors than large-scale loads. Furthermore, they also concluded that the forecast accuracy increase with the time horizon, i.e., long-term (yearly) and medium-term (monthly or quarterly) forecasts have smaller errors than short-term forecasts (from daily to sub-hourly). Naturally, demand is weather-driven, so by analyzing the studied models, the authors concluded that the forecast accuracy increase with the time scale being high to yearly forecasts than to hourly forecasts. However, considering hourly forecasts, the forecast accuracy increases how closer to real-time operation [26]. Koponen et al. presented a review of 12 models to forecast the short-term electrical energy load [27]. They considered six different scenarios to test these models. They concluded that the forecast errors decrease with an increase in the number of aggregated consumers, considering the normalized root mean square error (NRMSE). Furthermore, they indicated that their results do not support the use of specific criteria (such as MAPE or NRMSE) to compare methods. They also concluded that it should be used hybrid methods to compute demand forecasts. Algarvio and Lopes [28] presented a strategic bidding strategy for retailers considering hybrid forecast methodologies in spot day-ahead and intraday markets.

They concluded that the participation of retailers closer to real-time markets improves their forecast accuracy and their return from markets. It also has been concluded that retailers with larger and diversified portfolios have lower forecast errors.

Against this background, this paper focuses on upgrading the strategic bidding process for retailers in wholesale power markets presented in the previous work, considering its adaptation to CECs. It considers a new forecast methodology for the day-ahead market based on forced forecast and adapted the forecast methodology considered for the spot intraday market based on the short-run energy trends of the community, aiming at reducing forecast errors, and, consequently, the unbalances and penalties. Specifically, the purpose of the paper is threefold:

1. To use a model of management of the local members of the community;
2. To develop a strategic bidding process that aims at satisfying the energy needs of the community members, by submitting bids to wholesale markets with the goals of reducing forecast errors, unbalances and penalties, and the total cost of energy when compared to retail tariffs;
3. To develop a case study that tests the strategic bidding process, and compares its results with non-risk retail tariffs. The case study involves a community composed of 312 Portuguese consumers, considering their real consumption data from 2012 extrapolated to 2019, and the real Iberian market of electricity (MIBEL) and Portuguese IS costs from 2019.

The work presented here refines and extends the previous work on CECs composed of consumers [11], their agent-based management [13] and model, bilateral model [18], strategic bidding of retailers [28], and risk management [29,30]. The main novelty of the presented work consists of the equipment of the agent-based model of CECs with a new strategic bidding process that enables them to participate in wholesale electricity markets. Indeed, CECs have already been recognized by European legislation, and some CECs are already active in Portugal [11,14–16]. The main limitation of CECs is that they need to bid at least 1 MW of power to participate in spot markets. Therefore, CECs need to have a relevant weight not to need market intermediates.

The remainder of the paper is structured as follows. Section 2 presents an overview of electricity markets, considering spot, balancing, and IS markets. Section 3 introduces a model for strategic bidding of CECs. Section 4 presents a case study. Finally, concluding remarks are presented in Section 5.

2. Electricity Markets

Active market players have the option to trade electricity in five different markets: spots, continuous, derivatives (forwards, futures, swaps, and options), non-organized (private bilateral contracts), and ancillary services markets. In spot markets, agents can submit bids with a minimum of 1 MW to electricity pools based on day-ahead and intraday or real-time marginal auctions [3].

In Europe, day-ahead markets close at noon (CET time zone) of the day-ahead to real-time operation between 12–37 h before real-time commitment. European markets are coupled and use EUPHEMIA, a marginal pricing common algorithm used to solve power flows between different market zones with the goal of maximizing social welfare [31]. In Europe, it is also possible to trade energy in several intraday auctions a few hours ahead of real-time operation and in the continuous intraday market. In continuous intraday markets, players can negotiate 15 min-ahead of real-time operation based on the pay-as-bid scheme [3]. In derivatives markets, agents can sign standard financial and physical contracts on the exchanges (clearing houses) or over-the-counter (OTC) through electronic trading to reduce risk by hedging against spot price volatility and consumption uncertainty [17]. For non-standard agreements, agents can privately negotiate and set the terms and conditions of the contracts on non-organized markets [9]. These markets were designed for large dispatchable players, i.e., players that can comply with a programmed dispatch and have enough power to participate in these markets, which means that players like retailers,

CECs, and variable generation without storage capacity may have real-time deviations [4,5]. Real-time imbalances of BRPs concerning their final programmed dispatch may have to be balanced during real-time operation [6]. TSOs use balancing markets to guarantee the security of power systems by doing a real-time balance of demand and supply of energy. BRPs may have to pay/receive the down/up balancing costs, which normally results in penalties concerning spot markets [7].

2.1. European Balancing Markets

A variation in the kinetic energy, $qkin_t$, caused by different instantaneous powers of the rotating generators, ΔP_t^s, and/or motors, ΔP_t^d, from their defined set-point values in period t, may lead to deviations between supply and demand and cause frequency and/or voltage oscillations, as presented in the power equilibrium equation [32]:

$$\Delta P_t^s - \Delta P_t^d = \frac{dqkin_t}{dt} \tag{1}$$

In Europe, the maximum secure frequency oscillation in relation to the reference is 0.1%, being the maximum allowed oscillation of 0.5%. Frequency oscillations higher than 0.5% can lead to outages and to the division of connected control areas. When the frequency deviations achieve 0.1%, the balancing reserves are automatically activated to mitigate the deviations that originate this oscillation [33,34].

Traditionally, in Europe exist, four different mechanisms to balance power systems [6]:

- Frequency Containment Reserve (FCR);
- automatic-activated Frequency Containment Reserve (aFRR);
- manually-activated Frequency Containment Reserve (mFRR);
- Replacement Reserve (RR).

FCR is the fastest frequency reserve, being the first to be activated to solve frequency disturbances because of incidents or imbalances between production and consumption, which result in a frequency deviation in relation to the 50 Hz European programmed value. It has to be activated in a maximum of 15 s, and the disturbances need to be controlled in a few seconds. Power systems of the continental European synchronous grid have to reserve 3000 MW of their capacity to support FCR.

aFRR has to be activated in a maximum of 30 s and can stay active until a maximum of 15 min, replacing FCR. It also reestablishes the grid frequency to the scheduled value. Considering the programmed size of aFRR (power band), the TSO defines the band needs for every period. ENTSO-E suggests the minimum size of the symmetric power band [33].

mFRR is firstly used to free up and/or support aFRR and then to continue balancing long-term disturbances for long periods. The TSO is responsible for directly activating this reserve, which allows for solving medium and long-term active-power deviations originated by generators, loads, or other grid disturbances.

In the aFRR and mFRR products, TSOs typically define schedules for blocks of 15 min. In the corresponding markets, an auction for every hour of the day (or blocks of various hours) is carried out, and the technically capable generators are allowed to make bids. The auction criterion aims to determine the lowest capacity price (aFRR capacity market) and the lowest energy price (aFRR and mFRR energy markets), based on marginal pricing, pay-as-bid, or other pricing methods.

RRs are activated to solve long-term incidents that cannot be solved with the previous mechanisms. They are normally traded considering bilateral agreements between TSOs and providers. They can be activated in 15 min and can continue active for hours. This mechanism is activated considering the schedules of the programming dispatch agreed upon between TSOs and providers. While the other mechanisms can be directly activated and controlled by TSOs, in this mechanism, TSOs rely on providers to comply with the programmed dispatch.

Balancing reserves are directly traded between TSOs and providers. Providers of upward regulation will receive the up-regulation price of the reserves. On the contrary, providers of downward regulation will pay the down-regulation price. The costs or revenues of balancing markets are passed to BRPs that have deviations or need to be balanced according to the imbalance settlement mechanism. Normally, the prices of upward and downward regulation are higher and lower than spot prices, respectively, which originate the payment of penalties. Otherwise, BRPs that deviate from their schedules can be compensated or do not pay penalties.

In Europe, IS mechanisms strongly differ between countries. The following mechanisms are the most used [7]:

1. Only BRPs that deviate in the dominant balance direction may pay penalties;
2. Only BRPs who need to be balanced in the dominant direction may pay penalties;
3. All BRPs may pay penalties;
4. BRPs directly and equally pay/receive the balancing costs/revenues.

These mechanisms consider that BRPs will only pay for the balanced energy. The reserved capacity that guarantees the power system security is paid in the tariffs of end-use consumers.

The first two mechanisms are discriminatory, since only BRPs that contribute to deviations in the dominant direction may pay penalties. The second is more discriminatory because only BRPs that need to be balanced may pay penalties, but incentive BRPs to auto-regulate their set points, avoiding the payment of penalties. In these mechanisms, BRPs are not compensated, independently of the balancing prices, which may originate an economic surplus to TSOs. However, when the costs of balancing the system in the dominant direction are lower than in the non-dominant direction, TSOs may have an economic deficit. The third mechanism does not originate an economic deficit to TSOs, because all BRPs will pay penalties concerning their deviations. The fourth mechanism considers that all the balancing costs or revenues are passed to BRPs. This mechanism is fairer in the sense TSOs do not have an economic surplus or deficit. However, it does not incentive BRPs to balance themselves because they can be compensated for their imbalances.

Next, are going to be presented the details of the Portuguese balancing and IS markets.

Portuguese Balancing Markets

Portugal and Spain are members of the Iberian Market of Electricity (MIBEL). MIBEL only manages spot, derivatives, and bilateral markets. Ancillary services are independent for each country and managed by their local TSOs. However, some ancillary services can be traded between TSOs. For continuous balancing, Portugal considers the traditional European frequency reserves with the following specifications [6].

FCR is a mandatory and non-remunerated system service for all technically capable generators connected to the grid. They have to reserve 5% of their nominal power in stable conditions to support FCR. Portugal is part of the synchronous grid of continental Europe, contributing with its FCR reserved capacity to the required 3000 MW of positive and negative FCR ready to be activated in continental Europe.

The Portuguese TSO requires an asymmetrical aFRR power band where its up capacity doubles the down capacity. Historically, in Portugal, the aFRR power band is more used for up-regulation than down-regulation. Thus, concerning ENTSO-E suggestions, the Portuguese TSO upscales the up capacity of the aFRR until 60% and downscales its down capacity until 40%. In Portugal, the TSO allows the participation of all technically capable generators in hourly auctions of aFRR capacity. They are remunerated based on the marginal prices of the hourly auction. Generators have to be capable of providing both down-regulation and up-regulation, bidding an up capacity that has to double the down capacity. Due to the lack of competition, in Portugal are the combined cycle gas turbines that participate in aFRR markets, being the price of the energy they provide in aFRR defined by the regulator.

The energy of mFRR is obtained considering an hourly auction-based separate procurement of both upward and downward regulation on marginal markets. The problem with mFRR is that it is based on hourly auctions, so RRs shall be used for balancing long-term frequency deviations. RRs can be activated in 15 min and continue active for long time periods, based on bilateral contracts negotiated between TSOs and the participants.

2.2. Imbalance Settlement Mechanisms

The Portuguese mechanism considers that BRPs have to pay/receive the costs/revenues of all the energy used to balance the system [7]. Therefore, the TSO does not have an economic surplus or deficit concerning the energy used to balance the system. So, the TSO computes a single penalty, p_t^{pen}, and dual pricing, for period, t, considering the following formulations:

$$p_t^{pen} = \frac{\sum_{o=1}^{O}(p_{0,t} - p_{o,t}) \times q_{o,t}}{q_t^{dev}} \tag{2}$$

$$p_t^{up} = p_{0,t} + p_t^{pen} \tag{3}$$

$$p_t^{down} = -(p_{0,t} - p_t^{pen}) \tag{4}$$

where:

(i) $p_{0,t}$ is the spot price of the programmed energy;
(ii) $p_{o,t}$ is the price of the balancing mechanisms o, considering all balancing mechanisms O;
(iii) $q_{o,t}$ is the quantity of energy used by mechanism o to balance the system;
(iv) q_t^{dev} is all BRPs deviated quantity of energy;
(v) p_t^{up} is the upward imbalance price that all BRPs shall receive (if positive);
(vi) p_t^{down} is the downward imbalance price that all BRPs shall pay (if negative).

BRPs with upward deviations receive the sum of the spot price and the penalty. BRPs with downward deviations pay the subtraction of the penalty to the spot price. In the case of a positive penalty, BRPs are compensated because the prices of the ancillary services are lower when compared to spot markets. Otherwise, they are penalized. In the case of positive upward or downward imbalance prices, the TSO has to pay BRPs. Otherwise, are BRPs who pay to the TSO.

The Nordic and Spanish mechanisms compute the balance direction, and only the BRPs that originate those balance needs must directly pay the price of the energy used to balance the system [7,35]. Contrary to the Portuguese mechanism, this mechanism considers double penalty and single pricing, as presented in the following formulations:

$$penalties = \begin{cases} p_t^{up,pen} = 0 & \text{if } \sum_{o=1}^{O} q_{o,t}^{up} < \sum_{o=1}^{O} q_{o,t}^{down} \\ p_t^{up,pen} = min\left[\frac{\sum_{o=1}^{O}(p_{o,t}^{down} - p_{0,t}) \times q_{o,t}^{down}}{\sum_{o=1}^{O} q_{o,t}^{down}}, 0\right] & \text{if } \sum_{o=1}^{O} q_{o,t}^{up} \le \sum_{o=1}^{O} q_{o,t}^{down} \\ p_t^{down,pen} = 0 & \text{if } \sum_{o=1}^{O} q_{o,t}^{up} > q_{o,t}^{down} \\ p_t^{down,pen} = min\left[\frac{\sum_{o=1}^{O}(p_{0,t} - p_{o,t}^{up}) \times q_{o,t}^{up}}{\sum_{o=1}^{O} q_{o,t}^{up}}, 0\right] & \text{if } \sum_{o=1}^{O} q_{o,t}^{up} \ge \sum_{o=1}^{O} q_{o,t}^{down} \end{cases} \tag{5}$$

$$p_t^{up} = p_{0,t} + p_t^{up,pen} \tag{6}$$

$$p_t^{down} = -(p_{0,t} - p_t^{down,pen}) \tag{7}$$

where:

(i) $p_t^{up,pen}$ is the penalty of upward deviations;
(ii) $p_t^{down,pen}$ is the penalty of downward deviations;
(iii) $q_{o,t}^{up}$ is the quantity of energy used by mechanism o to upward balance;
(iv) $q_{o,t}^{down}$ is the quantity of energy used to downward balance;

Considering this mechanism, an upward penalty exists when the downward balancing needs are higher, penalizing BRPs with up deviation. On the contrary, are BRPs with down deviations who pay penalties when the upward balancing needs are higher. The problem with this mechanism is that only net deviations in the dominant direction are paid. It is an unfair system that highly penalizes the players that have to pay penalties. However, the Portuguese IS also does not incentive BRPs to be balanced.

The imbalance quantity, q_t^{dev}, assigned to a BRP is computed considering the difference between its final programmed dispatch, q_t^{prog}, and its real-time dispatch, q_t, in period T, as follows:

$$q_t^{dev} = q_t - q_t^{prog} = \int_{t=o}^{T} P_t - P_t^{prog} dt \tag{8}$$

where P_t and P_t^{prog} are the instantaneous powers of the final and programmed dispatch, respectively.

The next section presents the strategic bidding process of CECs able to reduce their imbalances.

3. Strategic Bidding in Wholesale Electricity Markets

Considering CECs with predefined members, as consumers or prosumers, they need to satisfy the energy needs of their members. CECs can enter into bilateral agreements to acquire energy with producers, retailers, or other sellers, and/or can submit bids to spot markets if they have the capability to trade the required minimum power. Bilateral contracts are a form of risk hedging against the volatility of spot prices, although they are subject to risk premiums. Normally, buyers of energy get worse prices in bilateral agreements. Thus, their risks are reduced to the consumption uncertainty of their portfolio and, in a smaller part, to the volatility of spot prices, since they could need to fix their energy quantities by submitting bids to spot markets, as the day-ahead market (DAM) and intraday market (IDM). The DAM is used to obtain/sell the need/excess of energy, that is expected not to be physically cleared by the members. Furthermore, each session of the IDM can be used to compensate for the expected short-run imbalances between all acquired and consumed electricity. Next, can be used the intraday continuous market 15-min ahead of a real-time operation to trade some of the close to real-time expected deviations [3]. Furthermore, as BRPs, CECs are responsible for their members' deviations. Thus, if they have imbalances in relation to their programmed dispatch, they could have to be penalized in balancing markets, paying/receiving the unbalanced down/up prices [6,7].

This section presents a process for strategic bidding in wholesale electricity markets, considering that CECs can also consider bilateral agreements to acquire electricity. The process uses different types of data. It uses historical data to forecast the next day's consumption in the DAM based on a forced forecast. It was selected from the database the most recent hour with an hourly consumption, h, according to the type of forecast day ($\mathcal{D}$): weekday ($\mathcal{W}$), Saturday ($\mathcal{S}$), Sunday ($\mathcal{U}$) or holiday ($\mathcal{H}$). Considering the database with the historical daily consumption data, $\mathcal{D} = \{\mathcal{W}, \mathcal{S}, \mathcal{U}, \mathcal{H}\}$, the formulation to obtain the forecast is:

$$\hat{q}_t = q_{t-h}, \forall h \in \mathcal{D} \tag{9}$$

subject to:

$$\min_{q_{t-h}} h, \{h | (\hat{q}_t \wedge q_{t-h}) \in (\mathcal{W} \vee \mathcal{S} \vee \mathcal{U} \vee \mathcal{H})\} \tag{10}$$

For every time period, CECs can have multiple contracts K, so the total quantity of electricity already guaranteed through bilateral contracts, $q_{c,t}$, is used to compute the bids to each period of the DAM, $q_{0,t}$.

$$q_{0,t} = \hat{q}_t - q_{c,t} \tag{11}$$

$$q_{c,t} = \sum_{k=1}^{K} q_{c_k,t} \tag{12}$$

For each intraday session, s, the forecast, $\hat{q}_{s,t}$, uses the most updated consumption information to forecast the consumption of the CEC and submit bids for the required electricity session. The intraday methodology has been adapted from a forecast methodology for retailers [28]. The computed quantity bid to an intraday session, $q_{s,t}$, to submit to every time period, t, of each intraday session, s, considering the short-run forecasts and all acquired electricity through bilateral contracts, $q_{c,t}$, the DAM, $q_{0,h}$ or the previous intraday session(s), $q_{i,t}$.

$$q_{s,t} = \hat{q}_{s,t} - q_{c,t} - q_{0,t} - \sum_{i=1}^{s-1} q_{i,t} \tag{13}$$

Then, the real-time imbalance, q_t^{dev}, of period t, is computed considering the difference between the real-time consumption of the CEC, q_t, and the final programmed dispatch, q_t^{prog}, respectively:

$$q_t^{dev} = q_t - q_{c,t} - q_{0,t} - \sum_{s=1}^{S} q_{s,t} = q_t - q_t^{prog} \tag{14}$$

Each time period balance responsibility of the CEC, C_t^{dev}, considering its deviations, q_t^{dev} and the prices of the excess or lack of electricity, in cases of up, P_t^{up}, or down, P_t^{down} deviations, respectively, are computed as follows:

$$\begin{cases} C_t^{dev} = q_t^{dev} P_t^{up}, & \text{for } q_t^{dev} > 0 \\ C_t^{dev} = \left| q_t^{dev} \right| P_t^{down}, & \text{for } q_t^{dev} < 0 \end{cases} \tag{15}$$

Each bilateral contract k has its own price, $p_{c_k,t}$, so, each time period cost, C_t, of the CEC is:

$$C_t = \sum_{k=1}^{K} p_{c_k,t} q_{c_k,t} + p_{0,t} q_{0,t} + \sum_{s=1}^{S} p_{s,t} q_{s,t} - C_t^{dev} \tag{16}$$

To evaluate the performance of the forecast techniques are used two different indicators, MAPE and NRMSE [28]:

$$MAPE = \frac{100\%}{T} \sum_{t=1}^{T} \left| \frac{q_t - \hat{q}_t}{q_t} \right| \tag{17}$$

$$NRMSE = 100\% \frac{\sqrt{\frac{1}{T} \sum_{t=1}^{T} (\hat{q}_t - q_t)^2}}{q_{max}} \tag{18}$$

where q_{max} is the maximum CEC's demand. The value of $\hat{q}_t$, depends on the time horizon of each market forecast, being equal to $q_{0,t} + q_{c,t}$ in the case of day-ahead forecasts and equal to q_t^{prog} in the case of intraday forecasts.

The following section presents a case study to test the strategic bidding process presented in this section when a CEC participates in the markets presented in the previous section.

4. Case Study

This section presents a case study that tests the process of strategic bidding on spot markets, considering a CEC composed of real-world consumers that want to be active market players.

The case study uses real-world data from 312 Portuguese consumers connected to the medium voltage of the transmission grid, representing around 5% of the national demand during the period from 2011 to 2013 [36]. The CEC is composed of 72 residential aggregations, 189 small commercial aggregations, 13 large commercial, 8 industrial, and 32 aggregations of diverse consumer types. They have a peak demand of 446 MW. Therefore, their consumption data from 2012 are extrapolated to 2019.

In 2019 the regulated energy tariff for medium voltage consumers was 111.93 €/MWh. From this tariff, 70.68 €/MWh is from the wholesale price of energy, 5.26 €/MWh is for retail commercialization, and the rest is for grid access and usage [16]. The last parcel includes the General Economic Interest Cost (GEIC), which results from economic incentives for renewable and thermal generation, with a value of 24.70 €/MWh. The Portuguese legislation highly incentives CECs and self-consumption. So, CECs and self-consumption have a discount of 50% in the GEIC, being the discount of CECs with self-consumption of 100%. Thus, CECs may only pay 23.64 €/MWh for grid access plus the wholesale cost of energy of their own trades, instead of the retail tariff (111.93 €/MWh). Against this background, the goal of this section is to test the strategic bidding process of CECs, considering its forecast accuracy and the market outcomes of the CEC, also considering the different IS mechanisms.

Considering the forecast accuracy, the DAM forecasts have a MAPE of 5.32% and a NRMSE of 4.6%. The IDM forecasts have a MAPE of 4.43% and a NRMSE of 3.62%. According to the literature, forecasts with a MAPE lower than 10% are considered highly accurate forecasts [25]. Comparing these results with the forecast accuracy of retailers when these consumers are part of their portfolios can be concluded that only one out of six retailers can obtain lower errors, and only in the IDM forecasts [28]. So, CECs can improve local forecast accuracy, but reducing the portfolios' diversification of retailers may decrease their national demand forecast accuracy. Thus, CECs can be relevant to balance power systems that consider local marginal pricing and balance, as in the USA and Australia. These values prove the strong accuracy of the employed forecast methodology, as can be seen in Figure 1.

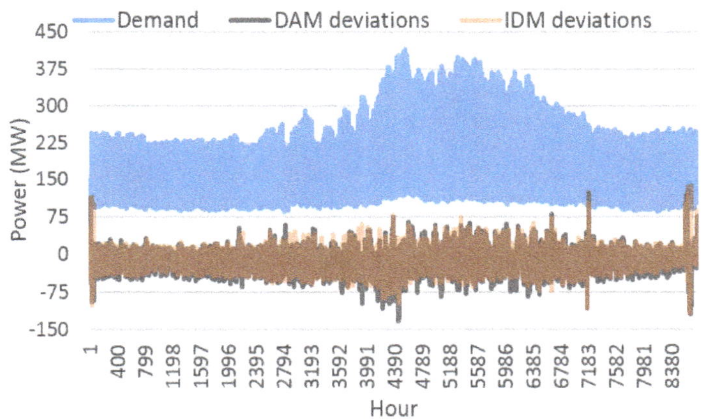

Figure 1. DAM and IDM deviations in relation to real consumption. Brown lines consider the merge between DAM and IDM deviations.

Analysing Figure 1 can be concluded that only a few hours during the year, IDM forecasts are worse than DAM forecasts. Analyzing the figure can be concluded that the CEC demand is higher during summer. This is true because in Portugal, during summer, cooling demand is satisfied by electric air conditioning, while during winter, heating demand is satisfied by natural gas, wood, and electricity. Furthermore, while cooling demand is satisfied during working hours, heating demand is satisfied during the night. Also, while the electrification of commercial buildings is advanced, residential consumers still use other sources of energy for heating demand. Moreover, the majority of the CEC participants are commercial consumers. Against this background, because of the high tourism rates and cooling demand during summer, the summer demand of the CEC is substantially higher than during other seasons. Concerning demand forecasts can be verified that during winter, deviations are higher, mainly at the beginning of January and during December, even considering lower demands when compared to summer. This may occur because of potentially uncertain cold waves that lead commercial consumers to use electrical heating against predictions. It was not detected significant differences in forecast accuracy according to the type of day (weekday, Saturday, Sunday, and holiday).

The main market outcomes of the CEC are presented in Table 1.

Table 1. Average hourly market outputs of the CEC on each market mechanism.

DAM €	IDM €	Portuguese IS €	Nordic IS €
−9579.59	94.05	−300.36	−284.45

From the results, it is possible to conclude that the DAM forecasts are overestimating the CEC consumptions, leading the CEC to sell part of its extra energy in the intraday market. Moreover, the average cost of the imbalances weighs around 3% of the total energy cost.

Table 2 presents the levelized cost of the CEC with energy on wholesale markets.

Table 2. Levelized energy costs of the wholesale market.

Levelized Cost €/MWh	Portuguese IS	Nordic IS
Total	48.89	48.81
IS	1.50	1.42

Analyzing Table 2, it is possible to conclude that consumers may reduce their costs in the energy part of the tariff from 70.68 €/MWh to values below 49.00 €/MWh by being an active market player, besides significant savings in all grid access costs for being part of a CEC. Also, the imbalance costs have a low weight when compared with the energy cost. Consumers may reduce their tariffs from 111.93 €/MWh to 72.53 €/MWh, a reduction of around 35%, by being part of a CEC and active market players. Furthermore, their cost of electrical energy may have a significant reduction in the case they invest in self-consumption.

The proposed strategic bidding already leads to high forecast accuracies and low imbalance costs. So, the CEC has no incentive to invest in storage capacity for self-control of its consumption. However, future power systems with majority penetrations of vRES may need the flexibility of demand players to guarantee the security of supply. Against this background, power systems shall design economically attractive demand response programs to incentive demand-side flexibility. However, in the case considering consumers with self-consumption (prosumers) and/or distributed generation as members of the CEC, the forecast accuracy of the methodology may decrease, which can increase the need for storage solutions or self-regulation of consumption to avoid the payment of high penalties. In the case of considering self-consumption, the CEC will not pay the GEIC costs, reducing their costs with grid access and usage from 28.90 €/MWh to 16.55 €/MWh.

The present study does not consider a change in each consumer behavior, which may be more conscious and active in the case of being part of a community. With increasing levels of distributed generation and local storage, such as solar photovoltaic and electric vehicles, the tendency is to increase the importance of the distribution grid and retire large-scale power plants of the transmission grid. To guarantee the security of supply and security standards in the energy dispatched to/from the transmission grid, local distribution system operators may rely on local consumption flexibility to avoid outages. In power systems with nearly 100% renewable generation, imbalances may be solved locally, avoiding the need for large-scale fossil fuel power plants providing reserves to balancing markets. So, CECs are important as BRPs of current and future power systems. The main problem of CECs is their lack of experience in participating in electricity markets. So, local consumers may be aggregated as a community, obtain bargaining power and then participate in the retail competition to avoid being divided throughout the portfolios of several retailers. However, retailers request substantial market premiums while negotiating long-term bilateral agreements [18]. CECs need to be more active as market players than as part of retailers' portfolios. So, the cost-benefit of being an active/passive consumer of an active/passive CEC may be considered.

In conclusion, it is economically beneficial for passive consumers to be part of an active CEC, considering savings of around 35% concerning retail tariffs, which may increase if consumers have self-consumption and flexibility.

5. Conclusions

This article has presented an overview of the European balancing and imbalance settlement markets. Furthermore, it has presented a strategic bidding process for Citizen Energy Communities (CECs) being active market players, by submitting bids on spot day-ahead (DAM) and intraday markets (IDMs).

The strategic bidding process uses two different hybrid forecast methodologies: a forced forecast for DAM bids and a short-run trend of the expected consumption behavior of the CEC members for IDM bids. The article has also presented a case study to evaluate the CECs' strategic bidding process in spot markets by using real data from the Iberian electricity market (MIBEL) in 2019 and from Portuguese consumers in 2012 but extrapolated for 2019. The model was tested by considering a CEC composed of 312 real medium voltage consumers. Results from the study confirm that large amounts of diversified aggregated demands conduct high forecast accuracies. Furthermore, it confirms that passive consumers economically benefit from being part of CECs, considering tariff incentives and lower wholesale market prices. Indeed, the study proved that consumers save 35% in electrical energy costs by being part of a CEC. Furthermore, their savings can increase if they invest in self-consumption. Moreover, the operation and outcomes of CECs can be improved in the case of having storage assets and flexible consumers, contributing to the local balance of the power system. Indeed, towards a carbon-neutral society, power systems may speed up the replacement of large-scale fossil fuel power plants by renewable distribution (small-scale) and transmission generation (large-scale) if consumers play an active role in the power system balance.

The main issues of CECs being active market players are the volatility of spot prices and the uncertain consumption of their members. They can mitigate the price risk by establishing medium to long-term bilateral agreements in wholesale markets. Furthermore, they can mitigate the quantity risk by signing demand response contracts with members and/or investing in storage solutions.

Future work is intended to study how the strategic bidding model can be adapted to prosumers and distributed generators as members of the CEC, and deal with flexibility considering demand response and storage assets. Moreover, are going to be analyzed the benefits of CECs being active market players or just part of retailers' portfolios.

Funding: This work has received funding from the EU Horizon 2020 research and innovation program under project TradeRES (grant agreement No 864276).

Institutional Review Board Statement: Not applicable.

Informed Consent Statement: Not applicable.

Data Availability Statement: The real consumption dataset of the consumers can be found in an online repository at https://archive.ics.uci.edu/ml/datasets/ElectricityLoadDiagrams20112014#. The market results of the Iberian market of electricity are available at https://www.omie.es/pt/market-results/daily/daily-market/daily-hourly-price. The market results of the Portuguese balancing markets and imbalance settlement are available at https://www.mercado.ren.pt/EN/Electr/MarketInfo/MarketResults/Pages/default.aspx. The Portuguese tariffs of electrical energy can be found at https://www.erse.pt/en/activities/market-regulation/tariffs-and-priceselectricity/. All data were accessed on 21 November 2022.

Conflicts of Interest: The author declares no conflict of interest.

Abbreviations

aFRR	automatic-activated Frequency Restoration Reserve
BRP	Balancing Responsible Party
CEC	Citizen Energy Community
CET	Central European Time
DAM	day-ahead market
FCR	Frequency Containment Reserve
GEIC	General Economic Interest Cost
IDM	intraday market
IS	Imbalance Settlement
MIBEL	Iberian market of electricity
MAPE	mean absolute percentage error
mFRR	manually-activated Frequency Restoration Reserve
NRMSE	normalized root mean square error
OTC	over-the-counter
RR	Replacement Reserve
TSO	Transmission System Operator
Indices	
k	contract number
K	number of contracts
$\mathcal{D}$	forecast day
$\mathcal{H}$	holidays set of days
h	hour
i	previous IDM session
o	balancing mechanism
O	number of balancing mechanisms
s	IDM session
S	number of IDM session
$\mathcal{S}$	Saturdays set
t	period
$\mathcal{T}$	number of periods
$\mathcal{U}$	Sundays set
$\mathcal{W}$	weekdays set
Variables	
C	energy cost
P	instantaneous power
P_t^{prog}	programmed power
$p_{0,t}$	DAM price
$p_{c_{ct},t}$	price of bilateral contract
$p_{s,t}$	IDM session price
p_t^{down}	downward imbalance price
p_t^{pen}	penalty price
p_t^{up}	upward imbalance price
q	quantity of energy
$\hat{q}$	forecasted energy
q_t^{dev}	deviated energy
$qkin$	kinetic energy

References

1. Hunt, S.; Shuttleworth, G. *Competition and Choice in Electricity*; Wiley: Chichester, UK, 1996.
2. Shahidehpour, M.; Yamin, H.; Li, Z. *Market Operations in Electric Power Systems*; Wiley: Chichester, UK, 2002.
3. Algarvio, H.; Lopes, F.; Couto, A.; Santana, J.; Estanqueiro, A. Effects of Regulating the European Internal Market on the integration of Variable Renewable Energy. *WIREs Energy Environ.* **2019**, *8*, e346. [CrossRef]
4. Strbac, G.; Papadaskalopoulos, D.; Chrysanthopoulos, N.; Estanqueiro, A.; Algarvio, H.; Lopes, F.; de Vries, L.; Morales-España, G.; Sijm, J.; Hernandez-Serna, R.; et al. Decarbonization of Electricity Systems in Europe: Market Design Challenges. *IEEE Power Energy Mag.* **2021**, *19*, 53–63. [CrossRef]

5. Algarvio, H.; Lopes, F.; Couto, A.; Estanqueiro, A.; Santana, J. Variable Renewable Energy and Market Design: New Market Products and a Real-world Study. *Energies* **2019**, *12*, 4576. [CrossRef]
6. Algarvio, H.; Lopes, F.; Couto, A.; Estanqueiro, A. Participation of wind power producers in day-ahead and balancing markets: An overview and a simulation-based study. *WIREs Energy Environ.* **2019**, *8*, e343. [CrossRef]
7. Frade, P.; Pereira, J.; Santana, J.; Catalão, J. Wind balancing costs in a power system with high wind penetration–Evidence from Portugal. *Energy Policy* **2019**, *132*, 702–713. [CrossRef]
8. Algarvio, H. Risk-Sharing Contracts and risk management of bilateral contracting in electricity markets. *Int. J. Electr. Power Energy Syst.* **2023**, *144*, 108579. [CrossRef]
9. Kirschen, D.; Strbac, G. *Fundamentals of Power System Economics*; Wiley: Chichester, UK, 2018.
10. Algarvio, H.; Lopes, F. Agent-based Retail Competition and Portfolio Optimization in Liberalized Electricity Markets: A Study Involving Real-World Consumers. *Int. J. Electr. Power Energy Syst.* **2022**, *137*, 107687. [CrossRef]
11. Algarvio, H. The Role of Local Citizen Energy Communities in the Road to Carbon-Neutral Power Systems: Outcomes from a Case Study in Portugal. *Smart Cities* **2021**, *4*, 840–863. [CrossRef]
12. Algarvio, H.; Lopes, F.; Sousa, J.; Lagarto, J. Multi-agent electricity markets: Retailer portfolio optimization using Markowitz theory. *Electr. Power Syst. Res.* **2017**, *148*, 282–294. [CrossRef]
13. Algarvio, H. Management of Local Citizen Energy Communities and Bilateral Contracting in Multi-Agent Electricity Markets. *Smart Cities* **2021**, *4*, 1437–1453. [CrossRef]
14. European Commission. Communication from the Commission to the European Parliament, the Council, the European Economic and Social Committee, the Committee of the Regions and the European Investment Bank. Clean Energy for All Europeans (COM/2016/0860 Final). Available online: https://eur-lex.europa.eu/legal-content/EN/TXT/?uri=CELEX%3A52016DC0860 (accessed on 21 November 2022).
15. European Commission. Regulation (EU) 2019/943 of the European Parliament and of the Council on the Internal Market for Electricity. 5 June 2019. Available online: http://data.europa.eu/eli/reg/2019/943/oj (accessed on 21 November 2022).
16. ERSE. Tariffs and Prices—Electricity. Available online: https://www.erse.pt/en/activities/market-regulation/tariffs-and-priceselectricity/ (accessed on 21 November 2022)
17. Algarvio, H. Multi-step optimization of the purchasing options of power retailers to feed their portfolios of consumers. *Int. J. Electr. Power Energy Syst.* **2022**, *142*, 108260. [CrossRef]
18. Algarvio, H. Agent-based model of citizen energy communities used to negotiate bilateral contracts in electricity markets. *Smart Cities* **2022**, *5*, 1039–1053. [CrossRef]
19. Astudillo, J.; De la Cruz, L. A joint multi-path and multi-channel protocol for traffic routing in smart grid neighborhood area networks. *Sensors* **2018**, *18*, 4052. [CrossRef] [PubMed]
20. Martín, P.; Moreno, G.; Rodríguez, F.; Jiménez, J.; Fernández, I. A Hybrid Approach to Short-Term Load Forecasting Aimed at Bad Data Detection in Secondary Substation Monitoring Equipment. *Sensors* **2018**, *18*, 3947. [CrossRef]
21. González-Briones, A.; Prieto, J.; De La Prieta, F.; Herrera-Viedma, E.; Corchado, J. Energy optimization using a case-based reasoning strategy. *Sensors* **2018**, *18*, 865. [CrossRef]
22. Lopes, F.; Algarvio, H. Demand Response in Electricity Markets: An Overview and a Study of the Price-Effect on the Iberian Daily Market. In *Electricity Markets with Increasing Levels of Renewable Generation: Structure, Operation, Agent-Based Simulation, and Emerging Designs, SSDC*; Lopes, F., Coelho, H., Eds.; Springer: Cham, Switzerland, 2018; Volume 144, pp. 265–303.
23. Wang, J.; Tse, N.; Poon, T.; Chan, J. A practical multi-sensor cooling demand estimation approach based on visual, indoor and outdoor information sensing. *Sensors* **2018**, *18*, 3591. [CrossRef]
24. Ayón, X.; Gruber, J.; Hayes, B.; Usaola, J.; Prodanovic, M. An optimal day-ahead load scheduling approach based on the flexibility of aggregate demands. *Appl. Energy* **2017**, *198*, 1–11. [CrossRef]
25. Wei, N.; Li, C.; Peng, X.; Zeng, F.; Lu, X. Conventional models and artificial intelligence-based models for energy consumption forecasting: A review. *J. Pet. Sci. Eng.* **2019**, *181*, 106187. [CrossRef]
26. Algarvio, H.; Couto, A.;Lopes, F.; Estanqueiro, A. Changing the day-ahead gate closure to wind power integration: A simulation-based study. *Energies* **2019**, *12*, 2765. [CrossRef]
27. Koponen, P.; Ikäheimo, J.; Koskela, J.; Brester, C.; Niska, H. Assessing and comparing short term load forecasting performance. *Energies* **2020**, *13*, 2054. [CrossRef]
28. Algarvio, H.; Lopes, F. Strategic Bidding of Retailers in Wholesale Energy Markets: A Model Using Hybrid Forecast Methods. In *Highlights in Practical Applications of Agents, Multi-Agent Systems, and Complex Systems Simulation. The PAAMS Collection. PAAMS 2022*; Communications in Computer and Information Science; Springer: Cham, Switzerland, 2022; Volume 1678.
29. Algarvio, H.; Lopes, F. Risk management and bilateral contracts in multi-agent electricity markets. In *Highlights of Practical Applications of Heterogeneous Multi-Agent Systems*; Springer: Cham, Switzerland, 2014; pp. 297–308.
30. Lopes, F.; Algarvio, H.; Santana, J. Agent-based simulation of electricity markets: Risk management and contracts for difference. In *Agent-Based Modeling of Sustainable Behaviors*; Springer: Cham, Switzerland, 2017; pp. 207–225.
31. Sleisz, A.; Sores, P.; Raisz, D. Algorithmic properties of the all-European day-ahead electricity market. In Proceedings of the 11th International Conference on the European Energy Market (EEM-14), Krakow, Poland, 28–30 May 2014; pp. 1–6.
32. Flynn, M.; Walsh, M.; O'Malley, M. Efficient use of generator resources in emerging electricity markets. *IEEE Trans. Power Syst.* **2000**, *15*, 241–249. [CrossRef]

33. ENTSO-E. ENTSO-E Network Code on Electricity Balancing. August 2014. Available online: https://www.entsoe.eu/Documents/Network%20codes%20documents/NC%20EB/140806_NCEB_Resubmission_to_ACER_v.03.PDF (accessed on 21 November 2022).
34. European Commission. Commission Regulation Establishing a Guideline on Electricity Balancing. March 2017. Available online: https://www.entsoe.eu/Documents/Network%20codes%20documents/NC%20EB/Informal_Service_Level_EBGL_16-03-2017_Final.pdf (accessed on 21 November 2022).
35. Khodadadi, A.; Herre, L.; Shinde, P.; Eriksson, R.; Söder, L.; Amelin, M. Nordic balancing markets: Overview of market rules. In Proceedings of the 17th International Conference on the European Energy Market (EEM-20), Stockholm, Sweden, 16–18 September 2020; pp. 1–6.
36. Rodrigues, F.; Trindade, A. Load forecasting through functional clustering and ensemble learning. *Knowl. Inf. Syst.* **2018**, *57*, 229–244. [CrossRef]

Disclaimer/Publisher's Note: The statements, opinions and data contained in all publications are solely those of the individual author(s) and contributor(s) and not of MDPI and/or the editor(s). MDPI and/or the editor(s) disclaim responsibility for any injury to people or property resulting from any ideas, methods, instructions or products referred to in the content.

eng

MDPI

Article

Experimental Design for the Propagation of Smoldering Fires in Corn Powder and Cornflour

Ana C. Rosa [1,2,*], Ivenio Teixeira [1], Ana M. Lacasta [3], Laia Haurie [3], Carlos A. P. Soares [4], Vivian W. Y. Tam [5] and Assed Haddad [6]

1 Programa de Engenharia Ambiental, Escola Politécnica da Universidade Federal do Rio de Janeiro, Av. Athos da Silveira Ramos, 149-Ilha do Fundão-Centro de Tecnologia-Bloco D, Rio de Janeiro 21941-909, Brazil
2 Departament d'Enginyeria Mecànica, Universitat Rovira i Virgili, Av. Paisos Catalans 26, 43007 Tarragona, Spain
3 Barcelona School of Building Construction, Universitat Politècnica de Catalunya, Av. Doctor Marañon 44, 08028 Barcelona, Spain
4 Pós-Graduação em Engenharia Civil, Universidade Federal Fluminense, Niterói 24210-240, Brazil
5 School of Engineering, Design and Built Environment, Western Sydney University, Locked Bag 1797, Penrith, NSW 2751, Australia
6 Departamento de Construção Civil, Escola Politécnica da Universidade Federal do Rio de Janeiro, Athos da Silveira Ramos, 149-Ilha do Fundão-Centro de Tecnologia-Bloco D, Sala 207, Rio de Janeiro 21941-909, Brazil
* Correspondence: carolinarosa@poli.ufrj.br

Abstract: Corn is an example of an agricultural grain with a specific combustibility level and can promote smoldering fires during storage. This paper conducts an experimental design to numerically evaluate how three parameters, namely particle size, moisture, and air ventilation, influence the smoldering velocity. The work methodology is based on Minitab's experimental design, which defined the number of experiments. First, a pile of corn is heated by a hot plate and a set of thermocouples registers all temperature variations. Then, a full-factorial experiment is implemented in Minitab to analyze the smoldering, which provides a mathematical equation to represent the smoldering velocity. The results indicate that particle size is the most influential factor in the reaction, with 35% and 45% variation between the dried and wet samples. Moreover, comparing the influence of moisture between corn flour and corn powder samples, a variation of 19% and 31% is observed; additionally, analyzing the ventilation as the only variant, we noticed variations of 15% and 17% for dried and wet corn flour, and 27% and 10% for dried and wet corn powder. Future studies may use the experimental design of this work to standardize the evaluation methodology and more effectively evaluate the relevant influencing factors.

Keywords: experimental design; corn; experiments; Minitab; smoldering velocity

Citation: Rosa, A.C.; Teixeira, I.; Lacasta, A.M.; Haurie, L.; Soares, C.A.P.; Tam, V.W.Y.; Haddad, A. Experimental Design for the Propagation of Smoldering Fires in Corn Powder and Cornflour. *Eng* **2023**, *4*, 15–30. https://doi.org/10.3390/eng4010002

Academic Editor: Antonio Gil Bravo

Received: 17 October 2022
Revised: 8 December 2022
Accepted: 19 December 2022
Published: 24 December 2022

Copyright: © 2022 by the authors. Licensee MDPI, Basel, Switzerland. This article is an open access article distributed under the terms and conditions of the Creative Commons Attribution (CC BY) license (https://creativecommons.org/licenses/by/4.0/).

1. Introduction

Smoldering is a term used to define the process of flameless burning within the material pores, with slow and low-temperature reactions [1], which is quite common in the storage of agricultural materials [2]. It can be defined as a process composed of two steps: pyrolysis and oxidation. The heat released by the oxidation step feeds the pyrolysis step, and if the pile height of the stored material is large enough not to dissipate heat and keep it stored, the reaction may be sustained for days or weeks. According to Ohlemiller [3], smoldering constitutes a severe fire hazard for two reasons: smoldering yields a higher fuel conversion to toxic compounds and allows a pathway to flaming combustion. A combustible powder or dust can react with oxygen and propagate the reaction without flaming, with velocities of mm/hour or cm/hour, evolving to glowing, flaming, or even explosive combustion [4]. Therefore, comprehensive knowledge of smoldering is essential to prevent facility accidents.

Some materials are susceptible to smoldering hazards triggered by self-heating or by an external source during storage [5]. Some grains, such as corn, have characteristics that may develop a smoldering reaction due to their level of combustibility. The physical and chemical features of the material and external conditions directly influence smoldering development [6]. The most evaluated characteristics of these materials are particle size and moisture; their variations may increase or decrease the propagation speed rate [2,7].

Although smoldering combustion is a common hazard in agricultural storage, there are few studies regarding the smoldering propagation phenomena and a lack of studies in agricultural materials, especially on corn grains. Most studies focus on other materials with a higher level of combustibility, such as sawdust, biomass, foam, and coal. This paper aims to fill this gap by developing an upward smoldering reaction in a corn pile. The novelty of this work lies in proposing a framework to develop a design of experiments (DoE) to understand the variables' impact and characterize the smoldering velocity inside the pile. Since the reaction propagation depends on material characteristics and external conditions, this work evaluates the influence of three factors in a full factorial experiment—particle size, moisture content, and air ventilation. The present study investigates how granularity, moisture, and ventilation affect smoldering fire propagation. In order to make the problem reliable and reproducible, DoE was designed and developed via Minitab software, which can offer a statistical representation for the extrapolation of the experiments and provide a mathematical equation with the three dependent variables. Following that, it allows the determination of smoldering velocity for different conditions.

2. Theoretical Foundation

Faced with the constant risk in industrial facilities, researchers have perceived the need to assess the hazard level of particulate materials igniting and developing smoldering combustion or flaming combustion. Furthermore, if accumulated on surfaces with high temperatures, even small layers of particulate material, such as powder or dust, may undergo thermal ignition, further leading to smoldering [8]. Over the years, this subject has been approached theoretically as well as experimentally. Palmer [9] was one of the pioneers and experimentally investigated the evolution of smoldering propagation within dust deposits up to 85 cm deep of cork dust, wood dust, and grass dust and measured the smoldering velocities and the temperatures along with the sample height. Another important study was done by Leish et al. [10], who investigated the spread of smoldering on grain, grain dust (corn and soybean), and wood dust under the influence of forced horizontal airflow.

As depicted by Ogle [11], smoldering requires four necessary conditions: a porous fuel that forms a solid char, an oxidizer, an ignition source with minimum ignition energy, and a minimum thickness for the deposit focusing on the thermal insulation to store the energy released by oxidation, which will increase the temperature within the fuel mass. Therefore, any changes in these four conditions or the external conditions will affect the smoldering propagation rate [12]. For this reason, many authors aimed to study the variations in the characteristics of both stored materials and external conditions. The essential features studied are particle size, moisture content, sample height, and oxygen supply [13–15]; their variations may increase or decrease the propagation speed. For example, El-Sayed and Abdel-Latif [4] investigated the critical temperature and critical flux for igniting a layer of corn flour dust and a mixture of wheat flour and corn flour (80% wheat flour + 20% corn flour) on a hot plate. In another work, El-Sayed e Khass [13] determined the minimum hot surface temperature for dust ignition, the ignition temperature of dust itself, and the ignition times and evaluated the effects of the dust particle size and the rice sample size on ignition parameters. Chunmiao et al. [16] in their studies proved that increasing the particle size of magnesium powder from 6 to 173 μm increased the minimum ignition temperature and increasing the thickness of the dust layer decreased this temperature as well. Sesseng et al. [17] performed some experiments with wood chips to evaluate how the granularity affects smoldering fire. All the aforementioned studies and other similar works elucidated the importance of studying the particle size and the minimum temperature at which dust layers or material piles ignite and can lead to smoldering fires.

Regarding the experiments assessing the spread progress, only a limited number of studies addressed the smoldering propagation in fuel beds, which can be divided according to the location of the ignition source and the direction of the spread rate: upward and downward. Palmer's study [9] concluded that sustained upward smoldering could be obtained inside dust deposits up to 85 cm deep. The propagation time was approximately proportional to the square of the depth of the dust. Torero and Fernandez-Pello [18] conducted an experimental study of upward smoldering of polyurethane foam and evaluated the smoldering velocity and reaction temperature as a function of the fuel height. He and Behrendt [19] compared theoretically natural upward and downward smoldering of piled sawdust char powder. Their results showed that upward smoldering was more than ten times faster, and the temperature inside the fuel bed was significantly higher lower than downward. Then, they experimentally investigated the natural smoldering of wood char granules in a packed bed and concluded that downward smoldering was stable.

In contrast, upward smoldering was affected by many factors like the fuel bed height, particle size, and ambient conditions [20]. He et al. experimentally investigated the effects of fuel type, moisture content, and particle size on the natural downward smoldering of biomass powder. Hagen et al. [21] studied the ignition temperature for smoldering in cotton for several densities both experimentally and theoretically, and showed that the ignition temperature decreases with increasing density.

Throughout the years, the mechanisms that affect the spread of smoldering fire have been studied. Agricultural materials are studied in smaller proportions when compared with these other materials because they are not considered as combustible. However, this fact does not negate the possibility of smoldering during storage. Therefore, the present study aims to experimentally evaluate a particular corn pile while varying three variables used in previous works with other materials. Two material variables were considered, particle size and moisture and an external variable referring to air ventilation on the sample. For the analysis, a full factorial experiment was carried out via Minitab (v. 17). Three influencing variables were chosen to evaluate the smoldering propagation velocity, and each factor was varied to identify the most influential factor.

3. Experimental Methodology

The methodology of this work consisted of five steps: planning, preliminary, performance, graph plots, and Minitab evaluation, as exemplified in Figure 1.

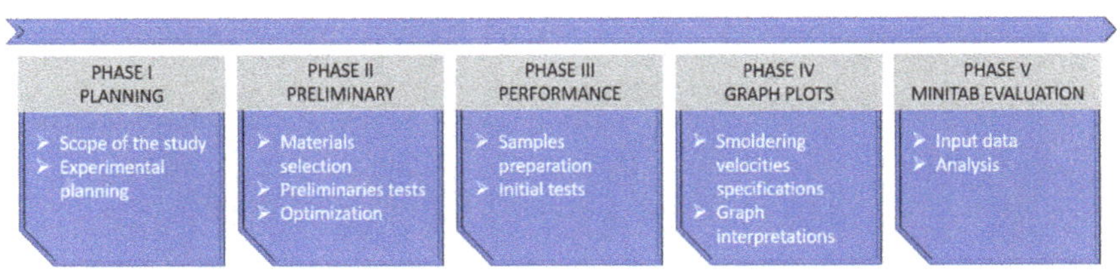

Figure 1. Work methodology.

3.1. Smoldering Schematic Setup

The experimental setup adopted in this study is based on previous works that investigated the upward smoldering process with other combustible materials on a similar experimental setup [22–24], depicted in Figure 2.

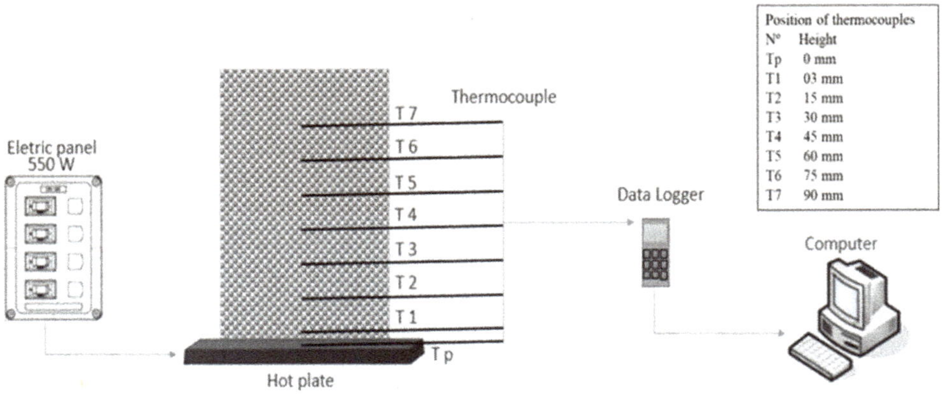

Figure 2. Schematic of the experimental setup.

The setup comprises an electric panel, a hot plate, a cylindrical reactor, eight thermocouples, and a data logger. The reactor consisted of a perforated cylindrical structure made of steel, 80 mm in diameter (D) and 120 mm in high (L), controlled by the electric panel. The panel is set to achieve a power of 550 W, and this can approximately elevate the hot plate temperature to 350 °C. Preliminary experiments that do not consider the use of the electric panel demonstrate that results could be impacted. Therefore, it ensures that there were no significant variations in hot plate temperature throughout the experiment.

During the tests, temperatures at different heights along the central axis of the reactor are measured with seven type K thermocouples, which are placed 3, 15, 30, 45, 60, 75, and 90 mm away from the hot plate. Furthermore, another thermocouple was positioned directly on the hot plate. After recording the temperatures, the smoldering velocities and the maximum temperature achieved in the burning process are determined.

3.2. Material Preparation

In order to perform the experimental planning, corn grains produced in Spain were used for these experiments, which were then crushed and sieved in different diameters. Thus, two samples were selected, and their particle diameter and particle distribution were estimated by laser diffraction (Table 1).

Table 1. Particle distribution of the corn materials.

Code	Sample	d10 (μm)	d50 (μm)	d90 (μm)
CF	Corn flour	18.62	181.6	408.7
CP	Corn powder	253.3	776.4	1199

Before the onset of the experiments, the material was separated into two environments to acquire two levels of moisture content. The first environment was a climatic chamber with a temperature of 16 °C that allowed moisture of 15%. The other one was a desiccator with a temperature of 100 °C that eliminated the moisture content of the corn samples. All samples were maintained at both environments for at least three days.

3.3. Design of Experiments

The objective of the experimental design is to perform a sequence of tests with some changes in the input variables of a process, which allows one to observe and identify corresponding changes in the output response. The core of this work was developed based on DoE defined by Minitab. As shown in Figure 3, its procedure is envisioned as a combination of influencing factors affecting a process and transforming an input material into an output product.

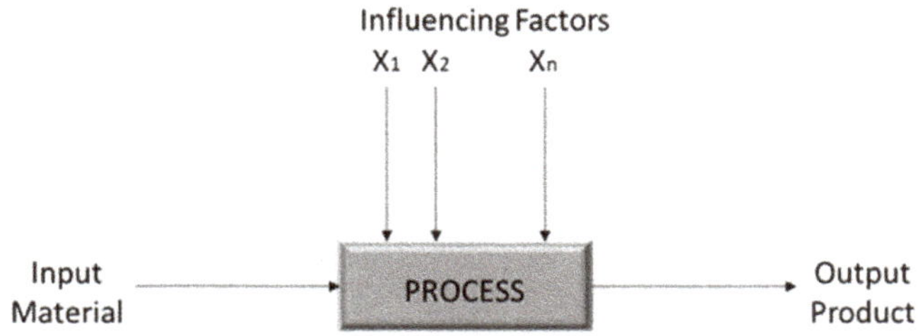

Figure 3. Schematic process of Minitab.

The sixteen experiments were performed, varying the three influencing factors chosen. For each particle diameter, two moisture content conditions were evaluated (named as dried and wet), and with and without a ventilation system. Table 2 shows the eight experiments that were conducted twice.

Table 2. Experiments performed.

Code	Sample	d50 (μm)	Humidity (%)	Ventilation (m/s)
CF-Dried	Corn flour	181.6	0	0
CF-Dried-V	Corn flour	181.6	0	0.1
CP-Dried	Corn powder	776.4	0	0
CP-Dried-V	Corn powder	776.4	0	0.1
CF-Wet	Corn flour	181.6	15	0
CF-Wet-V	Corn flour	181.6	15	0.1
CP-Wet	Corn powder	776.4	15	0
CP-Wet-V	Corn powder	776.4	15	0.1

4. Results and Discussion

4.1. Smoldering Combustion Analysis

Many specific behaviors of propagation development were noticed during the experiment, similar to those observed by other authors [19,22]. Figure 4 depicts a sequence of images taken during the test development of the corn flour sample. First, the fuel material was gently packed into the cylindrical reactor, avoiding the mass compaction located on the hot plate, and the thermocouples were connected to it (A). After the heating onset, some displacement, a contraction at the bottom and cracks on the top, indicated the drying step progress and the shrinkage of the material (B). Once the plate heating started, the heat generated by the plate began to heat the first layer by conduction, then the following layers were heated by the previous layers as the reaction front moved upwards. The occurrence of three basic steps can explain the smoldering process: pre-heating and drying of corn material, pyrolysis reaction transforming the fuel into char, and char oxidation reaction releasing more heat to the sample material. After 4 h, the electrical panel was turned off, and the heating source was interrupted. If the heat generated by the oxidation step is sufficient, the heat propagation continues, and the upward smoldering process remains active. However, the smoldering process ceases if the heat released is not higher than the heat absorbed by the heating and pyrolysis steps. As illustrated in some works [17], the appearance of a glowing mass may occur during the smoldering process, which was also observed during some experiments (C), especially experiments with dry corn flour. Additionally, the presence of smoke (D) was observed during the whole process. After the burning process, part of the material turned to ashes and char residue (E) or if the burning was incomplete, unburnt material.

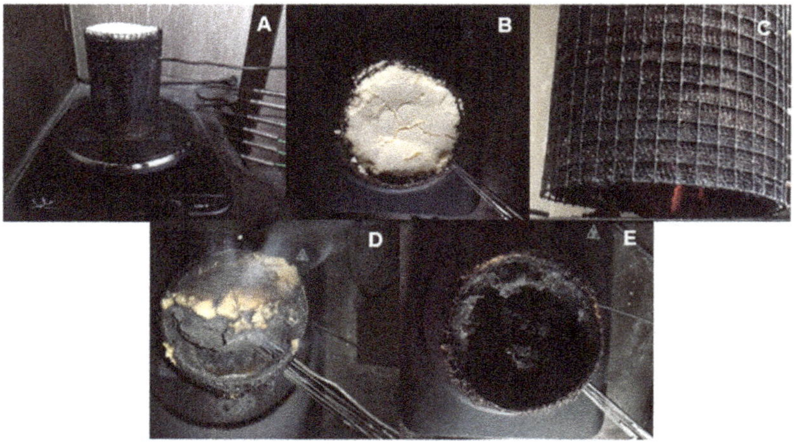

Figure 4. The sequence of images taken during the experiments.

As previously mentioned, seven thermocouples measured the evolution of the temperatures inside the corn bed. For each case, a graph was plotted indicating the temperature variation along the central axis of the reactor. As the heating, pyrolysis, and oxidation steps progressed, the heat was propagated toward the top of the sample, and temperatures at different heights of the corn bed were recorded. Subsequently, the graphs exhibited this temporal variation of the seven thermocouples within the sample and some differences and patterns could be observed.

Figure 5 displays four graphs showing the temporal evolutions registered in the tests with corn flour; each graph varied the moisture content and the ventilation condition. Figure 5A,B show the temporal evolutions of the temperature of the thermocouples obtained in an experiment with dried (CF-Dried) and wet corn flour (CF-Wet), while Figure 5C,D show the experiment with the air ventilation system for dried (CF-Dried-V) and wet corn flour (CF-Wet-V).

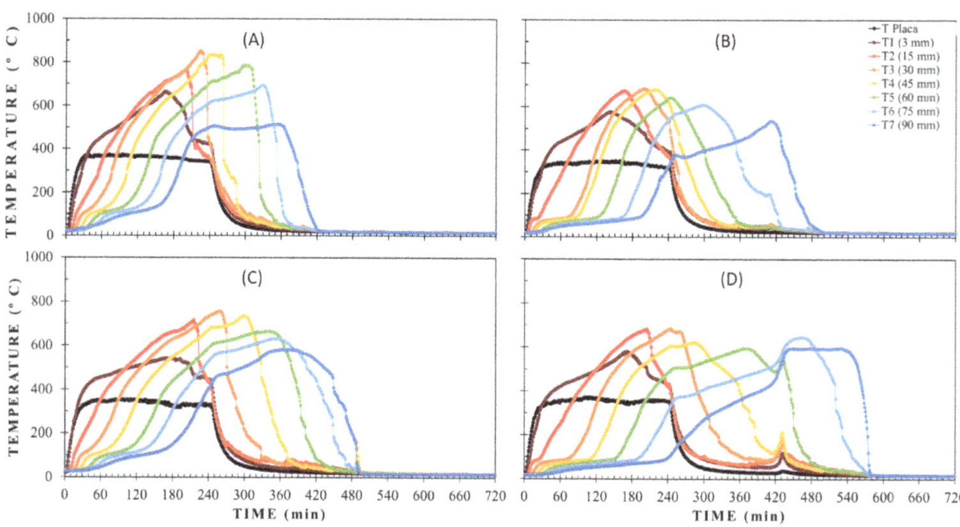

Figure 5. Temperature evolution of CF-Dried (**A**), CF-Wet (**B**), CF-Dried-V (**C**), and CF-Wet-V (**D**).

In all cases, it is possible to verify three specific zones. First, the pre-heating and drying zone is characterized by constant temperatures around 100 °C, followed by a temperature increase relative to the pyrolysis step. Finally, the sustained high temperatures due to the oxidation step ensure the smoldering propagation. Although comparing the wet and dried samples, this pre-heating zone is apparently shorter than the wet ones. This difference is more expressive in the layers of the corn flour bed closer to the top. After the pre-heating step, the propagated heat feeds the pyrolysis reaction in the lower layers. As the reaction rises toward the top, the layers underneath the top initiate the oxidation step. The temporal evolution of the temperatures registers a more significant development in the dried samples. However, the wet samples remain burning for longer.

Some researchers describe smoldering as a slow process, i.e., temperatures within a pile of particulate material take time to reach the temperature at which heat will be sufficient to self-sustain the propagation. For this reason, the corn flour layers packed in the cylindrical structure took a long time to reach the temperature to initiate the ignition and maintain the burning process. With the addition of the ventilation system to the samples with the two different moisture levels, a slight reduction in heat propagation is observed, which slows the temperature rise along the central axis of the fuel bed. At the initial pre-heating and drying phase, air ventilation influences the propagation time of the wet sample (CF-Wet-V), which extends the drying phase of the sample. Another critical point is the burning extension after switching off the heating plate. Ventilated samples increase the burning time at both moisture levels, approximately 1 h for dried corn flour (CF-Dried-V) and nearly 2 h for wet corn flour (CF-Wet-V).

At the sample's middle point (60 mm), it is observed that the temperature of the CF-Dried and CF-Wet samples at this point only reached the temperature of the hot plate at 2 h and 19 min, and 2 h and 42 min, respectively. With the addition of the ventilation, the values increase to 2 h and 38 min and 3 h and 21 min. The heat-dissipating effect is noted for this particle size and not the reaction acceleration due to more oxygen supplied. Another point that can be compared and evaluated is the point located 90 mm away from the heating plate, where the last thermocouple is placed. Regarding the influence of the ventilation system, there is a one-hour increase in the burning process for the dried corn flour sample and a one-hour and thirty-minute increase for the wet corn flour sample.

The graphs of corn flour exhibit differences in the maximum temperature achieved and the extension of the smoldering process. Although the CF-Dried (A) presents higher temperatures, CF-Wet-V (D) indicates the most extensive spread of smoldering propagation. After the hot plate shutdown, it can be noticed that smoldering propagation continued until the top of both samples, which indicates that all the fuel material is consumed.

Similar to what is presented and discussed above, Figure 6 also shows the temporal variation of temperature along the central axis of a corn powder bed, with a larger particle diameter than corn flour samples. Figure 6A,B show the temporal evolutions of the temperature of the thermocouples obtained in an experiment with dried (CP-Dried) and wet corn powder (CP-Wet), while Figure 6C,D show the experiment with the air ventilation system for dried (CP-Dried-V) and wet corn powder (CP-Wet-V).

These tests also present the pre-heating and drying, pyrolysis, and oxidation phases similar to the corn flour tests. The dried samples (CP-Dried and CP-Dried-V) show a less-defined drying phase than the wet samples (CP-Wet and CP-Wet-V). After the drying phase, unlike the corn flour samples, the temperatures of these corn powder samples slowly increase until they reach the hot plate temperature. Due to the larger particle size, the heat is not propagated at the same velocity as corn flour. In the case of dried corn powder, smoldering spread only up to 60 mm and 75 mm, while in the case of wet corn powder, the spread does not reach half of the fuel bed. This fact indicates that the upper layers of the fuel bed cannot sustain the reaction because the heat generated by the oxidation step is not sufficient to maintain the pyrolysis reaction in the upper layers. The addition of the air ventilation system does not extend the samples' burning time, which is not observed in the corn flour samples (CF-Dried-V and CF-Wet-V). However, the air ventilation affects the rise in temperatures in the CP-Dried-V case, which makes it difficult to reach the temperature of the hot plate as well, while in the CP-Wet-V case, there is very little influence of this factor. Therefore, the dissipative effect of the heat has more impact on the dried sample, being almost inexpressible in the wet sample.

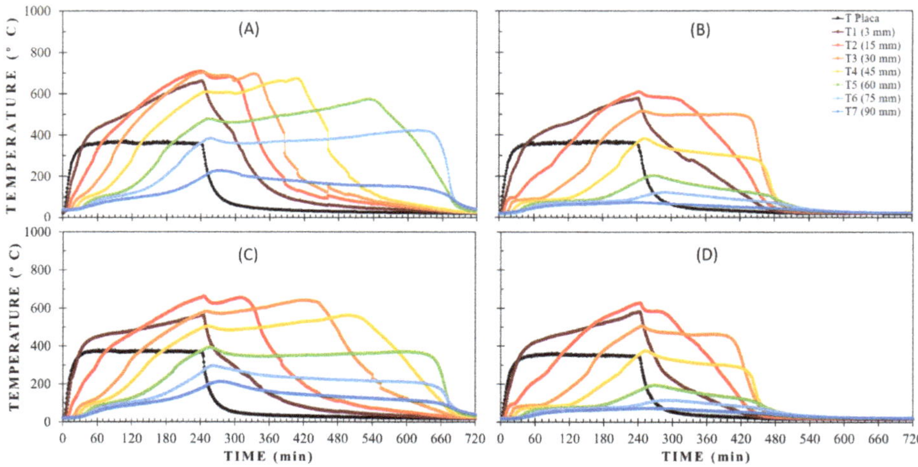

Figure 6. Temperature evolution of CP-Dried (**A**), CP-Wet (**B**), CP-Dried-V (**C**), and CP-Wet-V (**D**).

Evaluating the sample's middle-point for the corn powder graphs, the temperatures of CP-Dried and CP-Dried-V cases only reach the hot plate temperature at 3 h and 3 h and 47 min, respectively. However, the middle-point of the wet samples does not achieve the hot plate temperature, and they approximately reach 200 °C.

The graphs show that the smoldering ceases before it reaches the sample top. Comparing the two cases with the dried sample (CP-Dried and CP-Dried-V), we observe that the burning time is approximately the same, which ends almost 7 h after the hot plate shutdown. However, the height attained by smoldering propagation differs by at least 15 mm, i.e., the distance between thermocouples T5 and T6. In the cases with wet samples (CP-Wet and CP-Wet-V), the smoldering process extends to 45 mm upwards and remains burning for almost 4 h after the plate has been turned off.

These results exhibit similarities in propagation and extension of the smoldering, indicating the low influence of ventilation on this particle size. Although the CP-Dried sample (A) has the highest temperatures and longest burning time, the combustion process is incomplete, and the reaction does not entirely consume the corn powder.

The propagation rate of a smoldering process is more shallow than the rate of a flame combustion process. In the tests performed, the corn bed layers take a long time to reach the hot plate temperature, but they generate enough heat to continue feeding the process. Half of the sample takes more than 2 h to exceed the hot plate temperature, as with the CF-Dried sample, and more than 3 h with CP-Dried. In addition, the wet corn flour samples take a long time to reach high temperatures close to the hot plate's temperatures, as they remain longer in the drying phase. In contrast, wet corn powder samples do not develop smoldering in the layers located in the middle of the sample. Thus, the smoldering propagation is complete only for the corn flour samples, while the corn powder develops a partial smoldering.

The velocity at which smoldering propagates was defined according to the data gathered from the temperature evolution graphs. First, for each thermocouple positioned along the central axis of the fuel bed, the time taken to reach a temperature of 350 °C was verified. Then, with all the points of each thermocouple, a regression line was built, and the slope of the regression line quantifies the smoldering velocities. This process was repeated for each test, allowing us to achieve each condition's smoldering velocity.

Comparisons of the smoldering velocities acquired varying the three factors adopted to evaluate the process show that the smaller the particle size, the greater the smoldering velocity, as depicted in Figure 7. The corn flour propagation rates for dried and wet cases are 0.55 and 0.45 mm/min. In the meantime, the velocities of the corn powder samples achieve 0.36 and 0.25 mm/min, respectively. A comparison between the two-particle size samples with the same moisture content shows a variation of 35% for the dried samples and 45% for the wet samples. When comparing the influence of moisture, a variation of 19% for corn flour and 31% for corn powder is observed, indicating a more significant impact of this factor in corn powder samples. The air ventilation reduces the velocities of all samples. The corn flour samples, CF-Dried-V and CF-Wet-V, show velocity values of 0.47 and 0.37 mm/min, and the corn powder samples, CP-Dried-V and CP-Wet-V, registered 0.26 and 0.23 mm/min. This ventilation leads to 15% and 17% variations for dry and wet corn flour, and 27% and 10% for dry and wet corn powder. Although adding an airstream can provide more oxygen and thus promote a more intense combustion scenario, the tested cases indicate the opposite showing a slower propagation velocity.

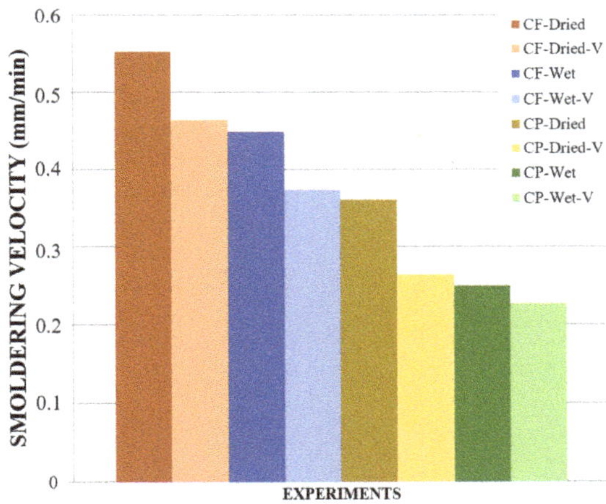

Figure 7. Smoldering velocities of the samples with different particle diameter, moisture, and ventilation system.

Rein [1] pointed out that a burning reaction usually spreads around 1 mm/min despite the variation between the chemical and physical properties of the fuel, which is utterly slower than flame propagation. The experiments performed in this work present slightly slower smoldering velocities between 0.23 mm/min and 0.55 mm/min. Although these values are lower than the value Rein pointed out, smoldering remains a hazard when storing corn materials.

As previously described in this work and by several authors, the smoldering process is known to have a slow burn propagation compared with flaming combustion. For this reason, temperatures along the central axis of the fuel bed take a long time to reach the hot plate temperature and then to sustain the reaction through heat-generating and storage. The temporal evolution of the temperatures inside the corn bed presents a behavior very similar to that shown by Hagen et al. in their experiments in cotton samples, which required an extended period for smoldering development [21]. In the first minutes of the experiments, the graphs have a drying step with little temperature change, followed by the pyrolysis and oxidation step with a rapid temperature increase sustained even after the hot plate shutdown. The temperature decay finally indicates the termination of the char oxidation.

Inside the corn fuel bed, the heat generated by the hot plate is responsible for the beginning of the drying of the sample material and then for the pyrolysis reaction in the first layers at the bottom of the sample, and finally followed by the oxidation reaction that releases more heat to the corn material. The smoldering process requires sufficient heat to turn all the material into reactive char for the oxidation step. As the reaction spreads upwards, the heat generated by the hot plate and pyrolysis step is stored, and thus the smoldering process continues to transfer heat to the upper layers. If the heat generated by the hot plate and the oxidation step of the lower layers is insufficient, the process ceases. Thus, part of the experiments conducted in this work develops total smoldering. Another part has partial smoldering because the heat stored in the material is insufficient to continue feeding the pyrolysis step.

4.2. Minitab Analysis

The tests of this work were elaborated according to DoE in Minitab to evaluate the smoldering velocity behavior in a corn bed. As previously mentioned, a full factorial experiment was selected, and three influencing factors were defined as the variables to evaluate the smoldering velocity. Each of these factors received two values. In addition, it was determined that each test would be reproduced twice. Therefore, the minimum number of tests required for this full factorial experiment was sixteen tests. Table 3 indicates the values of the selected factors, the smoldering velocity, and the smoldering level. During the tests and the graphs of the temporal evolution of the thermocouple temperatures, not all tests showed a complete smoldering level to the top of the corn bed. Only the corn flour samples developed a complete reaction. At the end of the tests, it could be visually verified that some samples were burned entirely, leaving only ash remaining. In contrast, others were partially burnt, with char and unburnt material as the residue.

Table 3. Experiment results.

Exp.	Type	Particle Diameter (μm)	Moisture	(%)	Vent. Condition	Ventilation (m/s)	Smoldering Velocity	Level of Smoldering
1	CP	776.4	Wet	15	Y	0.1	0.0207	Partial
2	CF	181.6	Wet	15	N	0	0.0465	Total
3	CP	776.4	Wet	15	N	0	0.0226	Partial
4	CP	776.4	Dry	0	Y	0.1	0.0292	Partial
5	CF	181.6	Wet	15	Y	0.1	0.0387	Total
6	CF	181.6	Wet	15	Y	0.1	0.0389	Total
7	CF	181.6	Wet	15	N	0	0.0451	Total
8	CP	776.4	Dry	0	Y	0.1	0.0291	Partial
9	CF	181.6	Dry	0	Y	0.1	0.0484	Total
10	CF	181.6	Dry	0	N	0	0.0548	Total
11	CP	776.4	Dry	0	N	0	0.0355	Partial
12	CP	776.4	Wet	15	N	0	0.0231	Partial
13	CP	776.4	Dry	0	N	0	0.036	Partial
14	CP	776.4	Wet	15	Y	0.1	0.021	Partial
15	CF	181.6	Dry	0	N	0	0.0554	Total
16	CF	181.6	Dry	0	Y	0.1	0.0473	Total

The tests were performed, varying the selected factors according to the values specified. After running the tests and calculating smoldering velocity in each case, all data were inserted in Minitab to continue with the DOE analysis. Afterward, it was possible to produce some graphs in this software to analyze the influencing factors and smoldering velocity.

The data set inserted in Minitab assisted in formulating a mathematical equation characterizing the smoldering phenomenon in corn material with the three variables chosen. However, as the linear regression model is not always appropriate for the specified data set, it is advisable to evaluate the model's suitability by examining residual plots. One of the first graphs produced in Minitab was the residual plots, which indicate the dataset's quality to specify smoldering velocity (Figure 8).

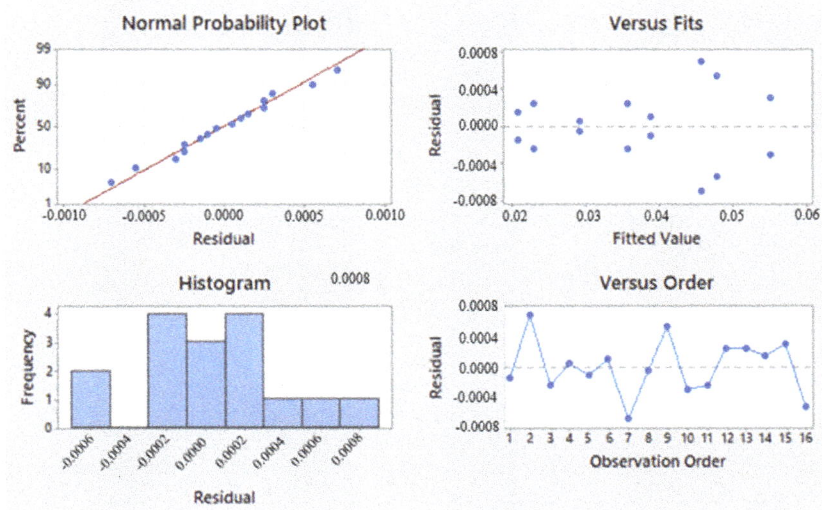

Figure 8. Residual plots for smoldering velocity.

The normal probability plot of these residuals is one of the four graphs presented in Figure 8. Most of the points in this plot fit precisely along the straight line, even though some do not fall precisely above the line, which indicates minor problems with the normality assumption. Still, no severe abnormality in the data set is suspected. Therefore, the data can be considered satisfactory for the analysis. Another graph related to residual plots is the histogram that shows a distribution similar to a Gaussian curve. This graph allows us to conclude that the number of tests reproduced is sufficient to guarantee that a normal distribution can represent the smoldering process. Finally, the last two graphs of residuals versus fitted values and residuals versus the observation order do not reveal any unusual or diagnostic pattern.

Figure 9 shows the Pareto Chart of the standardized effects that evaluate the effects of each influencing factor on smoldering velocity, which compares the relative magnitude and the statistical significance of both main and interaction effects. The standardized effect compares the t-statistic from each factor to the value corresponding to the error. This graph defines the particle diameter as factor A, the moisture content as factor B, and the air ventilation system as factor C. This graph evaluates the effects of each factor alone and the effects of more than one factor acting on the process. The factor with the most significant influence on smoldering propagation is the particle diameter, followed by the moisture content of the material. It is worth emphasizing that the impact of the material size is almost twice more significant than the influence of the moisture content. Additionally, according to the Pareto Chart, the effect of two or more factors together is not so relevant. This fact can also be verified in the set of interaction plots for the smoldering velocity (Figure 10). No factor presents an interaction between them.

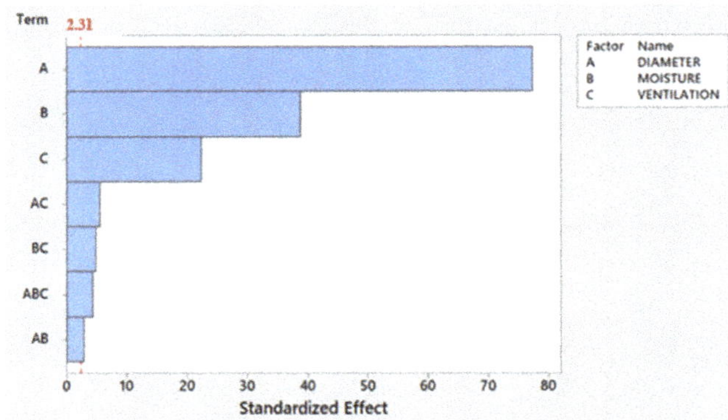

Figure 9. Pareto chart of the standardized effects.

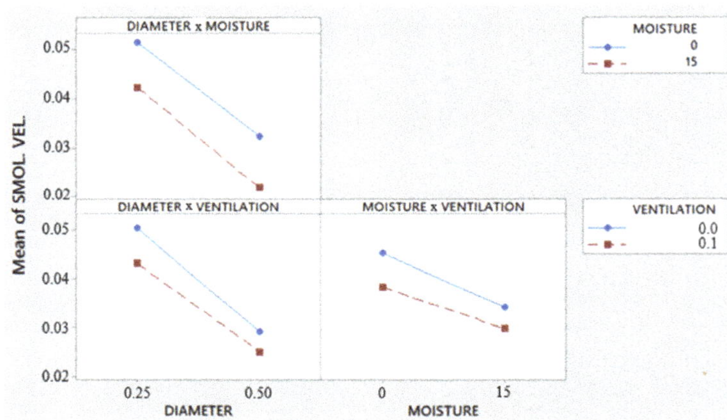

Figure 10. Interaction plot for smoldering velocity.

Figure 10 presents a set of three graphs with interactions between two factors (particle diameter versus moisture content, particle diameter versus air ventilation, and moisture content versus air ventilation). There is an absence of interactions between the factors that can be observed in the three plots. As shown in the temporal evolution of temperatures graphs, this factor interaction graph indicates that small particle diameters, low moisture content, and low air ventilation provided higher smoldering propagation velocities. Additionally, it can be observed that when the particle diameter or the moisture of the material achieves higher values, the two levels of ventilation system values tend to come closer and have closer smoldering velocities.

Figure 11 depicts the contour plots showing the relationship between two factors. Similar to the interaction plot, this graph also analyzes the interactions of two factors in the desired response, which is the smoldering velocity. Moreover, this set of plots presents a color degree, indicating ranges with different velocity values, which allows the evaluation of the extent of each range and even the specification of other cases between the limits of each level of the influencing factors. For example, the moisture content versus particle diameter plot shows more bands than the other two contour plots. It can be explained by the fact that these two factors are the most expressive in velocity.

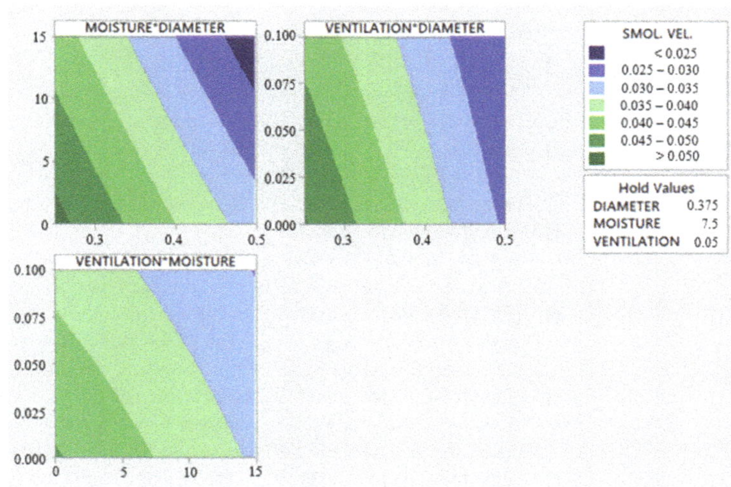

Figure 11. Contour plots of smoldering velocity.

These contour plots can analyze how variations between the factors' levels influence the smoldering and predict the best and worst cases where the smoldering propagation presents smaller values. The best case is in the dark blue band with greater diameter particles, higher moisture content, and higher air ventilation levels that exhibit bands of areas with lower velocities. Additionally, it is also possible to find out other values of the factors that fit in the same band. In the three graphs, we can see that slight variations in the axis regarding particle diameter lead to significant velocity changes and thus change the velocity bands. However, when evaluating the axes of the other two factors (moisture and air ventilation), it was observed that only a few velocity bands covered almost all values between the upper and lower levels of these two factors.

The moisture versus diameter plot has lower and higher velocity ranges, less than 0.25 mm/min and higher than 0.5 mm/min. In the meantime, the ventilation versus humidity plot indicates a smaller number of bands and a larger length than the other graphs. As it is possible to notice in the first contour plot, we can obtain the same velocity inferior to 0.25 mm/min if we have a level of moisture superior to 10% and a particle diameter slightly smaller than 0.5 mm. The following bands are broader and include moistures varying from 0% until 15% and a diameter greater than 0.300 mm. In the second and third contour plots, fewer bands appeared, and the blue bands are more vertical, i.e., for the same value on the x-axis, we get the same velocity if we change the values of the y-axis. The ventilation versus diameter plot shows that for a given range of diameter higher than 0.3 mm, ventilation may vary within the specified limits between 0 and 0.1 m/s, and the smoldering velocity will achieve the same value. The ventilation versus moisture plot indicates that with moisture higher than 7%, air ventilation can also change between the upper and lower limit levels. However, only the first allows the best factor combination to find the lowest smoldering velocity.

In addition to the plots, running the experimental design in Minitab provides a linear regression equation (Equation (1)) based on the data plugged into the software. The equation represents the contributions of the three influenced factors (Diameter—D, Moisture—M, Air ventilation—V) along with the three two-factor interactions (DM, MV, DV) and one three-factor interaction (DMV). Thus, this linear regression equation provided by the full-factorial experiment can be used to represent the smoldering process. Therefore, considering these factors, it will be possible to obtain the value of the smoldering rate for corn material from any set of values of these parameters. That can be helpful for future studies investigating the smoldering process with this material and similar conditions.

$$V = 7.445 \times 10^{-2} - 7.740 \times 10^{-2} \times D - 0.380 \times 10^{-3} \times M - 0.790 \times 10^{-1} \times V - 0.960 \times 10^{-3} \times D \times M + 2.60 \times 10^{-2} \times D \times V - 0.273 \times 10^{-2} \times M \times V + 1.160 \times 10^{-2} \times D \times M \times V \quad (1)$$

In this work, three factors are analyzed to evaluate the development of the smoldering process. First, the smaller the particle size, the larger the surface area and the greater the oxygen attack to the oxidation step. Therefore, the smoldering reaction in corn flour samples develops faster than in corn powder samples, which perfectly agrees with other literature experiments, demonstrating that granularity significantly affects the smoldering dynamics [17].

Second, regarding the material moisture, the dried samples do not need to absorb the initial heat of the drying stage. Instead, this heat was conducted to the next step (pyrolysis stage), which promoted higher temperatures throughout the corn bed. Correspondingly, the work performed by Huang et al. [14] also concluded that smoldering spread decreases with moisture content, and above a specific threshold, the experiment exhibited an incomplete burning reaction. Furthermore, another work also corroborates that increasing the moisture content reduces the propagation reaction during the drying step [23].

Third, the air ventilation system applied in the experiments had more dissipative heat than supplying oxygen to the oxidation reaction, reducing the velocity propagation. Experiments by Urban et al. [24] used an airflow of 0.5 m/s above the fuel bed. It demonstrated the importance of this factor and how it can affect the ignition process and establish a smoldering process.

The plots obtained by the experimental design indicate the influence and interactions of each adopted factor. Furthermore, it is possible to evaluate all the variations and impacts of the factors on the smoldering velocity. For example, the particle size is more influenceable on smoldering velocity than the other two factors. This work focused on applying an experimental design to investigate the behavior of smoldering combustion in corn grain. As a limitation and a suggestion for future works, one can consider different conditions and other factors to evaluate the smoldering velocity, which can be compared with the main outcomes of this paper.

5. Conclusions

This work proposed a new methodological framework evaluating the effects of the influencing factors on smoldering propagation—particle diameter, moisture content, and air ventilation, which was performed using the Design of Experiments in Minitab software. The experiments showed that upward propagation succeeded vertically from the base sample until insufficient heat exchange tried to sustain the process. The results plotted on the temperature evolution graphs and the graphs generated by the full factorial experiment indicated that the factor with the most significant influence on the propagation rate was the particle diameter, which represented a variation of 35% among the dried samples and 45% among the wet samples.

Moreover, comparing the influence of moisture between the corn flour and corn powder samples, a variation of 19% and 31% was observed, which indicated a more significant influence of this factor in corn powder samples. Regarding the ventilation influence as the only variant, 15% and 17% variations were noticed for dried and wet corn flour, and 27% and 10% for dried and wet corn powder. Additionally, the proposed framework considering the experimental planning developed a linear regression equation to represent the smoldering process in corn grain particles with the three influencing factors chosen, which can be used to extrapolate the results and obtain the smoldering velocities for other cases. Future work may use the proposed methodology to study other material properties that affect the propagation rate.

Author Contributions: Conceptualization, formal analysis, investigation, data curation, validation, writing—original draft, writing—review & editing: A.C.R., Writing—review & editing: I.T.; Writing—review & editing: A.M.L.; Writing—review & editing: L.H.; Writing—review & editing: C.A.P.S., Writing—review & editing: V.W.Y.T.; Formal analysis, investigation, data curation, writing—original draft, Writing—review & editing, validation, funding acquisition, project administration, supervision: A.H. All authors have read and agreed to the published version of the manuscript.

Funding: This research received financial support from CNPq (Brazilian National Council for Scientific and Technological Development) and CNE FAPERJ 2019-E-26/202.568/2019 (245653) Fundação de Amparo à Pesquisa do Estado do Rio de Janeiro.

Institutional Review Board Statement: Not applicable.

Informed Consent Statement: Not applicable.

Data Availability Statement: Some or all data used are proprietary or is confidential in nature and thus may only be provided with restrictions, stated clearly in the article.

Acknowledgments: The authors want to acknowledge the financial support from CNPq and CNE FAPERJ.

Conflicts of Interest: The authors declare no conflict of interest. The funders had no role in the design of the study; in the collection, analyses, or interpretation of data; in the writing of the manuscript, or in the decision to publish the results.

References

1. Rein, G. *SFPE Handbook of Fire Protection Engineering*, 5th ed.; Springer: Berlin/Heidelberg, Germany, 2016.
2. Wang, J.; Xing, W.; Huang, X.; Jin, X.; Yu, H.; Wang, J.; Song, L.; Zeng, W.; Hu, Y. Smoldering of Storage Rice: Effect of Moldy Degree and Moisture Content. *Combust. Sci. Technol.* **2022**, *194*, 1395–1407. [CrossRef]
3. Ohlemiller, T. *Smoldering Combustion*; Springer: New York, NY, USA, 1986; Volume 38.
4. El-Sayed, S.; Abdel-Latif, A. Smoldering combustion of dust layer on hot surface. *J. Loss Prev. Process Ind.* **2000**, *13*, 509–517. [CrossRef]
5. Liang, Z.; Lin, S.; Huang, X. Smoldering ignition and emission dynamics of wood under low irradiation. *Fire Mater.* **2022**, 1–11. [CrossRef]
6. Ross, A.; Blunck, D. The Influence of Particle Size in Influencing Smoldering Behavior Through Porous Woody Fuel Beds. *Combust. Sci. Technol.* **2022**, 1–23. [CrossRef]
7. Rosa, A.; Hammad, A.; Qualharini, E.; Vazquez, E.; Haddad, A. Smoldering fire propagation in corn grain: An experimental study. *Results Eng.* **2020**, *7*, 100151. [CrossRef]
8. Joshi, K. Factors Governing Spontaneous Ignition of Combustible Dusts. *Worcest. Polytech. Inst.* **2012**.
9. Palmer, K. Smouldering Combustion in Dusts and Fibrous Materials. *Combust. Flame* **1957**, *1*, 129–154. [CrossRef]
10. Leisch, S.; Kauffman, C.; Sichel, M. *Smoldering Combustion in Horizontal Dust Layers*; Elsevier: Amsterdam, The Netherlands, 1984; pp. 1601–1610.
11. Ogle, R. *Dust Explosion Dynamics*; Butterworth-Heinemann: Oxford, UK, 2016.
12. Cobian-Iñiguez, J.; Richter, F.; Carmignani, L.; Liveretou, C.; Xiong, H.; Stephens, S.; Finney, M.; Gollner, M.; Fernandez-Pello, C. Wind Effects on Smoldering Behavior of Simulated Wildland Fuels. *Combust. Sci. Technol.* **2022**, 1–18. [CrossRef]
13. El-Sayed, S.; Khass, T. Smoldering Combustion of Rice Husk Dusts on a Hot Surface. *Combust. Explos. Shock Waves* **2013**, *49*, 159–166. [CrossRef]
14. Huang, X.; Rein, G.; Chen, H. Computational Smoldering Combustion: Predicting the Roles of Moisture and Inert Contents in Peat Wildfires. *Proc. Combust. Inst.* **2015**, *35*, 2673–2681. [CrossRef]
15. Wu, D.; Schmidt, M.; Huang, X.; Verplaetsen, F. Self-Ignition and Smoldering Characteristics of Coal Dust Accumulations in O_2/N_2 and O_2/CO_2 Atmospheres. *Proc. Combust. Inst.* **2017**, *36*, 3195–3202. [CrossRef]
16. Chunmiao, Y.; Dezheng, H.; Chang, L.; Gang, L. Ignition behavior of magnesium powder layers on a plate heated at constant temperature. *J. Hazard. Mater.* **2013**, *246–247*, 283–290. [CrossRef] [PubMed]
17. Sesseng, C.; Reitan, N.; Storesund, K.; Mikalsen, R.; Hagen, B. Effect of particle granularity on smoldering fire in wood chips made from wood waste: An experimental study. *Fire Mater.* **2020**, *44*, 540–556. [CrossRef]
18. Torero, J.; Pello, A.F. Natural Convection Smolder of Polyurethane Foam, Upward Propagation. *Fire Saf. J.* **1995**, *24*, 35–52. [CrossRef]
19. He, F.; Behrendt, F. Comparison of natural upward and downward smoldering using the volume reaction method. *Energy Fuels.* **2009**, *23*, 5813–5820. [CrossRef]
20. He, F.; Behrendt, F. Experimental investigation of natural smoldering of char granules in a packed bed. *Fire Saf. J.* **2011**, *46*, 406–413. [CrossRef]
21. Hagen, B.; Frette, V.; Kleppe, G.; Arntzen, B. Onset of smoldering in cotton: Effects of density. *Fire Saf. J.* **2011**, *46*, 73–80. [CrossRef]

22. Ramírez, Á.; García-Torrent, J.; Tascón, A. Experimental determination of self-heating and self-ignition risks associated with the dusts of agricultural materials commonly stored in silos. *J. Hazard. Mater.* **2010**, *175*, 920–927. [CrossRef] [PubMed]
23. He, F.; Yi, W.; Li, Y.; Zha, J.; Luo, B. Effects of fuel properties on the natural downward smoldering of piled biomass powder Experimental investigation. *Biomass Bioenerg.* **2014**, *67*, 288–296. [CrossRef]
24. Urban, J.; Zak, C.; Song, J.; Fernandez-pello, C. Smoldering spot ignition of natural fuels by a hot metal particle. *Proc. Combust. Inst.* **2017**, *36*, 3211–3218. [CrossRef]

Disclaimer/Publisher's Note: The statements, opinions and data contained in all publications are solely those of the individual author(s) and contributor(s) and not of MDPI and/or the editor(s). MDPI and/or the editor(s) disclaim responsibility for any injury to people or property resulting from any ideas, methods, instructions or products referred to in the content.

eng MDPI

Article

Angle of the Perforation Line to Optimize Partitioning Efficiency on Toilet Papers

Joana Costa Vieira [1,*], André Costa Vieira [2], Marcelo L. Ribeiro [3], Paulo T. Fiadeiro [1] and Ana Paula Costa [1]

[1] Fiber Materials and Environmental Technologies (FibEnTech-UBI), Universidade da Beira Interior, R. Marquês D'Ávila e Bolama, 6201-001 Covilhã, Portugal
[2] Center for Mechanical and Aerospace Science and Technologies (C-MAST-UBI), Universidade da Beira Interior, R. Marquês D'Ávila e Bolama, 6201-001 Covilhã, Portugal
[3] Department of Aeronautical Engineering, University of São Paulo, Av. João Dagnone, 1100-Jardim Santa Angelina, São Carlos 13563-120, SP, Brazil
* Correspondence: joana.costa.vieira@ubi.pt

Abstract: Currently, tissue product producers try to meet consumers' requirements to retain their loyalty. In perforated products, such as toilet paper, these requirements involve the paper being portioned along the perforation line and not outside of it. Thus, it becomes necessary to enhance the behavior of the perforation line in perforated tissue papers. The current study aimed to verify if the perforation line for 0° (the solution found in commercial perforated products) is the best solution to maximize the perforation efficiency. A finite element (FE) simulation was used to validate the experimental data, where the deviations from the experiments were 5.2% for the case with a 4 mm perforation length and 8.8% for a perforation of 2 mm, and optimize the perforation efficiency using the genetic algorithm while considering two different cases. In the first case, the blank distance and the perforation line angle were varied, with the best configuration being achieved with a blank distance of 0.1 mm and an inclination angle of 0.56°. For the second case, the blank distance was fixed to 1.0 mm and the only variable to be optimized was the inclination angle of the perforation line. It was found that the best angle inclination was 0.67°. In both cases, it was verified that a slight inclination in the perforation line will favor partitioning and therefore the perforation efficiency.

Keywords: FE model; optimization; perforation efficiency; perforation line angle; tissue toilet paper

Citation: Vieira, J.C.; Vieira, A.C.; Ribeiro, M.L.; Fiadeiro, P.T.; Costa, A.P. Angle of the Perforation Line to Optimize Partitioning Efficiency on Toilet Papers. *Eng* **2023**, *4*, 80–91. https://doi.org/10.3390/eng4010005

Academic Editor: Antonio Gil Bravo

Received: 23 November 2022
Revised: 19 December 2022
Accepted: 20 December 2022
Published: 1 January 2023

Copyright: © 2023 by the authors. Licensee MDPI, Basel, Switzerland. This article is an open access article distributed under the terms and conditions of the Creative Commons Attribution (CC BY) license (https://creativecommons.org/licenses/by/4.0/).

1. Introduction

At the present time, there is a need for products that result in the use of less disposable material by environmentally conscious consumers. In the tissue paper converting industrial process, this has encouraged manufacturers to produce products with the ability to be partitioned [1].

In the production of finished tissue paper products, such as facial papers, paper towels and toilet papers, transversal perforation lines are used to facilitate the separation of the roll into individual "sheets" or services needed by the consumer. This feature of perforation allows the consumer to conveniently dispense a certain amount of the product according to their convenience [2]. Perforation takes place in the tissue paper converting machine when the sheet of paper passes through a nip between a stationary anvil and the perforator blades. These blades are usually mounted on a rotating cylinder and have alternately spaced teeth and notches. Both the anvil and the perforator are skewed in the machine direction (MD) to decrease the impact of the blade against the anvil by reducing vibration and keeping the cut line perpendicular to the MD of the tissue paper sheet. It is important that the perforator blades produce the desired cut in the finished product, so that consumer acceptance is as intended. The quality of the product cannot be affected by this operation due to poor distribution or the type of perforations. On the other hand, there has to be a balance between the number of cuts, the dimension of the cuts, the number of spacings, the dimension of the spacings and the number of plies, so that the partition of the paper roll partition by the consumer is neither easy nor too hard [3–5]. This balance is called the perforation efficiency and can be determined accordingly to the standard [6] by Equation (1):

$$E_p = 100\left[1 - \frac{\overline{S}_p}{\overline{S}_{np}}\right] \quad (1)$$

where E_p is the perforation efficiency (%), $\overline{S}_p$ is the average tensile strength of perforated papers (N/m) and $\overline{S}_{np}$ is the average tensile strength of unperforated papers (N/m).

During the tissue paper manufacturing process, raised up cellulosic fibers are found on the sheet surface, which help in consumer hygiene, but which in excess can form agglomerates, impairing the quality of the final product. To reduce the loss of cellulosic fibers on the paper surface, it is desirable that the perforation blade have relatively thin teeth [3,4]. Thus, the proper geometry of the blade must be considered. The perforator is also responsible for the visual appearance of the free edge of the remaining paper roll. The consumer wants an aesthetically pleasing free edge (smoother and less irregular between the cut and uncut areas) after tearing off the desired amount of paper [3,4].

The geometric discontinuity of the perforation line will affect the existing stress field in this area, thus affecting the stress concentration factor and consequently the final efficiency. The ratio between the highest value in a geometric discontinuity and the nominal stress in the minimum cross section is called the stress concentration factor [7]. In a previous work develop by Vieira et al. [8], they concluded that in toilet paper samples with a stress concentration factor above 0.11, a tear occurs at other locations away from the perforation line. On the other hand, toilet papers with a stress concentration factor below 0.11 tear along the perforation line. Another study carried out by Vieira et al. [9] showed that the perforation efficiency increases with an increase in the cut distance, stabilizing with a cut distance of 6 mm. The predicted differences of numerical simulations, when compared to experimental tests, decreases from 27% to 4% with a cutting distance ranging from 2 mm to 8 mm. However, the numerical simulations shown a trend in terms of the stabilization of the perforation efficiency for a cutting distance of 6 mm.

The current study aimed to verify if the perforation line at 0° is the best solution to maximize the perforation efficiency. To carry out this study, four commercial two-ply toilet papers were tested with the line of perforation at several angles. The perforation efficiency was evaluated at each angle. According to the authors' knowledge, there are limited studies on this subject.

2. Simulation–Materials and Methods

2.1. Optimization

The optimization of a constrained problem, using discrete variables, is better performed using the genetic algorithm (GA) [10] than using gradient-based methods, with the use of the GA avoiding the trap of local minima [11]. For this problem, the objective was to find the minimum force necessary to detach the toilet paper service by optimizing the angle α and the blank distance d of the paper cuts (see Figure 1), where the cut distance was maintained constant in all simulations (c = 3 mm). Additionally, a second optimization was performed regarding only the angle α by maintaining the blank distance d = 1.0 mm.

As usual, the design variables were coded as genes (coded as integer numbers) grouped into chromosomes (strings). The chromosomes were weighted as the fitness function (minimum force), representing the chromosome phenotype. Populations of possible optimal values were generated considering their probabilistic characteristics, which evolved over generations through reproductions. To avoid local minima, it is necessary to use enough search points within the design variables space [10]. The GA algorithm begins with a random population and assesses the fitness function. Reproduction is carried out by selecting the best individuals and generating the offspring. During reproduction, the genes can be exchanged by the crossovers [11].

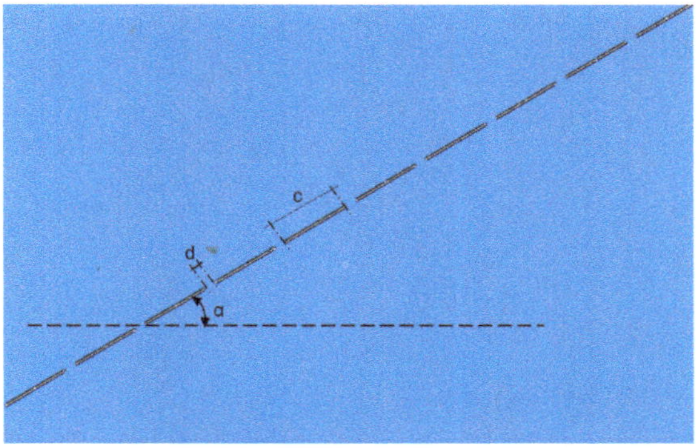

Figure 1. Design parameters.

The optimization parameters regard a population of 40 individuals (20 times the design parameters) and 150 generations (or as many generations as it takes for a convergence criterion to be reached), with 20% of mutation parameters and 50% crossover probability [12].

As mentioned before, the objective was to minimize the force to detach the toilet paper regarding specific design constrains, i.e., the angle, α, which ranged from 0° to 55°, and the blank distance, *d*, between the cuts, which ranged from 0.1 to 1.0 mm.

The GA created an angle, α, and a blank distance, *d*, population at random based on the angle range of interest. These parameters needed to be qualified according to how they may be more able than others to achieve the design objective.

When this was carried out by using the finite element (FE) model, population crossing could produce a new generation, which was again qualified by the FE model, and this process was repeated until the best generation was found, as shown by the flowchart in Figure 2. After each crossing, the algorithm made an elitism pre-definition, comparing the new generation with the previous one, and selecting the best members to compose the next generation to be crossed. For the genetic algorithm, the mutation probability is 1% and the crossover probability is 100%.

Regarding the optimization flowchart presented in Figure 2, four routines were developed separately:

i. a Python script to modify the FE model regarding the GA design parameters;
ii. a Python script to perform the FE results analyses (post-processing);
iii. a Fortran subroutine for the material model (more details in the section below);
iv. a MATLAB® script to control the FE analysis and GA.

The optimization process was controlled using the MATLAB® GA algorithm. The analysis started when MATLAB® GA generated the first generation of design parameters. Then, a Python script was called to modify the FE model regarding the design parameters. After that, the MATLAB® ran the FE analysis with the material model.

Die to the fact that explicit FE analyses can take a long time and the GA algorithm demands a considerable number of analyses, it was necessary to obtain the maximum force value and terminate the current analysis. This was performed by the MATLAB® code and a Python script that accesses the ABAQUS™ results several times until it detected a reduction of 20% in terms of the maximum force.

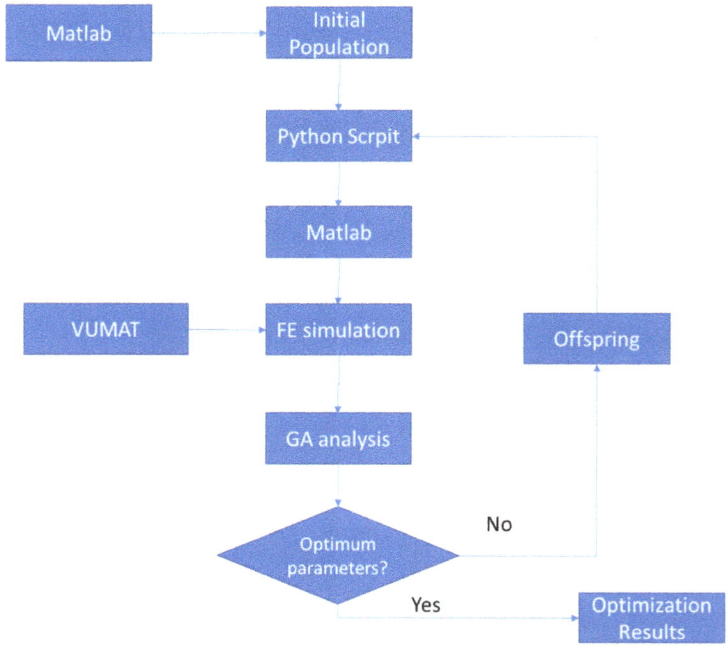

Figure 2. Analysis flowchart.

2.2. Material Model

It is not possible to adopt the isotropic behavior for tissue paper if the kind of paper has different behaviors in the machine and cross directions [8], and ABAQUS™ does not have a native constitutive law to model plasticity for orthotropic materials. Hence, a user material subroutine for explicit simulations (VUMAT) was implemented to simulate the orthotropic elastic–plastic behavior for the paper sheet. The material model, proposed by Mäkelä and Östlund [13], allows the paper anisotropic behavior to be accounted for, since the paper response is highly dependent on the fiber orientation. The model assumes the decomposition of the strain tensor into an elastic strain tensor and a plastic strain tensor (Equation (2)) while conserving the volume.

$$\varepsilon_{ij} = \varepsilon_{ij}^{e} + \varepsilon_{ij}^{p} \tag{2}$$

where ε_{ij} is the total strain, ε_{ij}^{e} is the elastic strain, and ε_{ij}^{p} is the plastic strain.

The material model adopts the concept of an isotropic plasticity equivalent material [14], a fictitious material that relates the orthotropic stress state to the isotropic stress state. Equation (3) gives the relation between the Cauchy stress tensor and the isotropic plasticity equivalent (IPE) deviatoric tensor.

$$s_{ij} = L_{ijkl}\sigma_{kl} \tag{3}$$

where s_{ij} is the deviatoric IPE stress tensor, σ_{kl} is the Cauchy stress and L_{ijkl} is the fourth order transformation tensor shown in Equation (4) for plane stress.

$$L = \begin{bmatrix} 2A & C-A-B & 0 \\ C-A-B & 2B & 0 \\ B-C-A & A-B-C & 0 \\ 0 & 0 & 3D \end{bmatrix} \tag{4}$$

where the parameters A, B, C and D are calibrated from the experimental results at 0° (MD—machine direction) and 90° (CD—cross direction) without perforation obtained in a previous work [15], using the following Equations (5)–(12) [11]:

$$A = \sqrt{1 - 12x^2} \tag{5}$$

$$B = 3(y - x) \tag{6}$$

$$C = 3(y + x) \tag{7}$$

$$D = \frac{K_{12}^{\frac{n}{(n+1)}}}{\sqrt{3}} \tag{8}$$

$$x = \sqrt{\frac{\alpha^2}{24(3\alpha^2 + \beta^2 - 4\beta + 4)}\left(\beta + 1 - \sqrt{6\beta - 3\alpha^2 - 3}\right)} \tag{9}$$

$$y = \frac{\alpha}{4x} - A \tag{10}$$

$$\alpha = K_{33}^{\frac{2n}{(n+1)}} - K_{22}^{\frac{2n}{(n+1)}} \tag{11}$$

$$\beta = K_{33}^{\frac{2n}{(n+1)}} + K_{22}^{\frac{2n}{(n+1)}} \tag{12}$$

The parameters K_{ii} and n are related to the curve fit of the tensile test applying the Ramberg–Osgood methodology. For the MD tensile test (see Equation (13)):

$$\varepsilon_{11} = \frac{\sigma_{11}}{E_{11}} + \left(\frac{\sigma_{11}}{E_0}\right)^n \tag{13}$$

For the CD (see Equation (14)):

$$\varepsilon_{kk} = \frac{\sigma_{kk}}{E_{kk}} + \left(\frac{K_{kk}E_{kk}}{E_0}\right)^n, k = 2,3 \tag{14}$$

Note that for Equation (13), the repeated indices do not mean the usual summation rule used in the indicial notation. Finally, the parameter K_{12} is obtained using Equation (15).

$$\gamma_{12} = \frac{\sigma_{12}}{G_{12}} + \left(\frac{K_{12}\sigma_{12}}{E_0}\right)^n \tag{15}$$

The Hooke's law for plane stress, small strain, linear elastic orthotropic material is given using Equation (16).

$$\sigma = C : \varepsilon^e \tag{16}$$

Where σ is the second order Cauchy stress tensor, C is the four-order plane stress, linear elastic, orthotropic constitutive law tensor and ε^e is the second order small strain elastic tensor using matrix notation.

2.3. Finite Element Model

The implementation of this model follows the well-known J_2 flow theory for isotropic materials using the backward Euler algorithm [11]. The explicit solver was used to overcome convergence issues that are common when using the implicit solver for this type of simulation. On the other hand, the stable time increment is very small, which increases the computational costs. Simulations were performed using a workstation with two intel Xeon E5-2630 8 cores (16 cores total with 32 threads) with 256 Gb ram.

The FE model dimensions, and boundary conditions are presented in Figure 3. The boundary conditions were imposed to represent a tensile test. Thus, all the displacement degrees of freedom are restricted (see Figure 3) in one side, and a prescribed displacement was applied on the reference point. A rigid link between the reference point and paper edge was used to connect the paper and the reference point.

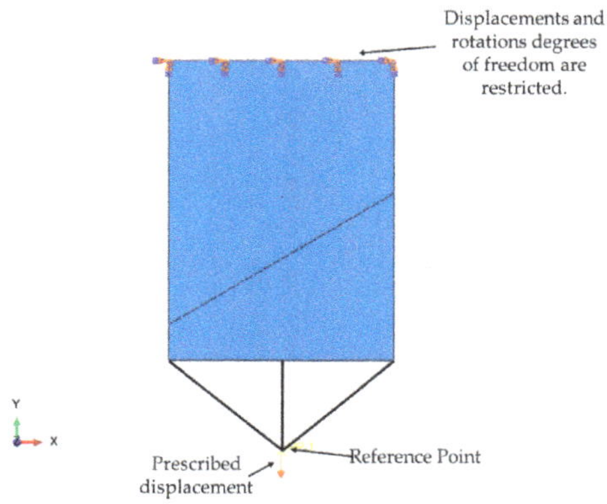

Figure 3. Finite element model and boundary conditions.

Modeling the tensile test using the reference point to apply the prescribed displacement was important for the post-processing once the number of procedures for the automatic results analysis had been reduced. This strategy does not affect the analysis results, as the resultant applied forces are the same for the case where a prescribed displacement is applied in each boundary node [8].

The paper was simulated using a four-node reduced integration membrane element (M3D4R). The model has a total of 11,086 elements and due to the cuts, a free mesh was used. It is important to mention that the mesh parameters did not change for all simulations. The material properties for the material model are: E_{11} = 13.89 MPa, E_{22} = E_{33} = 4.23 MPa, μ = 0.33 and G_{12} = 2.1 MPa. The parameters for the IPE model consider K_{22} = K_{33} since the mechanical behavior in the CD (direction 2) is similar to that in the thickness direction (direction 3). Thus, A = 1, B = 2.40, C = 2.40 and D = 1.38.

3. Experimental Tests–Materials and Methods

3.1. Materials

Four commercial two-ply toilet papers were selected. These toilet papers were identified A to D. It was previously verified that two of the two-ply papers tear off the perforation when loaded manually (toilet papers A and B). The other two papers tear on the perforation when loaded manually.

3.2. Methods

The grammage was determined accordingly with the standard ISO 12625-6:2005 [16] and defined as the mass per unit paper area (g/m^2). A Mettler Toledo PB303 Delta range analytical balance (Mettler Toledo, Columbus, OH, USA) was used to determine the paper sample weight. To determine the thickness, where a stack of sheets of paper or a sheet of paper were/was compressed at a given pressure between two parallel plates, a FRANK-PTI® Micrometer (FRANK-PTI GMBH, Birkenau, Germany) was used, in accordance to the standard ISO 12625-3:2014 [17]. According to this standard [17], the bulk, which is the inverse of density, can be determined by using the grammage and thickness previously determined.

According to the standard ISO 12625-12:2010 [5], the perforation line was evaluated. On a Thwing-Albert® VantageNX Universal Testing Machine, tensile tests were performed in the MD for all samples. For each paper, samples were prepared with the perforation in the center (0°) and with the line of perforation at different angles (20°, 30°, 37.5°, 41° and 45°). Other samples were also prepared, of each paper, with the length of a single "sheet" without perforation but with the orientation of the corresponding angle to annulate the fiber orientation contribution (see Figure 4)

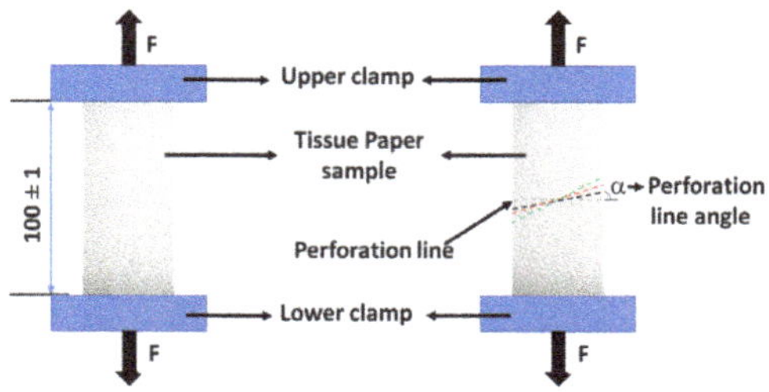

Figure 4. Experimental set-up to test non-perforated and perforated toilet papers. (F shows the force direction applied in the tensile test).

The cut and blank distances measurements were made with a paquimeter and repeated in 10 different perforations for each toilet paper sample.

4. Results and Discussion

Structural characterizations were carried out on the four commercial two-ply toilet papers samples, according to the above-referred standards. Table 1 shows the results in terms of grammage, thickness, bulk, cut and blank distances for all toilet paper samples.

Table 1. Physical characterization of the toilet papers: number of plies, grammage, thickness, bulk, cut and blank distance.

Toilet Paper ID	N° Plies	Grammage (g/m^2)		Thickness (µm)		Bulk (cm^3/g)		Cut Distance (mm)		Blank Distance (mm)	
		$\overline{x}$	σ	$\overline{x}$	σ	$\overline{x}$	σ	$\overline{x}$	σ	$\overline{x}$	σ
A	2	36.6	±0.64	374	±10.4	10.2	±0.36	1.5	±0.05	1.0	±0.05
B	2	35.4	±0.26	305	±12.4	8.6	±0.37	1.9	±0.05	1.2	±0.10
C	2	32.4	±0.42	611	±4.4	19.1	±0.41	4.0	±0.05	1.0	±0.05
D	2	44.9	±0.71	345	±8.7	7.7	±0.27	2.3	±0.05	1.0	±0.05

Looking at Table 1, the grammage shows values in the range of 32.4–44.9 g/m^2. Evaluating the outcomes for the thickness and bulk, values vary between 51% and 60%, respectively, due to the embossing process type.

Figure 5 shows the perforation efficiency behavior as function of the perforation line angle obtained for all toilet paper samples. Analyzing Figure 5, a decreasing trend in perforation efficiency can be observed with an increasing perforation line angle. Although the selected toilet papers present different characteristics, it was demonstrated that they present the same tendency in this regard. This fact is in line with what it was found by Vieira et al. [9], who stated that the perforation efficiency depends on the cut dimensions and not on the fibrous composition and/or the number of plies.

To validate the FE model, the perforation efficiency for papers B and C (Table 1), with a cut distance, *c*, of 1.9 mm and 4.0 mm, respectively, was simulated. The experimental and simulated results are compared in Figure 6. For these simulations, the FE model considered the same conditions (boundary conditions and fiber orientation) as the experiments with and without perforation.

There are some differences between the numerical and experimental results regarding the perforation efficiency (see Figure 6) that could be related to how the failure evolves in the FE model, resulting in higher failure loads (see Equation (1)). Despite these two cases, the FE model showed the same trend, and therefore optimization can be performed using this model (Figure 6). For the 4 mm perforation, the average error between the simulations and experiments was 5.2%, with the error being 8.8% for the 2 mm perforation.

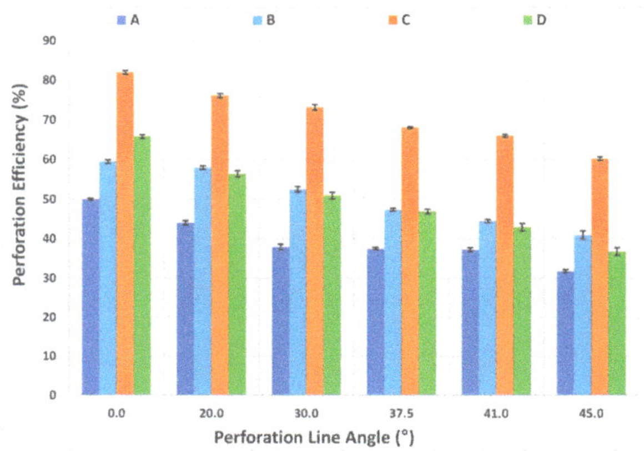

Figure 5. Perforation efficiency behavior as function of perforation line angle.

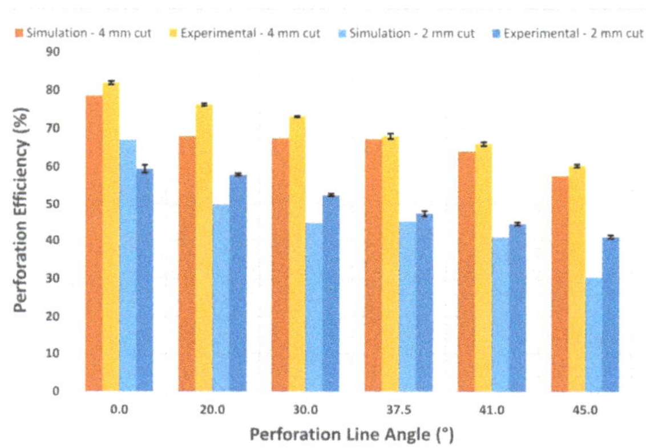

Figure 6. Experimental and theoretical perforation efficiency results as function of perforation line angle.

The first case considered the optimization of the two parameters, the blank distance, d, and the angle of the perforation line, according to Figure 1, to minimize the tear force. The parameter boundaries used in the GA were $0° \leq \alpha \leq 55°$ and $0.1 \leq d \leq 1.0$ mm. Regarding the upper boundary for the perforation line angle, α, the value of 55° was chosen to avoid the cut line cross of the upper or the lower edges of the paper model, where the displacement boundary conditions were applied.

For the case regarding the optimization of the perforation line angle and the blank distance, the optimum configuration was achieved after 51 generations, with the tear force being in the region of 0.064 N. In the configuration for the minimum tear force, the optimum angle was 0.56°, which corresponds to a perforation efficiency of 96.8% and, as expected, $d = 0.1$ mm. In comparison to a perforation efficiency of 0°, in the case of the optimal angle, an increase of 29.3% was obtained.

The GA's best value and mean value over the generations is presented in Figure 7. In this figure, the best value is almost equal through all generations and the mean value converges to the best value after 17 generations.

For the case where only the perforation line angle was the variable to be optimized (blank distance d was fixed and equal to 1 mm), the convergence occurred only after 66 generations, and the minimum tear force was 0.394 N. For this case, the angle for the minimum tear force was 0.67°, which corresponds to a perforation efficiency of 80.6%. Compared with a perforation efficiency at 0°, in the case of the optimal angle, an increase of 7.6% was obtained.

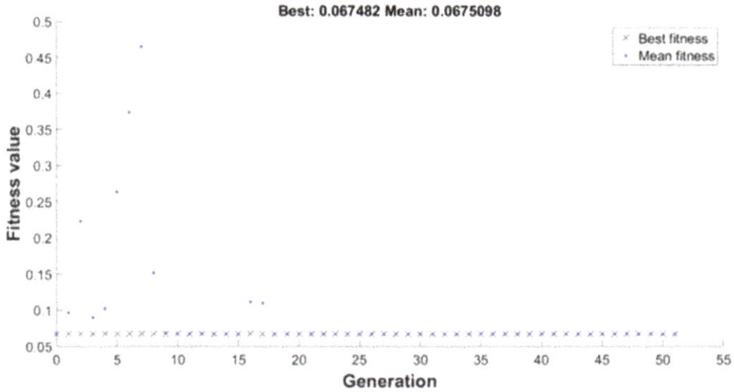

Figure 7. Optimization evolution of the best value and mean value.

As presented in Figure 8, the best value was almost constant after the 16th generation. On the other hand, the mean value did not converge. For this case, the stop criteria adopted was when the best value between generations was less than the MATLAB® default tolerance.

The stresses field for the optimum case, where the blank distance, *d*, and angle, α, were optimized, are presented in Figure 9, in the increment just before rupture.

The normal stress field in the MD (σ_{11} in *Y* direction), Figure 9a, shows a stress concentration between the cuts, as expected. As the distance between cuts are only 0.1 mm, the stress concentration is approximately K_t = 21 (regarding the stress in fiber direction, MD), justifying the low rupture force. The same behavior is detected for the other stresses (the CD (σ_{22} in *X* direction) in Figure 9b and shear stress (σ_{12}) in Figure 9c). Hence, cuttings affect the stress fields in the different directions of the paper plane. In this case, rupture begins at the center of the paper, moving fast towards the left and right edges (see Figure 9d), in the same way as it occurs experimentally in the laboratory.

Considering the other case, the optimization regarding only the inclination of the cuts, the stress fields are show in Figure 10. The stress concentration factor is approximately K_t = 4.1 for the MD stress (significantly lower as in the previous case). As in the previous case, the rupture starts at the center of the paper and moves towards the left and right edges (see Figure 10d).

Additionally, the simulations regarded the paper as a homogeneous media with no variations in fiber alignment or different concentrations throughout the model. This would not be the case in real paper, and such factors would have an influence on the paper rupture force. Figure 11a shows the MD stress field distribution (σ_{11} in *Y* direction) around the cuts with an orientation of 45°, and Figure 11b shows the same MD stress field distribution at the beginning of the paper rupture starting at the lower edge towards the center. Due to paper rupture (Figure 11b), stress flows in the non-ruptured region and hence stress is increased (darker green) in this region, while in the ruptured region stress field tends towards zero (darker blue).

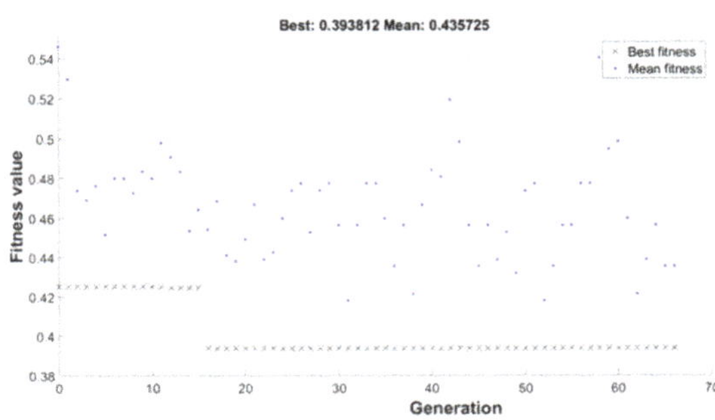

Figure 8. Optimization evolution of best and mean value for parameter *d*.

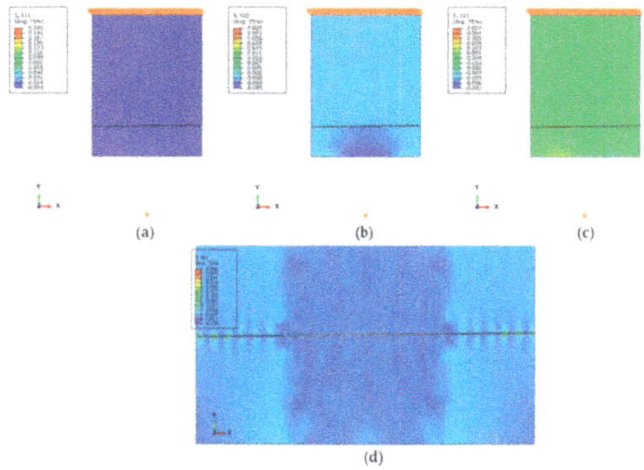

Figure 9. (a) Stress field MD; (b) stress field in CD; (c) shear stress; (d) rupture.

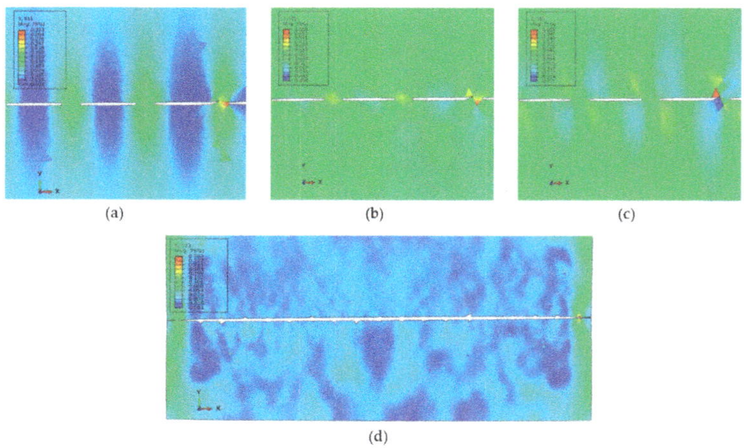

Figure 10. Stress field for the optimum orientation: (a) in fiber direction MD; (b) normal to fiber direction CD; (c) shear stress; (d) rupture for half of the model.

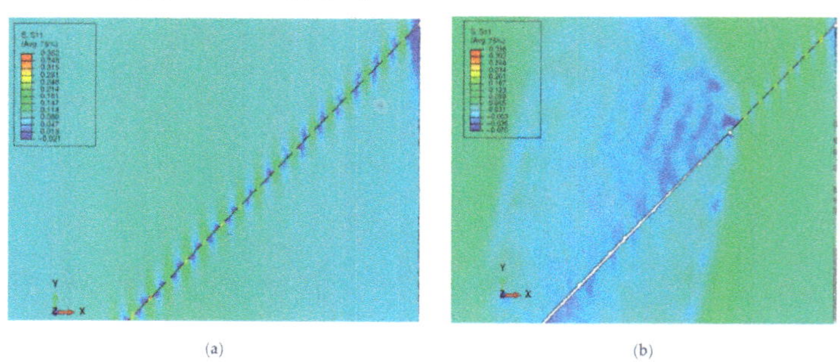

Figure 11. (a) Fiber direction stress field in MD for cuts at 45°; (b) rupture starting in the lower edge running to the center.

5. Conclusions

In general, the results of the FE model simulation analysis support the idea that the value of perforation efficiency tends to decrease with an increasing perforation line angle, in agreement with the experimental results.

A reduction in the tear force for the toilet paper was pursued using a genetic algorithm considering two different cases. In the first case, the blank distance and the angle of the cuts were the variables to be optimized and, for this case, the best configuration was achieved with a blank distance of 0.1 mm and a 0.56° inclination in terms of the perforation line, achieving an increase of 29.3% in perforation efficiency. Both the best and mean values converged for almost the same value for this case. For the case where the only variable to be optimized was the inclination of the cuts, with the blank distance fixed at 1.0 mm, the genetic algorithm found the best inclination angle to be 0.67°, achieving an increase of 7.6% in perforation efficiency, but the average values of the population did not converge. This was due to the complex failure mode of the paper and its kinematics as the damage evolved. Despite the complex failure behavior, the optimum configuration was achieved for both cases (with and without a blank distance fixed at 1.0 mm), and only a small inclination in the perforation line will reduce the tear force, regardless of the rupture progression along the perforation line.

Digital twining is an emergent simulation tool that will be commonly used in the near future because it will permit optimization in a digital environment and the subsequent transition to and application in the industrial environment, as proved with this work.

The main limitation of this work was that it considered the material to be homogeneous and orthotropic. In fact, the material used experimentally contained heterogeneously distributed fibers, preferentially oriented in the MD. But this macroscale model is accurate enough to simulate different geometries in terms of both the perforation line and the cut itself, such as waves, triangles, etc.

Author Contributions: J.C.V.: data acquisition and curation, investigation, writing—original draft, writing—review and editing. A.C.V.: FEM analysis, writing—original draft, simulation supervision, writing—review and editing. M.L.R.: FEM analysis, writing—original draft, simulation supervision, writing—review and editing. P.T.F.: supervision, writing—review and editing. A.P.C.: project supervisor, writing—review and editing. All authors have read and agreed to the published version of the manuscript.

Funding: The authors gratefully acknowledge the funding of this work that was granted under the Project InPaCTus—Innovative Products and Technologies from Eucalyptus, Project No. 21874 funded by Portugal 2020 through the European Regional Development Fund (ERDF) in the framework of COMPETE 2020 no. 246/AXIS II/2017. The authors are also very grateful for the support given by the research unit Fiber Materials and Environmental Technologies (FibEnTech-UBI), under the project reference UIDB/00195/2020, and by the Center for Mechanical and Aerospace Science and Technologies (C-MAST-UBI), under the project reference UIDB/00151/2020, both funded by the Fundação para a Ciência e a Tecnologia, IP/MCTES through national funds (PIDDAC).

Institutional Review Board Statement: Not applicable.

Informed Consent Statement: Not applicable.

Data Availability Statement: Not applicable.

Acknowledgments: The authors acknowledge the materials, access to equipment and installations, and all the general support given by The Navigator Company, RAIZ, the Optical Center, Department of Physics, Department of Textile Science and Technology, Department of Chemistry of the Universidade da Beira Interior.

Conflicts of Interest: The authors declare no conflict of interest.

References

1. Olson, S.R.; Hoadley, D.A.; Daul, T.A. Partitionable Paper Towel. U.S. Patent No. US20160345786A1, 1 December 2016.
2. Vieira, J.C.; Fiadeiro, P.T.; Costa, A.P. Converting Operations Impact on Tissue Paper Product Properties–A Review. *BioResources* **2023**, *18*, 24. [CrossRef]
3. Ogg, R.G.; Habel, M.A. Perforator Blade for Paper Products and Products Made Therefrom. U.S. Patent No. 5114771, 19 May 1992.
4. Schulz, G.; Gracyalny, D. Method and Apparatus for Pinch Perforating Multiply Web Material. U.S. Patent No. 5755654, 26 May 1998.
5. Hada, F.S.; Baggot, J.L.; Krautkramer, R.E. Method for Perforating Tissue Sheets. WO Patent No. WO 2010/076689 Al, 08 July 2010.
6. *ISO 12625-12:2010*; Tissue Paper and Tissue Products–Part 12: Determination of Tensile Strength of Perforated Lines–Calculation of Perforation Efficiency. International Organization for Standardization: Geneva, Switzerland, 2010.

7. Carvill, J. Mechanical Engineer's Data Handbook, Butterworth Heinemann, Oxford, United Kingdom. 2015. Available online: http://www.sciencedirect.com:5070/book/9780080511351/mechanical-engineers-data-handbook (accessed on 14 November 2022).
8. Vieira, J.C.; Vieira, A.C.; Mendes, A.d.O.; Carta, A.M.; Fiadeiro, P.T.; Costa, A.P. Mechanical behavior of toilet paper perforation. *BioResources* **2021**, *16*, 4846–4861. [CrossRef]
9. Vieira, J.C.; Vieira, A.C.; Mendes, A.d.O.; Carta, A.M.; Fiadeiro, P.T.; Costa, A.P. Toilet paper perforation efficiency. *BioResources* **2022**, *17*, 492–503. [CrossRef]
10. Almeida, J.H.S.; St-Pierre, L.; Wang, Z.; Ribeiro, M.L.; Tita, V.; Amico, S.C.; Castro, S.G.P. Design, Modeling, Optimization, Manufacturing and Testing of Variable-Angle Filament-Wound Cylinders. *Compos. Part B Eng.* **2021**, *225*, 109224. [CrossRef]
11. Kogiso, N.; Watson, L.T.; Gürdal, Z.; Haftka, R.T. Genetic Algorithms with Local Improvement for Composite Laminate Design. *Struct. Optim.* **1994**, *7*, 207–218. [CrossRef]
12. Goldberg, D.E. *Genetic Algorithms in Search, Optimization, and Machine Learning*; Addison-Wesley Pub. Co.: Reading, MA, USA, 1989; ISBN 978-0-201-15767-3.
13. Mäkelä, P.; Östlund, S. Orthotropic Elastic–Plastic Material Model for Paper Materials. *Int. J. Solids Struct.* **2003**, *40*, 5599–5620. [CrossRef]
14. Karafillis, A.P.; Boyce, M.C. A General Anisotropic Yield Criterion Using Bounds and a Transformation Weighting Tensor. *J. Mech. Phys. Solids* **1993**, *41*, 1859–1886. [CrossRef]
15. Vieira, J.C.; Mendes, A.d.O.; Ribeiro, M.L.; Vieira, A.C.; Carta, A.M.; Fiadeiro, P.T.; Costa, A.P. Embossing pressure effect on mechanical and softness properties of industrial base tissue papers with finite element method validation. *Materials* **2022**, *15*, 4324. [CrossRef] [PubMed]
16. *ISO 12625-6:2005*; Tissue Paper and Tissue Products—Part 6: Determination of Grammage. International Organization for Standardization: Geneva, Switzerland, 2005.
17. *ISO 12625-3:2014*; Tissue Paper and Tissue Products—Part 3: Determination of Thickness, Bulking Thickness and Apparent Bulk Density and Bulk. International Organization for Standardization: Geneva, Switzerland, 2014.

Disclaimer/Publisher's Note: The statements, opinions and data contained in all publications are solely those of the individual author(s) and contributor(s) and not of MDPI and/or the editor(s). MDPI and/or the editor(s) disclaim responsibility for any injury to people or property resulting from any ideas, methods, instructions or products referred to in the content.

eng MDPI

Article

Using ARIMA to Predict the Growth in the Subscriber Data Usage

Mike Nkongolo [1,2]

[1] Department of Informatics, Faculty of Engineering, Built Environment and Information Technology, University of Pretoria, Pretoria 0028, South Africa; u21629545@tuks.co.za

[2] Maven Systems Worx (Pty), Ltd. & NEC XON (Pty), Ltd., Centurion 0157, South Africa

Abstract: Telecommunication companies collect a deluge of subscriber data without retrieving substantial information. Exploratory analysis of this type of data will facilitate the prediction of varied information that can be geographical, demographic, financial, or any other. Prediction can therefore be an asset in the decision-making process of telecommunications companies, but only if the information retrieved follows a plan with strategic actions. The exploratory analysis of subscriber data was implemented in this research to predict subscriber usage trends based on historical time-stamped data. The predictive outcome was unknown but approximated using the data at hand. We have used 730 data points selected from the Insights Data Storage (IDS). These data points were collected from the hourly statistic traffic table and subjected to exploratory data analysis to predict the growth in subscriber data usage. The Auto-Regressive Integrated Moving Average (ARIMA) model was used to forecast. In addition, we used the normal Q-Q, correlogram, and standardized residual metrics to evaluate the model. This model showed a p-value of 0.007. This result supports our hypothesis predicting an increase in subscriber data growth. The ARIMA model predicted a growth of 3 Mbps with a maximum data usage growth of 14 Gbps. In the experimentation, ARIMA was compared to the Convolutional Neural Network (CNN) and achieved the best results with the UGRansome data. The ARIMA model performed better with execution speed by a factor of 43 for more than 80,000 rows. On average, it takes 0.0016 s for the ARIMA model to execute one row, and 0.069 s for the CNN to execute the same row, thus making the ARIMA 43$\times$ ($\frac{0.069}{0.0016}$) faster than the CNN model. These results provide a road map for predicting subscriber data usage so that telecommunication companies can be more productive in improving their Quality of Experience (QoE). This study provides a better understanding of the seasonality and stationarity involved in subscriber data usage's growth, exposing new network concerns and facilitating the development of novel predictive models.

Keywords: time series forecasting; subscriber data; seasonality; ARIMA; telecommunication; UGRansome; stationarity

Citation: Nkongolo, M. Using ARIMA to Predict the Growth in the Subscriber Data Usage. *Eng* **2023**, *4*, 92–120. https://doi.org/10.3390/eng4010006

Academic Editor: Antonio Gil Bravo

Received: 1 November 2022
Revised: 26 December 2022
Accepted: 27 December 2022
Published: 1 January 2023

Copyright: © 2023 by the author. Licensee MDPI, Basel, Switzerland. This article is an open access article distributed under the terms and conditions of the Creative Commons Attribution (CC BY) license (https://creativecommons.org/licenses/by/4.0/).

1. Introduction

The growth of competition in the telecommunications industry due to technological variety has facilitated the invention and expansion of new techniques for processing subscriber data to predict their behavior. Subscriber traffic represents all kinds of electronic data transmitted in the network [1]. This data is usually in the form of network flows passing from one node to another [2]. Furthermore, accurately predicting subscriber data can improve the Quality of Experience (QoE) to foresee and predict various anomalies, especially when the company faces revenue loss due to malicious activities. In addition, having the ability to forecast future data usage can be crucial for bandwidth sharing policy within the telecommunication business. Particularly, forecasting integrates a strong sense of seasonality towards data growth to enable management better predict potential revenue and anomalies. The above stems from a time series forecasting problem, and there is various research on different forecasting models [3–5]. Statistical models such as Auto-Regressive Integrated Moving Average (ARIMA) and Machine Learning (ML) models such as Long Short Term Memory (LSTM), gradient descent, and regression are popular techniques implemented within the time series forecasting space. In particular, the LSTM model has demonstrated great forecasting capabilities due to its ability to recall information and it is thus a strong contender against the traditional statistical ARIMA model. Execution speed is another factor to consider when selecting an appropriate model for subscriber data forecasting. This is because subscriber patterns contain petabytes made of historical time series data. It will thus be efficient to consider a model with fast execution speeds to enable faster decision-making. However, this research addresses the performance of the Convolutional Neural Network (CNN) and ARIMA models to forecast the growth of subscriber data usage. The research determines which algorithm is the most suitable in this scenario and aims to establish which of the two models performs better using speed and accuracy. In this article, we describe the advantages of using seasonality to examine changes in subscriber data. In a Time Series Analysis (TSA), seasonality is a characteristic of a time series in which the data experiences predictable and regular changes over a period [5]. Understanding seasonality in TSA can enhance the prediction performance of ML models. It can also assist in clearing the features by identifying the seasonality of time series samples and removing them from the original dataset. As a result, one can have a normalized dataset correlating input and output variables. The seasonality property can also provide more information about the seasonal component of the time series data that can provide insights to enhance predictors' performance [4]. Modeling seasonality ameliorates the data preparation and feature engineering steps. In each step, seasonal patterns can be extracted and modeled as input/output class labels with a supervised learning scheme. In adaptive computation, ARIMA is a class of time series forecasting models. Hence it is a special case of a class of regression models, not a class of classification models [6]. We have selected ARIMA as an adequate time series forecasting model to predict subscriber data usage and analyze the seasonality, trends, and cycles of features. The methodology was to use seasonality as the time series data property in the ARIMA model that implemented a distributed lag algorithm to forecast future subscriber data usage based on lagged parameters. This article implements a predictive ARIMA model using subscribers' data to study seasonality by predicting the growth in subscriber throughput.

1.1. Research Question

The main research question is as follows:

- Which forecasting model between ARIMA and CNN is effective in predicting subscriber data usage?

The research objective is to evaluate the two models using accuracy and computational speed.

1.2. Research Contribution

We propose the ARIMA model for subscriber data prediction using an unsupervised learning scheme. We have specifically implemented the ARIMA model with unlabelled features to predict the growth in subscriber data usage. In the model, the predictive layer forecasts the throughput rate fed into another layer that predicts the maximum usage growth. The remainder of the paper is structured as follows. Section 2 discusses related research works and Section 3 the research methodology. Section 4 presents the ARIMA results and the comparative analysis using the UGRansome dataset. Section 5 presents future research directions and concluding remarks.

2. Related Works

The section surveys the predictive techniques of TSA with attention to the proposed methodology.

2.1. Stationarity

Before delving deeper into the ARIMA model theory, it is important to understand the concept of stationarity. This is because, unlike the CNN, the ARIMA model does not perform well when the dataset is not stationary, and thus it needs to be made stationary before processing [7]. A dataset is said to be stationary if it conforms to all the following conditions [7]:

- The mean and standard deviation remain constant over time.
- The dataset exhibits no seasonal patterns. Seasonality refers to any predictable pattern or variation that repeats for a year [7].

There are various ways to make the data stationary. One commonly used approach is the differencing method where finite differencing is applied to the data points. A non-stationary trend is denoted by Y_t while the stationary trend is denoted by Z_t. We posit that Z_t will thus be equal to the difference between successive values of Y_t:

$$Z_t = Y_t - Y_{t-1} \tag{1}$$

However, should the nonstationary dataset also exhibit seasonal characteristics, it will therefore be recommended to apply seasonal differencing towards the dataset [8]:

$$Z_t = Y_t - Y_{t-m} \tag{2}$$

where m is the monthly timestamps. For instance, a 12 monthly differencing can be written as follows:

$$Z_t = Y_t - Y_{t-12} \tag{3}$$

The differencing method is very effective and thus recommended because in most cases, non-stationary data can easily be transformed to stationary after the first difference, and thus no further transformation would be required. This is not the case with other transformation methods where stationarity can be reached after multiple transformations along the same data points.

2.2. Background

In Ref. [9] a deep learning model to forecast a product usage of a given consumer based on historical data was developed. The authors adapted a CNN with auxiliary input to time-series data to demonstrate an improvement in the model accuracy which predicted future change. To improve the forecasting skills of aircraft in flight navigation systems, Ref. [10] undertook a study on weather forecasting comparing the predictive ability of LSTM and ARIMA models. The study found that the LSTM performs much better than the ARIMA with Root Mean Square Error (RMSE) values of 0.0007 for the LSTM and 0.948 for the ARIMA. A solution presented by [11] demonstrates that the LSTM outperforms the ARIMA model. The purpose of the study was to forecast a multi-step electricity load for Poland, Italy, and Great Britain. The RMSE values for each model were summarized, but the LSTM outperformed the ARIMA using the RMSE evaluation metric for predicting wind speed. In Ref. [12], a study to determine which forecasting time series techniques between ARIMA and LSTM produced the most accurate predictions with a minimalistic empirical error was undertaken. The LSTM outperformed the ARIMA for all stock markets prediction with an average RMSE of 64 dollars. Limestone is an important raw material in today's world. Around 10% of the sedimentary rocks on Earth are made up of limestone [13]. According to [13], over 25% of the world's population relies on limestone for drinking water, and about 50% of all known gas and oil reserves are encased in limestone rocks [13]. It is therefore crucial for various economies to accurately predict future prices of limestone. In Ref. [14] a study comparing the ARIMA and LSTM with regards to predicting future prices of limestone was conducted. The ARIMA performed slightly better than the LSTM with an accuracy of 95.7% compared to the LSTM's 91.8%. However, we argue that the probable reason for the LSTM model's subpar performance was due to manual tuning towards some of the model's hyper-parameters. For instance, the number of LSTM layers was manually tuned. In addition, the author did not disclose the exact units for their target variable. The authors in [15] used regression to learn the correlation between a time series and continuous variables. The approach was to detect the correct coefficients to forecast various attributes. The regression model predicted annual rainfall using historical temperature values [16] with Random Forest (RF) and Gradient Descent (GD) algorithms. The final results confirm the in-depth understanding of time series data to compute the optimal fitting algorithm. However, Ref. [17] attempted to predict respiratory rates using a sliding window that consists of three modules. The first module retrieved the signal of respiratory patterns; the second approximated the rates, and the third made various estimations. A Gaussian-based regression process extracted the respiratory features from the datasets. It also attempted to fit different Auto-Regressive (AR) algorithms to the retrieved signals. Unfortunately, the AR model failed to detect seasonality. In Ref. [18], Dynamic Time Warping (DTW) and K-Nearest Neighbor (KNN) used for time series forecasting exhibited a complexity time of 1-NN using the DTW that relied on the engineering of hand-crafted patterns. In Ref. [19], the CNN used on time-series data outperformed all other tested ML models. The author proposed a feature selection method to automate the learning from input variables. The learned patterns represent time series features with discriminatory layers. However, this technique relies on back-propagation that turns the NN into an adequate feature selector. According to [20], the juxtaposition of Recurrent Neural Networks (RNN) such as LSTM and CNN yielded enhanced accuracy for classification tasks with a range of 27% to 43% in comparison to other well-known ML models. The classification was also considered by [21] and assessed with J48, LSTM, RF, Support Vector Machine (SVM), and CNN. The LSTM-based CNN outperformed other models with three hidden layers. In Ref. [22], the authors used regression to allocate company resources. In addition, the authors undertook a substantial review of well-known ML models for time series data forecasting, but [23] used CNN to address multivariate time-series regression problems. The LSTM and Gated Recurrent Unit (GRU) portrayed transferable CNN units compared to other models. The research in [24] used LSTM with additional convolutional layers. The results provide a boost in predicting performance. Lastly, three CNN and four LSTM were implemented by [25] with an improved CNN execution time. Generally, regression models using CNN and LSTM are the most optimal ML techniques used in the literature for time series data forecasting (Table 1). The limitation of the discussed research relies on dataset misunderstanding, lack of feature engineering, non-seasonal patterns, computational biases, and time complexity. Classifiers such as SVM and Decision Trees (DT) are also prone to error in terms of time series prediction since they are not a better choice for forecasting (Table 1). The time series data forecasting solutions are also implemented in various fields such as weather, electricity, and price prediction (Table 1).

Table 1. Comparative analysis.

Source	Model	Limitation
[15]	RF & GD	Data understanding
[17]	Auto Regressive	Seasonality
[22]	CNN	Seasonality
[18]	DTW & KNN	Feature engineering
[19]	CNN	Back propagation
[20]	RNN & LSTM	Classification
[21]	LSTM & CNN	Classification
[24]	LSTM	Feature engineering
[25]	LSTM & CNN	Execution time
[23]	CNN	Biases

Table 1. *Cont.*

		Applicability
[10]	LSTM & ARIMA	Weather forecasting
[11]	LSTM & ARIMA	Electricity load forecasting
[12]	LSTM & ARIMA	Stock market prediction
[14]	LSTM & ARIMA	Limestone price prediction
[3]	ABC & ARIMA	Refineries
[9]	CNN	Subscriber usage

2.3. *Time Series Data Limitation*

Some attempts allow efficient computation of large-scale time series data. For instance, Ref. [26] implemented a Hadoop-based framework for accurate preprocessing of data which is important for feature selection. Unlike [26,27] concentrated on model selection by using MapReduce to compute the cross-validation that improved parallel rolling-window prediction using the training set of heterogeneous time series patterns. The predictive parameters computed the accuracy, but this technique could not tackle challenges associated with forecasting. In Refs. [28,29] multi-step forecasting was monitored by the ML models using the Spark environment. Specifically, Ref. [28] used H iterations to compute the multi-step prediction, while [29] implemented multivariate regression models using ML libraries. As a result, the H technique was not scalable for forecasting. With this, one can use a sample of patterns instead of the original data to predict. For example, Ref. [30] provides an overview of forecasting big data using time series traffic. The paper provides a premise for time series data forecasting, but it is still complicated to implement the proposed techniques to deal with subscriber data and forecast the future. Some researchers investigate the underlying intuition of parallel computing models using time series data. Unfortunately, these models resulted in expensive computational time complexity. For instance, Ref. [31] introduced a distributed approximator before the prediction calculation, requiring several iterations. Based on their frameworks, Refs. [32,33] proposed recursive techniques with Bayesian prediction while [34] refined the estimator computation of quantile regression model through various rounds of classification. Another well-known methodology is the alternation of eigenvectors for convex optimization of time series data. This technique blends the seasonality of time series data with the convergence properties of predictors [35], but the streams complicate the forecasting prediction. We argue that a one-shot averaging computation is a straightforward technique to compute the prediction. This method requires only a single computational round [36]. Various studies used distributed learning that split features in a specific frequency domain where the time series patterns are used in the splitting process [37]. These algorithms model successive refinements with a limitation that requires re-implementing each estimator scheme, but slow in terms of convergence accuracy compared to existing predictors designed for time series data forecasting [38]. For example, Ref. [39] analyzed cyclostationary properties of 0-day exploits with slow precision convergence. Boruta was the feature-based extraction method combined with Principal Components Analysis (PCA) to extract the most cyclostationary patterns from NSL-KDD, UGRansome, and KDD99 datasets. The RF and SVM were used to classify cyclostationary features. The supervised learning restricted the experiments, but our research implements an unsupervised learning scheme to study stationary prediction. Moreover, we have compared the ARIMA performance applied to the UGRansome [40] and subscriber datasets to assess the forecasting performance of stationary and time series data. The following section presents our methodology, Exploratory Data Analysis (EDA), and UGRansome dataset [41]. However, all mentioned articles in this section are crucial because they provide valuable recommendations regarding ML to forecast subscribers' usage data growth.

3. Materials and Methods

We have used subscriber data collected from a network database and analyzed the patterns to predict the growth in subscriber data usage. The Network Subscriber Data Management (NSDM) approach is thus the relevant aspect of this research as it stands at the core network layer and stores valuable data used by various subscribers. The NSDM extracts subscribers' patterns from the Insights Data Storage (IDS) and monitors all real-time traffic of subscriber data [42]. We have used the NSDM module that considers subscriber data in a centralized and secure environment having a scalable repository named IDS (Figure 1).

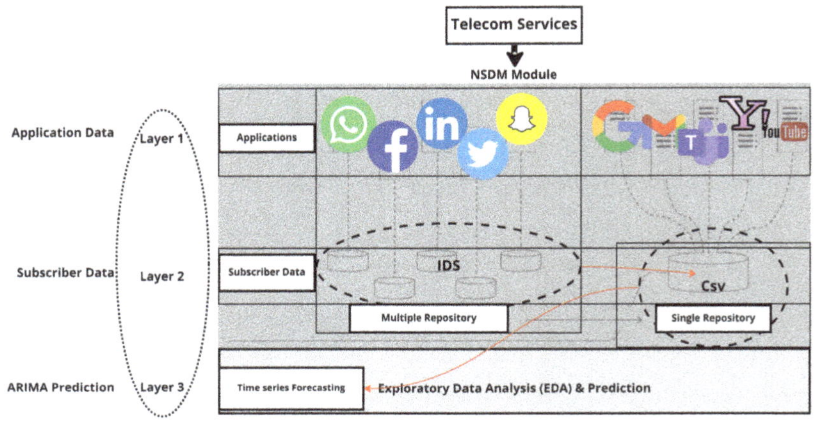

Figure 1. The NSDM architecture.

The IDS directory provides distributed and resilient subscriber patterns stored in a single repository. The ARIMA model was used on this repository to predict the growth in subscriber data usage (Figure 1).

3.1. Mathematical Formulation of ARIMA

An ARIMA model has a different Moving Average (MA), as well as AR components [43]. We use ARIMA(p, d, q) to denote an ARIMA model where the order of the AR module is (p, q) and d represents the number of differences needed for stationary series [43]. One can extend the ARIMA predictor to a Seasonal ARIMA (SARIMA) model by incorporating additional seasonal patterns to handle time series properties that exhibit a strong seasonal characteristic [43]. We can use ARIMA(p, d, q)(P, D, Q) to formulate a SARIMA model. Here, the uppercase Q, P, and D denote the order of the AR model, the number required for seasonal/stationary series, and the MA order. Similarly, the seasonality period is denoted by m [43,44]. An ARIMA(P, D, Q)$(p, d, q)_m$ model for time series $(y_t, t \in \mathcal{Z})$ has the following back-shift operator:

$$(1-\sum_{i=1}^{p}\theta_i B^i)-(1-\sum_{m=1}^{P}\alpha_i^m)(1-B)^d(1-B^m)^D y_t=(1+\sum_{i=1}^{q}\gamma_i)(1+\sum_{i=1}^{Q}\alpha_i B^m)\omega_t \quad (4)$$

where B denotes the backward shift function, ω_t the white noise, m the seasonality length, θ, and α represent the AR parameters, γ, and ω refer to the seasonal parameters of the MA. This mathematical formulation represents two major combinations of seasonal parameters P, D, Q, and p, d, q, where:

- the number of auto-regressive terms is p,
- nonseasonal differences denoted by d,
- and the number of lagged predictive biases denoted by q.

The variation of these ARIMA parameters can identify the most optimal set of features in obtaining precise predictive values [43–45].

3.2. Experimental Datasets

Figure 2 presents the research methodology where our framework provides subscriber data stored in the IDS module.

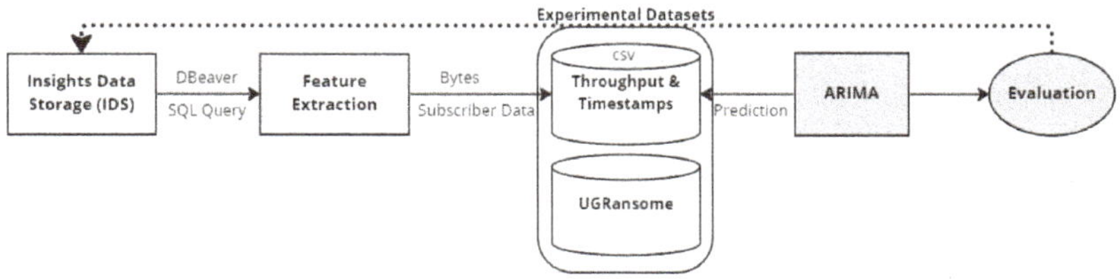

Figure 2. The experimental methodology.

The subscriber data was extracted from the real-time network traffic using a Structured Query Language (SQL). We pushed the features into a single comma-separated file and used EDA to visualize salient features of the network traffic. We have then obtained critical Key Performance Indicators (KPIs) that can support the prediction of data usage growth. The executed SQL retrieved the subscriber timestamps, incoming throughput, and outgoing throughout (Figure 3).

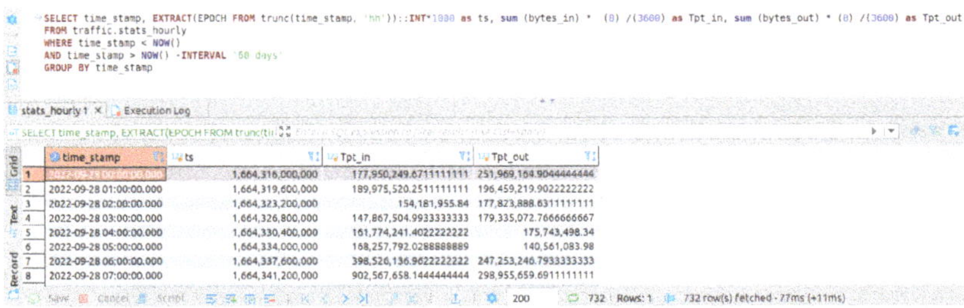

Figure 3. The subscriber data.

The query extracts the timestamps (ts) by truncating them into a human-readable format (Year-Month-Time). The incoming throughput was computed using the following Equation (5):

$$Tpt_{in} = \frac{sum(bytes_{in}) * (8)}{36{,}000} \tag{5}$$

The SQL in Figure 3 illustrates this process. Equation (6) denotes the outgoing throughput computation:

$$Tpt_{out} = \frac{sum(bytes_{out}) * (8)}{36{,}000} \tag{6}$$

It is hourly-based statistics retrieved from the traffic stats table of the IDS for 60 days (Figure 3). The 3600 represents an hour in seconds, and eight changed the bytes into bits. In addition, we grouped results by timestamps. Retrieved patterns were converted into Comma-Separated Values (CSV) (Figure 4).

The subscriber dataset has 730 entries with four attributes (human-readable timestamps, UNIX timestamps, incoming throughput (Tpt in), and outgoing throughput (Tpt out)). A timestamp represents the time when the subscriber traffic was collected [46]. The throughput is the flow that measures inputs/outputs movements within the network [46]. The following Figure 5 illustrates our research methodology.

time_stamp	ts	Tpt_in	Tpt_out
2022-09-28 0:00:00	1664316000000	177950249.7	251969164.9
2022-09-28 1:00:00	1664319600000	189975520.3	196459219.9
2022-09-28 2:00:00	1664323200000	154181955.8	177823888.6
2022-09-28 3:00:00	1664326800000	147867506	179335072.8
2022-09-28 4:00:00	1664330400000	161774241.4	175743498.3
2022-09-28 5:00:00	1664334000000	168257792	140561084
2022-09-28 6:00:00	1664337600000	398526137	247253246.8
2022-09-28 7:00:00	1664341200000	902567658.1	298955659.7
2022-09-28 8:00:00	1664344800000	2042488462	612111154.3
2022-09-28 9:00:00	1664348400000	1833562607	639776109.2

Figure 4. The CSV format of the subscriber data.

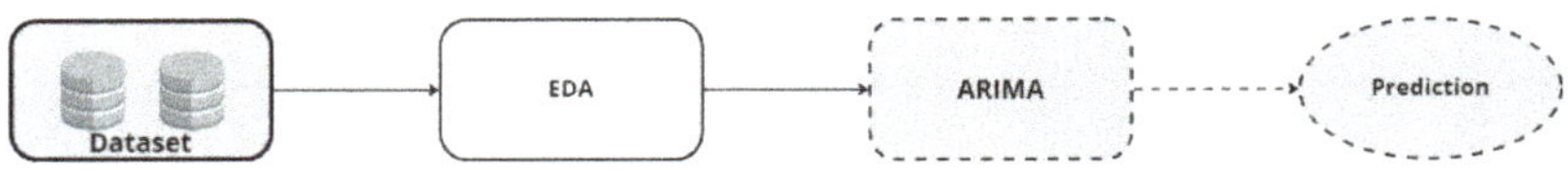

Figure 5. The research methodology.

The subscriber and UGRansome datasets are collected, and then the EDA is executed before the computation of the ARIMA model that predicts the growth in data usage based on the current timestamp. The techniques discussed in the literature train ML classifiers with human-labeled features, but this supervised learning method uses limited samples. We have used an unsupervised learning technique whereby we did not label the features. The ARIMA model attempted to use data points $x_1 \ldots x_n$ and assigned predicted values $\Theta_1 \ldots \Theta_n$ using predefined parameters.

3.3. Feature Engineering and Data Cleaning

Data cleaning is a method of mapping and transforming features from one-row data format to another, to make it more suitable and valuable for various downstream uses, such as time-series forecasting. One of the most important data cleaning processes is handling missing values [41]. Fortunately, concerning the subscriber data, the dataset contains no missing values. However, the data still needs to be transformed in other various ways for training and testing, and in this case, this will include:

1. Data normalization. The process of normalization is frequently used to prepare data for ML. The objective of normalization is to convert numerical columns to a common scale without losing information or distorting the ranges of values. This will reduce bias towards accurate prediction [40]. We have used Python Sklearn's MinMax Scaler function to normalize the throughput column down to values between 0 and 1.
2. Feature engineering. Also known as feature extraction, is a process of selecting and transforming the most important features from the data to utilize for developing predictive models using statistical or ML models [39]. Concerning the subscriber dataset, only the throughput and timestamp columns will be used to model the training and testing sets.
3. Train-test split. It is a method for validating models that enables one to simulate how a model would behave when presented with fresh untested data [47]. In our analysis, the training data will be split into k-fold cross-validation to avoid under/over-fitting. However, cross-validation is not necessary when the data is small.

3.4. *Stationarity of Data*

Two main methods can be used to determine the stationarity of a time series dataset:

1. Visual and graphical inspection. This is implemented by plotting the functions of the time-series dataset, and then inspecting visually whether the dataset is stationary or not, but the method is prone to inaccuracy.
2. Statistical Augmented Dickey-Fuller test. Named after famous statisticians David Dickey and Wayne Fuller, the Dickey-Fuller test is a more accurate stationarity test method that determines if a time series dataset is stationary by calculating the p-value to test the null hypothesis. The Dickey-Fuller null hypothesis is that the data is not stationary. If the p-value is more than 0.05, then there is strong support for the null hypothesis, and thus the time series dataset will be deemed to be non-stationary. Python's stats model library was used to perform this task by importing the fuller functionality.

In this research, the differencing method was applied to make the dataset stationary (Figure 6). The differencing and original data are distinguished in Figure 6.

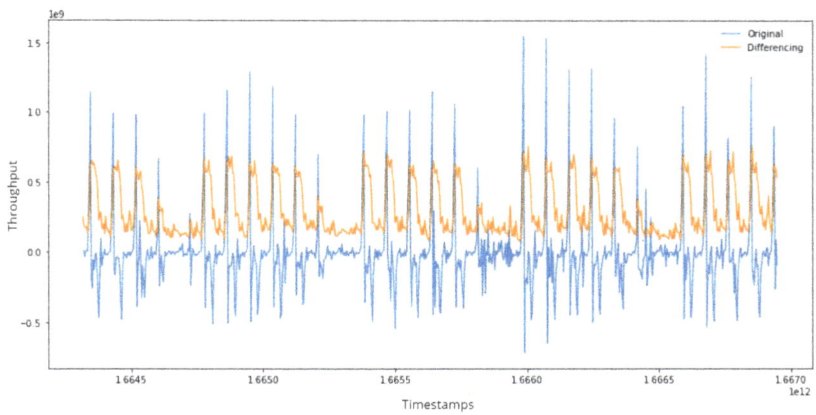

Figure 6. The original and differencing of subscriber data.

3.5. *The UGRansome Characteristics*

This dataset was created by extracting important features of two existing datasets (UGR'16 and ransomware) [41]. UGRansome is an anomaly detection dataset that includes normal and abnormal network activities [48]. The regular characteristic sequence makes up 41% of the dataset, whereas irregularity makes up 44%. The remaining 15% represents the predictive values of network attacks grouped into the signature (S), synthetic signature (SS), and anomalous (A) attacks (Figure 7).

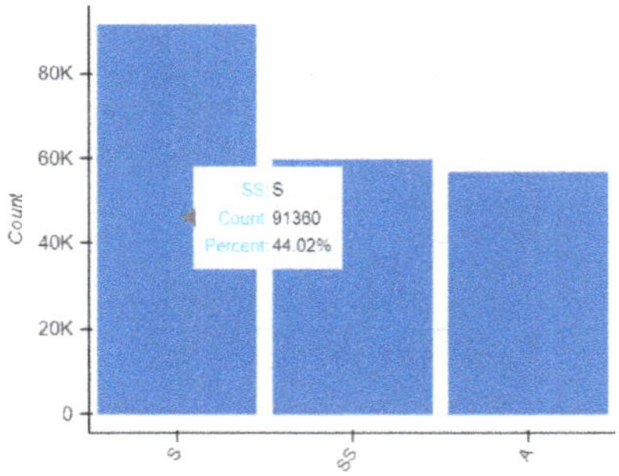

Figure 7. Distribution of network threats.

Figure 7 depicts the signature attacks having a proportion of 44.02% (synthetic signature 28.71%, and anomaly 27.27%). A significant proportion of signature traffic means that the UGRansome threatening concerns are detectable. Regular threats, like User Datagram Protocol (UDP) and Botnet, provide about 9% for the anomalous category. The Internet Protocol (IP) and ransomware addresses have a ratio of 1% [39]. In addition, a ratio of 2% exists between communication protocols and ransomware addresses [41]. According to Refs. [39,41] the significant distribution of the UGRansome could be summed up in the following Figure 8. However, UGRansome is more redundant compared to subscriber data and we removed 28.2% of duplicate records during the feature extraction phase (Figures 8 and 9).

Dataset Statistics

Number of Variables	14
Number of Rows	207533
Missing Cells	0
Missing Cells (%)	0.0%
Duplicate Rows	58491
Duplicate Rows (%)	28.2%
Total Size in Memory	106.9 MB
Average Row Size in Memory	540.2 B
Variable Types	Numerical: 4 Categorical: 9 GeoGraphy: 1

Figure 8. The UGRansome data summary.

Dataset Statistics

Number of Variables	4
Number of Rows	730
Missing Cells	0
Missing Cells (%)	0.0%
Duplicate Rows	0
Duplicate Rows (%)	0.0%
Total Size in Memory	74.3 KB
Average Row Size in Memory	104.2 B
Variable Types	Categorical: 1 Numerical: 3

Figure 9. The subscriber data summary.

3.6. Exploratory Techniques

The exploratory analysis provides a set of techniques to understand the dataset. The results produced by the EDA can assist in mastering the data structure [49], as well as the distribution of the features, detection of outliers, and correlation within the dataset. Some of the statistical metrics used to evaluate the ARIMA model are standard deviation, correlation, mean, standardized residual, normal Q-Q, correlogram, theoretical quantile, p-value, and accuracy:

- Standardized residual (r_i). It measures the strength of actual and predicted values and indicates the significance of features [50] (r_i facilitates the recognition of patterns that contribute the most to the predictive values):

$$r_i = \frac{e_i}{s(e_i)} = \frac{e_i}{RSE\sqrt{1-h_i}} \tag{7}$$

where e_i is the iTh residual, RSE is the standard error of the residual model, and h_i the iTh leverage observation.

- Normal Q-Q. The normal Q-Q means normal Quantile-Quantile. It is a plot that compares actual and theoretical quantiles [50]. The metric considers the range of random variables to plot normal Q-Q using a probabilistic computation. The x-axis represents the Z-score of the standardized normal distribution, but different formulations have been proposed in the literature to detect the plotting positions:

$$\frac{(k-a)}{(n+1-2a)}, \quad (8)$$

for some value between 0 and 1 [k, a]; which gives the following range (Equation (9)):

$$\frac{K}{(n+1)} \Longleftarrow \frac{(k-1)}{(n-1)}. \quad (9)$$

- Correlogram. It is a correlational and statistical chart used in TSA to plot the auto-correlations sample r_h versus the timestamp lags h to check for randomness [50]. The correlation is zero when randomness is detected. Equation (10) denotes the auto-correlation parameter at h lag:

$$r_h = \frac{c_h}{c_0}, \quad (10)$$

where c_h is the auto-covariance coefficient and c_0 the variance function.

- Augmented Dickey-Fuller (ADF) test. This statistical metric tests the stationarity of time series data [50] by using a unit root metric β that exists in a series of observations where $\alpha = 1$ as per the below Equation (11).

$$y_t \Longrightarrow \alpha_{t-1} + \beta x_e + \epsilon, \quad (11)$$

Here y_t represents the time series values at time t, but x_e is a separate time series variable.

- Theoretical quantile. The theoretical Q-Q explores the variable's deviation from theoretical distributions to visually evaluate if the ratio is significantly different for EDA purposes [50].
- Likelihood. The likelihood parameter maps $L : \Theta \Longrightarrow \mathbf{R}$ or $\mathbf{R} : \Theta \Longrightarrow \mathbf{L}$ given by $\mathbf{R}[\Theta]|y] \Longrightarrow f_y[x]|[\Theta]$ or $\mathbf{L}[\Theta]|[x] \Longrightarrow f_y[y]|[\Theta]$. This metric computes the most probable value assigned to a specific feature using Θ as the hypothesis in $\mathbf{R}$ and $\mathbf{L}$ spaces. Inputs x compute the predictive values y using a predefined Θ parameter. With this, the likelihood represents the quantile probability (Prob (Q)) of correlated features used for forecasting.
- Kurtosis. This metric evaluates the probability of the predicted variables by describing the probability proportion. There are various techniques to compute the theoretical distribution of Kurtosis, and there are subjective manners of approximating it with relevant samples [50]. With Kurtosis results, higher values determine the presence of outliers. The Kurtosis is as follows:

$$Kurtosis[x] = [(\frac{x-\mu}{\sigma})^n], \quad (12)$$

where μ is the random selection of inputs x using a standard deviation σ following the constraints:

$$\sum_{i=1}\sum_{j=1}\frac{\mu^i}{\sigma^j}. \quad (13)$$

- Jarque-Bera (JB) test. This metric uses a Lagrange multiplier to test for data normality. The JB value tests if the distribution is normal by testing the Kurtosis to determine if features have a normal distribution. A normal JB distribution will have symmetrical Kurtosis indicating the peaked in the distribution. We formulate the JB test as follows:

$$JB = n[\frac{\sqrt{b_1^2}}{6} + \frac{(b_2 - 3)^2}{24}], \quad (14)$$

where the sample size is n, $\sqrt{b_1}$ is the skewness sample, and b_2 is the Kurtosis coefficient.

- Heteroscedasticity. It checks the alternative hypothesis (H_A) versus the null hypothesis (H_0) [50]. With the alternative hypothesis, the empirical error is multiplying the function of various variables:

$$H_A : \sigma_1^2 = \sigma_2^2 * \ldots * \sigma_n^2. \quad (15)$$

However, a null hypothesis has equal error variances (homoscedasticity) [50]:

$$H_0 : \sigma_1^2 = \sigma_2^2 = \ldots = \sigma_n^2. \quad (16)$$

- Accuracy. The balanced accuracy B_A of the ARIMA model is calculated with the following mathematical formulation [47]:

$$B_A = \frac{((TP/TP + FN) + (TN/(TN + FP)))}{2} \quad (17)$$

where True Positive (TP) and True Negative (TN) denote correct classification, but misclassification is the False Positive (FP) and False Negative (FN) [50]. We used cross-validation rounds to build multiple training/testing subsets to decide which model is a suitable predictor of the growth in subscriber data (80% of the training set, 10% of the validation set, and 10% of the testing set).

3.7. Feature Extraction

ML models are used to address a range of prediction problems. The unsatisfactory prediction of ML classifiers originates from overfitting or underfitting features [7]. In this research, the removal of irrelevant patterns guaranteed improved performance of the ARIMA computation. PCA was utilized on the UGRansome data to extract relevant patterns. PCA is a feature extraction methodology of this research. We denote PCA as follows:

$$\mathbf{P} = \frac{1}{t-1} + \sum_{t=1}^{k}([x(t)x(t)^T]), \quad (18)$$

with stochastic $x(t)$ and $t = 1, 2 \ldots l$ with n-dimensional inputs x having a probability matrix $\mathbf{P}$ of zero mean. The PCA formulation uses the covariance given in Equation (19) with a linear calculation of $x(t)$ inputs into $y(t)$ outputs:

$$y(t) = Q^T x(t), \quad (19)$$

Q is an orthogonal n*n matrix type where i represents the columns viewed as eigenvectors computed as follows:

$$y_i(t) = Q^T x(t), \quad (20)$$

The range in Equation (20) starts from $1 \ldots$ n where y_i is the new component of the ith PCA. Table 2 depicts the PCA results using the UGRansome dataset. However, SQL was used as a feature extractor to extract subscriber data from the IDS (Figure 3).

Table 2. The PCA results using the UGRansome.

Attack	Feature	Total
Blacklist	Timestamp	2761
Spam	IP address	7425
Scan	Flag	1559
SSH	Prediction	7293
Botnet	Threats	4765
Total	-	23,803

3.8. Model Training and Testing

The parameters shown in Figure 10 will be used to train and test the ARIMA model.

	p	d	q
Subscriber data	**2**	**1**	**2**
UGRansome data	**1**	**2**	**1**

Figure 10. The ARIMA model parameters.

The choice of these parameters is justified because a model with only two AR terms would be specified as an ARIMA of order (2, 0, 0). A MA(2) model would be specified as an ARIMA of order (0, 0, 2). A model with one AR term, a first difference, and one MA term would have order (1, 1, 1). For the proposed model using the subscriber data, an ARIMA (2, 1, 2) model with two AR terms, the first difference, and two MA terms are being applied to improve the forecasting accuracy (Figure 10). One AR(1) term with two differences and one MA(1) term are used for the UGRansome data to account for a linear trend in the data (Figure 10). The differencing order refers to successive first differences. For instance, for a difference order = 2 the variable analyzed is $z_t = [x_t - x_{t-1}] - [x_{t-1} - x_{t-2}]$. This type of difference might account for a quadratic trend in the data. However, due to seasonality concerning the experimental data, the SARIMA model is used. The SARIMA model is an extension of the ARIMA with integrated seasonal components. The training set is used to train the model and thus used to predict the data usage for the 2022 year. The test set is also utilized to validate the final predictions for the year.

3.9. Model Tuning

In contrast to the ARIMA, the CNN has several parameters that are not estimated by the model (i.e., the p and q values), and thus the algorithm is trained by manually specifying a set of hyper-parameters using trial and error. The number of layers, neurons, batch sizes, and epochs utilized in the CNN, are some examples of hyper-parameters that need to be tuned manually. It should also be noted, that, unlike the ARIMA which required the dataset to be transformed to its stationary format, the CNN in this research will be trained against non-stationary time series patterns of the UGRansome dataset. Considering a single input of features as depicted in Figure 11, the CNN weights these features to enable the learning trend of particular observations. We have various observations, so we merge their outputs into a hidden layer.

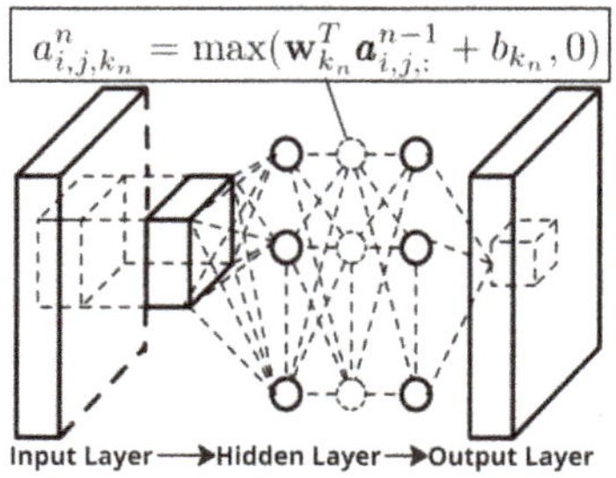

Figure 11. The CNN architecture.

Our CNN architecture uses binary convolutions (with 70 and 30 filters) and a densely connected layer of 50 neurons with the activation function (ReLU). This unit has six connected layers representing auxiliary outputs (Figure 11). Each layer predicts and passes the prediction value to the next layer which predicts growth in the subscriber data usage until the final layer produces long-term forecasting. Ideally, the convolution layer uses the ReLU computed as follows:

$$a_{i,k,j} = max[W_k^T x_{j,i} + b_k, 0] \tag{21}$$

where $a_{i,k,j}$ denote the activation value of the kth feature at location [j, i], and x_i, j is the input patch centered at location $[j, i]$. Here w_k and b_k are weighted vectors and bias terms of the kth filter. We use each layer to predict in advance some additional days and grid search to detect the optimal number of filters, convolutions, connected layers, and drop-out rate. For each layer $k \in [1, 2, 3, 4, 5, 6]$ we added an empirical error and loss function. Each layer k aims to produce future forecasting for more than 14 days ahead:

$$k_{loss} \Longrightarrow \frac{1}{n} \sum_i (y_i, 14_k - tanh(\hat{y}_i, 14_k))^2 \tag{22}$$

where $[y_i, 14_k]$ represents the features of time-series i, and $[\hat{y}_i, k]$ the k-th layer's forecasting the value. The function restricting the range of $[\hat{y}_i, k]$ to $[-1, 1]$ is denoted by tanh. With this, we can reformulate Equation (22) to the weighted sum and minimize the loss by decreasing λ_k values:

$$Min_{loss} \Longrightarrow \frac{1}{k} \sum_{k=1}^{K} (\frac{k_{loss}}{\lambda_{loss}}) \tag{23}$$

Table 3 shows a summary of the hyper-parameters used for the CNN model.

Table 3. CNN tuning parameters.

Hyper-Parameter	Value
Number of neurons	50
Activation function	Tanh
Number of dense layers	6
Optimizer	Adam
Batch size	2
Loss function	Mean Squared Error (MSE)
Number of epochs	50

3.10. ARIMA Predictor Model

Given a long period, $y_t[t = 1, 2, \ldots, T]$ of a spanning time series traffic, the aim is to come up with a new scheme that works well for predicting the future outcomes H. We define $S = [1, 2, \ldots, T]$ as the sequence of timestamp with time series y_t. The prediction of the problem can be written as $f[\Theta, \sum |y_t, t \in S]$, where the parameter is f, the global parameters Θ, and the covariance matrix $\sum$. However, the time series data is divided into different sub-series (k) having contiguous time intervals (Equation (24)):

$$S = \sum_{k=1}^{K} S_k, \tag{24}$$

where S_k extracts the kth sub-series timestamps and we posit $T = \sum_{k=1}^{K} T_k$. With this assumption, the predictor estimator of the sub-problem is shown in Equation (25):

$$f[\Theta, \sum |y_t, t \in S] = g[f_1, \sum_{1} |y_t, t \in S_1] \ldots [f_k, \sum_{k} |y_t, t \in S_k], \tag{25}$$

where f_k represents the estimator function for the kth sub-series and g the combination. The estimation was merged before the prediction. The idea is to use $g(.)$ as a single mean parameter, and our computational framework could be viewed as an averaging ML algorithm (Figure 12).

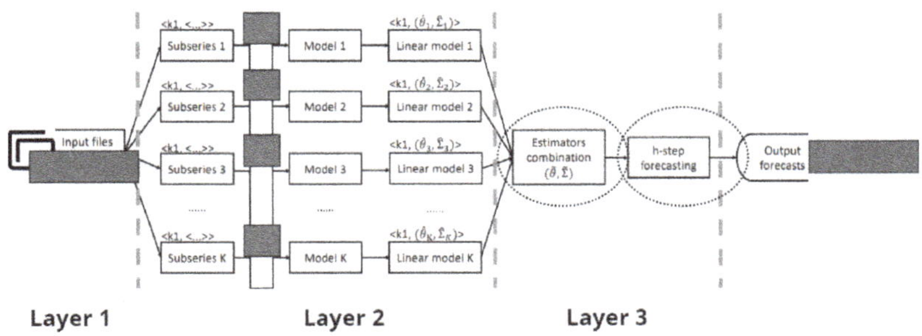

Figure 12. The ARIMA model.

Figure 12 outlines the proposed ARIMA model to forecast the growth in subscriber data. The timestamps of historical data were recorded in the IDS before being processed by the ARIMA. In simple terms, the proposed model consists of the following phases:

- Phase 1: Preprocessing. Subdivide the time series data into the K sub-series.
- Phase 2: Modeling. Train the algorithm using sub-series data by assuming that the IDS of the sub-series remains constant.
- Phase 3: Linear transformation. Translate the trained algorithm in phase 2 into linear representations k.
- Phase 4: Estimator combination. The obtained local estimator from phase 3 minimizes global losses parameters described in Section 3.1.
- Phase 5: Prediction. Predict the next observations H by utilizing the merged estimator's parameters presented in Equations (24) and (25).

We used available hourly-based timestamps to create a new set of timestamps (ts) used in the ARIMA prediction. The following formulation was used to predict new timestamps (Equation (26)):

$$Predicted_{ts} = LastEpochTimestamps + n * 3600 \tag{26}$$

where $n = range(1, 48)$. This computation provides new predicted timestamps on an hourly basis for the next hours. Figure 13 shows the computation of the needed inputs x using the current value, step size, and stop value. The x represents current hourly values starting at zero milliseconds (ms) and going up to 731. The computation of the predicted values used x, a step size of 360,000, and a stop value. The needed value represents the result of the prediction in which the current value of x was used as an index starting at zero and multiplied by the step size. This calculation stops at the 731st iteration, where 731 denotes the last index (Figure 13).

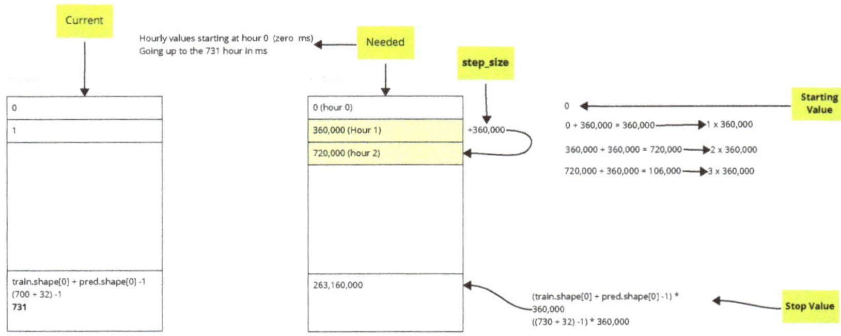

Figure 13. The predictive computation.

3.11. Computational Environment

The IDS used to build the subscriber data is installed on a DBeaver database. DBeaver is a database monitoring software that manipulates subscriber data. It can be used to build analytical dashboards from various data storage. Table 4 presents the computing environment and Table 5 the feature extraction results.

Table 4. Framework specification.

Node	Specification
RAM	39 GB
Service	Jupyter & DBeaver
ML algorithm	ARIMA & CNN
System	64-bits
Processor	2.60 GHz
Dataset	Subscriber data & UGRansome
Operating System (OS)	Windows & Linux
CPU	Intel i7-10
Language	Python & SQL

3.12. Feature Extraction

There are different reasons causing duplication in a dataset, among which are imperfections in the data collection process and the properties of patterns, but feature extraction solved redundancy dimensions. Features projected into a new space have lower dimensionality. Examples of such techniques include Linear Discriminant Analysis (LDA), Canonical Correlation Analysis (CCA), and PCA [40]. We have used PCA to extract relevant features of the UGRansome dataset. The PCA lowered computational complexity, built generalizable models, and optimized the storage space. To address the redundancy issue, the PCA selected a subset of relevant patterns from the original dataset based on their relevance.

We present the PCA results in Table 5 where the final dataset with the description of each attribute is presented. The prediction attribute facilitates the forecasting of any ML model to predict the category of novel intrusion. Our final dataset has 12 variables with 180,564 observations (Table 5). Consequently, if the deviation degree of a variable is high or low enough, it is considered an abnormality. However, we did not apply feature extraction on the subscriber data because it has no redundant patterns. The feature extraction using UGRansome leads to improved performance, higher prediction accuracy, minimized computational time, and efficient model interpretability.

Table 5. Extracted UGRansome features.

Number	Attribute	Description	Type
1	Timestamp	Traffic duration	Numeric
2	Protocol	Communication rule	Categorical
3	Flag	Network state	Categorical
4	IP address	Unique address	Categorical
5	Network traffic	Periodic network flow	Numeric
6	Threat	Novel malware	Categorical
7	Port	Communication port	Numeric
8	Expended address	Malware address	Categorical
9	Seed address	Malware address	Categorical
10	Cluster	Group assigned	Numeric
11	Ransomware	Novel malware	Categorical
12	Prediction	Novel malware class	Categorical

4. Results

This section is structured into the DFT, comparative results of ARIMA and CNN, execution speed test, and predictive results comparison using additional standard forecasting approaches such as BATS and TBATS. The section compares the ARIMA model performance using the subscriber and the UGRansome datasets. Figure 14 shows the format of the subscriber data following a stationarity distribution compared to the UGRansome timestamps.

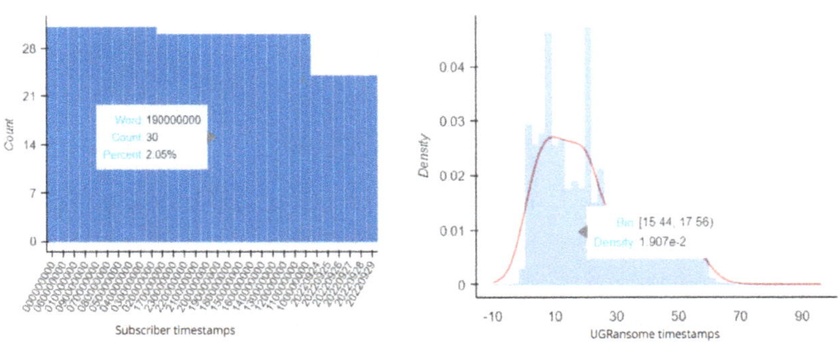

Figure 14. The timestamp and density comparison.

Figure 15 portrays a distribution of incoming and outgoing throughput of the subscriber data compared to the UGRansome port traffic (5066–5068).

Each attack flow is also depicted. The figure depicts NerisBonet threats with less traffic. This result reveals a time series forecasting property of the subscriber data but not the UGRansome. However, the UGRansome has more distributed or dependent variables (Figure 16).

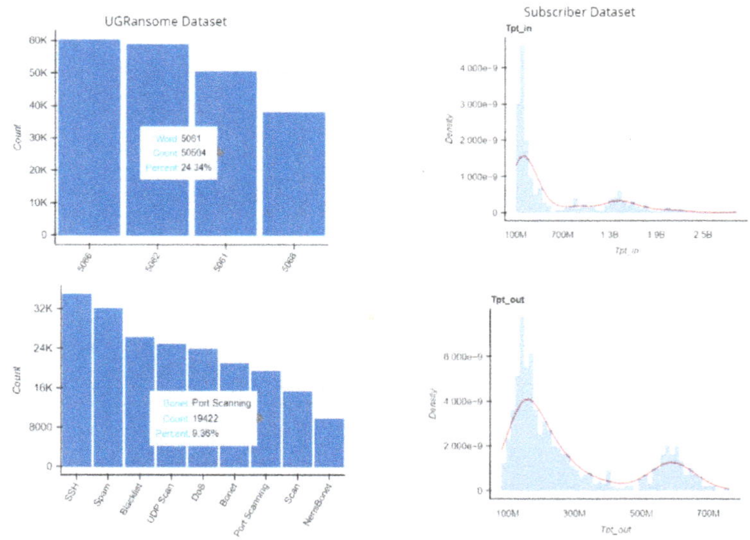

Figure 15. Additional features comparison.

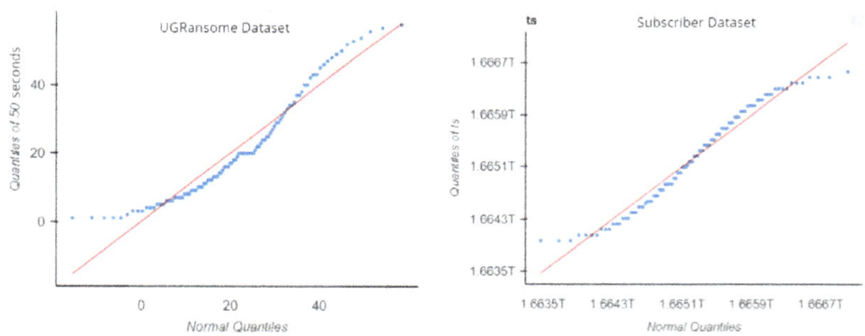

Figure 16. The normal Q-Q results.

The correlation of throughput is in Figure 17.

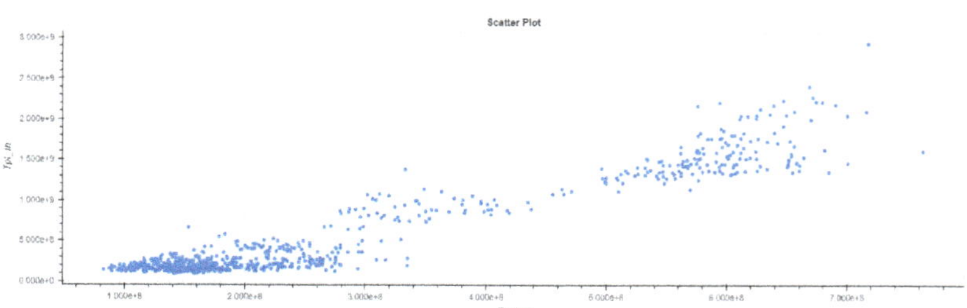

Figure 17. The throughput correlation.

This plot indicates the linear distribution of predicted values. The summary of the SARIMA model using the subscriber data is presented in Figure 18.

SARIMAX Results

Dep. Variable:	y	No. Observations:	732
Model:	SARIMAX(3, 0, 1)	Log Likelihood	-14399.470
Date:	Mon, 31 Oct 2022	AIC	28810.941
Time:	17:51:26	BIC	28838.515
Sample:	0	HQIC	28821.578
	- 732		
Covariance Type:	opg		

	coef	std err	z	P>|z|	[0.025	0.975]
intercept	4.899e+07	5.38e-09	9.1e+15	0.000	4.9e+07	4.9e+07
ar.L1	1.1487	0.103	11.154	0.000	0.947	1.351
ar.L2	-0.1228	0.135	-0.912	0.362	-0.387	0.141
ar.L3	-0.1855	0.051	-3.621	0.000	-0.286	-0.085
ma.L1	-0.1868	0.118	-1.588	0.112	-0.417	0.044
sigma2	7.181e+15	3.07e-17	2.34e+32	0.000	7.18e+15	7.18e+15

Ljung-Box (L1) (Q):	0.02	Jarque-Bera (JB):	376.22
Prob(Q):	0.88	Prob(JB):	0.00
Heteroskedasticity (H):	1.35	Skew:	1.22
Prob(H) (two-sided):	0.02	Kurtosis:	5.54

Figure 18. The SARIMA model summary.

The summary confirms that the prediction of subscriber data will have an increased mean or standard deviation given the likelihood, Prob(Q), and Kurtosis values. The SARIMA model reports no outliers for subscriber data given the Kurtosis value, and the JB test describes a normal distribution. Similarly, the level of heteroscedasticity supports our hypothesis with a degree of 1.35, a p-value of 0.112 having a likelihood value of $-14{,}399.47$. With these results, we reject the null hypothesis and accept the alternative. The former denies the growth of subscriber data, while the latter accepts. In the next section, we will compare the subscriber data with the UGRansome using the DFT results.

4.1. Dickey Fuller Test

The DFT results are in Table 6. A p-value of 0.007 with an accuracy of 90% was obtained. This result supports our hypothesis predicting an increase in subscriber data growth. As such, (i) the UGRansome used more iterations due to its size surpassing the subscriber dataset, (ii) the balanced accuracy of the DFT reached 81% of accuracy, and (iii) the dataset size has not to effect on the prediction performance. The residual and correlogram are in Figure 19 with the seasonality of the throughput.

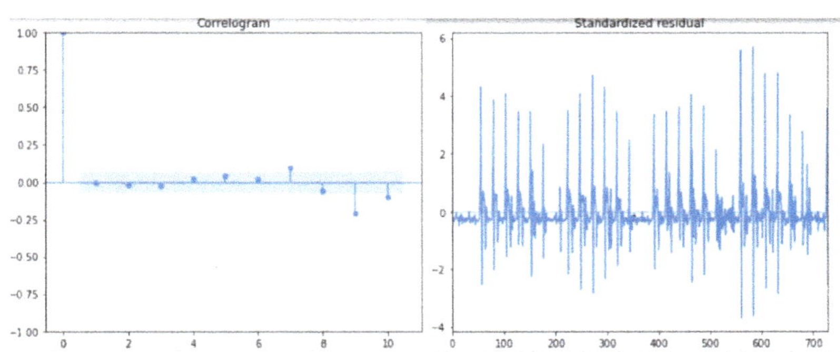

Figure 19. The standardized residual and correlogram.

Table 6. The DFT results.

Dataset	Test Statistic	*p*-Value	Iteration	Accuracy
Subscriber data	−3.537879	0.007066	20	90.567%
UGRansome data	−9.876982	0.0008044	342	90.456%
	Correlogram	**ADF**	**Q-Q**	
Subscriber training set	0.9	0.8	0.9	90.398%
UGRansome training set	0.8	0.9	0.7	89.453%
Subscriber testing set	0.8	0.9	0.9	91.348%
UGRansome testing set	0.8	0.8	0.9	88.298%
	Features Total	**Mean**	**Deviation**	
Subscriber data	700	54.23	22.45	92.351%
UGRansome data	8932	75.32	46.3	88.527%
Subscriber testing set	400	12.6	6.7	94%
UGRansome testing set	4765	26.87	39.65	88%
Balanced Accuracy	-	-	-	**81%**
Balanced Features	**3699**	-	-	-
Balanced Mean	-	**41.75**	-	-
Balanced Deviation	-	-	**28.25**	-

However, the data usage growth prediction is illustrated in Figure 20 using the ARIMA model that predicts a growth of 3 Mbps at a specific timestamp. UNIX timestamps of the subscriber data predicted the maximum data growth using the ARIMA model.

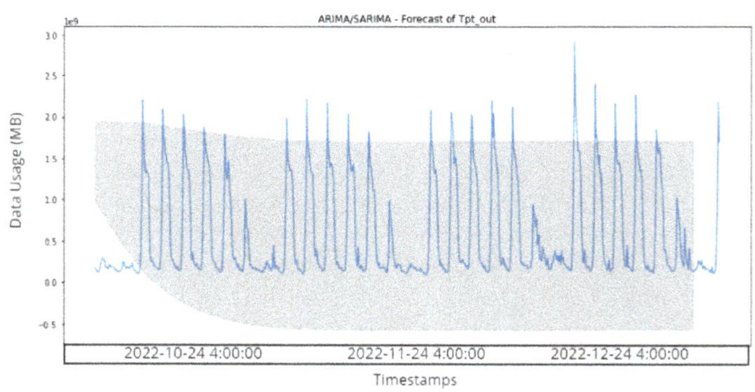

Figure 20. The ARIMA prediction.

In Figure 21, ARIMA predicts a maximal subscriber data usage growth of 14 Gbps (where blue denotes actual data, orange is the predicted ARIMA data, and green is the future predicted values).

The predicted mean and standard deviation are in Figure 22.

The original data represents current values, but ARIMA approximated a mean and standard deviation from these values. ARIMA predicted a maximum mean value of 5 Mbps with a standard deviation of 2 Mbps. The results show mean and standard deviation values lower than the original or current values. The scatter plot of the testing set is shown in orange with a testing size of 32 data points that have been plotted or predicted (Figure 23).

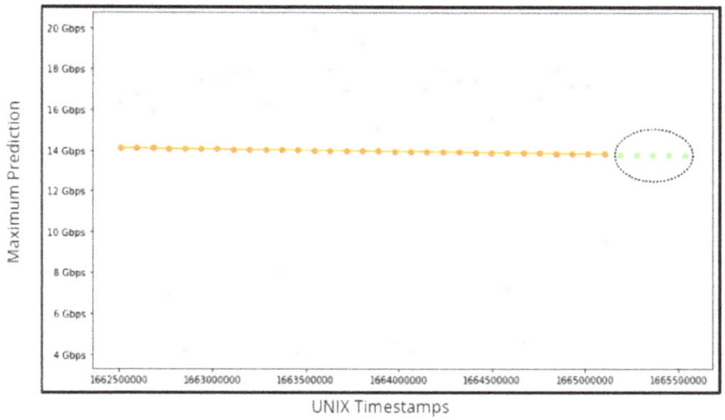

Figure 21. The maximum data usage prediction.

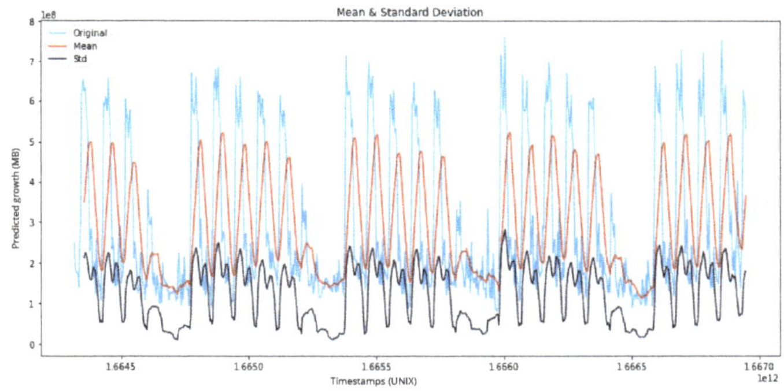

Figure 22. The prediction of the mean and standard deviation.

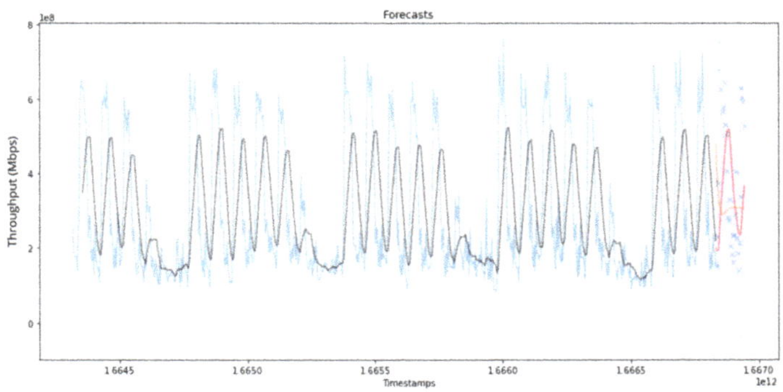

Figure 23. The throughput prediction.

The mean of the testing set is in red while the training set mean is black. The training set has more data points than the testing sample due to the cross-validation process. The prediction set has only 28 data points, and the forecasting is for 24 h ahead. On average, 3 Mbps is predicted for the next 24 h (Figure 23). In this study, the ARIMA code to generate the results is as follows:

```
def forecast(ARIMA model, periods = 0):
n periods = periods
fitted, interval = ARIMA model.predict(n periods = n periods, interval = True)
index = pd.date range(df.index[1], periods = n periods, freq = H)
fitted series = pd.Series(fitted, index = index)
lower series = pd.Series(interval[:, 0], index)
upper series = pd.Series(interval[:, 1], index)
plt.plot(fitted series)
plt.fill between(lower series.index,
lower series,
upper series,
alpha = 0.15)
plt.show()
forecast(ARIMA model, 730)
ARIMA model.summary( )
```

4.2. The Comparative Results of ARIMA and CNN

The CNN is compared to ARIMA using the subscriber and UGRansome datasets. The CNN depends on the predicted timestamps. In what follows, we present the CNN results compared to the results obtained by the ARIMA. The prediction includes 30 to 60 days. The comparative results of the ARIMA and CNN models are in Table 7.

Table 7. The CNN and ARIMA results.

Dataset	Features	*p*-Value	CNN	ARIMA
Subscriber data	450	0.006, 055	85.8%	**92.67%**
UGRansome data	120,000	0.0, 006, 043	88.9%	**91.65%**
Subscriber testing set	300	0.008	86.3%	**94.8%**
UGRansome testing set	60,500	0.007	88%	**95.3%**
Balance	45,312	0.005	87.25%	**93.605%**

The *p*-value is a number between 0 and 1 and can be interpreted as follows:

- A small *p*-value (typically <0.05) indicates strong evidence against the null hypothesis.
- A large *p*-value (>0.05) indicates weak evidence against the null hypothesis.
- The *p*-values very close to the cutoff (0.05) are considered to be marginal.

The reported *p*-values reject the null hypothesis stating a decrease in subscriber data usage. Moreover, the CNN is compared to ARIMA using experimental datasets. We used four samples: the first sample was the subscriber dataset, where the ARIMA model obtained 92% of accuracy and outperformed the CNN. The second sample was the UGRansome dataset containing more features, but the ARIMA model surpassed the CNN with 91% of accuracy. The third sample was the testing sample of the subscriber data where the ARIMA achieved 94%. In the last sample, the ARIMA accuracy outperformed the CNN with 95% of accuracy. Overall, the ARIMA model achieved the best results in all undertaken comparisons. The ARIMA model performed better with the UGRansome data, and this was due to the nature of seasonal network traffic. We computed our models on fewer features of the subscriber data without producing poor results. We believe this is due to time series data properties which improve the balanced accuracy with 93% of accuracy.

4.3. Execution Speed Test

The results showed a CNN not outperforming the ARIMA in terms of accuracy, whilst the ARIMA performed better than the CNN model in terms of execution speed by a factor of 43 for more than 80,000 rows. Table 8 summarises the speed test results for both models.

Table 8. Computational speed test results.

	ARIMA (s)	CNN (s)
0 rows	0	0
10 rows	0.44	4.39
100 rows	0.64	10.43
1000 rows	2.64	75.60
10,000 rows	18.24	685.76
100,000 rows	159.87	6951.91

A time Python package is used before and after each model's execution to measure how long it took to complete the entire procedure. The results are presented right after each execution is complete. The data point for each model is then plotted in batches of ten multiples (10 rows, 100 rows, 1000 rows, etc.). This gives a general notion of how each model's execution time grows as the number of data increases. Furthermore, this test is conducted using parameters for both the SARIMA and CNN models respectively. From the experimental results, it is evident that on average the ARIMA model outperforms the CNN in terms of accuracy. Table 8 showcases the speed test results for both models up to 100,000 rows. The ARIMA outperforms the CNN in terms of execution speed. In Table 8, which shows a linear growth for both models, the average rate of change can be determined by computing the following slope for each graph (Figure 24):

$$\therefore\ slope = \frac{\Delta y}{\Delta x}, \boxed{\text{ARIMA}}\ Slope = \frac{159.87 - 0}{100000 - 0} = 0.0016, \boxed{\text{CNN}}\ Slope = \frac{6951.91 - 0}{100000 - 0} = 0.069$$

Figure 24. The slope computation.

What can be deduced from the slopes is that on average, it takes 0.0016 s for the ARIMA model to execute one row, and 0.069 s for the CNN to execute the same row, thus making the ARIMA 43$\times$ $\frac{0.069}{0.0016}$ faster than the CNN model. Nevertheless, different parameters will yield different results. It can thus be argued that further tuning about the CNN would yield even better results. To put it into perspective, it took under three minutes for the ARIMA to successfully execute 100,000 rows of data as compared to the CNN which took nearly two hours to complete a similar task. This margin will only widen as more rows are introduced for training between the two models. In addition, both models must be fed a sufficient amount of data to successfully train and produce the best results. Furthermore, additional yearly data would provide findings that would boost the study's credibility in addition to pointing out parameter tuning issues for both models. Finding the ideal parameters was challenging in light of the above, especially with the CNN model. One option which was used in this research was to draw ideas from other models used in related contexts in addition to using the rule to fine-tune the model. Fortunately, this was not the case with the ARIMA model since the tuning process was relatively simple and was not based on trial and error as it was with the CNN. However, this process can further be improved for the ARIMA by a simple automated step-wise search using an Akaike Information Criterion (AIC). Developed by the Japanese statistician Hirotugu Akaike [51], AIC is used to evaluate various potential model parameters and choose the one that best fits the data. Fortunately, an AUTO ARIMA (.) function can be imported from Python's PMDARIMA library which can be used to compute the AIC for the ARIMA model. The main goal would be to develop ARIMA parameters with the lowest

AIC. For example, the ARIMA parameters for patterns in the experimental datasets were (2, 1, 2). However, if we were to perform an automated step-wise search using the AUTO ARIMA in Python (Figure 25), then the parameters would be (2, 0, 0) since it has the lowest AIC value at −103.947.

```
Performing stepwise search to minimize aic
 ARIMA(2,0,2)(0,0,0)[0] intercept   : AIC=-101.099, Time=0.11 sec
 ARIMA(0,0,0)(0,0,0)[0] intercept   : AIC=-99.831, Time=0.02 sec
 ARIMA(1,0,0)(0,0,0)[0] intercept   : AIC=-103.468, Time=0.03 sec
 ARIMA(0,0,1)(0,0,0)[0] intercept   : AIC=-101.532, Time=0.03 sec
 ARIMA(0,0,0)(0,0,0)[0]             : AIC=-55.837, Time=0.01 sec
 ARIMA(2,0,0)(0,0,0)[0] intercept   : AIC=-103.947, Time=0.05 sec
 ARIMA(3,0,0)(0,0,0)[0] intercept   : AIC=-101.948, Time=0.13 sec
 ARIMA(2,0,1)(0,0,0)[0] intercept   : AIC=-101.947, Time=0.07 sec
 ARIMA(1,0,1)(0,0,0)[0] intercept   : AIC=-103.001, Time=0.05 sec
 ARIMA(3,0,1)(0,0,0)[0] intercept   : AIC=-102.836, Time=0.15 sec
 ARIMA(2,0,0)(0,0,0)[0]             : AIC=-100.677, Time=0.03 sec

Best model:  ARIMA(2,0,0)(0,0,0)[0] intercept
Total fit time: 0.684 seconds
```

Figure 25. Step-wise AIC results using subscriber data.

It is thus safe to conclude that future ARIMA models can be tuned using the automated AIC step-wise search [51]. Due to the limited experience with constructing CNN models from scratch, the TensorFlow library was used to fully realize the model into practice. Unfortunately, due to the high level of abstraction of the library, it can be very challenging to realize full control over models built using TensorFlow. For example, one is unable to tune the weights of the model, thus resulting in mild changes over the final results every time the model is run due to TensorFlow's frequent changes concerning weights in the background, thus it is unclear precisely what type of network is being constructed in the background, which can result in uncertainty with regards to the performance of the model. Unfortunately, other libraries such as PyTorch and Theano exhibit similarly high levels of abstraction.

4.4. ARIMA, CNN, BATS, and TBATS Comparison

We have compared the obtained results with standard forecasting methods such as BATS and TBATS. BATS is an exponential smoothing technique that handles non-linear data. The advantage of using BATS is that it can treat non-linear patterns, resolve the autocorrelation issue, and account for multiple seasonality. However, BATS is computationally expensive with a large seasonal period. Hence, it is not suitable for hourly data. Thus, the TBATS model was developed to address this limitation. It represents each seasonal period as a trigonometric representation based on the Fourier series. This allows the model to fit large seasonal periods. It is thus a better choice when dealing with high-frequency data, and it usually fits faster than BATS. Figure 26 shows the comparative results with the testing sample and original subscriber data (baseline). The BATS and TBATS prediction are also illustrated in Figure 27. The TBATS outperformed the BATS with a Mean Absolute Percentage Error (MAPE) of 32.63 (Figure 27). This metric defines the accuracy of the forecasting method. The MAPE represents the average of the absolute percentage errors of each entry in a dataset to calculate how accurate the forecasted quantities were in comparison with the actual quantities. A maximum value of 6.5 KBS was predicted by the TBATS in terms of the network traffic volume (Figure 26). As such, the ARIMA model could still predict more subscriber data usage growth compared to the BATS and TBATS models. This is because ARIMA models have a fixed structure and are specifically built for time series or sequential features. It is also due to the non-modifications of predefined parameters before their implementation on time series data.

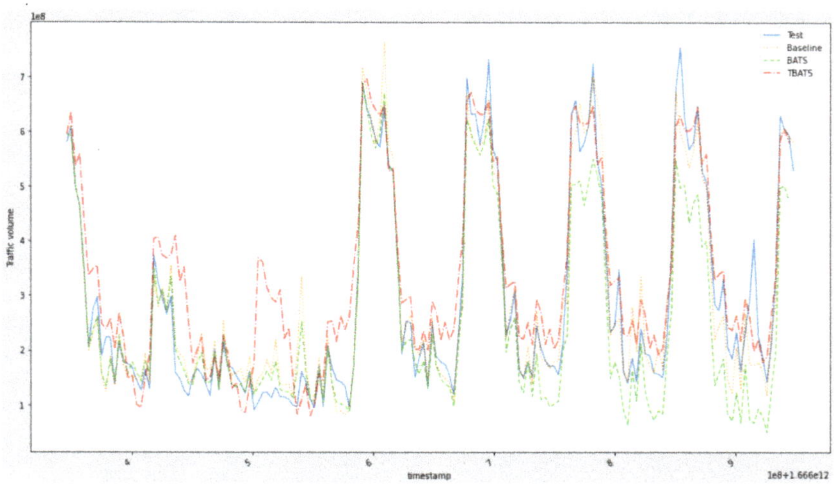

Figure 26. Standard forecasting approach comparison.

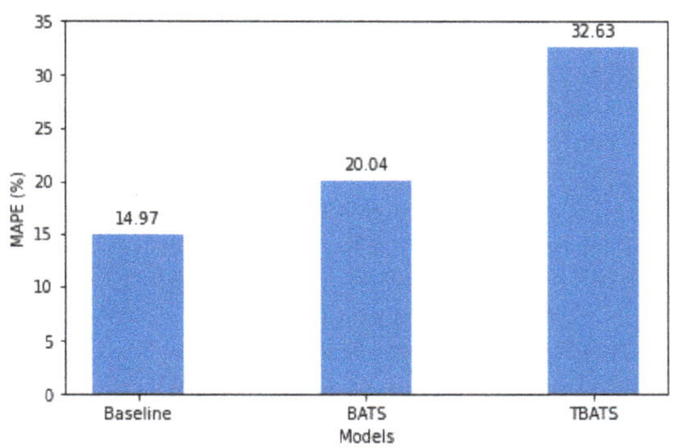

Figure 27. BATS and TBATS models comparison.

4.5. Recommendation

In general, the ARIMA model achieved the best predictive accuracy results on the subscriber data. With regards to model execution speed, it was posited that the ARIMA would perform better than the CNN due to the CNN's sequential weight computation for each hidden layer. However, it was not expected to outperform CNN by such a huge margin, where there was no need to validate future performance using the student *t*-test. Nonetheless, there are various ways to improve the execution time of CNN:

1. Python is unfortunately not the fastest language, and it is thus recommended to build ML models using low-level programming languages such as C, and C++ for faster processing times.
2. Running ML models locally via a Central Processing Unit (CPU) or Graphics Processing Unit (GPU) can slow down the learning process due to memory limits, and it is thus recommended to run models using cloud technologies such as Google Colab or Amazon Web Services (AWS) and SageMaker where memory can be defined before the model is run.
3. Decreasing the number of neurons that makes up the model can reduce the processing time. However, this will be at the expense of model accuracy since fewer neurons result in model underfitting with poor performance when new data is introduced.

4. Similar to decreasing the number of neurons, decreasing the number of epochs will also reduce the final model run time, however, at the expense of accuracy since reducing the number of epochs results in underfitting the model.

5. Conclusions and Discussion

Insights retrieval from subscriber data impacts the telecommunication landscape to facilitate information management and assist decision-makers in predicting the future using ML techniques. We explore time series forecasting analysis and predict subscriber usage trends on the network using the ARIMA model. The unknown forecasting value used by ARIMA relied on historical data. However, we used the data storage to build the subscriber dataset using hourly traffic statistics. We used various metrics to evaluate the ARIMA model. For instance, the normal Q-Q, standardized residual, theoretical quantile, correlogram, and accuracy. UGRansome was used to compare the obtained results that demonstrate similar accuracy values of 90% using the CNN model. The subscriber data was stationary but exhibited less seasonality. In the experimentation, ARIMA was compared to the CNN and achieved the best results with the UGRansome data. We have used an NSDM environment with subscriber data in a secure environment to retrieve relevant patterns such as timestamps and incoming/outgoing throughout to build the subscriber dataset. The variation of the auto-regressive and moving average components identified the most optimal features for obtaining precise predictive values. In addition, the subscriber data have normal distributions, but the UGRansome has more dependent variables. The ARIMA model predicted a growth of 3 Mbps with a maximum data usage growth of 14 Gbps. Furthermore, the performance concerning accuracy showed that ARIMA was superior to CNN. Concerning execution speed, the ARIMA outperforms the CNN by a 43:1 ratio for 100,000 rows. However, we recommend utilizing both methods depending on whether speed or accuracy is a priority. In the future, we will explore additional forecasting models by combining classical mathematical algorithms and compare the performance with Neural Networks, specifically Recurrent Neural Networks (RNN) using ensemble learning approaches. Lastly, it will be also better to explore other factors that affect subscriber's data usage such as multivariate forecasting.

Funding: The author wishes to extend his sincere appreciation and gratitude to the Editor-in-Chief for the invitation to contribute to this article entitled to a fee waiver discount.

Institutional Review Board Statement: Not applicable.

Informed Consent Statement: Not applicable.

Data Availability Statement: The dataset and code used can be obtained upon request or downloaded at https://www.researchgate.net/publication/342200905_An_Ensemble_Learning_Framework_for_Anomaly_Detection_in_the_Existing_Network_Intrusion_Detection_Landscape (Public Files/Ugransome.zip and subscriber data.csv). The code is under (Public Files/ARIMA). Accessed on 12 December 2022.

Acknowledgments: The author would like to thank his Ph.D. supervisor Jacobus Phillipus van Deventer from the University of Pretoria, Faculty of Informatics, who supervised this research, and Maven Systems Worx (Pty) Ltd for granting access to the subscriber data.

Conflicts of Interest: The author declares no conflict of interests.

Abbreviations

The following abbreviations are used in this manuscript:

ADFT	Augmented Dickey-Fuller Test
AWS	Amazon Web Services
ARIMA	Auto-Regressive Integrated Moving Average
AR	Auto-Regressive
CCA	Canonical Correlation Analysis
CPU	Central Processing Unit

CSV	Comma-Separated Values
CNN	Convolutional Neural Network
DL	Deep Learning
DPI	Deep Packet Inspection
DT	Decision Trees
DTW	Dynamic Time Warping
EDA	Exploratory Data Analysis
GRU	Gated Recurrent Unit
GD	Gradient Descent
GPU	Graphics Processing Unit
Tpt in	Incoming Throughput
IP	Internet Protocol
IDS	Insights Data Storage
KPI	Key Performance Indicator
KNN	K-Nearest Neighbor
LDA	Linear Discriminant Analysis
LSTM	Long Short Term Memory
ML	Machine Learning
MSE	Mean Squared Error
MA	Moving Average
MAPE	Mean Absolute Percentage Error
Ms	Milliseconds
NN	Neural Networks
NSDM	Network Subscriber Data Management
OS	Operating System
Tpt out	Outgoing Throughput
PCA	Principal Components Analysis
QoE	Quality of Experience
Prob (Q)	Quantile Probability
RF	Random Forest
ReLU	Rectified Linear Unit
RNN	Recurrent Neural Networks
RMSE	Root Mean Square Error
SARIMA	Seasonal ARIMA
SQL	Structured Query Language
SVM	Support Vector Machine
TSA	Time Series Analysis
Ts	Timestamps
UDP	User Datagram Protocol

References

1. Nkongolo, M.; van Deventer, J.P.; Kasongo, S.M.; van der Walt, W. Classifying Social Media Using Deep Packet Inspection Data. In *Inventive Communication and Computational Technologies*; Ranganathan, G., Fernando, X., Rocha, Á., Eds.; Springer: Singapore, 2023; pp. 543–557. [CrossRef]
2. Theodoridis, G.; Tsadiras, A. Applying machine learning techniques to predict and explain subscriber churn of an online drug information platform. *Neural Comput. Appl.* **2022**, *34*, 19501–19514. [CrossRef]
3. Kumar, R.; Kumar, P.; Kumar, Y. Multi-step time series analysis and forecasting strategy using ARIMA and evolutionary algorithms. *Int. J. Inf. Technol.* **2022**, *14*, 359–373. [CrossRef]
4. Li, X.; Petropoulos, F.; Kang, Y. Improving forecasting by subsampling seasonal time series. *Int. J. Prod. Res.* **2022**, 1–17. [CrossRef]
5. Jin, X.B.; Gong, W.T.; Kong, J.L.; Bai, Y.T.; Su, T.L. A variational Bayesian deep network with data self-screening layer for massive time-series data forecasting. *Entropy* **2022**, *24*, 335. [CrossRef] [PubMed]
6. Box, G.E.; Jenkins, G.M.; Reinsel, G.C.; Ljung, G.M. *Time Series Analysis: Forecasting and Control*; John Wiley & Sons: Hoboken, NJ, USA, 2015.
7. Adhikari, R.; Agrawal, R.K. An introductory study on time series modeling and forecasting. *arXiv* **2013**, arXiv:1302.6613.
8. Khashei, M.; Bijari, M. A novel hybridization of artificial neural networks and ARIMA models for time series forecasting. *Appl. Soft Comput.* **2011**, *11*, 2664–2675. [CrossRef]

9. Azaria, B.; Gottlieb, L.A. Predicting Subscriber Usage: Analyzing Multidimensional Time-Series Using Convolutional Neural Networks. In *Cyber Security, Cryptology, and Machine Learning*; Dolev, S., Katz, J., Meisels, A., Eds.; Springer: Cham, Switzerland, 2022; pp. 259–269.
10. Salman, A.G.; Heryadi, Y.; Abdurahman, E.; Suparta, W. Weather forecasting using merged long short-term memory model. *Bull. Electr. Eng. Inform.* **2018**, *7*, 377–385. [CrossRef]
11. Masum, S.; Liu, Y.; Chiverton, J. Multi-step time series forecasting of electric load using machine learning models. In Proceedings of the International Conference on Artificial Intelligence and Soft Computing, Zakopane, Poland, 3–7 June 2018; pp. 148–159.
12. Siami-Namini, S.; Tavakoli, N.; Namin, A.S. A comparison of ARIMA and LSTM in forecasting time series. In Proceedings of the 2018 17th IEEE International Conference on Machine Learning and Applications (ICMLA), Orlando, FL, USA, 17–20 December 2018; pp. 1394–1401.
13. Muhammad, U.L.; Musa, M.Y.; Usman, Y.; Nasir, A.B. Limestone as solid mineral to develop national economy. *Am. J. Phys. Chem.* **2018**, *7*, 23–28. [CrossRef]
14. Mbah, T.J.; Ye, H.; Zhang, J.; Long, M. Using LSTM and ARIMA to simulate and predict limestone Price variations. *Min. Metall. Explor.* **2021**, *38*, 913–926. [CrossRef]
15. Tan, C.W.; Bergmeir, C.; Petitjean, F.; Webb, G.I. Time series extrinsic regression. *arXiv* **2020**, arXiv:2006.12672.
16. Goldsmith, J.; Scheipl, F. Estimator selection and combination in scalar-on-function regression. *Comput. Stat. Data Anal.* **2014**, *70*, 362–372. [CrossRef]
17. Pimentel, M.A.; Charlton, P.H.; Clifton, D.A. Probabilistic estimation of respiratory rate from wearable sensors. In *Wearable Electronics Sensors*; Springer: Cham, Switzerland, 2015; pp. 241–262. [CrossRef]
18. Zheng, Y.; Liu, Q.; Chen, E.; Ge, Y.; Zhao, J.L. Time series classification using multi-channels deep convolutional neural networks. In *Web-Age Information Management*; Springer: Cham, Switzerland, 2014; pp. 298–310. [CrossRef]
19. Yang, J.; Nguyen, M.N.; San, P.P.; Li, X.L.; Krishnaswamy, S. Deep convolutional neural networks on multichannel time series for human activity recognition. In Proceedings of the Twenty-Fourth International Joint Conference on Artificial Intelligence, Buenos Aires, Argentina, 25–31 July 2015.
20. Okita, T.; Inoue, S. Recognition of multiple overlapping activities using compositional CNN-LSTM model. In Proceedings of the Proceedings of the 2017 ACM International Joint Conference on Pervasive and Ubiquitous Computing and Proceedings of the 2017 ACM International Symposium on Wearable Computers, Maui, HI, USA, 11–15 September 2017; pp. 165–168. [CrossRef]
21. Wang, J.; Long, Q.; Liu, K.; Xie, Y. Human action recognition on cellphone using compositional bidir-lstm-cnn networks. In Proceedings of the 2019 International Conference on Computer, Network, Communication and Information Systems (CNCI 2019), Qingdao, China, 27–29 March 2019; pp. 687–692.
22. Snow, D. AtsPy: Automated Time Series Forecasting in Python. 2020. Available online: https://papers.ssrn.com/sol3/papers.cfm?abstract_id=3580631 (accessed on 31 December 2022).
23. Mode, G.R.; Hoque, K.A. Adversarial examples in deep learning for multivariate time series regression. In Proceedings of the 2020 IEEE Applied Imagery Pattern Recognition Workshop (AIPR), Washington DC, USA, 13–15 October 2020; pp. 1–10. [CrossRef]
24. Antsfeld, L.; Chidlovskii, B.; Borisov, D. Magnetic sensor based indoor positioning by multi-channel deep regression. In Proceedings of the 18th Conference on Embedded Networked Sensor Systems, Virtual, 16–19 November 2020; pp. 707–708. [CrossRef]
25. Mehtab, S.; Sen, J.; Dasgupta, S. Robust analysis of stock price time series using CNN and LSTM-based deep learning models. In Proceedings of the 2020 4th International Conference on Electronics, Communication and Aerospace Technology (ICECA), Coimbatore, India, 5–7 November 2020; pp. 1481–1486. [CrossRef]
26. Mirko, K.; Kantelhardt, J.W. Hadoop. TS: Large-scale time-series processing. *Int. J. Comput. Appl.* **2013**, *74*, 1–8.
27. Li, L.; Noorian, F.; Moss, D.J.; Leong, P.H. Rolling window time series prediction using MapReduce. In Proceedings of the 2014 IEEE 15th international Conference on Information Reuse and Integration (IEEE IRI 2014), Redwood City, CA, USA, 13–15 August 2014; pp. 757–764. [CrossRef]
28. Talavera-Llames, R.; Pérez-Chacón, R.; Troncoso, A.; Martínez-Álvarez, F. Big data time series forecasting based on nearest neighbours distributed computing with Spark. *Knowl.-Based Syst.* **2018**, *161*, 12–25. [CrossRef]
29. Galicia, A.; Torres, J.F.; Martínez-Álvarez, F.; Troncoso, A. A novel Spark-based multi-step forecasting algorithm for big data time series. *Inf. Sci.* **2018**, *467*, 800–818. [CrossRef]
30. Petropoulos, F.; Apiletti, D.; Assimakopoulos, V.; Babai, M.Z.; Barrow, D.K.; Taieb, S.B.; Bergmeir, C.; Bessa, R.J.; Bijak, J.; Boylan, J.E.; et al. Forecasting: Theory and practice. *Int. J. Forecast.* **2022**, *38*, 705–871. [CrossRef]
31. Shamir, O.; Srebro, N.; Zhang, T. Communication-efficient distributed optimization using an approximate newton-type method. In Proceedings of the International Conference on Machine Learning, Beijing, China, 21–26 June 2014; pp. 1000–1008.
32. Wang, J.; Kolar, M.; Srebro, N.; Zhang, T. Efficient distributed learning with sparsity. In Proceedings of the International Conference on Machine Learning, Sydney, Australia, 6–11 August 2017; pp. 3636–3645.
33. Jordan, M.I.; Lee, J.D.; Yang, Y. Communication-efficient distributed statistical inference. *J. Am. Stat. Assoc.* **2018**, *114*, 668–681. [CrossRef]
34. Chen, X.; Liu, W.; Zhang, Y. Quantile regression under memory constraint. *Ann. Stat.* **2019**, *47*, 3244–3273. [CrossRef]
35. Ryu, E.K.; Yin, W. *Large-Scale Convex Optimization*; Cambridge University Press: Cambridge, MA, USA, 2022.

36. Challu, C.; Olivares, K.G.; Oreshkin, B.N.; Garza, F.; Mergenthaler, M.; Dubrawski, A. N-hits: Neural hierarchical interpolation for time series forecasting. *arXiv* **2022**, arXiv:2201.12886.
37. Fernández, J.D.; Menci, S.P.; Lee, C.M.; Rieger, A.; Fridgen, G. Privacy-preserving federated learning for residential short-term load forecasting. *Appl. Energy* **2022**, *326*, 119915. . [CrossRef]
38. Bennett, S.; Clarkson, J. Time series prediction under distribution shift using differentiable forgetting. *arXiv* **2022**, arXiv:2207.11486.
39. Nkongolo, M.; van Deventer, J.P.; Kasongo, S.M. The Application of Cyclostationary Malware Detection Using Boruta and PCA. In *Computer Networks and Inventive Communication Technologies*; Smys, S., Lafata, P., Palanisamy, R., Kamel, K.A., Eds.; Springer: Singapore, 2023; pp. 547–562. [CrossRef]
40. Nkongolo, M.; Van Deventer, J.P.; Kasongo, S.M.; Zahra, S.R.; Kipongo, J. A Cloud Based Optimization Method for Zero-Day Threats Detection Using Genetic Algorithm and Ensemble Learning. *Electronics* **2022**, *11*, 1749. [CrossRef]
41. Nkongolo, M.; van Deventer, J.P.; Kasongo, S.M. UGRansome1819: A Novel Dataset for Anomaly Detection and Zero-Day Threats. *Information* **2021**, *12*, 405. [CrossRef]
42. Ghaderi, A.; Movahedi, Z. Joint Latency and Energy-aware Data Management Layer for Industrial IoT. In Proceedings of the 2022 8th International Conference on Web Research (ICWR), Tehran, Iran, 11–12 May 2022; pp. 70–75. [CrossRef]
43. Mehdi, H.; Pooranian, Z.; Vinueza Naranjo, P.G. Cloud traffic prediction based on fuzzy ARIMA model with low dependence on historical data. *Trans. Emerg. Telecommun. Technol.* **2022**, *33*, e3731. [CrossRef]
44. Xiao, R.; Feng, Y.; Yan, L.; Ma, Y. Predict stock prices with ARIMA and LSTM. *arXiv* **2022**, arXiv:2209.02407.
45. Wang, X.; Kang, Y.; Hyndman, R.J.; Li, F. Distributed ARIMA models for ultra-long time series. *Int. J. Forecast.* 2022, *in press* . [CrossRef]
46. Chao, H.L.; Liao, W. Fair scheduling in mobile ad hoc networks with channel errors. *IEEE Trans. Wirel. Commun.* **2005**, *4*, 1254–1263. [CrossRef]
47. Nkongolo, M. Classifying search results using neural networks and anomaly detection. *Educor Multidiscip. J.* **2018**, *2*, 102–127.
48. Suthar, F.; Patel, N.; Khanna, S. A Signature-Based Botnet (Emotet) Detection Mechanism. *Int. J. Eng. Trends Technol.* **2022**, *70*, 185–193. [CrossRef]
49. Kotu, V.; Deshpande, B. Chapter 3—Data Exploration. In *Data Science*, 2nd ed.; Kotu, V., Deshpande, B., Eds.; Morgan Kaufmann: Burlington, MA, USA, 2019; pp. 39–64. . [CrossRef]
50. Ij, H. Statistics versus machine learning. *Nat Methods* **2018**, *15*, 233.
51. Akaike, H. A new look at the statistical model identification. *IEEE Trans. Autom. Control* **1974**, *19*, 716–723. [CrossRef]

Disclaimer/Publisher's Note: The statements, opinions and data contained in all publications are solely those of the individual author(s) and contributor(s) and not of MDPI and/or the editor(s). MDPI and/or the editor(s) disclaim responsibility for any injury to people or property resulting from any ideas, methods, instructions or products referred to in the content.

eng

Article

Investigating Metals and Metalloids in Soil at Micrometric Scale Using μ-XRF Spectroscopy—A Case Study

Sofia Barbosa [1,*], António Dias [2], Marta Pacheco [3], Sofia Pessanha [2] and J. António Almeida [1]

[1] GeoBioTec GeoBioSciences, GeoTechnologies and GeoEngineering & NOVA FCT (Department of Earth Sciences), Faculdade de Ciências e Tecnologia, 2829-516 Caparica, Portugal
[2] LIBPhys & NOVA FCT (Department of Physics), 2829-516 Caparica, Portugal
[3] NOVA FCT (Department of Earth Sciences), 2829-516 Caparica, Portugal
* Correspondence: svtb@fct.unl.pt; Tel.: +351-212-948-573

Abstract: Micrometric 2D mapping of distinct elements was performed in distinct soil grain-size fractions of a sample using the micro-X-ray Fluorescence (μ-XRF) technique. The sample was collected in the vicinity of São Domingos, an old mine of massive sulphide minerals located in the Portuguese Iberian Pyrite Belt. As expected, elemental high-grade concentrations of distinct metals and metalloids in the dependence of the existent natural geochemical anomaly were detected. Clustering and k-means statistical analysis were developed considering Red–Green–Blue (RGB) pixel proportions in the produced 2D micrometric image maps, allowing for the identification of elemental spatial distributions at 2D. The results evidence how elemental composition varies significantly at the micrometric scale per grain-size class, and how chemical elements present irregular spatial distributions in the direct dependence of distinct mineral spatial distributions. Due to this fact, elemental composition is more differentiated in coarser grain-size classes, whereas griding-milled fraction does not always represent the average of all partial grain-size fractions. Despite the complexity of the performed analysis, the achieved results evidence the suitability of μ-XRF to characterize natural, heterogeneous, granular soils samples at the micrometric scale, being a very promising investigation technique of high resolution.

Keywords: soil matrix; metal distribution per grain fraction; micro-X-ray elemental mapping; RGB clustering image analysis; k-means

Citation: Barbosa, S.; Dias, A.; Pacheco, M.; Pessanha, S.; Almeida, J.A. Investigating Metals and Metalloids in Soil at Micrometric Scale Using μ-XRF Spectroscopy—A Case Study. *Eng* **2023**, *4*, 136–150. https://doi.org/10.3390/eng4010008

Academic Editor: Antonio Gil Bravo

Received: 4 November 2022
Revised: 20 December 2022
Accepted: 21 December 2022
Published: 2 January 2023

Copyright: © 2023 by the authors. Licensee MDPI, Basel, Switzerland. This article is an open access article distributed under the terms and conditions of the Creative Commons Attribution (CC BY) license (https://creativecommons.org/licenses/by/4.0/).

1. Introduction

Quantification, imaging, and data processing of micro-X-ray Fluorescence (μ-XRF) outputs are presently an interesting but also very challenging area of investigation. To obtain the elemental distribution of a sample, specific instrumentation that provides precise positioning and good energy resolution must be used. Micro-XRF imaging spectrometers rely on scanning samples along the X and Y directions, with a micro-X-ray beam irradiating a region of interest (ROI), point by point [1]. Recent developments in μ-XRF consider quantitative analysis using fundamental parameter-based 'standardless' quantification algorithms [2,3].

The works developed by [2,4–6] evidence the suitability of this technique for various applications within the earth sciences. Further, 2D high-resolution chemical distribution maps can be used as qualitative multi-element maps or as semiquantitative single-element maps through which bulk and phase-specific geochemical data sets can be established [4].

In [2], the authors discuss the accuracy and precision of these quantitative analyses by using a simple-type calibration against a certified reference material of similar matrix and composition. μ-XRF is a non-destructive technique and leaves samples intact for other types of analyses, such as Raman spectroscopy or X-ray diffraction, which allow for the characterization of molecular components [7]. The use of μ-XRF in conjunction with these established methods of molecular analysis allows for a more complete characterization of grains and particles [2,8,9]. Heterogeneous samples, such as soils, are much harder to characterize. Both single particle as well as bulk analyses must be performed on sample specimens to ensure a full description by μ-XRF [8]. Its consideration to analyse bulk samples of soil implies, necessarily, a clear elemental identification and the distinction between different occurring grades [10]. Quantification of soil data by μ-XRF is still a topic of considerable investigation interest and has been reported only in a limited number of publications [11,12]. Recent research studies evidence how statistical and geostatistical techniques can be applied to co-relate distinct imaging results [13] and how it is already possible to generate 3D maps of chemical properties at the micrometric scale by combining 2D SEM-EDX data with 3D X-ray computed tomography images [14–16]. Effectiveness and potentialities that result from the integration of results of micro-X ray and SEM techniques are also well demonstrated by distinct researchers, even in the cases of very irregular, porous matrixes [14–16]. In fluorescence microscopy, colocalization refers to observation of the spatial overlap between two (or more) different fluorescent labels, each having a separate emission wavelength. Ref. [13] discussed co-localization analysis processes in the context of increasingly popular super-resolution imaging technique occurrence versus correlation, although this limits image pixel-based processing techniques. Ref. [15] developed a method to generate 3D maps of soil chemical properties at the microscale by combining 2D SEM-EDX data with 3D X-ray computed tomography images. The spatial correlation between the X-ray grayscale intensities and the chemical maps made it possible to use a regression-tree model as an initial step to predict 3D chemical composition.

Bulk-sample analysis is a test method used when individual particulate samples are not representative or are not obtained for a certain type of material. Particulate products, such as soils, granulated powders, dusts, or foodstuffs, are usually analysed through bulk-sampling principles [8]. The microscopic analysis of a heterogeneous matrix, such as bulk soil samples, with μ-XRF is complex but has unique potentialities.

The present work is an introductory study in which 2D image clustering analysis based on μ-XRF XY scanning maps of a soil sample was performed. The case study, a soil sample denominated as SD1, was collected at the former mine of São Domingos in Mértola, Portugal (Figure 1). São Domingos Mine is located at the Iberian Pyrite Belt (IPB). It is a world-renowned massive sulphide ore deposit, mainly exploited for its copper contents. High concentrations of As, Zn, and Pb area also found. Its exploitation started prior to the Roman occupation period, mainly for Au, Ag, and Pb. Due to the mine's extensive exploitation over the centuries, the area is filled with very heterogeneous mining waste. Natural gossan (iron caps) deposits and natural local mineralogy results in the generation of heterogeneous soils with high contents of several heavy metals and metalloids. At this mining site, the geology is dominated by greywackes and quartzwackes, quartzites, phyllites, schists, forming the "Baixo Alentejo" Flysch Group, turbidites, and a volcano–sedimentary complex. The lithostratigraphic units range mainly from the Devonian to the Carboniferous periods [17,18]. Due to its mining context and its local geology, the most common elements found in the soils around the mining area are, mainly, Al, Si, S, Ti, Mn, Cr, Fe, Cu, Zn, As, Ga, Pb, Sb, and Hg [19,20].

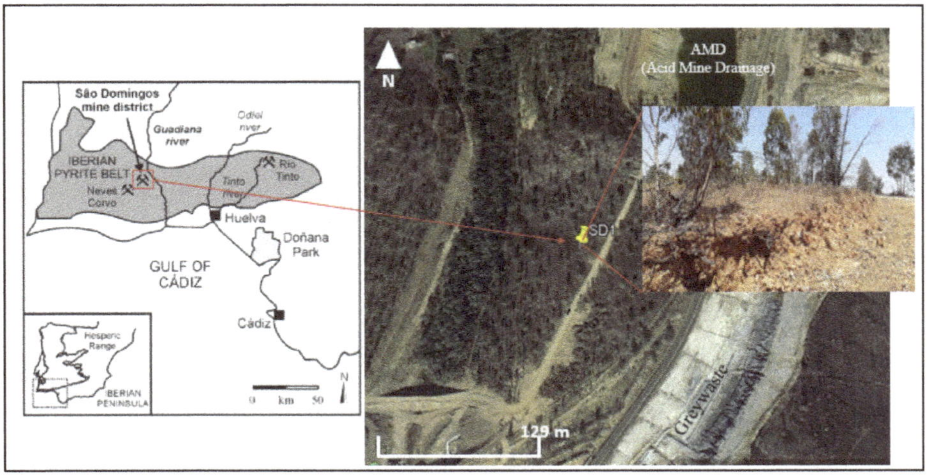

Figure 1. Location of São Domingos mine and location of the collected SD1 sample. Left figure is adapted from [17].

2. Materials and Methods

2.1. Sampling and Sample Preparation

The soil sample was collected with the aid of a small shovel, scooping the surface soil to a depth of about 10 to 20 cm. About 1.50 kg of material was collected, stored, and labelled adequately. SD1 consists of a reddish-brown soil with small to large particles (Figure 2). The sample was sieved into four classes of grain size, $\geq$2 to <3 mm, <2 mm to $\geq$500 µm, <500 µm to $\geq$250 µm, and <250 µm. A ground and milled bulk sample (TM, "Total Milled") was also prepared. Depending on the availability of the material, and using a manual benchtop press, two to five pellets were made from all the granulometry-size fractions and TM. Table 1 shows the number of pellets analysed by category. These pellets were analysed with a benchtop micro-XRF spectrometer, M4 TORNADO by Bruker (Billerica, MA, USA).

Figure 2. Pellet of an original SD1 sample of grain size fraction "<2 mm to $\geq$500 µm" (image source: Bruker's M4 TORNADO camera).

Table 1. Number of pellets by category.

Categories/Fraction	Number of Pellets (SD1)
TM	2
≥2 mm to <3 mm	2
<2 mm to ≥500 µm	5
<500 µm to ≥250 µm	3
<250 µm	3

2.2. Micro X-ray Fluorescence Multi-Point Measurements and 2D Image Mapping

The micro-X-ray fluorescence technique is applied by means of the energy dispersive spectrometer M4 TORNADO by Bruker. This instrument consists of a low-power X-ray tube with a Rh anode, which was operated in this case study at 50 kV and 300 uA. Placed after the X-ray tube, a poly-capillary lens focuses the beam to a spot size that can go down to 25 µm for Mo-Kα. This way, by selecting an area in the sample, point-by-point measurements can be performed and images of elemental distributions within the sample are generated.

In the case study, the pellets were analysed making use of an AlTiCu 100/50/25 µm filter composition. For elements emitting radiations from 5 to 35 keV, it is adequate to use filters that can lessen the effect of the Bremsstrahlung radiation that contribute to background radiation [21]. Therefore, for SD1, the two filters mentioned above were used due to the presence of elements with an atomic number (Z) superior to 21, i.e., from Titanium (Ti) to Yttrium (Y), which were identified in a primary analysis without filters.

The measurements were taken under 20 mbar vacuum conditions (to improve detection limits), with a step size of 15 µm and 10 ms acquisition per spectrum rendering for, on average, 1 h 30 min to ensure high-resolution 2D maps for each element.

Data treatment of micro-2D mapping was performed using the M4 TORNADO inbuilt software MQuant.

That is to say, only one pellet for each of the sample categories—TM, ≥2 mm to <3 mm, <2 mm to ≥500 µm, <500 µm to ≥250 µm, <250 µm—was chosen for 2D map surveys due to the big amount of data obtained.

2.3. Two-Dimensional Image Mapping Processing: Clustering RGB Pixel Analysis

µ-XRF 2D mapping outputs consisted of 2D image files. Possibilities related with the processing of these image files are mainly related with pixel quantification and statistical analysis of its distributions. In this case study, each image refers to a certain element spatial distribution for which its occurrence and concentration are locally represented by a certain intensity of a certain RGB (Red, Green, Blue) colour. The highest elemental concentrations are represented by the highest RGB light colour proportions (Figure 3).

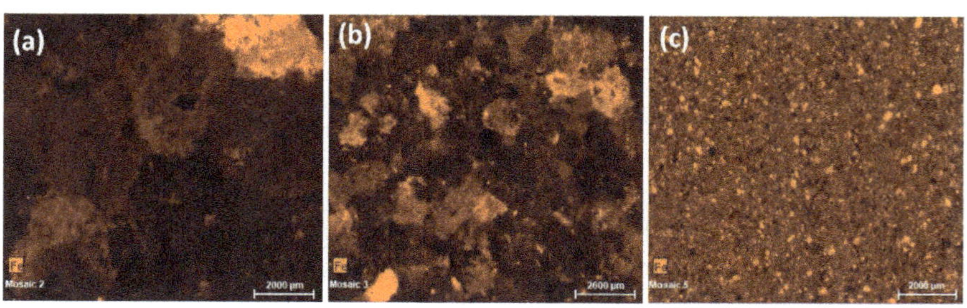

Figure 3. µ-XRF 2D mapping outputs for the element Iron (Fe). (**a**) Grain-size distribution: ≥2 mm to <3 mm; (**b**) <2 mm to ≥500 µm; (**c**) <250 µm.

Pixel proportion quantifications per distinct RGB colour intensity were established with R©Countcolors Package [22–25]). This package was developed originally with the aim of quantifying the area of white-nose syndrome infection of bat wings [25]. R©Countcolors Package allows users to quantify regions of an image by distinct colours. It is an R package that counts colours within specified colour ranges in image files and provides a masked version of the image with targeted pixels changed to a different selected colour by the utilizer. This package integrates techniques from image processing without using any machine learning, adaptive thresholding, or object-based detection, which make it reliable and easy to use but limited in terms of application.

The principle of the image processing analysis consisted of considering each RGB colour in three dimensions, where each colour is defined by its coordinates in R (red), G (green), and B (blue) axes. The range of each RGB colour is, thus, interpreted in a 3D space (Figure 4a). The quantitative RGB pixel analysis performed for each 2D image begins with the verification of the level of similarity of colour intensities according to its respective RGB code. Each RGB code represents a certain RGB cluster (and, thus, a certain colour intensity). RGB pixels per cluster are counted by samples of 10,000 pixels from the 2D image. For each RGB code representing a certain cluster, its respective frequencies are calculated (Figure 4b,c). Figure 4 presents an exemplification of the pixel-counting frequencies for six distinct colour clusters representing the concentrations of the element Fe. The pixels of more light-colour clusters represent the locations with highest concentrations on Fe. The number of clusters and the number of the sampling pixels are established by the user.

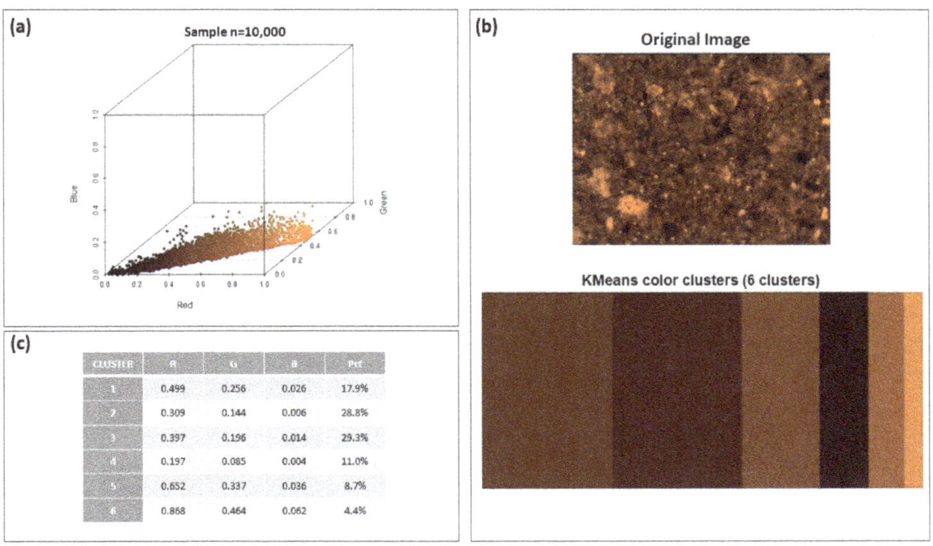

CLUSTER	R	G	B	Pct
1	0.499	0.256	0.026	17.9%
2	0.309	0.144	0.006	28.8%
3	0.397	0.196	0.014	29.3%
4	0.197	0.085	0.004	11.0%
5	0.652	0.337	0.036	8.7%
6	0.868	0.464	0.062	4.4%

Figure 4. RGB clustering analysis of a μ-XRF 2D map (element: Fe; pellet of a bulk sample). (**a**) RGB counting colours in three dimensions (sample size n = 10,000 pixels); (**b**) Pixel classification in 6 clusters; (**c**) RGB pixel proportions for each cluster.

One of the main objectives of this case study was to estimate the areas that are associated with a certain range of RGB pixels. The light-colour ranges that are associated with the highest colour intensities represent the highest elemental concentrations. In the adopted methodology, after selecting the colour clusters that are the most representative for a certain element occurrence, its respective areas are estimated. The images processed always integrate degrees of intensity of a unique colour, which relates to a certain element to be identified. The element occurrence is represented by the light-coloured clusters in each colour image. In [5], following the principles described in [23–25], the authors defined an analysis methodology based on two options: one that considers upper and lower limits for each colour range and where a box-shaped border is drawn around the region of that range (rectangular range) and a second option that considers the selection of a certain central colour and a search radius around it, were a "sphere" for the considered colour range is drawn (spherical range). Due to the possibilities of applying distinct criteria, estimated area calculations are referenced in terms of percentages of minimum and maximum probable areas (Figure 5). In fact, the calculated areas have distinct possibilities, directly dependent on the number of colour clusters and the search criteria, which are, in turn, user defined. Due to these distinct possibilities, it is more correct to suggest a range of probable estimated areas than to present only a specific estimated area. For this, the adopted methodology integrates the possibility of considering the search criteria to one, two, or three colour clusters simultaneously (Figure 5). When two or three colour clusters are to be considered, a search radius is applied to each colour. For minimum area calculations, it is advisable to consider "one colour cluster" with spherical or rectangular search criteria or "two colour clusters" procedures. To calculate possible maximum estimated areas, it is advisable to simultaneously consider "three colour clusters" for the estimations.

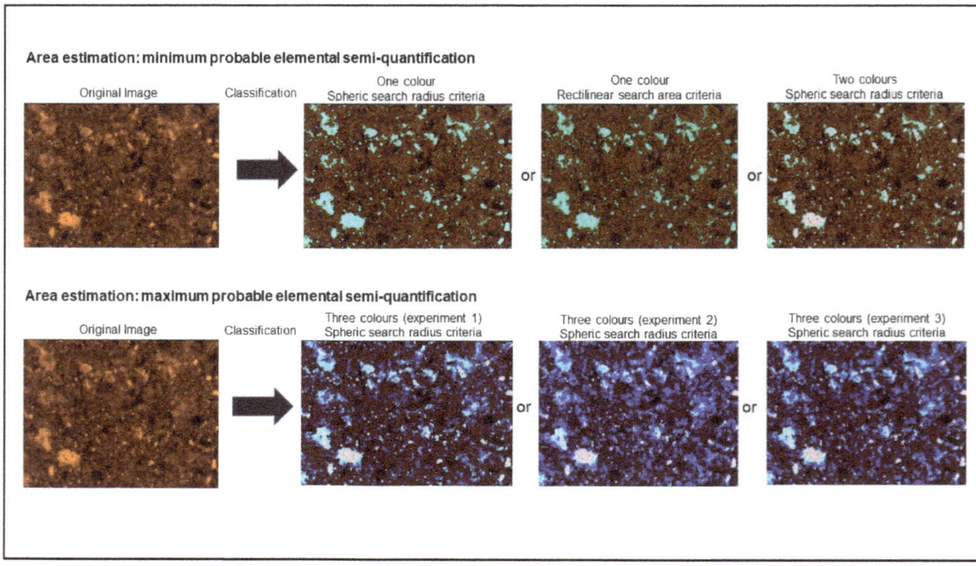

Figure 5. Methodology applied to estimate minimum and maximum probable elemental occurrence in a μ-XRF 2D map (example of element Fe in a bulk pellet sample).

This methodology allows one to accomplish a semi-quantitative analysis of the μ-XRF 2D mapping images. Uncertainty is mostly associated with the clustering classification and search criteria, which are user defined. The described methodology has already been applied to granular mining waste samples [5] and to a syenite nepheline rock sample in order to identify incompatible and scarce metals at the micrometric scale [5]. Results evidence the potentiality of this methodology to interpret elemental μ-XRF 2D mapping images of materials with heterogenous granular textures, such as soils and mining wastes, being also quite promising in elemental and mineral identification of distinct rock matrix [5,6].

3. Results—Elemental μ-2D Mapping Distributions

Through multi-point measurement analysis, it was possible to identify, in sample SD1, the following elements per size fraction class: aluminium (Al), silicon (Si), potassium (K), calcium (Ca), titanium (Ti), manganese (Mn), iron (Fe), nickel (Ni), copper (Cu), zinc (Zn), gallium (Ga), arsenic (As), rubidium (Rb), strontium (Sr), and yttrium (Y). Figure 6 presents the results achieved for the methodology applied for the case of the element Fe.

Estimations of minimum and maximum probable Fe occurrence in μ-XRF 2D maps are presented. Analogous results are presented for the elements Ca, Mn, Cu, Zn, and As in Appendix A.

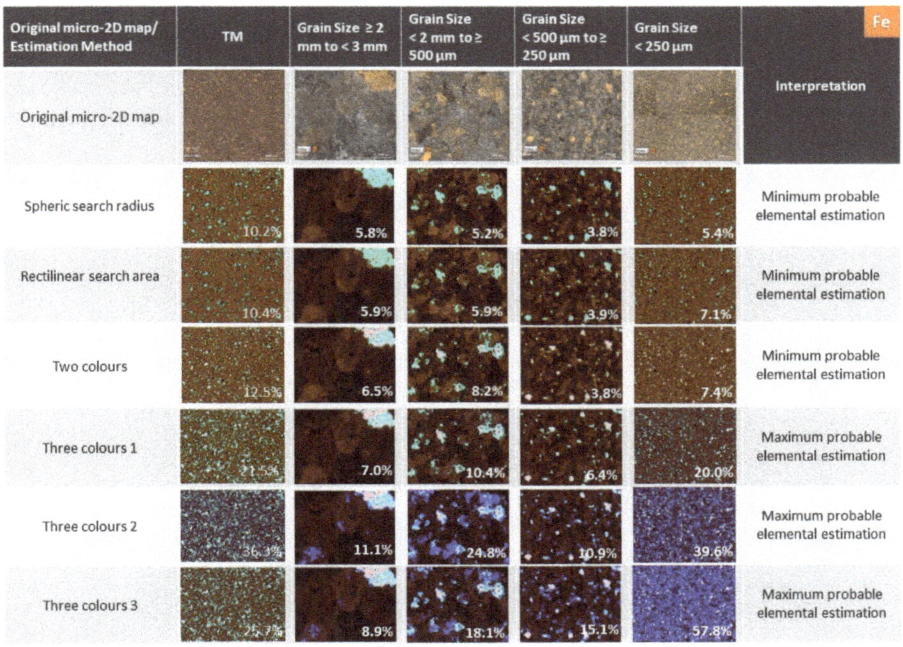

Figure 6. Minimum and maximum probable elemental occurrence in μ-XRF 2D map (percentage of area %) for Fe.

As can be observed, the difference in spatial distribution patterns and the estimated minimum and maximum elemental quantities is clear according to grain-size fractions. Further, patterns of TM (ground and milled) are more similar to grain-size fraction "<250 μm". This behavioural pattern can be observed in most of the analysed elements (Appendix A). Quantities per element are estimated in percentage (%) of the total mapped area and vary according to grinding, milling, and grain-size fraction (Figure 6, Appendix A and Figure 7). Bulk milled samples do not always represent the average between the distinct size fractions. In fact, for some elements, coarser gain-size fractions, such as "≥2 mm to <3 mm" and "<2 mm to ≥500 μm", tend to be present in distinct estimated quantities (Figure 6, Appendix A and Figure 7). These two facts are indicative of the occurrence of some elements in the direct dependence of the mineralogy and, in turn, in the dependence of its more representative granulometry. Table 2 includes a summary of the minimum and maximum elemental occurrence in the μ-XRF 2D map (percentage of area, %) of the elements Al, Si, K, Ca, Ti, Mn, Fe, Ni, Cu, Zn, Ga, As, Sr, and Y.

Elements presented in higher estimated percentages evidence the influence of the local geology in the soil's constitution [17,26–28]. Figure 8 presents some of the most representative results considering maximum estimated percentages of elemental occurrence area (%). The elements presented in this Figure, Si, Al, Cu, Zn, Ca, K, Ti, Fe, As, Ga, and Mn, are grouped according to their respective percentage of occurrence area (%). The results reflect not only the natural composition of soil (Si, Al, Ca, K, Ti) but also the presence of natural geochemical anomalies, which are related to the existence of massive sulphide ore deposit minerals, increasing the percentages of occurrence of Cu, Zn, Fe, As, Ga, and Mn among other elements. Apart from Si and Al, the elements Cu, Zn, Ca, K, Ti, Fe, As,

Ga, and Mn present specific spatial distribution patterns. For the case of Fe, As, Ga, and Mn, the dependence on coarser minerals is quite evident. Spatial overlap of the elements according to mineralogy is also possible to observe. In this context, the spatial overlap of Fe, As, and Ga is an example and is a consequence of the local geochemistry and mineralogy, which includes iron oxides and sulphides [17,26–28]. Simultaneously, the presence of As and Fe can be explained by the existence of arsenic-bearing sulfides, such as arsenopyrite or sulfosalts. The presence of Ga in the soil is usually connected with the occurrence of silty minerals. Ga tends to be sorbed by Fe(III) and Mn(III) oxides [29,30] and occurs as an impurity in iron oxides, hydroxides, and sphalerite minerals, which can explain the spatial correspondence between Fe and Ga in the SD1 sample.

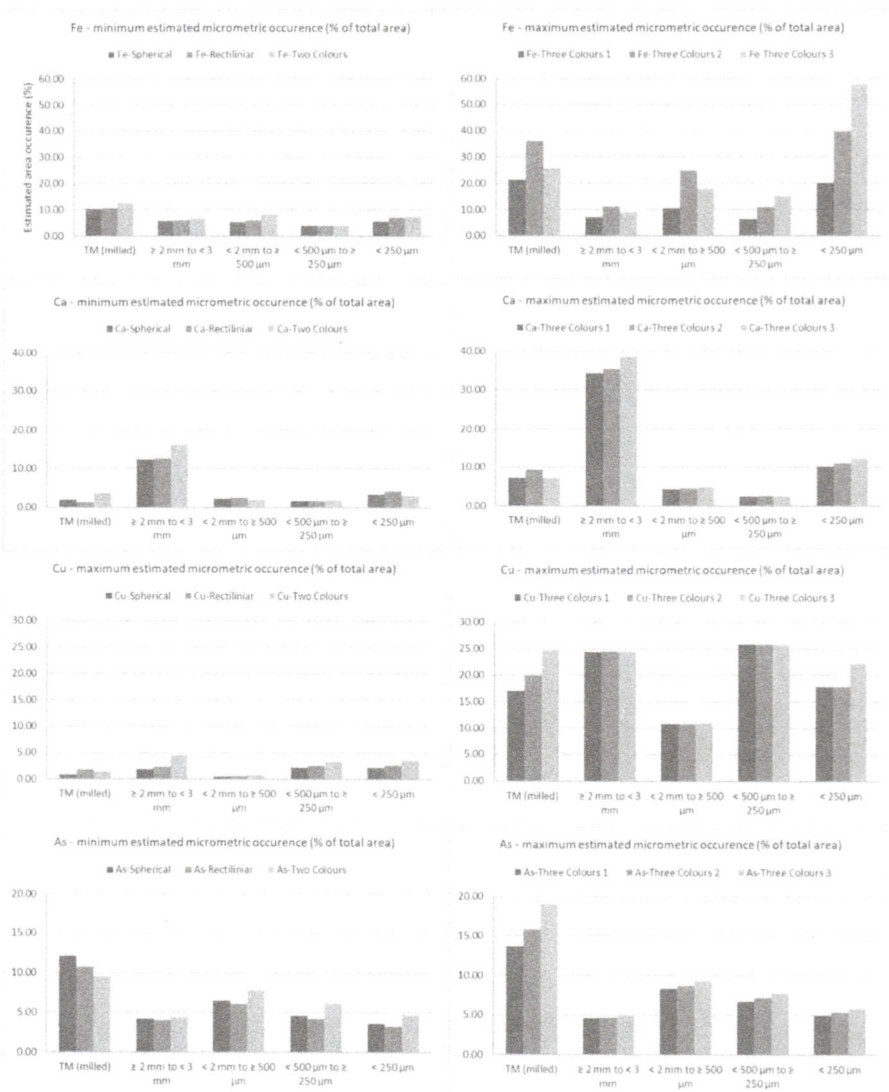

Figure 7. Minimum and maximum elemental occurrence in µ-XRF 2D map (percentage of area %) for Fe, Ca, Cu, and As.

Table 2. Synthesis of estimated minimum and maximum elemental occurrence (percentage of area %) for Al, Si, K, Ca, Ti, Mn, Fe, Ni, Cu, Zn, Ga, As, Sr, and Y.

Sample Fraction	TM (ground and milled)		Grain Size ≥2 mm to <3 mm		Grain Size <250 µm	
Element	Minimum probable	Maximum probable	Minimum probable	Maximum probable	Minimum probable	Maximum probable
Al	6.6–7.8	18.3–27.8	8.5–9.8	20.3–29.7	11.5–13.2	26.4–37.0
Si	3.8–4.4	5.4–5.5	14.8–19.1	24.5–24.9	26.1–32.1	41.6–42.2
K	5.8–7.8	35.8–44.7	2.3–2.8	10.4–13.3	5.4–8.8	35.9–44.2
Ca	2.0–3.6	7.3–9.4	12.3–16.3	34.1–38.6	3.1–4.3	10.1–12.3
Ti	2.1–4.6	3.3–3.5	7.3–9.6	10.0–14.0	5.9–6.2	5.6–6.9
Mn	2.4–12.7	19.8–21.5	0.9–1.4	1.7–1.8	1.9–9.3	13.5–14.8
Fe	10.2–12.5	21.5–36.3	5.8–6.5	7.0–11.1	5.4–7.4	20.0–57.8
Ni	3.2–7.5	15.3–24.4	3.8–8.8	17.7–27.9	9.3–16.0	28.0–39.3
Cu	0.8–1.7	17.0–24.7	1.9–4.5	24.4–24.5	2.2–3.6	13.5–14.8
Zn	11.2–12.9	49.4–59.7	11.0–19.6	30.9–37.4	7.0–18.1	33.2–41.9
Ga	4.8–9.4	10.6–16.2	3.0–4.4	4.7–6.8	7.0–15.7	17.4–24.2
As	9.5–12.1	13.7–19.0	4.0–4.4	4.6–4.9	3.3–4.7	5.0–5.8
Sr	13.8–19.6	42.8–49.5	22.2–27.3	41.9–45.4	28.0–36.1	54.0–54.1
Y	3.4–4.4	5.6–7.3	7.6–9.4	11.1–13.3	7.8–9.6	11.6–13.9

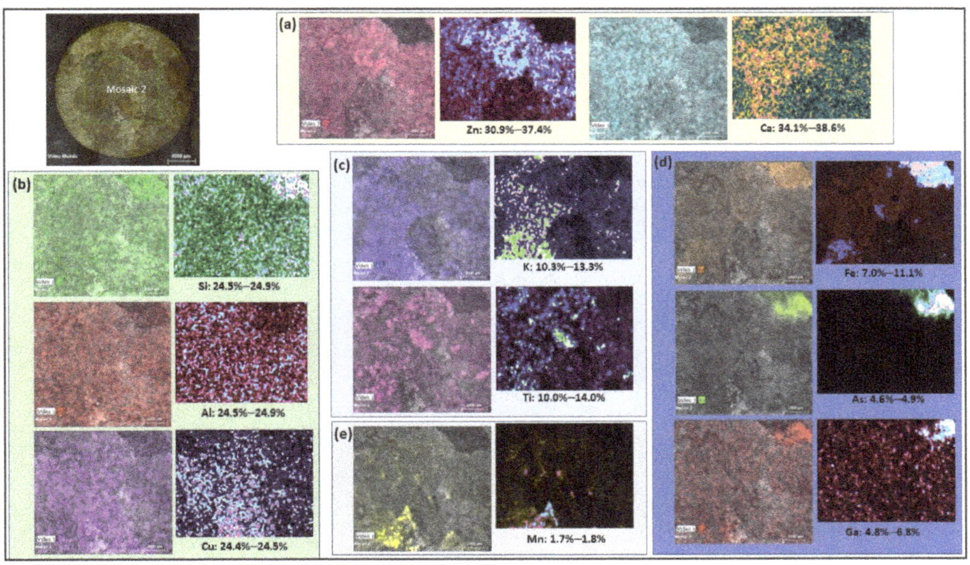

Figure 8. Image 2D micrometric maps of the elements (**a**) Zn, Ca (**b**) Si, Al, Cu (**c**) K, Ti (**d**) Fe, As, Ga (**e**) Mn in sample SD1, grain-size fraction "≥2 mm to <3 mm", and correspondent maximum estimated percentages of occurrence area (%).

4. Discussion and Conclusions

Elemental 2D spatial mapping through micro-XRF spectroscopy is a promising technique in the detailed study of granular heterogeneous samples, such as soils and mining wastes [31–33]. In this exploratory study, a clustering image analysis methodology was applied to detect elemental distribution at micrometric scale according to distinct colour intensities. The results present accurate information on the elemental distribution per grain fraction, offering clues of its geochemical occurrence (manly primary in coarse grain-size fraction and secondary in finer fractions). Results are more regular and similar between distinct fraction samples and milled samples when the element occurs at lower granulometries. The results showed that the elemental spatial patterns per grain-size fraction are not always coincident or similar to grinding and milled spatial pattern samples, showing that, for some cases, elemental distribution is dependent on specific mineralogy, which can have its own grain-size distribution pattern according to geochemical characteristics of the site. Some metals show distinctive percentages of occurrence according to grain-size fraction. Metal occurrence in milled fractions do not always correspond to the average of the grain-size fractions. Certain elements tend to be present in higher quantities in coarse fractions, mainly 2–3 mm, while other elements tend to present in smaller-size grain fractions (<250 μm). This will be dependent on the mineralogy and specific geochemical behaviour, especially mobility, of the elements. For sure, mobility and geochemical source of the element (primary or secondary) will dictate elemental specific spatial patterns at the micrometric scale.

In general, minimum and maximum elemental estimations from 2D maps show a tendency of greater discrepancies in results when the element is more abundant and widespread in the matrix. This is the example of element Si, K, Zn, Sr, and finer gain-size fractions of Fe. Major discrepancies in measurements are due to the higher difficulty in fixing the characteristic degree of colour intensity that marks the occurrence of the element, and distance between the distinct intensity colour degrees, which may make clustering classification difficult. In this context, the joint interpretation of 2D images to estimate 3D grades is currently an emerging research area [13,14,31–33] that will represent a quite interesting investigation upgrade.

The exploration of applicable data image analysis techniques able to identify elemental spatial overlaps in μ-XRF 2D map surveys and the estimation of grain-size distributions per element or per groups of elements are two promising areas for forward investigation in granular and heterogeneous samples, such as in the case of soil samples.

Author Contributions: Conceptualization, S.B. and A.D.; methodology, S.B. and A.D.; software, S.B. and M.P.; validation, S.B., A.D., S.P. and J.A.A.; investigation, S.B., A.D. and M.P.; writing—original draft preparation, S.B.; writing—review and editing, A.D., S.P. and J.A.A. All authors have read and agreed to the published version of the manuscript.

Funding: This research was funded by FCT-Fundação para a Ciência e a Tecnologia, Portugal, grants number UIDB/04035/2020, and UID/FIS/04559/2020.

Institutional Review Board Statement: Not applicable.

Informed Consent Statement: Not applicable.

Data Availability Statement: Not applicable.

Acknowledgments: The authors acknowledge the support of LIBPhys, GeoBiotec, Department of Physics and Department of Earth Sciences of Nova School of Science and Technology for the development of the laboratory work.

Conflicts of Interest: The authors declare no conflict of interest.

Appendix A

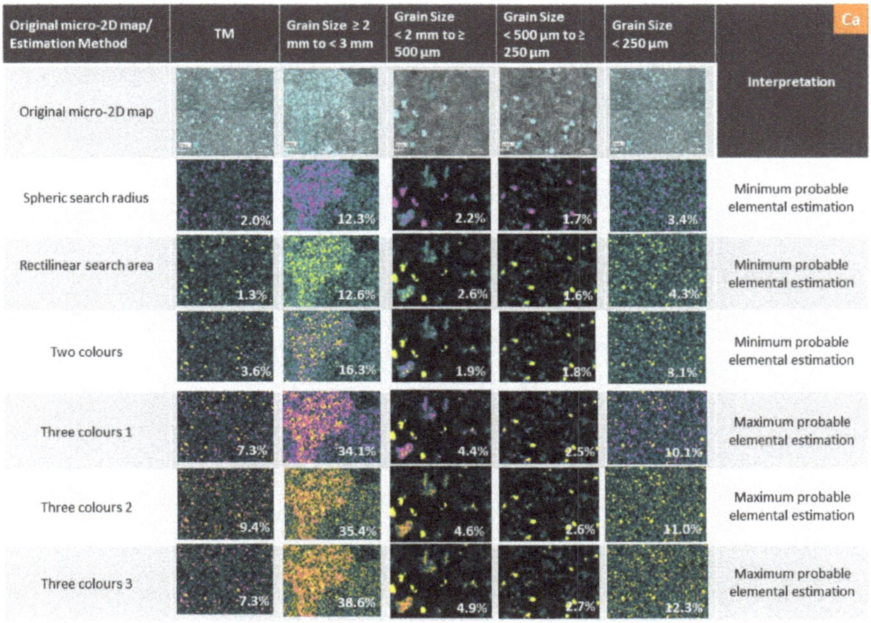

Figure A1. Minimum and maximum probable elemental occurrence in μ-XRF 2D map (percentage of area %) for Ca.

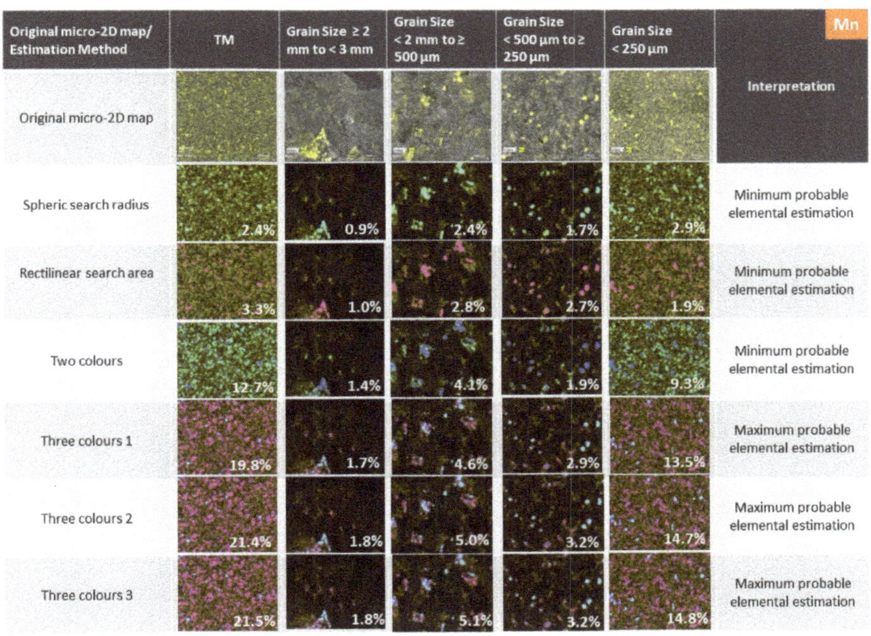

Figure A2. Minimum and maximum probable elemental occurrence in μ-XRF 2D map (percentage of area %) for Mn.

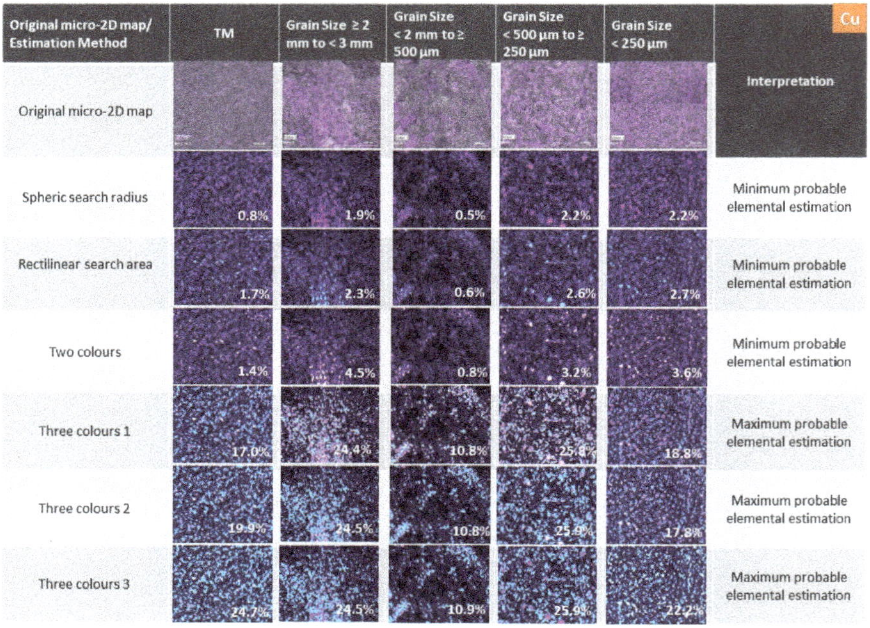

Figure A3. Minimum and maximum probable elemental occurrence in μ-XRF 2D map (percentage of area %) for Cu.

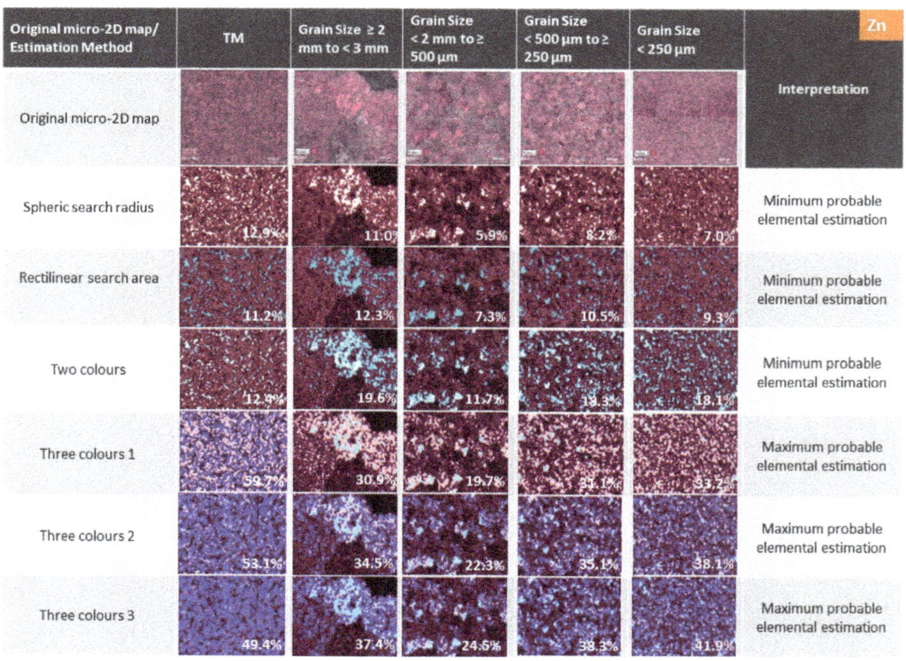

Figure A4. Minimum and maximum probable elemental occurrence in μ-XRF 2D map (percentage of area %) for Zn.

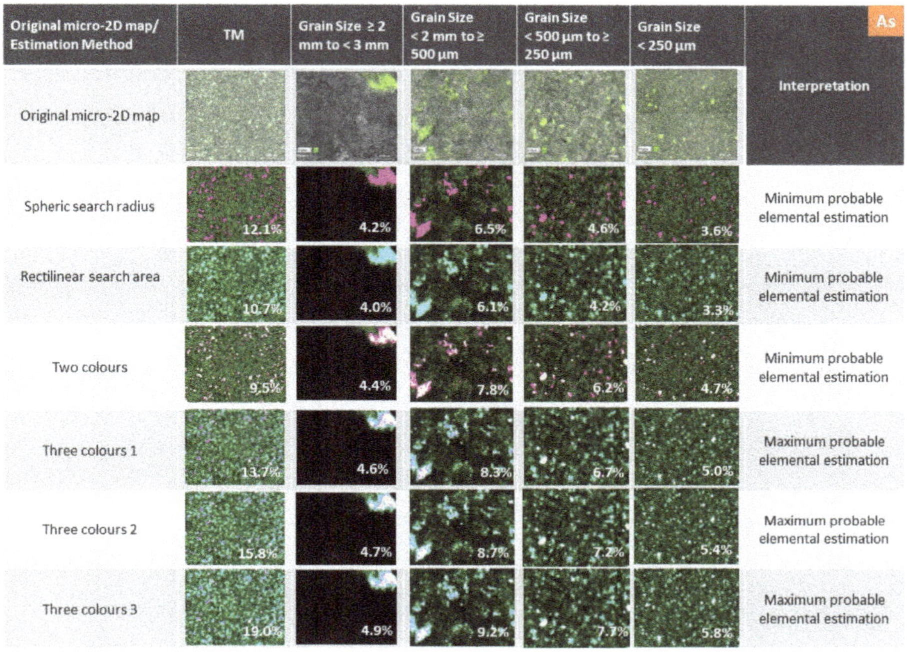

Figure A5. Minimum and maximum probable elemental occurrence in µ-XRF 2D map (percentage of area %) for As.

References

1. Carvalho, P.M.D.S.; Leite, F.; Silva, A.L.M.; Pessanha, S.; Carvalho, M.L.; Veloso, J.F.; Santos, J.P. Elemental mapping of Portuguese ceramic pieces with a full-field XRF scanner based on a 2D-THCOBRA detector. *Eur. Phys. J. Plus* **2021**, *136*, 423. [CrossRef]
2. Flude, S.; Haschke, M.; Storey, M.; Harvey, J. Application of benchtop micro-XRF to geological materials. *Mineral. Mag.* **2017**, *81*, 923–948. [CrossRef]
3. Pessanha, S.; Fonseca, C.; Santos, J.P.; Carvalho, M.L.; Dias, A.A. Comparison of standard-based and standardless methods of quantification used in X-ray fluorescence analysis: Application to the exoskeleton of clams. *X-ray Espectrometry* **2018**, *47*, 108–115. [CrossRef]
4. Kaskes, P.; Déhais, T.; de Graaff, S.J.; Goderis, S.; Claeys, P. Micro–X-ray Fluorescence (µXRF) Analysis of Proximal Impactites: High-Resolution Element Mapping, Digital Image Analysis, and Quantifications. In *Large Meteorite Impacts and Planetary Evolution VI*; Reimold, W.U., Koeberl, C., Eds.; Geological Society of America Special Paper: Boulder, CO, USA, 2021; Volume 550, pp. 171–206. [CrossRef]
5. Barbosa, S.; Dias, A.; Ferraz, A.; Amaro, S.; Brito, M.G.; Almeida, J.A.; Pessanha, S. The Dual Paradigm of Mining Waste: "From Ecotoxicological Sources to Potential Polymetallic Resources"—An Example from Iberian Pyrite Belt (Portugal). *Mater. Proc.* **2021**, *5*, 23. [CrossRef]
6. Barbosa, S.; Dias, A.; Durão, D.; Grilo, J.; Baptista, G.; Cagiza, J.; Pessanha, S.; Simão, J.; Almeida, J. Exploring High-Resolution Chemical Distribution Maps of Incompatible and Scarce Metals in a Nepheline Syenite from the Massif of "Serra de Monchique" (Portugal, Iberian Peninsula). *Minerals* **2022**, *12*, 1178. [CrossRef]
7. Le Gac, A.; Pessanha, S.; Longelin, S.; Guerra, M.; Frade, J.C.; Lourenço, F.; Serrano, M.C.; Manso, M.; Carvalho, M.L. New development on materials and techniques used in the heraldic designs of illuminated Manueline foral charters by multi-analytical methods. *Appl. Radiat. Isot.* **2013**, *82*, 242–257. [CrossRef] [PubMed]
8. Miller, T.C.; DeWitt, H.L.; Havrilla, G.J. Characterization of small particles by micro X-ray fluorescence. *Spectrochim. Acta Part B* **2005**, *60*, 1458–1467. [CrossRef]
9. Pessanha, S.; Costa, M.; Oliveira, M.I.; Jorge, M.E.M.; Carvalho, M.L. Nondestructive analysis of Portuguese "dinheiros" using XRF: Overcoming patina constraints. *Appl. Phys. A* **2015**, *119*, 1173–1178. [CrossRef]
10. Marguí, E.; Queralt, I.; de Almeida, E. X-ray fluorescence spectrometry for environmental analysis: Basic principles, instrumentation, applications and recent trends. *Chemosphere* **2022**, *303 Pt 1*, 135006. [CrossRef]

11. Colombo, F.; Bargalló, R.; Spalletti, L.A.; Enrique, P.; Queralt, I. Pumice clasts in cross stratified basalt-dominated sandstones and conglomerates. Characteristics and depositional significance: Huarenchenque Fm (Neuquén, Argentina). *J. Iber. Geol.* **2019**, *45*, 29–46. [CrossRef]
12. Pacheco, M.S.M. Variability Analysis and Spatial Elemental Distribution in Soils and Sediments Using Micro-XRF—A Case Study. Master's Thesis, Faculty of Science and Technology, NOVA University of Lisbon, Lisbon, Portugal, 2021; p. 308. Available online: https://run.unl.pt/handle/10362/113885 (accessed on 1 September 2022).
13. Aaron, J.S.; Taylor, A.B.; Chew, T.-L. Image co-localization—Co-occurrence versus correlation. *J. Cell Sci.* **2018**, *131*, jcs211847. [CrossRef] [PubMed]
14. Buchmann, M.; Borowski, N.; Leißner, T.; Heinig, T.; Reuter, M.A.; Friedrich, B.; Peuker, U.A. Evaluation of Recyclability of a WEEE Slag by Means of Integrative X-ray Computer Tomography and SEM-Based Image Analysis. *Minerals* **2020**, *10*, 309. [CrossRef]
15. Hapca, S.; Baveye, P.C.; Wilson, C.; Lark, R.M.; Otten, W. Three-Dimensional Mapping of Soil Chemical Characteristics at Micrometric Scale by Combining 2D SEM-EDX Data and 3D X-ray CT Images. Predictive Mapping of 3D Soil Chemical Composition at Micro-Scale. *PLoS ONE* **2015**, *10*, e0137205. [CrossRef]
16. Werner, F.; Mueller, C.W.; Thieme, J.; Gianoncelli, A.; Rivard, C.; Höschen, C.; Prietzel, J. Micro-scale heterogeneity of soil phosphorus depends on soil substrate and depth. *Nat. Sci. Rep.* **2017**, *7*, 3203. [CrossRef] [PubMed]
17. Álvarez-Valero, A.M.; Pérez-López, R.; Matos, J.; Capitán, M.A.; Nieto, J.M.; Sáez, R.; Delgado, J.; Caraballo, M. Potential environmental impact at São Domingos mining district (Iberian Pyrite Belt, SW Iberian Peninsula): Evidence from a chemical and mineralogical characterization. *Environ. Geol.* **2008**, *55*, 1797–1809. [CrossRef]
18. Oliveira, J.T.; Silva, J.B.; de Carvalho, D.; Van Den Boogaard, M.; Ribeiro, A. Notícia Explicativa da Folha 46-D Mértola. Departamento de Geologia, INETI, Lisboa. 2007. Available online: https://geoportal.lneg.pt/pt/dados_abertos/cartografia_geologica/cgp50k/46-D (accessed on 1 September 2022).
19. Santos, E.; Ferreira, M.; Abreu, M.M. Contribuição de Cistus Ladanifer L. e Cistus Salviifolius L. na Recuperação de Áreas Mineiras da Faixa Piritosa Ibérica. *Rev. De Ciências Agrárias* **2011**, *34*, 21–31. Available online: http://hdl.handle.net/10400.5/5002 (accessed on 1 September 2022).
20. Tavares, M.T.; Abreu, M.M.; Vairinho, M.M.; Sousa, A.J.; Quental, L. Comportamento geoquímico de alguns elementos vestigiais na envolvente das Minas de S. Domingos, Alentejo: Áreas da Tapada e do Telheiro. *Rev. De Ciências Agrárias* **2009**, *32*, 182–194. Available online: http://hdl.handle.net/10400.9/736 (accessed on 1 September 2022).
21. Pessanha, S.; Samouco, A.; Adão, R.; Carvalho, M.L.; Santos, J.P.; Amaro, P. Detection limits evaluation of a portable energy dispersive X-ray fluorescence setup using different filter combinations. *X-ray Spectrom.* **2017**, *46*, 102–106. [CrossRef]
22. R Core Team. *R: A Language and Environment for Statistical Computing*; R Foundation for Statistical Computing: Vienna, Austria, 2013; Available online: http://www.R-project.org/ (accessed on 1 September 2022).
23. Weller, H. Package 'Countcolors'. 2019. p. 10. Available online: https://cran.r-project.org/web/packages/countcolors/countcolors.pdf (accessed on 1 September 2022).
24. Weller, H. Introduction to Countcolors Package. 2019. Available online: https://cran.r-project.org/web/packages/countcolors/vignettes/Introduction.html (accessed on 1 September 2022).
25. Weller, H. Countcolors Package. 2019. Available online: https://rdocumentation.org/packages/countcolors/versions/0.9.1 (accessed on 1 September 2022).
26. Batista, M.J.; Matos, J.X.; Figueiredo, M.O.; de Oliveira, D.P.S.; Silva, T.; Santana, H.; Quental, L. Fingerprints for Mining Products and Wastes of the S. Domingos, Aljustrel and Neves Corvo Mines: A Sustainable Perspective. In Proceeding of the VIII Congresso Ibérico de Geoquímico/XVII Semana de Geoquímica, Castelo Branco, Portugal, 24–28 September 2011; p. 6.
27. Matos, J.X.; Pereira, Z.; Oliveira, V.; Oliveira, J.T. The Geological setting of the São Domingos pyrite orebody, Iberian Pyrite Belt. In Proceedings of the VII National Geology Congress, Estremoz, Portugal, 29 June–13 July 2006; pp. 283–286.
28. Matos, J.X.; Pereira, Z.; Batista, M.J.; de Oliveira, D.P.S. São Domingos Mining Site—Iberian Pyrite Belt. In Proceedings of the 9th International Symposium on Environmental Geochemistry, Aveiro, Portugal, 15–22 July 2012; pp. 7–12.
29. Négrel, P.; Ladenberger, A.; Reimann, C.; Birke, M.; Sadeghi, M. Distribution of Rb, Ga and Cs in agricultural land soils at European continental scale (GEMAS): Implications for weathering conditions and provenance. *Chem. Geol.* **2018**, *479*, 188–203. [CrossRef]
30. Połedniok, J. Speciation of scandium and gallium in soil. *Chemosphere* **2008**, *73*, 572–579. [CrossRef]
31. Hafez, I.T.; Sorrentino, G.; Faka, M.; Cuenca-García, C.; Makarona, C.; Charalambous, A.; Nys, K.; Hermon, S. Geochemical survey of soil samples from the archaeological site Dromolaxia-Vyzakia (Cyprus) by means of micro-XRF and statistical approaches. *J. Archaeol. Sci. Rep.* **2017**, *11*, 447–462. [CrossRef]

32. Kim, J.J.; Ling, F.T.; Plattenberger, D.A.; Clarens, A.F.; Lanzirotti, A.; Newville, M.; Peters, C.A. SMART mineral mapping: Synchrotron-based machine learning approach for 2D characterization with coupled micro XRF-XRD. *Comput. Geosci.* **2021**, *156*, 104898. [CrossRef]
33. Li, Q.; Hu, X.; Hao, J.; Chen, W.; Cai, P.; Huang, Q. Characterization of Cu distribution in clay-sized soil aggregates by NanoSIMS and micro-XRF. *Chemosphere* **2020**, *249*, 126143. [CrossRef] [PubMed]

Disclaimer/Publisher's Note: The statements, opinions and data contained in all publications are solely those of the individual author(s) and contributor(s) and not of MDPI and/or the editor(s). MDPI and/or the editor(s) disclaim responsibility for any injury to people or property resulting from any ideas, methods, instructions or products referred to in the content.

eng MDPI

Article

Network Pathway Extraction Focusing on Object Level

Ali Alqahtani [1,2]

[1] Department of Computer Science, King Khalid University, Abha 61421, Saudi Arabia; amosfer@kku.edu.sa
[2] Center for Artificial Intelligence (CAI), King Khalid University, Abha 61421, Saudi Arabia

Abstract: In this paper, I propose an efficient method of identifying important neurons that are related to an object's concepts by mainly considering the relationship between these neurons and their object concept or class. I first quantify the activation values among neurons, based on which histograms of each neuron are generated. Then, the obtained histograms are clustered to identify the neurons' importance. A network-wide holistic approach is also introduced to efficiently identify important neurons and their influential connections to reveal the pathway of a given class. The influential connections as well as their important neurons are carefully evaluated to reveal the sub-network of each object's concepts. The experimental results on the MNIST and Fashion MNIST datasets show the effectiveness of the proposed method.

Keywords: deep learning; network pathway extraction; neuron importance

Citation: Alqahtani, A. Network Pathway Extraction Focusing on Object Level. *Eng* **2023**, *4*, 151–158. https://doi.org/10.3390/eng4010009

Academic Editor: Antonio Gil Bravo

Received: 1 November 2022
Revised: 12 December 2022
Accepted: 27 December 2022
Published: 3 January 2023

Copyright: © 2023 by the author. Licensee MDPI, Basel, Switzerland. This article is an open access article distributed under the terms and conditions of the Creative Commons Attribution (CC BY) license (https://creativecommons.org/licenses/by/4.0/).

1. Introduction

Deep learning algorithms (e.g., NNs and CNNs) are often viewed as "black-box models" because of their vagueness and ambiguous working mechanisms [1]. Efforts have been made to investigate complex models, as well as to clarify and describe their work mechanisms and internal function, providing us a general understanding of how to handle and enhance such models. Different approaches have been developed to understand the importance of intermediate units in the neural networks, which consider substantial steps to gain insight into the characteristics of the latent representations, to understand how information is propagated through a network, and to evaluate the importance of a neuron by measuring the influence of hidden units. The established techniques have made an effort to visually interpret and understand the deep representations, mainly focusing on pixel-level annotations [2,3] and single-neuron properties via code inversion strategies (e.g., [4–8]) and activation maximization strategies (e.g., [9–12]) with regard to illustrating the learned representations of deep learning algorithms. Their interpretability has been applied to visually evaluate neurons' importance and to understand their properties. The major priority of these approaches is to clarify a model's predictions by looking for an explanation for specific activation and by analyzing individual neurons. However, it is still challenging to intuitively measure decision linkages and the sufficient associations between nodes with a massive number of connections. In this paper, I propose an efficient method to identify the important neurons that are related to object concepts and mainly consider the relationship between these neurons and their object concept or class. I first quantify the activation values among neurons, based on which histograms of each neuron are generated. Then, the obtained histograms are clustered to identify the neurons' importance. I then introduce a network-wide holistic approach that efficiently identifies important neurons and their influential connections to reveal the pathway of a given class. The influential connections as well as their important neurons are carefully evaluated to reveal the sub-network of each object's concepts.

The rest of the paper is organized as follows. In Section 2, I present related works, while I describe our proposed methodology in Section 3. In Section 4, I present our experimental results. Finally, concluding remarks are provided in Section 5.

2. Related Works

Considerable attention has been given to understanding the importance of internal units in neural networks. Understanding neuron properties and evaluating their importance raise awareness to the need to adopt the quantitative assessment of neurons' properties. Such manners are utilized to measure the activation importance of each node and to appoint a score to them. To determine the importance of hidden neurons, Dhamdhere et al. [13] apply integrated gradients by calculating a summation of the gradients of the prediction with respect to the input. Morcos et al. [14] explored the relationship between the output of individual neurons and the classification performance of neural networks to evaluate mutual information and class selectivity for each neuron's activation. Moreover, Na et al. [15] have recently used the highest mean activation to measure the importance of individual units on language tasks, showing that different units are selectively responsive to specific morphemes, words, and phrases. Despite the fact that most of the aforementioned techniques emphasize effective methods to identify important neurons, most of their concentration was on gaining the best understanding of the network's mechanism, with limited attention towards tracking the pathway of a given class and analyzing the network's behaviour at a sub-network level.

In an attempt to study the topic of cumulative network pruning, several approaches have been developed [16]. Frankle et al. [17] found that networks contain a sub-network that reaches a test accuracy that is comparable with the original network through an iterative pruning technique. Their core idea was to find a smaller, well-suited architecture to the target task at the training phase. Ashual et al. [18] proposed a composite network that focuses on extracting the sub-network for each class during the training process. These methods show the possibility of revealing the sub-network in an indirect way, whether through eliminating unimportant parts of the neural networks or through training multiple branches of the network, where different groups of branches denote different objects. Therefore, providing a way to measure the importance of different parts of the network, to detect the important neurons, and to identify relationships among neurons are worthwhile to extract the sub-network for a specific object or class.

In this paper, I propose an efficient method to identify the important neurons that are related to object concepts and mainly consider the relationship between these neurons and their object concept or class. I first quantify the activation values among neurons, based on which histograms of each neuron are generated. Then, the obtained histograms are clustered to identify the neurons' importance. I then introduce a network-wide holistic approach that efficiently identifies important neurons and their influential connections to reveal the pathway of a given class. The influential connections as well as their important neurons are carefully evaluated to reveal the sub-network of each object's concepts.

3. Method

Measuring the importance of different network parts always requires a more meticulous process. Most of the current techniques focus on providing efficient techniques to determine important neurons, with limited attention being paid to tracking the pathway of a given class and analyzing the network's behaviour at the sub-network level. My importance-measurement method introduces a novel way to reveal the sub-network; it estimates the importance of neurons in each layer and identifies a subset of their influential connections whose activation values are the most effective in identifying relationships among neurons. This section presents my overall proposed framework, which consists of two parts. First, the evaluation of neuron importance is discussed; this determines the importance of neurons in each layer. Then, I present a network-wide holistic method that efficiently identifies important neurons and their influential connections to reveal the pathway of a given class. The entire algorithm is provided in Algorithm 1. The details are provided below.

Algorithm 1: Network Pathway Extraction.

Input: a pre-trained model, training set (x, y);
Output: sub-network of each class;
for *each class* **do**
  compute the activation for each neuron Equation (1);
  generate histograms for each neuron;
  cluster the obtained histograms to identify the important neurons;
  identify the influential connections Equation (2);
  apply MV method to detect the most important connections for each neuron;
end

3.1. Analyzing the Importance of Individual Neuron

The aim was to reveal effective units in neural networks by estimating their activation. When the training data are fed through the network, a different representation is obtained for each example and has unique activation throughout all neurons in the network. A forward passing via a trained model is applied to derive the output of each unit. Different input examples can present more instances, and the output can be seen as random variables. I utilized a novel process to estimate the importance of units. In each layer, the weights are multiplied with an input sample, x, to deliver an output corresponding n activation. The activation at j-th unit is determined by summing the weights of the activations from all unit in the $(i-1)$-th layer. The production of the j-th unit in the i-th layer of the neural network is given by

$$t_j^{(i)}(x_n) = \sigma\left(b_j^{(i)} + \sum_p w_{p,j}^{(i-1)} t_p^{(i-1)}(x_n)\right), \tag{1}$$

where x_n is the n-th data sample at the input, σ denotes the activation function, b_j^i is the corresponding bias for the j-th unit in the i-th layer, and $w_{p,j}^{(i-1)}$ denotes the weight that connects p-th unit from the prior layer $(i-1)$ with the j-th unit in the i-th layer (current layer).

After obtaining a matrix of edge values for each example by Equation (1), histograms for each edge were generated (see Figure 1). Then, the obtained histograms were clustered into three different clusters (High, Medium, and Low). Therefore, I came up with a binary vector for every layer that demonstrates whether such units are critical, where 1 indicates that the neuron is essential and 0 otherwise.

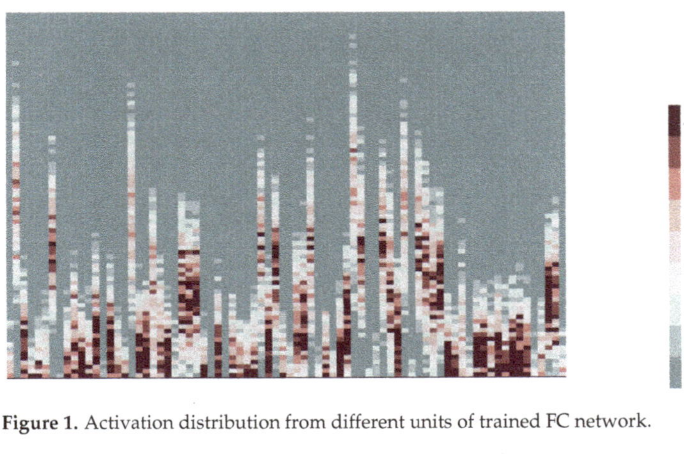

Figure 1. Activation distribution from different units of trained FC network.

3.2. From Individual Neuron to Sub-Network Analysis

The collection of important neurons at each layer in a network will only give localized feature descriptors, of which essential neurons for a particular class can be detected. By focusing on this assumption, I am able to justify how distinct neurons become the most representative for a given class, maintaining the class information within the input domain, consequently providing a way to detect how the activations from the nodes of the previous layer impact the activations of the current layer is required to extract a sub-network for a given class. A combination of neurons and their influential connections provides a novel way to reveal the sub-network.

As the activation of a neuron in the current layer is computed as the weighted sum of activations from neurons in the previous layer obtained by Equation (1), in fully connected layers, the influential connections $t_j^{(i-1)}(\hat{x}_n)$ of a single neuron are obtained by multiplying the weights of that neuron with its corresponding activations in the previous layer, as follows:

$$t_j^{(i-1)}(\hat{x}_n) = w_{p,j}^{(i-1)} t_p^{(i-1)}(x_n), \quad (2)$$

where x_n denotes the n-th data example at the input, $w_{p,j}^{(i-1)}$ is the weight that links p-th unit from the former layer $(i-1)$ with the j-th unit in the i-th layer (current layer), and $t_p^{(i-1)}(x_n)$ represents the corresponding activations in the previous layer $(i-1)$. After the values of the influential connections are obtained for each neuron, the majority voting (MV) method [19,20] is applied to detect the most important connections for each neuron.

4. Experiment and Discussion

I empirically investigated the performance of my proposed approach using two different datasets: MNIST [21] and MNIST-Fashion [22] through several models. The proposed method was implemented using Keras and TensorFlow [23] in Python. The specifications of the datasets and their architecture for the used models are presented in Table 1.

Table 1. Details of datasets and their architectures used in my experiments.

Dataset	Examples	Image Size	AutoEncoder Architecture	FC Architecture
MNIST [21]	70,000	$28 \times 28 \times 1$	784-1000-1000-1000-784	784-1000-1000-1000-10
MNIST-Fashion [22]	70,000	$28 \times 28 \times 1$	784-1000-1000-1000-784	784-1000-1000-1000-10

The first model was an auto-encoder model, which was optimized in an unsupervised manner. There are no fine-tuning and pre-training processes involved. The stochastic gradient descent was used, and each batch included 100 random shuffled examples. For both datasets, an initial learning rate of 0.006 with a momentum of 0.9 and weight decay of 0.0005 were utilized.

I carried out a visual assessment to evaluate the results of the proposed method. Some instances of actual inputs and reconstruction images generated by my model are presented in Figures 2 and 3. Two evaluation studies were adopted: an ablation study and an insertion study.

After the pathway was identified, I applied the ablation study by forcing the pathway of a particular class to be zero and performed the propagation of the forwarding pass. Three samples were fed to the network after identifying the class pathway: (different image (left), random noise image (middle), and same image (right)). The reconstruction images of such this assessment were visualized using the three samples (see Figures 2a,b and 3a,b). This ablation study is best suited for evaluating the effectiveness of extracting a particular class pathway. One clear example can be seen in Figure 2a; the model failed to reconstruct the image of digit 5 back to its original shape because the pathway of that digit was ablated. It is also worth noting that the reconstruction of the image of digit 7 obtains an optimal approximation of the underlying input data because the pathway of digit 7 was still available. These observations allow us to efficiently identify important neurons and their influential connections to reveal the pathway of a given class.

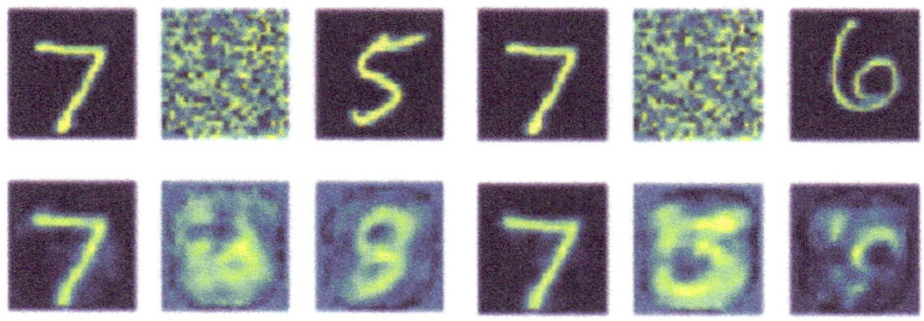

(**a**) Class (5) (**b**) Class (6)

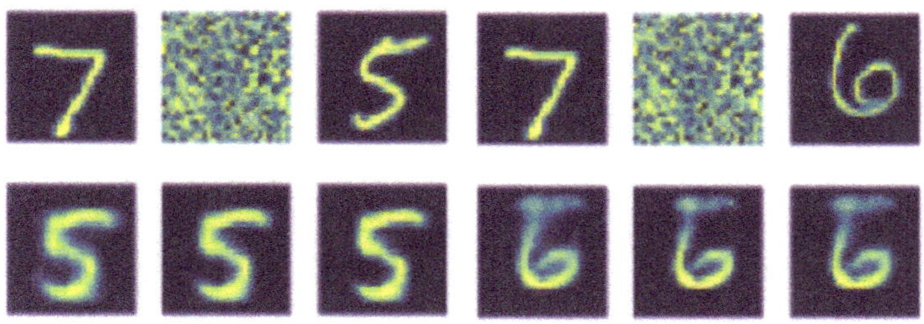

(**c**) Class (5) (**d**) Class (6)

Figure 2. An ablation study (**Top**) and an insertion study (**Bottom**) with different examples and different identification of class pathway on the MNIST dataset.

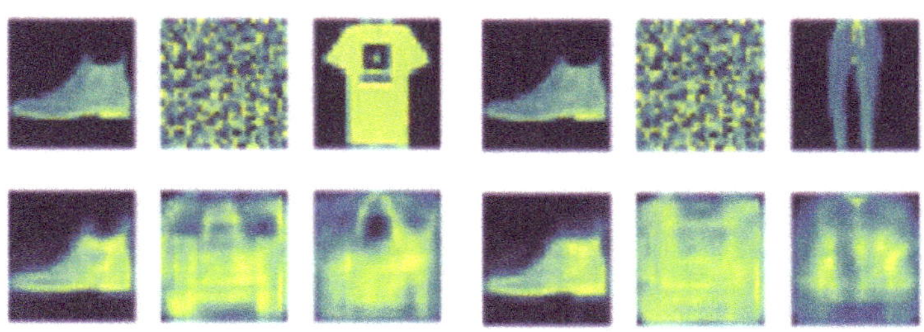

(**a**) Class (t-shirt) (**b**) Class (Trouser)

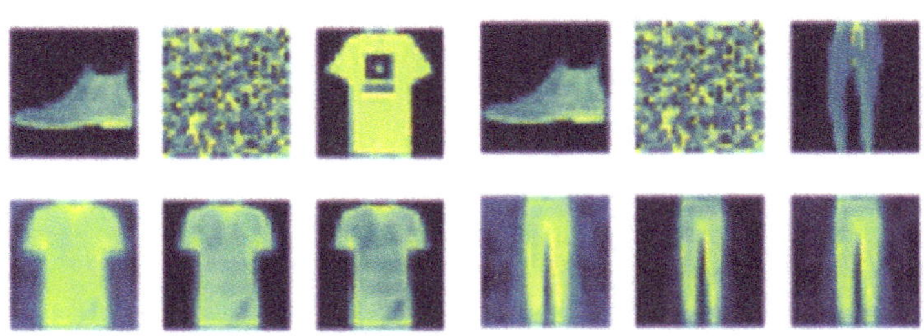

(**c**) Class (t-shirt) (**d**) Class (Trouser)

Figure 3. An ablation study (**Top**) and an insertion study (**Bottom**) with different examples and different identification of class pathway on the fashion MNIST dataset.

Figures 2 and 3 also show the results of the insertion study, where I forced the pathway of a particular class to be one and zero otherwise and performed the propagation of the forwarding pass (see Figures 2c,d and 3c,d). This study obviously showed that all input images ended up with the reconstruction of the same class of the identified pathway. The insertion study helped to evaluate the effectiveness of my method when revealing the pathway of a given class.

The second model was a classification model, which trained end-to-end in a supervised manner. Figure 4 experimentally analyzes the performance of my proposed method. Using my method, I identified the pathway of each class of trained fully-connected networks. My experiment showed that ablating a specific class pathway has no effect on other classes. One obvious explanation is that the proposed method succeeded in carefully identifying the pathway of each class. It is also crucial to note that because digit 6 has a comparable structure to that of digit 8, especially with regards to the bottom part of both digits (see Figure 4g), the model classified most examples of digit 6 into the class of digit 8. One reasonable justification is that the presented method was able to determine the homogeneous patterns for a particular digit, which leads to the identification of the pathway of the target class.

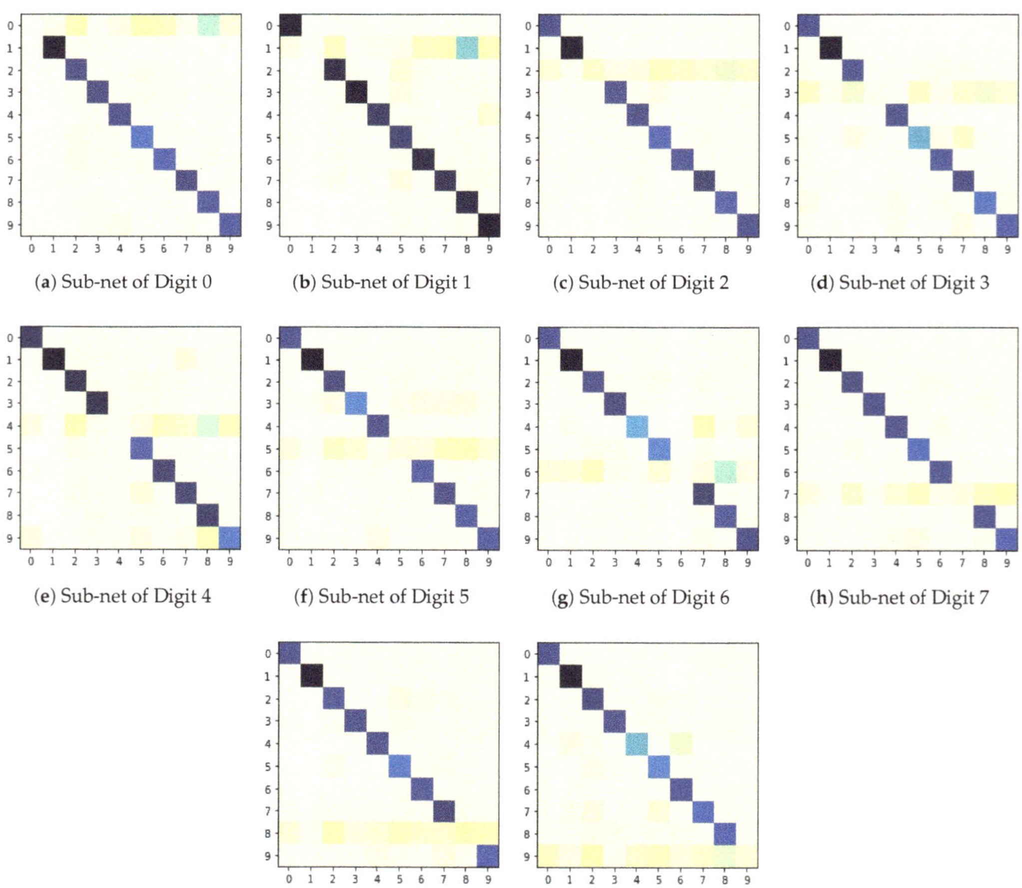

(**a**) Sub-net of Digit 0 (**b**) Sub-net of Digit 1 (**c**) Sub-net of Digit 2 (**d**) Sub-net of Digit 3

(**e**) Sub-net of Digit 4 (**f**) Sub-net of Digit 5 (**g**) Sub-net of Digit 6 (**h**) Sub-net of Digit 7

(**i**) Sub-net of Digit 8 (**j**) Sub-net of Digit 9

Figure 4. Results of different class pathways when applying an ablation study on the MNIST dataset.

5. Conclusions

In this paper, I proposed an efficient method to identify the important neurons, mainly considering the relationship between these neurons and their object concept or class. I introduce a network-wide holistic approach that efficiently identifies important neurons and their influential connections to reveal the pathway of a given class. The influential connections as well as their important neurons were carefully evaluated to reveal the sub-network of each object's concepts. I showed the effectiveness of the proposed method using two different datasets. Our potential future work is to expand the proposed framework to filters in CNNs and to investigate it with more difficult datasets. Although this procedure significantly identifies influential connections, the more theoretical analysis also requires further study to understand how such ideas can be generalized further. Moreover, I believe more investigation is needed to carefully study or gather evidence to determine whether the object's patterns can be a local receptive field in the FC networks, as connecting each neuron to only a local region of the input space might help to justify CNNs and to prove that the actual connections are local receptive fields. There are also several challenges and extensions we perceive as useful research directions. Extending the proposed framework and combining it to strengthen the discriminative features and improve the encoder's ability of deep clustering [24] is an important direction for future work.

Funding: This work was supported by the Deanship of Scientific Research, King Khalid University of Kingdom of Saudi Arabia under research grant number (RGP1/357/43).

Institutional Review Board Statement: Not applicable.

Informed Consent Statement: Not applicable.

Data Availability Statement: Not applicable.

Conflicts of Interest: The author declares no conflict of interest.

References

1. Bengio, Y.; Courville, A.; Vincent, P. Representation learning: A review and new perspectives. *IEEE Trans. Pattern Anal. Mach. Intell.* **2013**, *35*, 1798–1828. [CrossRef] [PubMed]
2. Bau, D.; Zhou, B.; Khosla, A.; Oliva, A.; Torralba, A. Network dissection: Quantifying interpretability of deep visual representations. In Proceedings of the IEEE Conference on Computer Vision and Pattern Recognition, Honolulu, HI, USA, 21–26 July 2017; pp. 6541–6549.
3. Bau, D.; Zhu, J.Y.; Strobelt, H.; Zhou, B.; Tenenbaum, J.B.; Freeman, W.T.; Torralba, A. Gan dissection: Visualizing and understanding generative adversarial networks. In Proceedings of the International Conference on Learning Representations, New Orleans, LA, USA, 6–9 May 2019.
4. Zeiler, M.D.; Fergus, R. Visualizing and understanding convolutional networks. In Proceedings of the European Conference on Computer Vision, Zurich, Switzerland, 6–12 September 2014; pp. 818–833.
5. Dosovitskiy, A.; Brox, T. Inverting visual representations with convolutional networks. In Proceedings of the IEEE Conference on Computer Vision and Pattern Recognition, Las Vegas, NV, USA, 27–30 June 2016; pp. 4829–4837.
6. Mahendran, A.; Vedaldi, A. Understanding deep image representations by inverting them. In Proceedings of the IEEE Conference on Computer Vision and Pattern Recognition, Boston, MA, USA, 7–12 June 2015; pp. 5188–5196.
7. Beguš, G.; Zhou, A. Interpreting intermediate convolutional layers of generative CNNs trained on waveforms. *IEEE/ACM Trans. Audio Speech Lang. Process.* **2022**, *30*, 3214–3229. [CrossRef]
8. Suganyadevi, S.; Seethalakshmi, V.; Balasamy, K. A review on deep learning in medical image analysis. *Int. J. Multimed. Inf. Retr.* **2022**, *11*, 19–38. [CrossRef] [PubMed]
9. Erhan, D.; Bengio, Y.; Courville, A.; Vincent, P. *Visualizing Higher-Layer Features of a Deep Network*; Technical Report; University of Montreal: Montreal, QC, Canada, 2009; Volume 1341.
10. Simonyan, K.; Vedaldi, A.; Zisserman, A. Deep inside convolutional networks: Visualising image classification models and saliency maps. *arXiv* **2013**, arXiv:1312.6034.
11. Yosinski, J.; Clune, J.; Nguyen, A.; Fuchs, T.; Lipson, H. Understanding neural networks through deep visualization. *arXiv* **2015**, arXiv:1506.06579.
12. Novakovsky, G.; Dexter, N.; Libbrecht, M.W.; Wasserman, W.W.; Mostafavi, S. Obtaining genetics insights from deep learning via explainable artificial intelligence. *Nat. Rev. Genet.* **2022**, 1–13. [CrossRef]
13. Dhamdhere, K.; Sundararajan, M.; Yan, Q. How important is a neuron? In Proceedings of the International Conference on Learning Representations, New Orleans, LA, USA, 6–9 May 2019.

14. Morcos, A.S.; Barrett, D.G.; Rabinowitz, N.C.; Botvinick, M. On the importance of single directions for generalization. In Proceedings of the International Conference on Learning Representations, Vancouver, BC, Canada, 30 April–3 May 2018.
15. Na, S.; Choe, Y.J.; Lee, D.H.; Kim, G. Discovery of Natural Language Concepts in Individual Units of CNNs. In Proceedings of the International Conference on Learning Representations, New Orleans, LA, USA, 6–9 May 2019.
16. Alqahtani, A.; Xie, X.; Jones, M.W. Literature Review of Deep Network Compression. *Informatics* **2021**, *8*, 77. [CrossRef]
17. Frankle, J.; Carbin, M. The lottery ticket hypothesis: Finding sparse, trainable neural networks. In Proceedings of the International Conference on Learning Representations, New Orleans, LA, USA, 6–9 May 2019.
18. Ashual, O.; Wolf, L. Specifying object attributes and relations in interactive scene generation. In Proceedings of the IEEE/CVF International Conference on Computer Vision, Seoul, Republic of Korea, 27 October–2 November 2019; pp. 4561–4569.
19. Alqahtani, A.; Xie, X.; Essa, E.; Jones, M.W. Neuron-based Network Pruning Based on Majority Voting. In Proceedings of the International Conference on Pattern Recognition, Milan, Italy, 10–15 January 2021; pp. 3090–3097.
20. Alqahtani, A.; Xie, X.; Jones, M.W.; Essa, E. Pruning CNN filters via quantifying the importance of deep visual representations. *Comput. Vis. Image Underst.* **2021**, *208*, 103220. [CrossRef]
21. Lecun, Y.; Bottou, L.; Bengio, Y.; Haffner, P. Gradient-based learning applied to document recognition. *Proc. IEEE* **1998**, *86*, 2278–2324. [CrossRef]
22. Xiao, H.; Rasul, K.; Vollgraf, R. Fashion-MNIST: A novel image dataset for benchmarking machine learning algorithms. *arXiv* **2017**, arXiv:1708.07747.
23. Abadi, M.; Barham, P.; Chen, J.; Chen, Z.; Davis, A.; Dean, J.; Devin, M.; Ghemawat, S.; Irving, G.; Isard, M.; et al. Tensorflow: A system for large-scale machine learning. In Proceedings of the Symposium on Operating Systems Design and Implementation, Savannah, GA, USA, 2–4 November 2016; pp. 265–283.
24. Alqahtani, A.; Xie, X.; Deng, J.; Jones, M.W. Learning discriminatory deep clustering models. In Proceedings of the International Conference on Computer Analysis of Images and Patterns, Salerno, Italy, 3–5 September 2019; pp. 224–233.

Disclaimer/Publisher's Note: The statements, opinions and data contained in all publications are solely those of the individual author(s) and contributor(s) and not of MDPI and/or the editor(s). MDPI and/or the editor(s) disclaim responsibility for any injury to people or property resulting from any ideas, methods, instructions or products referred to in the content.

eng

Article

Safety Occurrence Reporting amongst New Zealand Uncrewed Aircraft Users

Claire Natalie Walton * and Isaac Levi Henderson *

School of Aviation, Massey University, 47 Airport Drive, Palmerston North 4414, New Zealand
* Correspondence: clairenwalton@gmail.com (C.N.W.); i.l.henderson@massey.ac.nz (I.L.H.)

Abstract: Safety reporting has long been recognised as critical to reducing safety occurrences by identifying issues early enough that they can be remedied before an adverse outcome. This study examines safety occurrence reporting amongst a sample of 92 New Zealand civilian uncrewed aircraft users. An online survey was created to obtain the types of occurrences that these users have had, how (if at all) these are reported, and why participants did or did not report using particular systems. This study focussed on seven types of occurrences that have been highlighted by the Civil Aviation Authority of New Zealand as being reportable using a CA005RPAS form, the template for reporting to the authority for uncrewed aircraft occurrences. The number of each type of occurrence was recorded, as well as what percentage of occurrences were reported using a CA005RPAS form, an internal reporting system, or were non-reported. Qualitative questions were used to understand why participants did or did not report using particular systems. Categorical and numerical data were analysed using Chi-Squared Tests of Independence, Kruskal–Wallis H Tests, and Mann–Whitney U Tests. Qualitative data were analysed using thematic analysis. The findings reveal that 85.72% of reportable safety occurrences went unreported by pilots, with only 2.74% of occurrences being self-reported by pilots using a CA005RPAS form. The biggest reason for non-reporting was that the user did not perceive the occurrence as serious enough, with not being aware of reporting systems and not being legally required to report also being major themes. Significant differences were observed between user groups, providing policy implications to improve safety occurrence reporting, such as making reporting compulsory, setting minimum training standards, having an anonymous and non-punitive reporting system, and through working with member-based organisations.

Keywords: aviation safety; accident reporting; occurrence reporting; drones; unmanned aircraft; crewed aircraft; aviation regulation

Citation: Walton, C.N.; Henderson, I.L. Safety Occurrence Reporting amongst New Zealand Uncrewed Aircraft Users. *Eng* **2023**, *4*, 236–258. https://doi.org/10.3390/eng4010014

Academic Editor: Antonio Gil Bravo

Received: 12 December 2022
Revised: 4 January 2023
Accepted: 10 January 2023
Published: 12 January 2023

Copyright: © 2023 by the authors. Licensee MDPI, Basel, Switzerland. This article is an open access article distributed under the terms and conditions of the Creative Commons Attribution (CC BY) license (https://creativecommons.org/licenses/by/4.0/).

1. Introduction

While some types of uncrewed aircraft (UA) have been used within the civilian space for a long time (such as aeromodellers flying model aircraft), the proliferation of remotely-piloted aircraft and autonomous uncrewed aircraft (often called "drones") has accelerated in recent years [1], with the International Civil Aviation Organisation beginning to look at this issue as early as 2006 [2]. With applications ranging from aerial survey, mapping, aerial photography and video, along with inspection in the agricultural, security, energy, and construction industries, the benefits and range of uses continues to grow [3,4]. Worldwide, UA have become "more accessible, affordable, adaptable and more capable of anonymity" [5] p. 1. As UA utilisation expands, the potential for occurrences also increases. This highlights the importance of the communication of hazard information to mitigate risks and provide safety solutions. As technology progresses, UA are now performing Beyond Visual Line of Sight (BVLOS) operations and the amount of autonomy is also increasing, highlighting the need for safety systems that are fit for purpose [1]. While there is always a degree of inherent risk within aviation, the benefits of achieving a task must be perceived to outweigh any associated risks, particularly those toward crewed aircraft [6].

UA are inexpensive to purchase and because no one is aboard the aircraft, the user may take greater risks. Considering this, the importance of being proactive and identifying potential risks and vulnerabilities in advance ensures UA remain beneficial and do not become a danger to safety as the industry expands [7]. Occurrence reporting and monitoring of UA activities may be one method to reduce the risk of further occurrences [4].

With the development of UA technology and the likelihood of the UA industry expanding into shared airspace, along with increased numbers of UA operating BVLOS [1], the objective of this research is to examine the types of UA safety occurrences that users are having, how (if at all) these UA safety occurrences are being reported, and why they are being reported (or non-reported) using particular systems. An online survey was created to obtain this information from users. Our study was restricted to UA users in New Zealand, and only measured UA safety occurrences for each user between 2015 and 2022. This was because the current regulatory framework came into effect in 2015. With the data obtained, similarities and differences between users on their reporting (or non-reporting) of safety occurrences can be examined to ensure that risks are managed collectively, and safety improvements are made for the benefit of the industry.

The next section will present a literature review, which examines the applicable regulations in New Zealand and compares these with other jurisdictions and discusses relevant past literature in relation to safety occurrence reporting. Next, the methods are presented, including how the survey was created, how participants were recruited, our sample, and the how the data were analysed. The results are then presented, both quantitative (i.e., number of types of occurrences, number reported or non-reported using particular systems, and so on) and qualitative (i.e., the reasons for reporting or non-reporting using particular systems, as well as use of alternative safety performance measures). The results are then discussed, highlighting a number of potential strategies for improving safety reporting amongst UA users in New Zealand. Finally, the study is concluded and limitations and opportunities for future research are provided.

2. Literature Review

2.1. Terminology

Worldwide, several terms are used to describe UA. Colloquially, they are often referred to as drones, with more formal terms including Unmanned/Uncrewed Aerial Vehicles (UAVs), Unmanned/Uncrewed Aircraft Systems (UASs), Remotely Piloted Aircraft Systems (RPAS), and model aircraft [4,8]. Nuances exist between these terms; for example, UAVs refer to all uncrewed aircraft, while RPAS excludes autonomous aircraft, and model aircraft tends to refer to scale versions of crewed aircraft [4,8,9]. This study is interested in all types of UA, and henceforth we have consistently used the UA abbreviation throughout the paper.

2.2. Applicable Regulations

UA operations are primarily governed by two Civil Aviation Rule (CAR) Parts, CAR Parts 101 and 102. CAR Part 101 outlines a series of general operating rules, while CAR Part 102 outlines a process for certificating organisations that want to execute operations outside of those general operating rules [10,11]. Crewed aircraft incident and accident notification is regulated under CAR Part 12 but gives dispensation for UA to file occurrence reports [12]. According to CAR102.11, UA operators are required to produce an exposition to apply for Part 102 certification. The exposition must outline processes for a hazard register. The hazard register identifies risks and known hazards involved with the operation, the mitigation measures taken to avoid them along with details of the operation and equipment to be used [10]. While not being an explicit legal requirement to submit CA005RPAS forms to report occurrences, an advisory circular (outlining acceptable means of compliance) recommends that UA operators conducting operations under Part 102 should report occurrences in this manner [13]. This advisory circular lists seven events where reporting using a CA005RPAS form is recommended. These are " injury to persons; and loss of control; and fly-away; and motor or structural failure; and incidents involving manned aircraft; and incursion into airspace where not authorised; and damage to third party property" [13] p. 16. Nonetheless, the advisory circular does encourage UA operators to report any occurrences that they deem necessary, stating that regular recording of statistical data may help establish the reliability of UA and used when updating regulations [13]. However, with UA listed as a low-consequence and low-regulatory priority in the Civil Aviation Authority of New Zealand's (CAANZ) current safety strategy, these regulations may not be revised for some time [14].

New Zealand does not significantly differ in most respects with other jurisdictions with regard to regulation of UA [8]. However, internationally, there are differences with regard to the requirement for UA occurrence reporting. For example, Australia, the United States, the United Kingdom, and member states of the European Union Aviation Safety Agency (including the 27 European Union member countries, Switzerland, Iceland, Norway, and Liechtenstein) all provide very explicit thresholds for when UA occurrence reporting is required [15–19]. The specific thresholds nonetheless vary. In this respect, New Zealand diverges from these major jurisdictions by only providing an advisory circular, which, because it relates to CAR Part 102, implies it is not relevant to CAR Part 101 where more operations occur. Australia, the United States, and the European Union also provide confidential reporting systems, which are for all incidents, including those that do not reach the threshold for formal reporting [15,20,21]. These systems are aimed at collecting, analysing, and sharing safety information with users, industry, and regulators. They also emphasise that they are there to help learn from occurrences and will not be used to assign blame. There is no confidential reporting system available in New Zealand, and there is also no explicit statement on the CA005RPAS form to say that reports of UA occurrences will not be used to assign blame, though it does highlight the purpose of reporting is to help monitor risk, learn from incidents and accidents, and help to reduce the chances of accidents occurring [22].

2.3. Safety Occurrence Reporting

Reporting occurrences is an essential element of safety, providing information for the improvement and development of technology in addition to aiding UA pilots to develop their skills and knowledge to avoid further occurrences. Occurrences can be a forewarning system of a gap or weakness in the safety hazard management system. Incidents enable industry to develop learning to avoid more serious occurrences [23]. For an organisation to avoid occurrences leading to major accidents, it is crucial for occurrence information to be recorded and the information filtered to those affected, ensuring risk management strategies are adapted to avoid occurrences turning somewhat more serious [23]. For the sake of brevity, key findings and approaches from past studies related to safety occurrence reporting are summarised in Table 1.

Past research on occurrence reporting highlights some consistent themes. Firstly, occurrence reporting allows for learning and improvement of safety over time, not just for the affected organisation, but also for the industry as a whole when information is shared. Secondly, reporting has an attitudinal element, and organisational safety culture is often cited as having an influence on reporting, as well as the vice versa, where acting upon safety reports helps to reinforce a safety culture. Thirdly, underreporting is a common issue, with UA occurrence reporting estimated to be much lower than for crewed aircraft, and where only a minority of reports are self-reported by UA pilots. Lack of seriousness of an occurrence, knowledge of the legal system, usability of reporting systems, and fear of repercussions are cited as the key barriers to reporting. Fourthly, most past research has used reported occurrences as data, which is acknowledged by those studies as subject to bias because they are only the reported occurrences. This highlights the novel approach of this study by capturing participant occurrences, whether reported or unreported. Finally, no study appears to have asked UA users qualitative questions to understand why they are reporting or not, and if so, why they use particular systems.

Table 1. Abbreviated summary of past research relevant to UA safety occurrence reporting.

Study	Industry	Approach	Relevant Findings [1]
Cooke and Rohleder [23]	Non-specific	Simulating a model of a safety and incident learning system	Incident learning systems help to reinforce safety culture Organisations should implement rewards systems to encourage incident reporting When occurrence information is shared with industry, other organisations may learn and avoid the same occurrence
Cermelli et al. [24]	Petrochemical	Analysis of 5800 reported occurrences from the construction phase of a large petrochemical complex in the Middle East	Only 10–20% of near misses are self-reported
Nazeri et al. [25]	Crewed Aviation	Data mining and analysis of accidents and incidents pertaining to large aircraft commercial flights in the United States from 1995 to 2004	Incident reporting data were subject to bias and may be underreported A mixture of influences, when combined, create the situations that are more likely to result in an accident
Kyriakidis et al. [26]	Rail Transport	Precursor analysis from 18 major metros between 2002 and 2009 and a questionnaire to representatives in 11 metro organisations to assess safety maturity	More safety-conscious metros tend to record more accident precursors and top event and do so more rigorously Even metros that collect data likely do so only for more severe incidents Safety culture is appropriate to reduce accident precursors caused by human performance failures
Wild et al. [27]	Crewed and Uncrewed Aviation	Analysis of 152 UA accidents and incidents between 2006 and 2015 and comparison with crewed aircraft occurrences	UA occurrences were focussed on technology, whereas crewed aircraft occurrences were focussed on human factors
Ghasri and Maghrebi [4]	Uncrewed Aviation	Analysis of 138 UA accidents and incidents reported to the Australian Transport Safety Board between 2000 and 2018	Technology and equipment play a prevailing role in UA occurrences Data patterns from reported occurrences can be biased because of different attitudes among operators towards reporting occurrences
Kasprzyk and Konert [28]	Uncrewed Aviation	Content analysis of current European and US legislation covering UA accidents and serious incidents	The shortage of occurrence data does not provide sufficient information for improving safety and avoiding serious occurrences in the future
Konert and Kasprzyk [17]	Uncrewed Aviation	Content analysis of legislation covering incidents	There is a lack of consistent standards for reporting serious incidents and accidents involving UA UA incident reporting helps to improve safety policies and facilitate airspace integration
CAAUK [29]	Uncrewed Aviation	Survey of 32,933 UA users in the open and special categories	Only 25% of reports filed each month come from the pilots themselves UA occurrences are underreported by approximately ten times when compared with crewed aircraft occurrences Knowledge of the legal system, accessibility, the difficulty of filing reports, and the fear of repercussions were the key reasons why UA operators did not report occurrences
Henderson [30]	Uncrewed Aviation	Survey of 812 UA users to examine means of mitigating operational risks	UA users need to view risk more holistically as there is a tendency to focus only on the airworthiness of the UA and not other sources of risk

[1] We have been selective about which findings are immediately relevant to the study at hand—only a portion of the findings from any of these studies are reported here; for full results, readers will need to consult the sources themselves.

3. Methods

3.1. Materials

An online survey was created to examine what types of safety occurrences UA users are having, how (if at all) these are being reported, and why they are being reported (or non-reported) using particular reporting systems. To provide some clear guidelines about what should be reported and what systems are available, this survey was designed for

New Zealand UA users. Gender and age were collected for descriptive purposes. Users were categorised by user type (recreational, semi-professional, or professional), recency of flying, hours flown in the last 12 months, whether they had passed a course of UA operations, whether they had passed an Operational Competency Assessment (OCA) (the local term for a UA flight examination), whether they were a member of Model Flying New Zealand (MFNZ) (this is a nation-wide member organisation for aeromodellers), whether they were a member of UAVNZ and/or Aviation New Zealand (UAVNZ is an industry and professional body for UA operators, and Aviation New Zealand is its parent organisation representing the wider commercial aviation sector), which Rule Part they operated under (Part 101, Part 102, both, or unsure), and whether they pilot aircraft of 15 kg or more (as this is where some qualification or certification becomes required). These categorisations were used for later statistical analyses. Advisory Circular AC102-1 outlines seven types of occurrences that CAANZ would like reported using a CA005RPAS form, which are [13] p. 16:

1. Injury to persons (which includes the operator)
2. Loss-of-control incidents
3. Fly-aways
4. Motor and structural failures
5. Incidents involving manned aircraft
6. Incursion into airspace where not authorised
7. Damage to third party property

Because this advisory circular was published alongside the current regulatory framework in 2015, this study limited its scope to occurrences between 2015 and 2022. The survey asks participants to tick a box as to whether they have had any of each occurrence within this time period, and if they had, they were asked a follow-up question to obtain the number of each type of occurrence that they had within this timeframe. Participants who had no reportable occurrences during this time period were asked follow-up questions to see whether they had reported anything else, while participants who had a reportable occurrence were asked to provide percentages for how many of their occurrences had been reported using a CA005RPAS form, how many had been reported using an internal reporting system, how many had been reported using both the CA005RPAS form and an internal reporting system, and how many were not reported using either system. Qualitative questions were used to obtain the reasons why they did or did not report using particular systems and whether they had any alternative ways of measuring safety performance for their operations. There were two reasons why qualitative questions were used for this purpose. The first was to avoid the issue of self-generated validity where answers may be influenced by asking questions about measures that may not exist in long-term memory [31–34]. The second was to be consistent with a heterophenomenological epistemology, whereby it is important to recognise that each person lives in their own subjective reality, influenced by their life experiences and what they believe about those experiences [35,36]. By allowing participants to describe their reasoning in their own words, we build a better understanding of why they think they may behave in a particular manner [37], which is of interest when considering their behaviour in relation to safety occurrence reporting. The full list of questions in the survey, including display logic, is provided in Appendix A.

3.2. Procedure

An online survey hosted via Qualtrics was used to collect data. This was available from the 12 September 2022 until the 8 October 2022. Participants were recruited through posts on social media, through a link in an Aviation New Zealand weekly newsletter, and through encouraging participants to refer others onto the survey. Posts were made on the following Facebook forums to recruit participants: (1) Kiwi Pilots, (2) DJI Drones New Zealand Operators' Group, (3) Drone Fishing New Zealand, (4) Multirotors New Zealand, (5) New Zealand Drone Photography, (6) NZ Drone Photography, and (7) Drones on Farm NZ. A post was also made on LinkedIn using the personal account of the second author. Participants were asked how they found out about the survey before completing the survey.

When clicking on the link, participants were presented with an information sheet about the study. Three recruitment criteria were applied:

1. Participants had to reside in New Zealand
2. Participants had to have flown an unmanned aircraft at least once since 2015
3. Participants had to be 16 years or older (age to give consent to participate in New Zealand)

The use of convenience sampling was pragmatic—there are no reliable data about how many UA users there are in New Zealand, nor the split of different user types, though some estimations have been made [8]. Thus, this study simply aimed to make sure that there was a reasonable chance that different user types would be exposed to the recruitment materials posted across multiple forums. The "push-out" approach of social media recruitment (recruiting while they are engaged in other unrelated online activities) has been shown to provide demographically diverse samples [38]. Recruitment via Facebook has been shown to gather samples that are similarly representative to more traditional methods [39], with differences between Facebook data sets and comparison data sets being practically insignificant in their magnitude [40]. However, some researchers have found that it results in over-representation of young white women [41].

This project was peer-reviewed and deemed to be low risk. Consequently, it was not reviewed by one of the University's Human Ethics Committees but was registered as a low-risk study on the Massey University Human Ethics Database.

3.3. Sample

The survey obtained 110 responses during the study period. However, only 92 responses were complete enough to be useful (determined by completing at least 69% of the questions). Out of this sample of 92 participants, 83 (90.22%) were male, 6 (6.52%) were female, 1 (1.09%) was nonbinary, and 2 (2.17%) preferred not to say. The mean age was 42.78 (SD = 16.25), with one participant who did not provide age. All participants were current users of UA. Of these, 49 (53.26%) classified themselves as recreational users (primarily for enjoyment), 21 (22.83%) classified themselves as semi-professional (where less than 50% of the participants work time is spent on activities related to unmanned aircraft operations), and 22 (23.91%) classified themselves as professional users (where more than 50% of the participants work time is spent on activities related to unmanned aircraft operations).

To ensure that UA users from various groups were not over-represented, we asked participants how they found out about the survey. The majority of participants, 75 (81.52%), found out via social media, 14 (15.22%) found out from the Aviation New Zealand email, and 3 (3.26%) were referred by a friend.

Regarding flight recency, 62 (67.39%) had flown within the last month, 18 (19.57%) had flown within the last 6 months, 6 (6.52%) had flown within the last year, 6 (6.52%) had flown more than a year ago. With regards to flight currency, 18 (19.57%) had flown less than 5 h within the last 12 months, 19 (20.65%) had flown 5–10 h within the last 12 months, 16 (17.39%) had flown 10–25 h within the last 12 months, 13 (14.13%) had flown 25–50 h within the last 12 months, and 26 (28.26%) had flown more than 50 h within the last 12 months. There were 34 (36.96%) participants who had completed a course on UA operations, and 40 (43.48%) who had passed an operational competency assessment. In terms of member-based organisations, 21 (22.83%) were members of MFNZ, and 20 (21.74%) were members of UAVNZ and/or Aviation New Zealand.

Most of the participants (55, 59.78%) operated only under Part 101 of the CARs, while three (3.26%) operated only under a Part 102 Operator's Certificate, and 17 (18.48%) operated under both Part 101 and Part 102. Concerningly, 17 (18.48%) participants were unsure which set of CARs they were operating under. Only 6 (6.52%) participants operated UA with a mass of more than 15 kg.

3.4. Analysis

Participants were coded according to their user type, recency of flying, hours flown in last 12 months, whether they had completed a course or OCA, whether they were members of MFNZ or UAVNZ, and which Rule Part they used for operations. Chi-Squared Tests of Independence [42] were run to see whether occurrence reporting (both to CAANZ and internally) as well as non-reporting were associated with particular participant groups. Effect size is reported with Cramer's V [43]. The percentage of occurrences that were reported (using any means) was calculated for each participant as a numerical value. To see whether differences exist between the percentage of reported occurrences based upon user type, recency of flying, and hours flown, Kruskal–Wallis H tests [44] were run. Distributions were checked for similarity by visual inspection of a boxplot. Pairwise comparisons using Dunn's [45] procedure were performed and a Bonferroni [46] correction was applied. For other participant groupings (which only involve two groups), Mann–Whitney U tests [47] were used to see whether differences existed in the percentage of occurrences that were reported. Distributions were assessed to be similar based upon visual inspection, and directionality was assessed according to mean ranks and distributions using an exact sampling distribution for U [48].

For qualitative responses, a process consistent with Braun and Clarke's [49] fifteen-point checklist for a good thematic analysis was used. First, all responses to each qualitative question were collated. Next, they were divided into single units of thought, which this study has called "statements". This was important as sometimes a participant may outline multiple ideas within the same response, so they may have multiple statements for a particular question. Definitions for themes were created so that statements could be thematically classified. Definitions for themes did not overlap, so participants would only be grouped into multiple themes if they made statements that expressed ideas consistent with multiple themes.

4. Results

4.1. Descriptive Results

Out of the participants, 50 (54.35%) participants had a reportable occurrence in the period of 2015 to 2022. A summary of the types of occurrences is presented in Table 2, showing the number of participants having each occurrence, the total number of each occurrence observed, the mean number for each occurrence across the sample (n = 92), and the mean number for each occurrence across the sub-sample of only those users who had that occurrence type.

Table 2. Types of occurrences observed among users.

Type of Occurrence	No. Participants (%)	Observed Number	Mean (Sample)	Mean (Sub-Sample)
Injury to person	8 (8.70%)	9	$M = 0.10$, $SD = 0.33$	$M = 1.13$, $SD = 0.33$
Loss of control	34 (36.00%)	271	$M = 2.97$, $SD = 11.62$	$M = 7.97$, $SD = 18.05$
Fly-away	12 (13.04%)	34	$M = 0.36$, $SD = 1.65$	$M = 2.83$, $SD = 3.72$
Motor or structural failure	22 (23.91%)	122	$M = 1.33$, $SD = 5.54$	$M = 5.55$, $SD = 10.24$
Loss of separation with manned aircraft	3 (3.26%)	7	$M = 0.08$, $SD = 0.54$	$M = 2.33$, $SD = 1.89$
Airspace incursion	3 (3.26%)	7	$M = 0.08$, $SD = 0.54$	$M = 2.33$, $SD = 1.89$
Damage to third-party property	0	0	*N/A*	*N/A*
No reportable occurrences	42 (45.65%)	*N/A*	*N/A*	*N/A*
Total	50 [1] (54.35%)	450	$M = 4.89$, $SD = 18.79$	$M = 9.00$, $SD = 24.76$

[1] This number cannot be computed from the earlier numbers of participants for each occurrence type as a single participant may have had multiple types of occurrences.

Out of the 50 participants that indicated that they had reportable occurrences, 45 completed the follow-up questions about the percentage of occurrences that were reported using a CA005RPAS form (only), reported internally (only), reported both using a CA005RPAS form and internally, and the percentage that were not reported. Table 3 provides a summary of the prevalence of reporting systems to report occurrences.

Table 3. Prevalence of reporting systems to report occurrences.

Reporting System	Mean Percentage	No. Participants with >0%
CA005RPAS Form (only)	2.67% (*SD* = 14.97%)	2
Internal Reporting (only)	21.24% (*SD* = 38.63%)	12
Both CA005RPAS Form and Internal Reporting	4.96% (*SD* = 19.14%)	4
Not reported	71.13% (*SD* = 43.66%)	33
Reported to CAANZ [1]	7.62% (*SD* = 23.93%)	5
Reported Internally [2]	26.20% (*SD* = 41.84%)	12
Reported (any means)	28.87% (*SD* = 43.66%)	12

[1] Computed by adding percentages for CA005RPAS form (only) and those who used that form as well as internal reporting. [2] Computed by adding percentages for internal reporting (only) and those who reported internally and submitted a CA005RPAS form.

Using the total number of reportable occurrences for each of the 45 participants who answered the follow-up questions, we can now multiply the percentage reported to CAANZ, internally, and the percentage non-reported by the number of reportable occurrences. This is important as different users had different numbers of reportable occurrences; it is not just the average percentages that are important, but the percentage of actual reportable occurrences. The sample of 45 participants who answered the follow-up questions had a total of 427 reportable occurrences, Table 4 shows how those occurrences are divided amongst reporting systems.

Table 4. Number and percentage of occurrences that were reported using different systems.

Reporting System	No. Occurrences	% Occurrences (n = 427)
Reported to CAANZ [1]	11.69	2.74%
Reported Internally [2]	56.17	13.15%
Reported (any means)	60.97	14.28%
Non-Reported (any means)	366.03	85.72%

[1] Computed by adding percentages for CA005RPAS form (only) and those who used that form as well as internal reporting, and then multiplying by the user's observed number of occurrences. [2] Computed by adding percentages for internal reporting (only) and those who reported internally and submitted a CA005RPAS form, and then multiplying by the user's observed number of occurrences.

For this particular sample, despite the mean percentage for non-reporting being 71.13% across users, the percentage that were actually not reported was 85.72%. This was because some of the users with the highest number of occurrences had 0% reporting rates, skewing the percentage of occurrences upwards.

4.2. *Quantitative Results*

For the sake of brevity, statistically significant results from the Chi-Squared Tests of Independence, Kruskal–Wallis H Tests, and Mann–Whitney U Tests are reported in abbreviated form in Figure 1. In this figure, user groups are highlighted as being associated with either reporting to CAANZ via submitting a CA005RPAS Form, reporting using an internal process, or non-reporting of occurrences. These associations are based upon the results of the Chi-Squared Tests of Independence. Differences between groups in the overall percentage of occurrences that were reported (using either CA005RPAS Forms or an internal process) are shown in the bottom right corner. These are comparisons between groups determined using Kruskal–Wallis H Tests (for when there are more than two groups), or Mann–Whitney U Tests (for comparison between two groups). Full statistical reporting is provided in Appendix B. Effect sizes for statistically significant results are also reported in Appendix B.

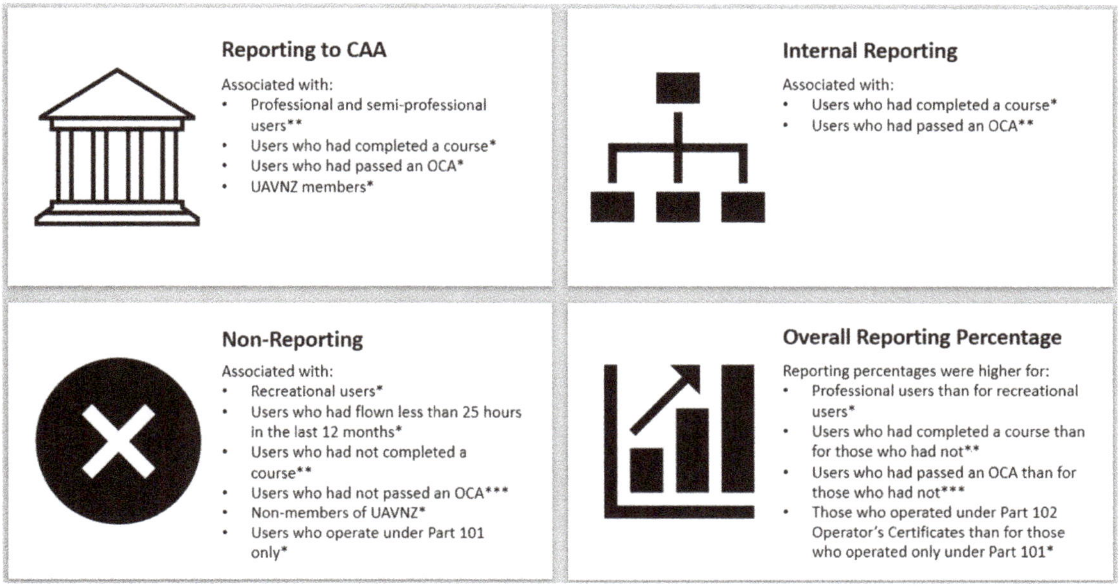

Figure 1. Abbreviated results of statistical analyses. Note: *, **, and *** indicate statistical significance at the $p < 0.05$, 0.01, and 0.001 levels, respectively.

4.3. *Qualitative Results*

To help explain the quantitative findings, participants were asked to explain why they reported using particular systems, and if they indicated that they had alternative ways of measuring safety performance, they were also probed on what alternative ways they were using. Thematic analyses were performed to make sense of the qualitative data and are reported in this section. Sub-sections divide the different thematic analyses, and an explanation is provided before each one about the number of participants who were asked the relevant question.

4.3.1. Non-Use of CA005RPAS Forms

Participants who did not report any occurrences using a CA005RPAS form were asked why they did not report any occurrences using this system. There were 40 participants who were presented with this question, however, only 31 participants provided an answer. These 31 participants made a total of 49 statements regarding why they did not use the CA005RPAS form to report any occurrences. Table 5 presents the themes from these qualitative answers.

4.3.2. Use of Both CA005RPAS Forms and Internal Processes

Participants who indicated that they reported some occurrences using both a CA005RPAS form and through an internal process were asked why they used both systems. Only four participants were presented with this question due to the lack of users who used both reporting systems for the same occurrence. All four of these participants provided answers, making a total of five statements. Table 6 presents the thematic analysis from their answers.

4.3.3. Use of Internal Reporting Instead of a CA005RPAS Form

Participants who indicated that they reported some occurrences internally, but not using a CA005RPAS form were asked why they reported only internally. Ten participants were presented with this question, with all ten participants providing answers, making a total of 13 statements. Table 7 presents the thematic analysis from their answers.

Table 5. Reasons for not using CA005RPAS forms to report any occurrences.

Theme	Explanation	Statements No.	Statements %	Participants No.	Participants % [1]	Example Quotes
Awareness	Statements that reflect not being aware about the form	8	16.33%	8	25.81%	"I haven't heard of it", "I don't know what the form is"
Own Property	Operations over own property	4	8.16%	4	12.90%	"... whilst flying on my own property", "A test flight of a new cinewhoop on my front lawn ... "
Rules-Based	Not required to do so under the CARs	9	18.37%	9	29.03%	"I didn't realise that I had violated any rule at the time", "Don't need to under Part 101 as far as I know"
Seriousness	Occurrence was perceived as not being serious enough to report	25	51.02%	14	45.16%	"Very minor incidents while learning to fly my first drone", "Nothing was damaged except my drone"
No Comment	Did not want to make a comment	1	2.04%	1	3.23%	"No comment"
Uncategorised	Statements that do not fit in any other theme	2	4.08%	2	6.45%	"I didn't want to waste the time of officials"

[1] Percentage calculated as of the number of participants who responded to the question, which was 31 (rather than the full sample of 92 participants).

Table 6. Reasons for using both a CA005RPAS form and an internal process.

Theme	Explanation	Statements No.	Statements %	Participants No.	Participants % [1]	Example Quotes
Organisational Policy	A procedure that the participant's organisation requires	2	40.00%	2	50%	"Internal reporting as per company SMS policies", "Organisational requirement"
Uncategorised	Statements that do not fit into any other theme	3	60.00%	3	75%	"To make it useful", "Reporting was required due to the failure that occurred"

[1] Percentage calculated as of the number of participants who responded to the question, which was 4 (rather than the full sample of 92 participants).

Table 7. Reasons for reporting only internally and not using a CA005RPAS form.

Theme	Explanation	Statements No.	Statements %	Participants No.	Participants % [1]	Example Quotes
Organisational Policy	A procedure that the participant's organisation requires	2	15.38%	2	20.00%	"We use a standard health and safety model ... So that the club level patterns can be seen"
Response to Occurrence	The organisation's response to the occurrence	3	23.08%	3	30.00%	"... rules and procedures refined to ensure minor incidents don't escalate", " ... Drone was not flown again until after software reinstall"
Rules-Based	Not required to do so under the CARs	3	23.08%	3	30.00%	"Reporting via 005 was not required in the regulation", "As both incidents were under 101 conditions there was no requirement to report them"
Seriousness	Occurrence was perceived as not being serious	3	23.08%	3	30.00%	"It had only gone less than 1 metre. I regained control and landed it"
Uncategorised	Statements that do not fit in any other theme	2	15.38%	2	20.00%	"Form didn't exist"

[1] Percentage calculated as of the number of participants who responded to the question, which was 10 (rather than the full sample of 92 participants).

4.3.4. Non-Reporting Using Either CA005RPAS Form or Internal Systems

Participants who indicated that they did not report some occurrences using either a CA005RPAS form or an internal system were asked why they did not report those occurrences using either system. There were 33 participants who were asked this question, with 27 providing answers, making a total of 42 statements. Table 8 presents the thematic analysis from their answers.

4.3.5. Alternative Ways of Monitoring Safety Performance

All participants were asked whether they used alternative ways of monitoring safety performance outside of occurrence reporting using CA005RPAS forms and internal reporting systems. Participants received different wording for this question based upon whether they were reporting occurrences or not. Eight participants out of the 13 that were reporting occurrences (including one who did not have a reportable occurrence but reported anyway) indicated that they also used alternative systems for monitoring safety performance. For those who were not reporting occurrences using CA005RPAS forms or using internal systems (whether those occurrences were reportable or not), 6 participants indicated that they used alternative systems for monitoring safety performance, while 61 participants indicated that they used no alternative systems. There were 12 participants that did not answer either version of the question. Of the 14 participants who indicated that they used alternative ways of monitoring safety performance, only 12 provided information about those alternative systems. Table 9 presents the thematic analysis based upon the responses of the 12 participants who were using alternative systems (whether in combination with CA005RPAS and internal reporting systems, or in lieu of those systems). These participants made a total of 15 statements.

Table 8. Reasons for not reporting occurrences using CA005RPAS forms or internal systems.

Theme	Explanation	Statements No.	Statements %	Participants No.	Participants % [1]	Example Quotes
Activity Type	Reporting was dependent upon the purpose of the flight	9	21.43%	8	29.63%	" … training flights to improve my flying skill … ", " … whilst engaged in recreational flying"
Awareness	Statements indicating lack of knowledge about occurrence reporting processes	6	14.29%	6	22.22%	" … didn't even know reporting incidents was a requirement in UAV flight or possible", "I have not even heard of the form"
No Internal process	Lack of organisational requirements for reporting	2	4.76%	2	7.41%	" … our internal process at work hadn't been set up", "Not a whole lot of internal processes in a sole trader company … "
Rules-Based	Not required to do so under the CARs	5	11.90%	5	18.52%	"Don't need to under part 101 as far as I know", "I am completely unaware of having to report to anyone"
Seriousness	Occurrence was perceived as not being serious	16	38.10%	11	40.74%	"Engine failure on model aircraft is a daily occurrence and does not require 005 reporting", " … no-one has ever been killed by the recreational use of multi-rotor drones"
Uncategorised	Statements that do not fit into any other theme	4	9.52%	3	11.11%	"Not worth the hassle", " … all safety precautions taken were always successful in mitigating any risk of injury or third party damage"

[1] Percentage calculated as of the number of participants who responded to the question, which was 27 (rather than the full sample of 92 participants).

Table 9. Alternative systems used by participants to monitor safety performance.

Theme	Explanation	Statements No.	Statements %	Participants No.	Participants % [1]	Example Quotes
Hazard Mitigation	Actions taken to eliminate or decrease the likelihood of an occurrence	5	33.33%	3	25.00%	"Test, test, test and test every aircraft and system prior to deploying … ", " … physical inspection of models, transmitter range checks before we fly as recommended by individual clubs, and model flying rules and requirements … "
Reporting systems	Other methods of incident reporting	5	33.33%	5	41.67%	" … MFNZ rules/procedures", "The main company (manned aviation) used Air Maestro for SMS, it made sense to also roll the the [sic] UAV division into this also"
Uncategorised	Statements that do not fit in any other theme	5	33.33%	5	41.67%	"DroneLogBook", " … we also have a safety officer … "

[1] Percentage calculated as of the number of participants who responded to the question, which was 12 (rather than the full sample of 92 participants).

5. Discussion

The results have indicated that a very large portion of reportable occurrences in New Zealand are going unreported, with each user on average reporting only 28.87% of their occurrences. Because of differences in the numbers of occurrences between users, the actual percentage of occurrences reported was only 14.28%. Reporting to the Civil Aviation Authority using a CA005RPAS form was particularly low, with an average of 7.62% of occurrences per user, and an observed rate of 2.74% of occurrences across this study's sample. Given the importance of safety occurrence reporting to improving organisational systems and for identifying common hazards across the sector and how these should be regulated, it is important to increase these percentages. The focus of this discussion section is on identifying ways that safety occurrence reporting might be improved, based upon the results of this study and also by examining the academic literature. It has been divided into five core areas of discussion: (1) the role of training and assessment, (2) working with member-based organisations, (3) seriousness of occurrences, (4) regulatory considerations, and (5) exploring confidential reporting systems.

5.1. The Role of Training and Assessment

One of the most useful findings from the statistical analyses was that having completed a course on UA operations and having passed an OCA both acted to improve reporting rates and decrease the likelihood of non-reporting occurrences. For the issuance of pilot licenses in crewed aviation, there are requirements in terms of passing both theory examinations and flight examinations [50]. Pilots operating under a Part 102 Operator's Certificate will need to have completed a theory course covering general aviation knowledge and UA-specific knowledge, as well as have passed an OCA [13]. However, no such requirement exists for operators under Part 101, whether flying commercially or not. As part of their *Enabling Drone Integration* discussion document of 2021, the Ministry of Transport in New Zealand are proposing to introduce a basic pilot qualification for all UA users to complete [51]. If this proposal is implemented, then there may be the opportunity to ensure that occurrence reporting is covered as part of this basic pilot qualification. Regardless of what approach is taken exactly, the results of this study suggest that the lack of any educational requirements to enter the aviation system as a UA user may be one of the drivers behind low reporting rates among users. A less formalised approach may also be to provide an online resource on CAANZ's website about occurrence reporting for UA users, which could provide clear guidelines on what should be reported and how this information will be used by the authority.

5.2. Working with Member-Based Organisations

This study found that members of UAVNZ/Aviation New Zealand were more likely to report an occurrence using the CA005RPAS form and were less likely to non-report occurrences. There were no statistically significant differences for MFNZ members. Both organisations have codes of conduct and can self-regulate their members. The difference in the observed results is likely because UAVNZ is a professional and industry body, meaning its members are commercial operators. MFNZ is an organisation for recreational operators who want to partake in aeromodelling. Unlike UAVNZ, MFNZ does have an accident reporting form and requires members to fill it out whenever someone is injured, when the model pilot files for insurance, or when the model has deviated into controlled airspace without permission [52]. However, this is also consistent with only reporting more serious occurrences (see later discussion).

One recent occurrence analysed on the UK's Confidential Human Factors Incident Reporting Programme (CHIRP) shows the significant amount of knowledge to be gained from occurrence reporting, analysing and discussion. In this report, a UA pilot and spotter had carried out a sizeable amount of hazard mitigation and planning prior to a training flight in a park. Regardless of this preparedness, plans were overthrown by an uninvolved person who walked across the UA's landing area and after a delay the UA ended up landing with minimal battery remaining. Lessons learnt and shared from this could help other new drone pilots in their hazard mitigations and preparations for training flights [53].

There is the opportunity for CAANZ to work more proactively alongside UAVNZ and MFNZ to ensure that their members report occurrences, even those that may not be strictly required by either organisation's internal policies. Working with organisations that have the ability to self-regulate may be an effective way of improving occurrence reporting, and member-based organisations have shown ability to improve performance in risk mitigation for UA operations [30]. Membership to MFNZ currently costs NZD 90 (USD 56.55) for an adult, while membership to UAVNZ currently costs NZD 253 (USD 158.98) for a company, with both organisations have other forms of membership available.

5.3. Seriousness of Occurrences

One reoccurring theme throughout this study was the perceived seriousness of the occurrence by participants. Although there is an argument for UA aircraft to have a safe place for training and currency practice where mistakes can be made, and lessons learnt, there is also an argument to be made for all occurrences being reported for statistical purposes. While these data could be recorded through a system similar to the Aviation Safety Reporting System run by NASA for the US aviation industry, CAANZ are currently the only organisation in New Zealand that collects occurrence information on the aviation industry. The importance of statistical data being available to CAANZ is highlighted in the organisation's Regulatory Safety and Security Strategy for the 2022 to the 2027 period, where the organisation's regulatory direction is towards an "intelligence-led" and "risk-based" assessment process, where CAANZ state "We rely in large part upon high-quality reporting by participants of occurrences" [14] p. 19. "In short, we rely on data and information to provide intelligence that informs the formation of our strategic and operational policies and plans . . . ." [14] p. 19. Without the data or with only limited data, the decisions made by the industry regulator may not be accurate to what the industry needs for safety and growth. The shortage of occurrence data may also limit the authority's ability to improve safety and avoid serious occurrences in the future [28].

Additionally, it is worth highlighting the importance of occurrence data being collected so that it can be shared within the industry so that other UA pilots and operators can avoid similar occurrences. The importance of understanding the UA modes, what could go wrong and the manual flight currency to recover prior to an occurrence was illustrated by the M600 Pro serious incident in 2019. In this occurrence the UA experienced a GPS-compass error where the UA reverted to a mode requiring manual control. However, both the pilot and observer did not recognise the error message or the loss of initial control as reason to take manual control and the UA continued drifting with the wind until it collided with a building [54]. The AAIB reported the operator had 10 s between the GPS-compass error and losing VLOS, and although CAAUK required the UA pilot to prove competence prior to carrying out commercial operations, there was no requirement at the time to maintain currency in emergency procedures. As a result of this serious incident the AAIB recommended UA pilots regularly practice manual emergency actions should automation ability be lost and raised the importance of recording minor occurrences where prior to this the UA had a minor similar occurrence in the weeks prior [54]. UA data and research are essential because UA are still in their infancy and the industry is growing [4].

5.4. Regulatory Considerations

A number of participants stated they do not submit occurrence reports because the current civil aviation regulations do not require an occurrence report to be submitted. This study also found those who operated under CAR Part 101, which does not require a hazard register, were associated with the non-reporting of occurrences. Those that operated under CAR Part 102, which includes a hazard register had reporting percentages higher than those who operated under CAR Part 101. Although the current New Zealand regulations may not specifically require occurrences to be reported, occurrence data are vital for regulations to evolve. A collision between a Robinson R44 helicopter and a Phantom 4 UA in Israel underlines the importance of reporting even when an occurrence happens within the law. In this accident, both operators were operating within the bounds of current regulations and were both approved to be in the areas of operation [55].

In 2019, an Alauda Airspeeder MK II entered controlled airspace affecting traffic in the Gatwick Airport area until its battery depleted resulting in an uncontrolled landing in a field. This shows the importance of report data for manufacturers to improve. During the resultant investigation, the AAIB found poor quality design and build contributed to the accident which saw parts within the kill-switch system detach from the UA circuit board [56]. The Australian-based organisation was compliant with Australian UA regulations and held a licence in accordance with CASA regulations. The UA was flying in the UK under an exemption approved by CAAUK for which no inspection by CAAUK was required [56]. On the day the exemption was issued, a test flight was carried out by the operator, without CAAUK present, where the UA experienced a heavy landing and damage to its landing gear. This was attributed to a power loss due to a battery fault. Although this previous day's occurrence was required under the exemption to be reported to CAAUK, no occurrence report was received to either the organisations base State authority CASA, the ATSB, or to CAAUK [56]. With UA's rapid advancement at the time, CAAUK team responsible for signing off on the exemption had little experience in the area of UA and insufficient resources [56]. This occurrence highlights how critical it is for information and in-depth knowledge of new technology for both operators and regulators, prior to approvals being granted and flights carried out. This occurrence also highlights the importance of the reporting and investigation of occurrences to develop knowledge and learning.

Like other jurisdictions have done, it would be beneficial for CAANZ to provide an explicit list of thresholds for when occurrence reporting is required, and for this to be on an accessible part of their website. Relying on users to come across AC102-1 means that those who do not have formal training are unlikely to know how to report using this system. Some of the thresholds that have been used internationally centre around the weight of the craft, airspace incursions, damage to property exceeding a certain amount, injury to uninvolved persons, loss of separation with crewed aircraft, and if the operation was commercial [15,16,18,19,57]. This study argues that any thresholds should be entirely risk-based and assessed using robust means such as the Joint Authorities for Rulemaking on Unmanned Systems' Specific Operations Risk Assessment (commonly called JARUS SORA) [58]. The current thresholds applied in AC102-1 could be used but need to be more accessible for users, such as through CAANZ's website, and clearly explained. A statement indicating that submitted CA005RPAS forms will not be used to assign blame or liability may also be appropriate if the purpose of data collection is to investigate and learn from occurrences, and would be more in line with the International Civil Aviation Organisation's Annex 13 requirements for air accident investigation [59].

5.5. *Exploring Confidential Reporting Systems*

This study discovered that CAANZ only received CA005RPAS forms from UA users for 2.74% of the occurrences within the sample over the study time period. This was similar to the CAAUK report which found only 25 percent of the 44 reports filed each month were by the UA pilots or operators themselves [60]. However, internal reporting was higher, with 13.15% of occurrences reported. The US confidential reporting system run by NASA, the ASRS, is aimed at collecting data and identifying deficiencies within the industry with the aim of improving safety. The ASRS system treats those that voluntarily report occurrences with immunity from prosecution so long as the act was not deliberate [20]. The EU and Australia have similar systems with ECCAIRS and REPCON, which are also aimed at improving safety.

6. Conclusions

This study presents the results obtained from a sample of 92 UA users in New Zealand regarding what sorts of occurrences they have had, how these have been reported (if at all), and why they chose to report using such systems (or non-report). There were 450 reportable safety occurrences from within the sample between 2015 and 2022, with the most common occurrences being loss of control and fly-aways. Concerningly, the average reporting rate (combining both internal reporting and the use of CA005RPAS forms) per user was only 28.87%, with the actual percentage of occurrences reported being even lower at 14.28%. This suggests that individual organisations and the industry as a whole are missing out on important safety information that may help to prevent occurrences escalating into accidents that result in injury, damage to property, or collisions with other aircraft. More so, reporting rates to CAANZ via CA005RPAS forms were particularly low, with the average reporting rate to CAANZ per user at only 7.62%, and only 2.74% of all occurrences being reported to CAANZ. This suggests that the regulator may be ill-informed about how to improve safety outcomes for the UA sector as it is not receiving sufficient data to lead evidence-based decisions. The statistical analyses and qualitative questions provide some potentially fruitful avenues for improving reporting rates amongst UA users in New Zealand. Namely, standards for training and assessment, working more effectively with member-based organisations, encouraging even non-serious occurrences to be reported, reconsideration of the current CARs, and the introduction of a confidential reporting system similar to what is used in the US, EU, and Australia. While these results are only directly applicable to New Zealand, the approach taken and the themes elucidated should also help guide other jurisdictions in the pursuit of improving safety occurrence reporting amongst UA users.

7. Limitations and Future Research

The key limitation of this research was the sample size, with only 92 valid responses. While understandable due to the topic area of the research and the number of qualitative questions that were presented to users, this does mean that the study only had the statistical power to find medium effect sizes. Small effect sizes may have been missed simply because of lack of statistical power. Nonetheless, the contribution of the qualitative themes may allow for a more structured questionnaire to be developed in the future and administered to a larger sample to check that the findings are indeed generalisable, and to provide enough statistical power to find small effect sizes. While we believe that this study has captured a useful and pragmatic cross-section of UA users in New Zealand, the convenience sampling method combined with anonymous responses means that we cannot be certain that the sample is representative. Nonetheless, we have described the recruitment methods in such a way that they could be replicated, and a similar result obtained.

Aside from generalisability, it would be valuable to have more in-depth discussions with UA users to understand why non-reporting of occurrences is so high. While some important conclusions can be made based upon this research, future research may benefit from focus groups or similar approaches being employed to properly understand why these conclusions hold true and to see how behaviours might be able to be changed in the future. That can then again be developed into a more structured approach, but with a solid foundation for making assumptions.

The final limitation is to highlight that we only observed self-reporting rates from the UA users themselves. In particular for CA005RPAS forms, it is possible for someone other than the UA user to file a report to CAANZ. In the United Kingdom, only a quarter of occurrence reports came from the UA pilots themselves [60]. Thus, it is possible that CAANZ is aware of some of the occurrences identified in this study from others reporting those occurrences. Reports from UA pilots are arguably more useful as these will contain operational and technical details that will not be able to be deduced from simply observing an occurrence. Regardless, it is important to highlight that our study is focussed only on UA users self-reporting the occurrences, rather than total occurrence reporting, including from third parties. Thus, reporting numbers inclusive of other parties may be higher than those reported in this study.

Author Contributions: Conceptualisation, C.N.W. and I.L.H.; methodology, C.N.W. and I.L.H.; formal analysis, C.N.W. and I.L.H.; investigation, C.N.W.; writing—original draft preparation, C.N.W.; writing—review and editing, I.L.H.; supervision, I.L.H. All authors have read and agreed to the published version of the manuscript.

Funding: This research received no external funding.

Institutional Review Board Statement: This study was peer-reviewed and deemed to be low risk. It was registered as such on the Massey University Human Ethics Database.

Informed Consent Statement: Informed consent was obtained from all subjects involved in the study.

Data Availability Statement: The full dataset that supports this study will be made available upon request to the corresponding author.

Conflicts of Interest: Claire Walton has been working on contract for the Civil Aviation Authority of New Zealand while completing this study, in the role of Regulatory Interventions Analyst. This job involves designing and managing safety initiatives for crewed aviation flight training. To manage any potential conflicts of interest with her research, she was not involved with any aspect of uncrewed aviation for the authority during her studies and has a signed conflict of interest with the authority. This research was completed in the capacity of her academic studies and none of the findings benefit her directly or indirectly. Isaac Henderson is the current Chair of UAVNZ, an industry and professional body representing the uncrewed aviation sector in New Zealand. He is also an active consultant in the uncrewed aviation industry. However, none of the findings directly or indirectly benefit him, and his primary involvement in this research was in the capacity of a supervisor.

Appendix A. Online Survey Questions

1. What is your gender?
    a. Male
    b. Female
    c. Other (please specify)
    d. Prefer not to say
2. What is your age?
3. How did you find out about this survey?
    a. Referred by a friend
    b. Sent a link from an organization that you are a member of
    c. Saw it on social media
    d. Other (please specify)
4. Which of the following best describes you?
    a. Not a current unmanned aircraft user
    b. Recreational unmanned aircraft user (primarily for enjoyment)
    c. Semi-professional unmanned aircraft user (where less than 50% of your work time is spent on activities related to unmanned aircraft operations)
    d. Professional unmanned aircraft user (where more than 50% of your work time is spent on activities related to unmanned aircraft operations)
5. When was the last time you flew an unmanned aircraft?
    a. Within the last month
    b. Within the last 6 months
    c. Within the last year
    d. More than a year ago
6. [Displayed unless participant selected d. for question 5] Within the last 12 months, roughly how many hours (flight time) have you spent flying unmanned aircraft?

a. Less than 5 hours
b. 5–10 hours
c. 10–25 hours
d. 25–50 hours
e. More than 50 hours

7. Have you ever done a course on unmanned aircraft operations?
   a. Yes
   b. No
8. Have you ever passed an operational competency assessment (also known as a flight examination) on an unmanned aircraft?
   a. Yes
   b. No
9. Are you a member of Model Flying New Zealand?
   a. Yes
   b. No
10. Are you or your organization a member of UAVNZ and/or Aviation New Zealand?
   a. Yes
   b. No
11. Under which set of Civil Aviation Rules do you conduct your unmanned aircraft operations?
   a. Under Part 101 of the Civil Aviation Rules
   b. Under a Part 102 Operator's Certificate
   c. Under both Part 101 and Part 102
   d. Unsure
12. Do you operate any unmanned aircraft with a mass of more than 15kg?
   a. Yes
   b. No
13. Since 2015, have you had any incidents while flying an unmanned aircraft that resulted in the following (please select all that apply, or none of the above)?
   a. Injury to persons (including yourself)
   b. Loss of control
   c. Fly-away
   d. Motor or structural failure
   e. Loss of separation with a manned aircraft
   f. Incursion into airspace where you were not authorised to fly
   g. Damage to third-party property
   h. None of the above
14. [Displayed if participant selected a. for question 13] Approximately how many incidents involving unmanned aircraft have you had that resulted in an injury to a person (including yourself) since 2015?
15. [Displayed if participant selected b. for question 13] Approximately how many incidents involving unmanned aircraft has you had that resulted in loss of control since 2015?
16. [Displayed if participant selected c. for question 13] Approximately how many incidents involving unmanned aircraft have you had that resulted in a fly-away since 2015?
17. [Displayed if participant selected d. for question 13] Approximately how many incidents have you had that resulted in motor or structural failure since 2015?
18. [Displayed if participant selected e. for question 13] Approximately how many incidents involving unmanned aircraft have you had that resulted in loss of separation with manned aircraft since 2015?

19. [Displayed if participant selected f. for question 13] Approximately how many incidents involving unmanned aircraft have you had that resulted in incursion into airspace where you were not authorised to fly since 2015?
20. [Displayed if participant selected g. for question 13] Approximately how many incidents involving unmanned aircraft have you had that resulted in damage to third-party property since 2015?
21. [Displayed if participant selected h. for question 13] Since 2015, have you ever reported any incidents involved an unmanned aircraft using a CA005RPAS form?
    a. Yes
    b. No
22. [Displayed if participant selected h. for question 13] Since 2015, have you ever reported any incidents involving an unmanned aircraft using an internal reporting process
    a. Yes
    b. No
23. [Displayed in answer to question 21 or question 22 is "yes"] Would you be willing to share your submitted CA005RPAS forms and/or internal incident reports with the researchers on the condition of anonymity and aggregation of data (so no individual or organisation could be identified)?
    a. Yes
    b. No
24. [Displayed if answer to question 23 is "yes"] Could you please enter your email address so that we can get in touch with you and provide a separate consent form to share your incident reports?

    Questions below this point were only asked to those who had at least one reportable accident since 2015
25. Out of the incidents that you had since 2015 (regardless of type), what percentage of them were reported using each of the following means?
    a. CA005 RPAS Form (only) [Slider from 0–100%]
    b. Internal reporting process (only) [Slider from 0–100%]
    c. Both the CA005RPAS form and an internal reporting process [Slider from 0–100%]
    d. Not reported using a CA005RPAS form or an internal reporting process [Slider from 0–100%] (Note that survey forced percentages to add up to 100%)
26. [Displayed if participant selected 0% for both options a. and c. in question 25] You have indicated that you did not report any incidents using the CA005RPAS form. Can you please explain why you did not report any incidents using a CA005RPAS specifically?
27. [Displayed if participant provided a value of greater than 0% for option c. in question 25] You have indicated that you reported some incidents using both the CA005RPAS form and an internal reporting process. Can you please explain why you used both reporting systems for those incidents?
28. [Displayed if participant provided a value of greater than 0% for option b. in question 25] You indicated that you reported some incidents using an internal reporting process instead of submitting a CA005RPAS form. Can you please explain why you chose an internal reporting process to report these incidents instead of using the CA005RPAS form?
29. [Displayed if participant provided a value of greater than 0% for option d. in question 25] You indicated that some incidents were not reported using either the CA005RPAS form or an internal reporting process. Can you explain why you did not report these incidents using either process?
30. Do you (or your organisation) use any alternative ways of monitoring safety performance for unmanned aircraft operations outside of using the CA005RPAS forms and/or and internal reporting process?

  a. Yes
  b. No

31. [Displayed if answer to question 30 is "yes"] Could you please outline the alternative ways of monitoring safety performance for unmanned aircraft operations that you (or your organisation) are using?
32. [Displayed if participant provided the value of 100% for option d. in question 25] You have indicated that none of the incidents you have had involving unmanned aircraft since 2015 were reporting using either CA005RPAS forms or an internal reporting process. Do you (or your organisation) have alternative ways of measuring safety performance for unmanned aircraft operations?
    a. Yes
    b. No
33. [Displayed if answer to question 32 is "yes"] Could you please outline the alternative ways of monitoring safety performance for unmanned aircraft operations that you (or your organisation) are using?
34. [Displayed if participant provided a value greater than 0% for option a., b., or c. in question 25] You have indicated that you have completed either CA005RPAS forms and/or internal incident reports since 2015. Would you be willing to share your submitting CA005RPAS forms and/or internal incident reports on condition of anonymity and aggregation of data (so that no individual or organisation could be identified)?
    a. Yes
    b. No
35. [Displayed if answer to question 34 is "yes"] Could you please enter your email address so that we can send you a consent form for you to share your incident reports?
36. Did you have any other comments you would like to add about safety reporting of unmanned aircraft incidents in New Zealand?

Appendix B. Full Statistical Reporting

Appendix B.1. Chi-Squared Tests of Independence

Results for Chi-Squared Tests of Independence are reported here and are divided by which demographic groupings were being tested for associations with use of particular reporting systems or non-reporting. Non-reported was tested in two manners: one as a categorical variable indicating that a participant had at least one non-reported occurrence, the other as a categorical variable indicating that a participant had not reported any occurrences.

Appendix B.1.1. User Type

- Reporting to the Civil Aviation Authority using a CA005RPAS form was associated with being professional or semi-professional users, $x^2(2) = 9.946$, $p = 0.007$, with a medium effect size, V = 0.470.
- Non-reporting of at least one accident was associated with being a recreational user or semi-professional user, $x^2(2) = 10.212$, $p = 0.006$, with a medium effect size, V = 0.476.
- Non-reporting of all accidents was associated with being a recreational user, $x^2(2) = 6.818$, $p = 0.033$, with a medium effect size, V = 0.389.

Appendix B.1.2. Recency of Flying UA

No associations were statistically significant.

Appendix B.1.3. Hours of Flying UA within Last 12 Months

- Non-reporting of at least one occurrence was associated with users in the less than 5 h, 5–10 h, and 10–25 h categorisations, $x^2 (4) = 11.700$, $p = 0.020$, with a large effect size, V = 0.510.

Appendix B.1.4. Completion of a Course on UA Operations

- Reporting to the Civil Aviation Authority using a CA005RPAS form was associated with having completed a course before, $x^2(1) = 6.429$, $p = 0.011$, with a medium effect size, V = 0.378.
- Reporting using an internal process was associated with having completed a course before, $x^2(1) = 5.278$, $p = 0.022$, with a medium effect size, V = 0.342.
- Non-reporting of at least one occurrence was associated with not having completed a course before, $x^2(1) = 8.839$, $p = 0.003$, with a medium effect size, V = 0.443.
- Non-reporting of all occurrences was associated with not having completed a course before, $x^2(1) = 10.045$, $p = 0.002$, with a medium effect size, V = 0.472.

Appendix B.1.5. Passing an OCA

- Reporting to the Civil Aviation Authority using a CA005RPAS form was associated with having passed an OCA before, $x^2(1) = 4.500$, $p = 0.034$, with a medium effect size, V = 0.316.
- Reporting using an internal process was associated with having passed an OCA before, $x^2(1) = 8.642$, $p = 0.003$, with a medium effect size, V = 0.438.
- Non-reporting of at least one occurrence was associated with not having passed an OCA before, $x^2(1) = 8.642$, $p = 0.003$, with a medium effect size, V = 0.438.
- Non-reporting of all occurrences was associated with not having passed an OCA before, $x^2(1) = 13.005$, $p < 0.001$, with a large effect size, V = 0.538.

Appendix B.1.6. MFNZ Membership

No associations were statistically significant.

Appendix B.1.7. UAVNZ/Aviation New Zealand Membership

- Reporting to the Civil Aviation Authority using a CA005RPAS form was associated with being a UAVNZ member, $x^2(1) = 10.864$, $p = 0.006$, with a medium effect size, V = 0.491.
- Non-reporting of all occurrences was associated with not being a UAVNZ member, $x^2(1) = 4.114$, $p = 0.043$, with a medium effect size, V = 0.302.

Appendix B.1.8. Rule Part Operated Under

- Non-reporting of at least one occurrence was associated with operating under Part 101 only, $x^2(1) = 4.752$, $p = 0.029$, with a medium effect size, V = 0.358.
- Non-reporting of all accidents was associated with operating under Part 101 only, $x^2(1) = 4.934$, $p = 0.026$, with a medium effect size, V = 0.365.

Appendix B.2. Kruskal–Wallis H Tests

Kruskal–Wallis H Tests revealed the following results:

- The percentage of occurrences that were reported were statistically significantly different between the user types, $x^2(2) = 7.935$, $p = 0.019$. Pairwise comparisons showed statistically significantly different reporting percentages between professional users (mean rank = 30.77) and recreational users (mean rank = 19.52) ($p = 0.015$). No other statistically significant differences between user groups existed.
- There were no statistically significant differences between users' reporting percentages based upon the recency of their flying.
- There were no statistically significant differences between users' reporting percentages based upon the number of hours they had flown in the last 12 months.

Appendix B.3. Mann–Whitney U Tests

Mann–Whitney U Tests revealed the following results:

- Reporting percentages for those who had completed a course before (mean rank = 28.50) were higher than those who had not completed a course before (mean rank = 18.19), $U = 367.500$, $z = 3.175$, $p = 0.001$.
- Reporting percentages for those who had passed an OCA before (mean rank = 28.04) were higher than those who had not passed an OCA before (mean rank = 16.70), $U = 376.000$, $z = 3.478$, $p < 0.001$.
- There was no statistically significant difference in reporting percentages based upon membership of MFNZ.
- There was no statistically significant difference in reporting percentages based upon membership of UAVNZ/Aviation New Zealand.
- Reporting percentages for those who operated under a Part 102 Operator's Certificate (mean rank = 24.75) were higher than those who operated only under Part 101 (mean rank = 16.87), $U = 192.500$, $z = 2.262$, $p = 0.048$.

References

1. Grote, M.; Pilko, A.; Scanlan, J.; Cherrett, T.; Dickinson, J.; Smith, A.; Oakey, A.; Marsden, G. Sharing airspace with Uncrewed Aerial Vehicles (UAVs): Views of the General Aviation (GA) community. *J. Air Transp. Manag.* **2022**, *102*, 102218. [CrossRef]
2. ICAO. Unmanned Aircraft Systems (UAS). 2011. Available online: https://www.icao.int/meetings/uas/documents/circular%20328_en.pdf (accessed on 30 September 2021).
3. New, W.K.; Leow, C.Y. Unmanned Aerial Vehicle (UAV) in future communication system. In Proceedings of the 26th IEEE Asia-Pacific Conference on Communications (APCC), Kuala Lumpur, Malaysia, 11–13 October 2021. [CrossRef]
4. Ghasri, M.; Maghrebi, M. Factors affecting unmanned aerial vehicles' safety: A post-occurrence exploratory data analysis of drones' accidents and incidents in Australia. *Saf. Sci.* **2021**, *139*, 105273. [CrossRef]
5. Bartsch, J. Unmanned and uncontrolled: The commingling theory and the legality of unmanned aircraft system operation. *J. Aeronaut. Aerosp. Eng.* **2015**, *4*, 100040. [CrossRef]
6. Clothier, R.A.; Walker, R.A. Safety risk management of unmanned aircraft systems. In *Handbook of Unmanned Aerial Vehicles*; Valavanis, K.P., Vachtsevanos, G.J., Eds.; Springer Science: Berlin/Heidelberg, Germany, 2014; pp. 2229–2275. [CrossRef]
7. Wild, G.; Gavin, K.; Murry, J.; Silva, J.M.; Baxter, G. A post-accident analysis of civil remotely-piloted aircraft system accidents and incidents. *J. Aerosp. Technol. Manag.* **2017**, *9*, 157–168. [CrossRef]
8. Henderson, I.L. Aviation safety regulations for unmanned aircraft operations: Perspectives from users. *Transp. Policy* **2022**, *125*, 192–206. [CrossRef]
9. Abdul Razak, A.S.B.; Henderson, I.L. Examining public-facing statements on airport websites related to unmanned aircraft. *Drone Syst. Appl.* **2022**, *10*, 199–234. [CrossRef]
10. CAANZ. Part 102: Unmanned Aircraft Operator Certification. 2015. Available online: https://www.aviation.govt.nz/assets/rules/consolidations/Part_102_Consolidation.pdf (accessed on 30 September 2021).
11. CAANZ. Part 101: Gyrogliders and Parasails, Unmanned Aircraft (Including Balloons), Kites, and Rockets—Operating Rules. 2021. Available online: https://www.aviation.govt.nz/assets/rules/consolidations/Part_101_Consolidation.pdf (accessed on 30 September 2021).
12. CAANZ. Part 12: Accidents, Incidents, and Statistics. 2020. Available online: https://www.aviation.govt.nz/assets/rules/consolidations/Part_012_Consolidation.pdf (accessed on 30 November 2022).
13. CAANZ. Unmanned Aircraft: Operator Certification [AC102-1]. 2015. Available online: https://www.caa.govt.nz/assets/legacy/Advisory_Circulars/AC102-1.pdf (accessed on 30 November 2022).
14. CAANZ. Regulatory Safety and Security Strategy 2022–2027. 2022. Available online: https://www.aviation.govt.nz/assets/publications/CAA-Regulatory-Strategy-2022-27.pdf (accessed on 30 November 2022).
15. ATSB. REPCON—Aviation Confidential Reporting Scheme. 2022. Available online: https://www.atsb.gov.au/voluntary/repcon_aviation (accessed on 30 November 2022).
16. CAAUK. Unmanned Aircraft System Operations in UK Airspace: Guidance. 2020. Available online: https://publicapps.caa.co.uk/docs/33/CAP722%20Edition8(p).pdf (accessed on 30 November 2022).
17. Konert, A.; Kasprzyk, P. UAS safety operation—Legal issues on reporting UAS incidents. *J. Intell. Robot. Syst.* **2021**, *103*, 51. [CrossRef]
18. FAA. Part 107—Small Unmanned Aircraft Systems. 2016. Available online: https://www.ecfr.gov/current/title-14/chapter-I/subchapter-F/part-107 (accessed on 17 January 2022).
19. The European Commission. Official Journal of the European Union L.243. 2015. Available online: https://eur-lex.europa.eu/legal-content/EN/TXT/PDF/?uri=CELEX:32015R1525&from=ES (accessed on 30 November 2015).
20. NASA. Aviaton Safety Reporting System. n.d. Available online: https://asrs.arc.nasa.gov/ (accessed on 30 November 2022).
21. ECCAIRS. ECCAIRS 2. 2020. Available online: https://aviationreporting.eu/en/homepage (accessed on 30 November 2022).
22. CAANZ. Occurrence Report: Remotely Piloted Aircraft Systems (RPAS/UAVs) [CA005RPAS]. 2019. Available online: https://www.aviation.govt.nz/assets/forms/CA005RPAS.pdf (accessed on 4 January 2023).

23. Cooke, D.L.; Rohleder, T.R. Learning from incidents: From normal accidents to high reliability. *Syst. Dyn. Rev.* **2006**, *22*, 213–239. [CrossRef]
24. Cermelli, D.; Pettinato, M.; Currò, F.; Fabiano, B. Major accident prevention: A construction site approach for pro-active management of unsafe conditions. *Chem. Eng. Trans.* **2019**, *74*, 1387–1392. [CrossRef]
25. Nazeri, Z.; Donohue, G.; Sherry, L. Analyzing relationships between aircraft accidents and incidents: A data mining approach. In Proceedings of the International Conference on Research in Air Transportation (ICRAT 2008), Fairfax, VA, USA, 1–4 June 2008.
26. Kyriakidis, M.; Hirsch, R.; Majumdar, A. Metro railway safety: An analysis of accident precursors. *Saf. Sci.* **2012**, *50*, 1535–1548. [CrossRef]
27. Wild, G.; Murray, J.; Baxter, G. Exploring civil drone accidents and incidents to help prevent potential air disasters. *Aerospace* **2016**, *3*, 22. [CrossRef]
28. Kasprzyk, P.J.; Konert, A. Reporting and investigation of Unmanned Aircraft Systems (UAS) accidents and serious incidents. Regulatory perspective. *J. Intell. Robot. Syst.* **2021**, *103*, 3. [CrossRef]
29. CAAUK. CAA RPAS Safety Reporting Project: Discovery Summary Report. 2022. Available online: https://publicapps.caa.co.uk/docs/33/CAA%20RPAS%20Safety%20Reporting%20Project%20(CAP2356).pdf (accessed on 30 November 2022).
30. Henderson, I.L. Examining New Zealand Unmanned Aircraft Users' Measures for Mitigating Operational Risks. *Drones* **2022**, *6*, 32. [CrossRef]
31. Feldman, J.M.; Lynch, J.G. Self-generated validity and other effects of measurement on belief, attitude, intention, and behavior. *J. Appl. Psychol.* **1988**, *73*, 421–435. [CrossRef]
32. Forbes, S.; Avis, M. Construct creation from research questions. *Eur. J. Mark.* **2020**, *54*, 1817–1838. [CrossRef]
33. Munnukka, J.; Järvi, P. The influence of purchase-related risk perceptions on relationship commitment. *Int. J. Retail Distrib. Manag.* **2015**, *43*, 92–108. [CrossRef]
34. Henderson, I.L.; Tsui, K.W.H.; Ngo, T.; Gilbey, A.; Avis, M. Airline brand choice in a duopolistic market: The case of New Zealand. *Transp. Res. Part A Policy Pract.* **2019**, *121*, 147–163. [CrossRef]
35. Dennett, D. *Consciousness Explained*; Little Brown: Boston, MA, USA, 1991.
36. Dennett, D.C. Heterophenomenology reconsidered. *Phenomenol. Cogn. Sci.* **2007**, *6*, 247–270. [CrossRef]
37. Deng, Q.; Henderson, I.L. Travel model choice for domestic intercity travel: A case study in Suzhou, China. *ASEAN J. Hosp. Tour.* **2022**, *20*, 1–26. [CrossRef]
38. Antoun, C.; Zhang, C.; Conrad, F.G.; Schober, M.F. Comparisons of Online Recruitment Strategies for Convenience Samples: Craigslist, Google AdWords, Facebook, and Amazon Mechanical Turk. *Field Methods* **2015**, *28*, 231–246. [CrossRef]
39. Thornton, L.; Batterham, P.J.; Fassnacht, D.B.; Kay-Lambkin, F.; Calear, A.L.; Hunt, S. Recruiting for health, medical or psychosocial research using Facebook: Systematic review. *Internet Interv.* **2016**, *4*, 72–81. [CrossRef]
40. Rife, S.C.; Cate, K.L.; Kosinski, M.; Stillwell, D. Participant recruitment and data collection through Facebook: The role of personality factors. *Int. J. Soc. Res. Methodol.* **2016**, *19*, 69–83. [CrossRef]
41. Whitaker, C.; Stevelink, S.; Fear, N. The Use of Facebook in Recruiting Participants for Health Research Purposes: A Systematic Review. *J. Med. Internet Res.* **2017**, *19*, e290. [CrossRef] [PubMed]
42. Agresti, A. *Categorical Data Analysis*, 3rd ed.; Wiley: Hoboken, NJ, USA, 2013.
43. Cohen, J. *Statistical Power Analysis for the Behavioral Sciences*, 2nd ed.; Lawrence Erlbaum Associates: Hillsdale, NJ, USA, 1988.
44. Kruskal, W.H.; Wallis, W.A. Use of ranks in one-criterion variance analysis. *J. Am. Stat. Assoc.* **1952**, *47*, 583–621. [CrossRef]
45. Dunn, O.J. Multiple comparisons using rank sums. *Technometrics* **1964**, *6*, 241–252. [CrossRef]
46. Bonferroni, C. Teoria statistica delle classi e calcolo delle probabilita. *Publ. Ist Super. Sci. Econ. Commer. Firenze* **1936**, *8*, 3–62.
47. Hart, A. Mann-Whitney test is not just a test of medians: Differences in spread can be important. *BMJ* **2001**, *323*, 391–393. [CrossRef]
48. Dinneen, L.C.; Blakesley, B.C. Algorithm AS 62: A generator for the sampling distribution of the Mann-Whitney U statistic. *J. R. Stat. Soc. Ser. C (Appl. Stat.)* **1973**, *22*, 269–273. [CrossRef]
49. Braun, V.; Clarke, V. Using thematic analysis in psychology. *Qual. Res. Psychol.* **2006**, *3*, 77–101. [CrossRef]
50. CAANZ. Advisory Circular AC61-1: Pilot Licenses and Ratings—General. 2021. Available online: https://www.aviation.govt.nz/assets/rules/advisory-circulars/ac061-1.pdf (accessed on 5 December 2022).
51. Ministry of Transport. Discussion Document: Enabling Drone Integration. 2021. Available online: https://www.transport.govt.nz//assets/Uploads/Discussion/EnablingDroneIntegration.pdf (accessed on 30 September 2021).
52. MFNZ. MFNZ Accident Form. n.d. Available online: https://docs.google.com/forms/d/e/1FAIpQLSf18WDp-mVsRxLq4hWVoKHsA-qKReM_RvnVx24iyHUOunhr7w/viewform (accessed on 5 December 2022).
53. Dent, R. CHIRP Confidential Human Factors Incident Reporting Programme: Landing Site Incursion—Initial Report. 2022. Available online: https://chirp.co.uk/report/duas13/ (accessed on 5 December 2022).
54. AAIB. AAIB Bulletin: 7/2020 DJI M600 Pro AAIB-26314. 2020. Available online: https://assets.publishing.service.gov.uk/media/5f1ae9f8d3bf7f596648297e/DJI_M600_Pro_UAS_reg_na_07-20.pdf (accessed on 30 November 2022).
55. State of Israel Ministry of Transport and Road Safety. Safety Investigation Information Report: Serious Incident File No. 81-18—Midair Collision. 2018. Available online: https://aviation-safety.net/reports/2018/20180814_R44_4X-BCR.pdf (accessed on 30 November 2022).

56. AAIB. AAIB Bulletin 3/2021: Alauda Airspeeder Mk II. 2021. Available online: https://assets.publishing.service.gov.uk/media/602bb22f8fa8f50388f9f000/Alauda_Airspeeder_Mk_II_UAS_reg_na_03-21.pdf (accessed on 8 October 2021).
57. CAAS. Advisory Circular: Permits for Unmanned Aircraft Operations AC 101-2-1(Rev 3). 2022. Available online: https://www.caas.gov.sg/docs/default-source/docs---srg/ac-anr101-2-1(3)-permits-for-ua-operations_7feb22.pdf (accessed on 30 November 2022).
58. JARUS. JARUS Guidelines on Specific Operations Risk Assessment (SORA). 2019. Available online: http://jarus-rpas.org/sites/jarus-rpas.org/files/jar_doc_06_jarus_sora_v2.0.pdf (accessed on 12 January 2022).
59. ICAO. *Annex 13 to the Convention on International Civil Aviation: Aircraft Accident and Incident Investigation*, 11th ed.; International Civil Aviation Organisation: Montreal, QC, Canada, 2016.
60. CAAUK. CAA RPAS Safety Reporting Project: Survey Summary and Results. 2022. Available online: https://publicapps.caa.co.uk/docs/33/CAA%20RPAS%20Safety%20Reporting%20Project%20Survey%20Summary%20and%20Results%20(CAP2357).pdf (accessed on 1 December 2022).

Disclaimer/Publisher's Note: The statements, opinions and data contained in all publications are solely those of the individual author(s) and contributor(s) and not of MDPI and/or the editor(s). MDPI and/or the editor(s) disclaim responsibility for any injury to people or property resulting from any ideas, methods, instructions or products referred to in the content.

eng

MDPI

Article

Application of Connected Vehicle Data to Assess Safety on Roadways

Mandar Khanal * and Nathaniel Edelmann

Department of Civil Engineering, Boise State University, Boise, ID 83725, USA
* Correspondence: mkhanal@boisestate.edu; Tel.: +1-208-426-1430

Abstract: Using surrogate safety measures is a common method to assess safety on roadways. Surrogate safety measures allow for proactive safety analysis; the analysis is performed prior to crashes occurring. This allows for safety improvements to be implemented proactively to prevent crashes and the associated injuries and property damage. Existing surrogate safety measures primarily rely on data generated by microsimulations, but the advent of connected vehicles has allowed for the incorporation of data from actual cars into safety analysis with surrogate safety measures. In this study, commercially available connected vehicle data are used to develop crash prediction models for crashes at intersections and segments in Salt Lake City, Utah. Harsh braking events are identified and counted within the influence areas of sixty study intersections and thirty segments and then used to develop crash prediction models. Other intersection characteristics are considered as regressor variables in the models, such as intersection geometric characteristics, connected vehicle volumes, and the presence of schools and bus stops in the vicinity. Statistically significant models are developed, and these models may be used as a surrogate safety measure to analyze intersection safety proactively. The findings are applicable to Salt Lake City, but similar research methods may be employed by researchers to determine whether these models are applicable in other cities and to determine how the effectiveness of this method endures through time.

Keywords: road safety; surrogate safety measure; crash; prediction; connected vehicle data; harsh braking

Citation: Khanal, M.; Edelmann, N. Application of Connected Vehicle Data to Assess Safety on Roadways. *Eng* **2023**, *4*, 259–275. https://doi.org/10.3390/eng4010015

Academic Editors: Sanjay Nimbalkar and Antonio Gil Bravo

Received: 25 November 2022
Revised: 28 December 2022
Accepted: 12 January 2023
Published: 14 January 2023

Copyright: © 2023 by the authors. Licensee MDPI, Basel, Switzerland. This article is an open access article distributed under the terms and conditions of the Creative Commons Attribution (CC BY) license (https://creativecommons.org/licenses/by/4.0/).

1. Introduction

Surrogate safety measures (SSMs) offer benefits over traditional safety analysis methods that use historical crash data. SSMs are a type of safety analysis that make use of data other than crash data, typically vehicle kinematic data. The first benefit of SSMs is that they use data which may be collected more rapidly than historical crash data. Crashes are rare events, and historical data may require years of accumulation to conduct a safety analysis. The second benefit is that SSM analysis is proactive, allowing for safety analysis prior to crashes occurring. An unsafe location may therefore be identified and improved before crashes occur, preventing injuries and property damage and possibly saving lives. The third benefit of SSMs is that the kinematic data used in a safety analysis with SSMs are much more voluminous, allowing for statistical methods to be more effective.

The kinematic data employed by SSMs may come from several sources. In the past, manual measurement at the study site was used. This method of data collection was problematic because it allowed for subjectivity and was difficult to perform accurately due to the fleeting nature of traffic interactions. Manual observation was replaced with video recordings which made it possible for traffic interactions to be replayed and offered the chance for multiple observers to analyze interactions, thus improving the problem of subjectivity. This problem has been further ameliorated with automated video data reduction with technology such as that offered by Transoft, Iteris, and similar companies. Additionally, microsimulation technology has allowed for simulation to be used as a source of kinematic data. This method eliminates subjectivity, as the computer running the simulation provides the data rather than human observers [1]. Microsimulation produces highly detailed and precise data and can produce large volumes of data with relatively little effort in comparison with manual collection. The fault of microsimulation lies in it being an abstraction rather than reality. While microsimulations are still highly useful, there has been research into the use of connected vehicle (CV) data with SSMs, meaning the use of data from the physical world rather than simulation.

CVs are a source of traffic data that allows for the high level of precision offered by microsimulation along with the realism of being generated by human drivers. CVs are automobiles sold to the public that include a transceiver which allows data to be collected regarding the vehicle's motion. For the sake of privacy, no individually identifiable information about the vehicle is visible. Vendors offer CV data to clients who wish to use the data for research and engineering projects. The main drawback of using data from CVs is that they currently comprise a small percentage of the total number of vehicles in the United States. A study from October 2021 found the median CV penetration rate to be approximately 4.5% [2]. Therefore, CVs do not offer a full picture of traffic. They are gradually becoming more common, though, as older vehicles are retired and replaced with new vehicles that are connected. Research into effective analysis methods with CV data will become more valuable as time goes on, speaking to the need for this research to take place now for a future increase in CVs.

One metric that is available from CVs is harsh braking event counts, which form the basis for the models developed in this study. Data points from CVs include information about braking and acceleration. The braking data may be filtered so that harsh braking events are identified and counted and then used as a regressor variable in a crash prediction model. This method is investigated in this paper. The significance of other regressor variables, such as CV volume and intersection geometric characteristics, was also investigated. The proposed crash prediction models may be used to estimate monthly counts of intersection-related crashes and offer all of the benefits of SSMs mentioned above.

The statistical models developed in this study show promise for use as a surrogate safety measure. Of the twelve statistical models developed in this study, ten possess a high level of statistical significance. Although connected vehicle penetration rates are too low at this time to depend upon models such as these, once these penetration rates increase, these models will offer an additional method of analysis.

1.1. Literature Review

Researchers have developed many SSMs which tend to fall into three categories. SSMs can be a time-based measure, a deceleration-based measure, or a safety index. Although most SSMs consider collisions involving two vehicles, it is possible to model single-vehicle crashes due to distraction or error [3]. SSMs operate upon the concept that events with greater risk tend to happen less frequently, with the riskiest and rarest events being the events that result in collision [4]. By analyzing less risky events that occur significantly more frequently, a safety analysis with SSMs can offer more insight into safety than an analysis with crash data alone.

1.1.1. Time-Based Measures

Time-based measures consider the kinematics of vehicles and how much of a time gap exists between vehicles. Time-to-collision (TTC), post-encroachment time (PET), and proportion of stopping distance (PSD) are time-based SSMs. TTC is a measure of the amount of time required for the space between two vehicles to close. TTC on its own is transient, but Minderhoud and Bovy developed aggregation methods in the form of their extended TTC measures, namely time-integrated TTC and time-exposed TTC [5]. Post-encroachment time is the difference in time between when an encroaching vehicle exits the path of travel and when a following vehicle first occupies the location where a collision would have occurred. A modified form of PET exists as initially attempted PET (IAPE). IAPE corrects the measure to account for the acceleration that commonly occurs when a driver determines that a conflict has ended [6]. PSD is a ratio between the distance a vehicle is from a potential collision location and the minimum stopping distance. These distances depend upon the velocity of the vehicles involved, making PSD a time-based measure.

There are both strengths and weaknesses associated with time-based SSMs. The strength of time-based SSMs lies in their simplicity and intuitiveness. TTC and PET may be implemented with kinematic data supplied by either on-site measurements or microsimulation. PSD also requires such kinematic data, but it also requires information on the vehicles' possible deceleration rates. This deceleration rate can be an established value or distribution of values or may be derived from environmental conditions. Drivers are aware of the importance of following distance and time headway, making these measures intuitive for researchers and practitioners alike. A weakness of time-based SSMs is the possibility of multiple encounters producing identical measures [7]. TTC may evaluate the same solution for both an encounter with a large speed differential between vehicles and a long following distance and another encounter with a small speed differential but a short following distance. This has made it difficult to establish particularly meaningful safety thresholds for these measures. Another weakness is the inability of time-based SSMs to evaluate the severity of a potential collision. In the encounters just described, which both result in an identical TTC, the severity of a resulting collision will be very different due to the differing speed differentials.

1.1.2. Deceleration-Based Measures

Deceleration-based measures consider braking action and the braking capacity of vehicles and are better equipped than time-based measures to evaluate potential crash severity. Additionally, this type of measure considers a driver's evasive action, an important component of traffic conflicts. Deceleration-based measures include braking applications and deceleration rate to avoid collision (DRAC). Brake applications have been found to be a poor SSM due to the variability in braking habits among drivers. Brake applications are such a common act, even in benign situations, that they are not highly indicative of a conflict [6]. Brake applications as an SSM fail to consider the severity of each particular braking action, something that DRAC and harsh braking are able to capture to their benefit. DRAC is a measure of the deceleration rate that a following vehicle would need to apply to avoid colliding with a leading vehicle. This measurement is compared to a safety threshold, commonly given as 3.35 m/s^2, to determine whether a conflict occurred [8].

Harsh braking events have also been suggested as an indicator of a conflict, which would also fall under the category of deceleration-based measures. A 2015 study found a high level of correlation between crash counts and harsh braking events, defined as events with a large absolute value of the first derivative of acceleration, known as jerk. These events were collected by vehicles with GPS units which collected data on the vehicles' location over time, allowing the jerk value to be computed. Mousavi found a threshold of -0.762 m/s^3 to be the most effective to define harsh braking but also noted that this threshold is lower than expected. Further investigation of a proper jerk threshold was recommended [9].

1.1.3. Safety Indices

Safety indices are the third category of SSM. These indices consider various factors and produce an indirect safety metric. Two examples are crash potential index (CPI) and the aggregated crash propensity metric (ACPM). CPI was developed to improve upon the drawbacks of the DRAC measure. While a constant safety threshold value is typically used with DRAC, the braking capacity of vehicles is variable for mechanical and environmental reasons. CPI considers this variability through the use of a maximum available deceleration rate (MADR) distribution. The probability that DRAC is greater than MADR is a term in the computation of CPI. ACPM also considers the MADR distribution in conjunction with a distribution of driver reaction times to compute the probability that each vehicle interaction will result in collision. These probabilities are aggregated to produce the ACPM [10]. CPI and ACPM indicate the safety level of a study location and time period without being a single measure of some observable quality.

Of the SSMs discussed, the analysis of harsh braking events holds potential due to its compatibility with CV data. Previous studies, such as Mousavi's thesis [9] and the work of Bagdadi and Varhelyi [11], have analyzed harsh braking data from GPS units due to the lack of availability of large-scale CV data when these studies were conducted. He et al. investigated the use of CV data for SSMs, using a safety pilot model dataset to compute TTC, DRAC, and a modified form of TTC [12]. Their study demonstrated the effectiveness of computing these measures with kinematic data from CVs. The development of a crash prediction model that uses harsh braking data from CVs would bridge the gap between these two studies and provide another tool for safety analysis.

2. Materials and Methods

The methods undertaken in this study include the three following phases: selection of study intersections, data collection, and statistical modeling. CV data collection was enabled by the automobile companies that manufactured the CVs. This study uses data within Salt Lake City, Utah for the months of March 2019, January 2021, and August 2021. These months were selected due to the availability of CV data for these particular months. A larger sample size in future studies would be preferable, but there were only three months of CV data due to budgetary restrictions.

2.1. Intersection Selection

The intersection selection process involved the collection of crash counts for all major intersections in Salt Lake City, amounting to 370 intersections. Crash counts for the three study months were obtained from the UDOT database and summed to find the total number of crashes for the intersections. The crashes within the UDOT system were filtered to include only those deemed to be intersection related by law enforcement. The sixty intersections with the most crashes were selected. The total monthly crashes ranged from zero to six. The sixty chosen study intersections included both signalized and unsignalized intersections.

2.1.1. Data Collection

The CV data interface comprises an interactive map and a control pane. The map displays waypoints that are produced by the CVs. When a CV is in motion, waypoints are produced once every three seconds. The waypoints are grouped by the overall trip of which it is a part by a journey ID number, making it possible to collect CV volumes. The waypoints also include data such as geographical location, a timestamp, speed, acceleration, jerk, heading, and information about the origin and destination of the trip that includes the particular waypoint. Harsh braking events were identified using the jerk values of these waypoints. Jerk is the first derivative of acceleration and is recorded for each of the waypoints. Jerk is a continuous measure for a vehicle, similar to speed or location. Each waypoint contains a value for jerk at the particular moment corresponding to the waypoint. This value is derived from the speed data. A geospatial filter was applied to limit the waypoints to those within the influence area of the study intersections, the main intersection square, and the legs of the intersection 250 ft behind the stop bar as displayed in Figure 1 [13]. Another filter was applied to limit waypoints to only those that possess a jerk value that is above the threshold that differentiates a regular braking event from a harsh braking event. This jerk threshold varied in this study to test the effectiveness of several harsh braking definitions. Thresholds tested varied between -0.15 m/s^3 and -3.2 m/s^3 in increments of 0.15 m/s^3. The query tool was used to obtain counts of harsh braking events for each of the jerk thresholds.

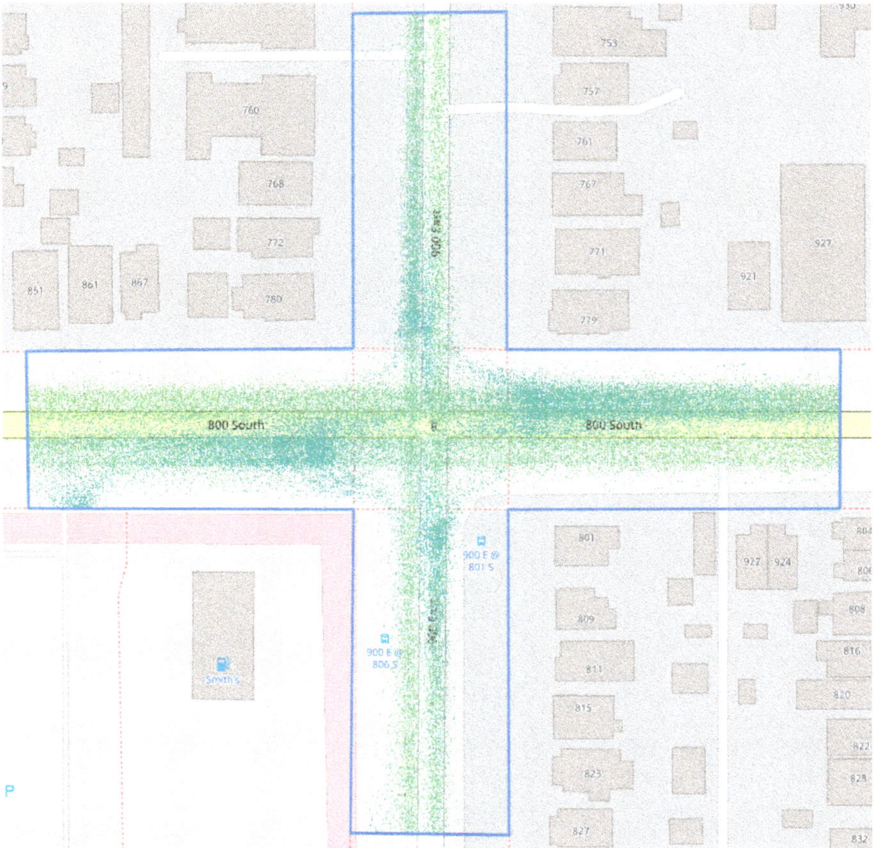

Figure 1. Intersection influence area with waypoints displayed.

Other metrics collected from the CV data included the CV volumes and the average jerk value for each of the intersections. The CV volumes were obtained by querying the unique count of the journey ID numbers. This counts the number of groups of waypoints that belong to trips that pass through the intersection. Thus, the volume of vehicles passing through the intersection is obtained. The total monthly CV volume was collected as was the total monthly volume that used the intersection between the hours of 7 AM and 9 AM and between the hours of 4 PM and 6 PM. The average jerk value among all waypoints within the intersection influence area was obtained on a monthly basis for each of the three study months for each of the intersections.

In addition to the crash data and CV data, information regarding the geometry and geography of each of the intersections was collected. The number of approaches with left turn lanes, the number of approaches with right turn lanes, and the maximum number of lanes that a pedestrian would have to cross were collected using Google Earth. Historical imagery was employed to ensure that these values were correct for the study months in question. ArcGIS Pro was used to determine the number of bus stops and the number of schools within a 305 m radius of the center point of each of the intersections. These metrics were included in this study because they are used in the safety performance functions within the Highway Safety Manual [14]. Table 1 is a summary of the dependent, exposure, and regressor variables collected for analysis in this study as organized per intersection or segment per month.

Table 1. Summary of variables.

Variable	Definition	Mean	SD	Min	Max
Monthly Crashes	Number of intersection-related crashes within the study month	0.7389	0.8347	0	6
Jerk1	Harsh braking events with the threshold being −0.15 m/s^3	93,813	80,218	1074	310,321
Jerk2	Harsh braking events with the threshold being −0.3 m/s^3	85,134	73,590	660	285,385
Jerk3	Harsh braking events with the threshold being −0.45 m/s^3	78,212	68,040	450	263,804
Jerk4	Threshold = −0.60 m/s^3	71,926	62,794	342	242,067
Jerk5	Threshold = −0.75 m/s^3	65,580	57,440	282	221,542
Jerk6	Threshold = −0.90 m/s^3	60,261	52,913	234	204,843
Jerk7	Threshold = −1.05 m/s^3	55,593	48,948	204	189,637
Jerk8	Threshold = −1.20 m/s^3	51,219	45,112	189	174,165
Jerk9	Threshold = −1.35 m/s^3	46,951	41,349	177	159,274
Jerk10	Threshold = −1.50 m/s^3	42,724	37,538	171	148,008
Jerk11	Threshold = −1.65 m/s^3	38,761	34,031	156	136,518
Jerk12	Threshold = −1.80 m/s^3	34,769	30,507	144	124,191
Jerk13	Threshold = −1.95 m/s^3	31,090	27,190	135	112,042
Jerk14	Threshold = −2.10 m/s^3	27,816	24,255	117	100,044
Jerk15	Threshold = −2.25 m/s^3	24,782	21,517	105	88,199
Jerk16	Threshold = −2.40 m/s^3	21,835	18,918	87	77,143
Jerk17	Threshold = −2.55 m/s^3	19,475	16,857	78	67,701
Jerk18	Threshold = −2.70 m/s^3	17,350	14,978	69	59,011
Jerk19	Threshold = −2.85 m/s^3	15,522	13,436	54	53,277
Jerk20	Threshold = −3.00 m/s^3	13,963	12,054	45	49,674
Jerk21	Threshold = −3.15 m/s^3	12,623	10,943	42	46,832
Jerk Avg	Average jerk value among all CV waypoints within the study month	−1.437	1.4076	N/A *	N/A
Monthly CVs	Number of unique CV trips through the intersection in the study month	9488	8041.9	187	29,481
Monthly AM CVs	Number of unique CV trips through the intersection in the study month between the hours of 7 AM and 9 AM	947.6	872.85	9	3412
Monthly PM CVs	Number of unique CV trips through the intersection in the study month between the hours of 4 PM and 6 PM	1425	1204.7	20	4720
Left-Turn Approaches	Number of intersection approaches with a designated left-turn lane	3.267	1.1987	0	4
Right-Turn Approaches	Number of intersection approaches with a designated right-turn lane	1.733	1.3684	0	4
Maximum Lanes Crossed by Ped	Maximum number of lanes a pedestrian must traverse to cross any of the intersection legs	6.383	1.7378	2	9
Bus Stops	Number of bus stops within a 305 m radius of the intersection center point	5.45	3.5709	0	13
Schools	Number of schools within a 305 m radius of the intersection center point	0.2667	0.5135	0	2

* N/A denotes not applicable.

2.1.2. Statistical Analysis

Once these data points were collected for each of the study intersections during each of the study months, a statistical regression analysis was performed to produce crash prediction models for Salt Lake City. Poisson regression, negative binomial regression, and generalized Poisson regression were considered in the analysis. Three statistical methods were used for the sake of producing a larger number of total models and investigating which of the regression methods performed best. Poisson regression requires that the mean and variance are equal for the dependent variable in the regression. The mean and variance of the monthly crashes at the intersections were approximately equal, making Poisson regression a viable option.

Poisson Regression

Poisson regression is applicable when the variable of interest is assumed to follow the Poisson distribution, which is a model of the probability that a particular number of events will occur. The dependent variable is the event count, which can be any of the nonnegative integers. Large counts are assumed to be uncommon, making Poisson regression similar to logistic regression, with a discrete response variable. Poisson regression, unlike logistic regression, does not limit the response variable to specific values. The Poisson distribution model takes the form given in Equation (1), in which Y is the dependent variable, y is a count from among the nonnegative integers, and μ is the mean incidence rate for an event per unit of exposure.

$$Pr(Y = y|\mu) = \frac{e^{-\mu}\mu^{y}}{y!} \quad (y = 0, 1, 2, \cdots) \tag{1}$$

If the Poisson incidence rate, μ, is assumed to be determined by a set of regressor variables, then Poisson regression is possible through the expression displayed in Equation (2) and the regression model displayed in Equation (3). In these equations, X is a regressor variable, β is a regression coefficient, and t is the exposure variable.

$$\mu = exp(\beta_1 X_1 + \beta_2 X_2 + \cdots + \beta_k X_k) \tag{2}$$

$$Pr(Y_i = y_i|\mu_i, t_i) = \frac{e^{-\mu_i t_i}(\mu_i t_i)^{y_i}}{y_i!} \tag{3}$$

The regression coefficients in Equation (2) may be estimated by maximizing the log-likelihood for the regression model. This is achieved by setting the derivative of the log-likelihood equal to zero to generate a system of nonlinear equations which may be solved with an iterative algorithm. The reweighted least squares iterative method is typically able to converge to a solution within six iterations [15].

Negative Binomial Regression

The negative binomial distribution is a generalization of the Poisson distribution that includes a gamma noise variable. This allows for negative binomial regression to be performed even if the dependent variable's mean and variance are not equal [16]. Negative binomial regression is commonly used for traffic safety applications because it has loosened restrictions in comparison to Poisson regression but is still capable of estimating an observed count, such as crash counts [17]. The negative binomial distribution takes the form presented in Equation (4), in which α is the reciprocal of the scale parameter of the gamma noise variable and other variables are as defined previously.

$$Pr(Y = y_i|\mu_i, \alpha) = \frac{\Gamma(y_i + \alpha^{-1})}{\Gamma(y_i + 1)\Gamma(\alpha^{-1})}\left(\frac{\alpha^{-1}}{\alpha^{-1} + \mu_i}\right)^{\alpha^{-1}}\left(\frac{\mu_i}{\alpha^{-1} + \mu_i}\right)^{y_i} \tag{4}$$

The mean of y in negative binomial regression depends upon the exposure variable and the regressor variables which are related by the expression displayed in Equation (5). Negative binomial regression is possible with the regression model displayed in Equation (6). In these equations, x is a regressor variable, and the other variables are as defined previously. As with Poisson regression, maximizing the log-likelihood may be used to estimate the regressor coefficients through an iterative algorithm [16].

$$\mu_i = exp(ln(t_i) + \beta_1 x_{1i} + \beta_2 x_{2i} + \ldots + \beta_k x_{ki}) \tag{5}$$

$$Pr(Y = y_i|\mu_i, \alpha) = \frac{\Gamma(y_i + \alpha^{-1})}{\Gamma(\alpha^{-1})\Gamma(y_i + 1)}\left(\frac{1}{1 + \alpha\mu_i}\right)^{\alpha^{-1}}\left(\frac{\alpha\mu_i}{1 + \alpha\mu_i}\right)^{y_i} \tag{6}$$

Generalized Poisson Regression

Generalized Poisson regression, like negative binomial regression, is applicable in a broader set of circumstances than Poisson regression. This is because it does not have the requirement that the mean and variance of the dependent variable in the regression be equal. There are two types of generalized Poisson regression models: Consul's generalized Poisson model and Famoye's restricted generalized Poisson regression model. Consul's model, also known as the Generalized Poisson-1 (GP-1) model, is the regression model that was employed in this study. The GP-1 model operates on the assumption that the dependent variable, y, is a random variable following the probability distribution presented in Equation (7), in which λ is the number of events per unit of time and α is the dispersion parameter which can be estimated using Equation (8) [18]. In Equation (8), N is the number of samples, k is the number of regression variables, y_i is the ith observed value, and $\hat{y}_i$ is the Poisson rate λ_i predicted for the ith sample [19].

$$Pr(Y = y_i) = \frac{e^{-(\lambda+\alpha*y_i)} * (\lambda + \alpha * y_i)^{y_i-1} * \lambda}{y_i!} \quad (7)$$

$$\alpha = \frac{\sum_{i=1}^{N} \left(\frac{|y_i - \hat{y}_i|}{\sqrt{\hat{y}_i}} - 1 \right)}{N - k - 1} \quad (8)$$

Poisson, negative binomial, and GP-1 regression techniques were explored by model generation in R. Models with many different combinations of regressor variables were created to find the model that performed best. In all models, the number of monthly crashes was used as the dependent variable, and the monthly CV volume was used as the exposure variable. The statistical models were evaluated based on the significance of the regressor variables used in the models, on the basis of the Akaike Information Criterion, and based on the residuals generated by the models. The best-performing models were selected and are summarized and discussed in the Results and Discussion Sections.

2.2. Segment Analysis

The preliminary results from the intersection study prompted interest in how the results of an intersection-based study would compare to the results of a segment-based study. To address this, a segment analysis was conducted. CV data were collected for thirty road segments in the Salt Lake City area. These segments include sections of interstate highway within the Salt Lake City limits and sections of interrupted state highway outside of the influence area of any intersections. The segments were all made to be approximately one-quarter mile in length to ensure that the segments had roughly equal exposure to crashes occurring. This prevented the need to determine a crash rate per unit length.

The segment CV data were collected in the same manner as the intersection CV data with a couple of key differences. First, the intersection CV data were all collected from within intersection influence areas. The segment CV data were all collected from areas entirely outside of intersection influence areas. Second, the geometric information and information related to schools and bus stops were not collected for the segments. Rather, the segment data included only harsh braking events for jerk thresholds ranging between -0.3 m/s^3 and -3.0 m/s^3 in increments of 0.3 m/s^3, as well as monthly CV counts, monthly CV counts between the hours of 7 AM and 9 AM, and monthly CV counts between the hours of 4 PM and 6 PM. As with the intersection analysis, crash data were collected for the segments from the UDOT database. The increment between successive jerk thresholds for segments differs from that which was used for intersections. This was done simply for the purpose of decreasing the amount of work needed for the analysis. More thresholds could have been tested, but the preliminary results from the intersection study indicated that the increment did not need to be as fine as 0.15 m/s^3. Table 2 is a summary of the variables collected for segments in this study.

Statistical analysis was conducted in the same manner as the intersection analysis, with Poisson, negative binomial, and generalized Poisson models generated and evaluated for the segment dataset. The best-performing models were selected and are summarized and discussed in the following sections.

Table 2. Summary of segment variables.

Variable	Definition	Mean	SD	Min	Max
Monthly Crashes	Number of intersection-related crashes within the study month	0.8222	1.2504	0	7
Jerk2	Harsh braking events with the threshold being -0.3 m/s^3	112,308	95,010	3906	401,183
Jerk4	Harsh braking events with the threshold being -0.6 m/s^3	42,168	38,254	993	148,024
Jerk6	Threshold = -0.9 m/s^3	14,109	16,790	213	94,961
Jerk8	Threshold = -1.2 m/s^3	8096	11,551	114	75,684
Jerk10	Threshold = -1.5 m/s^3	4687	8182	63	58,124
Jerk12	Threshold = -1.8 m/s^3	2939	6011	42	43,842
Jerk14	Threshold = -2.1 m/s^3	1894	4397	15	32,348
Jerk16	Threshold = -2.4 m/s^3	1288	3185	9	23,311
Jerk18	Threshold = -2.7 m/s^3	906	2313	3	16,746
Jerk20	Threshold = -3.0 m/s^3	649	1668	3	12,042
Monthly CVs	Number of unique CV trips through the segment in the study month	54,903	47,760	1327	185,293
Monthly AM CVs	Number of unique CV trips through the segment in the study month between the hours of 7 AM and 9 AM	6330	5247	132	20,226
Monthly PM CVs	Number of unique CV trips through the segment in the study month between the hours of 4 PM and 6 PM	8654	7433	177	27,639

3. Results

The collected intersection data were used for a statistical regression analysis, and the best regression model for each of the model families was found that had a high level of significance among the regressor variables and the intercept. The best Poisson model uses *Jerk18* and *Schools* from Table 1 as regressor variables. The best negative binomial model also uses *Jerk18* and *Schools* as regressor variables. The best generalized Poisson model uses *Jerk18* as a regressor variable. All of these models have a better than 0.1% significance level for their regressor variables and the intercept. In the case of the generalized Poisson model, both intercepts are significant at a better than 0.1% level. These models are summarized in Table 3.

Table 3. Summary of regression models for intersection analysis.

		Poisson Regression Model			
Parameter	Estimate	Std. Err.	Z-Score	Pr(>|z|)	Significance Level
Intercept	-8.576	0.2483	-34.544	$<2 \times 10^{-16}$	<0.1%
Jerk18	-4.056×10^{-5}	9.593×10^{-6}	-4.228	2.35×10^{-5}	<0.1%
Schools	1.103	0.3193	3.455	5.51×10^{-4}	<0.1%
Akaike Information Criterion		242.58			
Log Likelihood		-118.29			
RMSE		0.9468			
		Negative Binomial Regression Model			
Parameter	Estimate	Std. Err.	Z-Score	Pr(>|z|)	Significance Level
Intercept	-8.526	0.2664	-31.999	$<2 \times 10^{-16}$	<0.1%
Jerk18	-4.127×10^{-5}	1.037×10^{-5}	-3.981	6.87×10^{-5}	<0.1%
Schools	1.190	0.3566	3.337	8.46×10^{-4}	<0.1%
Akaike Information Criterion		242.67			
Log Likelihood		-117.337			

Table 3. *Cont.*

		Poisson Regression Model			
Theta		3.87			
RMSE		0.9627			
		Generalized Poisson Regression Model			
Parameter	Estimate	Std. Err.	Z-Score	Pr(>|z|)	Significance Level
Intercept 1	−8.264	0.2342	−35.288	<2 × 10^{-16}	<0.1%
Intercept 2	−11.81	1.741	−6.782	1.19 × 10^{-11}	<0.1%
Jerk18	−4.890 × 10^{-5}	1.019 × 10^{-5}	−4.797	1.61 × 10^{-6}	<0.1%
Log Likelihood		−122.8911			
Degrees of Freedom		197			
RMSE		0.9671			

The segment analysis also yielded three statistical models: a Poisson regression model, a negative binomial regression model, and a generalized Poisson regression model. The best Poisson, negative binomial, and generalized Poisson models identified use *Jerk2* as a regressor variable. All models have a better than 0.1% significance level for their regressor variable and intercept(s). These models are summarized in Table 4.

Table 4. Summary of regression models for segment analysis.

		Poisson Regression Model			
Parameter	Estimate	Std. Err.	Z-Score	Pr(>|z|)	Significance Level
Intercept	−8.881	0.2338	−37.991	<2 × 10^{-16}	<0.1%
Jerk2	−1.345 × 10^{-5}	1.661 × 10^{-6}	−8.098	5.57 × 10^{-16}	<0.1%
Akaike Information Criterion		199.32			
Log Likelihood		97.658			
RMSE		1.5102			
		Negative Binomial Regression Model			
Parameter	Estimate	Std. Err.	Z-Score	Pr(>|z|)	Significance Level
Intercept	−8.911	0.3329	−26.768	<2 × 10^{-16}	<0.1%
Jerk2	−1.212 × 10^{-5}	2.126 × 10^{-6}	−5.702	1.19 × 10^{-8}	<0.1%
Akaike Information Criterion		181.21			
Log Likelihood		−87.604			
Theta		0.880			
RMSE		1.5621			
		Generalized Poisson Regression Model			
Parameter	Estimate	Std. Err.	Z-Score	Pr(>|z|)	Significance Level
Intercept 1	−8.878	0.2416	−36.750	<2 × 10^{-16}	<0.1%
Intercept 2	−12.84	1.057	−12.149	<2 × 10^{-16}	<0.1%
Jerk2	−1.335 × 10^{-5}	1.799 × 10^{-6}	−7.425	1.13 × 10^{-13}	<0.1%
Log Likelihood		−96.9352			
Degrees of Freedom		117			
RMSE		1.5143			

The estimates for the coefficients of the harsh braking variable in each of these regression models (*Jerk18* and *Jerk2*) are all negative, indicating that an increase in hard braking events decreases the estimate for the number of crashes that will occur within the intersection area or along the segment in question. This suggests that hard braking events are an indication of safety. This is true at intersections as well as on segments away from the influence of intersections.

Tables 3 and 4 include the models with the best level of statistical significance, but there were numerous other models identified which also were statistically significant. A number of potential models could theoretically be used with similar results. The models display a gradual degradation in significance as the jerk variable used gets further away from the Jerk18 variable for intersections and the Jerk2 variable for segments.

Validation efforts conducted with the models produced the following graphs, displayed in Figures 2 and 3. These graphs display the expected monthly crash counts for each of the three models on the vertical axis. The horizontal axis represents the observed monthly crash counts that correspond to each of the expected crash counts. The "jitter" function in R has been used to generate these plots; hence, there is scatter around the integer counts of observed crashes.

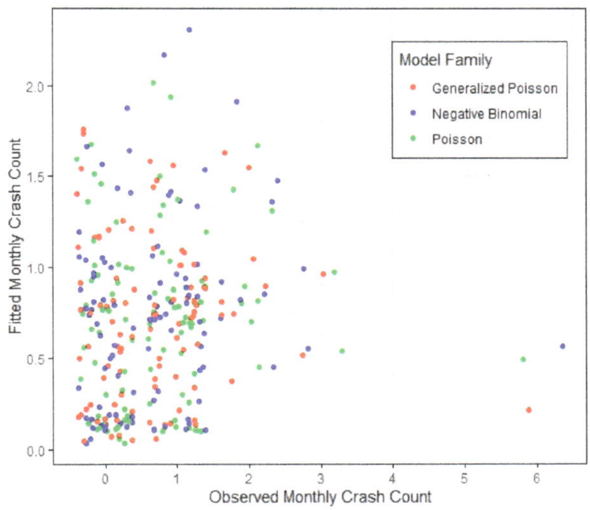

Figure 2. Intersection fitted crash counts versus observed crash counts.

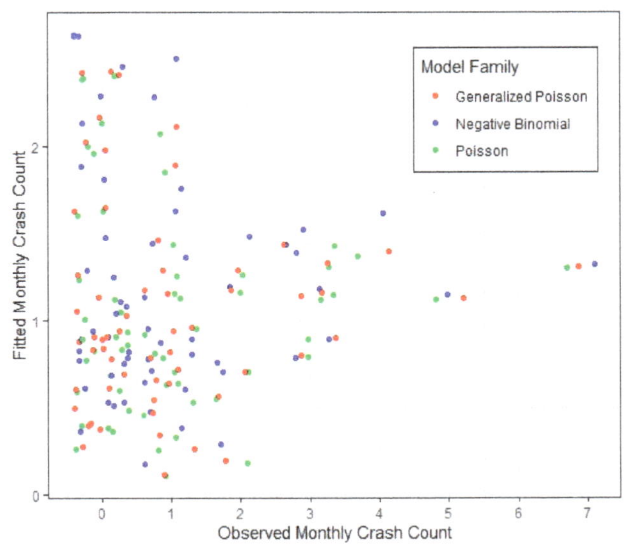

Figure 3. Segment fitted crash counts versus observed crash counts.

An additional analysis was conducted in the same manner as that which yielded the results presented up to this point, except with outlier crash counts removed from the intersection and segment datasets. The outliers were identified using boxplots generated for the observed crash counts. These boxplots are presented in Figure 4. The outliers are denoted as black points in Figure 4. The best identified Poisson, negative binomial, and generalized Poisson models are summarized in Tables 5 and 6.

Table 5. Summary of regression models for intersection analysis with outliers removed.

Poisson Regression Model					
Parameter	Estimate	Std. Err.	Z-Score	Pr(>∣z∣)	Significance Level
Intercept	−7.722	0.3456	−22.345	<2 × 10^{-16}	<0.1%
Jerk1	−9.144 × 10^{-6}	1.820 × 10^{-6}	−5.024	5.05 × 10^{-7}	<0.1%
Left-Turn Approaches	−0.2181	9.278 × 10^{-2}	−2.351	0.0187	<5%
Akaike Information Criterion		203.97			
Log Likelihood		−98.9848			
RMSE		0.6348			
Negative Binomial Regression Model					
Parameter	Estimate	Std. Err.	Z-Score	Pr(>∣z∣)	Significance Level
Intercept	−7.722	0.3456	−22.342	<2 × 10^{-16}	<0.1%
Jerk1	−9.145 × 10^{-6}	1.820 × 10^{-6}	−5.024	5.06 × 10^{-7}	<0.1%
Left-Turn Approaches	−0.2181	9.279 × 10^{-2}	−2.350	0.0188	<5%
Akaike Information Criterion		205.97			
Log Likelihood		−98.9875			
Theta		4676			
RMSE		0.6348			
Generalized Poisson Regression Model					
Parameter	Estimate	Std. Err.	Z-Score	Pr(>∣z∣)	Significance Level
Intercept 1	−7.722	0.3456	−22.345	<2 × 10^{-16}	<0.1%
Intercept 2	−38.95	7.415 × 10^{4}	−0.001	0.9996	None
Jerk1	−9.144 × 10^{-6}	1.820 × 10^{-6}	−5.024	5.05 × 10^{-7}	<0.1%
Left-Turn Approaches	−0.2181	9.278 × 10^{-2}	−2.351	0.0187	< 5%
Log Likelihood		−98.9848			
Degrees of Freedom		196			
RMSE		0.6348			

Table 6. Summary of regression models for segment analysis with outliers removed.

Poisson Regression Model					
Parameter	Estimate	Std. Err.	Z-Score	Pr(>∣z∣)	Significance Level
Intercept	−10.55	0.3682	−28.643	<2 × 10^{-16}	<0.1%
Jerk2	−6.322 × 10^{-6}	2.083 × 10^{-6}	−3.035	2.41 × 10^{-3}	<1%
Akaike Information Criterion		137.01			
Log Likelihood		−66.5050			
RMSE		0.7653			

Table 6. *Cont.*

Poisson Regression Model					
Negative Binomial Regression Model					
Parameter	Estimate	Std. Err.	Z-Score	Pr(>|z|)	Significance Level
Intercept	−10.47	0.3765	−27.812	$<2 \times 10^{-16}$	<0.1%
Jerk2	-6.638×10^{-6}	2.147×10^{-6}	−3.091	1.99×10^{-3}	<1%
Akaike Information Criterion		138.93			
Log Likelihood		−66.4645			
Theta		6.7			
RMSE		0.7730			
Generalized Poisson Regression Model					
Parameter	Estimate	Std. Err.	Z-Score	Pr(>|z|)	Significance Level
Intercept 1	−10.55	0.3682	−28.644	$<2 \times 10^{-16}$	<0.1%
Intercept 2	−38.46	9.572×10^{4}	0.000	0.99968	None
Jerk2	-6.322×10^{-6}	2.083×10^{-6}	−3.035	0.00241	<1%
Log Likelihood		−66.505			
Degrees of Freedom		117			
RMSE		0.7653			

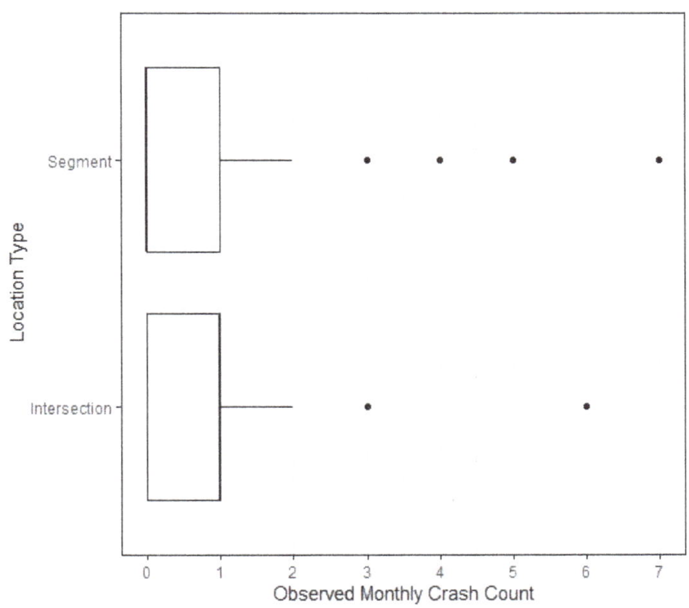

Figure 4. Boxplots of the observed monthly crash counts at intersections and segments.

Validation efforts were conducted for the models generated with outlier crash counts removed from the datasets. These validation efforts produced the graphs displayed in Figures 5 and 6. These graphs display the expected monthly crash counts for each of the three models on the vertical axis. The horizontal axis represents the observed monthly crash counts that correspond to each of the expected crash counts.

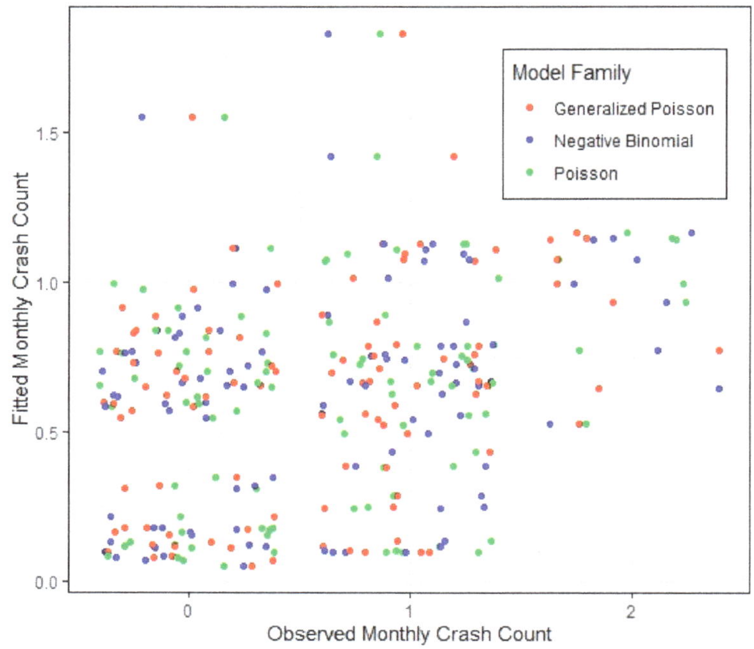

Figure 5. Intersection fitted crash counts versus observed crash counts with outliers removed.

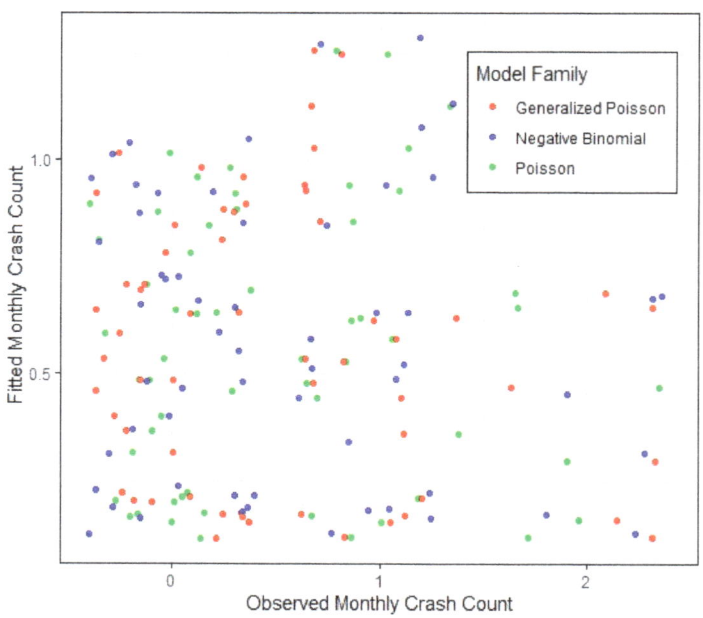

Figure 6. Segment fitted crash counts versus observed crash counts with outliers removed.

4. Discussion

This study demonstrates the effectiveness of using harsh braking data from CVs as a surrogate safety measure. For both intersections and segments, statistically significant models may be developed from multiple model families. Models such as these may be used to predict future crash rates for the purposes of prioritizing improvements and identifying risks to the public.

The results of this study reveal the jerk threshold for intersections and segments. For intersections, the jerk threshold is $-2.7\ \mathrm{m/s^3}$, corresponding to the regressor variable *Jerk18*. This threshold was identified to be the most effective for all three statistical model families. The jerk threshold for segments was found to be $-0.3\ \mathrm{m/s^3}$, corresponding to the variable *Jerk2*. A jerk threshold of $-2.7\ \mathrm{m/s^3}$ for intersections and $-0.3\ \mathrm{m/s^3}$ for segments indicates that intersections and segments operate differently in terms of safety. The jerk threshold is the value of jerk that differentiates ordinary events from harsh braking events. Events that do not meet the jerk threshold have little or no bearing on crash prediction. A larger absolute value of jerk threshold for intersections over segments indicates that braking must be more severe at an intersection to qualify a braking event as *harsh*. This could be due to different expectations of drivers in these differing contexts. At intersections, drivers expect to brake and are typically able to see the status of the traffic signal well in advance. Moderately hard braking, such as an event that generates a jerk value of $-1.5\ \mathrm{m/s^3}$, is expected and therefore ordinary. Such an event in a segment context, however, would be relatively unexpected and therefore extraordinary because segments are expected to have more uniform and smooth flow. This event would therefore qualify as a harsh braking event in a segment context but not in an intersection context.

The coefficient estimates for the harsh braking event count variables were found to be negative for all statistical models generated, indicating that an increase in harsh braking events is correlated with an increase in the frequency of zero crashes and a decrease in the frequency of one or more crashes. This means that harsh braking is correlated with increased safety on roads. The coefficient estimates for the jerk variable are small relative to other covariates, when the covariates are statistically significant. The small value for these coefficient estimates is due to the number of crashes being a small number relative to the high jerk event counts, as can be seen in Tables 1 and 2. To obtain a crash count estimate from a high jerk event count requires that the coefficient estimate be quite small. This study was predicated upon the notion that a harsh braking event corresponds to a traffic conflict and that traffic conflicts and collisions are related. That these models have negative estimates for the coefficients of the harsh braking event counts suggests that harsh braking events are indicative of the prevention of a traffic conflict which leads to the increase in the probability of zero crashes and to the increase in the probability of one or more crashes. Harsh braking events are events which might have been collisions but never were as a result of the evasive action of the drivers involved.

The statistical models presented in Tables 3 and 4 possess an excellent level of statistical significance at a level better than 0.1%. Poisson models are simpler than negative binomial models and generalized Poisson models, making them preferable if applicable. While the requirement that the mean and variance of the dependent variable be equal was approximately satisfied for the dataset used in this study, that may not be the case for other datasets. Therefore, negative binomial and generalized Poisson regression are recommended for crash prediction models based on harsh braking data.

As mentioned above, the presence of schools was found to increase crash frequency within intersection influence areas. This confirms the efficacy of the use of school presence in HSM safety analysis methodology. The estimated coefficients for the *Schools* variable are positive, indicating that the presence of a school or multiple schools nearby decreases the frequency of zero crashes and increases the frequency of one or more crashes. The presence of schools increases pedestrian activity and the presence of young drivers, which may help explain this increase.

The graphs presented in Figures 2 and 3 illustrate that these models fail to predict high crash counts while performing better at locations with lower numbers of observed crashes. This was not unexpected because the count models used in this study predict low probabilities for higher counts. The statistical significance of the regressor variables in the models, on the other hand, speaks to their overall strength. As CV penetration rates increase, allowing models based on CV data to be trained by a fuller picture of the activity on roads, models of this form will likely become more effective. Preliminary studies such as this, using CV data information in its technological infancy, set the stage for a future in which CVs become significantly more widespread and CV data capture a large portion if not a majority of roadway traffic. Figures 5 and 6, as well as the RMSE values presented in Tables 5 and 6 demonstrate that the models' predictive ability improves when outliers are removed. The RMSE values for intersections decreased from approximately 0.95 to 0.63 for intersections and from approximately 1.55 to 0.76 for segments. The decrease in RMSE indicates that the models produce more accurate crash count estimates when outliers are removed.

5. Conclusions

This study developed several statistical models which use harsh braking event counts from CV data in Salt Lake City as regressor variables and crash counts as the dependent variables. Both intersections and segments were considered separately in this study with models derived for each. Poisson, Negative Binomial, and Generalized Poisson models were developed and they revealed the jerk threshold for intersection influence areas to be -2.7 m/s^3 and the jerk threshold for segments to be -0.3 m/s^3. Additionally, the presence of schools within 305 m was found to be a statistically significant variable for intersection influence areas.

Crash prediction models such as these, based on harsh braking event counts, hold promise for agencies and industry as another tool for safety analysis. Agencies may investigate these models and tailor them to their jurisdictions for the purpose of adding such models to their established methodologies. Such tailored models may then be employed as a means of conducting comparative safety analysis for the purpose of identifying crash-prone locations and prioritizing improvements. Once a particular area is identified as being crash prone, further investigation into the cause of the safety hazard may commence. Employing harsh braking models such as those developed in this study requires less labor investment than existing methods, allowing for more frequent and widespread analyses to identify and characterize road hazards. It should be noted that the intersection models developed in this research are likely not applicable to sites with low crash activity, as intersections were selected to maximize the amount of historical crash activity.

Future research into SSMs that are based on harsh braking events could include the investigation of regional differences in models, the use of additional regressor variables in segment-based models, and harsh positive acceleration data from CVs. Regional differences may exist pertaining to the relationship between harsh braking and collisions. Harsh braking events were found to be positively correlated to crashes in a previous study in Louisiana which is contrary to the findings of this study [9]. While this may be due to the significant differences in the methods of data collection between these two studies, regional variations may also be a factor and ought to be investigated further. Additional regressor variables were not investigated in the segment-based models developed in this study to the degree to which they were investigated in the intersection-based models. The inclusion of such additional regressor variables for segments ought to be investigated more fully in a future study. These variables may include speed limits, curvature parameters, lane widths, or total number of lanes, among others. Finally, harsh positive acceleration data may be obtained in the same manner in which harsh braking data were collected in this study. Harsh acceleration may be an indicator of safety or the lack thereof because it can represent erratic driving behavior or situations in which a driver is attempting to clear a potential crash location rapidly. The consideration of harsh acceleration data may be performed separately from harsh braking data or in combination with harsh braking data. If attempts are successful, this would yield yet another tool for agencies and industry to employ for surrogate safety analysis.

Author Contributions: The authors confirm contribution to the paper as follows: study conception and design: M.K.; data collection: N.E.; analysis and interpretation of results: M.K. and N.E.;

manuscript preparation: N.E. and M.K. All authors have read and agreed to the published version of the manuscript.

Funding: This research was partially funded by Subaward No. UWSC9924 from the University of Washington to Boise State University from the U.S. Department of Transportation award to the University of Washington.

Data Availability Statement: Data used in this research are available by contacting the corresponding author at mkhanal@boisestate.edu.

Acknowledgments: The authors are grateful to the Boise State University Department of Civil Engineering for their support of this research. The authors would also like to express their gratitude to the support received from the PacTrans Region 10 University Transportation Center that made the procurement of the CV data possible.

Conflicts of Interest: The authors declare no conflict of interest.

References

1. Gettman, D.; Head, L. *Surrogate Safety Measures from Traffic Simulation Models Final Report*; U.S. Department of Transportation Federal Highway Administration Office of Research, Development, and Technology: Washington, DC, USA, 2003. [CrossRef]
2. Hunter, M.; Mathew, J.K.; Li, H.; Bullock, D.M. Estimation of Connected Vehicle Penetration on US Roads in Indiana, Ohio, and Pennsylvania. *J. Transp. Technol.* **2021**, *11*, 597–610. [CrossRef]
3. Astarita, V.; Caliendo, C.; Giofrè, V.P.; Russo, I. Surrogate Safety Measures from Traffic Simulation: Validation of Safety Indicators with Intersection Traffic Crash Data. *Sustainability* **2020**, *12*, 6974. [CrossRef]
4. Tarko, A.P. Use of Crash Surrogates and Exceedance Statistics to Estimate Road Safety. *Accid. Anal. Prev.* **2012**, *45*, 230–240. [CrossRef]
5. Minderhoud, M.M.; Bovy, P.H. Extended Time-to-Collision Measures for Road Traffic Safety Assessment. *Accid. Anal. Prev.* **2001**, *33*, 89–97. [CrossRef] [PubMed]
6. Allen, B.L.; Shin, B.T.; Cooper, D.J. Analysis of Traffic Conflicts and Collisions. *Transp. Res. Rec.* **1978**, *667*, 67–74.
7. Souza, J.Q.; Sasaki, M.W.; Cunto, F.J.C. Comparing Simulated Road Safety Performance to Observed Crash Frequency at Signalized Intersections. In Proceedings of the International Conference on Road Safety and Simulation, Indianapolis, IN, USA, 14–16 September 2011.
8. Guido, G.; Saccomanno, F.; Vitale, A.; Astarita, V.; Festa, D. Comparing Safety Performance Measures Obtained from Video Capture Data. *J. Transp. Eng.* **2010**, *137*, 481–491. [CrossRef]
9. Mousavi, S.M. Identifying High Crash Risk Roadways through Jerk-Cluster Analysis. Master's Thesis, Louisiana State University, Baton Rouge, LA, USA, 2015. Available online: https://digitalcommons.lsu.edu/gradschool_theses/159 (accessed on 15 February 2022).
10. Wang, C.; Stamatiadis, N. Surrogate Safety Measure for Simulation-Based Conflict Study. *Transp. Res. Rec.* **2013**, *2386*, 72–80. [CrossRef]
11. Bagdadi, O.; Várhelyi, A. Jerky driving—An indicator of accident proneness? *Accid. Anal. Prev.* **2011**, *43*, 1359–1363. [CrossRef] [PubMed]
12. He, Z.; Qin, X.; Liu, P.; Sayed, M.A. Assessing Surrogate Safety Measures Using a Safety Pilot Model Deplayment Dataset. *Transp. Res. Rec.* **2018**, *2672*, 1–11. [CrossRef]
13. *Highway Capacity Manual, Sixth Edition: A Guide for Multimodal Mobility Analysis*; Transportation Research Board, National Research Council: Washington, DC, USA, 2016.
14. *Highway Safety Manual*; American Association of State Highway and Transportation Officials: Washington, DC, USA, 2010.
15. Poisson Regression. NCSS Statistical Software. Available online: https://ncss-wpengine.netdna-ssl.com/wp-content/themes/ncss/pdf/Procedures/NCSS/Poisson_Regression.pdf (accessed on 5 May 2022).
16. Negative Binomial Regression. NCSS Statistical Software. Available online: https://ncss-wpengine.netdna-ssl.com/wp-content/themes/ncss/pdf/Procedures/NCSS/Negative_Binomial_Regression.pdf (accessed on 5 May 2022).
17. Wang, J.; Huang, H.; Zeng, Q. The effect of zonal factors in estimating crash risks by transportation modes: Motor vehicle, bicycle and pedestrian. *Accid. Anal. Prev.* **2017**, *98*, 223–231. [CrossRef] [PubMed]
18. Hilbe, J.M. *Negative Binomial Regression*, 2nd ed.; Cambridge University Press: Cambridge, UK, 2011. [CrossRef]
19. Date, S. Time Series Analysis, Regression and Forecasting: The Generalized Poisson Regression Model. Available online: https://timeseriesreasoning.com/contents/generalized-poisson-regression-model/ (accessed on 10 July 2022).

Disclaimer/Publisher's Note: The statements, opinions and data contained in all publications are solely those of the individual author(s) and contributor(s) and not of MDPI and/or the editor(s). MDPI and/or the editor(s) disclaim responsibility for any injury to people or property resulting from any ideas, methods, instructions or products referred to in the content.

eng

Article

An Approach to Quantifying the Influence of Particle Size Distribution on Buried Blast Loading

Ross Waddoups [1], Sam Clarke [1,*], Andrew Tyas [1,2], Sam Rigby [1], Matt Gant [3] and Ian Elgy [3]

[1] Department of Civil and Structural Engineering, University of Sheffield, Mappin Street, Sheffield S1 3JD, UK
[2] Blastech Ltd., The Innovation Centre, 217 Portobello, Sheffield S1 4DP, UK
[3] Defence Science and Technology Laboratory (Dstl), Porton Down, Salisbury, Wiltshire SP4 0JQ, UK
* Correspondence: sam.clarke@sheffield.ac.uk

Abstract: Buried charges pose a serious threat to both civilians and military personnel. It is well established that soil properties have a large influence on the magnitude and variability of loading from explosive blasts in buried conditions. In this study, work has been undertaken to improve techniques for processing pressure data from discrete measurement apparatus; this is performed through the testing of truncation methodologies and the area integration of impulses, accounting for the particle size distribution (PSD) of the soils used in testing. Two experimental techniques have been investigated to allow for a comparison between a global impulse capture method and an area-integration procedure from a Hopkinson Pressure Bar array. This paper explores an area-limiting approach, based on particle size distribution, as a possible approach to derive a better representation of the loading on the plate, thus demonstrating that the spatial distribution of loading over a target can be related to the PSD of the confining material.

Keywords: buried charges; impulse; particle size distribution; soil condition; landmine

Citation: Waddoups, R.; Clarke, S.; Tyas, A.; Rigby, S.; Gant, M.; Elgy, I. An Approach to Quantifying the Influence of Particle Size Distribution on Buried Blast Loading. *Eng* **2023**, *4*, 319–340. https://doi.org/10.3390/eng4010020

Academic Editor: Antonio Gil Bravo

Received: 16 December 2022
Revised: 17 January 2023
Accepted: 19 January 2023
Published: 28 January 2023

Copyright: © 2023 by the authors. Licensee MDPI, Basel, Switzerland. This article is an open access article distributed under the terms and conditions of the Creative Commons Attribution (CC BY) license (https://creativecommons.org/licenses/by/4.0/).

1. Introduction

Improvised Explosive Devices (IEDs) and landmines form a serious threat to life in military and civilian situations around the world. In 2020, over seven thousand people were killed or injured by landmines or 'Explosive Remnants of War' (ERWs) 80% of these were civilians [1]. A better understanding of the behaviour of these explosive charges can lead to better protection against them, thus saving lives and preventing injuries.

Much work has been performed to investigate the effects of explosive loading, especially for military applications, both in free air and with buried charges. The experiments are often conducted at a reduced scale [2] due to the high cost and difficulty of full-scale testing. Hopkinson [3]–Cranz [4] cube-root scaling is regularly used for this purpose. These experiments are supported by numerical modelling efforts [2], although these are often simplified models that do not incorporate soil-specific effects.

Studies have, in the past, failed to take sufficient account of soil conditions in experimentation and for the prediction of loading. Børvik et al. [5] and Kyner et al. [6] used ~200 µm glass microspheres as a synthetic soil to reduce the influence of variable soil conditions. McShane et al. [7] used compressed gas in place of explosives to reduce complexity and increase the ease of experimentation. This was found to be a suitable method of simulating sand-throw interactions with structures, although it only accounts for impulse transfer through said sand throw (ignoring blast-wave transfer).

It has been found that introducing/taking account of these complexities has wide-ranging effects on the loading generated and, thus, it is key that these are accounted for in future work. Hlady [8] used Concrete Fine Aggregate Sand (CFAS), a cohesionless well-graded sand, and compared this against Suffield Prairie Soil (PS), a fine-grained cohesive soil composed mainly of clay. It was found that CFAS led to a greater level of repeatability, alongside a much reduced level of labour required in preparation, compared to the cohesive soil. Fourney et al. [9] conducted small-scale experiments in a range of soil conditions, which showed that soil ejecta contributes the majority of impulsive loading from buried charges. Anderson et al. [10] varied plate and soil parameters in a plate-jump-height experimental setup, finding that increasing the moisture content (and bulk density) resulted in greater momentum transfer. Bergeron et al. [11] carried out a series of small-scale experiments, using high-speed imaging and flash X-ray to capture soil ejecta and air shock detachment at greater distances. This showed that the ejection velocity of the soil decreases with increasing overburden, as does the air shock propagation speed (with this being greatest in a soil surface flush-buried condition). Weckert and Resnyansky [12] also used flash X-ray to capture ejecta expansion in experiments utilising a range of soils of varying PSD for the validation of numerical modelling. Very good agreement was found between the numerical and experimental results for the ejecta-wave expansion rate and shape.

Clarke et al. [13,14] found that the use of well-graded cohesionless soils result in greater variability in total impulse between tests, when compared with uniform cohesionless soils, for all moisture contents and bulk densities. Although geotechnical conditions (such as moisture content and bulk density) could be controlled to a high level, the well-graded nature of 'Stanag' (an approximation of the sandy gravel defined by [15]) results in a wider spread of impulse values than in a uniform soil such as Leighton Buzzard Sand (LB). Comparing two LB fractions: 'Fraction B' (LB) and '25B Grit' (LBF), with respective C_u (coefficient of uniformity, defined in Equation (1)) values of 1.4 and 3.2, resulting in a higher spread of impulse for LBF by a factor of four, even though similar levels of control of geotechnical conditions were achieved [16], thus demonstrating that increased variability is to be expected with an increasingly well-graded soil. A comparision of the particle size distributions of the soils used in this and other studies is shown in Figure 1.

$$C_u = \frac{D_{60}}{D_{10}} \tag{1}$$

where D_{60} is the 60th percentile particle size by mass and D_{10} is the 10th percentile particle size.

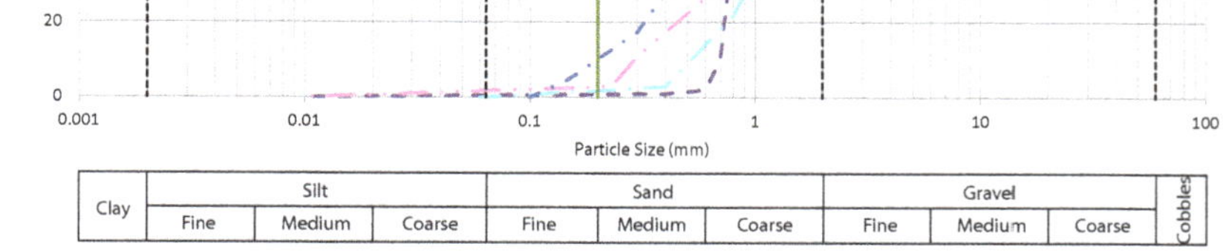

Figure 1. Particle Size Distribution (PSD) curves of a range of soils tested in the literature. LB, LBF, Clay [13]; Stanag [17]; CFAS (Centre of Allowable Bounds) [18]; and Glass Microspheres [5,6]

Computer programs have been used for decades for the prediction of blast loading from buried charges, with much based on earlier work by Westine et al. [19]. Tremblay [20] built on Westine's work to establish algebraic equations for impulsive loading; however, these do not account for moisture content as a separate influence from soil density. This is necessary as, for a constant bulk density, an increasing moisture content results in increasing impulse delivery [21].

Numerical modelling has begun to capture the specific loading characteristics associated with soil conditions. Børvik et al. [5] and Kyner et al. [6] used discrete-particle-based numerical models to simulate their small-scale soil-analogue experimental work. Grujicic et al. [22–24] developed material models for sand that take account of soil parameters including saturation and particle size. It is imperative that experimental results can be gathered and used to validate these models.

Research at the University of Sheffield has been conducted via two methods: 'Characterisation of Blast Loading' (CoBL) and 'Free-flying mass impulse capture apparatus' (FFM) [25]. FFM utilised a half-scale (of STANAG Threat Level 2, as defined by [15]) experimental setup wherein a deformable target plate and reaction mass captured the impulse from the buried charge, with the global impulse derived (as in Figure 2a). Hence, this method only captured the overall loading, without the spatial implications.

This spatial relationship has been determined previously using removable tapered plugs in the target plate [26], where their ejection velocity was measured using high-speed video. It was found that, as distance from the centre of the charge increases, the specific impulse decreases exponentially. A more accurate and repeatable experimental method has been developed for the CoBL setup [27] at quarter-scale, using 17 Hopkinson Pressure Bars (HPBs) of 10 mm diameter, arranged radially up to 100 mm from the charge centre in the face of a rigid target plate. Each HPB measures the axial strain, which is converted to stress with a specific impulse integrated for in time [27] and the global impulse interpolated over the instrumented area (as in Figure 2b).

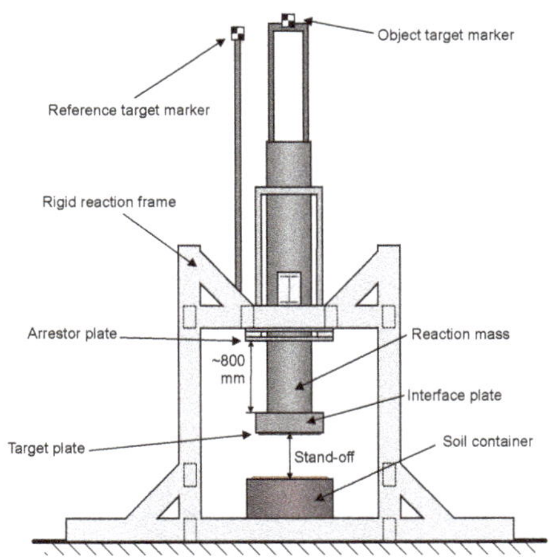

(a)

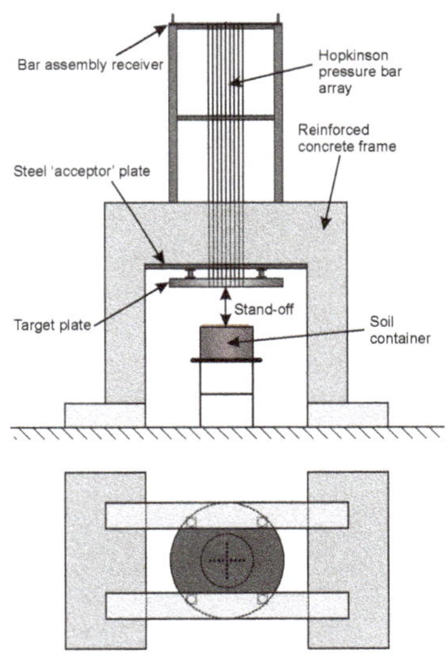

(b)

Figure 2. The two experimental setups at the University of Sheffield. (**a**) FFM (from Figure 3 of Clarke et al. [28]). (**b**) CoBL (from Figure 2 of Rigby et al. [29]).

In well-graded Stanag soil, individual particles can be over twice the size of the HPBs used in the CoBL setup. The total impulse values reported from FFM and CoBL testing are not in agreement for this soil type [17]; the impulse from CoBL is found to be much greater than that expected from scaling FFM, this is not the case for uniform soils. This suggests that the method of determining loading (a simple interpolation between discrete points) could be flawed for this well-graded soil. Hence, work is required to establish the relationship between a soil's PSD and the distribution and magnitude of impulsive loading.

2. Methodology

In order to address the disparity between the global impulse results for well-graded soil between CoBL and FFM experiments, alterations were required to the method of interpolation between the discrete measurement points (as laid out in Figure 3).

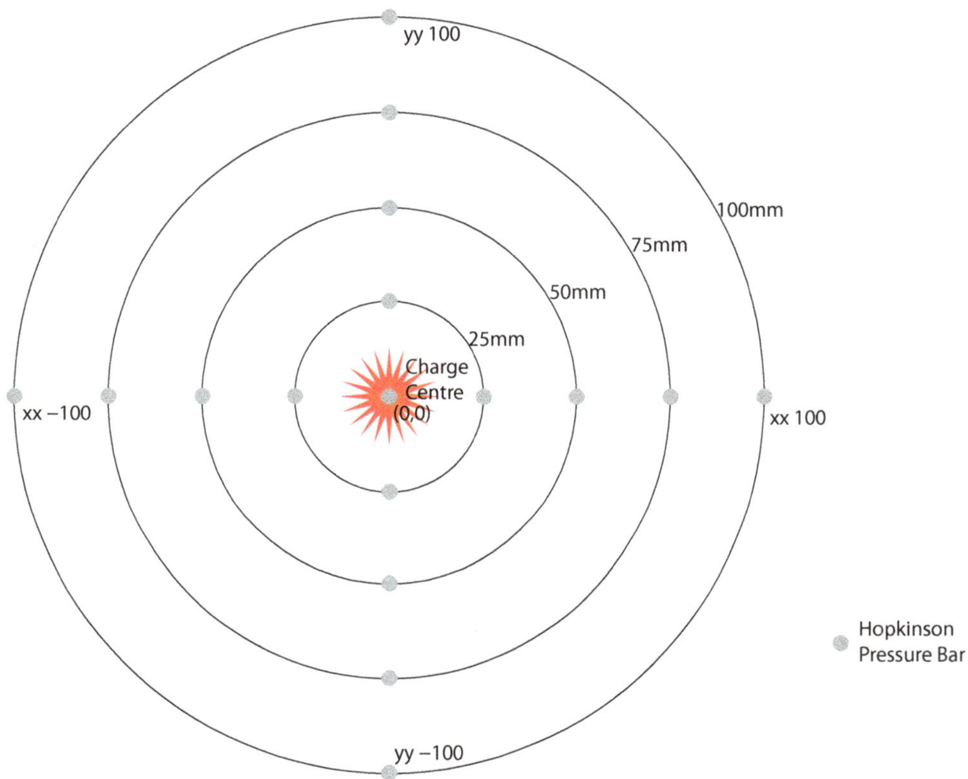

Figure 3. Layout of HPBs in the rigid target plate for the CoBL experimental setup

The previous data processing method (as used by Rigby et al. [17] and Clarke et al. [30] for all soil types, outlined in detail in [31]) operated by first importing the voltage signals from the experimental output, converting these to pressure signals then truncating them to a chosen length of time. A breakwire placed within the explosive charge was used to trigger the recording, so the truncation is applied after this time. Next, all of the pressure traces are aligned in time by their maximum pressure, so that, at any time after wave arrival, the value of the pressure can be interpolated between each HPB in the same axial direction (thus eliminating the temporal progression element and reducing the problem to a 1-dimensional interpolation). These four axes (positive xx, positive yy, negative xx, and negative yy) can then be interpolated between to populate the quadrants of a matrix with the expected pressure at every location (to a given mesh size). This occurs for the full test duration, after which the temporal wave-progression is reintroduced to allow the algorithm to represent both the temporal and spatial aspects of the loading. In the previous work, this temporal matrix of pressures over the plate was used to derive a specific impulse and global impulse over the whole plate. This methodology, along with the new interventions proposed herein, is outlined in the flowchart in Figure 4.

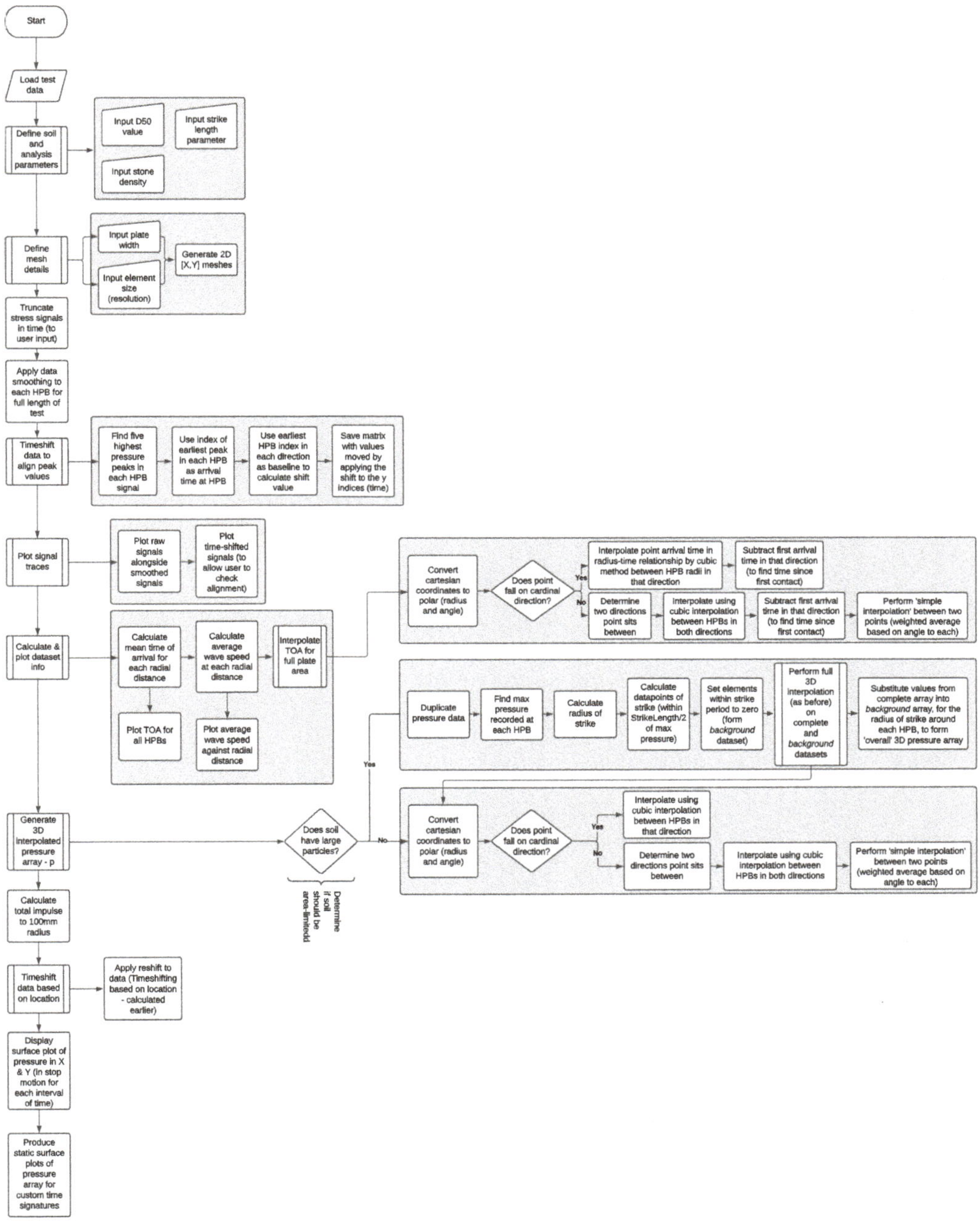

Figure 4. Flowchart of the methodology to convert discrete pressure measurements from HPBs to a full-plate dataset

2.1. Signal Truncation

Until now, it has been determined that an arbitrary cut-off, sufficiently later than the passing of the pressure wave, can be used to truncate input pressure signals. For the data presented by Clarke et al. [30], these results can be replicated by the use of a cut-off of 1.3 ms (1.2 ms after wave arrival), using peak global impulse as the reported values of impulse. Lide et al. [32] state that the speed of a wave in a narrow stainless steel rod is 5000 m/s, which means that over the 6m distance that the wave travels from the strain gauges to the end of the HPB and back, a 1.2 ms time period will have elapsed before the reflection will interfere with the pressure trace. This length of truncation has been found to have a significant effect on some test results due to the presence of pressure signal 'drift' after the loading has occurred. This drift is a phenomenon wherein, on certain tests, the gauge voltage (and, thus, the recorded pressure) does not return to zero after the loading period, even though the true pressure has returned to the ambient level. This can, with enough time to compound, lead to large increases or reductions in the global impulse derived (acceptable if there is a negative drift, as the peak impulse can be measured before the drift causes it to drop, leading to an increased reported impulse value for tests with a positive drift). This drift acts in opposite directions depending on the scope polarity during experimentation, this varied during the testing as it was assumed that it would not affect the results, with the systematic error only identified post-testing. As such, a reduced truncation time of 0.7 ms was introduced, as this has been found to reduce the influence of pressure drift whilst still capturing the full period of loading. The effects of this can be seen in Figure 5.

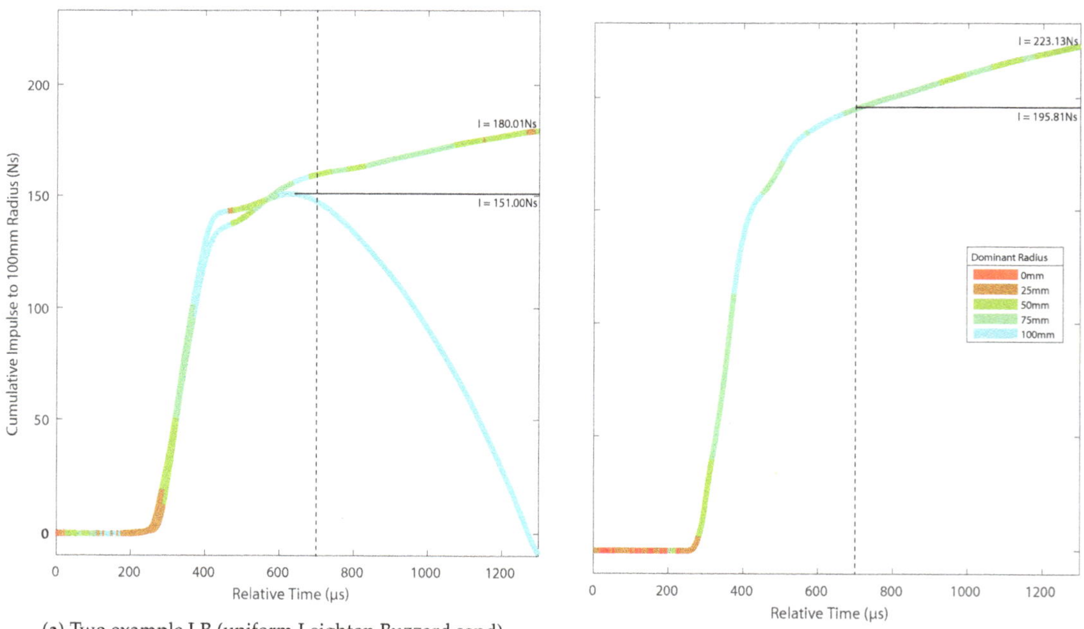

(**a**) Two example LB (uniform Leighton Buzzard sand) cumulative impulse plots with opposing positive and negative signal drift.

(**b**) Example Stanag (well-graded sandy gravel) cumulative impulse plot with positive signal drift.

Figure 5. 1.3 ms truncated signals with opposing drift directions, dashed line shows reduced 0.7 ms truncation period. 'Dominant radius' indicates where the majority of impulse contributions are occurring at that time.

2.2. *Signal Noise and Filtering*

In order to improve the readability of the pressure signals from the experimental apparatus, data smoothing was performed with the purpose of reducing signal noise without affecting the peak pressure and impulse values significantly. A Hampel (median) Filter [33], with various window widths, was trialled with limited success, with some especially noisy signals being improved marginally, though not to an adequate level. Savitzky–Golay filtering [34], as used by Pannell et al. [35] for the removal of noise from specific impulse data, was evaluated also. A first-order fit (moving average) with varying frame lengths was applied, with a frame length of 11 samples (equivalent to approximately 35 µs) selected as appropriate due to negligible reductions in peak pressure and global impulse whilst delivering a significantly 'cleaner' pressure signal.

To understand the origin of the noise in the signal, a Fast Fourier Transform (FFT) operation was undertaken to find the Discrete Fourier Transform (DFT) of the raw pressure signals, to determine if there were any dominant frequencies within the signal that could be attributed to physical causes such as electrical noise. Figure 6 shows the frequency spectrums resulting from the FFT process for three tests, showing that the majority of the signal is in the <100 kHz range. There is not a secondary peak in the frequency spectrum, suggesting that the noise cannot be attributed to a consistent external source.

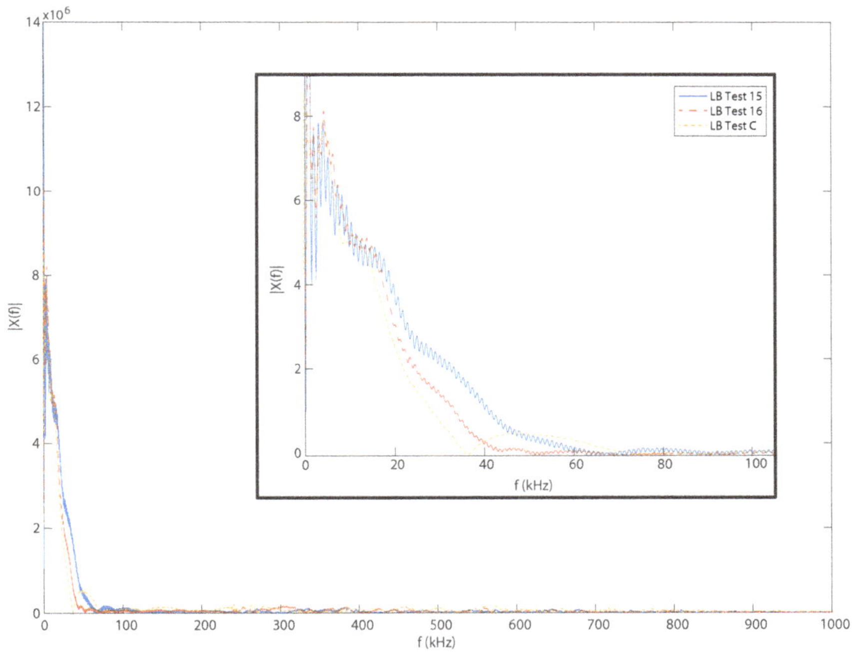

Figure 6. FFT analysis of LB experimental pressure signals. Numerical test IDs are consistent with [30], with alphabetical being previously unpublished. Figure inset is a zoomed in section of the same data.

Wang and Li [36] used FFTs to filter out high-frequency noise from signals in the Split Hopkinson Pressure Bar experiments, applying a low-pass filter to the FFT output, then performing an inverse FFT to retrieve a low-noise output. The same method was applied in this study. Tyas and Watson [37] state that, "the highest acceptable frequency component in a signal propagating in a steel bar [is limited] to approximately from $250/a$ to $500/a$ kHz", with a being the HPB radius in mm, in this case $a = 5$. Thus, the maximum theoretically acceptable frequency would be between 50 and 100 kHz. However, they go on to state that, in blast loading, the situation can be more complicated. They propose a dispersion correction method that results in a bar having "a bandwidth in excess of 1250/a kHz" [37]. Supported by Figure 6, 100 kHz was selected as the maximum frequency cut-off in the current analysis. A comparison between the original pressure signal, with specific impulse take-up for the selected bar, against the Savitzky–Golay filtered and Inverse FFT signals is presented in Figure 7. It can be seen that the Savitzky–Golay filtering is an effective method of reducing signal noise, whilst preserving the pressure peak and specific impulse. However, an inverse FFT method is not suitable as a noise-reduction method as the pressure peak and specific impulse takeup are not preserved. Thus, Savitzky–Golay filtering has been used to process each of the raw pressure signals before further analysis is performed.

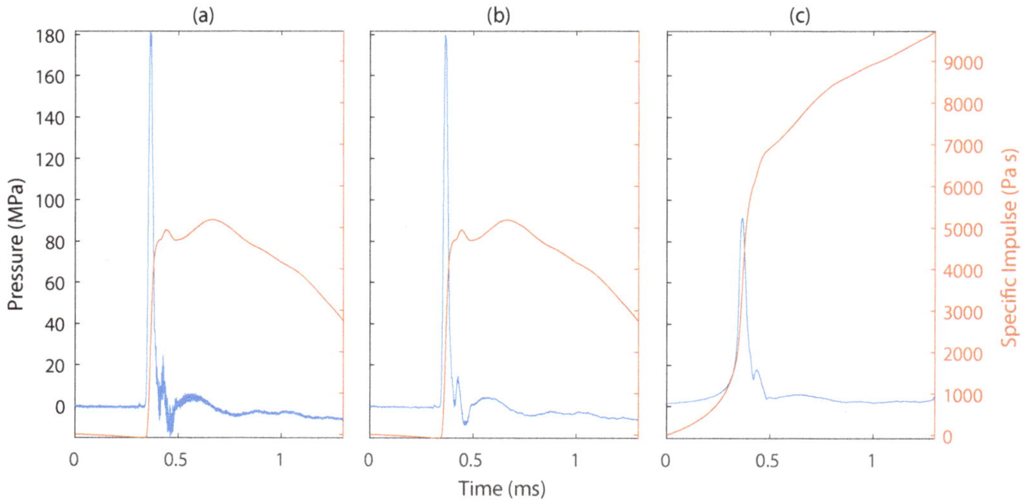

Figure 7. For a single HPB pressure trace: (**a**) Raw Signal vs. (**b**) Savitzky–Golay Filtered vs. (**c**) Inverse FFT Filtered. Blue = pressure, red = specific impulse.

2.3. Wave Arrival Time

The subsequent alignment of the pressure signals for interpolation from the maximum value of pressure in each individual signal is relatively good, with signals usually aligned within 50 μs of each other. However, this falls down when the arrival of the wave and the maximum pressure peak do not align (through the presence of a second peak slightly after the first, potentially from initial separation of the shock front and the soil ejecta wave). As such, this alignment was changed to operate by finding the n highest pressure peaks in the signal ($n = 5$ was selected as the optimum value), then using the time signature of the earliest of these as the arrival time of that pressure wave. This improved the signal alignment substantially and also allowed for the determination of wave expansion speed (across the plate), which can be compared with other experimental blast wave speed data. The transition from raw pressure signals to smoothed, arrival-time-aligned signals is demonstrated for an example LB test in Figure 8. It can be seen that many of these signals exhibit slight negative pressure drift, as outlined previously.

The time of arrival (TOA) of a pressure wave at each HPB for a single test is plotted against the radial distance from the centre of the plate in Figure 9a. A reasonable TOA curve can be seen to occur, growing somewhat exponentially with increasing radial distance, with behaviour exhibited at far-field distances in the air [38]. The corresponding average wave-expansion velocity (calculated from an interpolation of the change in TOA at each HPB over the horizontal distance, from the central HPB) is shown in Figure 9b. This velocity is that of the coupled soil ejecta and blast wave expansion, as the blast wave is found (through HSV) to usually detach from the soil ejecta only at greater distances [11,29]. It can be seen that, after an initial period of instability (indicated by the flat portion of the graph), the velocity reduces with the radial distance. The arrival times are typically variable within the 0–25 mm range, potentially due to experimental error in the centring of the sand bin and charge below the instrumented plate surface, as well as from the influence of sand plumes ejecting ahead of the main wave. The decision was made to limit analysis of the wave speed to radii ≥25 mm due to the unreliability of the data before that point, exacerbated by the presence of only one sensor at the centre rather than the four at every other instrumented distance.

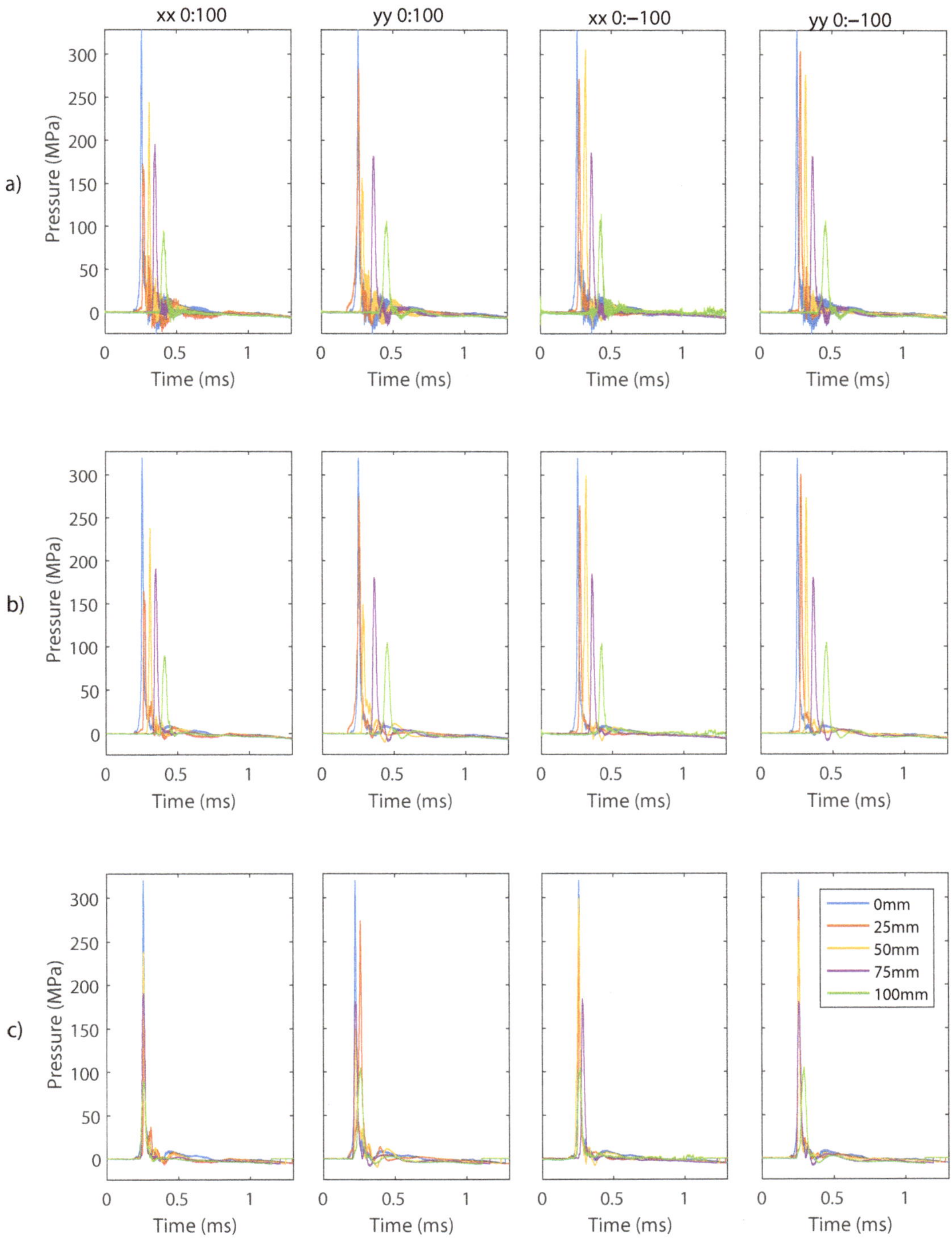

Figure 8. Full array of pressure signals for a single test: (**a**) raw pressure input, (**b**) smoothed, and (**c**) arrival-time-aligned.

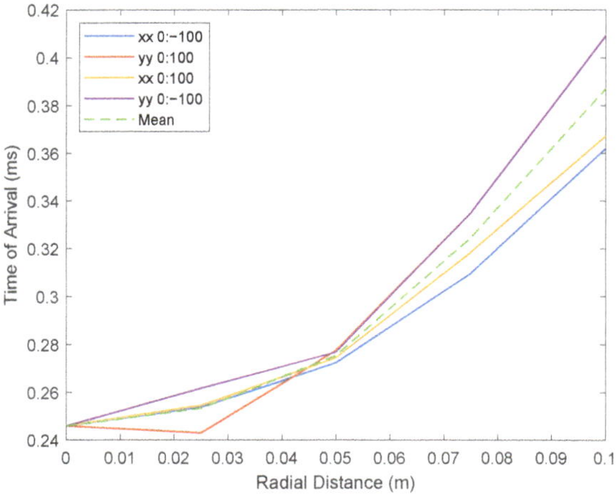

(a) TOA vs. radial distance from plate centre.

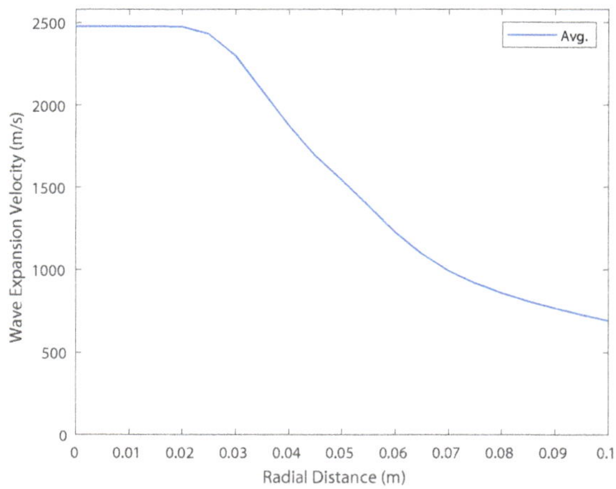

(b) Wave-expansion velocity vs. radial distance.

Figure 9. Time of arrival and wave-expansion velocity for an LB test.

2.4. Comparison of CoBL and FFM

So that the outputs of the modified interpolation algorithm could be reconciled against other data, global impulse values from the FFM tests were used as a comparison [28]. There is a disparity between the CoBL and FFM results for the Stanag tests (as discussed previously and in [17]), with the Stanag global impulse results being consistently higher than those of LB from CoBL testing but the reverse being true in FFM. There are two ways to reconcile the differences in these datasets. The first is that postulated in [17], wherein this higher impulse is caused by a more centralised loading in Stanag (due to a large number of discrete strikes directly above the charge, as well as the higher stiffness of Stanag), captured by the smaller relative instrumented area of CoBL. If this were the case, a greater peak deflection for the same impulse would be expected in plate-deflection experiments, such as that seen for testing performed with charges contained within a steel 'pot' (Minepot) in [28]. The peak deflections are marginally higher for Stanag when compared with LB (for the same impulse), but the extent and contribution of this loading centralisation is currently unknown. The second is that the existing area-integration of impulse inaccurately assumes a regular wave of soil expansion (through cubic interpolation) and occurs (disregarding the effects of discrete large particles), resulting in a consistent over-estimation of the global impulse. This study hypothesises that the particle strikes do not occur across the whole plate in this wave-like manner and, instead only occur at a limited number of locations, with the simple interpolation currently acting to skew the results by 'stretching' the increased pressure readings over an excessively large 'zone of influence'. This effect is shown in Figure 10, with strikes at the two HPBs. The interpolation algorithm required alteration so that it could correctly account for discrete particle strikes, on top of a background contiguous wave. This study is intended as a proof of concept to investigate whether it is plausible to correct the interpolation algorithm to account for the effect of discrete particle strikes.

It was determined that this alteration should use an area-limiting scheme, wherein the maximum area surrounding a HPB, over which a recorded pressure would be likely to have been applied, would be determined, as opposed to the existing algorithm that assumed a full 180 degrees of the plate could be influenced by this pressure spike. For example, the existing algorithm would assume that a strike effectively impacts an area of 25 mm by 236 mm at a 75 mm HPB, a total area of 5890 mm^2, when a typical Stanag particle (D_{50}) is around 10 mm in diameter (a cross-sectional area of 79 mm^2). As such, it was important to determine a way to limit this area of influence to the true limit that would be expected in the real, uninstrumented regions of the target plate (represented in simplified form for a single axis in Figure 11).

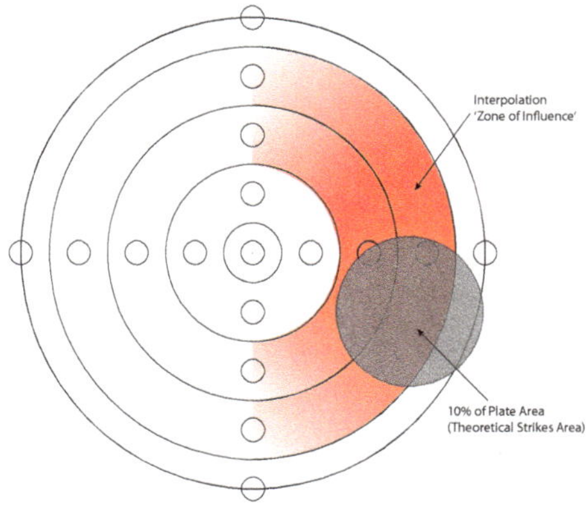

Figure 10. Demonstrating the difference between actual total discrete strike area (shown here as 10% of the plate area) versus the 'zone of influence' of the existing interpolation. It is assumed that if a proportion of the plate area is struck, the same proportion of the HPBs will be struck also, a strike area of 3142 mm^2 results in a 'zone of influence' of 9817 mm^2 in this example case.

This area limiting, as a consequence of the PSD, has been attempted as a possible method of understanding the behaviour of the soil in blast conditions. A number of approximations have been created, including assuming perfectly plastic collision behaviour and the straight-on impact of particles. Further work is required to establish the validity of these assumptions and to increase accuracy.

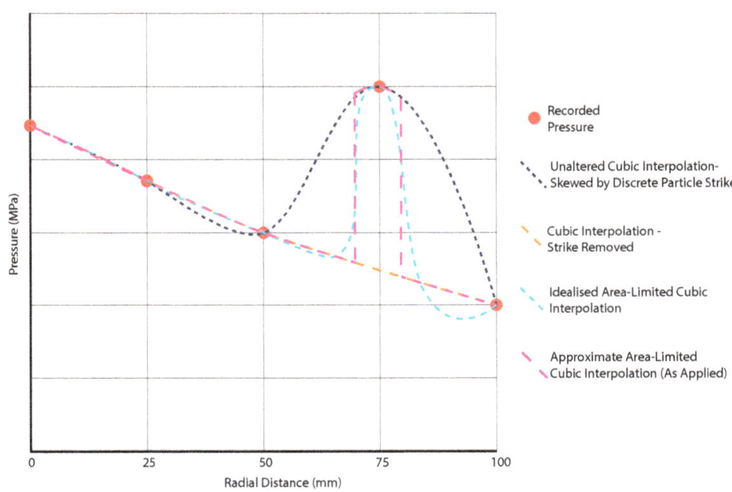

Figure 11. Diagram showing a simplified 1D approximation of the area-limiting process for a single cardinal direction with discrete particle strike at 75 mm.

2.5. Theoretical Particle Strike Area

In order to establish the area limit for any given particle strike at any HPB, it was first important to consider the known soil parameters that could be used to derive this, the D_{50} value (the median particle diameter) was selected for this purpose. From a particle size distribution graph, the D_{50} can be established, in order to gauge a typical value of the particle size that is representative of the soil as a whole.

This D_{50} value can be used, along with a stone density (assumed as a typical 2700 kg/m^3, in the calculation of a typical particle mass (assuming a spherical particle). When this is multiplied by the velocity of a strike (determined from average wave-expansion velocity at the relevant radial distance), the momentum of a typical particle strike can be established.

$$I_{\text{strike}} = F_{\text{typ.}} \cdot t_{\text{strike}} = m_{\text{particle}} \cdot v_{\text{avg.}} \tag{2}$$

This momentum transfer is equivalent to the impulse experienced if a plastic response is assumed (particle obliterated on impact, not usually the case and thus a simplification). The impulse would be greater if the particle were reflected elastically, up to a maximum of twice the incident momentum (for a perfectly elastic collision). This equivalence of recorded impulse to incoming momentum has been shown to be the case for a homogenous sand slug [39], but it is likely a simplification of the behaviour of discrete particle loading. However, the expected increase in impulse due to collision elasticity may be counteracted due to oblique or glancing impacts of particles on HPBs, reducing the incident force; for simplicity, these effects have been ignored in this study.

If divided by the time period of the strike (assumed as 25 µs from initial graphs), the impulse can be converted to a typical force value. This time period is supported by the findings of Liu et al. [40], who found that in a sand slug impact, the soil densification time (and thus strike length) is related to the column height of the soil (the overburden of 28 mm in Stanag testing), H, and the velocity of the soil (see Figure 12):

$$t = \frac{\bar{t} \cdot H}{v_0} \tag{3}$$

where, for an incompressible (rigid) target, $\bar{t} \approx 1$ when pressure drops to zero ($\bar{t}$ is a non-dimensional time).

Thus, for a 25 µs time period with H = 28 mm, a velocity of 1120 m/s would be expected, reasonable given the velocities of the soil in Stanag tests (peak average velocities: 864–2637 m/s and at 100 mm HPB: 636–1050 m/s).

$$\begin{aligned} F_{\text{typ.}} &= \frac{m_{\text{part.}} \cdot v_{\text{avg.}}}{t_{\text{strike}}} \\ &= \frac{\frac{4}{3} \cdot \pi \cdot \left(\frac{D_{50}}{2}\right)^3 \cdot \rho_{\text{stone}} \cdot v_{\text{avg.}}}{t_{\text{strike}}} \end{aligned} \tag{4}$$

If this typical force value is divided by the maximum actual recorded pressure at a given HPB, this will result in the area over which a particle strike should become effective, which can further be reduced to a radius of effect.

$$\begin{aligned} A_{\text{strike}} &= \pi \cdot r_{\text{strike}}{}^2 = \frac{F_{\text{typ.}}}{P_{\text{rec.}}} \\ r_{\text{strike}} &= \sqrt{\frac{A_{\text{strike}}}{\pi}} = \sqrt{\frac{\left(\frac{F_{\text{typ.}}}{P_{\text{rec.}}}\right)}{\pi}} \end{aligned} \tag{5}$$

$$r_{\text{strike}} = \sqrt{\frac{\frac{4}{3} \cdot \left(\frac{D_{50}}{2}\right)^3 \cdot \rho_{\text{stone}} \cdot v_{\text{avg.}}}{t_{\text{strike}} \cdot P_{\text{rec.}}}} \tag{6}$$

For example, the testing used Stanag soil with a D_{50} of 10 mm (extracted from Figure 1). For the HPB in test 34 at yy−50, the maximum recorded pressure was 353.7 MPa and the average wave speed at 50 mm radius was 1760.65 m/s. Therefore, the effective radius of this particular particle strike can be calculated as below:

$$r_{\text{strike}} = \sqrt{\frac{\frac{4}{3} \cdot \left(\frac{10 \times 10^{-3}}{2}\right)^3 \cdot 2700 \cdot 1760.65}{25 \times 10^{-6} \cdot 353.7 \times 10^6}}$$
$$= 0.0095\,\text{m}$$
$$= 9.5\,\text{mm}$$

The particle mass for this theoretical (spherical) D_{50} particle would be:

$$m = \frac{4}{3} \cdot \pi \cdot \left(\frac{D_{50}}{2}\right)^3 \cdot \rho_{\text{stone}}$$
$$= \frac{4}{3} \cdot \pi \cdot \left(\frac{10 \times 10^{-3}}{2}\right)^3 \cdot 2700$$
$$= 1.4\,\text{g}$$

The larger particles within the soil have much higher masses but are not representative of the soil as a whole because they are less likely to strike the HPBs due to their lower probability of occurrence. Some of these larger particles are pictured in Figure 13, with masses in the range from 9.70 g to 15.11 g. For reference, a spherical particle of diameter 20 mm has a mass of 11.3 g.

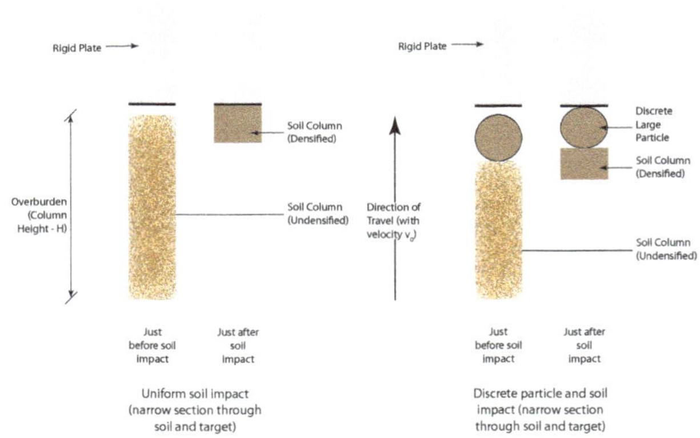

Figure 12. Densification of a soil column, without and with a discrete particle strike.

Figure 13. A selection of large particles extracted from a sample of Stanag soil, with masses of 10.41 g, 12.89 g, 9.70 g, and 15.11 g.

2.6. Application of Area-Limiting: Well-Graded Soil

To apply the area-limiting scheme to the pressure interpolation algorithm and, thus, garner a more accurate picture of the mechanisms occurring, a *background*-interpolated array was generated, onto which the discrete areas of an array inclusive of particle strikes would be superimposed.

This backgroundarray was generated by finding the time signature of the maximum pressure value at each HPB then removing a 25 μs section of data surrounding this from the signal, effectively removing the period of the particle strike. After performing this action on all 17 HPB signals, these were then interpolated using the original method to create a 3D pressure array.

This interpolation was also carried out on the unaltered data, inclusive of the particle strikes. The circular portions of this array (for the full time of the test), with the appropriate strike radii, were then applied to the background array to create an overall pressure array. This overall pressure array consists of data representing the standard wave of smaller particles expanding from the centre, with limited discrete pressure spikes from larger particle strikes (see Figure 14).

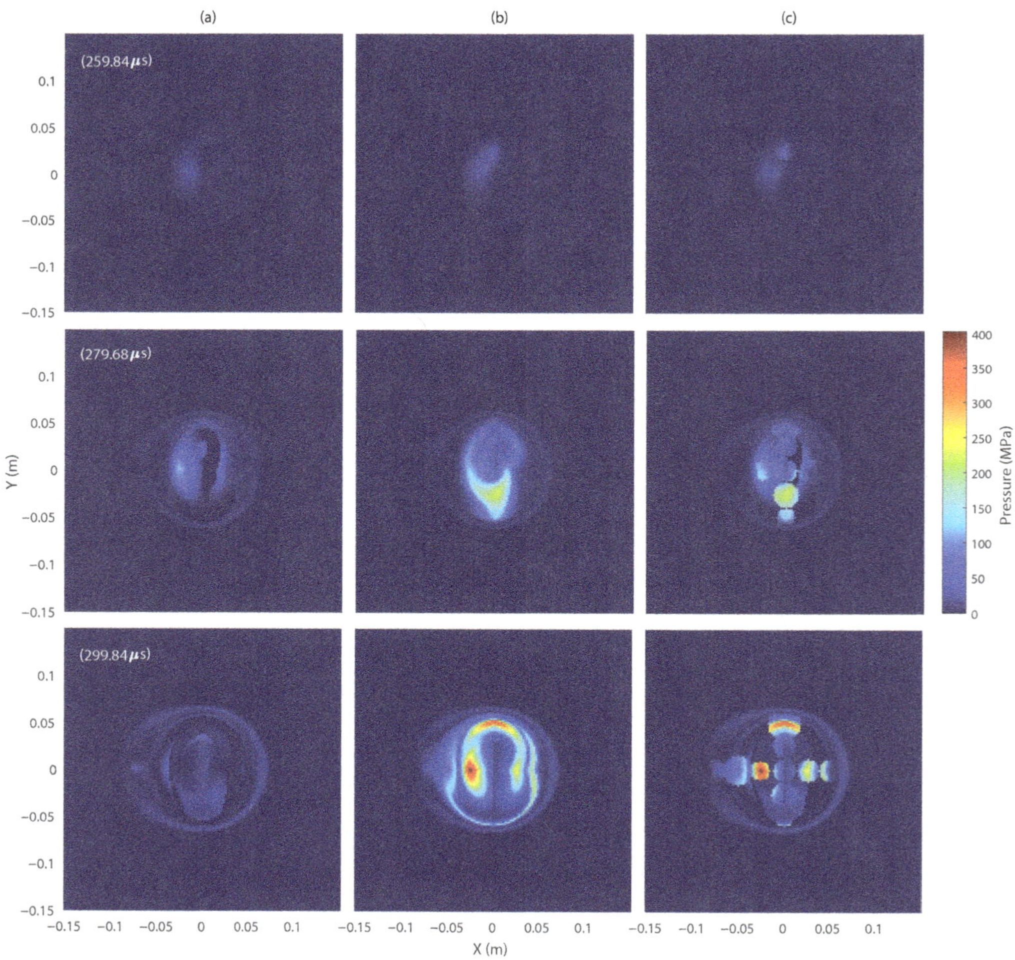

Figure 14. Test 34 (Stanag) pressure surfaces over a 40 μs period: (**a**) Background (strikes removed), (**b**) Original (skewed by particle strikes), and (**c**) Area-Limited.

2.7. Application of Area-Limiting: Uniform Soil

This new area-limiting algorithm was experimentally applied to the data from uniform soil (LB) tests in order to gauge its effectiveness. As can be seen from Figure 15, this failed to improve the interpolation of the data as the soil response consists of a contiguous pressure wave of similar-sized particles without discrete particle strikes. This meant that the algorithm removed a large proportion of real data from the array and, therefore, it performed much more poorly than the original method of interpolation.

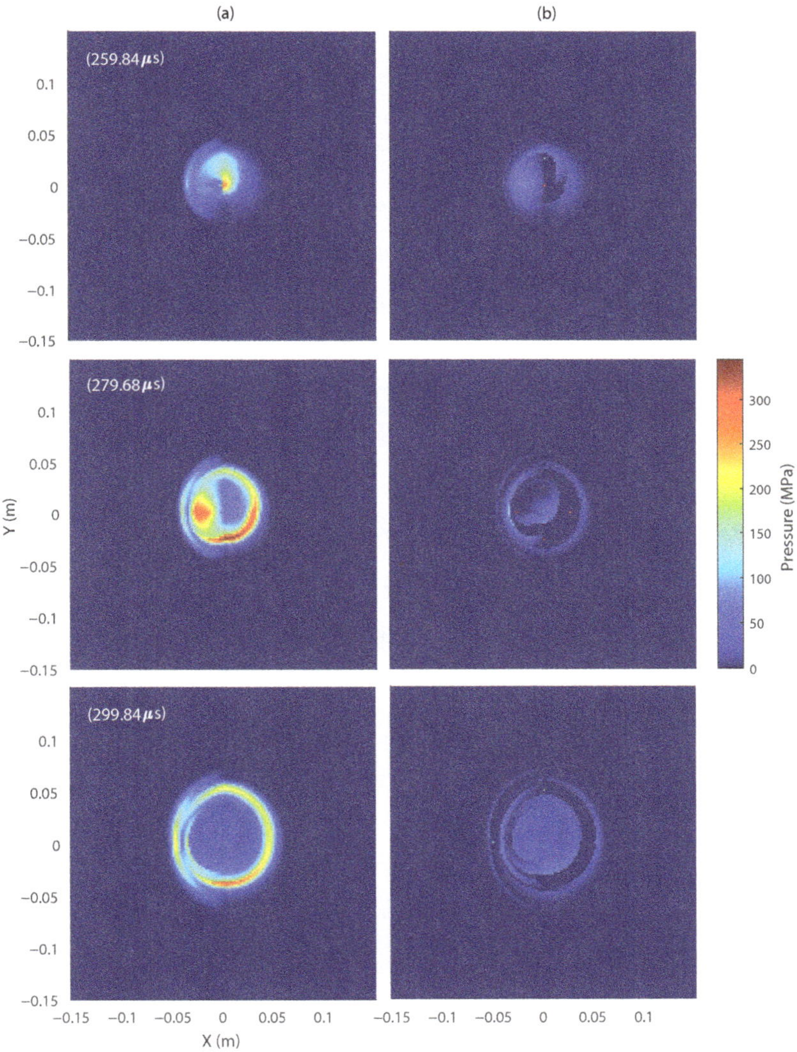

Figure 15. Test15 (LB) pressure surfaces over a 40 μs period: (**a**) Original and (**b**) Area-Limited

This mishandling of the data by the algorithm can be explained by calculating the radius of effect of an LB particle strike; thus, it can be seen that this is not an appropriate way to represent a pressure wave of uniform particles. For LB test 15 at the HPB at xx50, the D_{50} = 0.8 mm (from Figure 1), average wave speed at 50 mm radius was 769.8 m/s, and the maximum recorded pressure was 195.2 MPa.

$$
\begin{aligned}
r_{\text{strike}} &= \sqrt{\frac{\frac{4}{3} \cdot \left(\frac{0.8 \times 10^{-3}}{2}\right)^3 \cdot 2700 \cdot 769.8}{25 \times 10^{-6} \cdot 195.2 \times 10^6}} \\
&= 3.38 \times 10^{-4}\,\text{m} \\
&= 0.34\,\text{mm}
\end{aligned}
$$

Constricting the pressure wave peak to only a 0.34 mm radius around each HPB results in the area-limited pressure array represented in Figure 15, which clearly ignores the actual progression of the wave, limiting it to just the background pressure wave from outside of the 25 μs period of 'strike'.

Therefore, it was determined that this area-limiting process should only be applied to data from soils where the D_{50} particle size would result in a discrete impulse that forms a sizeable proportion of the global impulse experienced. A 1.4 g D_{50} Stanag particle, at a typical 1120 m/s (from the 25 μs strike length established earlier), would (assuming particle plasticity) result in an impulse of 1.57 N s. For a D_{50} LB particle, with a mass of 0.00072 g, at 1120 m/s, the impulse would be 0.81mN s, nearly 2000 times less. Given that a global impulse of the magnitude of hundreds of Newton–Seconds is to be expected from this testing, a single typical-particle strike (not accounting for the interpolation causing this error to spread) in Stanag soil could represent a >1% portion of the result, whereas, in LB, this would represent <0.001% (indicating that loading has to occur as part of a contiguous wave in LB soil).

3. Results and Discussion

The results from the CoBL experimental setup, presented herein, correspond to an over-burden (OB) of 28 mm and a stand-off distance (SOD) of 140 mm. A 78g 3:1 cylinder of PE4 explosive was utilised, buried within either LB (a uniform sand with a moisture content of 25% and a bulk density of 1990 kg/m^3) or Stanag (a well-graded sandy gravel with a moisture content of 14% and a bulk density of 2220 kg/m^3) both of which were fully saturated. Full saturation was achieved using the method outlined in [13]. The Test IDs used correspond to those in [30] if numerical and are previously unpublished data if alphabetical. The interpolation was carried out with an element size of 5 mm to ensure consistency between tests. It was found that increasing the mesh resolution had a minimal effect on the results.

3.1. Wave-Expansion Velocity

In order to validate the analytical methodology described previously, the wave-expansion velocities found in this study (used to calculate an area-of-effect of a particle strike) were compared against the wave speeds corresponding to arrival time data from work by Ehrgott [41]. In this work, Ehrgott used 2.27 kg C4 charges, buried in different soils with 100 mm OB, and measured the TOA at gauges suspended 500 mm above the soil surface, at varied horizontal distances from the charge centre. This charge size is equivalent to three times the scale of the CoBL tests using Hopkinson–Cranz cube root scaling and, as such, extrapolating the wave speed, on an exponential trendline, to a 300 mm radius of the centre should provide indicative values of expected wave speeds for CoBL data at 100 mm (with CoBL-scaled OB of 33 mm and SO of 167 mm) see Appendix A.

As can be seen in Table 1, for a poorly graded sandy soil, or a silty sand, wave speeds in the range 1028–1672 m/s are to be expected. Table 2 displays the wave expansion velocities at 100 mm radius for each CoBL test, where a similar over-burden (at an equivalent scale) in saturated Stanag soil results in speeds in the range 636–1050 m/s for a stand-off distance of 140 mm (compared with a 167 mm equivalent) and in saturated LB soil 523–694 m/s. These velocities are somewhat lower than those found by Ehrgott [41], likely a factor of a higher level of saturation causing increased soil throw volume due to a higher level of detonation product containment (reducing the expansion velocity). This is corroborated by a similar analysis of CoBL data from low-moisture-content (2.45%) LB resulting in velocities in the range 901–1611 m/s, which agrees much more closely with the range derived from Ehrgott [41].

Table 1. Wave speeds calculated from TOA data from Ehrgott [41], extrapolated to 300 mm for equivalence with CoBL data.

Distance from Blast Centre (mm)	Wave Speed (m/s)			
1060	Sandy Soil (Poorly Graded)	625	Silty Sand	606
705		754		793
537		895		1279
300 (Extrapolated)		1028		1672

Table 2. Wave speeds calculated from CoBL TOA data.

Soil Type	Test ID	Avg. Wave Speed at 100 mm (m/s)
Stanag, Saturated	Test 34	788.7
	Test 35	657.2
	Test 36	834.5
	Test 37	700.7
	Test A	1050.4
	Test B	636.1
LB, Saturated	Test 15	601.6
	Test 16	523.7
	Test 17	694.2
	Test C	690.7
LB, 2.45% M.C.	Test 7	901.2
	Test 8	1197.3
	Test 9	1610.8
	Test 10	1154.2

3.2. *CoBL Global Impulse*

3.2.1. Full Plate Integration (No Area Limiting)

The global impulse, used as an integration of pressure in time, from a full-plate cubic interpolation between the HPBs, is displayed in Table 3. The effects of the signal drift can be seen in the difference in the global impulses between the 0.7 ms and 1.3 ms signal truncation times. It can be seen that the Stanag results tend to increase with increased truncation time (due to a large net positive drift). On the other hand, the LB tests have much less of an increase (due to a much reduced net drift). This, if allowed to influence the results, would cause values from the Stanag tests to be inflated when compared to LB tests.

Considering the mean values of the impulse, it can be seen that the ratios of impulses for LB:Stanag is 1.00:1.29 for 0.7 ms truncation and 1.00:1.42 for 1.3 ms truncation (an effect of the higher positive drift), showing the possible effect of discrete particle strikes in skewing the interpolation, as it is expected that the saturated Stanag would have a lower value of impulse than saturated LB [17].

3.2.2. Area-Limiting Well-Graded Stanag

Application of the area-limiting process to the Stanag data results in the values of the global impulse displayed in Table 4. These results are for a 0.7ms truncation of each of the tests, to reduce the influence of signal drift.

Table 3. Global impulse results without area limiting for 0.7 ms and 1.3 ms truncations of signals from CoBL. * Test 17 data show no difference between truncation times as it was only recorded up to 0.5 ms.

Soil Type	Test ID	Global Impulse (Ns)		Truncation Diff.
		0.7 ms Trunc.	1.3 ms Trunc.	
Stanag, Saturated	Test 34	195.81	223.13	14%
	Test 35	182.00	209.84	15%
	Test 36	215.70	245.64	14%
	Test 37	199.33	227.73	14%
	Test A	172.04	188.49	10%
	Test B	222.62	262.40	18%
	Mean	211.48	245.47	16%
LB, Saturated	Test 15	147.09	151.00	3%
	Test 16	155.26	156.25	1%
	Test 17 *	150.46	150.46	0%
	Test C	159.90	180.01	13%
	Mean	153.18	159.43	4%

Table 4. Global impulse derived from integration of pressure over a 0.7 ms truncation time for an area-limited interpolation between HPBs (from CoBL).

Soil Type	Test ID	Global Impulse (Ns)
Stanag, Saturated	Test 34	125.58
	Test 35	131.59
	Test 36	152.49
	Test 37	132.57
	Test A	114.66
	Test B	152.45
	Mean	134.89

The mean value of global impulse, when this method has been applied, results in a ratio of impulse for LB:Stanag of 1.00:0.88, much closer to the 1.00:0.84 found in FFM testing by Rigby et al. [17].

3.3. Comparison to FFM

The global impulse results can be compared between CoBL and FFM experiments in order to validate the area-limiting method. If it is assumed that the LB total impulse results scale accurately, the ratio of I_{CoBL}:I_{FFM} from the mean LB values for each setup (1:40.53) can be used to derive an expected mean value of impulse for Stanag. For reference, geometric scaling alone, through projection of the FFM target plate to the CoBL SOD at the same Hopkinson–Cranz scale, results in a ratio of 1:30.07, however, this does not account for losses in the wave expansion through air. Utilising the LB-equivalency ratio results in an expected mean of 135.29 Ns for Stanag (only 0.3% more than the mean value of 134.89 Ns resulting from the area-limiting process). Given the number of assumptions made throughout this analysis, it is unfair to accept this accuracy as completely true, however, it is indicative that accounting for PSD is viable when analysing the spatial and temporal distributions of loading from buried charges. These assumptions require further investigation in the future. Each of the FFM test results, scaled by the 40.53 scale factor, are displayed in Table 5.

Table 5. Global impulse data from FFM testing, with impulse values scaled by the LB-equivalency scale factor (40.53). Results marked with * are from [28], with other data previously unpublished.

Soil Type	Total Impulse (Ns)	CoBL-Scaled Impulse (Ns)
Stanag, Saturated	4972.71 *	122.68
	5571.85 *	137.47
	5619.14	138.63
	5370.20	132.49
	5143.27	126.89
	5434.96	134.09
	5858.34	144.53
	5899.26	145.54
Mean	5483.71	135.29
LB, Saturated	6298.68 *	155.40
	6202.10 *	153.02
	6125.52 *	151.13
Mean	6208.77	153.18

4. Conclusions

This study has built on existing techniques to understand the spatial and temporal distributions of loading from soils in explosive blasts. Improvements have been made to pressure signal processing, including the use of data smoothing, alongside an improved arrival-time-finding algorithm. Further, a proof-of-concept method of area-limiting pressure spikes from discrete particle strikes was established, validated against another experimental setup to achieve similitude between results. The CoBL data (at one-quarter scale) has been directly compared against FFM (at one-half scale), with the area-limiting approximation allowing for agreement in the total impulsive loading for both well-graded and uniform soils.

From this, it can be understood that the spatial distribution of loading is largely impacted by the effects of a blast in well-graded soil, with high pressure strikes occurring over limited regions on a target. This leads to a lower level of global impulse than previously derived (thus bringing data from the CoBL experiment in line with expectations from FFM). This engenders a new understanding that the particle size distribution of a soil has not only a global effect on loading but also a discrete localised effect when larger particles are present, resulting in major implications for the design of protective structures and materials due to the presence of 'pockets' of much higher pressure (and thus specific impulse) loading within the overall wave.

Author Contributions: Conceptualization, S.C.; methodology, R.W.; formal analysis, R.W. and S.R.; writing, original draft preparation, R.W.; writing, review and editing, S.C., S.R., M.G. and I.E.; supervision, S.C. and A.T.; experimental management, A.T.; funding acquisition, M.G. and I.E. All authors have read and agreed to the published version of the manuscript.

Funding: This research was funded by a University of Sheffield Faculty Prize Scholarship. The original experimental work was funded by the Defence Science and Technology Laboratory under contract DSTLX-1000059883.

Data Availability Statement: The data presented in this study are available on request from the corresponding author.

Acknowledgments: The authors would like to recognise the work of the technical support staff at Blastech Ltd. without whom we would have never been able to have such excellent datasets to work with.

Conflicts of Interest: The authors declare no conflict of interest.

Abbreviations

The following abbreviations are used in this manuscript:

CoBL	Characterisation of Blast Loading (Experimental Setup);
FFM	Free Flying Mass (Experimental Setup);
FFT	Fast Fourier Transform;
HPB	Hopkinson Pressure Bar;
LB	Leighton Buzzard (Sand);
MC	Moisture Content;
OB	Over-Burden;
PSD	Particle Size Distribution;
SOD	Stand-Off Distance;
TOA	Time of Arrival.

Appendix A. Time of Arrival Data from Ehrgott [41]

Sandy Soil (100 mm OB)					1028 m/s (Speed at 300 mm)
		TOA (ms)			
Distance (inches)	**Distance (m)**	**1**	**2**	**Avg.**	**Wave Speed (m/s)**
41.74	1.06	1.68	1.71	1.70	625.48
27.77	0.71	0.95	0.92	0.94	754.39
21.15	0.54	0.6		0.60	895.35

Silty Sand (100 mm OB)					1672 m/s (Speed at 300 mm)
		TOA (ms)			
Distance (inches)	**Distance (m)**	**1**	**2**	**Avg.**	**Wave Speed (m/s)**
41.74	1.06	1.77	1.73	1.75	605.83
27.77	0.71	0.90	0.88	0.89	792.54
21.15	0.54	0.42		0.42	1279.07

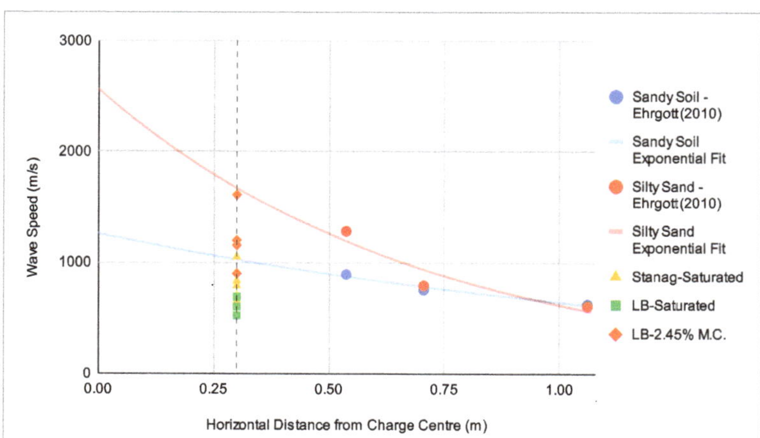

Arrival times for the soils most similar to those in use in this study were extracted from [41], converted to wave speeds, and then plotted with an exponential fit to find expected wave speeds at a CoBL-equivalent scale. The wave speeds found in this study have been plotted here also to allow for comparison.

References

1. International Campaign to Ban Landmines, Cluster Munition Coalition. Landmine Monitor 2021. Technical Report, ICBL-CMC. 2021. Available online: http://www.the-monitor.org/media/3318354/Landmine-Monitor-2021-Web.pdf (accessed on 16 January 2023).
2. Neuberger, A.; Peles, S.; Rittel, D. Scaling the response of circular plates subjected to large and close-range spherical explosions. Part II: Buried charges. *Int. J. Impact Eng.* **2007**, *34*, 874–882. [CrossRef]
3. Hopkinson, B. British ordnance board minutes 13565. *Natl. Arch. Kew UK* **1915**.
4. Cranz, C. *Lehrbuch der Ballistik*; Springer: Berlin/Heidelberg, Germany, 1925; Volume 1, p. 174.
5. Børvik, T.; Olovsson, L.; Hanssen, A.G.; Dharmasena, K.P.; Hansson, H.; Wadley, H.N. A discrete particle approach to simulate the combined effect of blast and sand impact loading of steel plates. *J. Mech. Phys. Solids* **2011**, *59*, 940–958. [CrossRef]
6. Kyner, A.; Dharmasena, K.; Williams, K.; Deshpande, V.; Wadley, H. High intensity impulsive loading by explosively accelerated granular matter. *Int. J. Impact Eng.* **2017**, *108*, 229–251. [CrossRef]
7. McShane, G.; Deshpande, V.; Fleck, N. A laboratory-scale buried charge simulator. *Int. J. Impact Eng.* **2013**, *62*, 210–218. [CrossRef]
8. Hlady, S. Effect of soil parameters on landmine blast. In Proceedings of the 18th International Symposium on the Military Aspects of Blast and Shock, Bad Reichenhall, Germany, 27 September–1 December 2004.
9. Fourney, W.; Leiste, U.; Bonenberger, R.; Goodings, D. Mechanism of loading on plates due to explosive detonation. *Fragblast* **2005**, *9*, 205–217. [CrossRef]
10. Anderson, C.E.; Behner, T.; Weiss, C.E. Mine blast loading experiments. *Int. J. Impact Eng.* **2011**, *38*, 697–706. [CrossRef]
11. Bergeron, D.; Walker, R.; Coffey, C. Detonation of 100-Gram Anti-Personnel Mine Surrogate Charges in Sand. Technical Report SR 668, Defence Research Establishment Suffield. 1998. Available online: https://cradpdf.drdc-rddc.gc.ca/PDFS/zbb68/p509935.pdf (accessed on 16 January 2023).
12. Weckert, S.A.; Resnyansky, A.D. Experiments and modelling for characterisation and validation of a two-phase constitutive model for describing sands under explosive loading. *Int. J. Impact Eng.* **2022**, *166*, 104234. [CrossRef]
13. Clarke, S.D.; Fay, S.D.; Tyas, A.; Warren, J.; Rigby, S.E.; Elgy, I.; Livesey, R. Repeatability of buried charge testing. In Proceedings of the 23rd International Symposium on the Military Aspects of Blast and Shock, Oxford, UK, 7–12 September 2014.
14. Clarke, S.D.; Fay, S.D.; Warren, J.A.; Tyas, A.; Rigby, S.E.; Reay, J.J.; Livesey, R.; Elgy, I. Geotechnical causes for variations in output measured from shallow buried charges. *Int. J. Impact Eng.* **2015**, *86*, 274–283. [CrossRef]
15. NATO. *AEP-55, Volume 2 (Edition 2)*; Procedures for Evaluating the Protection Level of Armoured Vehicles: Mine Threat. International Standard, NATO: Washington, DC, USA, 2011.
16. Clarke, S.D.; Warren, J.A.; Fay, S.D.; Rigby, S.E.; Tyas, A. The role of geotechnical parameters on the impulse generated by buried charges. In Proceedings of the 22nd International Symposium on the Military Aspects of Blast and Shock, Bourges, France, 4–9 November 2012.
17. Rigby, S.E.; Fay, S.D.; Tyas, A.; Clarke, S.D.; Reay, J.J.; Warren, J.A.; Gant, M.; Elgy, I. Influence of particle size distribution on the blast pressure profile from explosives buried in saturated soils. *Shock Waves* **2018**, *28*, 613–626. [CrossRef]
18. *ASTM C33/C33M-18*; Standard Specification for Concrete Aggregates. ASTM International: West Conshohocken, PA, USA, 2018.
19. Westine, P.S.; Morris, B.L.; Cox, P.A.; Polch, E. *Development of Computer Program for Floor Plate Response from Landmine Explosions*; Technical Report; Southwest Research Institute, Contract Report No. 1345; for US Army TACOM Research and Development Center: Detroit, MI, USA, 1985.
20. Tremblay, J. Impulse on blast deflectors from a landmine explosion. *Defence Research Establishment Valcartier Tech. Memo.DREV-TM-9814*. 1998. Available online: https://apps.dtic.mil/sti/pdfs/ADA482742.pdf (accessed on 16 January 2023).
21. Clarke, S.D.; Warren, J.A.; Tyas, A. The influence of soil density and moisture content on the impulse from shallow buried explosive charges. In Proceedings of the 14th International Symposium on Interaction of the Effects of Munitions with Structures, Seattle, WA, USA, 19–23 September 2011.
22. Grujicic, M.; Pandurangan, B.; Huang, Y.; Cheeseman, B.A.; Roy, W.N.; Skaggs, R.R. Impulse loading resulting from shallow buried explosives in water-saturated sand. *Proc. Inst. Mech. Eng. Part L J. Mater. Des. Appl.* **2007**, *221*, 21–35. [CrossRef]
23. Grujicic, M.; Pandurangan, B.; Mocko, G.M.; Hung, S.T.; Cheeseman, B.A.; Roy, W.N.; Skaggs, R.R. A combined multi-material Euler/Lagrange computational analysis of blast loading resulting from detonation of buried landmines. *Multidiscip. Model. Mat. Str* **2008**, *4*, 105–124. [CrossRef]
24. Grujicic, M.; Pandurangan, B.; Coutris, N.; Cheeseman, B.A.; Roy, W.N.; Skaggs, R.R. Computer-simulations based development of a high strain-rate, large-deformation, high-pressure material model for STANAG 4569 sandy gravel. *Soil Dyn. Earthq. Eng.* **2008**, *28*, 1045–1062. [CrossRef]
25. Rigby, S.E.; Clarke, S.D. Characterisation of blast loading: Current research at The University of Sheffield. *Off. J. Inst. Explos. Eng.* **2015**, 14–17.
26. Fox, D.M.; Huang, X.; Jung, D.; Fourney, W.L.; Leiste, U.; Lee, J.S. The response of small scale rigid targets to shallow buried explosive detonations. *Int. J. Impact Eng.* **2011**, *38*, 882–891. [CrossRef]
27. Clarke, S.D.; Fay, S.D.; Warren, J.A.; Tyas, A.; Rigby, S.E.; Elgy, I. A large scale experimental approach to the measurement of spatially and temporally localised loading from the detonation of shallow-buried explosives. *Meas. Sci. Technol.* **2015**, *26*, 015001. [CrossRef]

28. Clarke, S.D.; Fay, S.D.; Warren, J.A.; Tyas, A.; Rigby, S.E.; Reay, J.J.; Livesey, R.; Elgy, I. Predicting the role of geotechnical parameters on the output from shallow buried explosives. *Int. J. Impact Eng.* **2017**, *102*, 117–128. [CrossRef]
29. Rigby, S.E.; Fay, S.D.; Clarke, S.D.; Tyas, A.; Reay, J.J.; Warren, J.A.; Gant, M.; Elgy, I. Measuring spatial pressure distribution from explosives buried in dry Leighton Buzzard sand. *Int. J. Impact Eng.* **2016**, *96*, 89–104. [CrossRef]
30. Clarke, S.; Rigby, S.; Fay, S.; Barr, A.; Tyas, A.; Gant, M.; Elgy, I. Characterisation of buried blast loading. *Proc. R. Soc. A Math. Phys. Eng. Sci.* **2020**, *476:20190791*. [CrossRef]
31. Clarke, S.D.; Fay, S.D.; Rigby, S.E.; Tyas, A.; Warren, J.A.; Reay, J.J.; Fuller, B.J.; Gant, M.T.; Elgy, I.D. Blast quantification using hopkinson pressure bars. *J. Vis. Exp.* **2016**. *113, e53412*. [CrossRef]
32. Haynes, W.M.; Lide, D.R.; Bruno, T.J. Table: Speed of Sound in Solids at Room Temperature. In *CRC Handbook of Chemistry and Physics*; Internet Version; CRC Press: Boca Raton, FL, USA, 2005.
33. Pearson, R.; Neuvo, Y.; Astola, J.; Gabbouj, M. Generalized Hampel Filters. *EURASIP J. Adv. Signal Process.* **2016**. [CrossRef]
34. Savitzky, A.; Golay, M.J.E. Smoothing and Differentiation of Data by Simplified Least Squares Procedures. *Anal. Chem.* **1964**, *36*, 1627–1639.
35. Pannell, J.J.; Panoutsos, G.; Cooke, S.B.; Pope, D.J.; Rigby, S.E. Predicting specific impulse distributions for spherical explosives in the extreme near-field using a Gaussian function. *Int. J. Prot. Struct.* **2021**, *12*, 437–459. [CrossRef]
36. Wang, Z.; Li, P. Characterisation of dynamic behaviour of alumina ceramics: Evaluation of stress uniformity. *AIP Adv.* **2015**, *5*, 107224. [CrossRef]
37. Tyas, A.; Watson, A.J. An investigation of frequency domain dispersion correction of pressure bar signals. *Int. J. Impact Eng.* **2001**, *25*, 87–101. .: 10.1016/S0734-743X(00)00025-7. [CrossRef]
38. Farrimond, D.G.; Rigby, S.E.; Clarke, S.D.; Tyas, A. Time of arrival as a diagnostic for far-field high explosive blast waves. *Int. J. Prot. Struct.* **2022**, *14*, 379–402. [CrossRef]
39. Park, S.; Uth, T.; Fleck, N.; Wadley, H.; Deshpande, V. Sand column impact with a rigid target. *Int. J. Impact Eng.* **2013**, *62*, 229–242. [CrossRef]
40. Liu, T.; Wadley, H.; Deshpande, V. Dynamic compression of foam supported plates impacted by high velocity soil. *Int. J. Impact Eng.* **2014**, *63*, 88–105. . [CrossRef]
41. Ehrgott, J.Q., Jr. *Tactical Wheeled Vehicle Survivability: Results of Experiments to Quantify Aboveground Impulse*; Technical Report; Geotechnical and Structures Lab, Engineer Research and Development Center: Vicksburg, MS, USA, 2010.

Disclaimer/Publisher's Note: The statements, opinions and data contained in all publications are solely those of the individual author(s) and contributor(s) and not of MDPI and/or the editor(s). MDPI and/or the editor(s) disclaim responsibility for any injury to people or property resulting from any ideas, methods, instructions or products referred to in the content.

eng

MDPI

Article

Drone Detection Using YOLOv5

Burchan Aydin [1,*] and Subroto Singha [2]

[1] Department of Engineering and Technology, Texas A&M University-Commerce, Commerce, TX 75428, USA
[2] Department of Computer Science and Information Systems, Texas A&M University-Commerce, Commerce, TX 75428, USA
* Correspondence: burchan.aydin@tamuc.edu; Tel.: +1-903-886-5174

Abstract: The rapidly increasing number of drones in the national airspace, including those for recreational and commercial applications, has raised concerns regarding misuse. Autonomous drone detection systems offer a probable solution to overcoming the issue of potential drone misuse, such as drug smuggling, violating people's privacy, etc. Detecting drones can be difficult, due to similar objects in the sky, such as airplanes and birds. In addition, automated drone detection systems need to be trained with ample amounts of data to provide high accuracy. Real-time detection is also necessary, but this requires highly configured devices such as a graphical processing unit (GPU). The present study sought to overcome these challenges by proposing a one-shot detector called You Only Look Once version 5 (YOLOv5), which can train the proposed model using pre-trained weights and data augmentation. The trained model was evaluated using mean average precision (mAP) and recall measures. The model achieved a 90.40% mAP, a 21.57% improvement over our previous model that used You Only Look Once version 4 (YOLOv4) and was tested on the same dataset.

Keywords: YOLOv5; autonomous drone detection; image recognition; machine learning; mAP; unmanned aerial vehicle (UAV)

Citation: Aydin, B.; Singha, S. Drone Detection Using YOLOv5. *Eng* **2023**, 4, 416–433. https://doi.org/10.3390/eng4010025

Academic Editor: Antonio Gil Bravo

Received: 30 November 2022
Revised: 24 January 2023
Accepted: 28 January 2023
Published: 1 February 2023

Copyright: © 2023 by the authors. Licensee MDPI, Basel, Switzerland. This article is an open access article distributed under the terms and conditions of the Creative Commons Attribution (CC BY) license (https://creativecommons.org/licenses/by/4.0/).

1. Introduction

Drones are becoming increasingly popular. Most are inexpensive, flexible, and lightweight [1]. They are utilized in a variety of industries, including the military, construction, agriculture, real estate, manufacturing, photogrammetry, sports, and photography [2,3]. There were 865,505 drones registered as of 3 October 2022, with 538,172 of them being recreational [4]. Drones can take off and land autonomously, intelligently adapt to any environment, fly to great heights, and provide quick hovering ability and flexibility [5]. Increased usage of drones, on the other hand, poses a threat to public safety; for example, their capacity to carry explosives may be used to strike public locations, such as governmental and historical monuments [6]. Drones can also be used by drug smugglers and terrorists. Moreover, the increasing number of hobbyist drone pilots could result in interference with activities, such as firefighting, disaster response efforts, and so on [7]. A list of threats that drones currently pose and a discussion of how drones are being weaponized are offered in [8]. For instance, in April 2021, two police officers in Aguililla, Michoacan, Mexico were assaulted by drones *artillados* carrying explosive devices, resulting in multiple injuries [9]. Thirteen tiny drones attacked Russian soldiers in Syria, causing substantial damage [10]. Considering the possibility of drones being used as lethal weapons [11], authorities shut down the London Gatwick airport for 18 hours due to serious drone intrusion, causing 760 flights with over 120,000 people to be delayed [12].

Detecting drones may be difficult due to the presence of similar objects in the sky, such as aircrafts, birds, and so forth. The authors of [13] used a dataset made up of drones and birds. To create the dataset, they gathered drone and bird videos and extracted images using the MATLAB image processing tool. After gathering 712 photos to train the algorithms, they utilized an 80:20 train:test split to randomly choose the training and testing images.

They examined the accuracies of three different object detectors utilizing an Intel Core i5–4200M (2.5GHZ0), 2GB DDR3 L Memory, and 1TB HDD, reaching 93%, 88%, and 80% accuracy using the CNN, SVM, and KNN, respectively. The suggested technique examined included drone-like objects, i.e., birds in the dataset; however, it required 14 minutes and 28 seconds to attain 93% accuracy for just 80 epochs using the CNN methodology. As a result, their proposed approach was not feasible for real-time implementation.

Our previously proposed technique using fine-tuned YOLOv4 [14] overcame the speed, accuracy, and model overfitting issues. In that study, we collected 2395 images of birds and drones from public sources, such as Google, Kaggle, and others. We labeled the images and divided them into two categories: drones and birds. The YOLOv4 model was then trained on the Tesla K80 GPU using the Google deep learning VM. To test the detecting speed, we recorded two drone videos of our own drones at three different heights. The trained model obtained an FPS of 20.5 and 19.0. The mAP was 74.36%. In terms of speed and accuracy, YOLOv5 surpassed prior versions of YOLO [1]. In this study, we compared the performance increase using fine-tuned YOLOv5 for the same dataset used in [14] for drone detection using fine-tuned YOLOv4. YOLOv5 recently demonstrated improved performance in identifying drones. The authors of [1] presented a method for detecting drones flying in prohibited or restricted zones. Their deep learning-based technique outperformed earlier deep learning-based methodologies in terms of precision and recall.

Our key contributions to this study were the addition of a data augmentation technique to artificially overcome data scarcity difficulties, as well as the prevention of overfitting issues utilizing a random train:test split of 70:30, the fine-tuning of the original YOLOv5 based on our collected customized dataset, the testing of the model on a wide variety of backgrounds (dark, sunny), and the testing of different views of images. The model was tested on our own videos using two drones-DJI Mavic Pro, DJI Phantom; videos were taken at three common altitudes—60 ft, 40 ft, and 20 ft.

Paper Organization

The rest of the research study is structured as follows. Section 2 provides background for our research. Section 3 addresses the research materials and methodologies. Section 4 covers the findings of this study. Section 5 discusses the model's complexity and uncertainty. Section 6 depicts the performance improvement and gives an argumentative discussion. Section 7 brings our paper to a conclusion.

2. Background

In the past, various techniques, such as radar, were used to detect drones [15]. However, it is very difficult for radar to do so, due to the low levels of electromagnetic signals that drones transmit [16]. Similarly, other techniques, such as acoustic and radio frequency-based drone detection, are costly and inaccurate [17]. Recently, machine learning-based drone detectors, such as SVM and artificial neural network classifiers, have been used to detect drones, achieving better success than radar and acoustic drone detection systems [18]. The YOLO algorithm has outperformed competitor algorithms, such as the R-CNN and SSD algorithms, due to its complex feature-learning capability with fast detection [18]. In fact, the YOLO algorithm is now instrumental in object detection tasks [19]. Many computer vision tasks use YOLO due to its faster detection with high accuracy, which makes the algorithm feasible for real-time implementation [20]. One of the latest developments, YOLOv5, has greatly improved the algorithm's performance, offering a 90% improvement over YOLOv4 [21]. In the present research, we used YOLOv5 to build an automated drone detection system and compared the results against our previous system with the YOLOv4.

UAV detection systems are designed using various techniques. We have reviewed only those studies closely related to our methodology. UAV detection can be treated as an object detection problem in deep learning. Deep learning-based object detection techniques can be divided into one-stage and two-stage detection algorithms [22]. An example of a two-stage object detection technique is R-CNN [23]; examples of one-stage object detection techniques are YOLO [24], SSD [25], etc. The authors of [26] explained the mechanism of how object detectors work in general. Two-stage detectors use candidate object techniques, while one-stage detectors employ the sliding window technique. Thus, one-stage detectors are fast and operate in real-time [27]. YOLO is easy to train, faster, more accurate than its competitors, and can immediately train an entire image. Thus, YOLO is the most frequently used and reliable object detection algorithm [28]. It first divides an image into SXS grids and assigns a class probability with bounding boxes around the object [28]. It then uses a single convolutional network to perform the entire prediction. Conversely, R-CNNs begin by generating a large number of region proposals using a selective search method. Then, from each region proposal, a CNN is utilized to extract features. Finally, the R-CNN classifies and defines bounding boxes for distinct classes [28].

The authors of [28] used YOLOv2 to detect drones and birds, and achieved precision and recall scores above 90. The authors of [27] proposed a drone detection pipeline with three different models: faster R-CNN with ResNet–101, faster R-CNN with Inceptionv2, and SSD. After 60,000 iterations, they achieved mAP values of 0.49, 0.35, and 0.15, respectively. One example of an SSD object detector is MobileNet. MobileNetV2 was used as a classifier in [29]; the authors proposed a drone detection model where the methodology consisted of a moving object detector and a drone-bird-background classifier. The researchers trained the drone-vs-bird challenge dataset on the NVIDIA GeForce GT 1030 2GB GPU with a learning rate of 0.05. At an IoU of 0.5, their highest precision, recall, and F1 scores were 0.786, 0.910, and 0.801, respectively, after testing on three videos. The authors of [30] used YOLOv3 to detect and classify drones. The authors of [30] collected different types of drone images from the internet and videos to build a dataset. Images were annotated in the YOLO format in order to train a YOLOv3 model. An NVIDIA GeForce GTX 1050 Ti GPU was used to train the dataset with chosen parameter values, such as a learning rate of 0.0001, batch size of 64, and 150 total epochs. The best mAP value was 0.74. PyTorch, an open-source machine learning programming language, was used to train and test the YOLOv3 model.

The authors of [31] used YOLOv4 to automatically detect drones in order to integrate a trained model into a CCTV camera, thus reducing the need for manual monitoring. The authors collected their dataset from public resources such as Google images, open-source websites, etc. The images were converted into the YOLO format using free and paid image annotation tools. They fine-tuned the YOLOv4 architecture by customizing filters, max batches, subdivisions, batches, etc. After training the YOLOv4 model for 1300 iterations, the researchers achieved a mAP of 0.99. Though their mAP value was very high, they trained only 53 images and did not address model overfitting, resulting in a greater improvement scope.

The authors of [1] presented an approach based on YOLOv5. They utilized a dataset of 1359 drone images obtained from Kaggle. They fine-tuned the model on a local system with an 8 GB NVDIA RTX2070 GPU, 16 GB of RAM, and a 1.9 GHz CPU. They employed a 60:20:20 split of the dataset for training, testing, and validation. They trained the model on top of COCO pre-trained weights and obtained a precision of 94.70%, a recall of 92.50%, and a mAP of 94.1%.

3. Materials and Methods

In this research, we employed a recent version of the YOLO algorithm: YOLOv5 [32]. YOLOv5 is a high-performing and fast object detection algorithm that detects objects in real-time. Drones can fly at fast speeds; thus the detection speed also needs to be high. YOLOv5 has the ability to meet this requirement. The algorithm was developed using PyTorch, an open-source deep learning framework that has made training and testing easier for customized datasets and offers outstanding detection performance. YOLOv5 consists of three parts: the backbone, neck, and head [1].

The backbone is made of a CSPNet. The CSPNet reduces the model's complexity, resulting in fewer hyperparameters and FLOPS. At the same time, it resolves vanishing and exploding gradient issues, due to the depth of the neural networks. These improvements enhance inference speed and accuracy in object detection. Inside the CSPNet, there are several convolutional layers, four CSP bottlenecks with three convolutions, and spatial pyramid pooling. The CSPNet is responsible for extracting features from an input image and using convolutions and pooling to form a feature map that combines all extracted features. Thus, the backbone plays the role of feature extractor in YOLOv5.

The middle part of YOLOv5, often called the neck, is also known as the PANet. The PANet takes all the extracted features from the backbone and saves and sends them to the deep layers in order to perform feature fusions. These feature fusions are passed to the head so that high-level features are known to the output layer for final object detection.

The head of YOLOv5 is responsible for object detection. It consists of 1x1 convolutions that predict the class of an object, with bounding boxes around the target object and a class probability score. Figure 1 shows the overall architecture of YOLOv5.

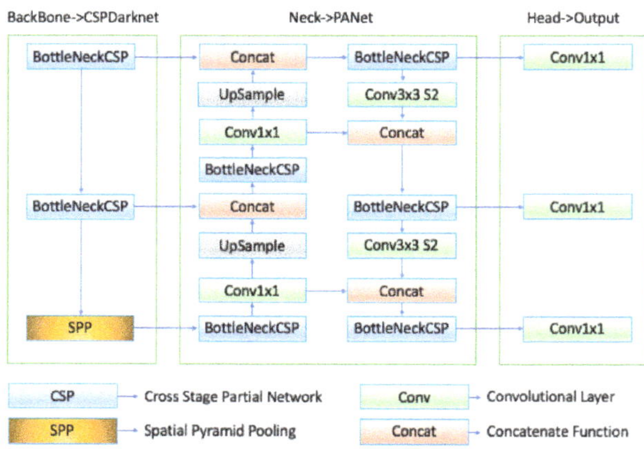

Figure 1. The YOLOv5 architecture.

The location of the bounding box is calculated using Equation (1):

$$\cup_x^y = \mathcal{P}_{x,y} * IOU_{predicted}^{ground\ truth} \quad (1)$$

In Equation (1), x and y are the y_{th} bounding box of the x_{th} grid. $\cup_x^y$ is the probability score for the y_{th} bounding box of the x_{th} grid. $\mathcal{P}_{x,y}$ equals 1 when there is a target and 0 when there is no target in the y_{th} bounding box. The IoU $IOU_{predicted}^{ground\ truth}$ is the IoU between the ground truth and the predicted class. Higher IoUs mean more accurately predicted bounding boxes.

The loss function of YOLOv5 is the combination of loss functions for the bounding box, classification, and confidence. Equation (2) represents the overall loss function of YOLOv5 [32]:

$$loss_{YOLOv5} = loss_{bounding\ box} + loss_{classification} + loss_{confidence} \quad (2)$$

$loss_{bounding\ box}$ is calculated using Equation (3):

$$loss_{bounding\ box} = \lambda_{\text{if}} \sum_{a=0}^{b^2} \sum_{c=0}^{d} E_{a,c}^{g} h_g (2 - K_a X n_a) \left[\left(x_a - x'^c_a\right)^2 + \left(y_a - y'^c_a\right)^2 + \left(w_a - w'^c_a\right)^2 + \left(h_a - h'^c_a\right)^2 \right] \quad (3)$$

In Equation (3), the width and height of the target object are denoted using $h\prime$ and $w\prime$. x_a and y_a indicate the coordinates of the target object in an image. Finally, the indicator function (λ_{if}) shows whether the bounding box contains the target object.

$loss_{classification}$ is calculated using Equation (4):

$$loss_{classification} = \lambda_{\text{classification}} \sum_{a=0}^{b^2} \sum_{c=0}^{d} E_{a,c}^{g} \sum_{C \in c_l} L_a(c) \log(LL_a(c)) \tag{4}$$

$loss_{confidence}$ is calculated using Equation (5):

$$loss_{confidence} = \lambda_{\text{confidence}} \sum_{a=0}^{b^2} \sum_{c=0}^{d} E_{a,c}^{confidence} (c_i - c_l)^2 + \lambda_g \sum_{a=0}^{b^2} \sum_{c=0}^{d} E_{a,c}^{g} (c_i - c_l)^2 \tag{5}$$

In Equations (4) and (5), $\lambda_{\text{confidence}}$ indicates the category loss coefficient, $\lambda_{\text{classification}}$ the classification loss coefficient, c_l the class, and c the confidence score.

Construction of the Experiment and Data Acquisition

We collected drone and bird images from public resources such as Google, Kaggle, Flicker, Instagram, etc. The drone images came from different altitudes, angles, backgrounds, and views, ensuring variability in the dataset. The bird images consisted of 300 different species. The entire dataset was formed using 479 bird images and 1916 drone images; altogether, the dataset consisted of 2395 images. We used a 70:30 train:test split to train and test the YOLOv5 model. The training dataset had 1677 images and the testing dataset had 718 images. We used data augmentation techniques to overcome data scarcity. In fact, using 3 variants of data augmentation, we generated a total of 5749 images. Using a freely available labeling tool, we annotated the images and divided them into two classes. Drone images were annotated as "first class" and bird images as "zero class." YOLO implementation requires that all images be saved in the. txt format, which has four coordinates for the object, including the class of 0 or 1.

We collected two videos of drones flying, using our own two drones: a DJI Mavic Pro and DJI Phantom III. We captured video shots at three different altitudes: 60 feet, 40 feet, and 20 feet. These are altitudes commonly used by drone pilots, especially drone hobbyists. At 60 feet, the drone looked almost like a bird. We captured the videos to evaluate the performance of the YOLOv5 model in terms of accuracy and speed, mainly at high altitudes.

We conducted the experiment using Google CoLab, a free cloud notebook in which we wrote the code, to implement YOLOv5. We fine-tuned the original YOLOv5 to train and test the model using our customized dataset. To accelerate and improve detection accuracy, we used a transfer learning technique. We employed the weights that were already available with the original YOLOv5 to implement transfer learning. We trained our customized model on top of the YOLOv5s.pt weight that was saved while training YOLOv5 on the COCO dataset. The original YOLOv5 was implemented using PyTorch. We also chose PyTorch. At the time we trained our model, Google CoLab allocated a Tesla T4 with a 15110MiB memory NVIDIA GPU. To fine-tune YOLOv5, we chose the values of the various hyperparameters suggested in the original. We used an lr of 0.01, momentum of 0.937, and decay of 0.0005. The model was optimized using stochastic gradient descent. The albumentations were Blur (p = 0.01, blur_limit = (3,7)), MedianBlur (p = 0.01, blur_limit = (3,7)), ToGray (p = 0.01), and CLAHE (p = 0.01, clip_limit = (1,4.0) title_grid_size = (8,8)).

We changed the number of classes from 80 to 2, since we had 2 classes: drone and bird. The model had 214 layers with 7,025,023 parameters, 7,025,023 gradients, and 16.0 GFLOPs. We used the Roboflow API to load and perform the data augmentation and preprocessing. We used the same dataset employed in our previous experiment. We performed auto-orient and modified classes as data preprocessing techniques. Auto-orient is an image processing technique that ensures that images match the source device orientation. Sometimes, the coordinates in various cameras may confuse (x,y) and (y,x). Auto-orient prevents bad data from being fed into YOLOv5. In addition to data preprocessing, we performed data augmentation using Roboflow API, which helped reduce the data shortage issue. We set the parameters as follows: flip: horizontal, hue: between −25 degrees and +25 degrees, cutout:

3 boxes with 10% size each, and mosaic: 1. Data augmentation ensured data variability, artificially generating 5749 total images, and randomly splitting the entire dataset into a 70:30 train:test split. Figure 2 shows the augmented dataset. We trained the model for 4000 iterations and saved the best weight to test the model using the testing images and videos. During training, we used %tensorboard to log the runs that autogenerated the learning curves in order to evaluate the model's performance beyond the evaluation metrics. Figure 3 shows a flowchart of the overall conducted experiment.

Figure 2. Augmented dataset.

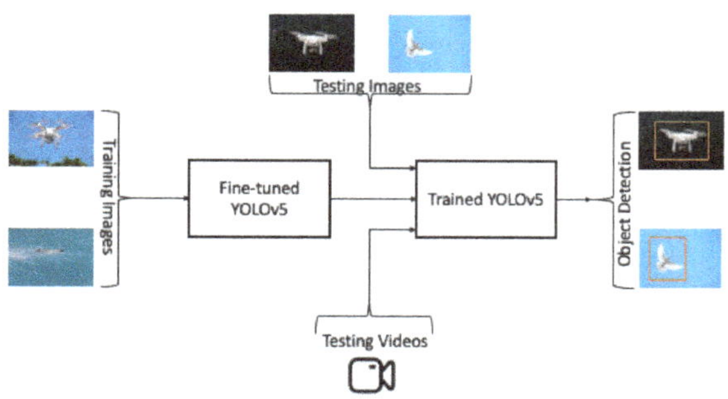

Figure 3. Overall conducted experiment flowchart.

4. Results

We evaluated the trained model using the mAP, precision, recall, and F1-scores. We used FPS as the evaluation metric to evaluate the speed of detection in the videos. Table 1 shows the mAP, precision, recall, and F1-scores. The model was evaluated on a testing dataset from a random train:test split. The testing images had data variabilities in terms of different backgrounds (e.g., bright, dark, blur, etc.) and weather conditions (e.g., cloudy, sunny, foggy, etc.), as well as images with multiple classes. To track the evaluation metrics, we plotted the values across iterations. Figure 4 shows the overall training summary of the model. The loss curves indicate a downward trend, meaning that during training, the losses were minimized both for training and validation. The metrics curves show upward trends, meaning the performance of the model improved over the iterations during training. We plotted the precision-recall curve to evaluate the model's prediction preciseness (see Figure 5). The curve tended towards the right top corner, meaning that the values were mostly close to one (i.e., the rate of misclassification was very low when using this model).

Table 1. Overall and individual evaluation metrics results.

Class	Precision	Recall	mAP50
All	0.918	0.875	0.904
Bird	0.860	0.766	0.820
Drone	0.975	0.985	0.987

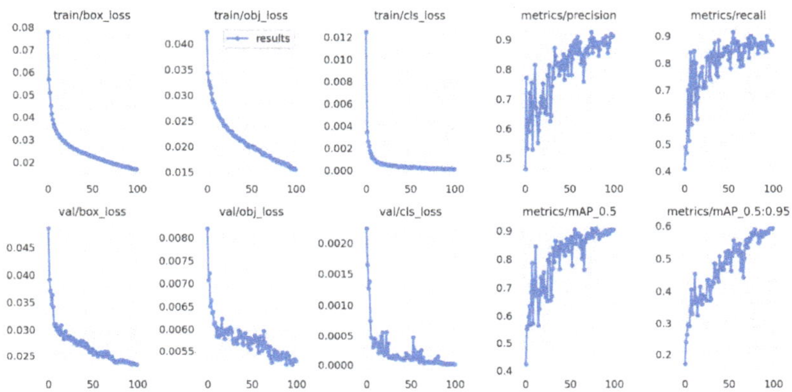

Figure 4. Overall summary of training.

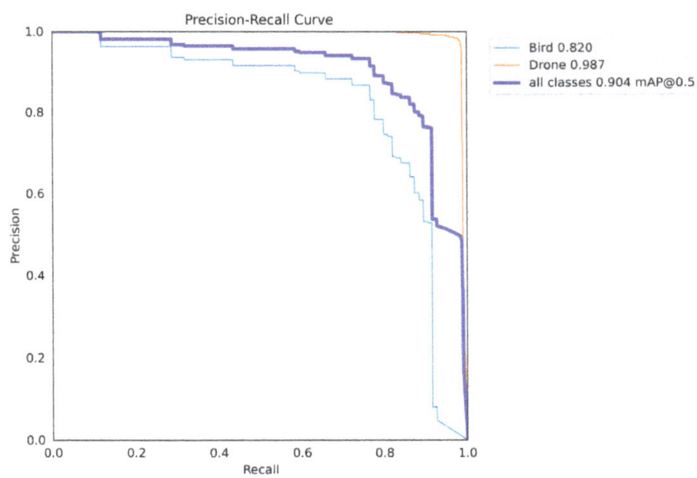

Figure 5. Precision-recall curve.

Finally, we show the model's evaluation metrics in Table 1, which offers an overall summary of the results regarding the trained model's performance when using the testing images. We achieved precision, recall, and mAP50 values of 0.918, 0.873, and 0.904 for all images, respectively. In addition, we calculated individual precision, recall, and mAP50 values for Classes 1 and 2 (see Table 1). Figure 6 shows the drones predicted by the model using randomly chosen testing images. Figures 7–12 show the predictions with bounding boxes and class scores at three different altitudes (20 ft, 40 ft, and 60 ft) in videos using two different types of drones (the DJI Mavic Pro and DJI Phantom III, Da-Jiang Innovations, Shenzhen, China).

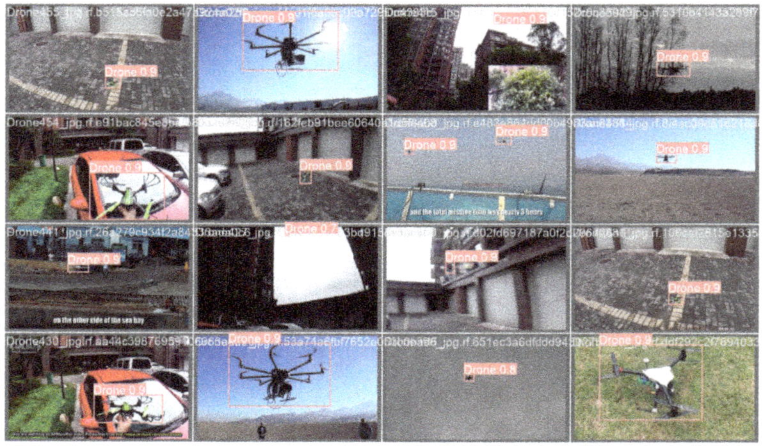

Figure 6. Drone predictions for test images.

Figure 7. At 20ft, DJI Mavic Pro.

Figure 8. At 40ft, DJI Mavic Pro.

Figure 9. At 60ft, DJI Mavic Pro.

Figure 10. At 20ft, DJI Phantom III.

Figure 11. At 40ft, DJI Phantom III.

Figure 12. At 60 ft, DJI Phantom III.

Appendix A contains more predictions based on the trained YOLOv5 model. While the model worked well on the majority of the test images, there were a few instances of misclassification. Figures A7 and A8 show two misclassifications in which the model misidentified certain drone-like objects as drones alongside correct predictions in these images. Blurred photos might be one of the causes of such misclassification. We can address this problem by employing more training photos, which is outside the scope of this study. The prediction confidence scores were poor, hovering around 10%. We might perhaps establish a confidence score threshold to avoid such misclassification while increasing the number of training images. There were just a few "bird" classes. Figure A3 depicts an example of correct "drone" and "bird" predictions. However, because of the uncertainty of both classes in a single video frame, we were unable to do any drone and bird detection in videos.

5. Model Complexity and Parameter Uncertainty

To do a quicker prediction, we employed YOLOv5, which mainly relies on GPU implementation. GPU implementation complicates CPU deployments. Data augmentation techniques such as rotation and flipping were used to artificially supplement the dataset for improved training and performance. The parameter uncertainty in our experiment included sampling errors, overfitting, and so forth. Too many classes from one class may create sampling error, whereas training a smaller number of images with higher parameters may result in overfitting. We used pre-trained model weights that were trained on the COCO dataset, and we trained our fine-tuned model on top of the pre-trained weights.

6. Discussion

Using deep learning for the detection of drones has become a common topic in the research community, due to the substantial importance of restricting drones in unauthorized regions; however, improvement is still needed. The authors of [30] proposed a drone detection methodology using deep learning, employing YOLOv3 to detect and classify drones. More than 10,000 different categories of drones were used to train the algorithm, and a mAP of 0.74 was achieved at the 150th epoch. Though they used a YOLO-based approach, their study did not consider testing the model using videos, different weather conditions, and backgrounds; most importantly, they did not test their model using images of objects like drones. The authors of [33] used deep learning-based techniques and Faster R-CNN on a dataset created from videos collected by the researchers. The following image augmentation techniques were employed: geometric transformation, illumination variation, and image quality. The researchers did not calculate the mAP values and instead plotted a precision-recall (AUC) curve to evaluate the performance. Using a synthetic dataset, their model achieved an overall AUC score of 0.93; for a real-world dataset, their model achieved an overall AUC score of 0.58. The dataset was trimmed from video sequences, and thus had no objects much of the time. In our previous research, we analyzed the performance of our proposed methodology using YOLOv4 and showed that the proposed methodology outperformed existing methodologies in terms of mAP, precision, recall, and F-1 scores. Using YOLOv4, we were able to achieve a mAP of 0.7436, precision of 0.95, and recall of 0.68. Most importantly, we included another evaluation metric, FPS, to evaluate the performance, achieving an average FPS of 20.5 for the DJI Phantom III videos and 19.0 FPS for the DJI Mavic Pro videos, all at three different high altitudes (i.e., 20 ft, 40 ft, and 60 ft). We tested the model using a highly variable dataset with different backgrounds (e.g., sunny, cloudy, dark, etc.), various drone angles (e.g., side view, top view, etc.), long-range drone images, and multiple objects in a single image. Our previous methodology achieved such an improvement due to the real-time detection capability of YOLOv4 acting as a single-stage detection process, and the various new features of YOLOv4 (e.g., CSP, CmBN, mish activation, etc.), which sped up detection. Furthermore, the default MOSAIC = 1 flag automatically performed the data augmentation. In this research, we employed Google CoLab and Google Deep Learning VM for parts of the training and testing. In addition to YOLOv4, YOLOv5 showed performance improvement, as shown in [1]. They obtained a precision of 0.9470, a recall of 0.9250, and a mAP of 0.9410. Although their evaluation metrics were higher than ours, our dataset was bigger. Furthermore, we had binary classes, whereas they just had a "drone" class. They did not employ data augmentation, whereas we used a data augmentation technique to build a collection of over 5700 images. As a result of the variability in the dataset and the addition of new classes with data augmentation, our suggested technique is resilient and scalable in real-world scenarios.

Our results for the present research outperformed our previous methodology, achieving a mAP of 0.904. Because of the lightweight design, YOLOv5 recognized objects faster than YOLOv4. YOLOv4 was created using darknet architecture; however, YOLOv5 is built with a PyTorch framework rather than a darknet framework. This is one of the reasons we obtained more accuracy and speed than earlier methodologies. In addition to the architecture itself, we fine-tuned the last layers of the original YOLOv5 architecture so that it performed better on our customized dataset. Other than the layer tuning, we customized the default values of the learning rate, momentum, batch size, etc. We trained the model for 100 iterations since we trained the custom dataset on top of the transferred weights for the COCO dataset. In addition to mAP, we achieved a precision of 0.918, recall of 0.875, and F-1 score of 0.896. In terms of F-1 score and recall, we also outperformed the previous model. We further tested the new model on two videos, using a Tesla T4 GPU. For the DJI Mavic Pro, we achieved a maximum FPS of 23.9 ms, and for the DJI Phantom III, a maximum FPS of 31.2 ms. Thus, in terms of inference speed, we also outperformed the previous model's performance. We achieved this improvement due to the new feature additions included in YOLOv5, such as the CSPDarknet53 backbone, which resolved the gradient issue using fewer parameters, and thus was more lightweight. Other helpful feature additions included the fine-tuning of YOLOv5 for our custom dataset, data augmentation performed to artificially increase the number of images, and data preprocessing to make training the model smoother and faster. The evaluation metric, F1 score, is a weighted sum of precision and recall. Precision is the accuracy of positive class prediction, whereas recall is the proportion of true positive classes. The greater the F1 score, the better the model in general. The correctness of bounding boxes in objects is measured by mAP, and the greater the value, the better. The speed of object detection is measured in frames per second (FPS). Table 2 compares the performance of the previous and proposed models in terms of four evaluation metrics: precision, recall, F1 score, mAP50, and FPS.

Table 2. Comparison between previous and proposed models' performance.

Models	Precision	Recall	F1 Score	mAP50		FPS
Previous YOLOv4 [30]	0.950	0.680	0.790	0.7436	DJI Mavic Pro DJI Phantom III	19.0 ms 20.5 ms
Proposed YOLOv5	0.918	**0.875**	**0.896**	**0.9040**	DJI Mavic Pro DJI Phantom III	**23.9 ms** **31.2 ms**

7. Conclusions

In this research, we compared the performance of one of the latest versions of YOLO, YOLOv5, to our previously proposed drone detection methodology that used YOLOv4. To make a fair comparison, we employed the same dataset and the same computing configurations (e.g., GPU). We first fine-tuned the original YOLOv5, as per our customized dataset that had two classes: bird and drone. We further tuned the values of the hyperparameters (e.g., learning rate, momentum, and decay) to improve the detection accuracy. In order to speed up the training, we used transfer learning, implementing the pre-trained weights provided with the original YOLOv5. The weights were trained on a popular and commonly used dataset called MS COCO. To address data scarcity and overfitting issues, we used data augmentation via Roboflow API and included data preprocessing techniques to smoothly train the model. To evaluate the model's performance, we calculated the evaluation metrics on a testing dataset. We used precision, recall, F-1 score, and mAP, achieving 0.918, 0.875, 0.896, and 0.904 values, respectively. We outperformed the previous model's performance by achieving higher recall, F-1 score, and mAP values (a 21.57% improvement in mAP). Furthermore, we tested the speed of detection on videos of two different drone models, the DJI Phantom III and the DJI Mavic Pro. We achieved maximum FPS values of 23.9 and 31.2, respectively, using an NVIDIA Tesla T4 GPU. The videos were taken at three altitudes—20 ft, 40 ft, and 60ft—to test the capability of the detector for objects at high altitudes. In future work, we will use different versions of YOLO and larger datasets. In addition, other algorithms for object detection will be included to compare the performance. Various drone-like objects such as airplanes will be added as classes alongside birds to improve the model's ability to distinguish among similar objects.

Author Contributions: Conceptualization, S.S. and B.A.; methodology, B.A. and S.S.; software, S.S.; validation, B.A. and S.S.; formal analysis, S.S and B.A.; investigation, B.A. and S.S.; resources, S.S.; data curation, S.S.; writing—original draft preparation, S.S.; writing—review and editing, B.A.; visualization, S.S.; supervision, B.A.; project administration, S.S. and B.A.; funding acquisition, B.A. All authors have read and agreed to the published version of the manuscript.

Funding: This research received no external funding.

Institutional Review Board Statement: Not applicable.

Informed Consent Statement: Not applicable.

Data Availability Statement: The datasets used or analyzed during the current study are available from the corresponding author upon reasonable request.

Conflicts of Interest: We declare that there is no conflict of interest.

Abbreviations

AUC	Area under the ROC Curve
CCTV	Closed-Circuit Television
CLAHE	Contrast Limited Adaptive Histogram Equalization
CNN	Convolutional Neural Network
COCO	Common Objects in Context
CPU	Central Processing Unit
CSPNet	Cross-Stage Partial Network

DJI	Da-Jiang Innovations
FLOPS	Floating-Point Operations per Second
FPS	Frames per Second
GFLOP	Giga Floating Point Operations per Second
Google CoLab	Google Colaboratory
GPU	Graphical Processing Unit
IoU	Intersection over Union
KNN	k-nearest neighbor
lr	Learning Rate
mAP	Mean Average Precision
MATLAB	Matrix Laboratory
PANet	Path Aggregation Network
R-CNN	Region-Based Convolutional Neural Network
ResNet	Residual Network
SSD	Single-Shot Multi-box Detector
SVM	Support Vector Machine
UAV	Unmanned Aerial Vehicle
VM	Virtual Machine
YOLO	You Only Look Once
YOLOv2	You Only Look Once version 2
YOLOv3	You Only Look Once version 3
YOLOv4	You Only Look Once version 4
YOLOv5	You Only Look Once version 5

Appendix A

In a variety of photos, our classifier effectively identified drone and bird objects. We evaluated images with intricate backgrounds and various climatic conditions. Here we have the detection results, where the images are displayed together with their corresponding class names and class probabilities. YOLOv5 generated the predictions in batches. Thus, predictions are shown all in one figure. Additionally, we tested images that have "drone" and "bird" in one image. In augmented training images, 0 refers to "bird" and 1 refers to "drone".

Figure A1. First batch prediction by YOLOv5.

Figure A2. Second batch prediction by YOLOv5.

Figure A3. Bird and drone in images predicted by YOLOv5.

Figure A4. First batch of augmented training image predicted by YOLOv5.

Figure A5. Second batch of augmented training image predicted by YOLOv5.

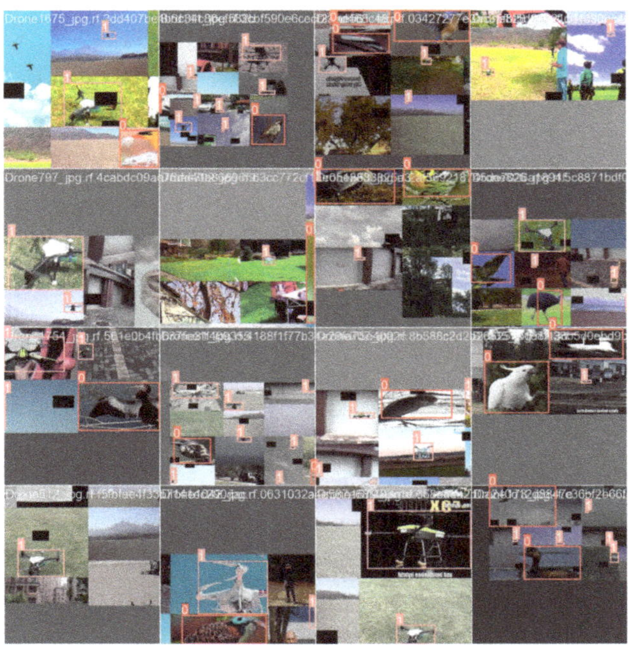

Figure A6. Third batch of augmented training image predicted by YOLOv5.

Figure A7. Instance1 of misclassified image predicted by YOLOv5.

Figure A8. Instance2 of misclassified image predicted by YOLOv5.

References

1. Al-Qubaydhi, N.; Alenezi, A.; Alanazi, T.; Senyor, A.; Alanezi, N.; Alotaibi, B.; Alotaibi, M.; Razaque, A.; Abdelhamid, A.A.; Alotaibi, A. Detection of Unauthorized Unmanned Aerial Vehicles Using YOLOv5 and Transfer Learning. *Electronics* **2022**, *11*, 2669. [CrossRef]
2. Jung, H.-K.; Choi, G.-S. Improved YOLOv5: Efficient Object Detection Using Drone Images under Various Conditions. *Appl. Sci.* **2022**, *12*, 7255. [CrossRef]
3. Aydin, B. Public acceptance of drones: Knowledge, attitudes, and practice. *Technol. Soc.* **2019**, *59*, 101180. [CrossRef]
4. Drones by the Numbers. Available online: https://www.faa.gov/uas/resources/by_the_numbers/ (accessed on 3 October 2022).
5. Liu, B.; Luo, H. An Improved Yolov5 for Multi-Rotor UAV Detection. *Electronics* **2022**, *11*, 2330. [CrossRef]
6. Long, T.; Ozger, M.; Cetinkaya, O.; Akan, O.B. Energy Neutral Internet of Drones. *IEEE Commun. Mag.* **2018**, *56*, 22–28. [CrossRef]
7. Singh, C.; Mishra, R.; Gupta, H.P.; Kumari, P. The Internet of Drones in Precision Agriculture: Challenges, Solutions, and Research Opportunities. *IEEE Internet Things Mag.* **2022**, *5*, 180–184. [CrossRef]
8. Ilijevski, I.; Dimovski, Z.; Babanoski, K. The Weaponisation of Drones—A Threat from Above Used for Terrorist Purposes. *J. Crim. Justice Secur.* **2021**, *3*, 336–349.
9. Mexican Cartel Tactical Note #49: Alleged CJNG Drone Attack in Aguililla, Michoacán Injures Two Police Officers | Small Wars Journal. 2021. Available online: https://smallwarsjournal.com/jrnl/art/mexican-cartel-tactical-note-49-alleged-cjng-drone-attack-aguililla-michoacan-injures-two (accessed on 27 January 2023).
10. Hambling, D. Swarm of drones attacks airbase. *New Sci.* **2018**, *237*, 12. [CrossRef]

11. Laksham, K.B. Unmanned aerial vehicle (drones) in public health: A SWOT analysis. *J. Fam. Med. Prim. Care* **2019**, *8*, 342–346. [CrossRef] [PubMed]
12. Xun, D.T.W.; Lim, Y.L.; Srigrarom, S. Drone detection using YOLOv3 with transfer learning on NVIDIA Jetson TX2. In Proceedings of the 2021 Second International Symposium on Instrumentation, Control, Artificial Intelligence, and Robotics (ICA-SYMP), Bangkok, Thailand, 20–22 January 2021; pp. 1–6.
13. Mahdavi, F.; Rajabi, R. Drone Detection Using Convolutional Neural Networks. In Proceedings of the 2020 6th Iranian Conference on Signal Processing and Intelligent Systems (ICSPIS), Mashhad, Iran, 23–24 December 2020; pp. 1–5. [CrossRef]
14. Singha, S.; Aydin, B. Automated Drone Detection Using YOLOv4. *Drones* **2021**, *5*, 95. [CrossRef]
15. Jian, M.; Lu, Z.; Chen, V.C. Drone detection and tracking based on phase-interferometric Doppler radar. In Proceedings of the 2018 IEEE Radar Conference (RadarConf18), Oklahoma City, OK, USA, 23–27 April 2018; pp. 1146–1149. [CrossRef]
16. Elsayed, M.; Reda, M.; Mashaly, A.S.; Amein, A.S. Review on Real-Time Drone Detection Based on Visual Band Electro-Optical (EO) Sensor. In Proceedings of the 2021 Tenth International Conference on Intelligent Computing and Information Systems (ICICIS), Cairo, Egypt, 5–7 December 2021; pp. 57–65. [CrossRef]
17. Shi, Q.; Li, J. Objects Detection of UAV for Anti-UAV Based on YOLOv4. In Proceedings of the 2020 IEEE 2nd International Conference on Civil Aviation Safety and Information Technology (ICCASIT), Weihai, China, 14–16 October 2020; pp. 1048–1052. [CrossRef]
18. Taha, B.; Shoufan, A. Machine Learning-Based Drone Detection and Classification: State-of-the-Art in Research. *IEEE Access* **2019**, *7*, 138669–138682. [CrossRef]
19. Zhu, X.; Lyu, S.; Wang, X.; Zhao, Q. TPH-YOLOv5: Improved YOLOv5 Based on Transformer Prediction Head for Object Detection on Drone-Captured Scenarios. In Proceedings of the IEEE/CVF International Conference on Computer Vision (ICCV) Workshops, Montreal, BC, Canada, 11–17 October 2021; pp. 2778–2788. Available online: https://openaccess.thecvf.com/content/ICCV2021W/VisDrone/html/Zhu_TPH-YOLOv5_Improved_YOLOv5_Based_on_Transformer_Prediction_Head_for_Object_ICCVW_2021_paper.html (accessed on 30 November 2022).
20. Quoc, H.N.; Hoang, V.T. Real-Time Human Ear Detection Based on the Joint of Yolo and RetinaFace. *Complexity* **2021**, *2021*, e7918165. [CrossRef]
21. Karthi, M.; Muthulakshmi, V.; Priscilla, R.; Praveen, P.; Vanisri, K. Evolution of YOLO-V5 Algorithm for Object Detection: Automated Detection of Library Books and Performace validation of Dataset. In Proceedings of the 2021 International Conference on Innovative Computing, Intelligent Communication and Smart Electrical Systems (ICSES), Chennai, India, 24–25 September 2021; pp. 1–6. [CrossRef]
22. Liu, L.; Ke, C.; Lin, H.; Xu, H. Research on Pedestrian Detection Algorithm Based on MobileNet-YoLo. *Comput. Intell. Neurosci.* **2022**, *2022*, e8924027. [CrossRef] [PubMed]
23. Girshick, R.; Donahue, J.; Darrell, T.; Malik, J. Rich Feature Hierarchies for Accurate Object Detection and Semantic Segmentation. In Proceedings of the IEEE Conference on Computer Vision and Pattern Recognition, Columbus, OH, USA, 23–28 June 2014; pp. 580–587. Available online: https://openaccess.thecvf.com/content_cvpr_2014/html/Girshick_Rich_Feature_Hierarchies_2014_CVPR_paper.html (accessed on 30 November 2022).
24. Redmon, J.; Farhadi, A. YOLO9000: Better, Faster, Stronger. In Proceedings of the 2017 IEEE Conference on Computer Vision and Pattern Recognition (CVPR), Honolulu, HI, USA, 21–26 July 2017; pp. 7263–7271. Available online: https://openaccess.thecvf.com/content_cvpr_2017/html/Redmon_YOLO9000_Better_Faster_CVPR_2017_paper.html (accessed on 30 November 2022).
25. Liu, W.; Anguelov, D.; Erhan, D.; Szegedy, C.; Reed, S.; Fu, C.-Y.; Berg, A.C. SSD: Single Shot MultiBox Detector. In Proceedings of the Computer Vision—ECCV 2016, Amsterdam, The Netherlands, 11–14 October 2016; Leibe, B., Matas, J., Sebe, N., Welling, M., Eds.; Springer International Publishing: Cham, Switzerland, 2016; pp. 21–37.
26. Saqib, M.; Khan, S.D.; Sharma, N.; Blumenstein, M. A study on detecting drones using deep convolutional neural networks. In Proceedings of the 2017 14th IEEE International Conference on Advanced Video and Signal Based Surveillance (AVSS), Lecce, Italy, 29 August–1 September 2017; pp. 1–5. [CrossRef]
27. Nalamati, M.; Kapoor, A.; Saqib, M.; Sharma, N.; Blumenstein, M. Drone Detection in Long-Range Surveillance Videos. In Proceedings of the 2019 16th IEEE International Conference on Advanced Video and Signal Based Surveillance (AVSS), Taipei, Taiwan, 18–21 September 2019; pp. 1–6. [CrossRef]
28. Aker, C.; Kalkan, S. Using deep networks for drone detection. In Proceedings of the 2017 14th IEEE International Conference on Advanced Video and Signal Based Surveillance (AVSS), Lecce, Italy, 29 August–1 September 2017; pp. 1–6. [CrossRef]
29. Seidaliyeva, U.; Akhmetov, D.; Ilipbayeva, L.; Matson, E.T. Real-Time and Accurate Drone Detection in a Video with a Static Background. *Sensors* **2020**, *20*, 3856. [CrossRef] [PubMed]
30. Behera, D.K.; Raj, A.B. Drone Detection and Classification using Deep Learning. In Proceedings of the 2020 4th International Conference on Intelligent Computing and Control Systems (ICICCS), Madurai, India, 13–15 May 2020; pp. 1012–1016. [CrossRef]
31. Mishra, A.; Panda, S. Drone Detection using YOLOV4 on Images and Videos. In Proceedings of the 2022 IEEE 7th International conference for Convergence in Technology (I2CT), Mumbai, India, 7–9 April 2022; pp. 1–4. [CrossRef]

32. Xu, Q.; Zhu, Z.; Ge, H.; Zhang, Z.; Zang, X. Effective Face Detector Based on YOLOv5 and Superresolution Reconstruction. *Comput. Math. Methods Med.* **2021**, *2021*, e7748350. [CrossRef] [PubMed]
33. Chen, Y.; Aggarwal, P.; Choi, J.; Kuo, C.-C.J. A deep learning approach to drone monitoring. In Proceedings of the 2017 Asia-Pacific Signal and Information Processing Association Annual Summit and Conference (APSIPA ASC), Kuala Lumpur, Malaysia, 12–15 December 2017; pp. 686–691. [CrossRef]

Disclaimer/Publisher's Note: The statements, opinions and data contained in all publications are solely those of the individual author(s) and contributor(s) and not of MDPI and/or the editor(s). MDPI and/or the editor(s) disclaim responsibility for any injury to people or property resulting from any ideas, methods, instructions or products referred to in the content.

eng

MDPI

Article

Image-Based Vehicle Classification by Synergizing Features from Supervised and Self-Supervised Learning Paradigms

Shihan Ma and Jidong J. Yang *

School of Environmental, Civil, Agricultural & Mechanical Engineering, University of Georgia, Athens, GA 30602, USA
* Correspondence: jidong.yang@uga.edu

Abstract: This paper introduces a novel approach to leveraging features learned from both supervised and self-supervised paradigms, to improve image classification tasks, specifically for vehicle classification. Two state-of-the-art self-supervised learning methods, DINO and data2vec, were evaluated and compared for their representation learning of vehicle images. The former contrasts local and global views while the latter uses masked prediction on multiple layered representations. In the latter case, supervised learning is employed to finetune a pretrained YOLOR object detector for detecting vehicle wheels, from which definitive wheel positional features are retrieved. The representations learned from these self-supervised learning methods were combined with the wheel positional features for the vehicle classification task. Particularly, a random wheel masking strategy was utilized to finetune the previously learned representations in harmony with the wheel positional features during the training of the classifier. Our experiments show that the data2vec-distilled representations, which are consistent with our wheel masking strategy, outperformed the DINO counterpart, resulting in a celebrated Top-1 classification accuracy of 97.2% for classifying the 13 vehicle classes defined by the Federal Highway Administration.

Keywords: vehicle classification; vision transformer; self-supervised learning; supervised learning; object detection

Citation: Ma, S.; Yang, J.J. Image-Based Vehicle Classification by Synergizing Features from Supervised and Self-Supervised Learning Paradigms. *Eng* **2023**, *4*, 444–456. https://doi.org/10.3390/eng4010027

Academic Editor: Antonio Gil Bravo

Received: 29 November 2022
Revised: 22 January 2023
Accepted: 30 January 2023
Published: 1 February 2023

Copyright: © 2023 by the authors. Licensee MDPI, Basel, Switzerland. This article is an open access article distributed under the terms and conditions of the Creative Commons Attribution (CC BY) license (https://creativecommons.org/licenses/by/4.0/).

1. Introduction

Vehicle classification is crucial information for highway infrastructure planning and design. In practice, a large quantity of sensors has been installed in state highway networks to collect vehicle information, such as weight, speed, class, and count of vehicles [1]. Many studies have been conducted to classify vehicle types based on sensor data. For example, Wu et al. (2019) used roadside LiDAR data for vehicle classification [2]. The study evaluated traditional machine learning methods (e.g., naive Bayes, k-nearest neighbors (KNN), random forest (RF), and support vector machine) for classifying eight vehicle categories and resulted in a best accuracy of 91.98%. In another study, Sarikan et al. (2017) employed KNN and decision trees for automated vehicle classification, where the inputs were extracted features from vehicle images. The method could distinguish all sedans and motorcycles in the test dataset [3]. Recent developments in vision-based deep learning, inspired by AlexNet [4], have made image-based vehicle classification a popular approach and continue to elevate the image classification benchmark. Zhou et al. (2016) demonstrated a 99.5% accuracy in distinguishing cars and vans and a 97.36% accuracy in distinguishing among sedans, vans, and taxis [5]. Similarly, Han et al. used YOLOv2 to extract vehicle images from videos and applied an autoencoder-based layer-wise unsupervised pretraining to a convolutional neural network (CNN) for classifying motorcycles, transporter vehicles, passenger vehicles, and others [6]. ResNet-based vehicle classification and localization methods were developed using real traffic surveillance recordings, containing 11 vehicle categories, and it obtained a 97.95% classification accuracy and 79.24% mean average precision (mAP) for the vehicle localization task [7]. To ensure the robustness of the models against weather and illumination variation, Butt et al. expanded a large dataset with six common vehicle classes considering adverse illuminous conditions and used it to finetune several pretrained CNN models (AlexNet, GoogleNet, Inception-v3, VGG, and ResNet) [8]. Among those, the finetuned ResNet was able to achieve 99.68% test accuracy. Regardless of the recent success in image-based vehicle classification, most of the models have been developed based on common vehicle categories that are not consistent with the vehicle classes established for engineering practice, such as the Federal Highway Administration (FHWA) vehicle classification, which defined 13 vehicle classes [9] with key axle information, as summarized in Table 1. The vehicle class details with illustrative pictures can be found in [10].

Table 1. This is a table. FHWA vehicle classification definitions.

Vehicle Class	Class Includes	Number of Axles	Vehicle Class	Class Includes	Number of Axles
1	Motorcycles	2	8	Four or fewer axle single-trailer trucks	3 or 4
2	All cars Cars with one- and two- axle trailers	2,3, or 4	9	Five-axle single-trailer trucks	5
3	Pick-ups and vans Pick-ups and vans with one- and two-axle trailers	2, 3, or 4	10	Six or more axle single-trailer trucks	6 or more
4	Buses	2 or 3	11	Five or fewer axle multi-trailer trucks	4 or 5
5	Two-Axle, six-Tire, single-unit trucks	2	12	Six-axle multi-trailer trucks	6
6	Three-axle single-unit trucks	3	13	Seven or more axle multi-trailer trucks	7 or more
7	Four or more axle single-unit trucks	4 or more			

Nonetheless, several vehicle classification studies have been conducted with respect to FHWA vehicle classes by focusing on truck classes. Given the detailed axle-based classification rules established by the FHWA, many researchers have used them explicitly for truck classification [11]. In statewide practice, weigh-in-motion (WIM) systems and advanced inductive loop detectors are typically utilized to collect data for truck classification based on the FHWA definition, from which high correct classification rates have been reported for both single-unit trucks and multi-unit trucks [12]. As mentioned previously, besides the traditional sensing technologies, vision-based models have recently been applied in truck classification. Similarly to [5], YOLO was adopted for truck detection. Then, CNNs were used to extract features of the truck components, such as truck size, trailers, and wheels, followed by decision trees to classify the trucks into three groups [13]. This work was continued by further introducing three discriminating features (shape, texture, and semantic information) to better identify the trailer types [14].

As noted in [14], some of the classes in the FHWA scheme only have subtle differences, and deep learning models have potential overfitting issues with their imbalanced datasets. It remains a challenge for vision-based models to successfully classify all 13 FHWA vehicle classes. The objective of this study is to leverage general representations distilled from the state-of-the-art self-supervised methods (DINO [15] and data2vec [16]) as well as specific wheel positional features extracted by YOLOR [17] to improve vehicle classification. Our results show that vehicle representations are the primary features for classifying different vehicle types while wheel positions are complementary features to help better distinguish similar vehicle classes, such as classes 8 and 9, where the only salient feature difference between them is the number of axles. To reinforce this feature complementarity, the general vehicle representations from self-supervised methods were further finetuned in a subsequent supervised classification task together with a random wheel masking strategy, which is compatible with the contextualized latent representations distilled by the data2vec method [16]. As a result, our method significantly improves the classification performance and achieved a Top-1 accuracy of 97.2% for classifying the 13 FHWA vehicle classes. This paper is organized into five sections. Section 2 describes the dataset, followed by our proposed method in Section 3, experiments in Section 4, and finally conclusions and discussions in Section 5.

2. Data Description

The dataset contains 7898 vehicle images collected from two sources: the Georgia Department of Transportation (GDOT) WIM sites and the ImageNet [18] opensource dataset. The GDOT data were collected by the cameras installed at selected WIM stations. A total of 6571 vehicles images was collected from the GDOT WIM sites, consisting mainly of the common classes, such as class 2 and class 9. The number of images across the 13 vehicle classes was not well balanced. Rare classes in the GDOT image dataset, such as classes 1, 4, 7, 10, and 13, contained a small number of images. In compensating for these low-frequency classes, an additional 1327 vehicle images were extracted from the ImageNet. Figure 1 summarizes the distribution of vehicle images across the 13 FHWA vehicle classes from both sources, the WIM sites and ImageNet. Exemplar images from each data source are shown in Figure 2.

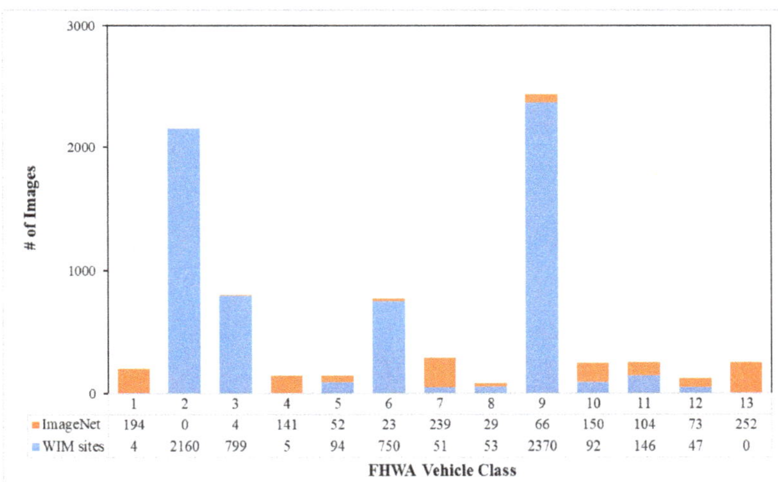

Figure 1. Distribution of image data among 13 FHWA classes from WIM sites and ImageNet.

Given the varying scale of vehicles relative to the image frame, the vehicles were cropped from the original images to remove the irrelevant background information. All models were trained based on the cropped vehicle images.

It should be noted that the number of vehicle axles is considered an important feature in FHWA vehicle class definition. For instance, class 8, class 9, and class 10 are both one-trailer trucks. The only difference among these classes is the number of axles (represented by the number of wheels in vehicle images). In addition, the relative locations of wheels are also different across vehicle categories. To extract these particular wheel positional features, a wheel detector was trained to locate all wheels in an image, as shown in Figure 3. With the locations of wheels being identified, the relative positions of all wheels were computed by dividing the wheel spacings (Di) by the maximum distance (i.e., distance between the center of the leftmost wheel and the center of the rightmost wheel) as depicted in Figure 4. This normalization process exists to remove the effect of different camera angles and unifying wheel positional information from different vehicle sizes and scales. The resulting normalized wheel positional features were a vector of the relative wheel positions, which were used to complement the features extracted by the vision transformer models for the downstream classification task.

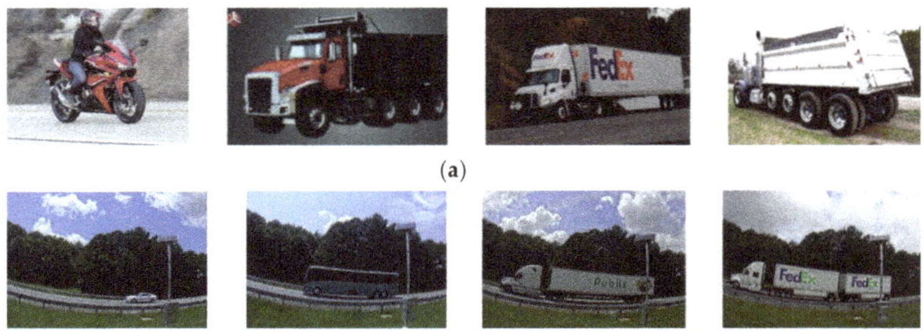

(a)

(b)

Figure 2. Examples of vehicle images in the dataset. (**a**) Vehicle images from the ImageNet. (**b**) Vehicle images collected at the GDOT WIM sites.

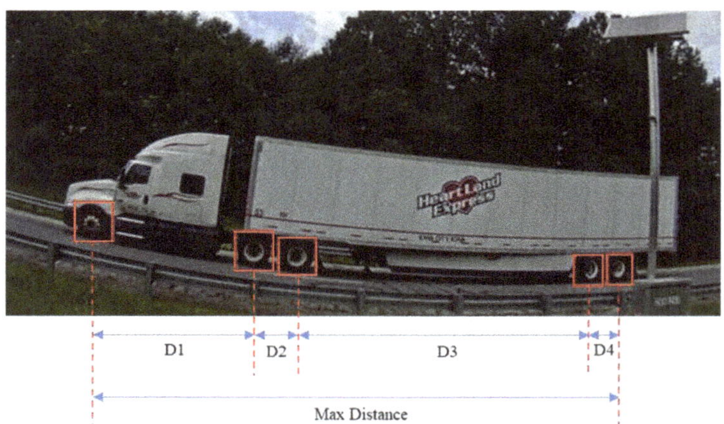

Figure 3. Illustration of wheel positional feature extraction (D1, D2, D3, and D4 are the center-to-center horizontal distances between two successive wheels from left to right).

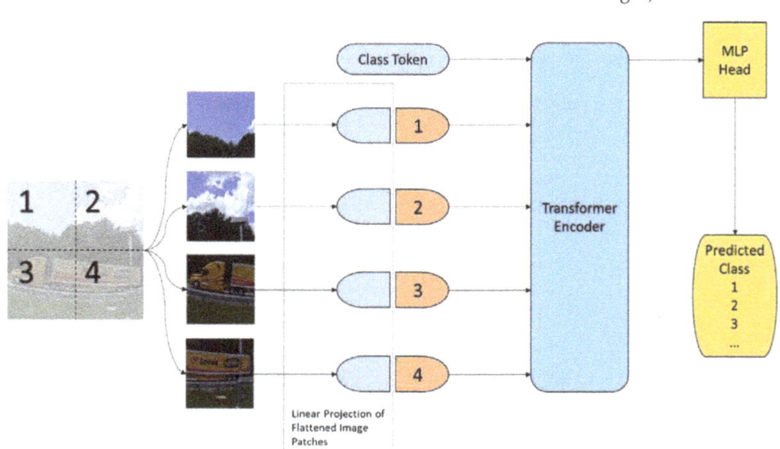

Figure 4. Illustration of a vision transformer.

3. Method

Artificial intelligence, especially deep learning, has grown dramatically over the past decade, fulfilling many real-world necessities. Many creative and influential models have been introduced especially in the cognitive computing, computer vision (CV), and natural language processing (NLP) areas. Some models have accomplished multidisciplinary success. One typical example is the transformer [19], which was first developed for NLP and has been successfully applied to vision tasks. The original transformer consists of multiple layers of encoder/decoder blocks, each of which has a combination of three key modules: a self-attention module, feedforward network modules, and layer normalization. Given its increasing popularity, the self-attention mechanism and adapted ViT architectures have been widely adopted across different fields (e.g., [20,21]). In this study, we leveraged the pretrained ViT encoders with the state-of-the-art self-supervised learning methods (i.e., DINO and data2vec) and complemented the ViT representations with wheel positional features retrieved from a finetuned object detection model (i.e., YOLOR). The two sets of features (i.e., ViT representations and wheel positional features) were harmonized by a wheel masking strategy during the classifier training. Our proposed method has shown dramatically improved classification performance when data2vec is used as the pretraining method and the ViT encoder is finetuned during the subsequent classifier training stage.

3.1. Vision Transformer

The transformer's architecture has recently been adapted to successfully handle vision tasks [22]. The ViT model has been demonstrated to achieve comparable or better image classification results than traditional CNNs [23–25]. Specifically, ViT leverages embeddings from the transformer encoder for image classification. As depicted in Figure 4, the input image is first divided into small image patches. Each patch is flattened and linearly projected to a latent vector dimension, which is then kept constant throughout all layers. The latent vector is learnable and referred to as patch embedding. A positional embedding is added to the patch embedding process to retain the spatial relationship among the image patches. The positional embedding process is illustrated using 2×2 patches in Figure 4. In practice, usually 7×7 or more patches are used. A class token is added and serves as a learnable embedding to the sequence of embedded patches. The learned representations from the encoder are passed to a multi-layer perception (MLP) for image classification. ViT has surpassed many popular CNN-based vision models, such as Resnet152 [22].

3.2. Self-Supervised Pretraining

To leverage the large number of unlabeled images (e.g., ImageNet [18]), self-supervised learning is adopted for pretraining a ViT encoder (base network). Two state-of-the-art methods: (1) self-distillation with no label (DINO) [15] and (2) data2vec [16] were evaluated in this setting. Figure 5 illustrates the structure of DINO, where input images are randomly cropped to form different views (global and local views) and fed to a teacher ViT and a student ViT, respectively. Therefore, only global view images are passed to the teacher ViT while both global and local view images are passed to the student ViT. In this way, the student model is able to extract multi-scale features. With the encodings from both ViTs, a "softmax" function is applied to produce two probability distributions, p_1 and p_2. A cross-entropy loss is then computed between p1 and p2. During the training, the parameters of the student ViT are updated by a stochastic gradient descent, while the parameters of the teacher ViT are updated from the exponential weighted average of the student ViT's parameters.

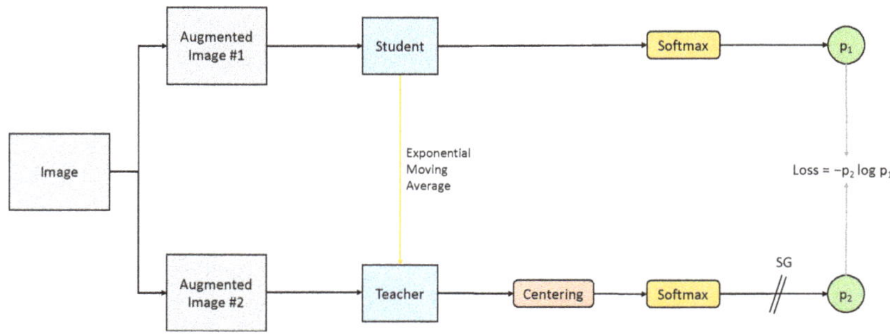

Figure 5. Illustration of DINO.

The data2vec was recently proposed by Baevski et al. [16], which represents a general self-supervised learning framework for speech, NLP, and computer vision tasks. The structure of data2vec is illustrated in Figure 6. Similar to DINO, data2vec also employs a teacher–student paradigm. The teacher generates representations from the original input image, while the student generates representations from the masked image. Different from DINO, data2vec predicts the masked latent representation and regressed multiple neural network layer representations instead of just the top layer. Instead of the cross-entropy loss in DINO, a smooth L_1 loss is used in data2vec.

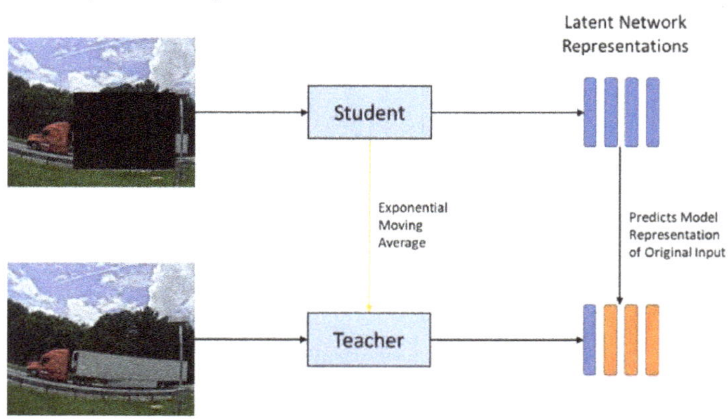

Figure 6. Illustration of data2vec.

3.3. Wheel Detection

As discussed previously, the relative wheel positions are critical features according to the FHWA vehicle classifications. Therefore, being able to detect all wheels in vehicle images provides more definitive features for vehicle classification. To leverage the existing object detection models, three real-time object detection architectures, i.e., Faster R-CNN [26], YOLOv4 [27], and YOLOR [17], were evaluated as potential wheel detectors.

Based on our experiments, both YOLOv4 and YOLOR achieved a mAP of 99.9%, slightly outperforming the Faster R-CNN model (mAPs = 99.0%). In light of the faster inference speed, YOLOR was chosen as the wheel detector in our study for extracting wheel positional features. In our experimental setting, the YOLOR model played dual roles: (1) detecting vehicles (with bounding boxes) so that they could be cropped out for further processing (as mentioned previously, all the models were trained and tested on cropped images rather than the original images) and (2) extracting wheel positional features, which were combined with the ViT features for the vehicle classification task. The fusion of these two sets of features was achieved by the end-to-end training of a composite model architecture as discussed in the following section.

3.4. Composite Model Architecture

A composite model architecture is proposed to improve vehicle classification by harnessing the features extracted from both ViT and wheel detection models, as shown in Figure 7. The input image is fed to YOLOR for vehicle and wheel detection. Then, the vehicle image is cropped based on the vehicle bounding boxes output from the YOLOR and resized to 224 × 224, which is the input size for the ViT encoder. Features extracted from the ViT encoder and the wheel positional features are concatenated and fed to a multi-layer perceptron (MLP) to classify the 13 FHWA vehicle classes. The wheel locations detected by YOLOR provide wheel positional features that are complementary to those features extracted by the ViT encoder since the former provides localized details on axle configuration while the latter emphasizes vehicle features at a coarser and larger scale (i.e., not necessarily attending to the wheel position details). To further reinforce this complementarity, one wheel was randomly masked when finetuning the ViT encoder. The vanilla ViT and pretrained ViTs by DINO and data2vec were all evaluated in this composite architecture setting. The experimental results are presented in the Experiments section.

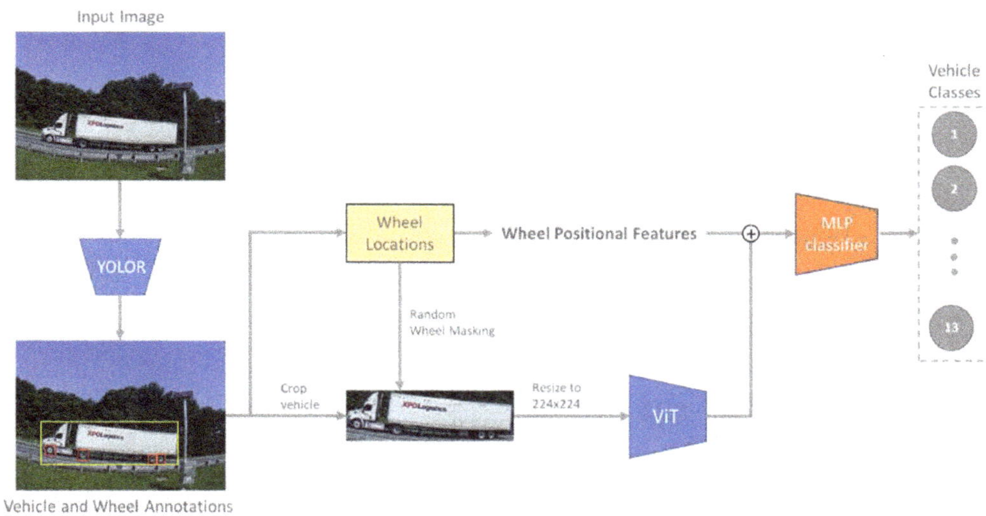

Figure 7. Structure of the composite model architecture.

4. Experiments

4.1. Effects of Self-Supervised Pretraining

The ViT model was trained with Adam [28] and a batch size of 64. The learning rate was initially set to 6×10^{-5}. For model development and evaluation, the dataset was split with 80% being relegated to training and 20% to testing. All cropped images were resized to 224 × 224 pixels and evenly divided into 14 × 14 patches. Two self-supervised methods, DINO and data2vec, were evaluated as a pretraining stage for the supervised classification task. An ImageNet-pretrained ViT with DINO and data2vec was utilized in this study. Our model training was conducted under two settings: (1) freezing the ViT backbone and only training the MLP classifier and (2) finetuning the ViT backbone while training the MLP classifier. For comparison purposes, the original ViT model was also trained end-to-end in a supervised fashion. The results are summarized in Table 2.

Table 2. Comparison of classification performance by different ViT training settings.

Network	Top-1 Acc. (%)	Weighted Avg. Precision (%)	Weighted Avg. Recall (%)
ViT	90.7	90.7	90.7
ViT + DINO (freeze-encoder)	94.6	94.6	94.6
ViT + DINO	95.6	95.7	95.6
ViT + data2vec (freeze-encoder)	93.5	93.5	93.5
ViT + data2vec	95.0	95.0	95.0

As shown in Table 2, the Top-1 accuracy, weighted average precision, and weighted average recall from the pretrained ViT models are significantly higher than those from the supervised ViT model regardless of whether the ViT backbone was frozen or not during the classifier training. There was a clear performance boost when the ViT encoder was finetuned during the classifier training. For the finetuned models, the ViT + DINO network performed slightly better than the ViT + data2vec.

For a detailed performance comparison across vehicle classes, the classification reports of the ViT, ViT + DINO, and ViT + data2vec are presented in Table 3. Overall, ViT performed well in the common classes (classes 2, 3, 6, and 9). However, for minority classes (classes 1, 4, 5, 7, 8, 10, 11, 12, and 13), pretrained ViTs reported much better precision, recall, and F1-scores.

Table 3. Comparison of classification reports of the ViT model with and without self-supervised pretraining.

	ViT without Pretraining			ViT Pretrained with DINO			ViT Pretrained with data2vec		
	Precision (%)	Recall (%)	F1-Score (%)	Precision (%)	Recall (%)	F1-Score (%)	Precision (%)	Recall (%)	F1-Score (%)
Class 1	88.0	55.0	67.7	100.0	100.0	100.0	100.0	100.0	100.0
Class 2	96.6	98.6	97.6	98.4	99.1	98.7	98.2	99.8	99.0
Class 3	93.5	89.4	91.4	97.5	95.7	96.6	99.4	95.0	97.1
Class 4	73.5	86.2	79.4	100.0	93.1	96.4	96.4	93.1	94.7
Class 5	90.0	62.1	73.5	88.5	79.3	83.6	88.5	79.3	83.6
Class 6	90.3	96.1	93.1	87.2	96.8	91.7	88.2	96.1	92.0
Class 7	66.2	74.1	69.9	95.8	79.3	86.8	84.9	77.6	81.1
Class 8	57.1	50.0	53.3	78.6	68.8	73.3	90.0	56.3	69.2
Class 9	96.2	98.2	97.2	96.8	98.4	97.6	97.0	98.4	97.7
Class 10	64.4	60.4	62.4	90.0	75.0	81.8	81.4	72.9	76.9
Class 11	78.9	82.0	80.4	97.8	90.0	93.8	94.0	94.0	94.0
Class 12	88.2	62.5	73.2	92.3	100.0	96.0	95.7	91.7	93.6
Class 13	68.6	68.6	68.6	90.6	94.1	92.3	80.4	80.4	80.4
Accuracy (%)		90.7			95.6			95.0	

4.2. *Performance of Composite Models*

As mentioned previously, the number of axles is a key factor in the FHWA vehicle classification rules. Therefore, we generated the wheel positional features from the wheel locations detected by YOLOR and fed these features to the classifier together with the ViT encodings. To assess the benefits of adding the wheel positional features, we evaluated two model scenarios: (1) ViT models, which did not include wheel positional features, and (2) composite models, which included the ViT models as well as the wheel positional features from YOLOR. Table 4 shows the results of the ViT models (upper part) and composite models (bottom part).

Table 4. Comparison of the supervised ViT model and the composite models under different training schemes.

Network	Top-1 Acc. (%)	Weighted Avg. Precision (%)	Weighted Avg. Recall (%)
ViT	90.7	90.7	90.7
ViT + DINO (freeze-encoder)	94.6	94.6	94.6
ViT + DINO	95.6	95.7	95.6
ViT + data2vec (freeze-encoder)	93.5	93.5	93.5
ViT + data2vec	95.0	95.0	95.0
ViT + YOLOR	91.4	91.6	91.4
ViT + DINO (freeze-encoder) + YOLOR	95.4	95.5	95.4
ViT + DINO + YOLOR	96.0	96.0	96.0
ViT + data2vec (freeze-encoder) + YOLOR	95.0	95.0	95.0
ViT + data2vec + YOLOR	95.3	95.2	95.3

The composite models, which fused the ViT encodings and wheel positional features, improved the classification accuracy by 0.3–1.5%. This confirms that specific wheel positional features are important for the vehicle classification task. The DINO-pretrained ViT models were slightly better than their data2vec counterparts. The best composite model was "ViT + DINO + YOLOR", which achieved an overall accuracy of 96%. The detailed performance metrics (precision, recall, and F1-score) across classes are included in Table 5 for both scenarios: with and without the wheel features.

Table 5. Comparison of classification reports of the pretrained model (ViT + DINO) with and without wheel features.

	DINO (Pretrained), without Wheel Features			DINO (Pretrained) + YOLOR, with Wheel Features		
	Precision (%)	Recall (%)	F1-Score (%)	Precision (%)	Recall (%)	F1-Score (%)
Class 1	100.0	100.0	100.0	100.0	100.0	100.0
Class 2	99.1	99.5	99.3	99.1	99.5	99.3
Class 3	98.7	97.5	98.1	98.7	97.5	98.1
Class 4	100.0	93.1	96.4	100.0	86.2	92.6
Class 5	88.5	79.3	83.6	80.7	86.2	83.3
Class 6	90.5	98.7	94.4	92.2	98.7	95.3
Class 7	91.4	91.4	91.4	92.7	87.9	90.3
Class 8	70.6	75.0	72.7	92.9	81.3	86.7
Class 9	98.2	98.0	98.1	97.6	98.6	98.1
Class 10	88.6	81.3	84.8	88.9	83.3	86.0
Class 11	86.2	100.0	92.6	88.9	96.0	92.3
Class 12	100.0	87.5	93.3	95.8	95.8	95.8
Class 13	95.1	76.5	84.8	100.0	80.4	89.1
Accuracy (%)		96.3			96.6	

The benefit of adding wheel features resulted in an obvious improvement in the F1-scores for classes 8 and 10. As indicated in Table 1, classes 8, 9, and 10 are all one-trailer trucks with minor differences (number of axles) and immensely imbalanced data distributions (with 82 images in class 8 but 2436 images in class 9). The model could have easily been confused among these classes. Adding the wheel positional features helped to better classify them, as well as classes 11, 12, and 13, which are all multi-trailer classes.

4.3. Random Wheel Masking Strategy

To regularize the learning process of the composite model, one of the YOLOR-detected wheels was randomly selected for masking when training the ViT. This could allow the learned ViT representations to adapt to the wheel noises being injected. The experiment results are summarized in Table 6.

Table 6. Comparison of model performance with and without random wheel masking.

Network	Without Wheel Masking			Randomly Masking One Wheel		
	Top-1 Acc. (%)	**WAP * (%)**	**WAR * (%)**	**Top-1 Acc. (%)**	**WAP (%)**	**WAR (%)**
ViT + YOLOR	91.4	91.6	91.4	91.7	91.6	91.7
ViT + DINO (freeze-encoder) + YOLOR	95.4	95.5	95.4	96.0	96.0	96.0
ViT + DINO + YOLOR	96.3	96.3	96.3	96.7	96.8	96.7
ViT + data2vec (freeze-encoder) +YOLOR	95.0	95.0	95.0	96.5	96.6	96.5
ViT + data2vec + YOLOR	95.3	95.2	95.3	97.2	97.2	97.2

* WAP, WAR: abbreviations for weighted average precision and weighted average recall, respectively.

As shown in Table 6, all models had improved performance when one wheel was randomly masked during the training of the classifier. Without wheel masking, the best composite model was DINO + ViT + YOLOR, which achieved an accuracy of 96.3%. After applying one wheel masking, its accuracy was raised up to 96.7%. In contrast, the "ViT + data2vec + YOLOR" model benefited tremendously from the random wheel masking. Its accuracy was boosted by 1.9% to 97.2%, surpassing the DINO + ViT + YOLOR. Table 7 shows the detailed classification results of the ViT + data2vec + YOLOR for both the with and without wheel masking settings.

Table 7. Comparison of classification reports of the composite model (ViT + data2vec + YOLOR) with and without wheel masking.

	ViT + Data2vec + YOLOR, without Wheel Masking			ViT + Data2vec + YOLOR, with Wheel Masking		
	Precision (%)	**Recall (%)**	**F1-Score (%)**	**Precision (%)**	**Recall (%)**	**F1-Score (%)**
Class 1	100.0	100.0	100.0	100.0	100.0	100.0
Class 2	98.2	99.8	99.0	99.3	99.8	99.5
Class 3	99.4	95.0	97.1	99.4	98.1	98.8
Class 4	93.3	96.6	94.9	96.6	96.6	96.6
Class 5	88.5	79.3	83.6	92.6	86.2	89.3
Class 6	88.6	95.5	91.9	93.2	97.4	95.3
Class 7	84.9	77.6	81.1	94.2	84.5	89.1
Class 8	100.0	62.5	76.9	100.0	81.3	89.7
Class 9	97.2	98.6	97.9	97.0	99.4	98.2
Class 10	81.8	75.0	78.3	97.6	83.3	89.9
Class 11	94.0	94.0	94.0	94.2	98.0	96.1
Class 12	95.8	95.8	95.8	88.5	95.8	92.0
Class 13	83.7	80.4	82.0	97.8	88.2	92.8
Accuracy (%)		95.3			97.2	

As indicated in Table 7, randomly masking one wheel increased the precision of classes 4, 5, 6, 7, 10, and 13 considerably. The F1-scores of most classes also improved, especially for the minority truck classes (5, 6, 7, 8, 10, and 13).

5. Conclusions and Discussions

The two self-supervised learning methods (DINO and data2vec) showed their superiority over the supervised ViT. The classification accuracies were further boosted after applying DINO or data2vec for pretraining. Finetuning the pretrained ViT encoders during the classifier training helped with the classification task. By adding additional wheel positional features, the models performed better than standalone ViTs. Additionally, the adoption of the random wheel masking strategy while finetuning the ViT encoder further improved the performance of the models, resulting in accuracies of 96.7 and 97.2%, respectively, for the DINO- and data2vec-pretrained models.

An important aspect to acknowledge is that classic supervised learning is largely constrained by limited annotated datasets. In contrast, self-supervised learning can take advantage of massive amounts of unlabeled data for representation learning and has become increasingly popular. Using self-supervised methods as a pretraining stage has been demonstrated to significantly improve the performance of vehicle classification. Between the two popular self-supervised learning methods, DINO and data2vec, there is an interesting finding: the DINO-pretrained ViT performed better than the data2vec-pretrained one, even with ViT finetuning and the addition of wheel positional features. However, by randomly masking a wheel during training, the data2vec-pretrained ViT outperformed the DINO-pretrained ViT. An arguable ratiocination is that during the pretraining stage, data2vec trains the ViT to predict the contextualized representations of masked image patches, which is consistent with our wheel masking strategy. This allows the data2vec-pretrained ViT to easily generalize over the masked wheel features, while for DINO, the ViT encoder learns from the cropped parts of input images and does not capture the contextual information, unlike with data2vec.

Although the ViT + data2vec + YOLOR model, coupled with the proposed strategy of random wheel masking, demonstrated an excellent performance in classifying 13 FHWA vehicle classes, there is still plenty of room for future improvement. Dataset imbalance issues can be further mitigated by acquiring more images of minority class vehicles. YOLOR was adopted as a wheel detector to extract wheel positional features, which increases the computational footprint since the two standalone models (ViT and YOLOR) are executed in parallel. A unified model architecture could be investigated to reduce computational costs for practical real-time applications. The work presented in this paper implicitly assumes that the full bodies of all vehicles are visible in the images while this may not be true in real-world settings, where vehicle occlusion and superimposition often occur during heavy traffic conditions, causing only parts of vehicles to be visible. This issue could be mitigated by purposely training the models to recognize vehicle classes with partially blocked images. In fact, the data2vec method learns general representations by predicting contextualized latent representations of a masked view of the input in a self-distillation setting. Thus, data2vec-distilled representations are robust in cases of partial blocking of vehicles in images. Other mitigation methods may consider leveraging multiple views from different cameras or even multimodal sensory inputs. For example, using thermal cameras and LiDAR could help to improve model performance under low light conditions (e.g., at night).

Author Contributions: Study conception and design, J.J.Y. and S.M.; data collection, S.M.; analysis and interpretation of results, S.M. and J.J.Y.; draft preparation, S.M. and J.J.Y. All authors have read and agreed to the published version of the manuscript.

Funding: This research was funded by the Georgia Department of Transportation, Grant Number: RP20-04.

Data Availability Statement: Some or all the data, models, and code that support the findings of this study are available from the corresponding author upon reasonable request.

Acknowledgments: The study presented in this paper was conducted by the University of Georgia under the auspices of the Georgia Department of Transportation (RP 20-04). The contents of this paper reflect the views of the authors, who are solely responsible for the facts and accuracy of the data, opinions, and conclusions presented herein. The contents may not reflect the views of the funding agency or other individuals.

Conflicts of Interest: The funder had no role in the design of the study; in the analyses or interpretation of data; in the writing of the manuscript; or in the decision to publish the results.

References

1. Cheung, S.Y.; Coleri, S.; Dundar, B.; Ganesh, S.; Tan, C.-W.; Varaiya, P. Traffic measurement and vehicle classification with single magnetic sensor. *Transp. Res. Rec.* **2005**, *1917*, 173–181. [CrossRef]
2. Wu, J.; Xu, H.; Zheng, Y.; Zhang, Y.; Lv, B.; Tian, Z. Automatic vehicle classification using roadside LiDAR data. *Transp. Res. Rec.* **2019**, *2673*, 153–164. [CrossRef]
3. Sarikan, S.S.; Ozbayoglu, A.M.; Zilci, O. Automated vehicle classification with image processing and computational intelligence. *Procedia Comput. Sci.* **2017**, *114*, 515–522. [CrossRef]
4. Krizhevsky, A.; Sutskever, I.; Hinton, G.E. Imagenet classification with deep convolutional neural networks. *Commun. ACM* **2017**, *60*, 84–90. [CrossRef]
5. Zhou, Y.; Nejati, H.; Do, T.-T.; Cheung, N.-M.; Cheah, L. Image-based vehicle analysis using deep neural network: A systematic study. In Proceedings of the 2016 IEEE International Conference on Digital Signal Processing (DSP), Beijing, China, 16–18 October 2016; pp. 276–280.
6. Han, Y.; Jiang, T.; Ma, Y.; Xu, C. Pretraining convolutional neural networks for image-based vehicle classification. *Adv. Multimed.* **2018**, *2018*, 3138278. [CrossRef]
7. Jung, H.; Choi, M.-K.; Jung, J.; Lee, J.-H.; Kwon, S.; Young Jung, W. ResNet-based vehicle classification and localization in traffic surveillance systems. In Proceedings of the IEEE Conference on Computer Vision and Pattern Recognition Workshops, Honolulu, HI, USA, 21–26 July 2017; pp. 61–67.
8. Butt, M.A.; Khattak, A.M.; Shafique, S.; Hayat, B.; Abid, S.; Kim, K.-I.; Ayub, M.W.; Sajid, A.; Adnan, A. Convolutional neural network based vehicle classification in adverse illuminous conditions for intelligent transportation systems. *Complexity* **2021**, *2021*, 6644861. [CrossRef]
9. Hallenbeck, M.E.; Selezneva, O.I.; Quinley, R. *Verification, Refinement, and Applicability of Long-Term Pavement Performance Vehicle Classification Rules*; United States. Federal Highway Administration. Office of Infrastructure: Washington, DC, USA, 2014.
10. Adm, F.H. *Traffic Monitoring Guide*; United States. Federal Highway Administration. Office of Highway Policy Information: Washington, DC, USA, 2001. Available online: https://rosap.ntl.bts.gov/view/dot/41607 (accessed on 22 January 2023).
11. Asborno, M.I.; Burris, C.G.; Hernandez, S. Truck body-type classification using single-beam LiDAR sensors. *Transp. Res. Rec.* **2019**, *2673*, 26–40. [CrossRef]
12. Hernandez, S.V.; Tok, A.; Ritchie, S.G. Integration of Weigh-in-Motion (WIM) and inductive signature data for truck body classification. *Transp. Res. Part C Emerg. Technol.* **2016**, *68*, 1–21. [CrossRef]
13. He, P.; Wu, A.; Huang, X.; Scott, J.; Rangarajan, A.; Ranka, S. Deep learning based geometric features for effective truck selection and classification from highway videos. In Proceedings of the 2019 IEEE Intelligent Transportation Systems Conference (ITSC), Auckland, New Zealand, 27–30 October 2019; pp. 824–830.
14. He, P.; Wu, A.; Huang, X.; Scott, J.; Rangarajan, A.; Ranka, S. Truck and trailer classification with deep learning based geometric features. *IEEE Trans. Intell. Transp. Syst.* **2020**, *22*, 7782–7791. [CrossRef]
15. Caron, M.; Touvron, H.; Misra, I.; Jégou, H.; Mairal, J.; Bojanowski, P.; Joulin, A. Emerging properties in self-supervised vision transformers. In Proceedings of the IEEE/CVF International Conference on Computer Vision, Seoul, Republic of Korea, 27 October–2 November 2019; pp. 9650–9660.
16. Baevski, A.; Hsu, W.-N.; Xu, Q.; Babu, A.; Gu, J.; Auli, M. Data2vec: A general framework for self-supervised learning in speech, vision and language. *arXiv* **2022**, arXiv:2202.03555.
17. Wang, C.-Y.; Yeh, I.-H.; Liao, H.-Y.M. You only learn one representation: Unified network for multiple tasks. *arXiv* **2021**, arXiv:2105.04206.
18. Deng, J.; Dong, W.; Socher, R.; Li, L.-J.; Li, K.; Fei-Fei, L. Imagenet: A large-scale hierarchical image database. In Proceedings of the 2009 IEEE Conference on Computer Vision and Pattern Recognition, Miami, FL, USA, 20–25 June 2009; pp. 248–255.
19. Vaswani, A.; Shazeer, N.; Parmar, N.; Uszkoreit, J.; Jones, L.; Gomez, A.N.; Kaiser, Ł.; Polosukhin, I. Attention is all you need. In Proceedings of the Thirty-First Annual Conference on Neural Information Processing Systems (NIPS), Long Beach, CA, USA, 4–9 December 2017.
20. Hu, H.; Lu, X.; Zhang, X.; Zhang, T.; Sun, G. Inheritance attention matrix-based universal adversarial perturbations on vision transformers. *IEEE Signal Process. Lett.* **2021**, *28*, 1923–1927. [CrossRef]
21. Zhou, X.; Bai, X.; Wang, L.; Zhou, F. Robust ISAR Target Recognition Based on ADRISAR-Net. *IEEE Trans. Aerosp. Electron. Syst.* **2022**, *58*, 5494–5505. [CrossRef]
22. Dosovitskiy, A.; Beyer, L.; Kolesnikov, A.; Weissenborn, D.; Zhai, X.; Unterthiner, T.; Dehghani, M.; Minderer, M.; Heigold, G.; Gelly, S. An image is worth 16x16 words: Transformers for image recognition at scale. *arXiv* **2020**, arXiv:2010.11929.
23. Wang, H.; Ji, Y.; Song, K.; Sun, M.; Lv, P.; Zhang, T. ViT-P: Classification of Genitourinary Syndrome of Menopause From OCT Images Based on Vision Transformer Models. *IEEE Trans. Instrum. Meas.* **2021**, *70*, 1–14. [CrossRef]
24. Cuenat, S.; Couturier, R. Convolutional Neural Network (CNN) vs Visual Transformer (ViT) for Digital Holography. *arXiv* **2021**, arXiv:2108.09147.
25. Yuan, L.; Chen, Y.; Wang, T.; Yu, W.; Shi, Y.; Jiang, Z.-H.; Tay, F.E.; Feng, J.; Yan, S. Tokens-to-token vit: Training vision transformers from scratch on imagenet. In Proceedings of the IEEE/CVF International Conference on Computer Vision, Seoul, Republic of Korea, 27 October–2 November 2019; pp. 558–567.

26. Ren, S.; He, K.; Girshick, R.; Sun, J. Faster r-cnn: Towards real-time object detection with region proposal networks. *IEEE Trans. Pattern Anal. Mach. Intell.* **2017**, *39*, 1137–1149. [CrossRef] [PubMed]
27. Bochkovskiy, A.; Wang, C.-Y.; Liao, H.-Y.M. Yolov4: Optimal speed and accuracy of object detection. *arXiv* **2020**, arXiv:2004.10934.
28. Kingma, D.P.; Ba, J. Adam: A method for stochastic optimization. *arXiv* **2014**, arXiv:1412.6980.

Disclaimer/Publisher's Note: The statements, opinions and data contained in all publications are solely those of the individual author(s) and contributor(s) and not of MDPI and/or the editor(s). MDPI and/or the editor(s) disclaim responsibility for any injury to people or property resulting from any ideas, methods, instructions or products referred to in the content.

eng

MDPI

Article

High-Performance Computation of the Number of Nested RNA Structures with 3D Parallel Tiled Code

Piotr Błaszyński *,†,‡ and Włodzimierz Bielecki †,‡

Faculty of Computer Science and Information Systems, West Pomeranian University of Technology in Szczecin, 70-310 Szczecin, Poland; wbielecki@zut.edu.pl

* Correspondence: pblaszynski@zut.edu.pl

† Current address: Faculty of Computer Science and Information Systems, West Pomeranian University of Technology in Szczecin, Żołnierska 49, 72-210 Szczecin, Poland.

‡ These authors contributed equally to this work.

Abstract: Many current bioinformatics algorithms have been implemented in parallel programming code. Some of them have already reached the limits imposed by Amdahl's law, but many can still be improved. In our paper, we present an approach allowing us to generate a high-performance code for calculating the number of RNA pairs. The approach allows us to generate parallel tiled code of the maximal dimension of tiles, which for the discussed algorithm is 3D. Experiments carried out by us on two modern multi-core computers, an Intel(R) Xeon(R) Gold 6326 (2.90 GHz, 2 physical units, 32 cores, 64 threads, 24 MB Cache) and Intel(R) i7(11700KF (3.6 GHz, 8 cores, 16 threads, 16 MB Cache), demonstrate a significant increase in performance and scalability of the generated parallel tiled code. For the Intel(R) Xeon(R) Gold 6326 and Intel(R) i7, target code speedup increases linearly with an increase in the number of threads. An approach presented in the paper to generate target code can be used by programmers to generate target parallel tiled code for other bioinformatics codes whose dependence patterns are similar to those of the code implementing the counting algorithm.

Keywords: bioinformatics; RNA folding; dynamic programming; tiled code generation; code parallelization; high-performance code

Citation: Błaszyński, P.; Bielecki, W. High-Performance Computation of the Number of Nested RNA Structures with 3D Parallel Tiled Code. *Eng* **2023**, *4*, 507–525. https://doi.org/10.3390/eng4010030

Academic Editor: Antonio Gil Bravo

Received: 21 December 2022
Revised: 28 January 2023
Accepted: 30 January 2023
Published: 3 February 2023

Copyright: © 2023 by the authors. Licensee MDPI, Basel, Switzerland. This article is an open access article distributed under the terms and conditions of the Creative Commons Attribution (CC BY) license (https://creativecommons.org/licenses/by/4.0/).

1. Introduction

Bioinformatics computing became one of the most important areas of science some time ago. Since the beginning, parallel versions of bioinformatics algorithms have been developed. Much of this work has led to significant speedup, some to new algorithms, but some remain in sequential versions. The approach presented in this paper used to modify the code implementing the algorithm for calculating the number of pairs in an RNA structure allows this algorithm to be parallelized and achieve significant target code speedup. The described method can also be used for a whole class of bioinformatics algorithms whose dependence patterns are similar to those of the examined code in this paper.

The serial code of bioinformatics algorithms subject to parallelization are Nussinov's, Zuker's, Smith–Waterman's algorithms, and some others. Typical dependency patterns available in such codes are non-uniform and generally presented with affine expressions. This makes transformations of such codes much more difficult in comparison with codes exposing only uniform dependences (all elements of dependence distance vectors are integers). Some bioinformatic algorithms are implemented in FPGA [1]. There are GPU-based solutions for both CUDA [2], and OpenCL [3], as well as Kokkos [4]. However, most implementations are realized by means of the OpenMP API [5] due to the popularity and simplicity of code development using this API.

In the Ref. [6], Smiths and Waterman described a mathematical analysis of a RNA secondary structure. In our paper, we use the implementation variant of the described algorithm. A description of this variant can be found in Raden's publications [7,8].

The counting algorithm computes the exact number of nested structures for a given RNA sequence. It populates matrix C using the following recursion:

$$C_{i,j} = C_{i,j-1} + \sum_{\substack{i \le k < (j-l) \\ S_k, S_j \text{ pair}}} C_{i,k-1} \cdot C_{k+1,j-1}, \tag{1}$$

where l is the minimal number of enclosed positions, and the entry $C_{i,j}$ provides the exact number of admissible structures for the sub-sequence from position i to j. The upper-right corner $C_{1,n}$ presents the overall number of admissible structures for the sequence. We choose value 1 for l. The minimal number of enclosed positions could also be 0, 2, or more. A value of l has an impact on results generated with the code, but it does not significantly change the execution time of the examined algorithm. Value 1 is the default in most experiments [7,8].

The C code implementing the counting algorithm is presented in Listing 1.

Listing 1. C code implementing the counting algorithm

```
for (int i = N - 2; i >= 1; i--) {
 for (int j = i + 2; j <= N; j++) {
  for (int k = i; k <= j - 1; k++) {
   c[i][j] += paired(k, j) ? c[i][k - 1] + c[k + 1][j - 1]: 0; //S0
  }
  c[i][j] = c[i][j] + c[i][j - 1]; //S1
 }
}
```

The counting algorithm requires high-performance computing for longer RNA sequences. Currently, high-performance computing is possible via the development of parallel tiled applications running on multi-core computers. Code parallelism allows for applying many cores, while tiling improves code locality and increases code granularity, that is crucial for achieving good code performance and scalability. To our best knowledge, there is no manual implementation (as parallel tiled code) of the counting algorithm. Parallel tiled code can be generated automatically by means of optimizing compilers. To generate such a code, we chose two optimizing compilers, Pluto and TRACO, and carried out experiments with codes generated with them. The results of experiments exposed the main drawback of the PLUTO and TRACO codes implementing the counting algorithm: insufficient code locality, which limits code performance and its scalability.

The problem statement is to derive an approach that allows for generation of parallel tiled code which is characterized by better code locality in comparison with that of PLUTO and TRACO code, to apply the approach to the source code implementing the counting algorithm to generate parallel tiled code, and carry out an experimental study to demonstrate the advantage of generated parallel target code implementing the counting algorithm.

A short description of the presented approach is the following. We discovered that the structure of dependences available in the source code implementing the counting algorithm prevents generation of 3D tiled code by means of PLUTO, PLUTO generates only 2D tiled code (the innermost loop is untiled). Increasing a tile dimension from 2D to 3D is crucial for enhancing tiled code locality. The reason is the following. A 2D tile is unbounded because it includes all the loop nest statement instances enumerated along the untiled innermost loop. In general, the upper bound of it is a parameter, so the number of statement instances enumerated with the innermost loop is parametric, that is, unbounded.

Thus, data associated with a single 2D tile cannot be held in cache that reduces code locality, whereas 3D tiles are bounded and choosing a proper tile size allows for keeping all the data of a single 3D tile in a cache that improves code locality.

We discovered that PLUTO generates 2D tiles (it fails to tile the innermost loop) instead of 3D ones because of complex dependences available in the source code implementing the counting algorithm. To improve the features of dependences, we suggest to apply to the source code scheduling according the data flow concept (DFC) that envisages that a loop nest statement instance can be executed when all its operands are ready (already calculated). That is, we suggest to generate new serial source code as a result of the transformation of the source one. For implementing such a transformation, DFC is derived and applied to the source code. We use a formal verification of the validity of derived schedules based on DFC. Then any compiler based on affine transformations can be applied to new source codes to generate parallel tiled codes. The crucial steps in the presented approach are deriving schedules based on DFC and verifying the legality of those schedules. The rest of the steps of the approach are easily implemented with open source tools.

Thus, the goal of the paper is to present an approach to generate high-performance 3D parallel tiled code on the basis of the code presented in Listing 1 and demonstrate the efficiency of that code on two modern multi-core platforms.

Loop tiling is discussed in the Refs. [9–12]. Let us illustrate loop tiling for the loop nest presented in Listing 2.

Listing 2. An illustrative example

```
for (int i = N - 2; i >= 1; i--) {
 for (int j = i + 2; j <= N; j++) {
  for (int k = i; k <= j - 1; k++) {
   c[i][j] += paired(k, j) ? c[i][k - 1] + c[k + 1][j - 1]: 0; //S0
  }
  c[i][j] = c[i][j] + c[i][j - 1]; //S1
 }
}
```

For this loop nest, the loop nest iteration space and the order of iteration execution is presented on the left side of Figure 1.

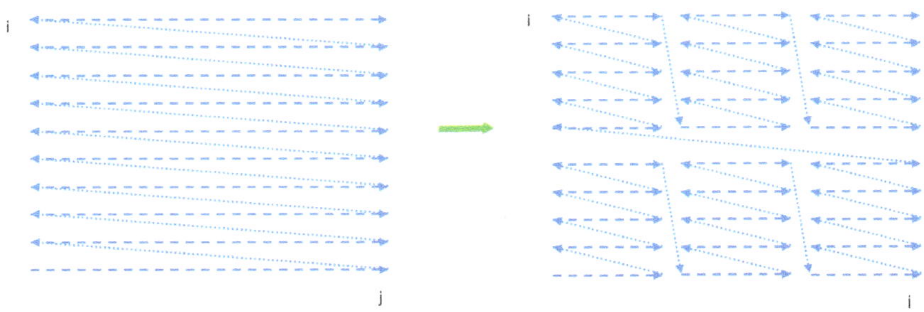

Figure 1. Loop transformation.

We may say that iteration execution takes place within a single tile (block). For a large problem size, it is impossible to hold all the data of such a tile in a cache that hampers code locality.

We may split the loop nest iteration space into tiles as shown on the right side of Figure 1. If a tile size is chosen properly, that is, all the data of a single tile can be held in the cache and those data occupy as much cache capacity as possible, we may considerably improve code locality.

The tiled code is presented in Listing 3. As we can see, the tiled code includes four loops. The two outermost loops enumerate tiles, while the two innermost loops enumerate iterations within a single tile. In general, it is difficult to build tiled code manually even for simple loop nests. Usually, optimizing compilers are used for automatic tiled code generation.

Automatically generated parallel tiled code is based on the polyhedral model [13]. For a given loop nest, this model envisages forming the following data: (i) a loop nest iteration space (a set of all the iterations executed with the loop nest), (ii) an original loop nest schedule in the global iteration space, and (iii) dependence relations. This model can be used for implementation of many transformations, for example, loop interchange, loop unrolling, loop fusion, loop fission, and register blocking. However, the most popular and effective transformations are loop parallelization and loop tiling. The polyhedral model is a basis of the affine transformation framework [13] and the correction approach [14], which allow for automatic parallel tiled code generation.

Listing 3. An illustrative example

```
for(TI=0; TI<N; TI+=16)
 for(TJ=0; TJ<N; TJ+=16)
  for(i=TI; i<min(TI+16,N); i++)
   for(j=TJ; j<min(TJ+16,N); j++)
    A[i][j] = B[j][i];
```

The main contributions of the paper are the following: (i) introducing an approach to generate 3D tiled parallel code for the counting algorithm; (ii) presenting an OpenMP parallel tiled code implementing the counting algorithm; (iii) presenting and discussing results of experiments on modern multi-core platforms.

The rest of the paper is organized as follows. Section 2.2 presents the background, our approach to generate 3D parallel tiled code, and closely related codes in C implementing the counting algorithm. Section 3 discusses results of experiments with examined codes on two modern multi-core machines. Section 4 concludes.

2. Materials and Methods

2.1. Background

Usually, to increase serial code performance, parallelization and loop tiling are applied to the source code. Parallelism allows us to use many threads to execute code, while loop tiling improves code locality and increases parallel code granularity that is crucial for improving multi-threaded code performance.

Loop tiling is a reordering loop transformation that allows data to be accessed in blocks (tiles), with the block size defined as a parameter of this transformation. Each loop is transformed in two loops: one iterating inside each block (intratile) and the other one iterating over the blocks (intertile).

As far as loop tiling is concerned, it is very important to generate target code with maximal tile dimension, which is defined with the maximal number of loops in the loop nest. If one or more the innermost loops remain un-tiled, resulting tiles are unbounded along those untiled loops. This makes tiles also unbounded that reduces tiled code locality because it is not possible to hold in cache all the data associated with a single unbounded tile [15,16]. If one or more of the outermost loops are untiled, they should be executed serially. This reduces target code parallelism and introduces additional synchronization events reducing target code performance [10,13].

Each iteration in the loop nest iteration space is represented with an iteration vector. All iteration vectors of a given loop statement form the iteration space of that statement.

Code can expose dependences among iterations in a code iteration space. A dependence is a situation when two different iterations access the same memory location and at least one of these accesses is written. Each dependence is represented by its source and destination.

To extract dependences and generate target code, we use PET [17] and the iscc calculator [18]. The iscc calculator is an interactive tool for manipulating sets and relations of integer tuples bounded by affine constraints over the set variables, parameters and existentially quantified variables. PET is a library for extracting a polyhedral model from a C source. Such a model consists of an iteration space, access relations, and a schedule, each of which is described using affine constraints. A PET schedule specifies the original execution order of loop nest statement instances.

PET extracts dependences in the form of relations, where the input tuple of each relation represents iteration vectors of dependence sources and the output tuple represents those of the corresponding dependence destinations; that is, the dependence relation, R, is presented in the following form:

$$R := [parameters] \rightarrow \{[input\ tuple] \rightarrow [output\ tuple] \mid constraints\},$$

where $[parameters]$ is the list of all parameters of affine $constraint$ imposed on $[input\ tuple]$ and $[output\ tuple]$.

For the dependence, a distance vector is the difference between the iteration vector of its destination and that of its source. Calculating such a difference is possible when both abovementioned vectors are of the same length. This is true for perfectly nested loops where all statements are surrounded with all loops. Otherwise, loops are imperfectly nested, that is, the dimensions of iteration spaces of loop nest statements are different and we cannot directly calculate a distance vector.

In such a case, to calculate distance vectors, we normalize the iteration space of each statement so that all the iteration spaces are of the same dimension. Normalization consists in applying a global schedule extracted with PET for each loop nest statement to an iteration space of the statement. The entire global schedule corresponds to the original execution order of a loop nest. As a result, the iteration spaces of each statement become of the same dimension in the global iteration space and we are able to calculate all distance vectors. We present details of normalization in the following subsection.

To tile and parallelize source codes, we should form time partition constraints [10] that state that if iteration I of statement $S1$ depends on iteration J of statement $S2$, then I must be assigned to a time partition that is executed no earlier than the partition containing J, that is, schedule $(I) \leq$ schedule (J), where schedule (I) and schedule (J) denote the discrete execution time of iterations I and J, respectively.

Linear independent solutions to time partition constraints are applied to generate schedules for statement instances of original code. Those affine schedules are used to parallelize and tile an original loop nest.

We strived to extract as many linear independent solutions to time partition constraints as possible because the number of those solutions defines the dimension of generated tiles [10].

The affine transformation framework comprises the above considerations and includes the following steps: (i) extracting dependence relations, (ii) forming time partition constraints on the basis of dependence relations, (iii) resolving the time partition constraints striving to find as many linearly independent solutions as possible, (iv) forming affine transformations on the basis of the independent solutions, and (v) generating parallel tiled code.

Details of the affine transformation framework can be found in the Ref. [13]. In the same paper, implementation details of the PLUTO compiler based on the affine transformation are presented.

An alternative approach to generate parallel tiled code is based on applying the transitive closure of a dependence graph. This approach is introduced in the Ref. [14]. It envisages the following steps: (i) extracting dependence relations, (ii) forming a dependence graph as the union of all dependence relations, (iii) calculating the transitive closure of the dependence graph, (iv) applying transitive closure to form valid tiles, and (v) generating parallel tiled code. This approach does not form and apply any affine transformation. The approach is implemented in the TRACO compiler.

2.2. 3D Tiled Code Generation

For the original loop nest in Listing 1, PET returns the following iteration spaces, $D0$ and $D1$, for statements $S0$ and $S1$, respectively.

$$D0 := N \rightarrow \{\, S_0(i,j,k) \mid i > 0 \wedge 2 + i \leq j \leq N \wedge i \leq k < j \,\},$$

$$D1 := N \rightarrow \{\, S_1(i,j) \mid i > 0 \wedge 2 + i \leq j \leq N \,\}.$$

As we can see, the dimensions of the iteration spaces of $S0$ and $S1$ are different, so we could not directly calculate distance vectors. To normalize the iteration spaces, we applied the following global schedules returned with PET:

$$M0 := N \rightarrow \{\, S0(i,j,k) \rightarrow (-i,j,0,k) \,\},$$

$$M1 := N \rightarrow \{\, S1(i,j) \rightarrow (-i,j,1,0) \,\}$$

to sets $D0$ and $D1$, respectively, and obtained the following global spaces:

$$D0' := N \rightarrow \{\, (-i,j,0,k) \mid i > 0 \wedge 2 + i \leq j \leq N \wedge i \leq k < j \,\},$$

$$D1' := N \rightarrow \{\, (-i,j,1,0) \mid i > 0 \wedge 2 + i \leq j \leq N \,\}.$$

It is worth noting that if a statement appears in a sequence of statements, PET extends the global schedule for these statements with a constant representing a global schedule dimension. The values of these dimensions correspond to the order of the statements in the sequence. If a statement appears as the body of a loop, then the schedule is extended with both an initial domain dimension and an initial range dimension. In schedules $M0$ and $M1$, in the output (right) tuples in the third positions, constants 0 and 1 are inserted, while in the fourth position of the output tuple of $M1$, constant 0 is inserted because statement $S1$ is not surrounded with iterator k.

In the same way, we transform dependence relations returned with PET and presented in original iteration spaces to dependence relations presented in the global iteration space, where all relation tuples are of the same dimension. Applying the deltas operator of the iscc calculator, which calculates the difference between the output and input tuples of dependence relations in the global iteration space, we obtained the following distance vectors represented with sets.

$$D1 := \{\, (0,i1,i2,i3) \mid i1 \geq 0 \wedge i1 \leq i2 \leq 1 \wedge i3 > i1 \,\},$$

$$D2 := \{\, (i0,1,0,i3) \mid i0 > 0 \wedge i3 < 0 \,\},$$

$$D3 := \{\, (0,i1,0,i3) \mid i1 \geq 2 \wedge i3 \geq 2 \,\},$$

$$D4 := \{\, (i0,1,-1,i3) \mid i0 > 0 \wedge i3 \leq -3 \,\},$$

$$D5 := \{\, (0,i1,-1,1) \mid i1 \geq 2 \,\},$$

$$D6 := \{\, (0,1,0,1) \,\}.$$

To simplify extracting affine transformations, we approximate the distance vectors above with a single distance vector, which represents each distance vector presented above: $D := \{ (i0, i1, i2, i3) \mid i0 \geq 0 \wedge (i1 = 1 \vee i1 \geq 2) \wedge (i2 = 0 \vee i2 = -1 \vee i1 \geq i2 \geq 1) \wedge (i3 > i1 \vee i3 < 0 \vee i3 > 2 \vee i3 \leq -3 \vee i3 = 1)\}$.

It is worth noting that the constraints of D are the logical conjunction of all the constraints of $D1, D2, D3, D4, D5$, and $D6$.

The time partition constraint formed on the basis of vector D is the following.

$$x_0 * i0 + x_1 * i1 + x_2 * i2 + x_3 * i3 \geq 0,\ constraints \quad (2)$$

where x_0, x_1, x_2, x_3 are unknowns, and *constraints* is the constraints of set D.

Because variable $i3$ is unbounded, that is, $-\infty \leq i3 \leq \infty$, we conclude that x_3 should be equal to 0 to satisfy constraint (2). We also conclude that unknown $x2$ should be 0 because variable $i2$ is not any loop iterator, it represents global schedule constants, which should not be transformed; they are used only to properly generate target code (correctly place loop statements in target code).

Taking into account the conclusions above, we consummate that there exist only two linearly independent solutions to constraint (2), for example, $(1,0,0,0)^T$ and $(0,1,0,0)^T$.

Thus, for the code in Listing 1, by means of affine transformations, we are able to generate only 2D tiles (see Background).

Next, we use the concept of a loop nest statement instance schedule, which specifies the order in which those instances are executed in target code. To improve the features of the dependences of the code presented in Listing 1, we suggest applying to the loop nest statements a schedule formed according to the data flow concept DFC): first, the readiness time for each operand of each statement should be defined, for example, if $t_1^i, t_2^i, \ldots, t_k^i$ are k discrete times of the readiness of k operands of statement i, then the schedule of statement i is defined as follows: $t_i = max(t_1^i, t_2^i, \ldots, t_k^i) + 1$. On the right of that formula, the first term defines the maximal time among all operand readiness times of statement i, and "+1" means that statement i can be executed at the next discrete time after all its operands are ready (already calculated).

DFC schedules should be defined and applied to all the statements of the source loop nest to generate a transformed serial loop.

Analyzing the operands $c[i][k-1]$ and $c[k+1][j-1]$ of statement $S0$ in Listing 1 as well as the bounds of loops i and j, we may conclude that their readiness times are $k-i-1$ and $j-k-2$, respectively. We also take into account that element $c[i][j]$ can be updated many times for different values of k, and the final value of $c[i][j]$ is formed in time $j-i-1$. Thus, according to DFC, statement $S0$ is to be executed at time $t = max(k-i-1, j-k-2) + 1$ for variables i and j satisfying the constraint $t <= j-i-1$. The last constraint means that element $c[i][j]$ formed with statement $S0$ can be updated many times at time t, satisfying the condition $t <= j-i-1$. Thus, taking into account the global schedule of statement $S0$ represented with relation $M0$, we obtain the following DFC schedule, $SCHED(S0)$.

$$SCHED(S0) := N \to \{ S_0(i,j,k) \to t = max(k-i-1, j-k-2) + 1, -i, j, 0, k) \mid t \leq j-i-1\}.$$

Analyzing statement $S1$, we may conclude that it should be executed when for given i and j, loop k is terminated, that is, the calculation of the value of element $c[i][j]$ is terminated, that is, at time $j-i-1$. Thus, we obtained the following schedule for statement $S1$ taking into account the global schedule for $S1$ presented with relation $M1$ above.

$$SCHED(S1) := N \to \{ S_0(i,j,k) \to (t = max(k-i-1, j-k-2) + 1, -i, j, 1, 0) \mid t = j-i-1\}.$$

The constraint $t = j-i-1$ means that statement $S1$ can be updated only when loop k is terminated. Constant 1 in the third position of the tuple $(t = max(k-i-1, j-k-2) + 1, -i, j, 1, 0)$ guarantees that statement $S1$ should be executed after terminating all the iterations of loop k.

Applying schedules $SCHED(S0)$ and $SCHED(S1)$ to statements $S0$ and $S1$, by means of the codegen iscc operator, we obtain the transformed code presented in Listing 4.

Listing 4. Transformed C code implementing the counting algorithm

```
for (int c0 = 1; c0 < N - 1; c0 += 1)
 for (int c1 = -N + c0 + 1; c1 < 0; c1 += 1)
  for (int c2 = c0 - c1 + 1; c2 <= min(N, 2 * c0 - c1 + 1); c2 += 1) {
   if (2 * c0 >= c1 + c2)
   {
    c[-c1][c2] += paired(-c0 + c2 - 1, c2) ? c[-c1][-c0 + c2 - 1 - 1] + c[-c0 + c2 - 1 + 1][c2 - 1] : 0;
   }
   c[-c1][c2] += paired(c0 - c1, c2) ? c[-c1][c0 - c1 - 1] + c[c0 - c1 + 1][c2 - 1] : 0;
   if (c1 + c2 == c0 + 1)
   {
    c[-c1][c0 - c1 + 1] = c[-c1][c0 - c1 + 1] + c[-c1][c0 - c1 + 1 - 1];
   }
  }
```

That code respects all dependences available in the code in Listing 1 due to the following reason. In the code in Listing 1, we distinguish two types of dependences: standard ones and reductions. If the loop nest statement uses an associative and commutative operation such as addition, we recognize the dependence between two references of this statement as a reduction dependence [19]. For example, in the code in Listing 1, statement $S0$

$$c[i][j] += paired(k,j)?c[i][k-1] + c[k+1][j-1] : 0; //S0$$

causes reduction dependences regarding to reads and writes of element $c[i][j]$.

We may allow them to be reordered provided that a new order is serial, that is, reduction dependences do not impose an ordering constraint; in the code in Listing 4, reduction dependences are respected due to the serial execution of loop nest statement instances.

Standard dependences available in the code in Listing 1 are respected via implementing the DFC concept.

To prove the validity of the applied schedules to generate the code in Listing 4 in a formal way, we use the schedule validation technique presented in the Ref. [20].

Given relation F representing all the dependences to be respected, schedule S is valid if the following inequality is true:

$$\Delta(S \circ F \circ S^{-1}) \succeq 0,$$

where Δ is the operator that maps a relation to the differences between image and domain elements.

The result of the composition $R = (S \circ F \circ S^{-1})$ is a relation where the input (left) and output (right) tuples represent dependence sources and destinations, respectively, in the transformed iteration space. A schedule is valid (respects all the dependences available in an original loop nest) if the vector whose elements are the differences between the image and domain elements of relation R is lexicographically non-negative ($\succeq 0$). In such a case, each standard dependence in the original loop nest is respected in the transformed loop nest.

To apply the schedule validity technique above, we extract dependence relations F by means of PET. Then we eliminate from F all reduction dependences, taking into account the fact that such dependences cause only statement $S0$. We present reduction dependences by means of the following relation:

$$\{\, S_0(i,j,k) \to S_0(i,j,k') \mid k' > k \,\}.$$

Next, applying the iscc calculator, we obtain a relation, R, as the result of the composition $(S \circ F \circ S^{-1})$, where S is the union of schedules $SCHED(S0)$ and $SCHED(S1)$ defined above to generate the target code.

To check whether each vector represented with set $C = \Delta(S \circ F \circ S^{-1})$ is lexicographically non-negative, we form the following set that represents all lexicographically negative vectors in the unbounded 5D space.

$LD5 := N \to \{\, (t, i0, i1, i2, i3) \mid t < 0 \,\} \cup N \to \{\, (0, i0, i1, i2, i3) \mid i0 < 0 \,\} \cup N \to \{\, (0, 0, i1, i2, i3) \mid i1 < 0 \,\} \cup N \to \{\, (0, 0, 0, i2, i3) \mid i2 < 0 \,\} \cup N \to \{\, (0, 0, 0, 0, i3) \mid i3 \le 0 \,\}$.

Then, we calculate the intersection of sets C and $LD5$. That intersection is the empty set, that means that all vectors of C are lexicographically non-negative. This proves the validity of schedules $SCHED(S0)$ and $SCHED(S1)$.

For the code presented in Listing 4, by means of PET and the iscc calculator, we obtained the following distance vectors.
$D1 := \{\, (i0, i1, 1, i3) \mid i0 \ge 2 \land -1 \le i3 \le 1 \land ((i1 \ge 2 + i0 \land i3 \ge 0) \lor (i1 > 0 \land i3 \le 0)) \,\}$,
$D2 := \{\, (2, 0, i2, -1) \mid i2 \ge 2 \,\}$,
$D3 := \{\, (i0, 0, i2, i3) \mid i3 \le 2 \land ((i0 \ge 3 \land i2 \ge 3 + i0 \land -2 \le i3 \le 0) \lor (i0 \ge 2 \land i2 \ge 2 \land 0 \le i3 \le 1) \lor (0 \le i2 \le 1 \land i2 \le i0 \land i3 \ge i2 \land i3 > -i0)) \,\}$,
$D4 := \{\, (2, i1, 1, -2) \mid i1 > 0 \,\}$,
$D5 := \{\, (1, 0, 1, 0) \,\}$,
$D6 := \{\, (i0, 0, 0, -1) \mid i0 > 0 \,\}$.

To simplify extracting affine transformations, we approximate the distance vectors above with a single distance vector, which represents each distance vector presented above. $D := \{\, (i0, i1, 1, i3) \mid i0 \ge 0 \land i1 \ge 0 \land i2 \ge 0 \land -2 \le i3 \le 2$. The time partitions constraint formed on the basis of the vector above is the following.

$$x_0 * i0 + x_1 * i1 + x_2 * i2 + x_3 * i3 \ge 0,\ constraints \tag{3}$$

where x_0, x_1, x_2, x_3 are unknowns, and *constraints* represent the constraints of set D above.

Taking into account that unknown $x3$ should be 0 because variable $i3$ is not any loop iterator, it represents global schedule constants and it should not be transformed. There exist three linearly independent solutions to constraint (3), for example, $(1,0,0,0)^T$, $(0,1,0,0)^T$, and $(0,0,1,0)^T$.

Applying the DAPT optimizing compiler [21], which automatically extracts and applies the affine transformations to the code in Listing 4 to tile and parallelize that code by means of the wave-front technique [22], we obtain the following target 3D tiled parallel code (Listing 5) with tiles of size 16 × 32 × 40. By means of experiments, this size was defined by us as the optimal one regarding tiled code performance.

In that code, the first three loops enumerate tiles, while the remaining three loops scan statement instances within each tile. The OpenMP [23] directive *#pragma omp parallel for* makes the loop *for (int h0=...)* parallel.

The Ref. [24] illustrates the advantage of 3D tiled codes in comparison with 2D tiled ones which implement RNA Nussinov's algorithm [25]. However, there are the following differences between the approaches used for code generation for the Nussinov problem [24] and for the counting problem considered in the current paper. The code for the Nussinov problem is derived on the idea of a calculation model based on systolic arrays (first figure in the Nussinov paper), while code for the counting problem is based on the data flow concept (DFC). The approach presented in the current paper uses the validity technique of the applied schedules to generate target code, while the approach presented in the Nussinov paper does not envisage any formal validation of applied schedules.

Listing 5. 3D tiled parallel code

```
for (int w0 = floord(-N + 34, 160) - 1; w0 < floord(7 * N - 10, 80)
    ; w0 += 1) {
  #pragma omp parallel for
  for (int h0 = max(max(0, w0 - (N + 40) / 40 + 2), w0 + floord(-4 *
       w0 - 3, 9) + 1); h0 <= min((N - 2) / 16, w0 + floord(N - 80 *
       w0 + 46, 240) + 1); h0 += 1) {
    for (int h1 = max(max(max(5 * w0 - 9 * h0 - 3, -((N + 29) / 32)),
         w0 - h0 - (N + 40) / 40 + 1), -((N - 16 * h0 + 30) / 32));
         h1 <= min(-1, 5 * w0 - 7 * h0 + 8); h1 += 1) {
      for (int i0 = max(max(1, 16 * h0), 20 * w0 - 20 * h0 - 4 * h1);
           i0 <= min(min(16 * h0 + 15, N + 32 * h1 + 30), 40 * w0 - 40
           * h0 - 8 * h1 + 69); i0 += 1) {
        for (int i1 = max(max(32 * h1, -40 * w0 + 40 * h0 + 40 * h1 +
             i0 - 38), -N + i0 + 1); i1 <= min(32 * h1 + 31, -40 * w0 +
             40 * h0 + 40 * h1 + 2 * i0 + 1); i1 += 1) {
          for (int i2 = max(40 * w0 - 40 * h0 - 40 * h1, i0 - i1 + 1);
               i2 <= min(min(N, 40 * w0 - 40 * h0 - 40 * h1 + 39), 2 * i0
               - i1 + 1); i2 += 1) {
            {
              if (2 * i0 >= i1 + i2) {
                c[-i1][i2] += (paired((-i0 + i2 - 1), (i2)) ? (c[-i1][-i0 +
                     i2 - 2] + c[-i0 + i2][i2 - 1]) : 0);
              }
              c[-i1][i2] += (paired((i0 - i1), (i2)) ? (c[-i1][i0 - i1 -
                   1] + c[i0 - i1 + 1][i2 - 1]) : 0);
              if (i1 + i2 == i0 + 1) {
                c[-i1][i0 - i1 + 1] = (c[-i1][i0 - i1 + 1] + c[-i1][i0 - i1
                     ]);
              }
            }
          }
        }
      }
    }
  }
}
```

2.3. Related Codes

To our best knowledge, there is no manual implementation (as parallel tiled code) of the counting algorithm. To generate such a code, we chose two optimizing compilers, PLUTO and TRACO. They are open source codes and allow for automatic code optimization (tiling and parallelization). We did not consider other optimizing compilers because they do not satisfy one or more of the following demands: the compiler must be a source-to-source translator, it should be able to tile and parallelize source code, be based on polyhedral techniques, be currently maintained and have no building problems, and be well-documented. Applying PLUTO to counting source code allows us to generate only 2D tiled parallel code while applying TRACO results in the generation of codes with irregular tiles.

The usefulness of generated tiled parallel code presented in this paper is the following. In relation to the 2D tiled PLUTO code, the generated code by means of the presented approach is 3D. This allows us to increase code locality and, as a consequence, improve target code performance. With regard to the PLUTO code, 3D tiled code enumerates regular bounded tiles that allows for improving code locality in comparison with the PLUTO code, and this results in improving code performance. The tile regularity of 3D tiled code also provides better code locality in comparison with the TRACO code because a tile size is limited and there is a possibility to choose a tile size such that all the data associated with a single tile can be held in cache.

Listing 6. PLUTO [13] code implementing the counting algorithm

```
if (N >= 3)
{
 for (t1 = 3; t1 <= N; t1++)
 {
  lbp = 0;
  ubp = floord(t1 - 2, 32);
#pragma omp parallel for private(lbv, ubv, t3, t4, t5)
  for (t2 = lbp; t2 <= ubp; t2++)
  {
   for (t3 = t2; t3 <= floord(t1, 32); t3++)
   {
    if ((t1 >= 32 * t3 + 1) && (t1 <= 32 * t3 + 31))
    {
     for (t4 = max(1, 32 * t2); t4 <= min(t1 - 2, 32 * t2 + 31); t4++)
     {
      for (t5 = max(32 * t3, t4); t5 <= t1 - 1; t5++)
      {
       c[t4][t1] += paired(t5, t1) ? c[t4][t5 - 1] + c[t5 + 1][t1 - 1] : 0;
      }
      c[t4][t1] = c[t4][t1] + c[t4][t1 - 1];
     }
    }
    if (t1 >= 32 * t3 + 32)
    {
     for (t4 = max(1, 32 * t2); t4 <= min(t1 - 2, 32 * t2 + 31); t4++)
     {
      for (t5 = max(32 * t3, t4); t5 <= 32 * t3 + 31; t5++)
      {
       c[t4][t1] += paired(t5, t1) ? c[t4][t5 - 1] + c[t5 + 1][t1 - 1] : 0;
      }
     }
    }
    if (t1 == 32 * t3)
    {
     for (t4 = max(1, 32 * t2); t4 <= min(t1 - 2, 32 * t2 + 31); t4++)
     {
      if (t1 % 32 == 0)
      {
       c[t4][t1] = c[t4][t1] + c[t4][t1 - 1];
      }
     }
    }
   }
  }
 }
}
```

The codes used for comparison are presented below. These codes were obtained from the code in Listing 1. The code in Listing 6 is generated by the PLUTO parallel compiler [13] and the code in Listing 7 is generated by TRACO [14].

The code in Listing 6 enumerates 2D tiles of size 32×32, which was established as the optimal one by us via experiments. PLUTO implements the affine transformation framework, and as we demonstrate in Section 2.2, there exist only two linearly independent solutions for the time partition constraints formed for the code in Listing 1. Thus, the maximal dimension of the tiles in the code in Listing 6 is 2D.

Traco generates 3D tiles of size $8 \times 127 \times 16$ (defined by us as the optimal one by means of experiments), but tiles are irregular, some of them are unbounded, hampering thread load balance and reducing code locality because not all the data associated with a single unbounded tile can be held in the cache.

Listing 7. TRACO [14] code implementing the counting algorithm

```
for (c1 = 0; c1 < N + floord(N - 3, 128) - 2; c1 += 1)
#pragma omp parallel for
 for (c3 = max(0, -N + c1 + 3); c3 <= c1 / 129; c3 += 1)
  for (c4 = 0; c4 <= 1; c4 += 1)
  {
   if (c4 == 1)
   {
    for (c9 = N - c1 + 129 * c3; c9 <= min(N, N - c1 + 129 * c3 + 127); c9 += 1)
     for (c10 = max(0, -c1 + 64 * c3 - c9 + (N + c1 + c3 + c9 + 1) / 2 + 1); c10 <= 1; c10 += 1)
     {
      if (c10 == 1)
      {
       c[(N - c1 + c3 - 2)][c9] = c[(N - c1 + c3 - 2)][c9] + c[(N - c1 + c3 - 2)][c9 - 1];
      }
      else
      {
       for (c11 = N - c1 + 129 * c3 + 1; c11 < c9; c11 += 1)
        c[(N - c1 + c3 - 2)][c9] += paired(c11, c9) ? c[(N - c1 + c3 - 2)][c11 - 1] + c[c11 + 1][c9 - 1] : 0;
      }
     }
   }
   else
   {
    for (c5 = 0; c5 <= 8 * c3; c5 += 1)
     for (c9 = N - c1 + 129 * c3; c9 <= min(N, N - c1 + 129 * c3 + 127); c9 += 1)
      for (c11 = N - c1 + c3 + 16 * c5 - 2; c11 <= min(min(N - c1 + 129 * c3, N - c1 + c3 + 16 * c5 + 13), c9 - 1); c11 += 1)
       c[(N - c1 + c3 - 2)][c9] += paired(c11, c9) + c[(N - c1 + c3 - 2)][c11 - 1] + c[c11 + 1][c9 - 1] + 0;
   }
}
```

3. Results

First we show impact of the number of threads on performace in Figures 2 and 3. To carry out the experiments, we used two machines: (i) a processor Intel(R) Xeon(R) Gold 6326 (2.90GHz, 2 physical units, 32 cores, 64 threads, 24 MB Cache) (results obtained on that machine are presented in Figures 4 and 5) and (ii) a processor Intel(R) i7-11700KF (3.6GHz,

8 cores, 16 threads, 16MB Cache), where results achieved on that machine are depicted in Figures 6 and 7. All examined codes were compiled by means of the gcc 11.3 compiler with the –O3 flag of optimization. The reason for using this option was to generate optimal serial and parallel executable code. Codes without this option run much longer for both sequential and parallel code.

Experiments were carried out for 24 RNA randomly generated sequence lengths of the problem defined with parameter N from 500 to 12,000. The results presented in the Ref. [26,27] show that cache-efficient code performance does not change based on the strings themselves, but it depends on the size of a string. We also performed a study with a variable number of threads, with sequence length 12,000 on Intel Xeon and sequence length 10,000 on Intel i7. We used a range from 1 to 32 threads with step 1 for both machines. Results are shown in Figures 2 and 3. We compared the performance of the 3D tiled code generated with the presented approach with that of the following codes:

1. Listing 6—PLUTO parallel tiled code (based on affine transformations) (PLUTO in charts) [13].
2. Listing 7—Tiled code based on the correction technique (TRACO in charts) [14].
3. Listing 1—Original code of counting algorithm (ORIGINAL in charts) [7,8].

All source codes used for carrying out experiments as well as a program allowing us to run each parallel program for a random or real RNA strand can be found in the Data Availability section.

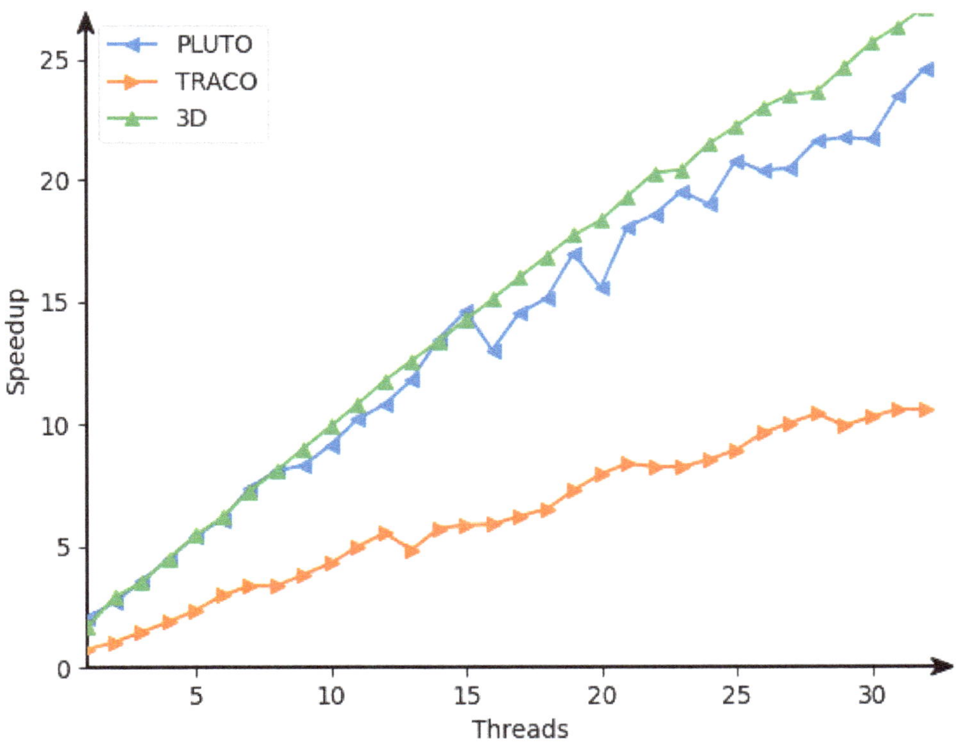

Figure 2. Speedup for different thread numbers. Intel Xeon—sequence length 12,000.

3.1. Impact of the Number of Threads on Code Performance

3.1.1. Intel Xeon

Figure 2 shows the speedup of the three examined parallel codes—the ratio of the serial code execution time to that of the parallel one. The 3D code speedup demonstrates nearly linear speedup. From this chart, it can be assumed that the 3D code is well-scalable, that is, it is possible to increase code parallelism further when increasing the number of threads. The 3D code can be run on a machine with a large number of threads without any code modification.

3.1.2. Intel i7

Figure 3 shows the speedup of the same parallel codes on a i7 processor. The important part of this chart is between 12 and 16 threads (16 threads is the maximum for this unit). From this chart, it can be assumed that the 3D code works very well at the maximum level of threads, and it is also possible to increase code parallelism further when increasing the number of cores and threads.

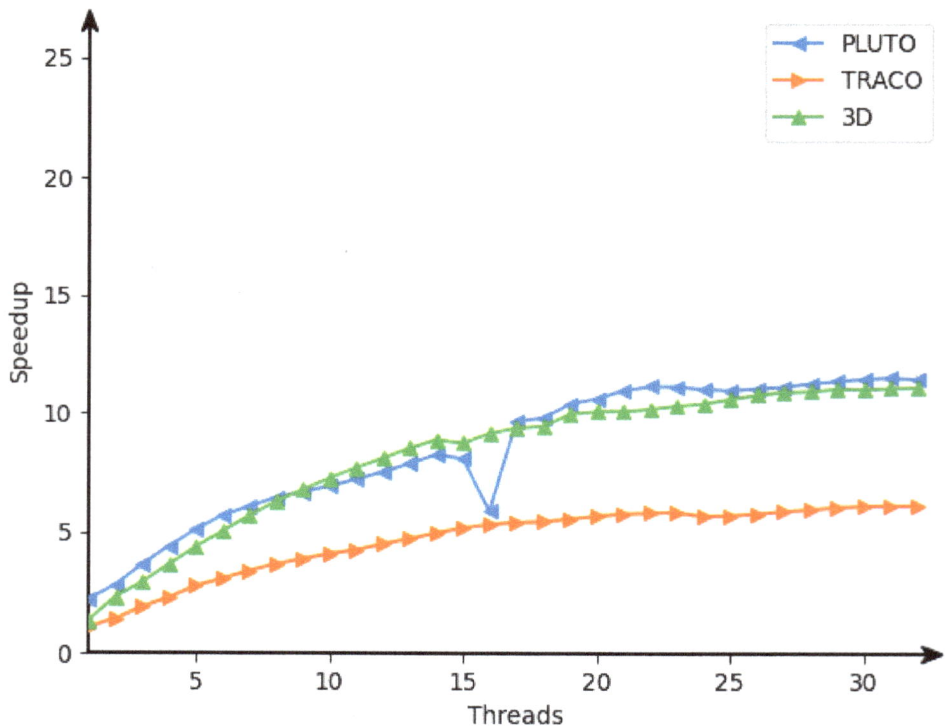

Figure 3. Speedup for different thread numbers. Intel i7—sequence length 10,000.

3.2. Impact of the Problem Size on Code Performance

3.2.1. Intel Xeon

For the results depicted in Figures 4 and 5, one can see a clear advantage of the 3D code for larger problem sizes. For smaller problem sizes, taking into account how the PLUTO code is simpler than 3D code (it comprises 5 loops while 3D code includes 6 loops), that is, PLUTO executes less iterations than 3D code does, it demonstrates better performance than that of the 3D code, while for larger problem sizes, better code locality of 3D code in comparison with that of PLUTO one outweighs the benefits of PLUTO code simplicity.

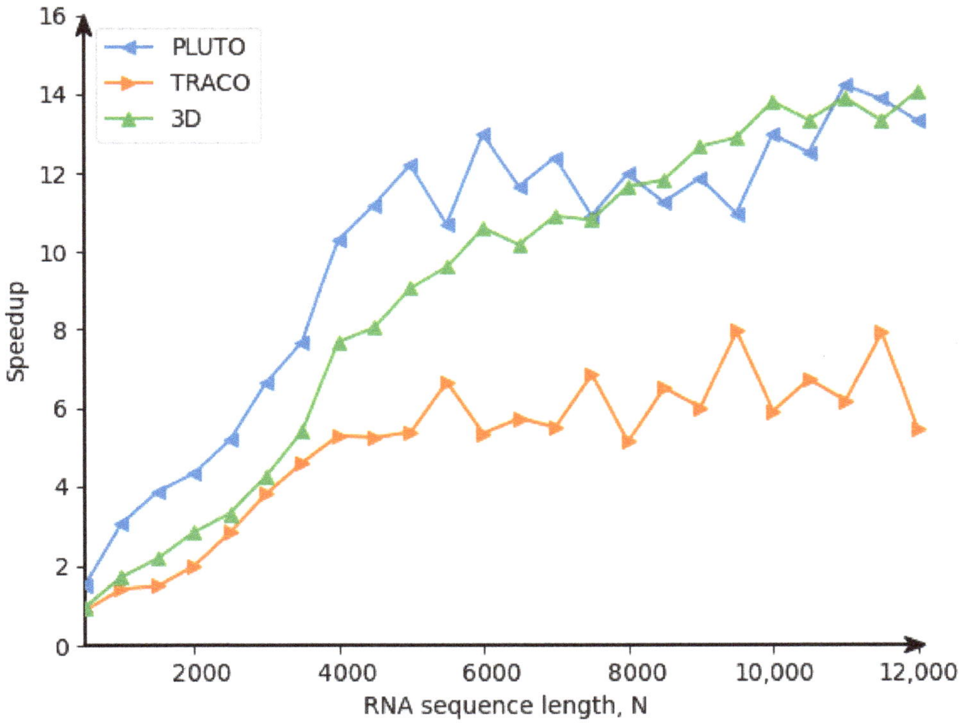

Figure 4. Speedup for different sequence lengths. Intel Xeon—16 threads.

3.2.2. Intel i7

The code in Figure 6 shows that with a larger problem size, it is possible to calculate faster for the 3D code. In addition to that, the 3D code is characterized by much greater stability of performance (no spikes in speedup for the both processors tested).

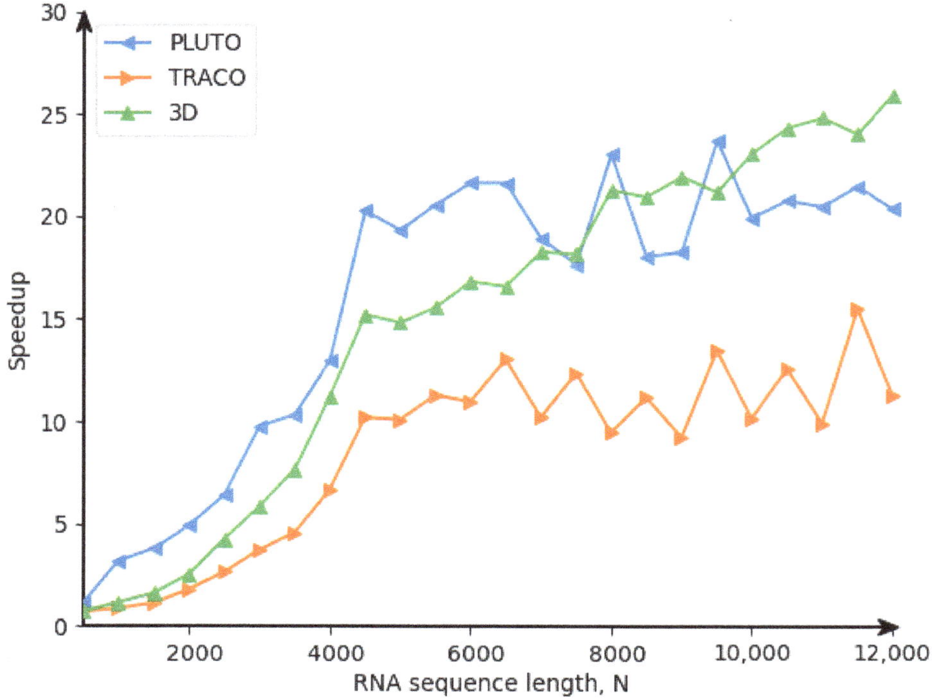

Figure 5. Speedup for different sequence lengths. Intel Xeon—32 threads.

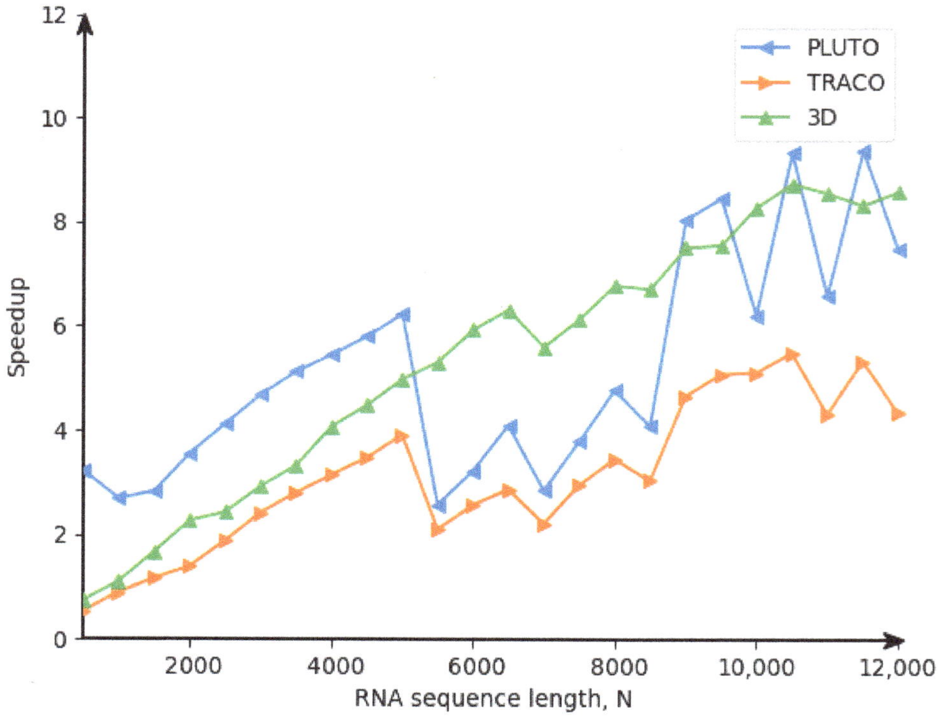

Figure 6. Speedup for different sequence lengths. Intel i7—16 threads.

In addition to that, Figure 7 shows that for the 3D code, it is possible to utilize the available threads fully. However, with more threads than the number of threads available on the processor, the operation of this code is not the fastest one.

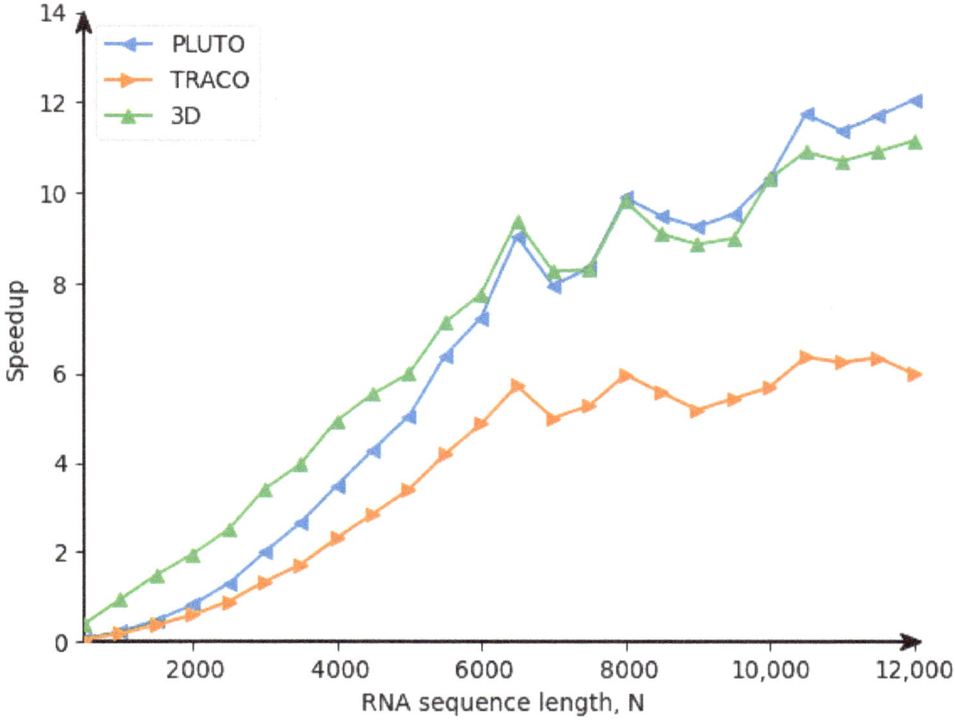

Figure 7. Speedup for different sequence lengths. Intel i7—32 threads.

4. Discussion

The approach presented in this paper allows for parallel tiled code generation for the counting algorithm. Target code demonstrates a significant increase in code performance, largely over original sequential code. For larger problem sizes, it outperforms related parallel tiled codes and exposes better scalability. The experimental results carried out by us show that the 3D parallel tiled code, implementing the counting algorithm, utilises computational capabilities of modern processor cores very well. The advantages of the obtained 3D code are more obvious for a large problem size. We plan to apply the presented approach to other bioinformatics codes whose dependence patterns are similar to those available in the code implementing the counting algorithm. This allows for increasing the tile dimension as a consequence of increasing the performance and scalability of target codes. We also intend to fully automate the process of target code generation and implement it in an optimizing compiler.

Author Contributions: Conceptualization and methodology, W.B. and P.B.; software, P.B.; validation, W.B., P.B.; data curation, P.B.; original draft preparation, P.B.; writing—review and editing, W.B. and P.B.; visualization, P.B. All authors have read and agreed to the published version of the manuscript.

Funding: This research received no external funding.

Institutional Review Board Statement: Not applicable.

Informed Consent Statement: Not applicable.

Data Availability Statement: Source codes to reproduce all the results described in this article can be found at: https://github.com/piotrbla/counting3d. The iscc script (validitycheck.iscc) carrying out the calculations above is presented at https://github.com/piotrbla/counting3d/blob/main/validity_check.iscc (accessed on 12 January 2023).

Conflicts of Interest: The authors declare no conflict of interest.

Abbreviations

The following abbreviations are used in this manuscript:

RNA	RiboNucleic Acid
TRACO	compiler based on the TRAnsitive ClOsure of dependence graphs

References

1. Nawaz, Z.; Nadeem, M.; van Someren H.; Bertels K. A parallel FPGA design of the Smith-Waterman traceback. In Proceedings of the 2010 International Conference on Field-Programmable Technology, Beijing, China, 8–10 December 2010; Volume 18, pp. 454–459.
2. Manavski, S.A.; Valle, G. CUDA compatible GPU cards as efficient hardware accelerators for Smith-Waterman sequence alignment. *BMC Bioinform.* **2008**, *9*, S10. [CrossRef] [PubMed]
3. Gruzewski, M.; Palkowski, M. RNA Folding Codes Optimization Using the Intel SDK for OpenCL. In Proceedings of the Artificial Intelligence and Soft Computing: 20th International Conference, ICAISC 2021, Virtual Event, 21–23 June 2021; Springer: Cham, Switzerland, 2021; Volume 12855.
4. Gruzewski, M., Palkowski, M. Implementation of Nussinov's RNA Folding Using the Kokkos Library. In *Progress in Image Processing, Pattern Recognition and Communication Systems, Proceedings of the Conference (CORES, IP&C, ACS), Virtual Event, 28–30 June 2021*; Springer: Cham, Switzerland, 2022; Volume 255, pp. 15–30.
5. Palkowski, M.; Bielecki, W. Tiling Nussinov's RNA folding loop nest with a space-time approach. *BMC Bioinform.* **2019**, *20*, 208. [CrossRef]
6. Smith, T.F.; Waterman, M.S. Identification of common molecular subsequences. *J. Mol. Biol.* **1981**, *147*, 195–197. [CrossRef] [PubMed]
7. Raden, M.; Mohamed, M.M.; Ali Syed, M.; Backofen, R. Interactive implementations of thermodynamics-based RNA structure and RNA-RNA interaction prediction approaches for example-driven teaching. *PLoS Comput. Biol.* **2018**, *14*, e1006341. [CrossRef]
8. Raden, M.; Ali, S.M.; Alkhnbashi, O.S.; Busch, A.; Costa, F.; Davis, J.A.; Eggenhofer, F.; Gelhausen, R.; Georg, J.; Heyne, S.; et al. Freiburg RNA tools: A central online resource for RNA-focused research and teaching. *Nucleic Acids Res.* **2018**, *46*, W25–W29. [CrossRef] [PubMed]
9. Bondhugula, U.; Baskaran, M.; Krishnamoorthy, S.; Ramanujam, J.; Rountev, A.; Sadayappan, P. Automatic Transformations for Communication-Minimized Parallelization and Locality Optimization in the Polyhedral Model. In *Compiler Construction, Proceedings of the 17th International Conference, CC 2008, Held as Part of the Joint European Conferences on Theory and Practice of Software, ETAPS 2008, Budapest, Hungary, 29 March–6 April 2008*; Springer: Berlin/Heidelberg, Germany, 2008; pp. 132–146.
10. Lim A.; Cheong G.L.; Lam M.S. An Affine Partitioning Algorithm to Maximize Parallelism and Minimize Communication. In Proceedings of the 13th International Conference on Supercomputing, Rhodes, Greece, 20–25 June 1999; pp. 228–237.
11. Wolf, M.E.; Lam, M.S. A data locality optimizing algorithm. In Proceedings of the ACM SIGPLAN 1991 Conference on Programming Language Design and Implementation, Toronto, ON, Canada, 26–28 June 1991; pp. 30–44.
12. Xue, J. *Loop Tiling for Parallelism*; Springer: Berlin/Heidelberg, Germany, 2000, Volume 575.
13. Bondhugula, U.; Hartono, A.; Ramanujam, J.; Sadayappan, P. Pluto: A practical and fully automatic polyhedral program optimization system. In Proceedings of the ACM SIGPLAN 2008 Conference on Programming Language Design and Implementation (PLDI 08), Tucson, AZ, USA, 7–13 June 2008; pp. 101–113.
14. Palkowski, M.; Bielecki, W. TRACO parallelizing compiler. In *Soft Computing in Computer and Information Science*; Springer: Cham, Switzerland, 2015; pp. 409–421.
15. Mullapudi, R.T.; Bondhugula, U. Tiling for dynamic scheduling. In Proceedings of the 4th International Workshop on Polyhedral Compilation Techniques, Vienna, Austria, 20 January 2014; Volume 20.
16. Palkowski, M.; Bielecki, W. Tuning iteration space slicing based tiled multi-core code implementing Nussinov's RNA folding. *BMC Bioinform.* **2018**, *19*, 12. [CrossRef] [PubMed]
17. Verdoolaege, S. Counting affine calculator and applications. In Proceedings of the First International Workshop on Polyhedral Compilation Techniques (IMPACT'11), Chamonix, France, 3 April 2011.
18. Verdoolaege, S.; Grosser, T Polyhedral extraction tool. In Proceedings of the First Second International Workshop on Polyhedral Compilation Techniques (IMPACT'12), Paris, France, 23 January 2012.
19. Pugh, W.; Wonnacott, D. Static analysis of upper and lower bounds on dependences and parallelism. *ACM Trans. Program. Lang. Syst. (TOPLAS)* **1994**, *16*, 1248–1278. [CrossRef]

20. Verdoolaege, S.; Carlos Juega, J.; Cohen, A.; Ignacio Gomez, J.; Tenllado, C.; Catthoor, F. Polyhedral parallel code generation for CUDA. *ACM Trans. Archit. Code Optim. (TACO)* **2013**, *9*, 1–23. [CrossRef]
21. Bielecki, W.; Poliwoda, M. Automatic parallel tiled code generation based on dependence approximation. In *International Conference on Parallel Computing Technologies*; Springer: Berlin/Heidelberg, Germany, 2021; pp. 260–275.
22. Kennedy K., Allen J. R. *Optimizing Compilers for Modern Architectures: A Dependence-Based Approach*; Morgan Kaufmann Publishers Inc.: San Francisco, CA, USA, 2001.
23. Van der Pas, R.; Stotzer, E.; Terboven, C. *Using OpenMP# The Next Step: Affinity, Accelerators, Tasking, and SIMD*; MIT Press: Cambridge, MA, USA, 2017.
24. Bielecki, W.; Błaszyński, P.; Pałkowski, M. 3D Tiled Code Generation for Nussinov's Algorithm. *Appl. Sci.* **2022**, *12*, 5898. [CrossRef]
25. Nussinov, R.; Pieczenik, G.; Griggs, J.R.; Kleitman, D.J. Algorithms for loop matchings. *SIAM J. Appl. Math.* **1978**, *35*, 68–82. [CrossRef]
26. Li, J.; Ranka, S.; Sahni, S. Multicore and GPU algorithms for Nussinov RNA folding. *BMC Bioinform.* **2014**, *15*, S1. [CrossRef]
27. Zhao, C.; Sahni, S. Cache and energy efficient algorithms for Nussinov's RNA folding. *BMC Bioinform.* **2017**, *18*, 15–30. [CrossRef]

Disclaimer/Publisher's Note: The statements, opinions and data contained in all publications are solely those of the individual author(s) and contributor(s) and not of MDPI and/or the editor(s). MDPI and/or the editor(s) disclaim responsibility for any injury to people or property resulting from any ideas, methods, instructions or products referred to in the content.

eng

MDPI

Article

Covering Arrays ML HPO for Static Malware Detection

Fahad T. ALGorain *,† and **John A. Clark** *,†

Department of Computer Science, University of Sheffield, Sheffield S10 2TN, UK
* Correspondence: ftalgorain1@sheffield.ac.uk (F.T.A.); john.clark@sheffield.ac.uk (J.A.C.)
† These authors contributed equally to this work.

Abstract: Malware classification is a well-known problem in computer security. Hyper-parameter optimisation (HPO) using covering arrays (CAs) is a novel approach that can enhance machine learning classifier accuracy. The tuning of machine learning (ML) classifiers to increase classification accuracy is needed nowadays, especially with newly evolving malware. Four machine learning techniques were tuned using cAgen, a tool for generating covering arrays. The results show that cAgen is an efficient approach to achieve the optimal parameter choices for ML techniques. Moreover, the covering array shows a significant promise, especially cAgen with regard to the ML hyper-parameter optimisation community, malware detectors community and overall security testing. This research will aid in adding better classifiers for static PE malware detection.

Keywords: cAgen; combinatorial testing; covering arrays; machine learning; static PE malware detection; hyper-parameter optimisation; grid search

1. Introduction

1.1. Malware and Its Detection

Malicious software is any programme that can be executed that is intended to cause harm. Academic and commercial research and development into malware detection has been a constant focus for some time now [1] and malware remains one of the most important concerns in contemporary cybersecurity. There are three approaches to malware detection: static, dynamic, and hybrid detection. Static malware detection analyses malicious binary files without executing them; this is the focus of this paper. Dynamic malware detection uses the features of run-time execution behaviour to identify malware. Hybrid detection combines the two previous approaches. Many companies and universities have significantly invested in developing new methods for identifying malware and many researchers have looked into the possibility of using machine learning (ML) to detect it.

1.2. Ml-Based Static Malware Detection Related Literature

Windows Portable Execution (PE) malware is one of the most common forms of encountered malware. Several works have explored the use of machine learning for PE malware detection, e.g., [2–4]. In [5] the authors provided a dataset (usually referred to as the Ember dataset) accompanied by various Python routines to facilitate access. They also provided baseline applications of various ML techniques to their dataset. In [6], the authors considered imbalanced dataset issues and model training duration. They also applied a static detection method using a gradient-boosting decision tree algorithm. Their model achieved better performance than the baseline model with less training time. (They used feature reduction based on the recommendation of the authors of [5].) Another approach used a subset of the Ember dataset for their work and compared different ML models [7]. Their goal was to identify malware families and their work was mainly concerned with scalability and efficiency. The proposed random forest model achieved a slightly better performance than the baseline model. In [8], the authors used a hybrid of two datasets, Ember (version 2017) and another dataset from the security partner of Meraz' 18 techno

Citation: ALGorain, F.T.; Clark, J.A. Covering Arrays ML HPO for Static Malware Detection. *Eng* **2023**, *4*, 543–554. https://doi.org/10.3390/eng4010032

Academic Editor: Antonio Gil Bravo

Received: 9 December 2022
Revised: 16 January 2023
Accepted: 7 February 2023
Published: 9 February 2023

Copyright: © 2023 by the authors. Licensee MDPI, Basel, Switzerland. This article is an open access article distributed under the terms and conditions of the Creative Commons Attribution (CC BY) license (https://creativecommons.org/licenses/by/4.0/).

cultural festival (IIT Behali). A feature selection method—Fast Correlation-based Feature Selection (FCBF)—was used to improve their model's performance. Thirteen features (with high variance) were selected. Several ML models (Decision Trees, Random Forest, Gradient boost, AdaBoost, Gaussian Naive Bayes) were introduced and trained. The Random Forest approach achieved the highest accuracy (99.9%) [9]. The study used the same dataset as this paper. It proposed an ensemble learning-based method for malware detection. A stacked ensemble of fully connected, one-dimensional convolutional neural networks (CNNs) performs the initial stage classification, while a machine learning algorithm handles the final stage classification. They evaluated 15 different machine learning classifiers in order to create a meta-learner. Several machine learning techniques were utilised for this comparison: Naive Bayes, Decision Tree, Random Forests, GB, K-Nearest Neighbours, Stochastic Gradient Descent and Neural Nets. The evaluation was conducted on the Windows Portable Executable (PE) malware dataset. An ensemble of seven neural networks with the ExtraTrees classifier as the last-stage classifier performed the best, achieving perfect accuracy. The model parameters were not stated.

Determining the full detection capabilities of the various methods is a tricky business, particularly when such methods are ML-based. Parameter selections for ML algorithms, for instance, are typically crucial to their performance and yet specific choices in the literature often lack convincing (or sometimes any) rationale. In this paper, we explore how to optimise the parameters of such algorithms, a process known as hyper-parameter optimisation (HPO). We specifically investigate the use of covering arrays as a way to combat the curse of dimensionality that results from Grid Search, which is the most common systematic approach used.

1.3. Grid Search and the Curse of Dimensionality

Grid Search is a powerful and widely used means of searching a parameter space to seek sets of values that give the best performance. Grid Search applies the full combinatoric evaluation of the cross-product of discretised parameter domains. A discretised domain is a set of 'representative' elements that 'span' the domain in some way. For example, the set $0, 5, ..., 95, 100$ can be considered to span the set comprising the integers 0..100. The real interval $[0, 1]$ can be spanned for some purposes by the set with the elements 0.0, 0.25, 0.5, 0.75, and 1.0.

The total number of combinations for Grid Search is the product of the (assumed finite) cardinalities of the individual discretised domains D_i.

$$totalCombinations = \prod_{i=1}^{n} card(D_i)$$

Grid Search can obviously give a thorough exploration of the parameter space, assuming that the individual domains are suitably discretised. However, in some areas of engineering, it is found that full combinatorial evaluation can be wasteful. For example, in software testing, a particular sub-combination coverage of parameter values can provide a very high fault detection capability. However, we do not know in advance the specific sub-combinations that will be the most revealing. Some effective means of exploring the combinatorial space is needed so that we do not incur the costs of a full grid search.

Covering arrays provide one such mechanism. Furthermore, the concept can be applied at different 'strengths', allowing flexibility in the thoroughness of the exploration of the search space at hand. Each discretised parameter domain has a set of values. A covering array is defined over the cross-product of the discretised domains $D = D_1 \times D_2 \times ... \times D_n$. The rows of the array denote specific tests. The columns of the array denote specific parameters. The (i, j) element of the array is the value of parameter j in test i. The Cartesian product of all parameter sets defines complete combinatorial coverage. The rows of a covering array provide a subset of that with a particular strength t. In a CA with strength t, then for any subset of t parameters, each possible t-tuple of values occurs in at least one row (test). This is often called t-way testing. Orthogonal arrays (OAs) are the optimal version of

CAs where each t-tuple occurs *exactly* once (rather than at least once). For some problems, an OA may not actually exist. Pairwise testing ($t = 2$) is widely used. Furthermore, it has been found more generally that small values of t can actually give a strong performance in fault-finding. As t increases, the size of the covering array also increases. The test set reduction achieved by covering arrays compared with a full combinatorial grid search may be very significant. The concept of strength is illustrated below.

We will generally denote a covering array with cross-product D as above and with strength t by CA_t^D. Consider a combinatorial search space with the (discretised) parameter domains $A = \{0,1\}$, $B = \{0,1\}$, and $C = \{0,1\}$, i.e., with a cross-product $D_{abc} = A \times B \times C$. A $CA_1^{D_{abc}}$ provides a suite of cases where each value of each domain occurs at least once. This is easily achieved by an array with just two rows (i.e., two cases) as shown below. $A = 0$ occurs in the first row and $A = 1$ occurs in the second. This is similar for B and C.

A	B	C
0	0	0
1	1	1

If we had, say, 26 binary domains $A, B, ..., Z$, then a similar covering array, i.e., a $CA_1^{D_{abc..z}}$ with two rows would satisfy the $t = 1$ strength requirement, i.e., with rows as shown below.

A	B	C	..	X	Y	Z
0	0	0	..	0	0	0
1	1	1	..	1	1	1

A CA_1^D clearly gives a rather weak coverage (exploration) of the domain space for most purposes. In the $A, B, ..., Z$ example, only 2 from 2^{26} possible row values are sampled. For a CA_2^D, each combination of values from any two ($t = 2$) domains is present in the array. A $CA_2^{D_{abc}}$ for the A, B, and C example is given below.

A	B	C
0	0	0
0	1	1
1	0	1
1	1	0

We can see that the four possible values of (A, B) are present, i.e., (A, B) =$(0, 0)$ in row 0, $(0, 1)$ in row 1, $(1, 0)$ in row 2, and $(1, 1)$ in row 3. Similarly, we can see that four possible values of (A, C) and the four values of (B, C) are also present. Thus, all pairs of values from any two domains from A, B, and C are present and so the given array is indeed a $CA_2^{D_{abc}}$. The simplest $CA_3^{D_{abc}}$ array would give full combinatorial coverage, i.e., with all eight (A, B, C) combinations, with the usual binary enumeration of 0–7 for the rows, i.e., $[0, 0, 0]$ through to $[1, 1, 1]$.

1.4. Generating Covering Arrays

The actual generation of arrays is not our focus. A good deal of theoretical and practical work has been carried out into algorithms to do so. Our motivation to use covering arrays was inspired by their use in software testing. The in-parameter order (IPO) method for generating CA_2^D arrays for test suites is given in [10]. As they state, "For a system with two or more input parameters, the IPO strategy generates a pairwise test set for the first two parameters, extends the test set to generate a pairwise test set for the first three parameters, and continues to do so for each additional parameter."

CAs have become widely used in the combinatorial testing field where they provide a means of reducing the number of tests needed in comparison to exhaustive combinatorial testing. This has led to an increased use of a specific instance of the IPO strategy called in-parameter-order-general (IPOG). IPOG can be used to generate covering arrays of arbitrary strengths [11]. It is a form of greedy algorithm and might not yield the test suites of minimal size. It has been noted that providing an optimal covering array is an NP-complete problem [12].

The IPOG strategy has gained traction in the software testing field. This is due to the competitive test suites that are yielded by the covering arrays it generates in comparison with other approaches for generating test suites. Additionally, it exhibits a lower generation time than other algorithms. The main goal of IPOG is to minimise the generated test suite size. This is a significant area to explore, especially when the cost of testing is very high. The duration of test cycles will be reduced with fewer tests. However, there are some cases when the test execution is very fast and does not impact the overall testing time. Instead, optimising test suites can be very costly as the test generation time can become dominating [13,14]. Optimisation of the IPOG family was introduced by [15].

There are many problems for CAs [16], where the construction of optimal values is known to be the hardest [17]. Various methods for generating covering arrays have been proposed. These include the automatic efficient test generator (AETG) system [18], deterministic density algorithm (DDA) [19,20], in-parameter order [21], and the advanced combinatorial testing system (ACTS) [22], each with its own advantages and disadvantages. Interested readers are referred to [18–21,23], respectively, for more information. The in-parameter-order (IPO) strategy grows the covering array column by column, adding rows as needed to ensure full t-way coverage. Various kinds of research on improving the covering array generation with the in-parameter-order strategy have been conducted. The original aim of the strategy was the generalisability of generating covering arrays of arbitrary strength [11] resulting in the in-parameter-order-general (IPOG) algorithm. In [10] a modification to IPOG resulted in smaller covering arrays in some instances and faster generation times. In [24], a combination of IPOG with a recursive construction method was proposed that reduces the number of combinations to be enumerated. In [25], the use of graph-colouring schemes was proposed to reduce the size of the covering arrays. In [26], IPOG was modified with additional optimisations aimed at reducing don't-care values in order to have a smaller number of rows. Most of these presented works primarily aimed to reduce the size of generated covering arrays. The FIPOG technique was shown to outperform the IPOG implementation of ACTs in all benchmarks and improved test generation times by up to a factor of 146 [15].

In this paper, we use an implementation of FIPOG provided by the cAgen tool. We show that the use of FIPOG's covering arrays can achieve excellent results for the hyper-parameter optimisation of ML-based classifiers (and better than using the default parameters) far more quickly than when using full grid search. Below, we describe the cAgen tool [27], which implements the FIPOG technique.

1.5. The Cagen Toolset

The cAgen toolset provides a means of generating the covering arrays and is available online [27]. This allows the user to specify parameters and sets of associated values. For technical reasons that are concerned with our specific approach to the use of covering arrays, we will assume that a discretised parameter domain with R elements is indexed by values 0, 1, (R-1). Figure 1 shows a completed specification for the (A, B, C) example above.

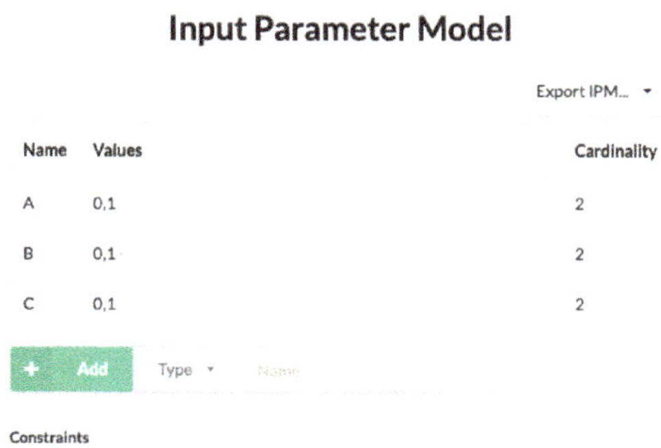

Figure 1. Full parameter specification for the ABC example (no constraints)

Having specified the parameters, we can invoke the generation capability of cAgen, as Figure 2 shows the array generation stage for the A, B, and C examples, where a value of t = 2 was selected. If we wanted each pair to occur multiple times, we could specify a larger value of *lambda*. Several generation algorithms are available. Figure 2 shows that we have chosen FIPOG for better performance and fast generation [27]. The array can then be stored in a variety of formats. We chose to use the CSV format throughout.

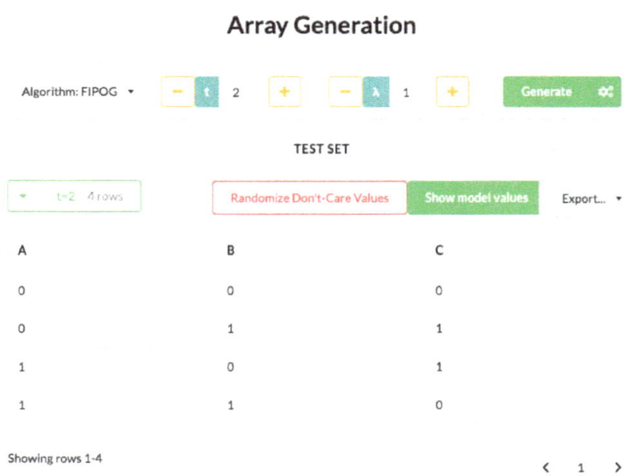

Figure 2. Array generation for ABC example above with t = 2.

1.6. Array Indexing

We use lists to represent parameter spaces. A list's elements will be either actual parameter values or else a list representing a subdomain. The values $0, 1, \ldots, (R-1)$ are interpreted as indices to the corresponding elements in the discretised domain list. For example, $MAX\_DEPTH = [5, 10, 15, 20, None]$ would be a simple list with four specific integer values and a 'None' value. $LEARNING\_RATE = [0.001, 0.01, 0.1, 0.2]$ is a simple list of four real values. $MAX\_LEAVES = [[1,2,3],[4,5,6],[7,8,9]]$ is a list of lists of values. Here, the list cardinalities are given by $card(MAX\_DEPTH) = 5$, $card(LEARNING\_RATE) = 4$ and $card(MAX\_LEAVES) = 3$. Thus, for the list of lists, the cardinality is the cardinality of the highest-level list. Covering array values are indices to the top-level list.

Python lists allow us to include different types of elements. Thus, in MAX_DEPTH, we see that integer values, as well as a 'None' parameter value, can be specified. 'None' typically means that the algorithm can proceed as it sees fit, with no direction from the user for this parameter. ScikitLearn's ML algorithms often have such parameters as defaults. Where a parameter is represented by a simple list, then the covering array value for the parameter is used to index the specific element of the list. Thus, an array value of 2 in the covering array column corresponding to MAX_DEPTH corresponds to a parameter value of 15, i.e., $MAX\_DEPTH[2] = 15$. The indexed array element may be a list. Thus, a covering array value of 2 for MAX_LEAVES, gives $MAX\_LEAVES[2] = [7,8,9]$. In such a case, a value is randomly selected from the indexed list $[7,8,9]$. Thus, each of the values 7, 8, and 9 are now selected with a probability of 0.333. In practice, we represent regular integer ranges of integer values more compactly, via the use of low, high, and increment indicators. Thus, we will typically represent the list $[1,2,3,4,5]$ by $[low, high, incr] = [1,5,1]$. We adopt the convention of both low and high being included in the denoted range. We distinguish between simple lists with three elements and 'compact' lists with the same three elements (selection is resolved by different routines, determined at the set-up time by the user).

1.7. Structure of the Paper

In this paper, we investigate whether the clear efficiency benefits a covering array approach can be brought to bear on the ML-based static malware detection problem. Section 2 describes our methodology. Section 2.2 details the performed experiments. The results are presented in Sections 3 and 4 concludes our paper.

2. Methodology

2.1. Overall Approach

We apply a variety of ML techniques and specify suitable domains for the parameters we wish to experiment with (other parameters assume defaults). We evaluate over the full combinatorial domain (for Grid Search) and over all rows of the covering arrays of interest (for t = 2, 3, 4). A full combinatorial evaluation or a full covering array evaluation (i.e., all rows evaluated) will be referred to as a 'run' or 'iteration'. We carry out 30 runs for each array of interest and for the full combinatorial case. We do this in order to gain insight into the distribution of outcomes from the technique. Some runs may give better results than others, even if the same array has been used as the basis for the run. This is due to the stochastic selection of elements within selected ranges as indicated above. Pooling the results from the 30 runs provides a means of determining an accurate and useful distribution for the approach. In practice, a user may simply use one run of a covering array search, if they are confident that it will give good enough results. Our evaluation activities aim to determine whether such confidence is justified.

2.2. Experimental Details

Our work uses two powerful toolkits: scikitLearn [28] and the cAgen tool [27]. The experiments are carried out using the Windows OS 11, with 11 Gen Intel Core i7-11800H, with a 2.30 GHz processor, and 16 GB RAM. The work uses a dataset [29] built using PE files from [30]. The dataset has 19,611 labelled malicious and benign samples from different repositories (such as VirusShare). The samples have 75 features. The dataset is split into a training dataset and a testing dataset (80% training, 20% testing) and can be found in [29]. All results are obtained using Jupyter Notebook version 6.1.0 and Python version 3.6.0.

A small amount of pre-processing is carried out on the malware dataset. The 'Malware' feature records the label for the supervised learning. From the remaining (i.e., input) feature columns, we restrict ourselves to binary and numerical features and so drop the 'Name', 'Machine', and 'TimeDateStamp' features. The filtered input features are then subject to scaling via scikitLearn's StandardScalar fit_transform function. The same approach is taken for all the ML approaches considered. No further feature engineering is performed. This is deliberate. Our aim is investigate ML model hyper-parameters; we wish to keep other factors constant (researchers whose focus is any of the specific ML approaches are free to engage in further optimisations should they so wish).

We evaluate a covering-array-based hyper-parametrisation on three well-established ML approaches (Decision Trees (DTs) [28], xgboost [31], and Random Forest (RF) [28,32]) together with a state-of-the-art approach (LightGBM [33]). Table 1 shows the implementation details for all four ML models using the cAgen tool. In particular, the hyper-parameter ranges of interest are shown for each technique, together with the corresponding IPM values (giving possible indices into the top level array list). The ML evaluation metric is *accuracy* as implemented by scikitLearn [34]. All three-element lists in our experiments are compact lists. The results are processed using SciPy's 'descriptive statistics' method [35].

Table 1. ML Models cAgen configurations.

ML Algorithms	Hyper-Parameters	Hyper-Parameter IPM Values	T-Strengths Values	IPM Values	Number of Iterations
RF	n_estimators, max_depth, criterion, min_samples_split, min_samples_leaf, max_features	[[100, 300, 50], [350, 550, 50], [600, 800, 50]] [[1, 10, 1], [11, 15, 1], [16, 20, 1],None] ['entropy', 'gini'] [[5, 25, 5], [30, 50, 5]] [[5, 25, 5], [30, 50, 5]] ['auto', 'sqrt', 'log2', 'None']	T-2, 3, 4	0, 1, 2 0, 1, 2, 3 0, 1 0, 1 0, 1 0, 1, 2, 3	30
LightGBM	num_leaves, boosting_type, Subsample_for_bin, Is_unbalance, max_depth	[[20, 80, 20], [100, 160, 20]] ['GBDT', 'GOSS'] [[1000, 5000, 1000], [6000, 10000, 1000], [11000, 15000, 1000]] [True, False] [1, 5, 10, 15, 20, 25]	T-2, 3, 4	0, 1 0, 1 0, 1, 2 0, 1 0, 1, 2, 3, 4, 5	30
Xgboost	Min_child_weight gamma max_leaves reg_alpha max_depth	[1, 2, 4, 6, 8, 10, 12, 14] [[1, 4, 1], [5, 8, 1]] [2, 4, 6, 8, 10, 12] [0.01, 0.1, 0.2, 0.3, 0.4, 0.5] [1, 5, 10, 15, 20, 25]	T-2, 3, 4	0, 1, 2, 3, 4, 5, 6, 7 0, 1 0, 1, 2, 3, 4, 5 0, 1, 2, 3, 4, 5 0, 1, 2, 3, 4, 5	30
DT	max_depth, criterion, min_samples_split, min_samples_leaf, max_features	[[1, 10, 1], [11, 15, 1], [16, 20, 1], None] ['entropy', 'gini'] [[5, 25, 5], [30, 50, 5]] [[5, 25, 5], [30, 50, 5]] ['auto', 'sqrt', 'log2', 'None']	T-2, 3, 4	0, 1, 2, 3 0, 1 0, 1 0, 1 0, 1, 2, 3	30

3. Results

The results of hyper-parameter optimisation based on covering arrays (with strengths of 2, 3, or 4) and grid search are shown in the following tables. The best-performing parameter values are given, together with the time taken to complete the corresponding search, coverage (number of evaluations), and summary accuracy data. Tables 2–5 show results for RF, LightGBM, Xgboost, and DT, respectively. The results for Grid Search (over the same discretised parameter ranges) are also shown in each table. In the tables "No. of evaluations" is equal to the number of rows (i.e., combinations) in the covering array multiplied by the number of iterations (30).

Table 2. RF Model cAgen Results Comparison.

ML Algorithms	Optimal Values	T-Values/ Grid Search	Time to Complete	No. of Evaluations	Score (min/max Accuracy and Mean)
RF	400 14 entropy 5 10 None	T2	2 h 35 min 21 s	480	minmax = (0.9031205384458495, 0.9904140322251682) mean = 0.9743610662231913
	700 None entropy 5 10 None	T3	7 h 31 min 17 s	1500	minmax = (0.8916989598205181, 0.9902100754640016), mean = 0.9760138690597593
	150 18 entropy 5 10 None	T4	11 h 58 min 4 s	2880	minmax = (0.8949622679991842, 0.9902100754640016), mean = 0.975666825299703
	650 19 entropy 5 5 None	Full Grid Search	2 Days, 23 h, 58 min and 18 s	11520	minmax = (0.8853763002243524, **0.9906179889863349**), mean = 0.9755531264871848

Table 3. LightGBM Model cAgen Results Comparison.

ML Algorithms	Optimal Values	T-Values/ Grid Search	Time to Complete	No. of Evaluations	Score (min/max Accuracy and Mean)
LightGBM	60 gbdt 9000 False 25	T2	1 h 41 min 18 s	540	minmax = (0.9726697940036713 0.9908219457475015), mean = 0.9858956345699156
	60 gbdt 15000 False 25	T3	3 h 4 min 23 s	1080	min,max = (0.9702223128696716 0.9908219457475015) mean = 0.985933215491649
	40 gbdt 1000 False 15	T4	5 h 36 min 55 s	2160	min,max = (0.9704262696308382, **0.9912298592698348**), mean = 0.9859165023681646
	140 goss 1000 False 20	Full Grid Search	7 h 57 min 14 s	4320	min,max = (0.9696104425861717, 0.9910259025086682), mean = 0.985890630075313

Table 4. Xgboost Model cAgen Results Comparison (Kaggle dataset).

ML Algorithms	Optimal Values	T-Values/ Grid Search	Time to Complete	No. of Evaluations	Score (min/max Accuracy and Mean)
Xgboost	1 1 10 0.4 20	T2	1 h 9 min 21 s	1440	min,max = (0.9763410157046706, 0.9902100754640016), mean = 0.9856683549754118
	2 1 2 0.01 25	T3	5 h 37 min 32 s	8640	min,max = (0.9763410157046706, 0.9902100754640016), mean = 0.985969474471412
	2 1 12 0.1 25	T4	23 h 12 min 33 s	51840	min,max = (0.9763410157046706, 0.9906179889863349), mean = 0.9859859987460435
	1 1 10 0.01 15	Full Grid Search	1 d 12 h 21 min 51 s	103680	min,max = (0.9763410157046706, **0.9906179889863349**), mean = 0.9856382316161938

Table 5. DT Model cAgen Results Comparison.

ML Algorithms	Optimal Values	T-Values/ Grid Search	Time to Complete	No. of Evaluations	Score (min/max Accuracy and Mean)
DT	Entropy 20 None 5 25	T2	42.5 s	480	min,max = (0.744034264735876 0.9836834591066694) mean = 0.9656014174994901
	Entropy 11 None 5 15	T3	1 min 28 s	960	minmax = (0.7352641240057108, 0.9840913726290027), mean = 0.9662880719287512
	Entropy None None 5 10	T4	2 min 46 s	1920	min,max = (0.7352641240057108, 0.9840913426290027), mean = 0.9662736249915017
	Gini None None 10 20	Full Grid Search	5 min 43 s	3840	min,max = (0.7352641240057108 **0.9849071996736691**) mean = 0.9660357285505473

We can see that the DT classifier in Table 5 is the fastest of all ML models. Even if we look at the grid search, it is still efficient with this particular technique, taking only 5 min and 43 s to finish 3840 evaluations. However, cAgen is much more efficient with only 42.5 s to finish. Although only 480 evaluations with $t = 2$ were made, this achieves the same accuracy as Grid Search but with less time and effort. The second fastest ML model after DT was LightGBM, which highlights covering the array capability even more. Table 3 shows a huge disparity in time between the Grid Search and cAgen runs. The cAgen approach is faster than the Grid Search with only 1 h, 41 min and 18 s taken to complete the search, while the latter took 7 h, 57 min, and 14 s. Both reached excellent values for finding hyper-parameter choices while having higher accuracy. Both strength values $t = 2$ and $t = 3$ in LightGBM, even though they have almost the same results obtained, reached a score with different hyper-parameter values. cAgen is more efficient than Grid Search, using less time. The third ML model was RF, where cAgen runs reached the highest performing choices for $t = 2$ with 2 h and 35 min. In contrast, Grid Search took 2 days, 23 h and 58 min to complete the search. The difference between cAgen and Grid Search in Table 2 is significant evidence of the usefulness of covering arrays for hyper-parameter optimisation. Xgboost was the slowest of all models to achieve the best values. It took more computational time than the other techniques to achieve the best values for strengths $t = 3$ and $t = 4$, and even Grid Search. The figures below Figures 3–5 compare the accuracy results between the selected models ($t = 2$, $t = 3$ and $t = 4$) in a histogram. (These histograms are not normalised between techniques, i.e., the total counts may vary between techniques. However, the general distributions can be compared.)

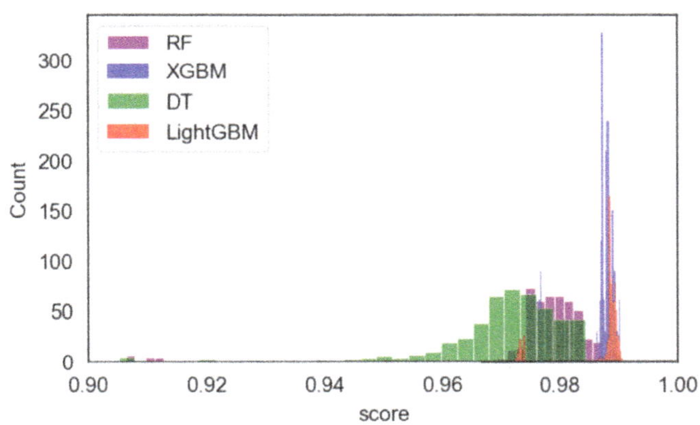

Figure 3. cAgen ML Models Results Comparison for Strength $t = 2$.

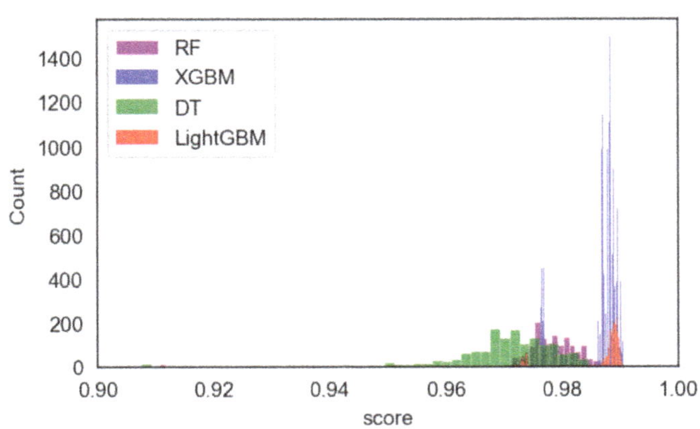

Figure 4. cAgen ML Models Results Comparison for Strength $t = 3$.

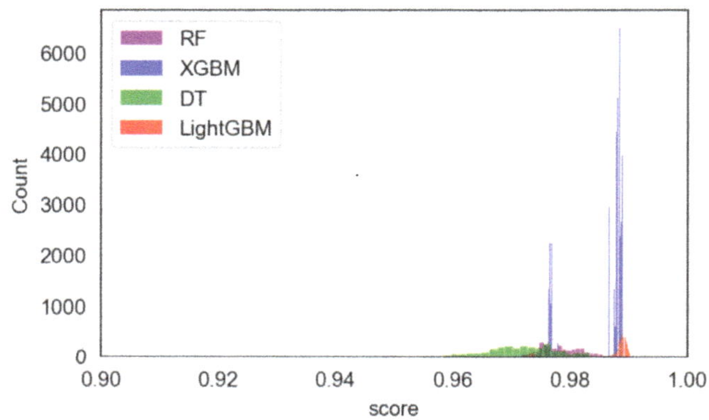

Figure 5. cAgen ML Models Results Comparison for Strength $t = 4$.

The authors in [9] benchmarked several ML models' performances using ensemble learning with 10-fold cross-validation. For DT and RF, the accuracy results were 0.989 and 0.984, respectively. Our model achieved 0.9849 and 0.9904. However, the main aim of our paper was to evaluate various coverage strategies, and not necessarily achieving an optimal value for each ML technique application. If explicit optima are the target, then further optimisations should be considered (see below).

4. Discussion

cAgen, a covering array approach with various strengths, was used to find high-performing hyper-parameters for targeted ML models. It was compared to Grid Search. Our results show that the systematic coverage offered by covering arrays can be both highly effective and efficient. The covering arrays produced by cAgen produced superior results to Grid Search across all four ML models. We highly recommend the covering arrays approach for ML researchers and the community overall. Although our work focused on improving the attained accuracy of malware classification, other security tasks may benefit from such an approach (particularly ML-based classification tasks).

For future work, we would like to assess the feasibility of adding more ML models techniques and hyper-parameters, increasing the complexity of search space/test sets, comparing different settings within the workspaces itself (e.g., FIPOG-F and FIPOG-2F), and increasing the complexity of t-way testing by adding constraints. We also believe that there is merit in considering hierarchical approaches to hyper-parametrisation, i.e., using the best values to come out of a set of runs (or even a single run) as identifying a reduced space to be systematically searched (e.g., using another covering array).

We note that we generally seek only excellent results. There is no guarantee of optimality from any of our tested approaches. An optimal result may well be given at a point that is simply not present in the cross-product of discretised domains because the discretisation process only defines *representative points* to span the domain. Furthermore, for each ML model considered, we presented what we believe are *plausible* discretised parameters ranges as the basis for our experiments. The specific choices made may affect the results. We acknowledge that other ranges are possible.

Furthermore, although one might legitimately expect higher strengths of a covering array to give rise to improved results, this is not actually guaranteed. Furthermore, after building up experience with the approaches for a specific system, one might accept that a low-strength array gives highly acceptable results very quickly, and so choose to use such arrays for all subsequent runs when training data are updated.

Author Contributions: Writing—original draft, F.T.A. and J.A.C.; Writing—review & editing, F.T.A. and J.A.C. All authors have read and agreed to the published version of the manuscript.

Funding: This research received no external funding.

Institutional Review Board Statement: Not applicable.

Informed Consent Statement: Not applicable.

Data Availability Statement: Dataset can be found in https://www.kaggle.com/datasets/amauricio/pe-files-malwares.

Conflicts of Interest: The authors declare no conflict of interest.

References

1. Pandey, A.K.; Tripathi, A.K.; Kapil, G.; Singh, V.; Khan, M.W.; Agrawal, A.; Kumar, R.; Khan, R.A. Trends in Malware Attacks: Identification and Mitigation Strategies. In *Critical Concepts, Standards, and Techniques in Cyber Forensics*; IGI Global: Hershey, PA, USA, 2020; pp. 47–60.
2. Schultz, M.G.; Eskin, E.; Zadok, F.; Stolfo, S.J. Data mining methods for detection of new malicious executables. In Proceedings of the Proceedings 2001 IEEE Symposium on Security and Privacy. S&P 2001, Oakland, CA, USA, 14–16 May 2000 ; pp. 38–49.
3. Kolter, J.Z.; Maloof, M.A. Learning to detect and classify malicious executables in the wild. *J. Mach. Learn. Res.* **2006**, *7*, 470–478.
4. Raff, E.; Barker, J.; Sylvester, J.; Brandon, R.; Catanzaro, B.; Nicholas, C.K. Malware detection by eating a whole exe. In Proceedings of the Workshops at the Thirty-Second AAAI Conference on Artificial Intelligence, New Orleans, LA, USA, 2–7 February 2018.
5. Anderson, H.S.; Roth, P. elastic/ember. Available online: https://github.com/elastic/ember/blob/master/README.md (accessed on 8 December 2022).
6. Pham, H.D.; Le, T.D.; Vu, T.N. Static PE malware detection using gradient boosting decision trees algorithm. In *Proceedings of the International Conference on Future Data and Security Engineering*; Springer: Berlin/Heidelberg, Germany, 2018; pp. 228–236.
7. Fawcett, C.; Hoos, H.H. Analysing differences between algorithm configurations through ablation. *J. Heuristics* **2016**, *22*, 431–458. [CrossRef]
8. Malik, K.; Kumar, M.; Sony, M.K.; Mukhraiya, R.; Girdhar, P.; Sharma, B. *Static Malware Detection and Analysis Using Machine Learning Methods*; 2022 Available online: https://www.mililink.com/upload/article/793128668aams_vol_217_may_2022_a47_p4183-4196_kartik_malik_et_al..pdf (accessed on 8 December 2022).
9. Azeez, N.A.; Odufuwa, O.E.; Misra, S.; Oluranti, J.; Damaševičius, R. Windows PE Malware Detection Using Ensemble Learning. *Informatics* **2021**, *8*, 10. [CrossRef]
10. Forbes, M.; Lawrence, J.; Lei, Y.; Kacker, R.N.; Kuhn, D.R. Refining the in-parameter-order strategy for constructing covering arrays. *J. Res. Natl. Inst. Stand. Technol.* **2008**, *113*, 287. [CrossRef] [PubMed]
11. Lei, Y.; Kacker, R.; Kuhn, D.R.; Okun, V.; Lawrence, J. IPOG: A general strategy for t-way software testing. In Proceedings of the 14th Annual IEEE International Conference and Workshops on the Engineering of Computer-Based Systems (ECBS'07), Tucson, AZ, USA, 26–29 March 2007; pp. 549–556.
12. Seroussi, G.; Bshouty, N.H. Vector sets for exhaustive testing of logic circuits. *IEEE Trans. Inf. Theory* **1988**, *34*, 513–522. [CrossRef]
13. Kitsos, P.; Simos, D.E.; Torres-Jimenez, J.; Voyiatzis, A.G. Exciting FPGA cryptographic Trojans using combinatorial testing. In Proceedings of the 2015 IEEE 26th International Symposium on Software Reliability Engineering (ISSRE), Washington, DC, USA, 2–5 November 2015; pp. 69–76.
14. Kleine, K.; Simos, D.E. Coveringcerts: Combinatorial methods for X. 509 certificate testing. In Proceedings of the 2017 IEEE International Conference on Software Testing, Verification and Validation (ICST), Toyko, Japan, 13–17 March 2017; pp. 69–79.
15. Kleine, K.; Simos, D.E. An efficient design and implementation of the in-parameter-order algorithm. *Math. Comput. Sci.* **2018**, *12*, 51–67. [CrossRef]
16. Hartman, A.; Raskin, L. Problems and algorithms for covering arrays. *Discret. Math.* **2004**, *284*, 149–156. [CrossRef]
17. Colbourn, C.J.; Dinitz, J.H. Part VI: Other Combinatorial Designs. In *Handbook of Combinatorial Designs*; Chapman and Hall/CRC: Boca Raton, FL, USA, 2006; pp. 349–350.
18. Cohen, D.M.; Dalal, S.R.; Fredman, M.L.; Patton, G.C. The AETG system: An approach to testing based on combinatorial design. *IEEE Trans. Softw. Eng.* **1997**, *23*, 437–444. [CrossRef]
19. Bryce, R.C.; Colbourn, C.J. The density algorithm for pairwise interaction testing. *Softw. Test. Verif. Reliab.* **2007**, *17*, 159–182. [CrossRef]
20. Bryce, R.C.; Colbourn, C.J. A density-based greedy algorithm for higher strength covering arrays. *Softw. Testing, Verif. Reliab.* **2009**, *19*, 37–53. [CrossRef]
21. Lei, Y.; Tai, K.C. In-parameter-order: A test generation strategy for pairwise testing. In Proceedings of the Proceedings Third IEEE International High-Assurance Systems Engineering Symposium (Cat. No. 98EX231), Washington, DC, USA, 13–14 November 1998; pp. 254–261.

22. Yu, L.; Lei, Y.; Kacker, R.N.; Kuhn, D.R. Acts: A combinatorial test generation tool. In Proceedings of the 2013 IEEE Sixth International Conference on Software Testing, Verification and Validation, Luxembourg, 18–22 March 2013; pp. 370–375.
23. Torres-Jimenez, J.; Izquierdo-Marquez, I. Survey of covering arrays. In Proceedings of the 2013 15th International Symposium on Symbolic and Numeric Algorithms for Scientific Computing, Washington, DC, USA, 23–26 September 2013; pp. 20–27.
24. Lei, Y.; Kacker, R.; Kuhn, D.R.; Okun, V.; Lawrence, J. IPOG/IPOG-D: Efficient test generation for multi-way combinatorial testing. *Softw. Testing, Verif. Reliab.* **2008**, *18*, 125–148. [CrossRef]
25. Duan, F.; Lei, Y.; Yu, L.; Kacker, R.N.; Kuhn, D.R. Improving IPOG's vertical growth based on a graph coloring scheme. In Proceedings of the 2015 IEEE Eighth International Conference on Software Testing, Verification and Validation Workshops (ICSTW), Graz, Austria, 13–17 April 2015; pp. 1–8.
26. Younis, M.I.; Zamli, K.Z. MIPOG-an efficient t-way minimization strategy for combinatorial testing. *Int. J. Comput. Theory Eng.* **2011**, *3*, 388. [CrossRef]
27. Group, M.R. Covering Array Generation. Available online: https://matris.sba-research.org/tools/cagen/#/about (accessed on 21 July 2022).
28. Pedregosa, F.; Varoquaux, G.; Gramfort, A.; Michel, V.; Thirion, B.; Grisel, O.; Blondel, M.; Prettenhofer, P.; Weiss, R.; Dubourg, V.; et al. Scikit-learn: Machine learning in Python. *J. Mach. Learn. Res.* **2011**, *12*, 2825–2830.
29. Mauricio. Benign Malicious. Available online: https://www.kaggle.com/amauricio/pe-files-malwares (accessed on 10 November 2022).
30. Carrera, E. pefile. 2022. Available online: https://github.com/erocarrera/pefile (accessed on 15 January 2022).
31. Chen, T.; Guestrin, C. Xgboost: A scalable tree boosting system. In Proceedings of the 22nd ACM Sigkdd International Conference on Knowledge Discovery and Data Mining, San Francisco, CA, USA, 13–17 August 2016; pp. 785–794.
32. Buitinck, L.; Louppe, G.; Blondel, M.; Pedregosa, F.; Mueller, A.; Grisel, O.; Niculae, V.; Prettenhofer, P.; Gramfort, A.; Grobler, J.; et al. API design for machine learning software: experiences from the scikit-learn project. *arXiv* **2013**, arXiv:1309.0238.
33. LightGBM documentation. Available online: https://lightgbm.readthedocs.io/en/latest (accessed on 20 August 2021).
34. Sklearn. Sklearn-Accuracy-Metrics. Available online: https://scikit-learn.org/stable/modules/generated/sklearn.metrics.accuracy_score.html (accessed on 21 July 2022).
35. Virtanen, P.; Gommers, R.; Oliphant, T.E.; Haberland, M.; Reddy, T.; Cournapeau, D.; Burovski, E.; Peterson, P.; Weckesser, W.; Bright, J.; et al. SciPy 1.0: Fundamental Algorithms for Scientific Computing in Python. *Nat. Methods* **2020**, *17*, 261–272. [CrossRef] [PubMed]

Disclaimer/Publisher's Note: The statements, opinions and data contained in all publications are solely those of the individual author(s) and contributor(s) and not of MDPI and/or the editor(s). MDPI and/or the editor(s) disclaim responsibility for any injury to people or property resulting from any ideas, methods, instructions or products referred to in the content.

eng

MDPI

Article

Similarity of Musical Timbres Using FFT-Acoustic Descriptor Analysis and Machine Learning

Yubiry Gonzalez * and **Ronaldo C. Prati**

Center of Mathematics, Computer Science, and Cognition, Federal University of ABC, Av. Dos Estados, 5001, Santo André 09210-580, SP, Brazil
* Correspondence: yubiry.gonzalez.17@gmail.com

Abstract: Musical timbre is a phenomenon of auditory perception that allows the recognition of musical sounds. The recognition of musical timbre is a challenging task because the timbre of a musical instrument or sound source is a complex and multifaceted phenomenon that is influenced by a variety of factors, including the physical properties of the instrument or sound source, the way it is played or produced, and the recording and processing techniques used. In this paper, we explore an abstract space with 7 dimensions formed by the fundamental frequency and FFT-Acoustic Descriptors in 240 monophonic sounds from the Tinysol and Good-Sounds databases, corresponding to the fourth octave of the transverse flute and clarinet. This approach allows us to unequivocally define a collection of points and, therefore, a timbral space (Category Theory) that allows different sounds of any type of musical instrument with its respective dynamics to be represented as a single characteristic vector. The geometric distance would allow studying the timbral similarity between audios of different sounds and instruments or between different musical dynamics and datasets. Additionally, a Machine-Learning algorithm that evaluates timbral similarities through Euclidean distances in the abstract space of 7 dimensions was proposed. We conclude that the study of timbral similarity through geometric distances allowed us to distinguish between audio categories of different sounds and musical instruments, between the same type of sound and an instrument with different relative dynamics, and between different datasets.

Keywords: musical timbre; FFT; musical instruments; acoustic descriptors; machine learning; data analysis; tinysol; goodsounds

Citation: Gonzalez, Y.; Prati, R.C. Similarity of Musical Timbres Using FFT-Acoustic Descriptor Analysis and Machine Learning. *Eng* **2023**, *4*, 555–568. https://doi.org/10.3390/eng4010033

Academic Editor: Antonio Gil Bravo

Received: 25 December 2022
Revised: 4 February 2023
Accepted: 6 February 2023
Published: 9 February 2023

Copyright: © 2023 by the authors. Licensee MDPI, Basel, Switzerland. This article is an open access article distributed under the terms and conditions of the Creative Commons Attribution (CC BY) license (https://creativecommons.org/licenses/by/4.0/).

1. Introduction

Musical timbre is a multidimensional attribute of musical instruments and of music in general, which, as a first approximation, allows one to differentiate one sound from another when they have the same intensity, duration, and pitch. It is well known that the complexity of musical timbre is not only associated with the identification of a musical instrument. We can find musical sounds with more similar timbre characteristics between acoustically different instruments than those of instruments with the same acoustic characteristics, considering the same pitch and dynamics.

Since musical timbre is a phenomenon of auditory perception, many of the investigations were developed in line with psychoacoustics with the aim of evaluating verbal descriptors that reveal measurable attributes of musical timbre [1–4]. The attributes of color vision and the perception of musical timbre were revealed through experiments on the subjective evaluation of perception [5]. Other more recent studies focused on similarities in the perception of images and the perception of timbre in various types of musical instruments, with models that represent timbre through linguistic-cognitive variables in a two-dimensional space [6,7]. Although the psychoacoustic perception of the musical timbre cannot be ignored, it must be recognized that the main timbral characteristics must be inscribed somehow within the Fast Fourier Transform (FFT) that enables the recording and subsequent reproduction of musical sound.

For the sake of argument, suppose that there are significant timbral characteristics that are not contained in the FFT performed on a musical audio record. In this scenario, the deconvolved audio (inverse convolution) of the reproduced digital record cannot be distinguished timbrally. However, this does not occur in musical digitization, as we are able to distinguish timbral aspects from deconvolved audios. Therefore, the FFT contains all of the significant timbral characteristics. If monophonic audio recordings of constant frequency (separate musical notes), equal intensity, and duration are considered, then the FFT will account for the timbre differences. Although psychoacoustic aspects are important, under these considerations, their effect on audio records does not affect the differences and timbral similarities of comparisons between the various audio records.

The characterization of musical timbre from the analysis of the spectrum contained in the Fast Fourier Transform has been one of the research topics of recent years in the fields of Musical Information Retrieval (MIR), Automatic Music Transcription (AMT), and performances of electro-acoustic music, among others. Recent developments in Signal Processing for Music Analysis [8–10] have allowed important applications in audio synthesis and in the deconvolution of polyphonic musical signals using spectrograms. However, it remains to be quantified which of the minimal descriptors of musical timbre characteristics, when present in the FFT of the audio records, are responsible for the acoustic stimulus, which allows the auditory identification of the sound source.

To extract information from the frequency spectrum of musical records, one must define the magnitudes, functions, or coefficients that describe or characterize a certain spectrum, which is generically called the acoustic descriptors. These provide quantitative measures that describe the set of amplitudes and frequencies of the FFTs of the audio records. Many researchers [11–23] focused on the presentation of an exhaustive collection of timbre descriptors (Timbre ToolBox, Librosa, etc.) that can be computationally extracted from a statistical analysis of the spectrum (FFT). Several other spectral descriptors appear in the literature, although there is no consensus on which or how many acoustic descriptors are necessary to characterize musical timbre. However, it is recognized that many of them are derivatives or combinations of others and that, in general, they are correlated with each other [12].

The use of the FFT and its representation in the frequency domain could be a way to study the physical characteristics of the musical timbre, thus, having a collection of well-bounded, discrete, and measurable pairs of computable numbers that represent the frequencies and amplitudes of the components of the Fourier analysis. In previous work, the authors [24,25] presented a minimum set of six dimensionless descriptors, motivated by musical acoustics and using the spectra obtained by the FFT, which allows for the description of the timbre of wooden aerophones (Bassoon, Clarinet, Transverse Flute, and Oboe) using individual sound recordings of the musical tempered scale. We show that these descriptors are sufficient to describe the timbral characteristics in the aerophones studied, allowing for the recognition of the musical instrument by means of the acoustic spectral signature. Also, Gonzalez & Prati [26] studied the timbral-variation dynamics (pianissimo, mezzo-forte, and fortissimo) in wooden aerophones using this set of six timbral descriptors in the Principal Component Analysis (PCA) of the TinySol audio library [27] and considering the common tessitura.

The goal of the present communication is to use the FFT-timbral coefficients to decrypt the similarity of musical timbres of different instruments. To this end, it is necessary to establish categories and build a space that classifies certain structures by applying Machine-Learning techniques. In Section 2 we used the timbre descriptors for defining a point for each musical sound of frequency f_0, each dynamic, and each instrument in an abstract timbral space of seven dimensions; then, the set of points is represented as a moduli space, and, therefore, the classification of the similarity problem of musical sound can be approached using Category Theory [28,29]. Section 3 presents an algorithm that is based on a data table corresponding to the fundamental frequencies and timbral coefficients for classifying each sound in terms of Euclidean distances. Further, in Section 4, we present the preliminary results of variations arising as a function of the musical instrument, the dynamics, and the audio database used. Finally, the conclusions are presented in the last Section.

2. Acoustic Descriptors and Timbral Representation

It should be noted that, unlike the timbral study of speech and environmental sounds, musical frequencies make up a finite, countable, and discrete set of only 12 different values in each musical octave for a total of 96 possible fundamental frequencies, and their integer multiples are in the audible range: from 20 Hz to 20 kHz. Therefore, the musical timbre can be characterized by a limited set of timbral coefficients, which are dimensionless quantities related to the frequencies and amplitudes in the Fourier spectrum of the audio records. Motivated by musical acoustics, these coefficients are tonal descriptors and, in essence, functionally describe the discrete distribution of normalized frequencies and amplitudes. As the amplitudes of the spectra of the FFTs are normalized (using the quotient of the amplitude of each partial frequency with respect to the greatest amplitude measured in each spectrum), it is possible to compare the relative amplitudes among them. They can be grouped into descriptors of the fundamental frequency (musical scale, 96 possible frequencies) and descriptors of the rest of the partial frequencies that arise when performing the FFT of the audio under analysis (descriptors of the shape of the distribution and statistical-frequency distribution). These proposed descriptors are dimensionless coefficients.

The FFT values are essentially a discrete collection of pairs of different amplitudes and frequencies; therefore, they can be summarized by the following six dimensionless parameters, see [24,26] for further details.

2.1. Fundamental Frequency Descriptors

The measurement of the fundamental frequency in relation to the average frequency (Affinity A) is as follows:

$$A \equiv \frac{\sum_{i=1}^{N} a_i f_i}{f_0 \sum_{i=1}^{N} a_i} \tag{1}$$

The quantification of the amplitude of the fundamental frequency with respect to the collection of amplitudes (Sharpness S) follows below, where f_0 and a_0 represent the fundamental frequency and their amplitude, and f_i and a_i denote the frequency and amplitude of the ith FFT peak.

$$S \equiv \frac{a_0}{\sum_{i=1}^{N} a_i} \tag{2}$$

2.2. Distribution Statistics

A descriptor of how close the secondary pulses are to being integer multiples of the fundamental frequency (Harmonicity H) is as follows:

$$H \equiv \sum_{j=1}^{N} \left(\frac{f_j}{f_0} - \left[\frac{f_j}{f_0} \right] \right) \tag{3}$$

where the [ ] denotes the integer part.

The envelope descriptor through the average slope in the collection of pulses (Monotony M) follows:

$$M \equiv \frac{f_0}{N} \sum_{j=1}^{N} \left(\frac{a_{j+1} - a_j}{f_{j+1} - f_j} \right) \tag{4}$$

2.3. Descriptors of the Frequency Distribution

The measurement of the frequency distribution with respect to the average frequency (Mean Affinity MA) is:

$$MA \equiv \frac{\sum_{i=1}^{N}\left|f_i - \bar{f}\right|}{N f_0} \quad (5)$$

The quantification of the average amplitude of the pulse collection (Mean Contrast MC) is:

$$MC \equiv \frac{\sum_{j=1}^{N}\left|a_0 - a_j\right|}{N} \quad (6)$$

These dimensionless timbral coefficients, together with the fundamental frequency, form a vector $(f_0, A, S, H, M, MA, MC)$ in an abstract or seven-dimensional configurational space for each monophonic audio record, which could represent the musical timbral space. Then, given a certain musical instrument, there will be only 96 possible sounds in western music (12 semitones in 8 octaves), with each one represented by a unique septuple. The set of points is represented as a Moduli space [29] or equivalently as a vector space.

As a potential representation of the timbres, Grey [30] proposed a three-dimensional timbre space based on the dissimilarity between pairs of sounds of musical instruments. Stimulus-neighboring points are represented in evolution points by their physical representations in terms of amplitude, time, and frequency. McAdams [31] found two dimensions for the set of wind/string musical instruments that qualitatively included the spectral and temporal envelopes, those for the set of percussion instruments including the temporal envelope, and either the spectral density or pitch clarity/noisine of the sound. The combined set had all three perceptual dimensions. Peeters et al. [12] calculated several measures on various sets of sounds and found that many of the descriptors correlated quite strongly with each other. Using a hierarchical cluster analysis of correlations between timbral descriptors, they concluded that there were only about ten classes of independent descriptors.

The problem of timbral representation is very similar to that of color–space representations. In both cases, the perceptions (audio, color) need to be defined operationally in abstract spaces for their computation and operational management. Thus, there are 256 digital colors (0–255) represented in an RGB configuration. By analogy, one could think of an analogous representation of the 96 monophonic musical sounds. This assumption is formally justified through Category theory in abstract mathematics [29]. The color and audio categories form groupoids, where the colors (timbres) are objects, and the color variations (timbre variations) are morphisms. The functors between them are induced in the continuous maps [30]. Hence, if the sounds constitute groupoids, all of their morphisms or forms of representation are equivalent, and consequently, the categories of musical sounds admit a representation through a vector space where the functors are linear transformations and a Euclidean metric could be defined for the distances between points in this abstract space.

3. Timbre Similarities in Musical Instruments

Two databases, Tinysol [32] and Good-Sounds [33], were used for the study of timbral similarities. The first dataset contained 2478 samples in the WAV audio format, sampled at 44.1 kHz, with a single channel (mono), at a bit depth of 16, each containing a single musical note from 14 different instruments, played in the so-called "ordinary" style and in the absence of a mute. The second dataset (Good-Sounds) contained monophonic recordings of two kinds of exercises: single notes and scales, from 12 different instruments and four different microphones. For the instruments, the entire set of playable semitones in the instrument was recorded several times with different tonal characteristics: "good-sound", "bad", "scale-good", and "scale-bad", see [33] for details.

For this study, only monophonic sounds were analyzed using the FFT of the audio records for the two common woodwind instruments in the databases Tinysol and Good-Sounds: the Transverse Flute and the Clarinet. The analysis presented includes only the fourth octave of the equal temperament scale. These are the most typical types of musical scales in Western music culture and are also the ones used by the audio recordings of the datasets used in this work. We used the following nomenclature for each sound of that octave: C4, C#4, D4, D#4, E4, F4, F#4, G4, G#4, A4, A#4, and B4. From the Good-Sounds database, only single-note sound recordings labeled "Good-Sounds" were used, and records in the database were called "AKG" and "Neumann" in all of the dynamics differences (*p*, *mf*, *f*).

The general procedure is summarized in Figure 1. First, the fundamental frequencies and their corresponding timbral coefficients were obtained for each of the 240 sounds analyzed from the 2 databases, namely, List 1, in Figure 1. With this data, a general dataframe was built. After the mean value of the timbral coefficients was listed, the data was grouped by instrument, note, and dynamics for List 2. The mean of the standardized data was calculated (namely, list 2), grouping the data by instrument, note, and dynamics. Subsequently, the Euclidean distance for each sound was calculated considering the data in List 3, which was grouped by musical sound and the dynamics of the instrument, specifically, by Flute and Clarinet, 12 sounds of the fourth octave, and 3 dynamics (*p*, *mf*, *f*) for a total of 72 types of audio in 244 records of the data set. When the test audio was incorporated, the software obtained the timbral vector "b" of that audio and identified it using List 2. The characteristic value of the vector "a" corresponds to the said instrument, sound, and dynamics. We proceeded to calculate the Euclidean distance between both (d). If, statistically, the distance is probably significant (less than 2.4 times SD), the audio test was considered as corresponding to the sound, instrument, and dynamics of List 3, in the *i*th position. Finally, the new audio was incorporated into the database. Otherwise, the software indicated the Euclidean distance, such as the similarity weighting and the timbre characteristics associated with the said audio.

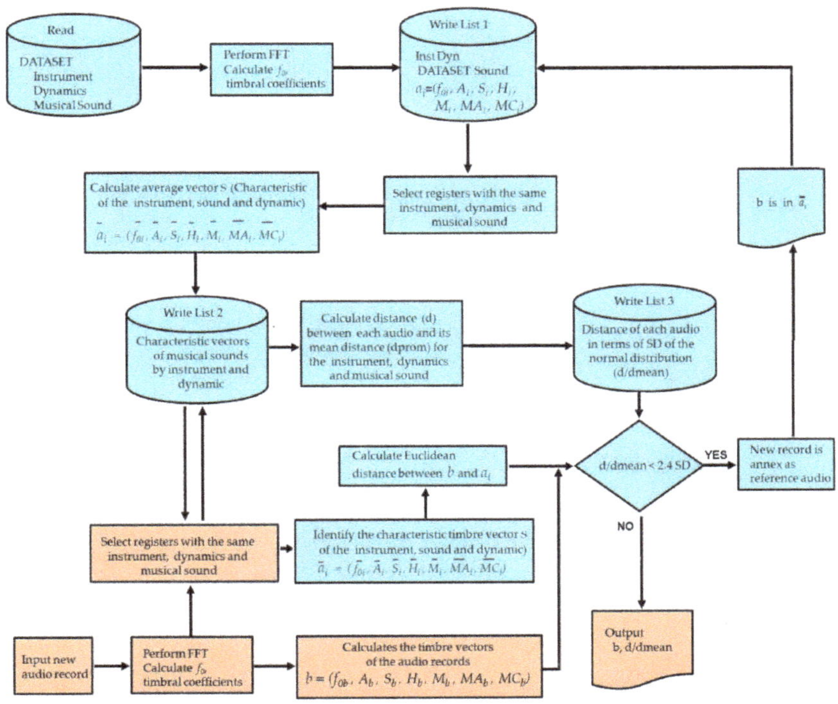

Figure 1. Flowchart diagram of the algorithm to calculate distances and relations of timbral similarity.

4. Results

4.1. Variations Due to the Musical Instrument

Following the previous procedure, for the Clarinet and Flute reference audios of the 4th octave and a dynamics of mezzo-forte for both databases, we obtained the average distances in the seven-dimensional timbral space between the positions of each sound with respect to all of the others (Figure 2). It was observed that the minimum distance occurs precisely for the correspondence between the sounds (diagonal elements), and in all cases, it is statistically discernible for a normal distribution (less than 2.4 times the standard deviation). In addition, the distance of any sound of the Clarinet with respect to those of the Flute, and reciprocally any of the sounds of the Flute with respect to those of the Clarinet (matrix sub-blocks without color), is greater than those corresponding to the distance between sounds of the same instrument (matrix sub-blocks in color green and violet).

The representation of the audio records by means of the timbral-coefficients vector allows the representation of a timbral space, where the distance between points is a measure of their timbral proximity. Then, the distances between any two audio records can be represented in a matrix (Figure 2). To facilitate its reading, a color scale is included, highlighting the distances that are statistically significant (blue) with those that are not (red), using as a criterion the value of 2.4 times the standard deviation in a normal distribution.

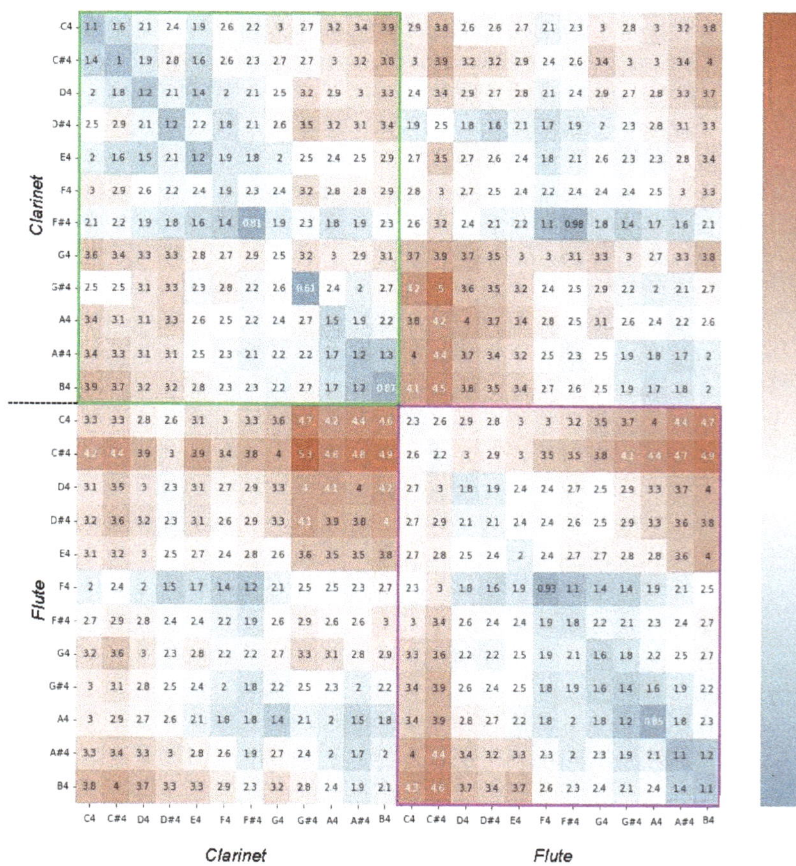

Figure 2. Patterned distance between musical sounds for Clarinet and Flute in mezzo-forte for the 4 ta octave, reference sounds in the data set.

The Good-Sounds database contains several registers considered "spurious" (called P 1 and P 2 in Figure 3); in addition to sounds selected as standards for each instrument and dynamics (called Neu1 and AKG1 in Figure 3), they present variations of the ratio d/d_mean that are significantly higher with respect to the reference audios, in all the musical sounds of the 4th Octave: This variation of the P 1 and P 2 audios occurs randomly in both series and in both instruments. The procedure outlined in Figure 1, when going through the records of these audios, incorporates the Neu1 and AKG1 audios into the database, while the audios P 1 and P 2 are discarded because they are incompatible with the registered standard values (see Figure 3).

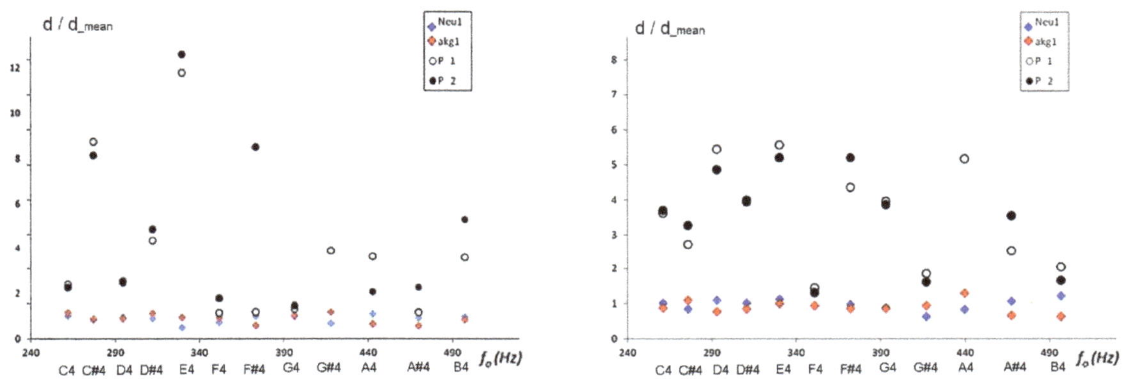

(a) (b)

Figure 3. Patterned distance between musical sounds in mezzo—forte for the 4 ta octave, reference sounds in the dataset: (**a**) Clarinet (**b**) Flute.

4.2. *Variations Due Musical Dynamics*

When variations of the dynamics are considered for the sounds of the fourth octave, we observed that the minimum distance occurs precisely for the correspondence between the sounds (diagonal elements). In Figure 4(up) Clarinet and Figure 4(down) Flute, it was observed that, in each row, the minimum distance occurs for the corresponding sound on the musical and dynamic scale. We also noted that the dynamics of mezzo-forte are always less than those of the adjacent sounds (by rows or columns) for all musical notes and in both instruments.

There does not seem to be a d/d_mean behavior for the various dynamics in the Clarinet. The dynamic of fortissimo in the Flute (Figure 4) is always close to the value of d_mean for each musical sound. This may be due to the fact that the monophonic sounds of the Flute are well-defined by the performer when the pressure of air is at its maximum within the resonant cavity and by the relative ease to play of this dynamic. So, for the sounds of the flute in fortissimo, the execution is very similar in different interpreters, and the dispersion of distance values in the registers is smaller. That is, the standard deviation of the sample is very small (d/d_mean was less than a tenth of the standard deviation).

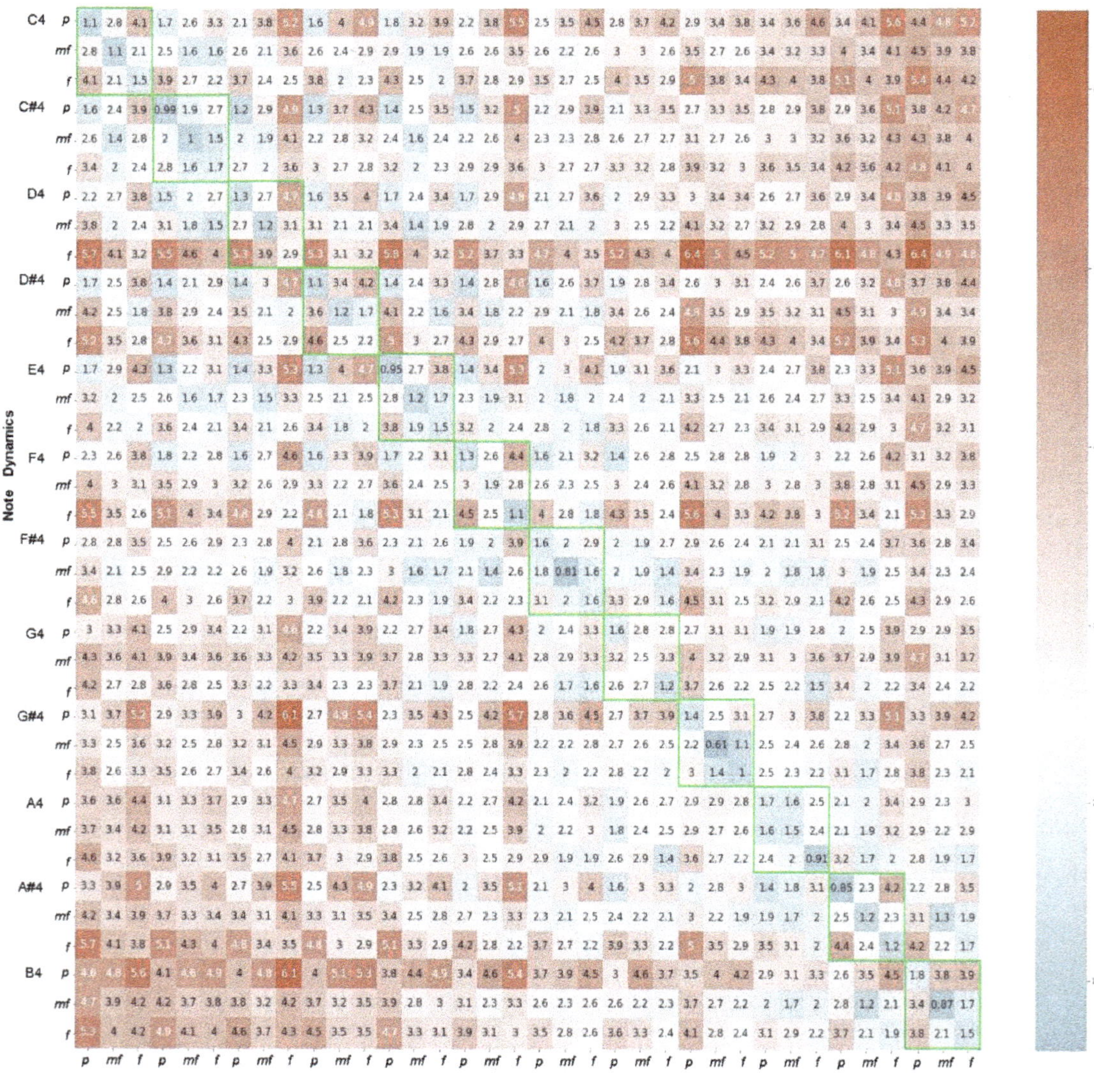

Figure 4. *Cont.*

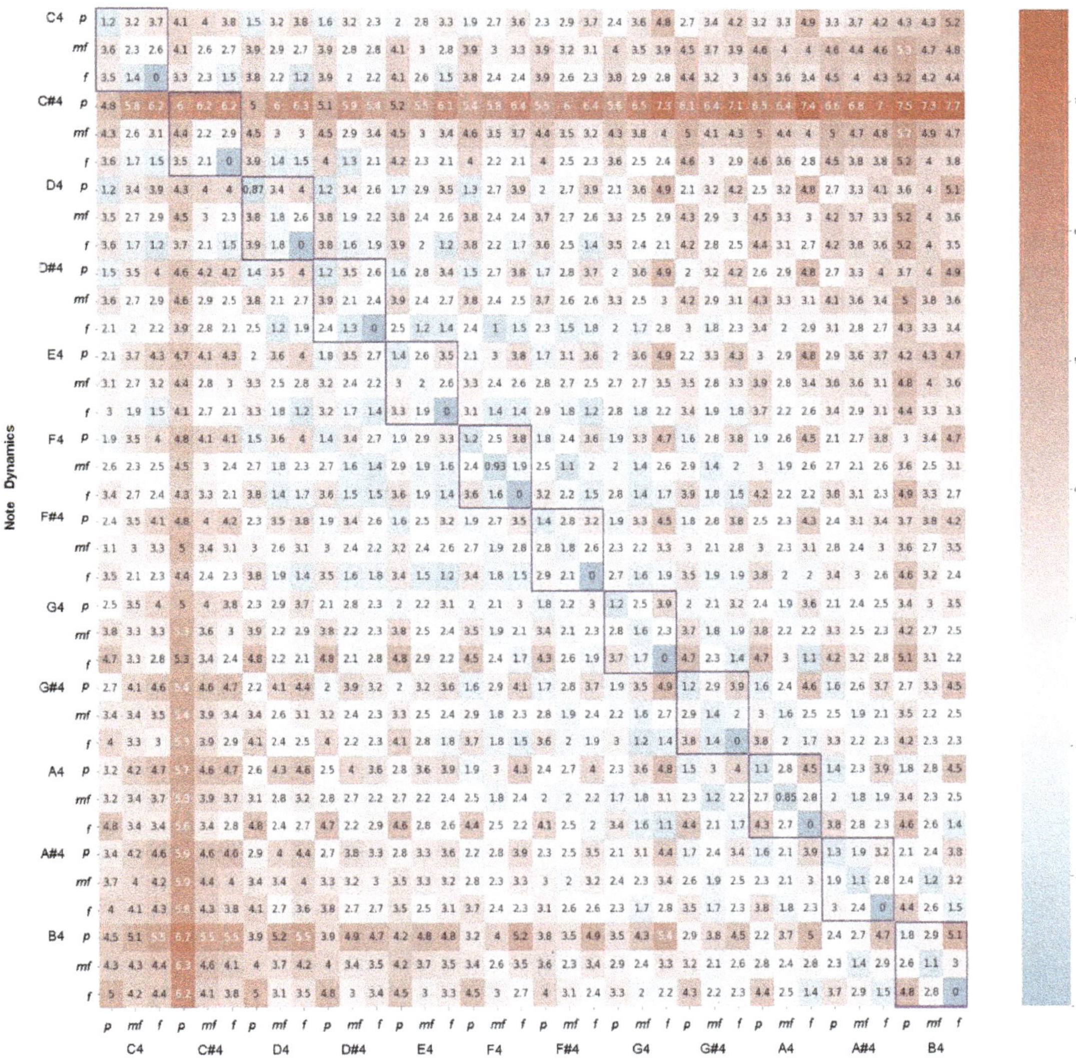

Figure 4. Patterned distance between musical sounds for clarinet (**up**) and Transverse Flute (**down**) in several dynamics (p, mf, f) for the 4 ta octave, reference sounds in the dataset.

4.3. Variations Due to the Tinysol and GoodSounds Database

The classification of the timbre of musical instruments in the proposed seven-dimensional space critically depends on the standardization of the real audios taken as reference. For this reason, it is important to ensure the robustness and reliability of the reference audio records. The databases used, as already mentioned, are reliable [32,33]; the standardized distances of the records for Clarinet and Flute are shown in Figures 5 and 6, respectively, and are grouped by dataset type (Tinysol, Goodsounds-Neumann, and Goodsounds-AKG). Figures 5 and 6 show that the audio records are within the radius of reliability radius of the normal distribution, with the separation from the mean value less than 2.2 times the standard deviation.

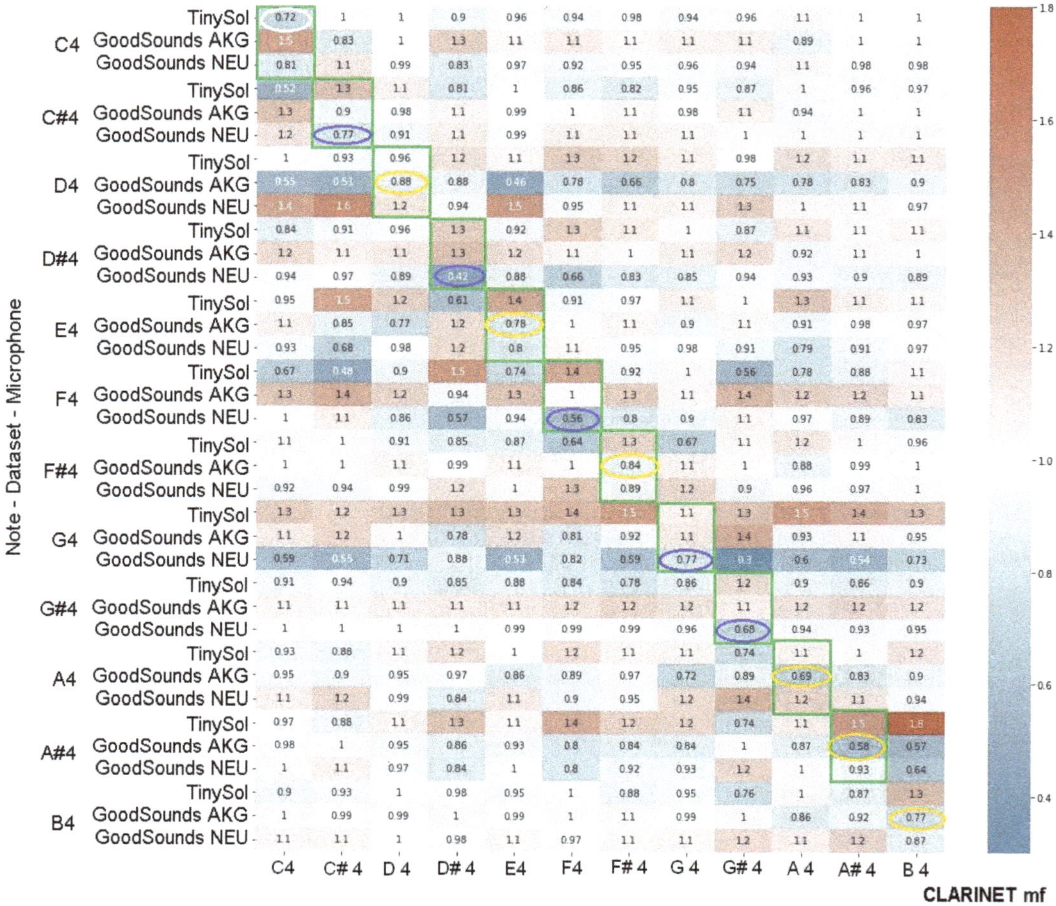

Figure 5. Patterned distance between musical sounds for Clarinet to reference sounds, according to several data sets. The circles of colors represent the shortest distance between the three dynamics.

The diagonal sub-blocks of the matrix indicate the correspondence with the expected values for the ratio between d/d_mean. The minimum value in each data set has been highlighted. For the Clarinet, the GoodSounds Neumann audio recordings are closer to the mean value and better discriminate the smallest distance in relation to the other distances of the other sounds (minimum value for each row, highlighted in dotted ovals). Figure 6 shows the comparison of the Transverse Flute data sets. The GoodSounds database provides two sets of Neumann-type and AKG-type records.

For some sounds, the distance closest to the mean value belongs to different recordings with no apparent systematic variation. However, the mode with which a given database provides a weighted distance (d/d_mean) closest to the mean value could be used as a quantitative evaluation criterion for various sound libraries.

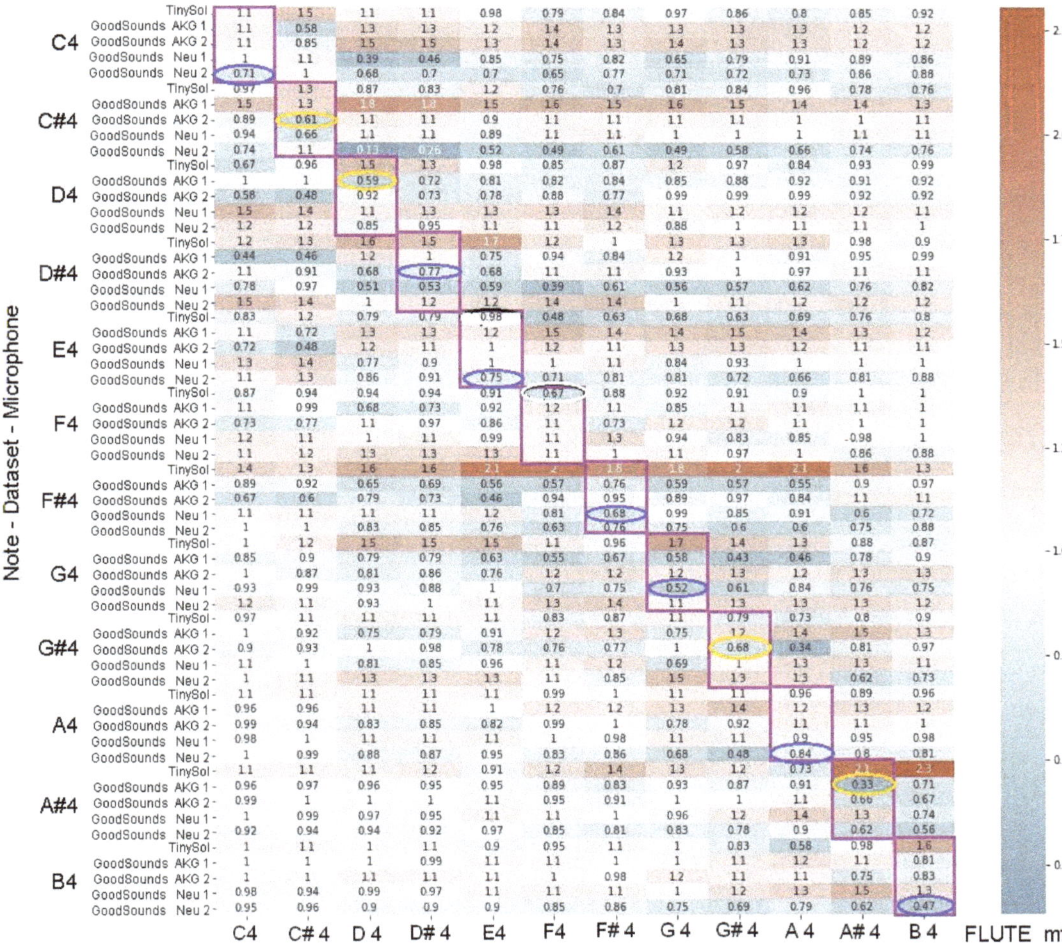

Figure 6. Patterned distance between musical sounds for Flute to reference sounds, according to several data sets. The circles of colors represent the shortest distance between the sound libraries.

4.4. Timbral Similarity between Clarinet and Transverse Flute

For the common tessiture between Clarinet and Flute in mezzo-forte dynamics, the results of timbral distances (Figure 2) can be used to find those sounds generated by different instruments such that the separation in timbral space is statistically significant (less than 2.4 times the average of the distance), in accordance with the machine-learning algorithm presented in Figure 1. Table 1 shows the distance values between similar sounds for the fourth octave. Note that the minimum distance corresponds to the diagonal elements, as expected. However, there are sounds between different instruments that are also significant because their distance is less than 2.4 times of the average distance. This suggests that they are timbrically related, that is to say that the FFTs of these sounds should be similar in terms of the number of harmonics, envelopes, and distribution of partial frequencies. Indeed, in Figure 7 the FFTs are shown where it is possible to see their similarity.

Table 1. Average distance in the timbral space of the sounds A#4 and B4 for Flute and Clarinet.

	ClBb A#4	ClBb B4	Fl A#4	Fl B4
ClBb A#4	1.198	1.314	1.667	1.985
ClBb B4	1.184	0.867	1.807	2.015
Fl A#4	1.665	2.026	1.091	1.247
Fl B4	1.936	2.141	1.362	1.127

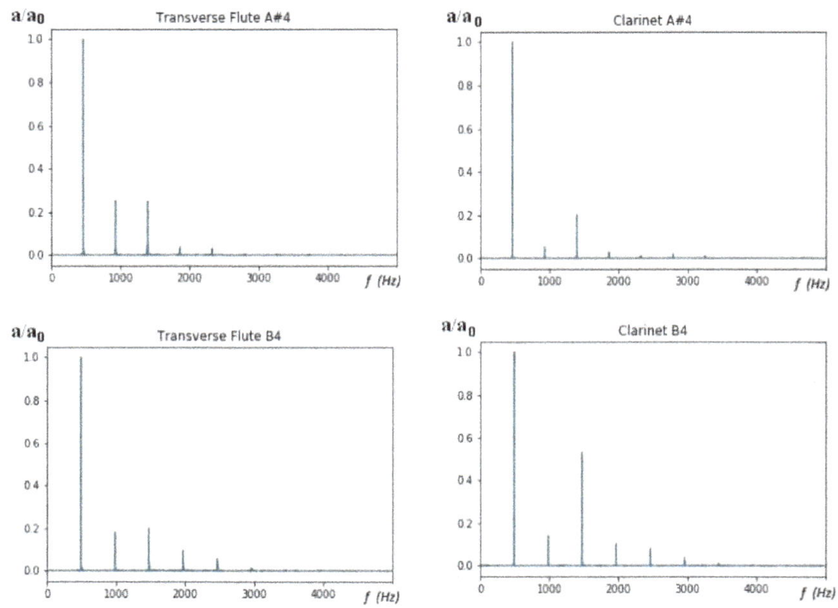

Figure 7. Comparison of the normalized FFTs of the sounds A#4 and B4 (rows) for Flute and Clarinet (columns). The intensities are normalized with respect to the amplitude (a_0) of the fundamental frequency.

5. Conclusions

The septuple made up of the fundamental frequency and the six timbral coefficients of each musical sound unambiguously define a collection of points and, therefore, formally (Category Theory), a timbral space can be devised to represent the sounds. In such a space, the subsets of musical sounds are groupoids and are related to each other by morphisms. This suggests that for each musical instrument, the dynamics and musical sound would be represented by a single characteristic vector (f_0, A, S, H, M, AM, CM) containing significant timbral characteristics. The real audio recordings would constitute statistical variations due to randomness in the execution of the musical sound by the interpreter and to specificities of the musical instrument with which the audio was made (model and manufacturer, quality of the same, materials used, imperfections of its acoustics, etc.), or even the recording equipment. Then, the audio sets for a type of musical instrument, specific dynamics, and a specific sound, will cover a spatial region in addition to the characteristic timbral vector.

In this work, we were able to determine the timbral variations of the following audio categories:

- Audios of different sounds and different instruments (Section 4.1).
- Audios of the same type of sound and instrument with different relative musical dynamics (Section 4.2).
- Audios of different databases (Section 4.3).

We find that for all the case studies, the smallest distances always occur between the elements of the diagonal. In the case of timbral variations by dynamics, we found that for most of the sounds, the dynamics of pianissimo and mezzo-forte have the smallest distances. This is related to the acoustic properties of the instrument and the difficulty of the air control for the dynamics of fortissimo by the performers. Regarding the analyzed databases, we found that the GoodSounds—Neumann database was the one with the lowest distance values. This suggests that it is a more reliable database for analysis of timbral properties of instruments.

For the study of the timbral similarities between the audio recordings, we proposed an algorithm (Figure 1) that evaluates such timbral similarities through Euclidean distances in the abstract space of 7 dimensions. This allowed us to find which FFTs are similar across different instruments (Section 4.4). For the two instruments in this study, statistically significant sounds were found because their distance is less than 2.4 times of the average distance. This suggests that these sounds (Table 1) are timbrically related, that is, that the FFTs of these sounds are similar in terms of the number of harmonics, envelope, and distribution of partial frequencies.

We plan to investigate different machine learning algorithms for future work, as well as different measures of distance.

Author Contributions: Conceptualization, Y.G. and R.C.P.; methodology, Y.G. and R.C.P.; software, Y.G. and R.C.P.; validation, Y.G. and R.C.P.; formal analysis, Y.G. and R.C.P.; investigation, Y.G. and R.C.P.; resources, Y.G. and R.C.P.; data curation, Y.G. and R.C.P.; writing—original draft preparation, Y.G.; writing—review and editing, Y.G. and R.C.P.; visualization, Y.G.; supervision, R.C.P.; project administration, Y.G. and R.C.P.; funding acquisition, Y.G. and R.C.P. All authors have read and agreed to the published version of the manuscript.

Funding: This study was financed in part by the Coordenação de Aperfeiçoamento de Pessoal deNível Superior—Brasil (CAPES)—Finance Code 001.

Institutional Review Board Statement: Not applicable.

Informed Consent Statement: Not applicable.

Data Availability Statement: The sounds used in this work are available at the following link: https://zenodo.org/record/3685367#.XnFp5i2h1IU%22, accessed on 15 May 2022.

Conflicts of Interest: The authors declare no conflict of interest.

References

1. Jiang, W.; Liu, J.; Zhang, X.; Wang, S.; Jiang, Y. Analysis and Modeling of Timbre Perception Features in Musical Sounds. *Appl. Sci.* **2020**, *10*, 789. [CrossRef]
2. Guven, E.; Ozbayoglu, A.M. Note and Timbre Classification by Local Features of Spectrogram. *Procedia Comput. Sci.* **2012**, *12*, 182–187. [CrossRef]
3. Fourer, D.; Rouas, J.L.; Hanna, P.; Robine, M. Automatic timbre classification of ethnomusicological audio recordings. In Proceedings of the International Society for Music Information Retrieval Conference (ISMIR 2014), Taipei, Taiwan, 27–31 October 2013.
4. McAdams, S. The perceptual representation of timbre. In *Timbre: Acoustics, Perception, and Cognition*; Springer: Cham, Switzerland, 2019; pp. 23–57.
5. Liu, J.; Zhao, A.; Wang, S.; Li, Y.; Ren, H. Research on the Correlation Between the Timbre Attributes of Musical Sound and Visual Color. *IEEE Access* **2021**, *9*, 97855–97877.
6. Reymore, L.; Huron, D. Using auditory imagery tasks to map the cognitive linguistic dimensions of musical instrument timbre qualia. *Psychomusicol. Music. Mind Brain* **2020**, *30*, 124–144. [CrossRef]
7. Reymore, L. Characterizing prototypical musical instrument timbres with Timbre Trait Profiles. *Musicae Sci.* **2022**, *26*, 648–674. [CrossRef]
8. Muller, M.; Ewert, S.; Kreuzer, S. Making chroma features more robust to timbre changes. In Proceedings of the 2009 IEEE International Conference on Acoustics, Speech and Signal Processing, Taipei, Taiwan, 19–24 April 2009; pp. 1877–1880. [CrossRef]
9. Muller, M. *Fundamentals of Music Processing*; Springer: Erlangen, Germany, 2021.
10. Muller, M.; Ewert, S. Towards Timbre-Invariant Audio Features for Harmony-Based Music. *IEEE Trans. Audio Speech Lang. Process.* **2010**, *18*, 649–662. [CrossRef]
11. Lartillot, O.; Toiviainen, P.; Eerola, T. A Matlab Toolbox for music information retrieval. In *Data Analysis, Machine Learning and Applications*; Springer: Berlin/Heidelberg, Germany, 2008; pp. 261–268.

12. Peeters, G.; Giordano, B.L.; Susini, P.; Misdariis, N.; McAdams, S. The timbre toolbox: Extracting audio descriptors from musical signals. *J. Acoust. Soc. Am.* **2011**, *130*, 2902–2916. [PubMed]
13. Barbedo, J.G.; Tzanetakis, G. Musical instrument classification using individual partials. In *IEEE Transactions on Audio, Speech, and Language Processing*; IEEE: New York, NY, USA, 2010; Volume 19, pp. 111–122.
14. Joshi, S.; Chitre, A. Identification of Indian musical instruments by feature analysis with different classifiers. In Proceedings of the Sixth International Conference on Computer and Communication Technology 2015, Allahabad, India, 25–27 September 2015; pp. 110–114. [CrossRef]
15. Ezzaidi, H.; Bahoura, M.; Hall, G.E. Towards a characterization of musical timbre based on chroma contours. In Proceedings of the International Conference on Advanced Machine Learning Technologies and Applications, Cairo, Egypt, 8–10 December 2012; pp. 162–171.
16. Böck, S.; Korzeniowski, F.; Schlüter, J.; Krebs, F.; Widmer, G. Madmom: A new python audio and music signal processing library. In Proceedings of the 24th ACM International Conference on Multimedia, Santa Barbara, CA, USA, 23–27 October 2016; pp. 1174–1178.
17. McFee, B.; Raffel, C.; Liang, D.; Ellis, D.P.; McVicar, M.; Battenberg, E.; Nieto, O. Librosa: Audio and music signal analysis in python. In Proceedings of the 14th Python in Science Conference, Austin, TX, USA, 6–12 July 2015; Volume 8, pp. 18–25.
18. Krimphoff, J.; McAdams, S.; Winsberg, S. Characterization of the timbre of complex sounds. II Acoustical analysis and psychophysical quantification. *J. Phys.* **1994**, *4*, 625–628.
19. Johnston, J. Transform coding of audio signals using perceptual noise criteria. *IEEE J. Sel. Areas Commun.* **1988**, *6*, 314–323.
20. Gaikwad, S.; Chitre, A.V.; Dandawate, Y.H. Classification of Indian Classical Instruments using Spectral and Principal Component Analysis based Cepstrum Features. In Proceedings of the IEEE International Conference on Electronic Systems, Signal Processing and Computing Technologies (ICESC), Nagpur, India, 9–11 January 2014.
21. Joder, C.; Slim Essid, S. Temporal Integration for Audio Classification with Application to Musical Classification. In *IEEE Transaction on Speech and Audio Processing*; IEEE: New York, NY, USA, 2009; Volume 17, pp. 174–186.
22. Pollard, H.F.; Jansson, E.V. A tristimulus method for the specification of musical timbre. *Acta Acust. United Acust.* **1982**, *51*, 162–171.
23. Burred, J.J.; Röbel, A.; Sikora, T. Dynamic spectral envelope modeling for timbre analysis of musical instrument sounds. *IEEE Trans. Audio Speech Lang. Process.* **2010**, *18*, 663–674.
24. Gonzalez, Y.; Prati, R.C. Acoustic Descriptors for Characterization of Musical Timbre Using the Fast Fourier Transform. *Electronics* **2022**, *11*, 1405. [CrossRef]
25. Gonzalez, Y.; Prati, R.C. Applications of FFT for timbral characterization in woodwind instruments. In Proceedings of the Brazilian Symposia On Computer Music (SBCM), Recipe, PE, Brazil, 24–27 October 2021. [CrossRef]
26. Gonzalez, Y.; Prati, R.C. Acoustic Analysis of Musical Timbre of Wooden Aerophones. *Rom. J. Acoust. Vib.* 2022, *in press*.
27. Cella, C.E.; Ghisi, D.; Lostanlen, V.; Lévy, F.; Fineberg, J.; Maresz, Y. OrchideaSOL: A dataset of extended instrumental techniques for computer-aided orchestration. *arXiv* **2020**, arXiv:2007.00763.
28. Awodey, S. *Category Theory*; Oxford University Press: Oxford, UK, 2010.
29. Mannone, M.; Arias-Valero, J.S. Some Mathematical and Computational Relations Between Timbre and Color. In *Mathematics and Computation in Music. MCM 2022. Lecture Notes in Computer Science*; Montiel, M., Agustín-Aquino, O.A., Gómez, F., Kastine, J., Lluis-Puebla, E., Milam, B., Eds.; Springer: Cham, Switzerland, 2022; Volume 13267, pp. 127–139.
30. Grey, J.M. Multidimensional perceptual scaling of musical timbres. *J. Acoust. Soc. Am.* **1977**, *61*, 1270–1277. [PubMed]
31. McAdams, S. Perception et Cognition de la Musique. 2015. Available online: https://www.erudit.org/en/journals/sqrm/2016-v17-n2-sqrm03970/1052743ar/ (accessed on 10 November 2022).
32. Carmine, E.; Ghisi, D.; Lostanlen, V.; Lévy, F.; Fineberg, J.; Maresz, Y. TinySOL: An Audio Dataset of Isolated Musical Notes. Zenodo 2020. Available online: https://zenodo.org/record/3632193#.Y-QrSnbMLIU (accessed on 15 May 2022).
33. Romaní Picas, O.; Parra-Rodriguez, H.; Dabiri, D.; Tokuda, H.; Hariya, W.; Oishi, K.; Serra, X. A real-time system for measuring sound goodness in instrumental sounds. In Proceedings of the 138th Audio Engineering Society Convention, AES 2015, Warsaw, Poland, 7–10 May 2015; Audio Engineering Society: New York, NY, USA, 2015; pp. 1106–1111.

Disclaimer/Publisher's Note: The statements, opinions and data contained in all publications are solely those of the individual author(s) and contributor(s) and not of MDPI and/or the editor(s). MDPI and/or the editor(s) disclaim responsibility for any injury to people or property resulting from any ideas, methods, instructions or products referred to in the content.

eng

MDPI

Article

Tensor CSRMT System with Horizontal Electrical Dipole Sources and Prospects of Its Application in Arctic Permafrost Regions

Alexander K. Saraev *, Arseny A. Shlykov and Nikita Yu. Bobrov

Institute of Earth Sciences, St. Petersburg State University, 199034 St. Petersburg, Russia
* Correspondence: a.saraev@spbu.ru; Tel.: +7-921-799-43-16

Abstract: When studying horizontally-inhomogeneous media, it is necessary to apply tensor modifications of electromagnetic soundings. Use of tensor measurements is of particular relevance in near-surface electrical prospecting because the upper part of the geological section is usually more heterogeneous than the deep strata. In the Enviro-MT system designed for the controlled-source radiomagnetotelluric (CSRMT) sounding method, two mutually perpendicular horizontal magnetic dipoles (two vertical loops) are used for tensor measurements. We propose a variant of the CSRMT method with two horizontal electrical dipole sources (two transmitter lines). The advantage of such sources is an extended frequency range of 1–1000 kHz in comparison with 1–12 kHz of the Enviro-MT system, greater operational distance (up to 3–4 km compared to 600–800 m), and the ability to measure the signal at the fundamental frequency and its subharmonics. To implement tensor measurements with the equipment of the CSRMT method described in the paper, a technique of creating a time-varying polarization of the electromagnetic field (rotating field) has been developed based on the use of two transmitters with slightly different current frequencies and two mutually-perpendicular transmitter lines grounded at the ends. In this way, we made it possible to change the direction of the electrical and magnetic field polarization continuously. This approach allows realization of the technique of tensor measurements using the new modification of the CSRMT method. In permafrost areas, the hydrogenic taliks are widespread. These local objects are important in the context of study of environmental changes in the Arctic and can be successfully explored by the tensor CSRMT method. For the numerical modeling, a 2D model of the talik was used. Results of the interpretation of synthetic data showed the advantage of bimodal inversion using CSRMT curves of both TM and TE modes compared to separate inversion of TM and TE curves. These new data demonstrate the prospects of the tensor CSRMT method in the study of permafrost regions. The problems that can be solved using the CSRMT method in the Arctic permafrost regions are discussed.

Keywords: electromagnetic soundings; controlled source; radio magnetotellurics; polarization of electromagnetic field; tensor measurements; permafrost; hydrogenic taliks

Citation: Saraev, A.K.; Shlykov, A.A.; Bobrov, N.Y. Tensor CSRMT System with Horizontal Electrical Dipole Sources and Prospects of Its Application in Arctic Permafrost Regions. *Eng* **2023**, *4*, 569–580. https://doi.org/10.3390/eng4010034

Academic Editor: Antonio Gil Bravo

Received: 24 November 2022
Revised: 1 February 2023
Accepted: 7 February 2023
Published: 9 February 2023

Copyright: © 2023 by the authors. Licensee MDPI, Basel, Switzerland. This article is an open access article distributed under the terms and conditions of the Creative Commons Attribution (CC BY) license (https://creativecommons.org/licenses/by/4.0/).

1. Introduction

Horizontally-inhomogeneous media are challenging objects for electrical prospecting. To obtain reliable data, it is necessary to use tensor modifications of the prospecting methods. Methods for measuring and interpreting tensor data are developed in detail for the magnetotelluric (MT) sounding method, operating in the frequency range 0.001–10 Hz, and for the audiomagnetotelluric (AMT) sounding method, frequency range 1–10,000 Hz, both based on the use of natural electromagnetic fields. Using multidirectional measuring arrays, recording the horizontal components of the electrical field Ex and Ey and the magnetic field Hx and Hy, as well as the vertical component of the magnetic field Hz, and using the tensor processing and interpretation of sounding data, reliable information can be obtained about the structure of horizontally inhomogeneous media [1,2].

The tensor version of the controlled-source audio magnetotelluric (CSAMT) sounding method, frequency range 0.1–10,000 Hz, is used in areas with complex geology. Two differently-directed transmitter lines (placed nearby or separated) are used, and the measurements of electrical and magnetic field components are done in the same way as in MT and AMT methods [3]. If information about the predominant strike of geological structures at the studied site is available, orienting the transmitter lines parallel and perpendicular to the strike simplifies the interpretation, since measurement data are directly related to the transverse-electric (TE) and transverse-magnetic (TM) modes of the electromagnetic field. The CSAMT method is more efficient than MT and AMT methods when the level of industrial noise is high, as well as when it is necessary to obtain high-quality data in the MT "dead band" (frequency range 0.5–7 Hz) and AMT "dead band" (frequency range 800–5000 Hz) [4].

The features of horizontally-inhomogeneous media are of particular relevance when interpreting data produced by near-surface electrical prospecting methods, since the upper part of a section is usually more heterogeneous than the deep horizons. This is especially true for permafrost regions. The structure of permafrost areas is characterized by a significant heterogeneity, including rock properties changing dramatically both vertically and horizontally. Areas of thawed rocks alternate with ice-rich sediments. In this case, the cryogenic boundaries in a section may not coincide with the geological boundaries. Underground ice and cryopegs create large contrasts in electrical properties.

Currently, processes of permafrost degradation occur in large areas, leading to an increase in the heterogeneity of frozen strata. The production of mineral resources in the Arctic regions and the construction of infrastructure require considerable year-round engineering and geophysical research. Electrical-prospecting methods are widely used in permafrost regions, since the electrical properties of frozen sediments are a sensitive indicator of their state. The direct current (DC) methods, such as vertical electrical sounding (VES) and electrical resistivity tomography (ERT), are the most commonly used in these studies [5–9]. Taking into account the horizontally-heterogeneous structure of frozen strata, for engineering surveys in the permafrost regions, the VES method is often used not in a standard version with a single array, but in the modification of two-components (MTC) with two differently-directed lines [10]. However, one of the main disadvantages of direct-current methods is the necessity of using grounded electrodes, which imposes seasonal restrictions on fieldwork and significantly slows the survey speed [8].

Electromagnetic methods are more promising for year-round surveys in the permafrost zone, but they are not as widely used as direct-current methods, despite having a number of advantages: higher survey speed, the ability to work at any season of the year with ungrounded electrical sensors, and the ability to study geological sections under high-resistivity screens. Among electromagnetic methods, transient electromagnetics (TEM) [11–15] and ground penetrating radar (GPR) [16–20] are most often used on permafrost. TEM has always been considered as a relatively deep geophysical technique compared to DC methods. In particular, it was proposed to be used for mapping the base of permafrost [21]. When studying the near-surface part of the section, TEM has a limitation regarding the minimum depth of research ("dead zone") associated with the influence of interfering transients when the current is turned off in the transmitter loop [22–24]. GPR is successfully used on glaciers and solves a number of problems in the study of near-surface horizons in permafrost areas. However, the penetration depth of GPR reduces to the uppermost few meters when clay sediments are present in a section. In addition, GPR data can be adversely influenced by multiple reflections from ice lenses and other heterogeneities in the upper part of the section.

For shallow-depth studies down to 30–50 m, the tensor measurement technique is implemented in radio-magnetotelluric (RMT) sounding, based on the registration of radio-transmitter signals in the frequency range of 10 kHz to 250–1000 kHz [25], which covers very low frequency 10–30 kHz (VLF), low frequency 30–300 kHz (LF), and medium frequency 300–1000 kHz (MF) signals. However, in remote polar regions, the spectra of the fields of LF and MF radio transmitters are quite limited. For these conditions, a tensor variant of the controlled-source radiomagnetotelluric (CSRMT) sounding method has been developed, which expands the capabilities of the RMT method. The first Enviro-MT system implementing this method was built at the University of Uppsala, Sweden [26]. Two mutually-perpendicular horizontal magnetic dipoles (HMD, two vertical loops) are used as controlled sources. Enviro-MT equipment has been widely applied to a variety of problems in near-surface geophysics [27–30]. A disadvantage of this technology is that it is difficult to generate large moments (the product of the current strength and the area of the loop) of the vertical loop source, reducing the amplitude of the response signal and thus limiting the area of survey without moving the source.

St. Petersburg State University, Russia, and the University of Cologne, Germany, have developed a variant of CSRMT method when two mutually-perpendicular horizontal electrical dipoles (HED, two transmitter lines) are used as sources for tensor measurements [31–33]. It is much easier to generate a large moment with such a source, which in this case is the current strength multiplied by the line length.

One of the problems of current interest for the CSRMT method in permafrost regions is detection and contouring of hydrogenic taliks. These objects (unfrozen rock mass surrounded by permafrost) are usually formed as a result of the warming effect of water reservoirs and streams on the underlying ground [34]. Despite their wide distribution in the permafrost, taliks are still relatively poorly studied. Often there is no reliable information even about their thickness [14,35]. Meanwhile, taliks are of interest as indicators of climate change in the Arctic and may serve as a potential source for the water supply in the permafrost areas.

Here, we describe features of the tensor version of the CSRMT method using the developed equipment with the HED sources, and present the results of numerical modeling applied to the problem of identifying and contouring hydrogenic taliks in the permafrost zone.

2. Radiomagnetotelluric and Controlled Source Radiomagnetotelluric Methods

The RMT method uses the electromagnetic fields of remote radio transmitters. The primary field of a radio transmitter is a linearly-polarized wave, in which, at the earth-air interface, the horizontal components of the electrical and magnetic fields are mutually perpendicular. Values of the complex surface impedance (Z) are determined by measuring horizontal components of the electrical (E_X) and magnetic (Hy) fields:

$$Z = E_X/Hy. \quad (1)$$

Z is then transformed into the apparent resistivity

$$\rho_a = (1/\omega\mu_0)\cdot|Z|^2, \quad (2)$$

where $\omega = 2\pi f$, f is the frequency in Hz, $\mu_0 = 4\pi\cdot10^{-7}$ H/m is the vacuum magnetic permeability (magnetic constant).

The impedance phase (phase difference between E_X and Hy components) is determined as:

$$\varphi_Z = \varphi_{E_X} - \varphi_{Hy}. \quad (3)$$

The sounding curves are frequency dependences of ρ_a and φ_Z; inversion of the latter yields the resistivity section at the observation point. At a distance of several kilometers from a radio transmitter, the measured impedance coincides with the impedance of the vertically-incident plane wave, which depends only on the structure and properties of the underlying half-space. For the plane-wave model, interpretation methods have been developed in detail to ensure reliable sounding results [1,2].

The RMT sounding method is most effectively used in populated regions, where it is possible to measure the signals of VLF, LF, and MF radio transmitters in the full frequency range from 10 to 1000 kHz. Usually, the signals of 20–30 radio transmitters are measured with confidence, allowing both the acquisition of sounding curves suitable for the inversion and the derivation of resistivity sections [31]. In remote regions, such as permafrost areas in the Arctic, only VLF radio transmitter signals can be recorded, allowing use of the profiling technique only. In addition, existing radio transmitters operate at frequencies above 10 kHz, which limits the depth of investigation to 30–50 m, depending on the resistivity of the rocks.

To overcome limitations of the RMT method, the controlled source of the electromagnetic field is used in addition to measuring the signals from remote radio transmitters. Concerning the Enviro-MT equipment developed at the University of Uppsala, the frequency range was extended down to 1 kHz, increasing the depth of investigation to 100–150 m. Two mutually-perpendicular HMDs (two vertical loops) were used as controlled sources [26]. These sources have a number of advantages: absence of grounding, compactness and ease of installation, and the possibility of implementing tensor measurements with two differently-oriented HMDs. However, the working area of these sources is limited, measurements being possible at a distance from the source not exceeding 600–800 m. This leads to the need to move the source frequently. Furthermore, the Enviro-MT equipment measures the controlled-source field only in a narrow frequency range from 1 to 12 kHz. The need to tune the source in resonance with a load at each emitted frequency limits the survey productivity. The possibility of using subharmonics of the fundamental frequency has not been studied.

St. Petersburg State University, Russia, and the University of Cologne, Germany, jointly with St. Petersburg small enterprises Microkor LLC, Tenzor LLC, Magnetic Devices LLC, and with the Russian Institute for Power Radioengineering, have developed CSRMT equipment that includes a recorder and transmitter allowing operation of the CSRMT method in an extended frequency range of 1 kHz–1 MHz. The frequency range is extended by using as a source a transmitter line several hundred meters long grounded at the ends. This source has a wider working area than an HMD source. Measurements are possible at a distance from the source of up to 3–4 km. Along with the measurements of the signal at the fundamental frequency, signals of subharmonics are measured, increasing the survey productivity. On the other hand, the HMD source does not require grounding, which may be an advantage, particularly during fieldwork in winter.

First surveys by the CSRMT method with an HED source were carried out using the scalar technique. Significant experience has been gained in solving various problems of near-surface geophysics [31,32,36–39]. In recent years, the tensor variant of the CSRMT method, using two mutually perpendicular HEDs as electromagnetic field sources, has been developed. Examples of bimodal inversion of field data using the tensor modification of the CSRMT method have been reported in our papers [33,40].

3. Equipment for the Controlled Source Radiomagnetotellurics

The CSRMT hardware-software complex (Figure 1) includes a recorder with receiving electrical and magnetic antennas, a transmitter with electrical dipole sources, and data processing and interpretation software tools. The RMT-5 recorder [31] has five channels for synchronous measurements, with 16-bit ADCs in each channel (two electrical and three magnetic channels). The recorder frequency range is 1–1000 kHz, the built-in memory is 8 GB. The built-in display and keypad allow autonomous field work without having to connect to an external PC, and the built-in power supply provides an operation time of 6–8 h. Measured data are transferred to an external computer via an Ethernet channel. The GPS receiver records the survey coordinates and time. During operation, time series of magnetic and electrical field signals are recorded and stored in the built-in memory. The recorder operates in four frequency ranges: D1 (1–10 kHz, signal sampling frequency fs = 39 kHz), D2 (10–100 kHz, fs = 312 kHz), D3 (10–300 kHz, fs = 832 kHz), and D4 (100–1000 kHz, fs = 2496 kHz).

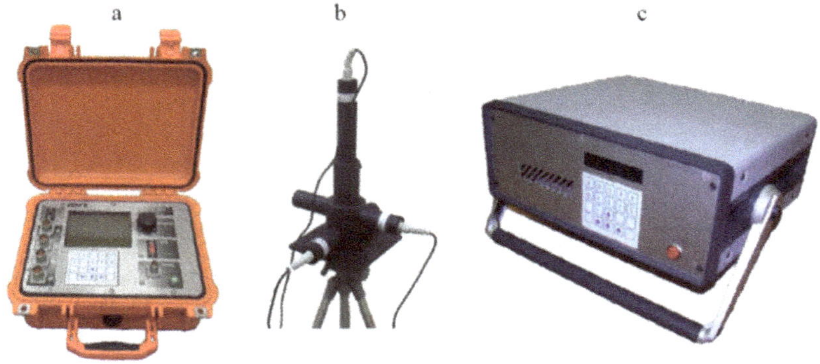

Figure 1. Recorder (**a**), magnetic sensors (**b**), and transmitter (**c**) of the CSRMT equipment.

Magnetic antennas have a frequency range of 1–1000 kHz, a self-noise level of 25 $fT/\sqrt{Hz}$, and a transfer factor of the magnetic-induction antenna into a signal voltage of 20 mV/nT. Electrical field measurements can be carried out with grounded and ungrounded (capacitive) receiving lines (2 × 10 m long wires), which allow one to work both in summer and in winter with ice and snow cover, as well as in conditions unfavorable for the grounding of electrical lines (asphalt, concrete, gravel). The compact size of the equipment allows for its use in restrictive areas.

After measurements, a fast Fourier transformation of time series of electrical field E_X and Ey (V/m) and magnetic field H_X, Hy and H_Z (A/m) components is performed, and auto-spectra and mutual spectra of the electrical and magnetic fields and their coherence are calculated. At a coherence level > 0.8, the data are considered suitable for further processing and are used to calculate the apparent resistivity and impedance phase.

The GTS-1 transmitter for the CSRMT method is designed to excite rectangular bipolar pulses in the frequency range of 0.1 Hz–1 MHz, with an adjustable pulse-time ratio to a load with a resistance of 10–1000 Ω. The supply voltage is 220 V; power frequency is 50 Hz. Output voltage is up to 300 V, output current is from 100 mA to 7.5 A, and output power at a load of 100 Ω is up to 1 kW. The transmitter is operated from a control panel or remotely from an external computer.

4. Measurement Technique

When conducting the survey using the CSRMT method, we use an HED source, which is a cable with a length of 400 to 1000 m, grounded at the ends. As noted above, this source is more efficient for the CSRMT method than an HMD source having larger range of coverage, wider frequency range, and the ability to register the main harmonic of the emitted signal and its subharmonics in a wide frequency range.

A finite-length cable source has long been used in the CSAMT method [3]. The high efficiency of this source is confirmed by many years of experience in applying the CSAMT method in different regions and with equipment from different manufacturers. The experience of using HED in the CSRMT method confirms the efficiency of this type of source.

In the surveys using the CSRMT method with HED, the measurement performance is significantly increased by measuring the signals at the main frequencies and their odd subharmonics. As shown in Figure 2, at the main signal frequency of 1 kHz, nine odd subharmonics with a coherence level higher than 0.8 are visible in the signal spectra for the electrical and magnetic channels. The spectra shown in this figure were obtained at a distance of 1 km from the HED source, with transmitter-cable length of 200 m. To cover the full frequency range of 1–1000 kHz, three fundamental frequencies are usually used, each being accompanied by 8–12 subharmonics. As a result, high productivity of measurements, about 70–80 sounding stations per day is achieved.

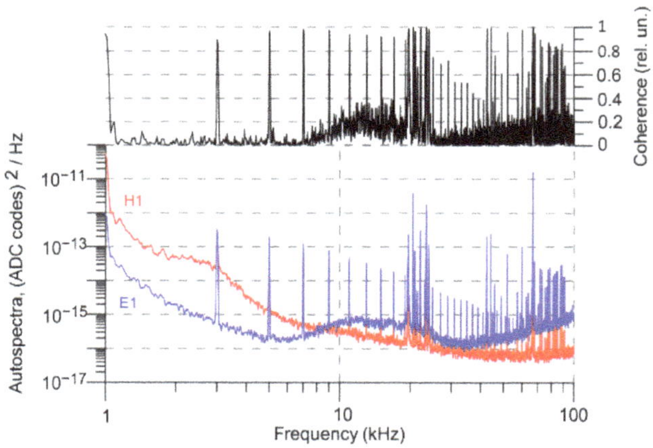

Figure 2. Autospectra of electrical E1 and magnetic H1 field signals from an HED source and radio transmitters in the frequency range of 1–100 kHz. Directions of E1 and H1 are mutually perpendicular.

Using the CSRMT method with electrical transmitter lines, it is necessary to use electrodes for the groundings. The installation of the transmitter lines is usually performed once per several days or weeks of surveying, when it is possible to measure electromagnetic fields of the HED source, due to the large operational distance (up to 3–4 km). In the winter season, we use for each grounding several (up to 10–15) electrodes, and the grounding resistance usually does not exceed 100–200 Ω.

For controlled sources including HED, three electromagnetic field zones are commonly introduced: near-field, transition, and far-field zone [3]. The near-field zone corresponds to the condition $|k|r << 1$, where k is the wave number of the earth, and r is the distance from the source to the observation point. For the case of low frequency (quasi-stationary approximation), the near-field zone is introduced via the skin layer thickness d:

$$d \approx 503\,(\rho/f)^{1/2} \tag{4}$$

where ρ is the resistivity in Ω·m, f is the frequency in Hz.

In this case, the above condition for the near-field zone corresponds to the ratio $r/d < 0.5$. In the near-field zone, the alternating electromagnetic field behaves like the DC field. Components of the electrical field depend on the resistivity of the rocks, but do not depend on the frequency. Components of the magnetic field do not depend on either the frequency or the resistivity of rocks. Therefore, near-field impedance measurements cannot be used for frequency soundings. In the transition zone at $|k|r \approx 1$ or $r/d \approx 1$, the components of the electromagnetic field depend on both the frequency and the coordinates of the observation point. In the far-field zone, with $|k|r >> 1$ or $r/d >$ 4–5, the components of the electromagnetic field correspond to the model of a vertically-incident plane wave [3]. The components depend only on the field frequency and do not depend on the coordinates of the observation point.

Surveys by the CSRMT method are carried out both in the far-field and in the transition zones of the controlled source. It should be noted that measurements in the transition zone make it possible to determine the anisotropy parameters of rocks, such as horizontal and vertical resistivity and anisotropy coefficient [38].

5. Tensor Measurements

Tensor measurements require registration of an electromagnetic field of different polarizations [1,2]. To generate a field with time-varying polarization (rotating field), two alternately-operating and differently-oriented sources are commonly used [26,41]. In this case, the transmitter is connected in turn to one then to the other transmitter line. The field registered by the recorder has a different orientation at different times. Time series obtained at the same frequency but with different sources are processed as a single data set.

Apart from sequential connection of sources with the same current frequency, field rotation can be achieved by using currents with slightly different frequencies in each of two transmitters operating simultaneously. In this case, the period of change in the direction of the total field will be proportional to the difference between the periods of current in each transmitter. The total field (both electrical and magnetic) will be elliptically polarized, and the direction of polarization will continuously change in time.

This approach is effective at frequencies below 10 kHz. At high frequencies (above 10 kHz) used in the CSRMT method, the wavelength of the current in the transmitter line hundreds of meters long becomes comparable to or smaller than the length of the line itself. In this case, the amplitude and phase of the recorded field will be affected by the current distribution along the wire, which in turn will be determined by the distributed electrical parameters of the wire, such as linear resistance, capacitance, and inductance. The wave effects that arise in the wire will influence the direction of polarization of the electromagnetic field.

The layout of the field experiment conducted to analyze the polarization of the rotating field is presented in Figure 3. We have used two transmitters and two mutually-perpendicular lines grounded at the ends. In the first experiment, each line was 200 m long, and in the second experiment the lines were 600 m long.

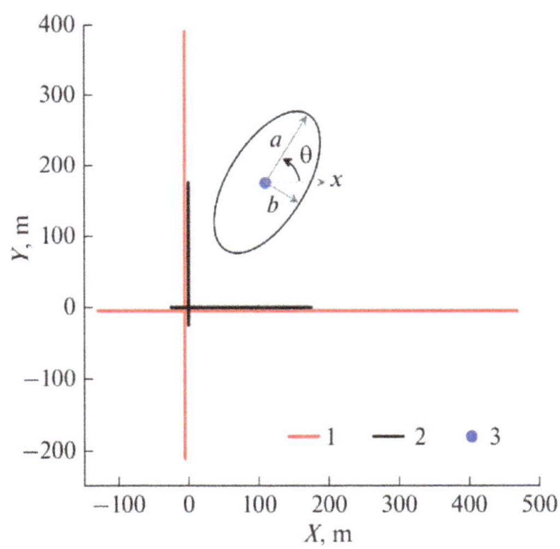

Figure 3. Layout of the field experiment: 1—600 m long lines; 2—200 m long lines; 3—measurement point. Polarization ellipse elements are schematically shown around the measurement point: *a*—major semi axis, *b*—minor semi axis, θ—angle of rotation of the semi-major axis.

Figure 4 shows dynamic spectra of the parameter $\Delta\theta_E$, which is a measure of the deviation of the polarization direction of the electrical field, characterized by angle θ, at a particular moment of time from the average direction of polarization over the entire measurement time. We measured the time series and estimated the directions of electrical field polarization for short realizations. Then we estimated an averaged direction for each transmitter frequency (odd harmonics of the main frequency) over the entire measurement time.

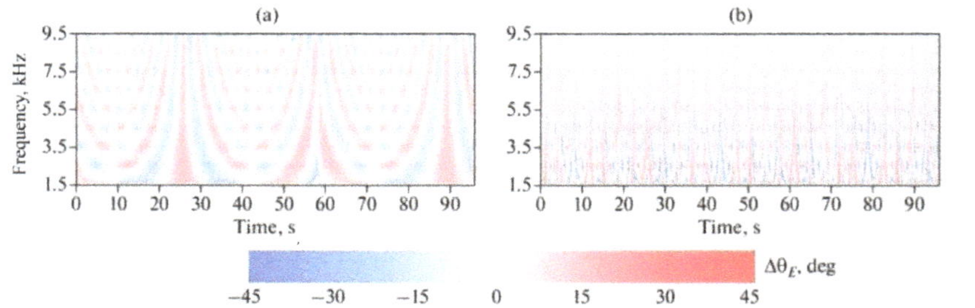

Figure 4. Dynamic spectrum of the $\Delta\theta_E$ parameter for frequencies with a small shift in different transmitter lines: (**a**) 200 m long lines; (**b**) 600 m long lines. Base frequencies are 0.5000 kHz in the first transmitter and 0.5001 kHz in the second transmitter, respectively.

This figure shows how much directions of the horizontal electrical field polarization deviate from a certain averaged direction during the measuring time, i.e., the spatial precession of the direction of horizontal electrical field for different frequencies over time. A small frequency shift between two transmitters, 0.02% for a base frequency of 0.5 kHz (0.5000 and 0.5001 kHz), appears to produce oscillations in the orientation of the horizontal electrical field with an amplitude of ±15–20°. Period of oscillations for a 200-m transmitter lines varies from about 20 s at 1.5 kHz to about 4 s at 9.5 kHz (Figure 4a). For a 600-m transmitter line, the main oscillation period is about 3 s for all frequencies in the considered range (Figure 4b). The same properties apply to the magnetic field. This is sufficient for the use in tensor measurements.

6. Modeling of Taliks in Permafrost Regions

We present here an assessment of the possibility of detecting and contouring hydrogenic taliks based on the data of numerical modeling. A two-dimensional model of a sub-lake talik is assumed (Figure 5). Under the lake with depth ranging from 2 to 6.5 m and water resistivity of 120 Ω·m, there is a zone of thawed sedimentary rocks with a maximum thickness up to 18 m and resistivity of 30 Ω·m. Surrounding frozen rocks have resistivity of 1000 Ω·m. This model is representative for arctic thermokarst lakes [42–45].

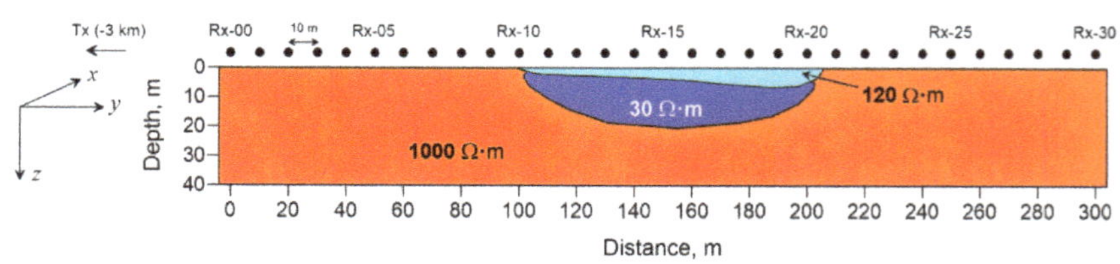

Figure 5. Two-dimensional model of a sub-lake talik. The direction to the field generation system (Tx) is shown by an arrow in the upper left corner. The measurement stations (Rx) are marked by black dots along the surface.

The synthetic field generation system consists of two perpendicular HEDs. The first dipole is directed along the x axis, and the position of the measurements profile corresponds to the equatorial area of the source. The second dipole is directed along the y axis, and in this case the profile is in the axial area of the source.

The sounding stations are located along the profile with a separation of 10 m. The nearest station, with number 00, is located at a distance of 3 km from the source. In total, there are 31 sounding stations in a 300 m long profile.

The modeling has been performed for 18 frequencies evenly distributed on a logarithmic scale in the range of 1–1000 kHz. For the selected talik model and measurement geometry, the field of the dipole directed along the x-axis corresponds to the TE mode, and the field of the dipole directed along the y-axis corresponds to the TM mode. At the lowest frequency of 1 kHz, the thickness of the skin layer for a half-space with a resistivity of 1000 Ω·m is approximately 500 m. Therefore, we can assume that the position of all sounding stations on the profile corresponds to the ratio r/d = 6 or, in other words, to the source far-field zone [3].

To obtain synthetic 2D data in the source far-field zone, the MARE2DEM software [46] has been used. Two-dimensional inversion of synthetic curves of apparent resistivity and impedance phase has been carried out using the ZondMT2D software [47]. When only curves of the TM mode are used, an anomalous object, the talik, is identified, but its contour is not determined exactly (Figure 6a). From the inversion of the TE mode, the contour of the talik has been determined more accurately; however, its shape relative to the model is somewhat distorted and resistivity of the host medium under the talik is overestimated up to 5000–7000 Ω·m (Figure 6b). The most reliable result with good correspondence of the shape and properties of the anomalous object to the original model was obtained by bimodal inversion using CSRMT curves of TM and TE modes (Figure 6c).

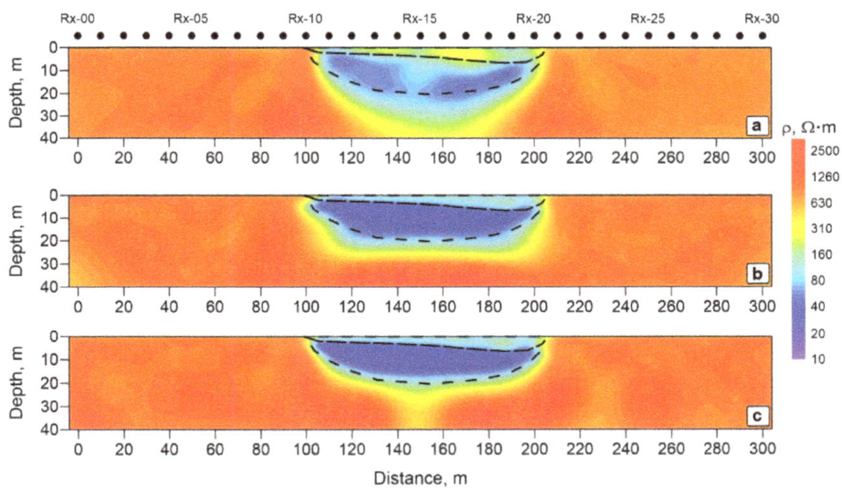

Figure 6. Results of the 2D inversion of the synthetic CSRMT data in the plane-wave approximation. (**a**)—inversion of the TM mode; (**b**)—inversion of the TE mode; (**c**)—bimodal inversion. The black dashed line indicates the contour of the talik and the water body assumed in the model.

From the analysis of the experience in application of different electrical and electromagnetic methods and results of numerical modeling, we can conclude that the CSRMT method has good prospects for application in the Arctic regions to study the structure and properties of frozen rocks, determine the depth of seasonal thawing, monitor the dynamics of permafrost degradation, identify and contour lenses of underground ice and taliks, map ice wedges and cryopegs, and control soil and groundwater pollution in the sensitive permafrost ecosystem. Frozen strata are characterized by high spatial heterogeneity that can be better resolved by tensor soundings using both TM and TE modes of the electromagnetic field. Data obtained by the CSRMT method can be used during planning and construction of oil and gas facilities in permafrost regions, of the infrastructure of the ports at the Northern seas, and for tracing linear industrial objects in the North (oil and gas pipelines, power lines, railways and roads).

7. Conclusions

The features of the CSRMT tensor system using two differently directed HEDs (transmitter lines) as sources are considered. Compared to a previous variant of the CSRMT method with two multidirectional HMDs (vertical loops), realized in the Enviro-MT system, the proposed technique has a number of advantages. It has an extended frequency range of 1–1000 kHz compared to 1–12 kHz of the old system, greater operational distance, (up to 3–4 km, compared to 600–800 m), and the ability to measure the signal of the main frequency and its subharmonics. Nevertheless, HMD still has its advantages in that loop sources do not require grounding, which is especially important during winter fieldwork.

Technical parameters of the developed CSRMT system, which is used for realization of tensor measurements, are described. To implement tensor measurements, a technique of creating time-varying polarization of electromagnetic field (rotating field) has been developed. It is based on the use of two mutually perpendicular HEDs and two transmitters simultaneously operating with a slight frequency shift (below 10 kHz) or at the same frequency (above 10 kHz). This technique makes it possible to obtain, within a shorter measurement time, results similar to the results of tensor measurements based on a sequential switching of two perpendicular HEDs, and to process data as a single time series.

The numerical modelling was performed for one of the common objects in permafrost areas—a sub-lake hydrogenic talik, which can be studied using the new tensor CSRMT system. We considered a 2D model of talik and obtained synthetic CSRMT curves for TM and TE modes. Upon inversion of TM curves, we found that the talik was identified, but its borders were not determined exactly. The inversion of TE curves allows the creation of a more accurate contour of the talik; however, its shape compared to the model was somewhat distorted, and resistivity of the host medium under the talik was overestimated. The most reliable result with good correspondence of the shape and properties of the talik to the initial model was obtained by bimodal inversion using curves for both TM and TE modes. The numerical modelling confirmed the expected conclusion about the necessity and relevance of CSRMT tensor measurements in permafrost areas.

According to the analysis of applications of different electrical and electromagnetic methods in permafrost regions, the prospects for the use of the CSRMT method are highlighted. The possibility of working with a bimodal electromagnetic field is an important advantage of CSRMT compared to other electrical prospecting methods employed in the Arctic.

Author Contributions: Conceptualization, A.K.S.; methodology, A.K.S.; software, A.A.S.; validation, A.K.S., A.A.S. and N.Y.B.; formal analysis, A.A.S.; investigation, A.K.S., A.A.S. and N.Y.B.; resources, A.K.S. and A.A.S.; data curation, A.A.S.; writing—original draft preparation, A.K.S.; writing review and editing, N.Y.B.; visualization, A.K.S. and A.A.S.; supervision, A.K.S.; project administration, A.K.S. and N.Y.B.; funding acquisition, A.K.S. All authors have read and agreed to the published version of the manuscript.

Funding: This research was funded by the Russian Science Foundation, project No. 21-47-04401.

Institutional Review Board Statement: Not applicable.

Informed Consent Statement: Not applicable.

Data Availability Statement: The datasets generated and analyzed in the current study are available from the corresponding author on reasonable request.

Acknowledgments: The presented results were obtained with the support of the Russian Science Foundation, project No. 21-47-04401, and the Research park of St. Petersburg State University "Center for Geo-Environmental Research and Modeling (GEOMODEL)".

Conflicts of Interest: The authors declare no conflict of interest.

References

1. Berdichevsky, M.N.; Dmitriev, V.I. *Models and Methods of Magnetotellurics*; Scientific World: Moscow, Russia, 2009; 680p. (In Russian)
2. Chave, A.D.; Jones, A.G. (Eds.) *The Magnetotelluric Method: Theory and Practice*; Cambridge University Press: Cambridge, UK, 2012; 570p. [CrossRef]
3. Zonge, K.L.; Hughes, L.J. Controlled-source audio-frequency magnetotellurics. In *Electromagnetic Methods in Applied Geophysics*; Nabighian, M.N., Ed.; Society of Exploration Geophysicists: Houston, TX, USA, 1991; Volume 2, pp. 713–809. [CrossRef]

4. Milsom, J.; Eriksen, A. *Field Geophysics*, 4th ed.; Wiley Publication: New York, NY, USA, 2011; 287p.
5. Yakupov, V.S. *Study of Frozen Strata by Geophysical Methods*; Syberian Branch of RAS, Yakutsk Subsidiary: Yakutsk, Russia, 2000; 336p. (In Russian)
6. Zykov, Y.D. *Geophysical Methods of Permafrost Investigations*; Moscow State University: Moscow, Russia, 2007; 264p. (In Russian)
7. Hauck, C.; Kneisel, C. *Applied Geophysics in Periglacial Environments*; Cambridge University Press: Cambridge, UK, 2008; 240p. [CrossRef]
8. Hauck, C. New Concepts in Geophysical Surveying and Data Interpretation for Permafrost Terrain. *Permafr. Periglac. Process.* **2013**, *24*, 131–137. [CrossRef]
9. Briggs, M.A.; Campbell, S.; Nolan, J.; Walvoord, M.A.; Ntarlagiannis, D.; Day-Lewis, F.D.; Lane, J.W. Surface geophysical methods for characterising frozen ground in transitional permafrost landscapes. *Permafr. Periglac. Process.* **2017**, *28*, 52–65. [CrossRef]
10. *Guide to the Interpretation of VES and MTC Curves*; PNIIS; Stroyizdat: Moscow, Russia, 1984; 200p. (In Russian)
11. Harada, K.; Wada, K.; Fukuda, M. Permafrost mapping by transient electromagnetic method. *Permafr. Periglac. Process.* **2000**, *11*, 71–84. [CrossRef]
12. Hauck, C.; Guglielmin, M.; Isaksen, K.; Mühll, D.V. Applicability of frequency-domain and time-domain electromagnetic methods for mountain permafrost studies. *Permafr. Periglac. Process.* **2001**, *12*, 39–52. [CrossRef]
13. Ageev, V.; Ageev, D. Solving of hydrogeological problems in permafrost zone conditions of the polar part of Western Siberia by the TEM method. In Proceedings of the Near Surface Geoscience Conference & Exhibition 2019, Hague, The Netherlands, 8–12 September 2019.
14. Creighton, A.L.; Parsekian, A.D.; Angelopoulos, M.; Jones, B.M.; Bondurant, A.; Engram, M.; Lenz, J.; Overduin, P.P.; Grosse, G.; Babcock, E.; et al. Transient electromagnetic surveys for the determination of talik depth and geometry beneath thermokarst lakes. *J. Geophys. Res. Solid Earth* **2018**, *123*, 9310–9323. [CrossRef]
15. Buddo, I.; Sharlov, M.; Shelokhov, I.; Misyurkeeva, N.; Seminsky, I.; Selyaev, V.; Agafonov, Y. Applicability of transient electromagnetic surveys to permafrost imaging in Arctic West Siberia. *Energies* **2022**, *15*, 1816. [CrossRef]
16. Stevens, C.W.; Moorman, B.J.; Solomon, S.M.; Hugenholtz, C.H. Mapping subsurface conditions within the near-shore zone of an Arctic delta using ground penetrating radar. *Cold Reg. Sci. Technol.* **2009**, *56*, 30–38. [CrossRef]
17. Watanabe, T.; Matsuoka, N.; Christiansen, H. Ice- and soil-wedge dynamics in the Kapp Linné Area, Svalbard, investigated by two- and three-dimensional GPR and ground thermal and acceleration regimes. *Permafr. Periglac. Process.* **2013**, *24*, 39–55. [CrossRef]
18. Shennen, S.; Tronicke, J.; Wetterich, S.; Allrogen, N.; Schwamborn, G.; Schirrmeister, L. 3D ground-penetrating radar imaging of ice complex deposits in northern East Siberia. *Geophysics* **2016**, *81*, WA185–WA192. [CrossRef]
19. Jafarov, E.E.; Parsekian, A.D.; Schaefer, K.; Liu, L.; Chen, A.C.; Panda, S.K.; Zhang, T. Estimating active layer thickness and volumetric water content from ground penetrating radar measurements in Barrow, Alaska. *Geosci. Data J.* **2017**, *4*, 72–79. [CrossRef]
20. Tregubov, O.; Kraev, G.; Maslakov, A. Hazards of activation of cryogenic processes in the Arctic Community: A geopenetrating radar study in Lorino, Chukotka, Russia. *Geosciences* **2020**, *10*, 57. [CrossRef]
21. Buddo, I.; Misyurkeeva, N.; Shelokhov, I.; Chuvilin, E.; Chernikh, A.; Smirnov, A. Imaging Arctic Permafrost: Modeling for Choice of Geophysical Methods. *Geosciences* **2022**, *12*, 389. [CrossRef]
22. Plotnikov, A.E. Evaluation of limitations of the transient electromagnetic method in shallow-depth studies: Numerical experiment. *Russ. Geol. Geophys.* **2014**, *55*, 907–914. [CrossRef]
23. Kozhevnikov, N.O.; Sharlov, M.V. Early-time and late-time limitations on the performance of near-surface TEM measuring systems. In Proceedings of the Near Surface Geoscience Conference & Exhibition 2019, Hague, The Netherlands, 8–12 September 2019.
24. Antonov, E.Y.; Mogilatov, V.S.; Epov, M.I. Effect of transmitter current waveform on transient electromagnetic responses. *Russ. Geol. Geophys.* **2019**, *60*, 492–499. [CrossRef]
25. Tezkan, B. Radiomagnetotellurics. In *Groundwater Geophysics: A Tool for Hydrogeology*; Kirsch, R., Ed.; Springer: Berlin/Heidelberg, Germany, 2008; pp. 295–318. [CrossRef]
26. Bastani, M. Enviro-MT—A New Controlled Source/Radio Magnetotelluric System. Ph.D. Thesis, Acta Universitatis Upsaliensis, Uppsala, Sweden, 2001.
27. Bastani, M.; Malehmir, A.; Ismail, N.; Pedersen, L.B.; Hedjazi, F. Delineating hydrothermal stockwork copper deposits using controlled-source and radio-magnetotelluric methods: A case study from northeast Iran. *Geophysics* **2009**, *45*, B167–B181. [CrossRef]
28. Bastani, M.; Savvaidis, A.; Pedersen, L.B.; Kalscheuer, T. CSRMT measurements in the frequency range of 1–250 kHz to map a normal fault in the Volvi basin, Greece. *J. Appl. Geophys.* **2011**, *75*, 180–195. [CrossRef]
29. Shan, C.; Bastani, M.; Malehmir, A.; Persson, L.; Lundberg, E. Integration of controlled-source and radio magnetotellurics, electric resistivity tomography, and reflection seismics to delineate 3D structures of a quick-clay landslide site. *Geophysics* **2016**, *81*, B13–B29. [CrossRef]
30. Wang, S.; Bastani, M.; Constable, S.; Kalscheuer, T.; Malehmir, A. Boat-towed radio-magnetotelluric and controlled source audio-magnetotelluric study to resolve fracture zones at Äspö Hard Rock Laboratory site, Sweden. *Geophys. J. Int.* **2019**, *218*, 1008–1031. [CrossRef]

31. Saraev, A.; Simakov, A.; Shlykov, A.; Tezkan, B. Controlled source radiomagnetotellurics: A tool for near surface investigations in remote regions. *J. Appl. Geophys.* **2017**, *146*, 228–237. [CrossRef]
32. Tezkan, B.; Muttaqien, I.; Saraev, A. Mapping of buried faults using the 2D modelling of far-field controlled source radiomagnetotelluric data. *Pure Appl. Geophys.* **2019**, *176*, 751–766. [CrossRef]
33. Shlykov, A.; Saraev, A.; Agrahari, S.; Tezkan, B.; Singh, A. One-dimensional laterally constrained joint anisotropic inversion of CSRMT and ERT data. *J. Environ. Eng. Geoph.* **2021**, *26*, 35–48. [CrossRef]
34. Romanovsky, N.N. Taliks in the area of permafrost and the scheme of their subdivision. *Moscow Univ. Geol. Bull.* **1972**, *1*, 23–34. (In Russian)
35. O'Neill, H.B.; Roy-Leveillee, P.; Lebedeva, L.; Ling, F. Recent advances (2010–2019) in the study of taliks. *Permafr. Periglac. Process.* **2020**, *31*, 346–357. [CrossRef]
36. Saraev, A.; Simakov, A.; Tezkan, B.; Tokarev, I.; Shlykov, A. On the study of industrial waste sites on the Karelian Isthmus/Russia using the RMT and CSRMT methods. *J. Appl. Geophys.* **2020**, *175*, 103993. [CrossRef]
37. Saraev, A.K.; Shlykov, A.A.; Tezkan, B. Application of the Controlled Source Radiomagnetotellurics (CSRMT) in the Study of Rocks Overlying Kimberlite Pipes in Yakutia/Siberia. *Geosciences* **2022**, *12*, 34. [CrossRef]
38. Shlykov, A.; Saraev, A.; Agrahari, S. Study of the anisotropy of horizontally layered section using data of the controlled source radiomagnetotelluric. *Geophysica* **2019**, *54*, 3–21.
39. Shlykov, A.; Saraev, A.; Tezkan, B. Study of a permafrost area in the northern part of Siberia using controlled source radiomagnetotellurics. *Pure Appl. Geophys.* **2020**, *177*, 5845–5859. [CrossRef]
40. Smirnova, M.; Shlykov, A.; Asghari, S.F.; Tezkan, B.; Saraev, A.; Yogeshwar, P.; Smirnov, M. Three-dimensional controlled-source electromagnetic inversion in the radio-frequency band. *Geophysics* **2022**, *88*, E1. [CrossRef]
41. Wannamaker, P.E. Tensor CSAMT survey over the Sulphur Springs thermal area, Valles Caldera, New Mexico, United States of America, Part I: Implications for structure of the western caldera. *Geophysics* **1997**, *62*, 451–465. [CrossRef]
42. Sjoberg, Y.; Marklund, P.; Pettersson, R.; Lyon, S.W. Geophysical mapping of palsa peatland permafrost. *Cryosphere* **2015**, *9*, 465–478. [CrossRef]
43. Fedorova, I.; Bobrov, N.; Pankova, D.; Konosavskiy, P.; Alekseeva, N. Modeling of thermic process of the Arctic ecosystems. In Proceedings of the 19th International Multidisciplinary Scientific GeoConference SGEM, Albena, Bulgaria, 30 June–6 July 2019; Book 3.1. pp. 401–410. [CrossRef]
44. Shein, A.N.; Olenchenko, V.V.; Kamnev, Y.K.; Sinitskiy, A.I. Structure of freezing talik under lake at the Parisento field station (Gydan Peninsula) according to electrical resistivity tomography. *Interexpo Geo-Sibyr* **2019**, *2*, 103–110. (In Russian) [CrossRef]
45. Rangel, R.C.; Parsekian, A.D.; Farquharson, L.M.; Jones, B.M.; Ohara, N.; Creighton, A.L.; Gaglioti, B.V.; Kanevskiy, M.; Breen, A.L.; Bergstedt, H.; et al. Geophysical observations of taliks below drained lake basins on the Arctic Coastal Plain of Alaska. *J. Geophys. Res. Solid Earth* **2021**, *126*, e2020JB020889. [CrossRef]
46. Key, K. MARE2DEM: A 2-D inversion code for controlled-source electromagnetic and magnetotelluric data. *Geophys. J. Int.* **2016**, *207*, 571–588. [CrossRef]
47. Zond Software. Available online: http://zond-geo.com/english/ (accessed on 24 November 2022).

Disclaimer/Publisher's Note: The statements, opinions and data contained in all publications are solely those of the individual author(s) and contributor(s) and not of MDPI and/or the editor(s). MDPI and/or the editor(s) disclaim responsibility for any injury to people or property resulting from any ideas, methods, instructions or products referred to in the content.

eng

MDPI

Article

Use of the Analytic Hierarchy Process Method in the Variety Selection Process for Sugarcane Planting

Luiza L. P. Schiavon *, Pedro A. B. Lima *, Antonio F. Crepaldi and Enzo B. Mariano

Department of Production Engineering, School of Engineering of Bauru, Campus Bauru, São Paulo State University (UNESP), Bauru 17033-360, Brazil
* Correspondence: luiza.schiavon@unesp.br (L.L.P.S.); pedro.ab.lima@unesp.br (P.A.B.L.)

Abstract: The sugar and alcohol sectors are dynamic as a result of climate alterations, the introduction of sugarcane varieties, and new technologies. Despite these factors, Brazil stands out as the main producer of sugarcane worldwide, being responsible for 45% of the production of fuel ethanol. Several varieties of sugarcane have been developed in the past few years to improve features of the plant. This, however, led to the challenge of which variety producers should choose to plant on their property. In order to support this process, this research aims to test the application of the analytic hierarchy process (AHP) method to support producers to select which sugarcane variety to plant on their property. To achieve this goal, the research relied on a single case study performed on a rural property located inland of São Paulo state, the main producer state in Brazil. The results demonstrate the feasibility of the approach used, specifically owing to the adaptability capacity of the AHP method.

Keywords: multicriteria method; decision-making process; sugarcane; analytic hierarchy process

1. Introduction

Sugarcane is an important commodity for several developing countries' economies, such as China [1], India [2], Belize [3], and Brazil—the main sugarcane producer in the world. Brazil started producing sugarcane in the 14th century, still in the colonial age, and around the 17th century, the country became the major sugarcane producer worldwide [4]. Nowadays, Brazil is responsible for producing 45% of the ethanol used for fuel in the world. The country is also one of the major exporters of sugar. São Paulo state is the main producer in the country, producing 53.7% of the Brazilian sugarcane in the 2019/2020 harvest, producing 29.03 million tons of sugar and 35.5 billion liters of ethanol. Moreover, the sector represented 26.89% (U.S. $4.07 billion) of the state exportation [5]. The northeastern region of the state received the greatest increase in sugarcane production, which took place in areas that previously held cattle and other agricultural production [6]. Other states in the south-central region of the country also experienced a similar pattern to São Paulo, although of a lower magnitude [7].

Among the reasons for the expansion of Brazilian production of ethanol from sugarcane was the introduction of flex-fuel engines in the internal market, which provides the ability to use any amount of ethanol and gasoline combination in vehicles [6,8]. The global demand for less environmentally harmful fuels played a significant part in this process [8], motivating research on several alternatives for replacing fossil fuels in the agriculture field, e.g., producing biomass from agricultural residues [9]. Brazilian governmental and industrial stakeholders took advantage of these positive scenarios and worked for the development of sugarcane production in the country. The substantial production of sugarcane in the São Paulo state is related to several factors, such as the presence of a large amount of land with appropriate quality for sugarcane, the best infrastructure in the country, and a regional system of innovation to support the development of production [10]. An expansive picture of Brazilian sugarcane-based ethanol production can be found in [11].

Citation: Schiavon, L.L.P.; Lima, P.A.B.; Crepaldi, A.F.; Mariano, E.B. Use of the Analytic Hierarchy Process Method in the Variety Selection Process for Sugarcane Planting. *Eng* **2023**, *4*, 602–614. https://doi.org/10.3390/eng4010036

Academic Editor: Antonio Gil Bravo

Received: 29 November 2022
Revised: 10 February 2023
Accepted: 13 February 2023
Published: 15 February 2023

Copyright: © 2023 by the authors. Licensee MDPI, Basel, Switzerland. This article is an open access article distributed under the terms and conditions of the Creative Commons Attribution (CC BY) license (https://creativecommons.org/licenses/by/4.0/).

Sugarcane cultivation and processing plants have a significant impact on the Brazilian socio-economic structure, being a source of employment and income generation for several municipalities [12]. With the significant reduction in the burning technique previous to the harvesting [6], bioethanol production from sugarcane can also have positive outcomes for environmental sustainability, especially related to reducing CO_2 emission compared with using fossil fuels (see [13] for a review about sugarcane production and sustainability in Brazil). Recent studies show new opportunities to increase environmental sustainability in the sector, such as with circular economy practices [14], which is a concept that can contribute to achieving sustainable and human development [15].

Different varieties of sugarcane present different features that affect the products made from sugarcane as well as the sugarcane growth and production itself [16]. Brazil has been experiencing a significant increase in the number of sugarcane varieties [17]. New varieties of sugarcane are useful for achieving higher efficiency (e.g., lower costs and higher productivity) according to different factors [10], such as the new intensive mechanization (as well as other management aspects) and new environments for plantation (including soil and climatic conditions) [18]. Notwithstanding all the apparent benefits, a massive number of options cannot be easily processed, resulting in difficulty in selecting the best choice [19]. The bounded rationality of individuals affects the efficiency of the decision-makers, including in the agriculture sector [20]. In other words, the higher the number of sugarcane varieties, the more complex the process to select the best sugarcane variety [21].

A decision-making process should be structured based on rules, methods, and procedures and its goal should be the selection of the best-performing option, best expectation, or best evaluation among all the available choices [22]. Within the strategies for decision-making, multicriteria techniques are among the main approaches used, as they consider several factors—different conflicting criteria—related to the decision [23,24].

The use of multicriteria methods is a common approach in agriculture-related literature [14,25]. Within these applications, there is a branch of studies focused on variety selection using the analytical hierarchical process (AHP) method. AHP is a suitable approach to indicate preferences for different objectives [26]. For example, AHP has already been applied to support crop selection considering oilseed crops [27], as well as to select the best grape option for organic viticulture [28]. However, little research has relied on AHP to support the selection process of sugarcane variety (with the exception of [29]). Approaches to support variety selection can be quite useful for producers, especially in locations with low resources [30]. Thus, this research aims to test the application of the AHP method to support producers to select which sugarcane variety to plant on their property. In order to achieve this goal, the research relied on a single case study performed in a rural property located in São Paulo state, Brazil. Therefore, while the generalization of the approach used in this research can be made to other contexts, the specific outcomes (i.e., the variety selected) and the specific variables in the model (i.e., the varieties and their features) should be understood considering this case study. This is because of the external aspects that influence the sugarcane varieties, such as soil and climate. It is expected that the approach presented in this research can be especially useful for the decision-making processes related to variety selection faced by small and medium-sized rural producers.

After this introduction, Section 2 presents the materials and methods used in this research, describing the AHP method and the case study of this research. Next, Section 3 presents the results and discussions of the research's findings. Finally, Section 4 presents the conclusions.

2. Materials and Methods

2.1. Analytic Hierarchy Process

Analytic hierarchy process (AHP) was developed in the 1970s by Tomas L. Saaty, and consists of a multicriteria method used and known to support decision-making in problems with multiple criteria. It is based on the Newtonian and Cartesian method that seeks to solve a problem by decomposing it into factors, which can be decomposed into new factors up to the lowest level [31].

AHP is a method commonly applied to define weights for different criteria, being used in a different set of problems and fields, e.g., [27,28,32], mainly thanks to its robustness and simplicity [33]. These applications can be suitable for supporting the solution of simple issues involving only one person or extremely complex situations related to several variables [33]. Generally, the AHP method follows four main steps: problem modeling, pairwise comparison, judgment scale, and priorities' derivation [34].

The hierarchical structure begins with an overall objective that descends to criteria and then alternatives [35]. The first level of the structure presents the general objective to be achieved. The second level indicates the criteria that contribute to reaching the general objective. The third level contains the decision alternatives for the problem [36] (Figure 1).

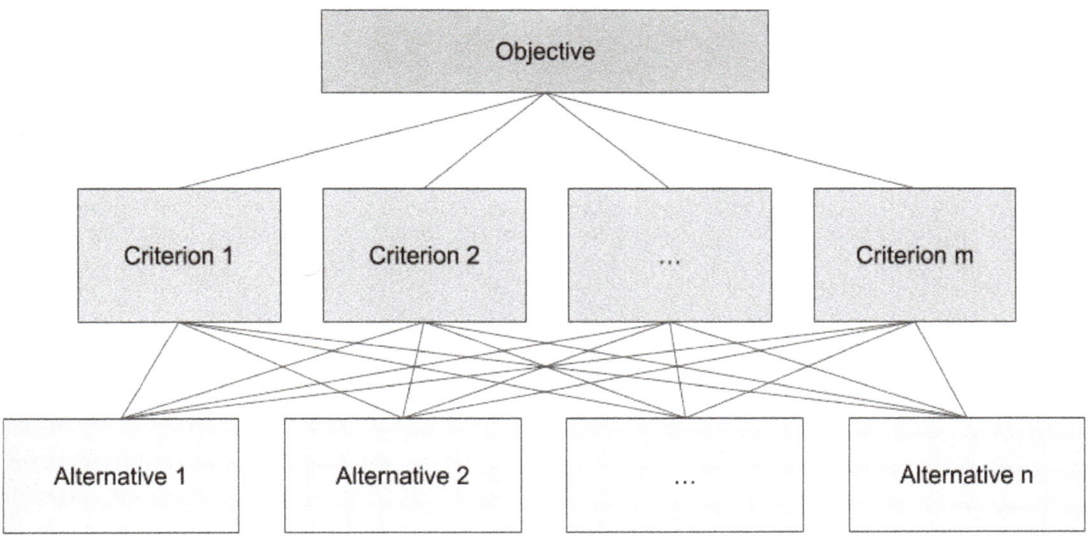

Figure 1. Representation of the basic hierarchical structure of the AHP method. Source: adapted from [37].

After the hierarchical model is developed to address the issue, the next step is the pairwise comparisons for each level of the model by the decision-makers participating in the research. This step aims to achieve a weight factor for each element on the level considering the element right on the next higher level. This process provides a measure of the relative importance of the considered element [36].

The priority definition should be based on the ability of the individuals to perceive the relationship between the objects, comparing the pairs in relation to a criterion or judgment. It is necessary to apply the following steps to achieve this [35,38]:

- Elaborate on the problem to be solved;
- Consider the objectives and results of the problem;
- Identify the criteria that influence the behavior;
- Structure the problem in a hierarchy of different levels, criteria, sub-criteria, and alternatives;

- Parity judgment: to judge pair-by-pair the elements in the hierarchy in relation to each element in the superior level, compounding a matrix of judgment A, using the scale presented in Table 1. The quantity of judgments for the construction of a matrix A is n × (n − 1), where n is the number of elements.

Table 1. Saaty's fundamental scale for the AHP method.

Intensity of Importance on an Absolute Scale	Definition	Explanation
1	Equal importance	Two options contribute equally to the objective
3	Moderate importance of one over another	Experience and judgment strongly favor one activity over another
5	Essential or strong importance	Experience and judgment strongly favor one activity over another
7	Very strong importance	An activity is strongly favored and its dominance demonstrated in practice
9	Extreme importance	The evidence favoring one activity over another is of the highest possible order of affirmation
2, 4, 6, 8	Intermediate values between the two adjacent judgments	Applied when compromise is needed

Source: adapted from [35].

- Normalization of the judgment matrices: through the sum of the elements of each column of the judgment matrix, their normalized values are obtained. After that, it is necessary to divide each element of these matrices for the summation of the values of the respective column;
- Calculus of the global priorities: it is necessary to identify a global priorities vector that can store the priority associated with each alternative in relation to the main focus;
- Logical consistency: this method calculates the consistency ratio of the judgment, being $CR \doteq CI/IR$, where IR is the random consistency index obtained for a reciprocal matrix of order n, with non-negative elements and automatically generated. In order to be considered consistent, it is necessary that $CR \leq 0.10$.

2.2. AHP Application

In this research, the AHP method was used to support two medium producers of sugarcane located in the inland of São Paulo state, Brazil, to select varieties of sugarcane to be planted. Both are the owners and are responsible for the decision-making in the property. Thus, the application of the AHP method can support them to select the best option according to their preferences regarding the sugarcane's features. The land has purple oxisol soil, also known as "purple land", a kind of reddish soil that is very fertile. The climate is high-altitude tropical, which is characterized by the concentration of rains in the summer and temperatures below 18 °C in the winter.

The São Paulo state (Figure 2) has an estimated population of 44.7 million inhabitants, being the most populous Brazilian state, representing 21% of Brazil's population. In 2019, the GDP achieved by the state was approximately USD 582.18 billion, representing 31% of the national GDP [39]. The state also has among the country's highest levels of human development and municipal environmental management practices [40].

Figure 2. Brazil and São Paulo state location.

For decision applications, the AHP method was carried out in two phases: (1) the hierarchic design and (2) the evaluation [33]. One of the ways to develop the hierarchic design—phase (1)—is by reaching a consensus in a group, with the presence of individuals with knowledge and experience in the analyzed field being recommended [33]. The hierarchical model should be "complex enough to capture the situation, but small and nimble enough to be sensitive to changes" [31] (p. 163). Thus, in order to decompose the problem into hierarchy elements and to establish the criteria to be evaluated, the first author arranged a meeting with an agronomist engineer. The interview with the agronomist engineer was used to understand the most relevant criteria to be considered for AHP and the best varieties of sugarcane to be included in the model. This study considered the following possible varieties as options to be selected: RB867515, RB966928, CT9001, and RB855156, which are the most cultivated varieties in the regions according to the agronomist engineer. The selected criteria were as follows:

1. Potential for sucrose accumulation: this is the sugarcane's capacity that determines the agricultural production. The values vary according to the time of the year and support the steps that compound the industrialization process of the sugarcane [41];
2. Ratoon sprouting: physiological processes that encompass the period from plantation to the beginning of tillering, after the second cut [42];
3. Ton per hectare: mass of sugarcane produced in one hectare, where 1 ton/hectare is equivalent to 0.1 kg/m^2;
4. Longevity: this is the life expectancy of the cane field, that is, the number of cuts between cane field renovation cycles. As planting is one of the most important stages of sugarcane, a variety of sugarcane that has great longevity has a direct impact on production costs and economic return [43];
5. Soil requirement: this consists of the nutrients required by plants for proper growth [44].

Figure 3 represents the hierarchy constructed in this research.

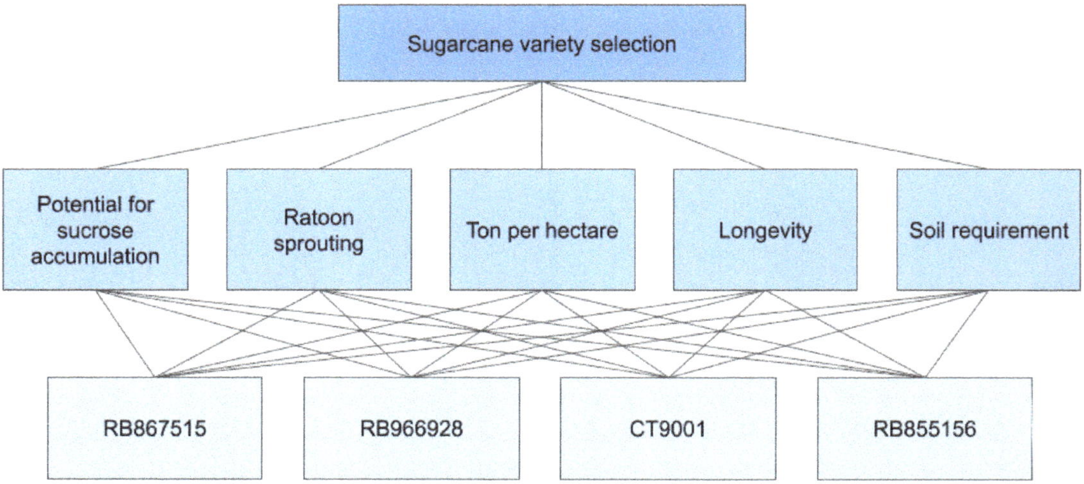

Figure 3. Representation of the basic hierarchical structure of sugarcane used in this research.

There is not a pre-established number of individuals that should be interviewed for the AHP method. In the agriculture-related literature, this number has ranged from large numbers in complex issues, such as 60 [45] and 144 [46], to small numbers when related to decision-making problems for farmers to apply in their work context, such as 1 decision-maker [36] and 3 decision-makers [28]. This last context is the case of this research, as the main goal of the AHP application was to support the farmers in selecting which sugarcane variety to choose on their farm. Thus, for the evaluation phase (phase 2), the first author arranged individual meetings with two local farmers who are co-owners of a farm located in São Paulo's inland and have more than 50 years of experience in agriculture production. From now on, they will be called decision-maker 1 and decision-maker 2. This step was performed in order to conduct the paired comparison of each element on the hierarchical level, creating a matrix of quadratic decisions. The paired comparison used the Saaty's fundamental scale for AHP (Table 1). Next, the authors determined the degrees of preference for each criterion, developing five matrices that compare the degrees of intensity for pairs as a function of each characteristic, referencing the five criteria adopted. With the comparing matrices fulfilled, the authors created an algorithm in C language and in MatLab R2015a to implement the AHP method. The results were validated with the application of the Super Decision Software. Next, the authors evaluated the consistency ratio (CR) of all hierarchies, dividing the consistency index (CI) by the random consistency index (RCI) obtained for one matrix of order n, with non-negative elements and randomly generated. The RCI of the hierarchy must be inferior or equal to 10%. The flowchart in Figure 4 presents all of the research steps.

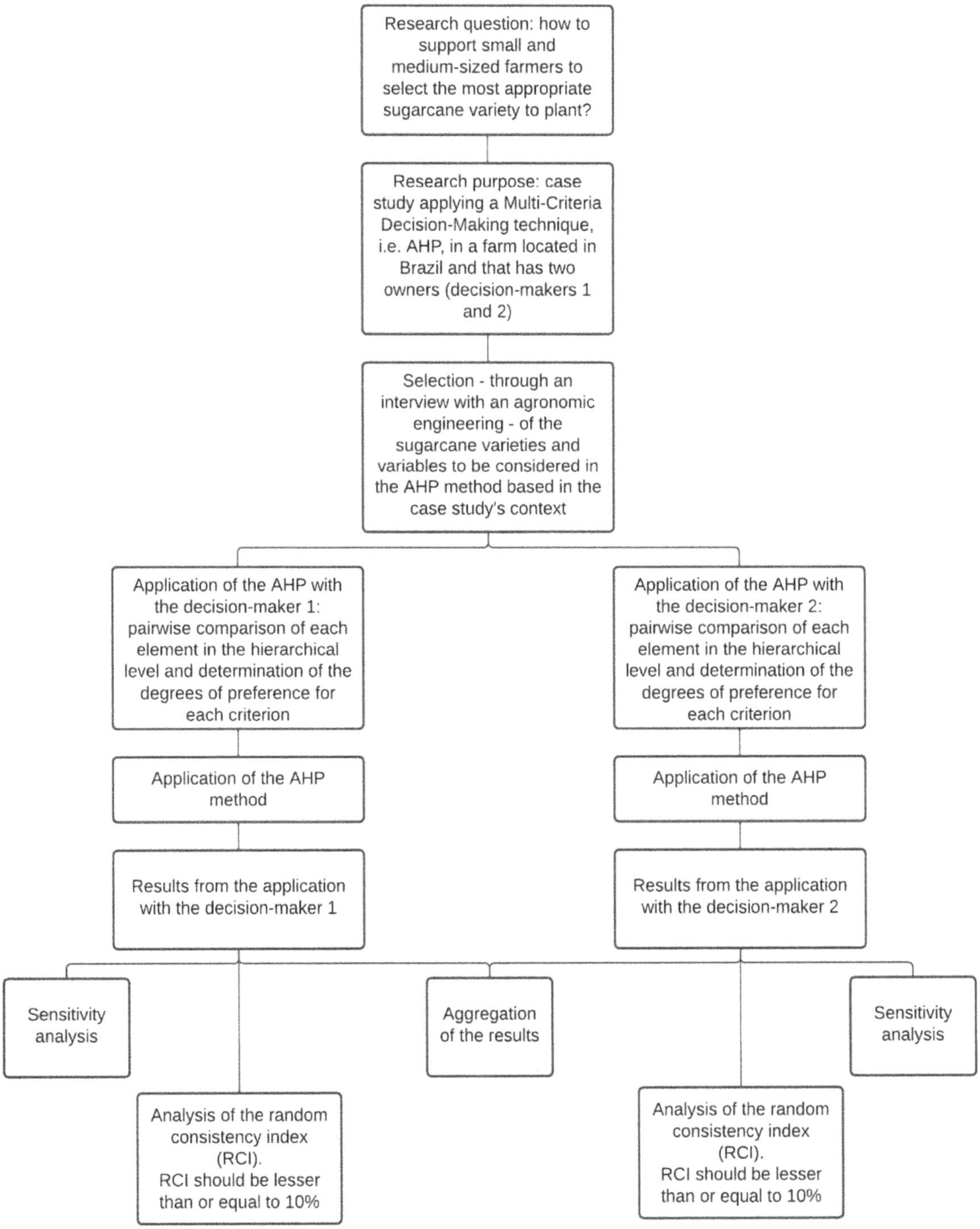

Figure 4. Flowchart describing all of the steps of the AHP method applied in this research.

3. Results

Table 2 presents the prioritization of each alternative (variety of sugarcane) related to each criterion and the main focus of decision-maker 1 and decision-maker 2.

Table 2. Prioritization of each alternative of decision-maker 1 and decision-maker 2 for the variables considered in the AHP application.

Criterion	Decision-Maker	Sugarcane Variety	RB867515	RB966928	CT9001	RB855156
Potential for sucrose accumulation	1	RB867515	1	1/9	1/5	1/3
		RB966928	9	1	5	9
		CT9001	5	1/5	1	3
		RB855156	3	1/9	1/3	1
	2	RB867515	1	1/7	1/3	1/5
		RB966928	7	1	5	3
		CT9001	3	1/5	1	1/5
		RB855156	5	1/3	5	1
Ratoon sprouting	1	RB867515	1	1/9	1/5	1/3
		RB966928	9	1	3	1
		CT9001	5	1/3	1	1
		RB855156	3	1	1	1
	2	RB867515	1	1/7	3	1/3
		RB966928	7	1	7	5
		CT9001	1/3	1/7	1	1/5
		RB855156	3	1/5	5	1
Ton per hectare	1	RB867515	1	1/3	1/5	1/3
		RB966928	3	1	1/3	3
		CT9001	5	3	1	5
		RB855156	3	1/3	1/5	1
	2	RB867515	1	1/5	1/3	1
		RB966928	5	1	5	7
		CT9001	3	1/5	1	3
		RB855156	1	1/7	1/3	1
Longevity	1	RB867515	1	1/5	1	1/9
		RB966928	5	1	5	1/7
		CT9001	1	1/5	1	1/9
		RB855156	9	7	9	1
	2	RB867515	1	1/5	3	1/5
		RB966928	5	1	7	3
		CT9001	1/3	1/7	1	1/3
		RB855156	5	1/3	3	1
Soil requirement	1	RB867515	1	1/5	1/5	1/9
		RB966928	5	1	1/3	1/5
		CT9001	5	3	1	1
		RB855156	9	5	1	1
	2	RB867515	1	1/7	1/5	1/3
		RB966928	7	1	5	7
		CT9001	5	1/5	1	3
		RB855156	3	1/7	1/3	1

With the application of the AHP method, after performing the matrices' normalization and the calculation of the average of each criterion, it was possible to determine the preference matrices, as presented in Table 3 for decision-maker 1 and decision-maker 2.

Table 3. Preference matrix of each variable and sugarcane variety for both decision-makers.

Decision-Maker	Sugarcane Variety	Potential for Sucrose Accumulation	Ratoon Sprouting	Ton per Hectare	Longevity	Soil Requirement
1	RB867515	0.047321	0.059865	0.076463	0.057550	0.049690
	RB966928	0.660862	0.446506	0.244503	0.212166	0.144201
	CT9001	0.199117	0.226610	0.543046	0.057550	0.350881
	RB855156	0.092700	0.267019	0.135988	0.372734	0.455229
2	RB867515	0.05385091	0.099166143	0.088053939	0.104356652	0.050389565
	RB966928	0.551807627	0.638159813	0.6302765	0.567573554	0.638159813
	CT9001	0.101457857	0.050389565	0.200719575	0.062934631	0.21228448
	RB855156	0.292883606	0.21228448	0.080949986	0.265135162	0.099166143

Regarding the sucrose accumulation criterion, both decision-makers indicated a preference for RB966928. However, while decision-maker 1 preferred CT9001 as the second-best variety, decision-maker 2 opted for the RB55156 variety. Regarding the criterion of ratoon sprouting, both decision-makers agreed that the best variety was RB966928 and the second preference was RB855156. Considering the criterion ton per hectare, decision-maker 1 elected CT9001 as the best choice, while decision-maker 2 opted for RB855156. For the longevity criterion, decision-maker 1 considered RB855156 as the best variety, followed by RB966928; decision-maker 2, however, considered the RB966928 variety as the best and RB855156 as the second preference. Finally, considering the soil requirement criterion, decision-maker 1 chose the RB855156 variety, while decision-maker 2 preferred the RB966928 variety.

Table 4 presents a criteria comparison matrix for decision-maker 1 and decision-maker 2. Both decision-makers presented similar options regarding the comparison of the criteria; the majority of comparisons were different in two points of intensity importance. The main difference is that, while decision-maker 1 considers that longevity has a moderate importance over ton per hectare, decision-maker 2 considers that ton per hectare has a strong importance over longevity. In other words, when comparing only these two criteria, decision-maker 1 considers longevity more important while decision-maker 2 considers ton per hectare more important. There was also a difference when comparing the sucrose accumulation criterion with the longevity criterion. While decision-maker 1 considered longevity with very strong importance over the sucrose accumulation criterion, decision-maker 2 considered longevity with moderate importance over sucrose accumulation.

Table 4. Criterion comparison matrix of each variable and sugarcane variety for both decision-makers.

Decision-Maker	Sugarcane Variety	Potential for Sucrose Accumulation	Ratoon Sprouting	Ton per Hectare	Longevity	Soil Requirement
1	Potential for sucrose accumulation	1	1/9	1/5	1/7	1/3
	Ratoon sprouting	9	1	5	5	7
	Ton per hectare	5	1/5	1	1/3	3
	Longevity	7	1/5	3	1	5
	Soil requirement	3	1/7	1/3	1/5	1
2	Potential for sucrose accumulation	1	1/7	1/7	1/3	1/5
	Ratoon sprouting	7	1	3	7	9
	Ton per hectare	7	1/3	1	5	5
	Longevity	3	1/7	1/5	1	3
	Soil requirement	5	1/9	1/5	1/3	1

Next, the authors normalized the comparison matrix of the criteria and calculated the average in order to achieve the final result, which is displayed in Table 5.

Table 5. Result of the AHP application with the preferences of both decision-makers for each sugarcane variety.

Variety	Decision-Maker 1	Decision-Maker 2
RB867515	5.8477466%	9.1871988%
RB966928	35.7146925%	62.6347183%
CT9001	24.3115487%	10.6052233%
RB855156	34.1260123%	17.5728596%

The results in Table 5 present the final quantification of each alternative according to the answers provided by decision-maker 1 and decision-maker 2. Considering decision-maker 1, 5.84% of the quantification was for selecting variety RB867515, 35.71% for choosing variety RB966928, 24.31% for variety CT9001, and 34.12% for variety RB855156. For decision-maker 2, 9.18% of the quantification was for selecting variety RB867515, 62.63% was for selecting variety RB966928, 10.60% was for selecting variety CT9001, and 17.57% was for selecting variety RB855156.

Comparing the final results of both decision-makers, it was possible to notice that, even though there were differences in the percentage for each variety, they presented the same ranking order. In this way, for both decision-makers, the best choice was the RB966928 variety, the second-best choice was RB855156, the third was CT9001, and the last was RB867515.

In order to verify the method's validity, the authors calculated the consistency ratio in the research's matrices, that is, they compared the consistency index with the random consistency index corresponding to the dimension of each matrix. As the consistency ratio of the hierarchy of all matrices was lower than 10%, the method can be considered valid. Therefore, the sugarcane variety RB966928, with numeric results of 35.71% and 62.63% for decision-maker 1 and 2, respectively, should be selected considering the pairwise comparisons provided and the verification of the matrix's coherence. After achieving the final results, the first author presented them to both producers, which reported that the variety RB966928 is usually the one that they prefer and the one they were thinking of planting in the next season. Therefore, the method presented in this research can also be applied to evaluate whether the variety selected by the producers is indeed the one that the producers believe is the best option. Future studies should include longitudinal and economic data in order to verify whether the selected option is indeed the one that presents the best economic outcomes for the producers.

The aggregation of individual judgments (AIJ) method was used in order to aggregate the results of decision-makers into a single group [47] (Table 6). It is noticeable that the decision-makers' ranking was maintained, that is, the selected variety was RB966928 with 49.28%, followed by the RB855156 variety with 27.57%, next was the CT9001 variety with 15.19%, and finally the RB867515 variety with 7.96%.

Table 6. Aggregated results for both decision-makers 1 and 2.

Variety	Aggregated Results
RB867515	7.9622971%
RB966928	49.2800982%
CT9001	15.1858805%
RB855156	27.5717243%

The final priority of the alternatives is mainly determined by the weights assigned to the main criteria. Therefore, small changes in the relative weights can lead to large changes in the final ranking. In this context, sensitivity analysis can be performed based on the scenarios they reflect, increasing or decreasing the weight of individual criteria, resulting in changes in priority and rank [48].

As a final analysis, the authors performed a sensitivity analysis of the chosen criteria. For decision-maker 1, when the weight of the potential for sucrose accumulation criterion was changed, the alternative to be chosen remained RB855156. When the weight of the ratoon sprouting criterion was below 53.05%, the selected variety was RB855156; when it was above 53.05%, then the selected variety changed to RB966928. For the ton per hectare criterion, when its weight was at 0%, the selected varieties were RB855156 and CT9001; as the weight increased, the selected variety became RB966928. The variety chosen was CT9001 when the weight was 36.65%. The chosen alternative was RB966928 when the longevity criterion had a weight lower than 22.59%; when it was above this value, the selected variety was RB855156. On the other hand, when the weight of the soil requirement criterion was below 10.73%, the alternative chosen was RB966928; when it was above this level, the RB855156 variety was chosen. This same analysis was performed for decision-maker 2; however, regardless of the weight of the criteria, the variety chosen was RB966928. These analyses indicate that, for decision-maker 1, there is a system instability that varies, mainly, between the RB855156 and RB966928 varieties, which is justifiable because, as shown in Table 5, the variation between the choice of these varieties is only 1.58866802%. On the other hand, decision-maker 2′s system is stable, making it possible to change the relative importance levels of the criteria without affecting the choice of sugarcane variety, proving to be a robust choice, allowing the decision-maker greater security in relation to his choice.

4. Conclusions

The application of the analytic hierarchy process (AHP) to support a decision enables the analysis of all of the criteria and alternatives in light of each criterion. It can be considered that the goal to select the best variety of sugarcane was accomplished. The method used in this research is a supporting tool to the decision-making process, which does not diminish the farmer's role in it; he/she remains the decision element and source of information for judging the value and construction of the hierarchical model. Besides, the objective of the tool is to deal with the selection process scientifically and to model the subjectivity inherent to the decision-making process, not removing its subjectivity [29].

This research's importance is highlighted by applying a relatively simple method that can support farmers in selecting the best variety of sugarcane to plant. Another application of the method is to analyze whether farmers are selecting the best choice for their farms. Future studies could rely on a bigger sample in order to compare the varieties that farmers are planting and the choice that they reached as the best one with the AHP method. Considering theoretical implications, this research is important to increase the knowledge of AHP usefulness in the agricultural field.

Considering that the agricultural environment is very dynamic owing to environmental changes and the introduction of new technologies and new varieties of sugarcane into the market, the application of AHP proved to be adequate owing to its dynamic capacity of adaptation. Although there is no methodological issue in applying the AHP method with a sample of only two individuals, this can be considered one limitation of this research. Another limitation is that both farmers are from the same region of the state; therefore, future studies could compare the results of farmers in different locations and kinds of farms to better understand the issue. Finally, future studies should include longitudinal and economic data in order to perform further analysis and increase the outcomes of the method.

Author Contributions: Conceptualization, L.L.P.S. and P.A.B.L.; methodology, L.L.P.S.; software, L.L.P.S.; validation, A.F.C. and E.B.M.; formal analysis, L.L.P.S.; investigation, L.L.P.S.; resources, L.L.P.S.; data curation, L.L.P.S.; writing—original draft preparation, L.L.P.S. and P.A.B.L.; writing—review and editing, L.L.P.S., P.A.B.L., A.F.C. and E.B.M.; visualization, L.L.P.S. and P.A.B.L.; supervision, L.L.P.S. and P.A.B.L.; project administration, L.L.P.S. and P.A.B.L. All authors have read and agreed to the published version of the manuscript.

Funding: This research received no external funding.

Institutional Review Board Statement: Not applicable.

Informed Consent Statement: Not applicable.

Data Availability Statement: Not applicable.

Conflicts of Interest: The authors declare no conflict of interest.

References

1. Li, Y.R.; Yang, L.T. Sugarcane agriculture and sugar industry in China. *Sugar Tech* **2015**, *17*, 1–8. [CrossRef]
2. Solomon, S. Sugarcane production and development of sugar industry in India. *Sugar Tech* **2016**, *18*, 588–602. [CrossRef]
3. Gongora, A.; Villafranco, D. Sugarcane bagasse cogeneration in Belize: A review. *Renew. Sustain. Energy Rev.* **2018**, *96*, 58–63. [CrossRef]
4. Rodrigues, D.; Ortiz, L. *Em Direção à Sustentabilidade da Produção de Etanol de Cana de Açúcar no Brasil*; Amigos da Terra Brasil: Porto Alegre, Brasil, 2006.
5. IEA—Instituto de Economia Agrícola. Cana-de-Açúcar: Produção e Processamento em 2019. 2020. Available online: http://www.iea.sp.gov.br/out/TerTexto.php?codTexto=14767 (accessed on 29 May 2021).
6. Rudorff, B.F.T.; de Aguiar, D.A.; da Silva, W.F.; Sugawara, L.M.; Adami, M.; Moreira, M.A. Studies on the rapid expansion of sugarcane for ethanol production in São Paulo State (Brazil) using Landsat data. *Remote Sens.* **2010**, *2*, 1057–1076. [CrossRef]
7. Adami, M.; Rudorff, B.F.T.; Freitas, R.M.; Aguiar, D.A.; Sugawara, L.M.; Mello, M.P. Remote sensing time series to evaluate direct land use change of recent expanded sugarcane crop in Brazil. *Sustainability* **2012**, *4*, 574–585. [CrossRef]
8. Dias, M.O.S.; Maciel Filho, R.; Mantelatto, P.E.; Cavalett, O.; Rossell, C.E.V.; Bonomi, A.; Leal, M.R.L.V. Sugarcane processing for ethanol and sugar in Brazil. *Environ. Dev.* **2015**, *15*, 35–51. [CrossRef]
9. Voultsos, I.; Katsourinis, D.; Giannopoulos, D.; Founti, M. Integrating LCA with Process Modeling for the Energetic and Environmental Assessment of a CHP Biomass Gasification Plant: A Case Study in Thessaly, Greece. *Eng* **2020**, *1*, 2. [CrossRef]
10. Furtado, A.T.; Scandiffio, M.I.G.; Cortez, L.A.B. The Brazilian sugarcane innovation system. *Energy Policy* **2011**, *39*, 156–166. [CrossRef]
11. Moraes, M.A.F.D.; Zilberman, D. *Production of Ethanol from Sugarcane in Brazil: From State Intervention to a Free Market*; Springer Science & Business Media: Berlin/Heidelberg, Germany, 2014; Volume 43.
12. Gilio, L.; de Moraes, M.A.F.D. Sugarcane industry's socioeconomic impact in São Paulo, Brazil: A spatial dynamic panel approach. *Energy Econ.* **2016**, *58*, 27–37. [CrossRef]
13. Bordonal, R.D.O.; Carvalho, J.L.N.; Lal, R.; de Figueiredo, E.B.; de Oliveira, B.G.; La Scala, N. Sustainability of sugarcane production in Brazil. A review. *Agron. Sustain. Dev.* **2018**, *38*, 1–23. [CrossRef]
14. Jesus, G.M.K.; Jugend, D.; Paes, L.A.B.; Siqueira, R.M.; Leandrin, M.A. Barriers to the adoption of the circular economy in the Brazilian sugarcane ethanol sector. *Clean Technol. Environ. Policy* **2021**, 1–15. [CrossRef]
15. Lima, P.A.B.; Jesus, G.M.K.; Ortiz, C.R.; Frascareli, F.C.O.; Souza, F.B.; Mariano, E.B. Sustainable Development as Freedom: Trends and Opportunities for the Circular Economy in the Human Development Literature. *Sustainability* **2021**, *13*, 13407. [CrossRef]
16. Jane, S.A.; Fernandes, F.A.; Silva, E.M.; Muniz, J.A.; Fernandes, T.J.; Pimentel, G.V. Adjusting the growth curve of sugarcane varieties using nonlinear models. *Cienc. Rural* **2020**, *50*. [CrossRef]
17. Dal-Bianco, M.; Carneiro, M.S.; Hotta, C.T.; Chapola, R.G.; Hoffmann, H.P.; Garcia, A.A.F.; Souza, G.M. Sugarcane improvement: How far can we go? *Curr. Opin. Biotechnol.* **2012**, *23*, 265–270. [CrossRef]
18. Dias, H.B.; Inman-Bamber, G.; Everingham, Y.; Sentelhas, P.C.; Bermejo, R.; Christodoulou, D. Traits for canopy development and light interception by twenty-seven Brazilian sugarcane varieties. *Field Crops Res.* **2020**, *249*, 107716. [CrossRef]
19. Schwartz, B. *The Paradox of Choice: Why More Is Less*; Ecco Press: New York, NY, USA, 2004.
20. Musshoff, O.; Hirschauer, N. A behavioral economic analysis of bounded rationality in farm financing decisions: First empirical evidence. *Agric. Financ. Rev.* **2011**, *71*, 62–83. [CrossRef]
21. Ramburan, S.; Paraskevopoulos, A.; Saville, G.; Jones, M. A decision support system for sugarcane variety selection in South Africa based on genotype-by-environment analyses. *Exp. Agric.* **2010**, *46*, 243–257. [CrossRef]
22. Choo, C.W. The knowing organization: How organizations use information to construct meaning, create knowledge and make decisions. *Int. J. Inf. Manag.* **1996**, *16*, 329–340. [CrossRef]
23. Gebre, S.L.; Cattrysse, D.; Alemayehu, E.; Van Orshoven, J. Multi-criteria decision making methods to address rural land allocation problems: A systematic review. *Int. Soil Water Conserv. Res.* **2021**, *9*, 490–501. [CrossRef]
24. Tervonen, T.; Figueira, J.R. A survey on stochastic multicriteria acceptability analysis methods. *J. Multi-Criteria Decis. Anal.* **2008**, *15*, 1–14. [CrossRef]
25. Kaim, A.; Cord, A.F.; Volk, M. A review of multi-criteria optimization techniques for agricultural land use allocation. *Environ. Model. Softw.* **2018**, *105*, 79–93. [CrossRef]
26. Salas-Molina, F.; Pla-Santamaria, D.; Garcia-Bernabeu, A.; Reig-Mullor, J. A compact representation of preferences in multiple criteria optimization problems. *Mathematics* **2019**, *7*, 1092. [CrossRef]

27. Dekamin, M.; Barmaki, M.; Kanooni, A. Selecting the best environmental friendly oilseed crop by using Life Cycle Assessment, water footprint and analytic hierarchy process methods. *J. Clean. Prod.* **2018**, *198*, 1239–1250. [CrossRef]
28. Dragincic, J.; Korac, N.; Blagojevic, B. Group multi-criteria decision making (GMCDM) approach for selecting the most suitable table grape variety intended for organic viticulture. *Comput. Electron. Agric.* **2015**, *111*, 194–202. [CrossRef]
29. Costa, H.G.; Moll, R.N. Emprego do método de análise hierárquica (AHP) na seleção de variedades para o plantio de cana-de-açúcar. *Gest. Prod.* **1999**, *6*, 243–256. [CrossRef]
30. Shanthy, T.R. Participatory varietal selection in sugarcane. *Sugar Tech* **2010**, *12*, 1–4. [CrossRef]
31. Saaty, R.W. The analytic hierarchy process—What it is and how it is used. *Math. Model.* **1987**, *9*, 161–176. [CrossRef]
32. Aguiar, C.R.D.; Nuernberg, J.K.; Leonardi, T.C. Multicriteria GIS-Based Approach in Priority Areas Analysis for Sustainable Urban Drainage Practices: A Case Study of Pato Branco, Brazil. *Eng* **2020**, *1*, 6. [CrossRef]
33. Vargas, L.G. An overview of the analytic hierarchy process and its applications. *Eur. J. Oper. Res.* **1990**, *48*, 2–8. [CrossRef]
34. Ishizaka, A.; Labib, A. Review of the main developments in the analytic hierarchy process. *Expert Syst. Appl.* **2011**, *38*, 14336–14345. [CrossRef]
35. Saaty, T.L. How to make a decision: The analytic hierarchy process. *Eur. J. Oper. Res.* **1990**, *48*, 9–26. [CrossRef]
36. Alphonce, C.B. Application of the analytic hierarchy process in agriculture in developing countries. *Agric. Syst.* **1997**, *53*, 97–112. [CrossRef]
37. Zarghami, M.; Szidarovszky, F. Introduction to Multicriteria Decision Analysis. In *Multicriteria Analysis*; Springer: Berlin/Heidelberg, Germany, 2011; pp. 1–12.
38. Vaidya, O.S.; Kumar, S. Analytic hierarchy process: An overview of applications. *Eur. J. Oper. Res.* **2006**, *169*, 1–29. [CrossRef]
39. SEADE—Fundação Sistema Estadual de Análise de Dados. Home Page. Available online: https://www.seade.gov.br/# (accessed on 15 January 2021).
40. Lima, P.A.B.; Paião Júnior, G.D.; Santos, T.L.; Furlan, M.; Battistelle, R.A.G.; Silva, G.H.R.; Ferraz, D.; Mariano, E.B. Sustainable Human Development at the Municipal Level: A Data Envelopment Analysis Index. *Infrastructures* **2022**, *7*, 12. [CrossRef]
41. Sachdeva, M.; Bhatia, S.; Batta, S.K. Sucrose accumulation in sugarcane: A potential target for crop improvement. *Acta Physiol. Plant.* **2011**, *33*, 1571–1583. [CrossRef]
42. Silva, M.A.; Véliz, J.G.E.; Sartori, M.M.P.; Santos, H.L. Glyphosate applied at a hormetic dose improves ripening without impairing sugarcane productivity and ratoon sprouting. *Sci. Total Environ.* **2022**, *806*, 150503. [CrossRef]
43. Xu, F.; Wang, Z.; Lu, G.; Zeng, R.; Que, Y. Sugarcane ratooning ability: Research status, shortcomings, and prospects. *Biology* **2021**, *10*, 1052. [CrossRef]
44. Mariano, E.; Otto, R.; Montezano, Z.F.; Cantarella, H.; Trivelin, P.C. Soil nitrogen availability indices as predictors of sugarcane nitrogen requirements. *Eur. J. Agron.* **2017**, *89*, 25–37. [CrossRef]
45. Veisi, H.; Deihimfard, R.; Shahmohammadi, A.; Hydarzadeh, Y. Application of the analytic hierarchy process (AHP) in a multi-criteria selection of agricultural irrigation systems. *Agric. Water Manag.* **2022**, *267*, 107619. [CrossRef]
46. Cay, T.; Uyan, M. Evaluation of reallocation criteria in land consolidation studies using the Analytic Hierarchy Process (AHP). *Land Use Policy* **2013**, *30*, 541–548. [CrossRef]
47. Forman, E.; Peniwati, K. Aggregating individual judgments and priorities with the analytic hierarchy process. *Eur. J. Oper. Res.* **1998**, *108*, 165–169. [CrossRef]
48. Chang, C.W.; Wu, C.R.; Lin, C.T.; Chen, H.C. An application of AHP and sensitivity analysis for selecting the best slicing machine. *Compu. Ind. Eng.* **2007**, *52*, 296–307. [CrossRef]

Disclaimer/Publisher's Note: The statements, opinions and data contained in all publications are solely those of the individual author(s) and contributor(s) and not of MDPI and/or the editor(s). MDPI and/or the editor(s) disclaim responsibility for any injury to people or property resulting from any ideas, methods, instructions or products referred to in the content.

eng

MDPI

Article

Formalising Autonomous Construction Sites with the Help of Abstract Mathematics

Dmitrii Legatiuk [1] and Daniel Luckey [2,*]

[1] Chair of Mathematics, Universität Erfurt, 99089 Erfurt, Germany
[2] Chair of Computing in Civil Engineering, Bauhaus-Universität Weimar, 99423 Weimar, Germany
* Correspondence: daniel.luckey@uni-weimar.de

Abstract: With the rapid development of modern technologies, autonomous or robotic construction sites are becoming a new reality in civil engineering. Despite various potential benefits of the automation of construction sites, there is still a lack of understanding of their complex nature combining physical and cyber components in one system. A typical approach to describing complex system structures is to use tools of abstract mathematics, which provide a high level of abstraction, allowing a formal description of the entire system while omitting non-essential details. Therefore, in this paper, autonomous construction is formalised using categorical ontology logs enhanced by abstract definitions of individual components of an autonomous construction system. In this context, followed by a brief introduction to category theory and ologs, exemplary algebraic definitions are given as a basis for the olog-based conceptual modelling of autonomous construction systems. As a result, any automated construction system can be described without providing exhausting detailed definitions of the system components. Existing ologs can be extended, contracted or revised to fit the given system or situation. To illustrate the descriptive capacity of ologs, a lattice of representations is presented. The main advantage of using the conceptual modelling approach presented in this paper is that any given real-world or engineering problem could be modelled with a mathematically sound background.

Keywords: modelling; abstract approach; formalisation; category theory; ontology logs; robotic construction; autonomous construction; conceptual modelling

Citation: Legatiuk, D.; Luckey, D. Formalising Autonomous Construction Sites with the Help of Abstract Mathematics. *Eng* **2023**, *4*, 799–815. https://doi.org/10.3390/eng4010048

Academic Editors: Antonio Gil Bravo and Jingzheng Ren

Received: 7 December 2022
Revised: 21 January 2023
Accepted: 24 February 2023
Published: 1 March 2023

Copyright: © 2023 by the authors. Licensee MDPI, Basel, Switzerland. This article is an open access article distributed under the terms and conditions of the Creative Commons Attribution (CC BY) license (https://creativecommons.org/licenses/by/4.0/).

1. Introduction

Civil engineering is widely considered very traditional, especially in comparison to other engineering disciplines, such as mechanical engineering, mainly because of the unique character of each structure. Even for standard residential buildings, the particular conditions on each construction site may require changes in the design and modelling of the building, which may affect the entire construction process. This one-of-a-kind-production has been a major obstacle on the way to integrating modern technologies and automation in the field of civil engineering for a long time. Furthermore, the availability of cheap manual labour and the small and medium-sized enterprise structure of the construction sector have hindered advances in research and development. Recently, the development of integrating affordable yet highly flexible industrial robots into the digital design flow of architecture and construction has prompted a surge in robotic systems in construction [1].

The spectrum of automation in construction ranges from industrial and on-site prefabrication to autonomous on-site robots. However, for mobile in-situ construction robots, only prototypes have been presented so far; see, for example, works [2–4]. This is mainly because industrial robots are designed for repetitive tasks in controlled environments and are now used in manufacturing systems for different types of materials, for example, timber [5], masonry [6], and concrete [7]. In this regard, the transition to autonomous mobile on-site robots requires modifications to the robotic system, possibly adding wheels for movement, tracking or scanning devices for orientation and perception of obstacles, other robots or human co-workers. Furthermore, each construction site is unique, implying that a robot must adapt to sudden changes in environmental conditions.

As the success of robotic construction evidently depends on the interaction of robotic systems with the environment, it has been stated in [8] that design and construction systems have to be aligned to the capabilities of the robot arm and tolerances of the material, i.e., weight, friction, rigidity, as well as the robotic placement and potential connection systems. In addition to the one-of-a-kind character, structures are an aggregation of possibly hundreds of work steps for each individual part of the structure. From structural work to interior fitting, robotic systems have to be able to handle each step of the way and, therefore, need to interact or collaborate with the environment and other robots on the same construction site. Therefore, advanced feedback and localisation systems, e.g., external or integrated sensors, cameras or laser scanners, are required to cope with tolerances, uncertainties and possibly human interaction. To overcome these obstacles, a workflow and prototypes for autonomous construction sites have been presented in [4,9].

However, summarising current results related to robotic construction sites, it is noticeable that researchers solely present particular solutions, aiming to address specific tasks. This approach results in creating somehow similar yet different workflows, each highlighting specific aspects of robotic construction sites relevant to a particular task. Furthermore, each prototype uses different tools as well as system components and is based on different types of robotic manipulators from various manufacturers programmed by different types of code to control the robots, emphasising the lack of a general approach to modelling robotic construction sites that would be applicable to different tasks. Therefore, this paper aims to provide a fundamental basis for the conceptual modelling of robotic construction sites based on mathematical abstractions and, in particular, category theory.

In recent years, several results related to the formal modelling of engineering systems have been presented. In particular, abstract approaches based on graph theory [10], abstract Hilbert spaces [11,12], relational algebra [13,14], predicate logic [15,16], type theory [17,18], and category theory [19–22] have been proposed. However, direct use of these results in the context of autonomous construction sites is difficult because of the coupled system robot-construction site, which requires conceptual modelling not only of a robot itself but also its surrounding and, in particular, kinematic constraints on robot movements. Therefore, to overcome this difficulty, the use of *categorical ontology logs*, or simply *ologs*, combined with an abstract algebraic approach is proposed in this paper.

Ologs were introduced by Spivak and Kent in [23] and are based on category theory, implying that ologs have a strong mathematical basis while providing the flexibility of general-purpose ontologies. In particular, ologs provide two distinct features making them very attractive for practical use:

(i) Ologs follow the ideas of "lattice of theories" presented in [24], implying, simply speaking, that the same system can be represented by ologs with different levels of details, thus constituting a *lattice* of representations. This point of view can be adapted to the conceptual modelling of autonomous construction sites or engineering systems in general. A system could be modelled on a very general level at first and, after that, by using specific movements along the lattice, such as contraction, expansion, revision, and analogy, specific parts of the system can be "zoomed in".

(ii) The categorical foundation of ologs provides a clear formal procedure for relating two different ologs. This procedure is based on the concept of *common ground*, which is represented by a third olog related to the two other ologs. Practically, it implies that different ologs can be created for individual parts of an engineering system and then coupled together in one system of ologs.

Although ologs have the obvious advantages discussed above, they also share an obstacle typical for all general-purpose ontologies: a subjective worldview of the ontology creator. Therefore, to overcome this obstacle, in this paper, we propose a slight modification of the concept of common ground presented in [23]. This modification is based on a two-step procedure: at first, formal definitions of individual components of an autonomous construction site, based on an abstract algebraic approach presented in [13], are introduced; after that, the abstract definitions are used as a common ground for all ologs describing an autonomous construction site. In this case, the subjectivity of the ologs' creator worldview can be overcome, and thus, a formally sound lattice of ologs, representing an autonomous construction site, will be obtained.

This paper aims to provide a basis to overcome the previously mentioned issues of conceptual modelling of automated construction sites by coupling ologs with abstract algebraic definitions. In this context, algebraic definitions are used as a common basis for olog-based conceptual modelling of autonomous construction systems. As a result, any automated construction system can be described, without providing exhausting detailed definitions of the system components. Existing ologs can easily be extended, contracted or revised to fit the given system or situation. With these operations, e.g., revision, precise translation terminologies are provided. To illustrate the capacity of ologs, a lattice of representations for automated construction sites is presented. The main advantage of using the conceptual modelling approach presented in this paper is that any given real-world scenario or engineering problem could be modelled with a mathematically sound background.

The paper is organised as follows: Section 2 provides a few basic facts about category theory and ologs; Section 3 introduces an abstract description of autonomous construction sites, used as the common ground for olog-based representation of advanced structures presented in Section 4; finally, a discussion on the results of the paper and remarks on further applications are provided in Section 5.

2. Fundamentals of Category Theory and Ologs

In this section, the concept of ologs for the purpose of conceptual modelling of real-life scenarios is described, following a basic introduction to category theory as mathematical basis of ologs.

2.1. Basics of Category Theory

Ologs are based on category theory, and therefore, to support the reader in the upcoming discussion, a few basic definitions of category theory are provided in this section. Generally speaking, category theory can be seen as an abstract theory of functions studying different mathematical structures (objects) and relations between them [25]. A *category* is introduced via the following definition:

Definition 1 ([25]). *A category consists of the following data:*

(i) *Objects:* $A, B, C, \ldots$

(ii) *Arrows:* $f, g, h, \ldots$

(iii) *For each arrow* f*, there are given objects* $\mathrm{dom}(f)$, $\mathrm{cod}(f)$ *called the domain and codomain of* f*. We write* $f\colon A \longrightarrow B$ *to indicate that* $A = \mathrm{dom}(f)$ *and* $B = \mathrm{cod}(f)$*.*

(iv) *Given arrows* $f\colon A \longrightarrow B$ *and* $g\colon B \longrightarrow C$*, i.e., with* $\mathrm{cod}(f) = \mathrm{dom}(g)$*, there is given an arrow* $g \circ f\colon A \longrightarrow C$ *called the composite of* f *and* g*.*

(v) *For each object* A*, there is given an arrow* $1_A\colon A \longrightarrow A$ *called the identity arrow of* A*.*

These data are required to satisfy the following laws: $h \circ (g \circ f) = (h \circ g) \circ f$ *and* $f \circ 1_A = f = 1_B \circ f$*.*

A category is everything satisfying this definition, and therefore, very general objects can be put together to form a category by specifying relations between objects via the arrows, which are sometimes called morphisms. This generality is the starting point for introducing ologs, as it will be shown later. Mappings between different categories are introduced by the notion of a *functor*:

Definition 2 ([25]). *A functor* $F\colon \mathbf{C} \longrightarrow \mathbf{D}$ *between categories* $\mathbf{C}$ *and* $\mathbf{D}$ *is a mapping of objects to objects and arrows to arrows in such a way that:*

(i) $F(f\colon A \longrightarrow B) = F(f)\colon F(A) \longrightarrow F(B)$;
(ii) $F(1_A) = 1_{F(A)}$;
(iii) $F(g \circ f) = F(g) \circ F(f)$.

That is, F respects domains and codomains, identity arrows, and composition. In other words, functors are structure-preserving mappings between categories.

2.2. Introduction to Ologs

Ologs, in general, as first introduced in [23], are intended to provide a framework for knowledge representation, in order to organise data and results, to make them comprehensible and comparable to other scientists. As stated by the name, **o**ntology **logs** are closely related to ontologies, which focus on defining what entities exist, thus consequently categorising entities and defining relationships between these categories. In engineering applications, ontologies are used to develop models of reality. Subsequently, ologs are intended to structure and represent the results of defining entities and modelling relationships between categories by recording them in a structure based on category theory.

However, as for every model, the structure is highly dependent on the subjective worldview of the creator(s). When creating ontologies, subjectivity should be eliminated as far as possible, as it may lead to information not being perceived by readers as intended by the creators. Hence, ologs are aware that the views of the creators and readers may not correspond. Therefore, ologs do not attempt to accurately reflect reality but to be structurally sound and accurate in correspondence with the views of the creator. However, discrepancies in the views of different creators do not prevent ologs from being aligned and connected. Because of the strong mathematical basis provided by category theory, ologs can be linked and precisely connected by functors, as the main advantage for conceptual modelling.

Functors allow ologs to be referenceable by other authors and, in addition, extendable since any model, respectively olog, needs to be extended in order to correctly represent new developments, features or different views. Moreover, the mapping of ologs by functors allows the generation of precise translation terminologies between models. Thus, as well as being represented as graphs, ologs can serve as database schemas that provide a human-readable interface, with the components of ologs representing tables and attributes to translate one system of tables into another. Therefore, the basic components and the respective graphic representation of ologs are presented in the following.

To keep the presentation short, the detailed discussion on the construction of ologs and their structure from [23,26] is compressed in the form of one definition. More advanced concepts from ologs theory will be discussed at the places of their direct use for the olog-based description of engineering systems. Additionally, to support the reader, the definition of ologs is placed in the engineering context. The following definition introduces ologs:

Definition 3. *An olog is a category, which has types as objects, aspects as arrows, and facts as commutative diagrams. The types, aspects, and facts are defined as follows:*

- *A **type** is an abstract concept represented as a box containing a singular indefinite noun phrase. Types are allowed to have compound structure, i.e., being composed of smaller units. The following boxes are types:*

a structure	*an engineer e and a building b such that e constructed b*	*a triple (w, t, m) where w is a wall, t is a thickness of w, and m is a material of w*

- *Aspects are functional relationships between the types represented by labelled arrows in ologs. Consider a functional relationship called f between types X and Y, which can be denoted $f\colon X \rightarrow Y$, then X is called the domain of definition for the aspect f, and Y is called the set of result values of f. Here are two examples of using aspects:*

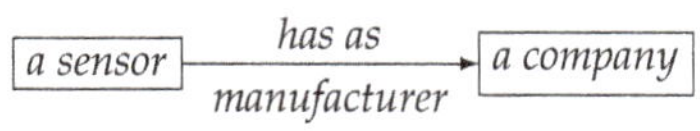

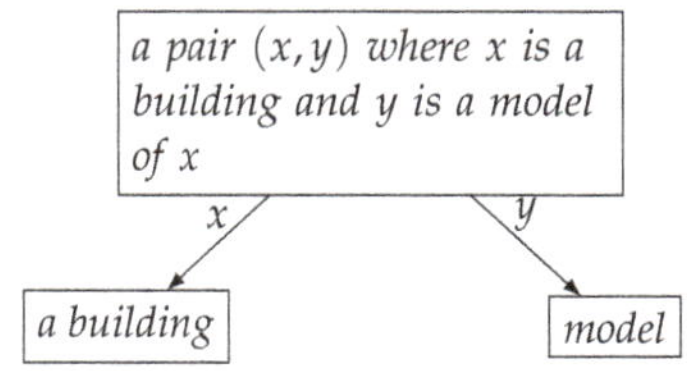

- *Facts are commutative diagrams, i.e., graphs with some declared path equivalences, in ologs. Facts are constructed by composing several aspects and types.*

Facts, represented by commutative diagrams, have a crucial role in practical applications of ologs, because facts can be straightforwardly converted into databases of knowledge; see [23,26] again for a detailed discussion. Thus, ologs provide a general framework for knowledge representation supporting an easy integration into the engineering modelling process via the link to databases.

With the definition of types, aspects, and facts, the main components of ologs have been introduced. However, as shown in Definition 3, the construction of ologs as graphs follows several rules, in order to keep the system readable, such as the declaration of types should begin with "a" or "an" and aspects with a verb. Detailed information on the construction of ologs can be found in [23].

3. Abstract Description of Autonomous Construction Sites

For enhancing ologs with more objective constructions, it is necessary to provide a formal common ground for the development of ologs of an autonomous construction site. Therefore, this section provides an abstract algebraic description of essential parts definitions, such as robot and robotic environment, constituting an autonomous construction site. As a result of this section, an abstract framework for describing autonomous construction is created.

It is important to remark that existing definitions of a robot attempt to find a balance between being too vague and too specific, with a valid general definition regarding robots seemingly missing or still subject to debate given the sheer amount of robot variations. An overview of several varying robot definitions is given in [27]. Although this section takes steps in this direction, it is not the aim to claim that the definitions provided below should be used as an industry standard. Additionally, it is worth remarking that it is certainly possible to connect existing definitions of a construction robot to the abstract constructions presented in this paper. However, this connection goes beyond the scope of the current paper and is therefore kept for future work.

Because of the predominant role in research and development as stated above, robots in the following context are considered industrial robots, with their components, structure, and operation described in [28]. The aim of this work is to illustrate how the coupling of the abstract algebraic approach and ologs can improve the conceptual modelling of autonomous construction sites.

In terms of abstract constructions, it is possible to follow either the top-to-bottom approach by first defining an autonomous construction site and then scaling it down to its components, or the bottom-to-top approach, by defining the components and then scaling them up to the autonomous construction site. For the purpose of this article, the first approach will be used. Therefore, we start with the following definition:

Definition 4 (Autonomous construction site)**.** *An* ***autonomous construction site*** *or a* ***robotic environment*** *is the object* $\mathfrak{A} = \langle \mathfrak{T}, \mathfrak{R}, \mathfrak{H}, \mathfrak{E}, \mathfrak{G}, \mathfrak{B} \rangle)$*, where*

- $\mathfrak{T}$ *is a task to be solved by robots on a construction site;*
- $\mathfrak{R} = (R_1, R_2, \ldots, n)$ *is an n-tuple of robots;*

- *$\mathfrak{H}$ is the object describing human–robot interaction on a construction site;*
- *$\mathfrak{E}$ is a set of pairs representing environmental conditions on a construction site;*
- *$\mathfrak{G}$ is a 4-tuple of GPS information for important parts of a construction site;*
- *$\mathfrak{B}$ is a base station controlling the autonomous construction site.*

Let us discuss the role of each component from Definition 4 in more detail:

- The $\mathfrak{T}$ is represented by a tuple $\mathfrak{T} = (\mathcal{V}, \mathcal{A})$, where $\mathcal{V}$ is a natural language sentence formulating the task, and $\mathcal{A}$ is a formalisation of $\mathcal{V}$ in terms of a sequence of control signals controlling cyber components of the autonomous construction site.
- The n-tuple of robots $\mathfrak{R}$ evidently contains information about all robots used on the construction site. A precise definition of a robot in the framework of the abstract approach presented in this paper is provided in Definition 5.
- Considering that autonomous construction sites naturally combine human workers and robots, it is necessary to address the question of human–robot interaction [29]. However, an abstract definition of such an interaction goes beyond the scope of the current paper. Therefore, we address the point of human–robot interaction simply by placing a specific object $\mathfrak{H}$ for it, which can still be defined later without the need to change any other definition presented in this paper.
- The role of set $\mathfrak{E} = (E_1, E_2)$ is to provide information about environmental conditions on a construction site. In this way, this information is formalised in terms of a denotation E_1 and the corresponding value E_2.
- For integration of a robotic system in the construction progress, it is necessary to provide information on the positioning of the robotic system, as well as all essential parts of the construction site. For that purpose, the 4-tuple $\mathfrak{G} = (O, x_1, x_2, x_3)$ is introduced, where O denotes the object, and x_1, x_2, x_3 are object coordinates.
- Finally, cyber parts of the autonomous construction site must be controlled, and therefore, the base station $\mathfrak{B}$ needs to be included in the definition.

In summary, Definition 4 provides an abstract point of view on autonomous construction sites. This abstract point of view helps to "sieve out"all details that are not critical for the first stage of planning and designing an autonomous construction site.

Next, an abstract description of a robot needs to be introduced. It is also necessary to take into account that a robotic system can generally be subdivided into two parts: a physical part (physical components of the systems) and a logical part (control and communication signals). Hence, an abstract definition must also reflect this coupled nature of a robot. Therefore, the following definition is proposed:

Definition 5 (Robot). *A **robot** is the object $\mathfrak{R} = \langle \mathfrak{C}, \mathfrak{K}, \mathfrak{P}, \mathfrak{S}, \mathfrak{Ac} \rangle$, where*

- *$\mathfrak{C}$ is a robotic controller generating control signals;*
- *$\mathfrak{K}$ is a finite set of kinematic properties of a robot;*
- *$\mathfrak{P}$ is a k-tuple of physical properties of a robot;*
- *$\mathfrak{S} = (\mathcal{S}_1, \mathcal{S}_2, \ldots, \mathcal{S}_n)$ is an n-tuple of sensors installed on a robot;*
- *$\mathfrak{Ac} = (\mathcal{A}_1, \mathcal{A}_2, \ldots, \mathcal{A}_m)$ is an m-tuple of actuators installed on a robot.*

For providing a clear practical interpretation of this definition, let us now discuss the robot components individually:

- The robotic controller $\mathfrak{C}$ is needed for a communication with a base station $\mathfrak{B}$ introduced in Definition 4, and in general, it sends a sequence of control signals for operating the robot.
- A set of kinematic properties $\mathfrak{K}$ represents physical constraints limiting the possible movements of a robot. In practice, $\mathfrak{K}$ is determined by the kinematic chain that is formed by the series of manipulators, connected by joints, and may differ in specifications and movement, providing the (internal) axes of the robot. Furthermore, a robot system may consist of external axes, e.g., track systems. The degrees of freedom of

the robotic system is the combination of internal and external axes determined by the kinematic chain. Based on the kinematic properties representing specific constraints, the robotic controller $\mathfrak{C}$ is able to generate control signals for the robot to reach target coordinates in the determined work area. Additionally, it is important to notice that for making $\mathfrak{K}$ consistent from the point of view of set theory, it is assumed that all kinematic constraints are formalised in terms of equations and inequalities, i.e., mathematical expressions.
- The tuple $\mathfrak{P}$ contains robot specification information (e.g. type, manufacturer, or information about a motor driving the system), which includes information. Physical properties have to be also known for generating control signals by the robotic controller $\mathfrak{C}$.
- Evidently, various sensors might be installed on a robot for measuring environmental conditions, as well as important physical quantities of a robot itself, e.g., the temperature of individual parts. These sensors are combined in an n-tuple $\mathfrak{S}$.
- Similar to sensors, various actuators need to be installed on a robot and are activated via control signals. These actuators are combined an m-tuple $\mathfrak{Ac}$.

For completing basic abstract definitions related to autonomous construction sites, it is necessary to introduce abstract descriptions of sensors and actuators. Abstract definitions for sensors and sensor networks have already been introduced in [13], and a sensor is then defined as follows:

Definition 6 (Sensor, [13]). *A* ***sensor*** *is the object* $\mathcal{S} = \langle \mathcal{I}, \mathcal{Y}, \mathcal{T} \rangle$, *where*
- $\mathcal{I} = (I_1, I_2, \ldots, I_n)$ *is an n-tuple of finite index sets;*
- $\mathcal{Y} = (Y_1, Y_2, \ldots, Y_n)$ *is an n-tuple of measurements with* $Y_i \in \mathbb{R}^{N_i}$, $i = 1, \ldots, n$;
- $\mathcal{T}$ *is a k-tuple of specifications (type information).*

By this definition, sensors are allowed to measure several physical quantities, and not just one. Moreover, for simplicity, we assume that card $I_i = N_i \, \forall i$. Nonetheless, it is important to remark that the case card $I_i < N_i \, \forall i$ is also of practical interest for further use of measurements, i.e., data and signal analysis, since it underlines that not all measured data can be used, but only a subset, which corresponds to the idea of frame analysis and sparse representations [30].

Further, the following definition of a sensor cluster has been presented in [13]:

Definition 7 (Sensor cluster, [13]). *A sensor cluster is the object* $\mathcal{S}_\mathcal{C} = \langle \mathfrak{B}, \mathfrak{S}, \mathfrak{R} \rangle$, *where*
- $\mathfrak{B}$ *is a sensor node or a base station controlling the sensor cluster;*
- $\mathfrak{S} = (\mathcal{S}_1, \mathcal{S}_2, \ldots, \mathcal{S}_n)$ *is an n-tuple of sensors, introduced in Definition 6;*
- $\mathfrak{R} = (R_1, R_2, \ldots, R_m)$ *is an m-tuple of relations.*

In this definition, the m-tuple of relation $\mathfrak{R}$ specifies the rules of communication between sensors, which are specified during the sensor network design; see [14,31] for specific examples of relations and practical meanings of the relations in wireless sensor network modelling.

Taking into account Definition 7, it is also possible to change Definition 5 of a robot by replacing the n-tuple of sensors $\mathfrak{S}$ with the sensor cluster $\mathcal{S}_\mathcal{C}$. However, this approach might be a bit inconsistent because typically robots have built-in sensors, which are directly controlled by the robotic controller $\mathfrak{C}$ and not by a separated sensor node $\mathfrak{B}$, as required by Definition 7. Therefore, the current form of Definition 5 is preferred. Moreover, if extra sensors need to be installed on a robot, then it is always possible to combine both definitions via the composition

$$\mathfrak{R} \circ \mathcal{S}_\mathcal{C},$$

where the composition $\circ$ represents communication rules between the sensor node and the robotic controller. Hence, the separation of two definitions provides more flexibility in terms of the descriptive capabilities of the whole abstract framework.

Finally, following Definition 6, let us now introduce a definition for an actuator:

Definition 8 (Actuator). *An actuator is the object $\mathcal{A} = \langle \mathfrak{B}, \mathcal{A}_{\mathcal{S}}, \mathcal{T} \rangle$, where*

- *$\mathfrak{B}$ is a sensor node or a base station controlling the actuator;*
- *$\mathcal{A}_{\mathcal{S}}$ is an actuation signal;*
- *$\mathcal{T}$ is an k-tuple of specifications (typing information).*

This definition is based on the fact that each actuator has a sensor node attached to it, controlling the actuation process. The control of the actuation process is realised via the corresponding control model embedded into the sensor node, which is abstracted here in terms of the actuation signal $\mathcal{A}_{\mathcal{S}}$. The k-tuple of specification information $\mathcal{T}$ represents standard information about the actuator (e.g., type, manufacturer).

Further, if necessary, a definition of an actuator cluster, similar to a sensor cluster introduced in Definition 7, can be provided. In this case, actuators have to be combined in a tuple, and another tuple of relation, specifying communications between various actuators and base stations, must be introduced. For the purpose of this paper, a definition of an actuator cluster is omitted. Instead, Definition 8 shall conclude by defining a common ground for an autonomous construction site, which shall subsequently be used for creating ologs.

4. Olog Representations of Robotic Construction Sites

In this section, the abstract definitions of autonomous construction sites introduced in Section 3 will be used as a common ground for creating olog representations of autonomous construction sites. In particular, the concept of the lattice of representations will be discussed and illustrated by examples.

Ologs reflect the idea of lattice of theories by a lattice of representations, see again [23], as mentioned in Section 1. Formally, the lattice of representations is represented by an entailment pre-order as part of the global category of specifications. Practically, it means that it is possible to move between different ologs by using four operations/mapping, see Figure 1: contraction **C**, expansion **E**, revision **R**, and analogy **A**.

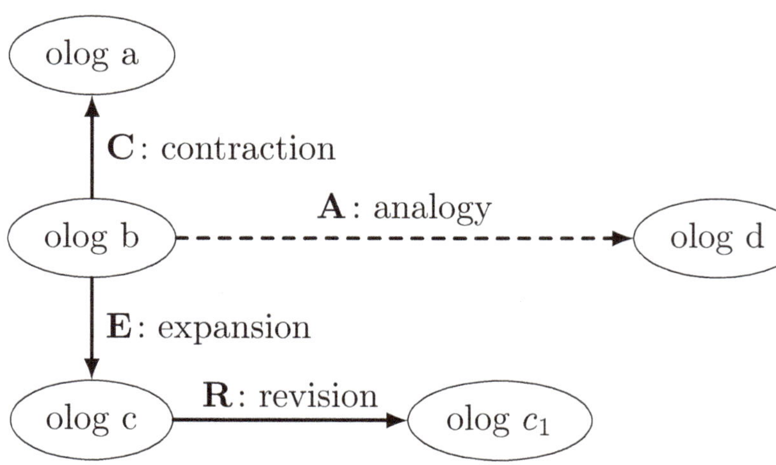

Figure 1. A general idea of the lattice of representation concept.

More general or detailed ologs are created, by moving upwardes or downwards, respectively, between ologs, i.e., by contraction **C** or expansion **E**, as it is indicated in Figure 1. In addition, ologs may be revised to update or remove details, which is realised via revision **R**, or translated into another olog, using a morphism.

As mentioned in Section 3, a top-to-bottom approach was chosen for olog development. Therefore, we start our lattice of representations by developing olog $\mathcal{O}_1$ for describing an autonomous construction site based on Definition 4. The olog starts at type *A*, the autonomous construction site itself. Based on Definition 4, an autonomous construction site consists of a tuple of tasks $\mathfrak{T}$, an n-tuple of robots $\mathfrak{R}$, a human–robot interaction $\mathfrak{H}$, a set of pairs of environmental conditions $\mathfrak{E}$, a 4-tuple of GPS information $\mathfrak{G}$, and a base station $\mathfrak{B}$. Every object of the autonomous construction site is represented by a type in the corresponding olog illustrated in Figure 2.

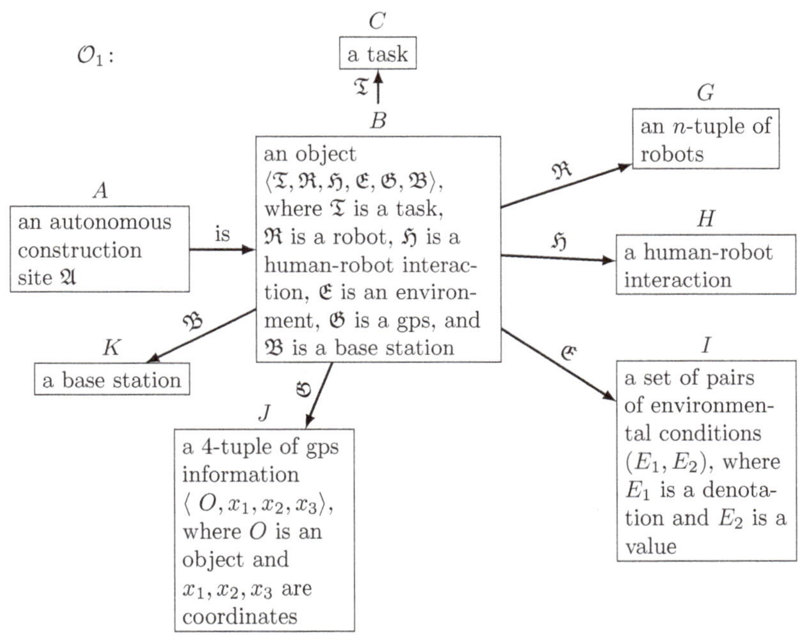

Figure 2. Olog representation of an autonomous construction site.

In conformity with Section 2.2, every type begins with a or an. An aspect is connecting a type, representing all possible objects of that type, called *domain*, with another type, called *codomain*, representing a subset of possible results. The olog presented in Figure 2 provides a very general description of an autonomous construction site. To provide a better overview, several arrows connecting types have been omitted. These connecting arrows constitute facts about an autonomous construction site. For example, by connecting type *A* and type *K* via an arrow labelled as *has*, we would obtain the following fact: an autonomous construction site $\mathfrak{A}$ has a base station. Similarly, other facts can be deduced from olog $\mathcal{O}_1$.

Next, let us illustrate how expansion works on the example of olog $\mathcal{O}_1$. We formally apply an expansion mapping **E** to $\mathcal{O}_1$, which results in adding more types and connecting arrows to the original olog of an autonomous construction site. Figure 3 presents the results of this expansion, a new olog $\mathbf{E}\mathcal{O}_1$.

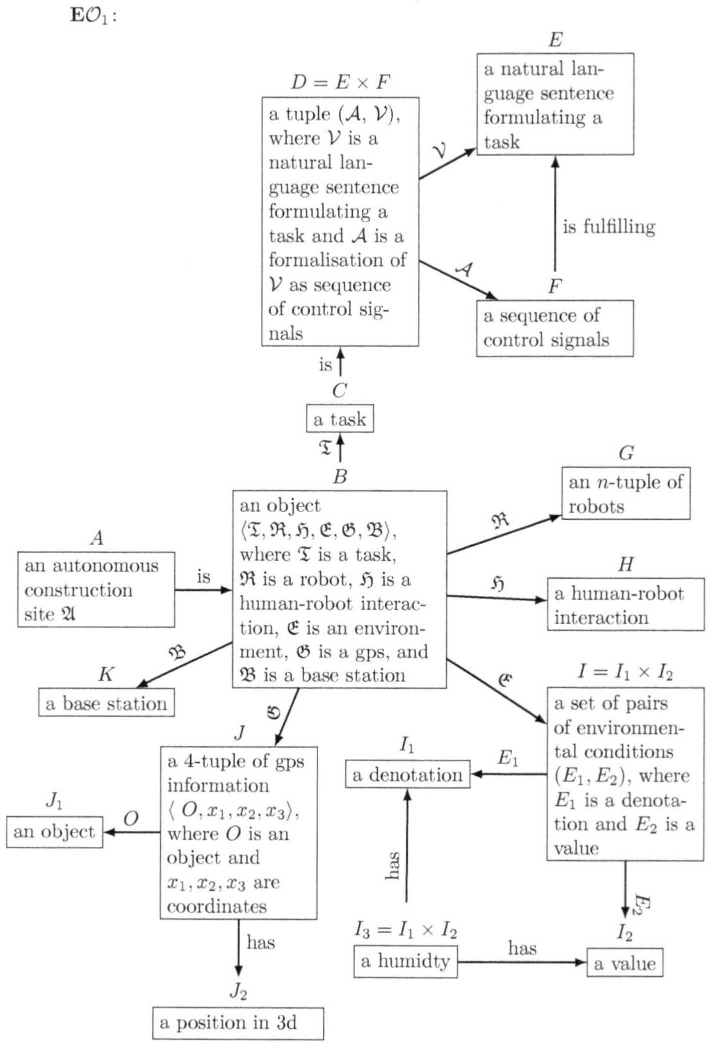

Figure 3. Olog representation of an autonomous construction site after the application of an expansion mapping **E**.

Evidently, olog $\mathbf{E}\mathcal{O}_1$ has been expanded with more types and more facts provided by commutative paths, for example, the triangle DEF. Additionally, for illustrative purposes, we have added type I_3 as humidity, which is a particular instance of an environmental condition. This shows how concrete data can be added to an abstract olog, implying that an olog can be directly translated into a database of knowledge about an autonomous construction site.

Further, let us illustrate how contraction works by the example of olog $\mathbf{E}\mathcal{O}_1$. It is important to underline that $\mathbf{C} = \mathbf{E}^{-1}$ is not required, meaning that by contracting an expanded olog, we do not need to obtain the original olog. Figure 4 illustrates a possible (one of many) output(s) of applying a contraction mapping $\mathbf{C}$ to olog $\mathbf{E}\mathcal{O}_1$, where some details of Definition 4 have been omitted.

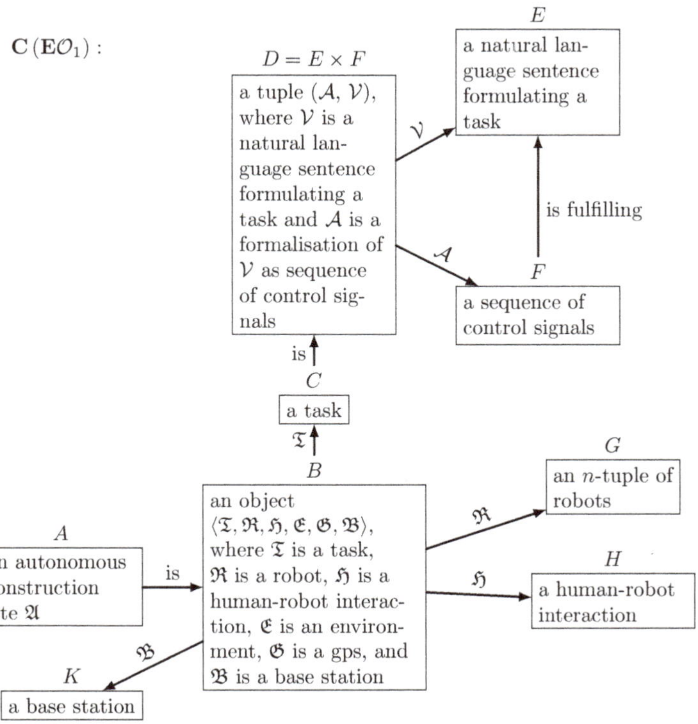

Figure 4. Olog representation of an autonomous construction site after the application of a contraction mapping **C** to the expanded olog $\mathbf{E}\mathcal{O}_1$.

Thus, we have the following diagram on the level of ologs:

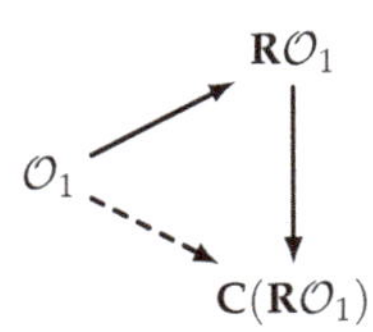

where the dashed arrow indicates that we cannot arrive at olog $\mathbf{C}(\mathbf{R}\mathcal{O}_1)$ from the olog $\mathcal{O}_1$ in one step, and a combination of contractions and expansions is required. Hence, we obtain a lattice of representations containing several ologs that are convertible between each other and represent different levels of details about an autonomous construction site.

Similar to the construction of olog $\mathcal{O}_1$, ologs for Definitions 5–8 can be established. For illustrative purposes, only the olog for a robot is presented, while the other ologs will only be denoted. For keeping consistency with the order of definition presented in Section 3, let us denote by $\mathcal{O}_2$ an olog for a robot, by $\mathcal{O}_3$ an olog for a sensor, by $\mathcal{O}_4$ an olog for a sensor cluster, and by $\mathcal{O}_5$ an olog for an actuator. Figure 5 presents olog $\mathcal{O}_2$, which is based on Definition 5. Similar to olog $\mathcal{O}_1$, olog $\mathcal{O}_2$ might be expanded or contracted in different ways, as well as some arrows making olog $\mathcal{O}_2$ commute, and facts could be easily added.

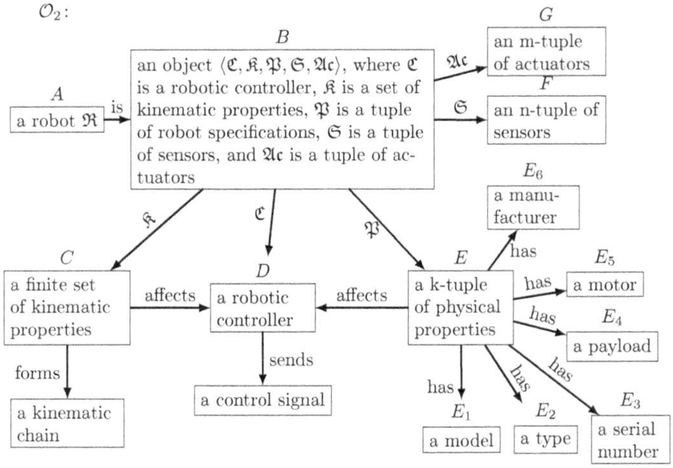

Figure 5. Olog representation of a robot based on Definition 5.

In addition, a potential connection between ologs $\mathcal{O}_1$ and $\mathcal{O}_2$ is worth discussing. In general, there are two possibilities to formally connect these ologs:

(i) Olog $\mathcal{O}_1$ can be expanded to include olog $\mathcal{O}_2$ as a sub-part.
(ii) Olog $\mathcal{O}_1$ can be expanded to include olog $\mathcal{O}_2$, and then the resulting olog should be contracted to olog $\mathcal{O}_2$.

From the point of view of constructing a lattice of theories (or representations), the first approach is preferable, while an olog containing more information and, thus, in this sense, a more general olog can be generated from $\mathcal{O}_1$. Following this, let us further introduce formal denotations:

- $\mathcal{O}_1 \cup \mathcal{O}_2$ denotes the olog obtained by expanding the autonomous construction site olog by adding the robot olog $\mathcal{O}_2$ to it;
- $\mathcal{O}_1 \cup \mathcal{O}_2 \cup \mathcal{O}_3$ denotes the olog obtained by expanding the olog $\mathcal{O}_1 \cup \mathcal{O}_2$ by adding the sensor olog $\mathcal{O}_3$ to it;
- $\mathcal{O}_1 \cup \mathcal{O}_2 \cup \mathcal{O}_3 \cup \mathcal{O}_5$ denotes the olog obtained by expanding the olog $\mathcal{O}_1 \cup \mathcal{O}_2 \cup \mathcal{O}_3$ by adding the actuator olog $\mathcal{O}_5$ to it.

Thus, we obtain the following lattice of representations:

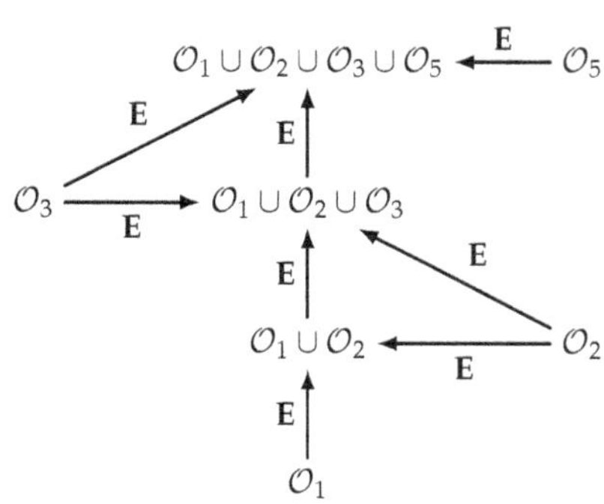

Evidently, all arrows can be reverted and, hence, turned into a contraction of ologs. The diagram above provides a clear structure of how different parts of a system "autonomous construction site" are connected. This structure underlines advantages of working with abstract definitions introduced in Section 3 in combination with ologs:

(i) The resulting ontological description can be easily extended by adding new definitions and ologs without a need for changing previous results;
(ii) The lattice of representation can even be created at first, serving as a guideline for creating ologs and definitions;
(iii) Each olog can be directly converted into a database, see again [23], and, thus, used as a basis for practical implementations of ontologies and formal representations.

It is also worth underlining that the composition of expansion followed by contraction can be viewed as a "zoom-in" operation on an olog. This operation can be seen as a special kind of revision **R**, when a type of the original olog is expanded and then everything except this expansion is removed. This procedure corresponds to the second alternative on connecting ologs $\mathcal{O}_1$ and $\mathcal{O}_2$, as discussed above.

Next, let us briefly illustrate a revision of an olog. According to [23], a revision is a composite, which uses a contraction to discard irrelevant details, followed by an expansion to add new facts. Referring back to the discussion around Definition 7, it is possible to replace an n-tuple of sensors in the definition of a robot with a sensor cluster $\mathcal{S}_\mathcal{C}$. For olog $\mathcal{O}_2$, it means that revision $\mathbf{R} = \mathbf{E} \circ \mathbf{C}$ is applied. Figure 6 shows the resulting olog $\mathbf{R}\mathcal{O}_2$. It is worth noting that this revision of ologs, as well as of the definition, can be easily performed within the abstract approach proposed in this paper.

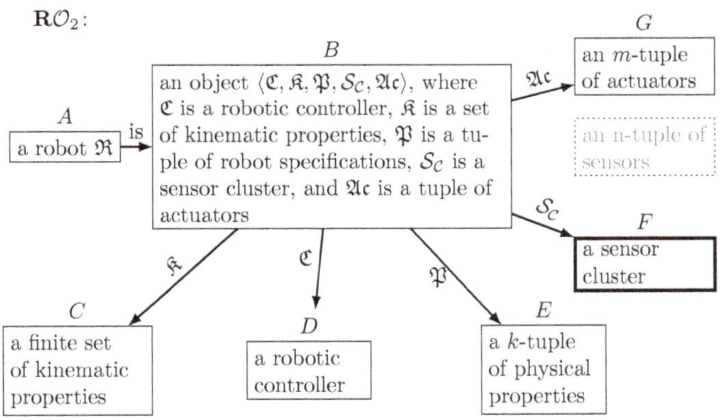

Figure 6. Illustration of olog revision on the example the robot-olog $\mathcal{O}_2$.

Finally, let us briefly discuss how analogy mapping works. An analogy is obtained by systematically renaming all types and aspects of an olog to describe/model a different real-world situation. For example, a mobile unmanned aerial vehicle can be seen as a system, which is similar to a robot. In this case, an analogy between olog $\mathcal{O}_2$ and olog $\mathcal{O}_{UAV}$, describing a mobile unmanned aerial vehicle, can be created. This analogy is formally represented by the diagram

$$\mathcal{O}_2 \xrightarrow{\mathbf{A}} \mathcal{O}_{UAV}$$

Exemplarily, an analogy could mean that the types for kinematic properties need to be renamed or reorganised, or the type for a robotic controller needs to be replaced by a remote controller device. Evidently, Definition 5 needs to be adapted then as well, which can be easily accomplished within the abstract approach proposed in this paper.

5. Discussion and Conclusions

In this paper, a conceptual modelling approach for autonomous construction based on categorical ontology logs coupled with abstract algebraic definitions was presented. The motivation for this coupling is twofold: first, introducing abstract definitions of individual components of an autonomous construction system allows removing subjectivity, which is typical for ontology-based representations; and second, these abstract definitions serve as a common ground for ologs making the whole framework easily extendable and interpretable. Therefore, after introducing abstract definitions of individual components of an autonomous construction system, several ologs for these definitions have been developed. Moreover, basic operations, i.e., contraction, expansion, revision, and analogy, have been discussed.

Let us now summarise and discuss the main points of the paper:

- **Abstract description of autonomous construction sites**
  Several abstract definitions formalising autonomous construction sites have been introduced in Section 3. The idea of these definitions is to provide a common ground for an olog-based description of autonomous construction. A top-to-bottom approach for conceptual modelling of autonomous construction sites has been chosen. Hence, starting with an autonomous construction site, definitions of its more detailed components have been added step-by-step. The main advantage of this approach is that the resulting conceptual modelling framework is scaleable and extendable with more details, if necessary. Any of the Definitions 4–8 can be revised or updated without the need for a general restructuring of the complete framework presented in this paper.
  It is also important to underline that the field of robotic construction still misses generally accepted "standard" definitions. Therefore, the results presented in Section 3 should not be understood in the way of the definitions to become an industrial standard but rather as an approach on how to address practical engineering problems on a more abstract level sieving out all concrete details.
- **Olog-based representations of autonomous construction**
  An olog-based representation of autonomous construction sites has been presented in Section 4. As described in Section 2.2, ologs are designed to handle the subjectivism of the creator of the abstract model. This point has been further strengthened by coupling ologs with abstract definitions introduced in Section 3. This coupling makes the relation and comparison, as well as the translation of ologs, even more mathematically sound and formal. Hence, the ologs presented in this paper can be straightforwardly implemented in the form of databases, as well as the extension/contraction rules. Further, if more details are desired in a concrete application, these details can be easily added via revision of existing ologs, as has been demonstrated in the paper.
- **Lattice of representations**
  Finally, Section 4 presents a lattice of representations, which is developed by extending and revising existing ologs. Arguably, the concept of the lattice of representations is the most powerful tool of olog-based description of engineering systems. First, the lattice can be easily extended without the need for changing previous results. In this case, a new olog is simply added to the lattice, and the corresponding extension is then formally defined. Second, the lattice of representation can even be created first and, hence, provide a guideline for creating ologs and missing definitions.

It is also beneficial to provide a few comments on practical applications of the conceptual modelling framework presented in this paper:

1. The first step should be the formal creation of a lattice of representations, where, of course, instead of ologs, only names of important parts to be described are written. In this step, it is important to decide what should be the least detailed olog and how many different parts need to be modelled.
2. Collect/create definitions of all parts to be described by ologs. In this step, it is important to keep the balance between the number of details and the level of abstractions. This balance is generally to be defined by the modeller and the objective of the work. Evidently, existing definitions, for example, industry standards, can be used, or new definitions can be developed, as has been done in this paper.

3. Create ologs for each part and fit them into the lattice of representations defined in Step 1. Further, if necessary, ologs can be converted into databases and connected to other conceptual models, if available.

In summary, the results presented in this paper indicate that a coupling of ologs and abstract algebraic definitions provides a high degree of flexibility to the resulting framework. Moreover, as it has been shown in some examples, the abstract framework can be easily extended with new definitions and, hence, with new ologs. Therefore, ologs are proposed to overcome the issues of incomparable prototypes and isolated solutions of systems for autonomous construction. As a result, any automated construction system can be described without providing exhausting detailed definitions of the system components, as existing ologs can be extended, contracted or revised to fit the given system or situation. To illustrate the capacity of ologs, an exemplary lattice of representations for autonomous construction sites has been presented. Additionally, the results obtained for autonomous construction can be transferred to other fields of engineering by using analogy operations on two levels: adapting ologs and translating the respective definitions. Thus, the results presented in this paper can be seen not only as an attempt to formalise an autonomous construction but as a general approach to formalising engineering problems.

For future work, since the definitions for a robot and an autonomous construction site are only exemplary, the detailed description of a complete system of autonomous construction would be of relevance to determine the exact ramifications and parameters of describing such a complex system by means of ologs. Subsequently, the process of how an existing olog representation of an autonomous construction system can be translated or revised into another system needs to be examined. Furthermore, the investigation of different systems would allow the identification of matching parameters in order to identify possible system-inherent properties in order to approach a general system definition, if required.

Further direction of future work could be related to using the abstract definitions presented in Section 3 in concrete engineering applications. In particular, using these definitions in the context of path optimisation on graphs with the help of Clifford operator calculus, as it has been presented in [31], to further underline the advantages of coupling abstract mathematics and engineering.

Author Contributions: Conceptualisation, D.L. (Daniel Luckey) and D.L. (Dmitrii Legatiuk); methodology, D.L. (Daniel Luckey) and D.L. (Dmitrii Legatiuk); writing—original draft preparation, D.L. (Daniel Luckey); writing—review and editing, D.L. (Daniel Luckey) and D.L. (Dmitrii Legatiuk); funding acquisition, D.L. (Dmitrii Legatiuk). All authors have read and agreed to the published version of the manuscript.

Funding: This research is supported by the German Research Foundation (DFG) through grant LE 3955/4-1.

Institutional Review Board Statement: Not applicable.

Informed Consent Statement: Not applicable.

Data Availability Statement: Not applicable.

Conflicts of Interest: The authors declare no conflict of interest. The funders had no role in the writing of the manuscript.

Abbreviations

The following abbreviations are used in this manuscript:

Olog	Ontology log
UAV	Unmanned aerial vehicle

References

1. Bock, T. The future of construction automation: Technological disruption and the upcoming ubiquity of robotics. *Autom. Constr.* **2015**, *59*, 113–121. [CrossRef]
2. Keating, S.; Lel, J.; Cai, L.; Oxman, N. Toward site-specific and self-sufficient robotic fabrication on architectural scales. *Sci. Robot.* **2017**, *2*, eaam8986. [CrossRef] [PubMed]
3. Hack, N.; Dörfler, K.; Walzer, A.N.; Wangler, T.; Mata-Falcón, J.; Kumar, N.; Buchli, J.; Kaufmann, W.; Flatt, R.J.; Gramazio, F.; et al. Structural stay-in-place formwork for robotic in situ fabrication of non-standard concrete structures: A real scale architectural demonstrator. *Autom. Constr.* **2020**, *115*, 103197. [CrossRef]
4. Wagner, H.J.; Alvarez, M.; Kyjanek, O.; Bhiri, Z.; Buck, M.; Menges, A. Flexible and transportable robotic timber construction platform—TIM. *Autom. Constr.* **2020**, *120*, 103400. [CrossRef]
5. Willmann, J.; Knauss, M.; Bonwetsch, T.; Apolinarska, A.A.; Gramazio, F.; Kohler, M. Robotic timber construction—Expanding additive fabrication to new dimensions. *Autom. Constr.* **2016**, *61*, 16–23. [CrossRef]
6. Dörfler, K.; Sandy, T.; Giftthaler, M.; Gramazio, F.; Kohler, M.; Buchli, J. Mobile Robotic Brickwork. In *Robotic Fabrication in Architecture, Art and Design 2016*; Reinhardt, D., Saunders, R., Burry, J., Eds.; Springer: Cham, Switzerland, 2016; pp. 204–217.
7. Lindemann, H.; Gerbers, R.; Ibrahim, S.; Dietrich, F.; Herrmann, E.; Dröder, K.; Raatz, A.; Kloft, H. Development of a Shotcrete 3D-Printing (SC3DP) Technology for Additive Manufacturing of Reinforced Freeform Concrete Structures. In *First RILEM International Conference on Concrete and Digital Fabrication—Digital Concrete 2018. DC 2018. RILEM Bookseries*; Wangler, T., Flatt, R., Eds.; Springer: Cham, Switzerland, 2018; Volume 19, pp. 204–217.
8. Helm, V.; Willmann, J.; Gramazio, F.; Kohler, M. In-Situ Robotic Fabrication: Advanced Digital Manufacturing Beyond the Laboratory. In *Gearing up and Accelerating Cross-fertilization between Academic and Industrial Robotics Research in Europe: Technology Transfer Experiments from the ECHORD Project. Springer Tracts in Advanced Robotics*; Röhrbein, F., Veiga, G., Natale, C., Eds.; Springer: Cham, Switzerland, 2014; Volume 94, pp. 63–83.
9. Chai, H.; Wagner, H.J.; Guo, Y.; Qi, Y.; Menges, A.; Yuan, P.F. Computational design and on-site mobile robotic construction of an adaptive reinforcement beam network for cross-laminated timber slab panels. *Autom. Constr.* **2022**, *142*, 104536. [CrossRef]
10. Keitel, H.; Karaki, G.; Lahmer, T.; Nikulla, S.; Zabel, V. Evaluation of coupled partial models in structural engineering using graph theory and sensitivity analysis. *Eng. Struct.* **2011**, *33*, 3726–3736. [CrossRef]
11. Dutailly, J.C. *Hilbert Spaces in Modelling of Systems*; 2014; 47p. Available online: https://hal.archives-ouvertes.fr/hal-00974251 (accessed on 14 August 2021).
12. Dutailly, J.C. *Common Structures in Scientific Theories*; 2014; 34p. Available online: https://hal.archives-ouvertes.fr/hal-01003869 (accessed on 14 August 2021).
13. Legatiuk, D.; Smarsly, K. An abstract approach towards modeling intelligent structural systems. In Proceedings of the 9th European Workshop on Structural Health Monitoring, Manchester, UK, 10–13 July 2018.
14. Nefzi, B.; Schott, R.; Song, Y.Q.; Staples, G.S.; Tsiontsiou, E. An operator calculus approach for multi-constrained routing in wireless sensor networks. In Proceedings of the 16th ACM International Symposium on Mobile Ad Hoc Networking and Computing, New York, NY, USA, 22–25 June 2015.
15. Vassilyev, S.N. Method of reduction and qualitative analysis of dynamic systems: I. *J. Comput. Syst. Int.* **2006**, 17–25. [CrossRef]
16. Vassilyev, S.N.; Davydov, A.V.; Zherlov, A.K. Intelligent control via new efficient logics. In Proceedings of the 17th World Congress The International Federation of Automatic Control, Seoul, Republic of Korea, 6–11 July 2008.
17. Gürlebeck, K.; Nilsson, H.; Legatiuk, D.; Smarsly, K. Conceptual modelling: Towards detecting modelling errors in engineering applications. *Math. Methods Appl. Sci.* **2020**, *43*, 1243–1252. [CrossRef]
18. Legatiuk, D.; Nilsson, H. Abstract modelling: Towards a typed declarative language for the conceptual modelling phase. In Proceedings of the 8th International Workshop on Equation-Based Object-Oriented Modeling Languages and Tools, Weßling, Germany, 1 December 2017.
19. Foley, J.D.; Breiner, S.; Subrahmanian, E.; Dusel, J.M. Operands for complex system design specification, analysis and synthesis. *Proc. R. Soc.* **2021**, *477*. https://arxiv.org/abs/2101.11115.
20. Gürlebeck, K.; Hofmann, D.; Legatiuk, D. Categorical approach to modelling and to coupling of models. *Math. Methods Appl. Sci.* **2017**, *40*, 523–534. [CrossRef]
21. Kavrakov, I.; Legatiuk, D.; Gürlebeck, K.; Morgenthal, G. A categorical perspective towards aerodynamic models for aeroelastic analyses of bridges. *R. Soc. Open Sci.* **2019**, *6*, 181848. [CrossRef] [PubMed]
22. Legatiuk, D. Mathematical modelling by help of category theory: Models and relations between them. *Mathematics* **2022**, *9*, 1946. [CrossRef]
23. Spivak, D.; Kent, R. Ologs: A categorical framework for knowledge representation. *PLoS ONE* **2012**, *7*, e24274. [CrossRef] [PubMed]
24. Sowa, J. *Knowledge Representation: Logical, Philosophical, and Computational Foundations*; Brooks/Cole: Pacific Grove, CA, USA, 2000.
25. Awodey, S. *Category Theory*; Oxford University Press Inc.: New York, NY, USA, 2010.
26. Spivak, D. *Category Theory for Scientists*; MIT Press: Cambridge, MA, USA, 2014.
27. Haidegger, T. Taxonomy and Standards in Robotics. In *Encyclopedia of Robotics*; Ang, M.H., Khatib, O., Siciliano, B., Eds.; Springer: Berlin/Heidelberg, Germany, 2021.
28. Lynch, K.M.; Park, F.C. *Modern Robotics: Mechanics, Planning, and Control*; Cambridge University Press: Cambridge, UK, 2017.

29. Brosque, C.; Galbally, E.; Khatib, O.; Fischer, M. Human-Robot Collaboration in Construction: Opportunities and Challenges. In Proceedings of the 2020 International Congress on Human-Computer Interaction, Optimization and Robotic Applications (HORA), Ankara, Turkey, 26–27 June 2020; pp. 1–8.
30. Christensen, O. *An Introduction to Frames and Riesz Bases*; Springer International Publishing: Heidelberg, Germany, 2016.
31. Syarif, A.; Abouaissa, A.; Idoumghar, L.; Lorenz, P.; Schott, R.; Staples, S.G. New path centrality based on operator calculus approach for wireless sensor network deployment. *IEEE Trans. Emerg. Top. Comput.* **2019**, *7*, 162–173. [CrossRef]

Disclaimer/Publisher's Note: The statements, opinions and data contained in all publications are solely those of the individual author(s) and contributor(s) and not of MDPI and/or the editor(s). MDPI and/or the editor(s) disclaim responsibility for any injury to people or property resulting from any ideas, methods, instructions or products referred to in the content.

eng

Article

Bending and Torsional Stress Factors in Hypotrochoidal H-Profiled Shafts Standardised According to DIN 3689-1

Masoud Ziaei

Department of Mechanical and Automotive Engineering, Institute for Machin Development, Westsächsische Hochschule Zwickau, D-08056 Zwickau, Germany; masoud.ziaei@fh-zwickau.de

Abstract: Hypotrochoidal profile contours have been produced in industrial applications in recent years using two-spindle processes, and they are considered effective high-quality solutions for form-fit shaft and hub connections. This study mainly concerns analytical approaches to determine the stresses and deformations in hypotrochoidal profile shafts due to pure bending loads. The formulation was developed according to bending principles using the mathematical theory of elasticity and conformal mappings. The loading was further used to investigate the rotating bending behaviour. The stress factors for the classical calculation of maximum bending stresses were also determined for all those profiles presented and compiled in the German standard DIN3689-1 for practical applications. The results were also compared with the corresponding numerical and experimental results, and very good agreement was observed. Additionally, based on previous work, the stress factor was determined for the case of torsional loading to calculate the maximum torsional stresses in the standardised profiles, and the results are listed in a table. This study contributes to the further refinement of the current DIN3689 standard.

Keywords: hypotrochoidal profile shafts; DIN3689 H-profiles; bending stress; rotating bending loads in profiled shafts; flexure; torsional stress in profiled shafts; noncircular shafts; bending stress factor; torsional stress factor

Citation: Ziaei, M. Bending and Torsional Stress Factors in Hypotrochoidal H-Profiled Shafts Standardised According to DIN 3689-1. *Eng* **2023**, *4*, 829–842. https://doi.org/10.3390/eng4010050

Academic Editor: Antonio Gil Bravo

Received: 14 December 2022
Revised: 8 February 2023
Accepted: 1 March 2023
Published: 6 March 2023

Copyright: © 2023 by the author. Licensee MDPI, Basel, Switzerland. This article is an open access article distributed under the terms and conditions of the Creative Commons Attribution (CC BY) license (https://creativecommons.org/licenses/by/4.0/).

1. Introduction

In the field of modern drive technology, there is an increasing demand for higher power transmission in a smaller construction space. A necessary and important component in drive trains is the form-fit shaft and hub connections. Thereby, a widely used standard solution is the key-fit connection according to DIN 6885 [1]. However, this technique is reaching its mechanical limitations, which is why industry focus has been increasingly on form-fit connections with polygon profiles in the past few years. With the hypotrochoidal polygonal connection (H-profiles in Figure 1), a polygonal contour has been the new standard according to DIN 3689-1 [2] since November 2021. The great advantages of H-profiles via key-fit connections were studied in [3]. These investigations display a significant reduction of around 50% in the fatigue notch factor.

Additionally, a significant advantage of hypotrochoidal profiles (H-profiles) is their manufacturability through two-spindle turning [4,5] (Figure 2) and oscillating–turning [6] processes, as well as roller milling [7] (Figure 3). This allows time-efficient production.

Despite the excellent manufacturability described above and the great mechanical advantages of H-profiles, there is currently no reliable and cost-effective calculation method for the dimensioning of such profiles. The determination of the strength limit of H-profiles is still performed by means of extensive numerical investigations.

DIN 3689-1 refers to geometric specifications for H-profiles. Design guidelines are compiled in Part 2 of the standard. This paper represents an analytical solution for purely bending-loaded H-profile shafts in general and specifically for all standardised H-profiles for the first time. Furthermore, the author uses the analytical solution developed in another paper [8] for all standard profiles for torsional stresses and puts them together for practical and industrial applications.

The results can be used for a reliable and cost-effective calculation method of H-profile shafts with a simple pocket calculator for pure bending as well as torsional loads.

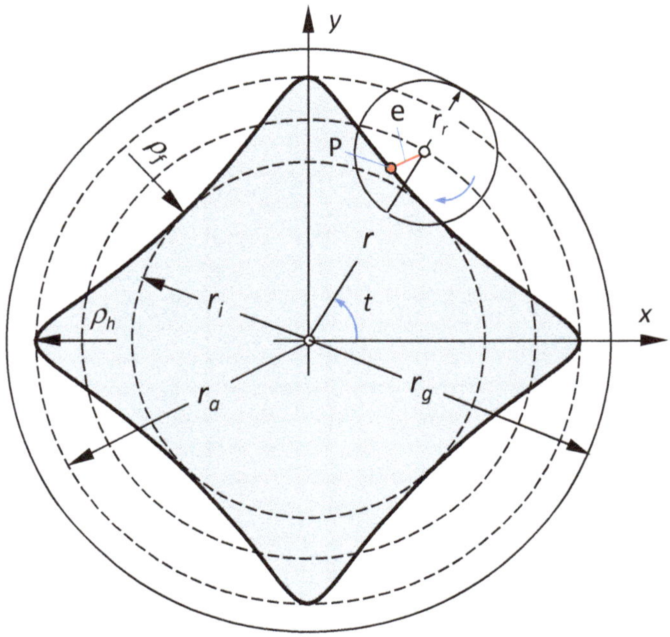

Figure 1. Description of exemplary hypotrochoid (H-profile) with four concave sides. A detailed explanation of the parameters is given below in Section 2.

Figure 2. Some H-profiles manufactured by two-spindle process, Iprotec GmbH, © Guido Kochsiek, www.iprotec.de, Zwiesel, Germany [5].

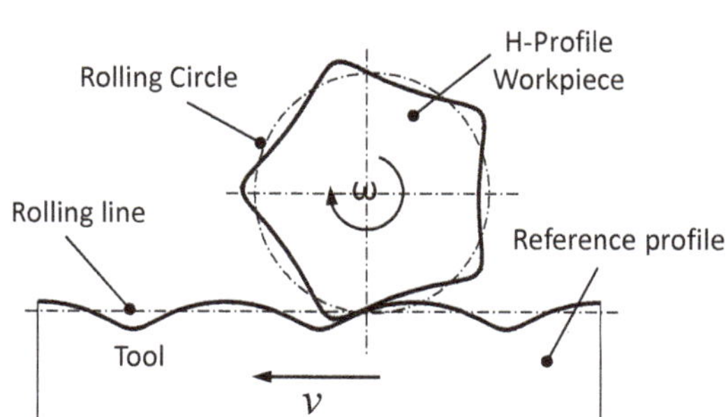

Figure 3. Roller milling manufacturing for H-profile [7].

2. Geometry of H-Profiles

A hypotrochoid (H-profile) is created by rolling a circle with radius r_r (called a rolling circle) on the inside of a guiding circle with radius r_g with no slippage (see, for instance, [9]). The distance between the centre point of the rolling circle and the generating point P is defined as eccentricity (Figure 1). Depending on the diameter ratios of the two circles and the location of the generating point P in the rolling circle, different H-profiles may be formed.

The diameter ratio (r_g/r_r) defines the number of sides "n" and should be an integer $(n > 2)$ to obtain a closed curve without intersection. The coordinates of the generated point P describe the parameter equations for the hypotrochoid (H-profile) as follows:

$$\begin{aligned} x(t) &= r \cdot \cos(t) + e \cdot \cos[(n-1) \cdot t] \\ y(t) &= r \cdot \sin(t) - e \cdot \sin[(n-1) \cdot t] \text{ with } 0^\circ \le t \le 360^\circ. \end{aligned} \tag{1}$$

The overlapping of the profile contour starts from the limit eccentricities of $e_{lim} = \frac{r}{n-1}$ and, accordingly, the limit relative eccentricity of $\varepsilon_{lim} = \frac{e_{lim}}{r} = \frac{1}{n-1}$.

Figure 4 shows some examples of the H-profiles obtained for different numbers of sides (n) and eccentricities.

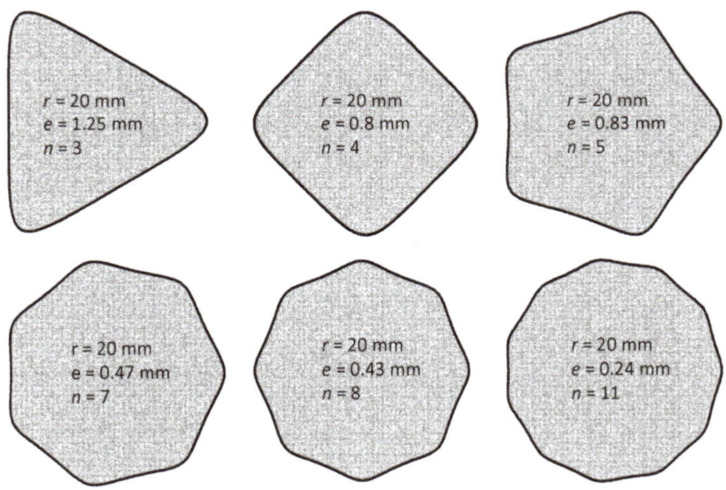

Figure 4. Examples of H-profiles with different numbers of sides (n) and eccentricities.

If a rolling circle rolls on the outside of a guiding circle, the profile generated is called an epicycloid (E-profile).

2.1. Geometric Properties

Area

Starting from the parameter representation (1) for the hypotrochoidal contours, the following complex mapping function is formulated as follows:

$$\omega(\zeta) = r \cdot \zeta + \frac{e}{\zeta^{n-1}} \tag{2}$$

This function conformally maps the perimeter of a unit circle to the contour of a H-profile. However, when the area enclosed by the polygon was mapped, multiple poles were formed at the corners of the contour. A complete conformal mapping is not essential for the determination of bending stresses. However, for shear force bending, a complete mapping of profile cross-section is necessary (analogue to torsion problem [8]).

By substituting mapping (2) into the equation for the area [10,11]:

$$A = \frac{1}{2}\int_0^{2\pi} Im\left[\bar{\omega}(\zeta) \cdot \dot{\omega}(\zeta)\right]dt \tag{3}$$

the following relationship can be derived for the area enclosed by an H-profile for any number of flanks n and eccentricity e:

$$A = A_a - \pi \cdot e \cdot \left[d_a + e \cdot (n-2)\right] \tag{4}$$

where $\dot{\omega} = d\omega/dt$ is the first derivative of the mapping function, t defines the parameter angle, and $A_a = \frac{\pi}{4} \cdot d_a^2$ is the area of the head circle (with $d_a = 2 \cdot r_a$).

2.2. Radius of Curvature at Profile Corners and Flanks

From a manufacturing point of view, the radius of the curvature of the contour at profile corners (on the head circle) plays an important role. Using the equation presented in [11], the radius of curvature can be determined:

$$\rho = 2i \cdot \frac{(\dot{\omega} \cdot \bar{\dot{\omega}})^{\frac{3}{2}}}{\bar{\dot{\omega}} \cdot \ddot{\omega} - \dot{\omega} \cdot \bar{\ddot{\omega}}} = \frac{|\dot{\omega}|^3}{Im(\bar{\dot{\omega}} \cdot \ddot{\omega})} \tag{5}$$

The second derivative of the mapping function in (5) is defined as $\ddot{\omega} = \frac{d^2\omega}{dt^2}$.

The radius of curvature at profile corners (on the head circle in Figure 1) can be determined by substituting mapping function (2) into Equation (5) for $t = 0$ as follows:

$$\rho_a = \frac{(d_a - 2 \cdot e \cdot n)^2}{2 \cdot \left[d_a + 2 \cdot e \cdot n \cdot (n-2)\right]} \tag{6}$$

The radius of curvature at profile corners ρ_a is important in connection with the minimum tool diameter regarding the manufacturability of the profile.

The radius of curvature of the profile in the profile flank ρ_f (Figure 1) can also be determined using Equation (5) for $t = \pi/n$:

$$\rho_f = \frac{\left[d_a + 2 \cdot e \cdot (n-2)\right]^2}{2 \cdot \left[d_a - 2 \cdot e \cdot (n^2 - 2 \cdot n + 2)\right]} \tag{7}$$

The radius of curvature in the flank area ρ_f is a measure of the degree of the form closure of profile contours.

2.3. Bending Stresses

In many practical applications, a failure may occur in the profiled shaft outside of the connection due to the excessive stresses. For these cases, the following analytical approach based on [12] is used to solve the bending problem.

It is assumed that the cross-sections remain flat (without warping) after bending. The following relationships are valid for the stresses:

$$\begin{gathered} \sigma_x = \sigma_y = \tau_{xy} = \tau_{yz} = \tau_{xz} = 0 \\ \sigma_z = -\frac{M_b}{I_y} \cdot x, \end{gathered} \tag{8}$$

where I_y denotes the moment of inertia for profile cross-section relative to the y-axis (Figure 5).

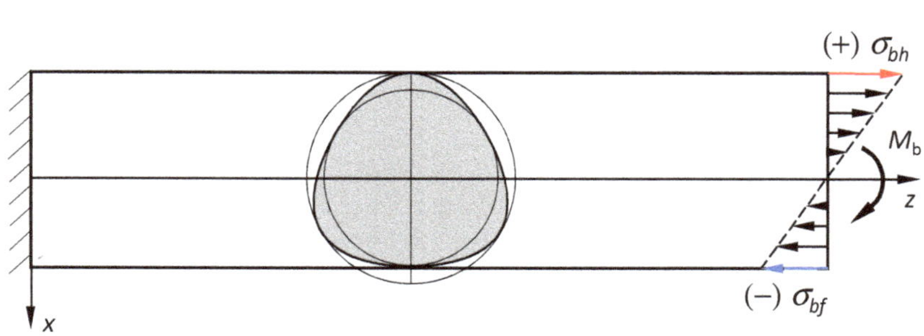

Figure 5. The bending coordinate system for a loaded profile shaft.

2.4. Bending Deformations

Displacement is determined using Hooke's law, and the corresponding correlation between displacements and the strain is as follows (see [12,13]):

$$u_x = \frac{M_b}{2 \cdot E \cdot I_y} \cdot \left[z^2 + \nu \cdot \left(y^2 - x^2 \right) \right] \tag{9}$$

2.5. Moments of Inertia

The moments of inertia involve a double integral over the profile's cross-section, but this can be reduced to a simple curvilinear integral over the profile contour using Green's theorem, as follows:

$$\begin{aligned} I_x &= -\frac{1}{3}\int_\gamma y^3 dx \\ I_y &= \frac{1}{3}\int_\gamma x^3 dy \\ I_{xy} &= \frac{1}{2}\int_\gamma x^2 y dy. \end{aligned} \tag{10}$$

The contour description according to Equation (2) is also advantageous here. For the contour of the profile's cross-section, the following coordinates apply:

$$\begin{aligned} x &= \frac{\omega(\lambda) + \omega(\lambda)}{2} \\ y &= \frac{\omega(\lambda) - \omega(\lambda)}{2 \cdot i}. \end{aligned} \tag{11}$$

By substituting Equation (11) in (10), I_y, I_y, I_{xy} can be determined as such:

$$\begin{aligned} I_x &= -\tfrac{i}{48}\int_\gamma \left(\omega(\lambda) - \omega(\lambda)\right)^3 d\left(\omega(\lambda) + \omega(\lambda)\right) \\ I_y &= \tfrac{i}{48}\int_\gamma \left(\omega(\lambda) + \omega(\lambda)\right)^3 d\left(\omega(\lambda) - \omega(\lambda)\right) \\ I_{xy} &= -\tfrac{1}{32}\int_\gamma \left(\omega(\lambda) + \omega(\lambda)\right)^2 \left(\omega(\lambda) - \omega(\lambda)\right) d\left(\omega(\lambda) - \omega(\lambda)\right), \end{aligned} \tag{12}$$

where $\lambda = e^{it}$. Function (12) facilitates the determination of moment of inertia with the assistance of Equation (2).

The moment of inertia I_y is necessary for the calculation of the bending stress σ_z as well as for the determination of bending deformation u_x (Equations (8) and (9)).

Inserting the mapping function from (2) into Equation (12) for I_y, the following relationship is determined for the bending moment of inertia for an arbitrary number of flanks n and eccentricity e:

$$I_y = \frac{\pi}{4} \cdot \left(r^4 - 2e^2(n-2)r^2 - e^4(n-1) \right) \tag{13}$$

If one substitutes $x(t)$ from (1) and I_y from (13) into Equation (8), the distribution of the bending stress on the lateral surface of the profile can be determined as follows:

$$\sigma_b(t) = \frac{4M_b}{\pi} \cdot \frac{r\cos(t) + e\cos((n-1)t)}{r^4 - 2e^2(n-2)r^2 - e^4(n-1)} \tag{14}$$

The maximum bending stress on the tension side occurs at $x = r + e$ (on the profile head, Figure 5), and therefore the following equation can be obtained:

$$\sigma_{bh} = \frac{4M_b}{\pi} \cdot \frac{r + e}{r^4 - 2e^2(n-2)r^2 - e^4(n-1)} \tag{15}$$

The bending stress on the pressure side occurs at $x = r - e$ in the middle of a profile flank (on the profile foot, Figure 5) can also be determined as follows:

$$\sigma_{bf} = \frac{4M_b}{\pi} \cdot \frac{r - e}{r^4 - 2e^2(n-2)r^2 - e^4(n-1)} \tag{16}$$

2.6. Example

An H-profile from DIN 3689-1 [2] with three sides, a head circle diameter of 40 mm and eccentricity $e = 1.818$ mm ($r = 18.18$ mm; related eccentricity $\varepsilon = 0.1$) was chosen as the object of investigation. The bending load was chosen as M_b = 500 Nm.

In order to compare the analytical results, numerical investigations were carried out using FE analyses, and the MSC-Marc programme system was used.

Figure 6 shows the mesh structure and the corresponding boundary conditions. The shaft is fixed on the right side. A bending moment is applied on the left side of the shaft via a reference node using REB2s. Bending stresses were evaluated at an adequate distance (l_b) from the loading point. The FE mesh in Figure 6 contains hexahedral elements with full integration, type 7 according to the Marc Element Library [14].

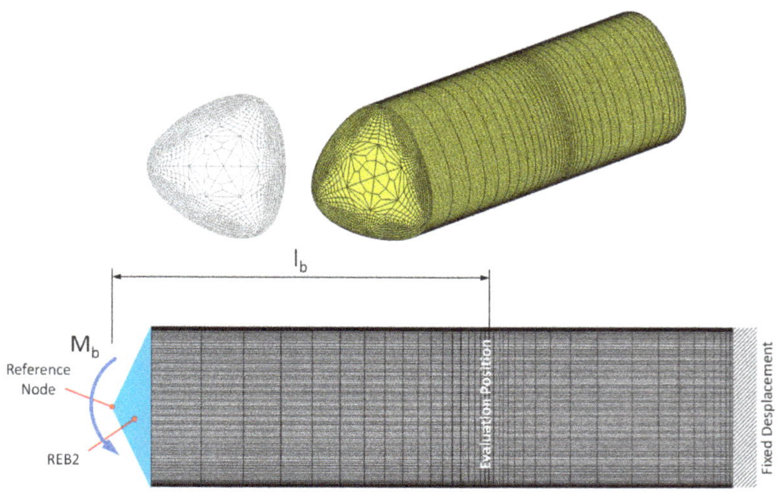

Figure 6. FE mesh and boundary conditions for the H-profile with n = 3 according to DIN 3689-1.

FE structures are generated by employing software written in Python language at the Chair of Machine Elements at West Saxon University of Zwickau, Germany. The FE meshes were then transferred to MSC-Marc program system and integrated into pre-processing.

Figure 7 displays the distribution of bending stress on the circumference of the profile according to Equation (14) and its comparison with the numerical result. A good agreement between the results was observed.

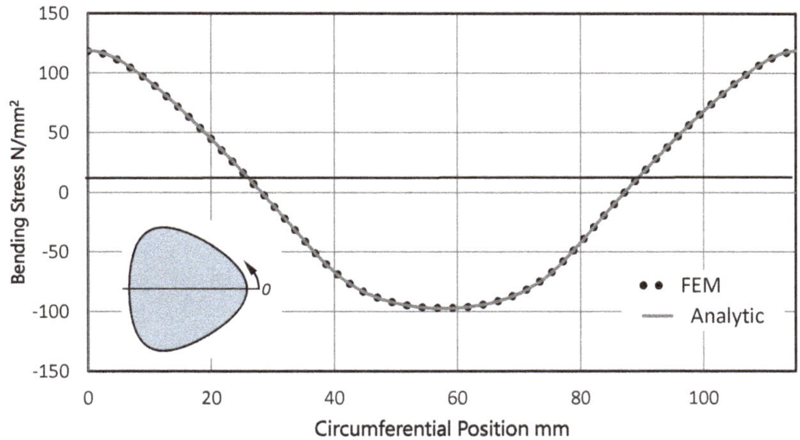

Figure 7. Circumferential distribution of the bending stress on the lateral surface of a standardised H3 profile.

Additionally, bending stresses were experimentally determined for the profile head and foot areas. Figure 8 shows the test bench for bending load.

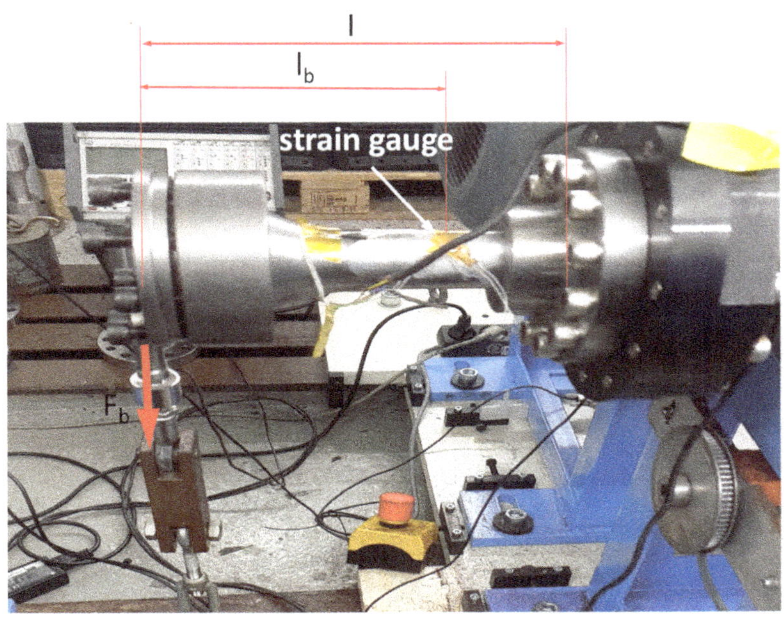

Figure 8. Bending loads test bench (Machine Elements Laboratory at West Saxon University of Zwickau).

Experimental results for head and foot areas are compared with Equations (15) and (16) in Figure 9, where a good agreement of the results is evident.

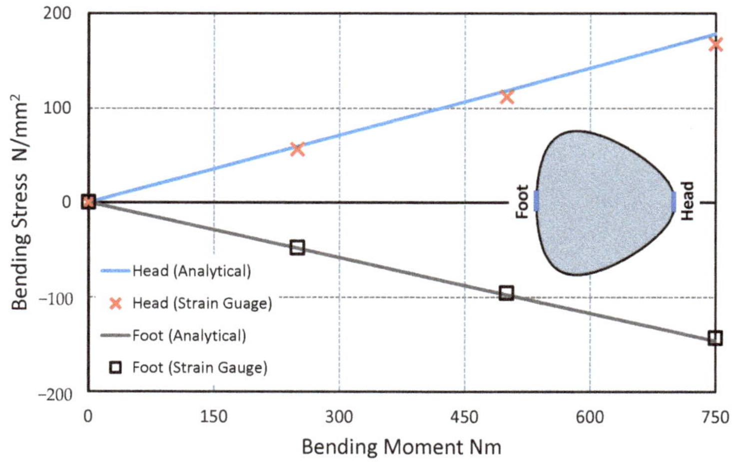

Figure 9. Comparison of the experimental results with the analytical solutions.

2.7. Stress Factor for Bending Loads

The stress factor is defined as the ratio of bending stress in a profile shaft to a corresponding reference stress for a round cross-section with radius r (nomial radius of the profile):

$$\alpha_b = \frac{\sigma_b}{\sigma_{b,ref}} \quad \text{with: } \sigma_{b,ref} = \frac{M_b \cdot r}{I_{y,ref}} \text{ and } I_{y,ref} = \frac{\pi}{4} \cdot r^4. \tag{17}$$

For the head of the profile, the stress factor is determined as follows:

$$\alpha_{bh} = \frac{1 + e}{1 - 2e^2(n-2) - e^4(n-1)} \tag{18}$$

Figure 10 shows the curves for the stress factor α_{bh} as a function of the relative eccentricity ε for different numbers of sides n. It can be recognised that the stress factor rises with an increase in eccentricity and or the number of sides.

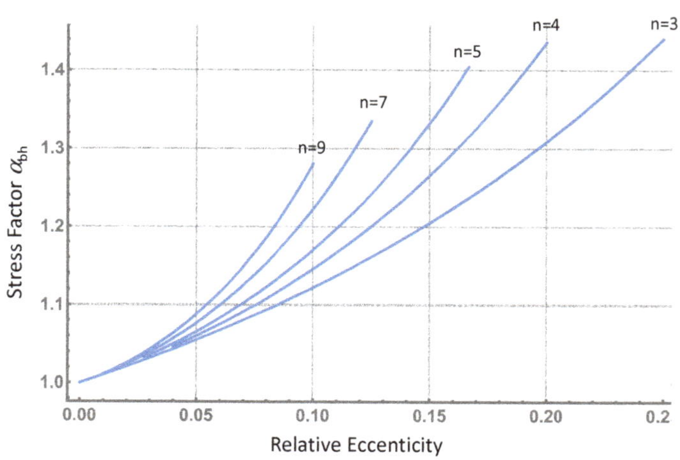

Figure 10. Stress factors for the bending stress at the profile head (Equation (18)) with varying relative eccentricity and number of sides.

For the profile base (foot), the following stress factor is analogously obtained:

$$\alpha_{bf} = \frac{1-e}{1-2e^2(n-2)-e^4(n-1)} \tag{19}$$

2.8. Rotating Bending Stress

During power transmission, the gear shaft always shows rotational movement. Therefore, the rotating bending was also investigated.

Figure 11 schematically represents the rotated position of an H-profile with three flanks according to the Cartesian coordinates.

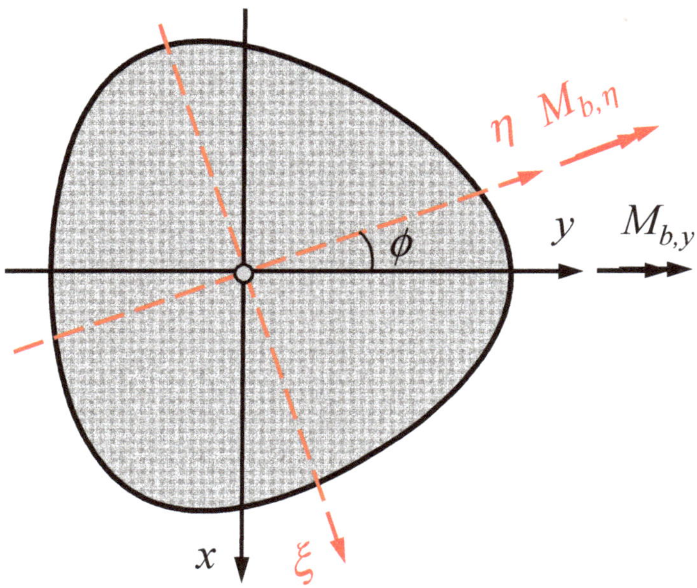

Figure 11. Rotated coordinate system for determining the bending moment of inertia.

The moment of inertia remains invariant due to the periodic symmetry of the cross-section of the H-profile presented based on Equation (2). Therefore, the following relationships are valid from Equation (12):

$$I_x = I_y \text{ and } I_{xy} = 0. \tag{20}$$

From Equation (20) and the use of Mohr's circle, it can be proven that the moment of inertia is independent of the rotation angle ϕ (see also [10]):

$$\begin{aligned} I_\xi = I_\eta (= I_x = I_y) \\ I_{\xi\eta} = I_{xy} = 0. \end{aligned} \tag{21}$$

In order to obtain the general solution of the bending stress according to Equation (8) for an arbitrary angle of rotation, the perpendicular distance ξ is to be calculated in the rotated coordinate system:

$$\xi(\phi) = y\cos(\phi) - x\sin(\phi) \tag{22}$$

where ϕ denotes the angle of rotation. If the values for x and y from (1) are inserted into the relationship (22), the following equation results for the perpendicular distance in the rotated coordinate system ($0 \leq t \leq 2\pi$):

$$\xi(\phi, t) = r\sin(t - \phi) - e\sin((n-1)t + \phi) \tag{23}$$

The distribution of bending stress on the profile contour may be determined by using (23) in the relation of bending stress as follows:

$$\sigma_b(\phi, t) = -\frac{M_b}{I_\eta} \cdot \xi(\phi, t) = \frac{4M_b}{\pi} \cdot \frac{r\sin(t - \phi) - e\sin((n-1)t + \phi)}{r^4 - 2e^2(n-2)r^2 - e^4(n-1)} \tag{24}$$

Figure 12 shows the distributions of the bending stresses on the profile contour for different angles of rotation, which were determined using Equation (24). As expected, the maximum stress occurred at the profile head.

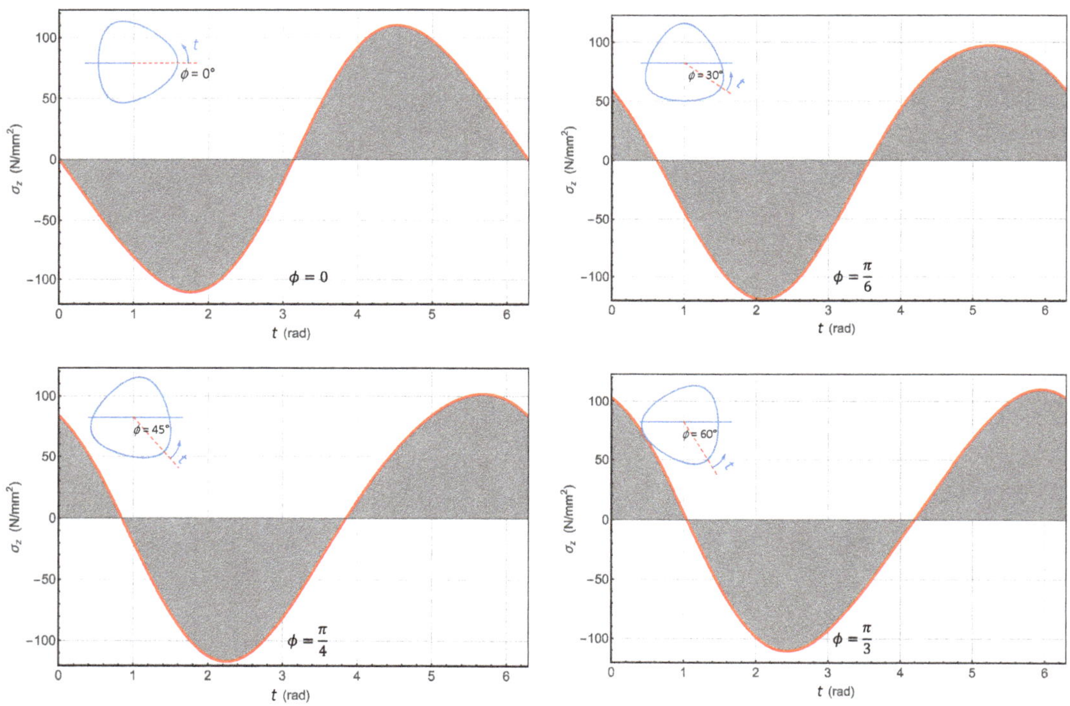

Figure 12. Distributions of the bending stresses on the profile contour for different angles of rotation ϕ, with $r = 18.18$ mm, $n = 3$, $e = 1.818$ mm, and M_b = 500 Nm.

2.9. Deflection

The deflection of the profile shaft can also be determined with the help of the bending moment of inertia I_y. As explained above, this is independent of the angular position of the cross-section (Equation (21)).

The deflection of the neutral axis is determined from Equation (9) for $x = y = 0$ as follows:

$$\delta_x = \frac{M_b}{2 \cdot E \cdot I_y} \cdot z^2 \tag{25}$$

Substituting (13) in (25), the deflection can be determined as

$$\delta_x = \frac{2M_b}{\pi E} \cdot \frac{z^2}{r^4 - 2e^2(n-2)r^2 - e^4(n-1)} \tag{26}$$

2.10. Example

Figure 13 shows the deflection for an H-profile shaft with three flanks according to DIN3689-1 with d_a = 40 mm (H3-40 × 32.73 with ε = 0.1) and a length of 160 mm made of steel (E = 210,000 $\frac{\mathrm{N}}{\mathrm{mm}^2}$). The comparison with FE analysis shows very good agreement with Equation (26), as can also be seen in Figure 13. The bending load was chosen as M_b = 500 Nm.

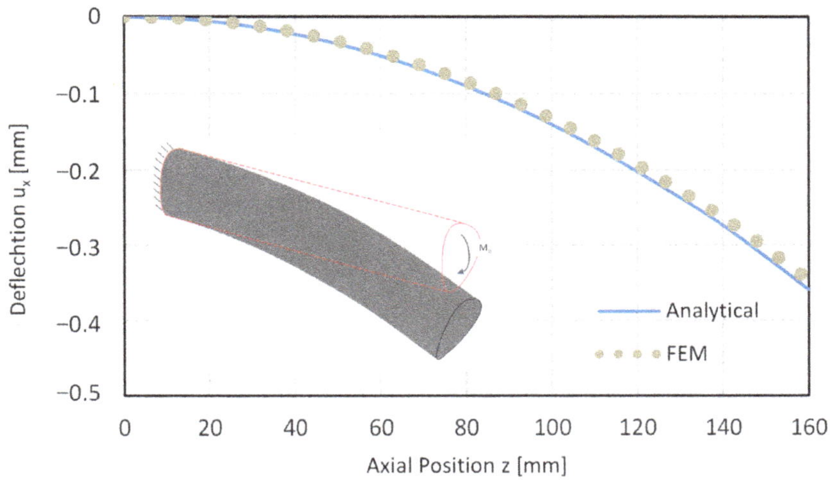

Figure 13. Deflection in a DIN3689-H3-40 × 32.73 profile shaft.

2.11. H-Profiles According to DIN3689-1

DIN3689-1 is a new standard that was published for the first time in November 2021. It describes the geometric properties of 18 specified H-profiles in two series. Series A is based on the head diameter, and series B involves the foot diameter as the nominal size of the profile. The respective corresponding profiles are geometrically similar. Each series contains 48 nominal sizes, which remain geometrically similar amongst themselves. Consequently, all standardised profiles are limited to 18 variants. This facilitates the processing of a generally valid design concept.

2.12. Stress Factor for Bending

The maximum bending stresses at the head and foot of the profile are important from a technical point of view for the design of a profile shaft subject to bending. Therefore, in this section, the two stress factors α_{bh} and α_{bf} for all the 18 standard profile series were determined using Equations (18) and (19).

2.13. Stress Factor for Torsion

The stress concentration factor for torsion α_t is defined as the ratio of the maximum torsional stress $\tau_{t,max}$ (occurring in the middle of the profile flank) and the torsional stress in a round reference shaft with radius r:

$$\alpha_t = \frac{\tau_{t,max}}{\tau_{t,ref}} \quad \text{with: } \tau_{t,ref} = \frac{M_t \cdot r}{I_{t,ref}} \text{ and } I_{t,ref} = \frac{\pi}{2} \cdot r^4. \tag{27}$$

In [15], purely numerical investigations were carried out on the torsional stresses in H-profile shafts to calculate the stress factor.

The analytical solution for torsion may be performed using the approach of Muskhelishvili [12]. However, this requires a conformal mapping of the unit circle onto the polygon's cross-section. For H-profiles, the mapping function derived from the parametric equation, Equation (1), cannot be directly used to solve the torsional stresses due to the multiple poles. The authors of [16] employed an elaborate computational process to determine the polynomials required for the description of the mappings of H-profiles. In [8,17–19], successive methods according to Kantorovich [20] were used to develop a suitable mapping function in the form of a series converging to the profile contour. The convergence quality and limit were examined and presented depending on the number of terms in the series developed in [8], calculating the torsional deformations for all standardised profiles. In the presented work, this method, accompanied by FEA, was used for all the 18 standardised profile geometries of DIN3689-1 to determine the maximum torsional stresses, which occur in the middle of the profile flank at the profile foot. A stress concentration factor for torsional loading α_t was also determined analogously to that defined for the case of bending load.

For practical applications, the results for the bending and torsional stress factors are compiled in Table 1. Using the relative eccentricity, no dependence on the shaft diameter appears. Table 1 lists the results obtained for the bending and torsional stress factors for all standardised profile geometries according to DIN3689-1 (rounded to two decimal places).

Table 1. Stress factors for bending and torsional loads for the H-profiles standardised according to DIN3689-1.

n	ε	α_{bh}	α_{bf}	I_y/I_0	α_t
3	0.100	1.12	0.92	0.98	1.23
4	0.056	1.07	0.96	0.99	1.17
4	0.111	1.17	0.94	0.95	1.37
5	0.031	1.04	0.97	0.99	1.12
5	0.062	1.09	0.96	0.98	1.24
5	0.094	1.16	0.96	0.95	1.38
6	0.020	1.02	0.98	1.00	1.10
6	0.040	1.05	0.97	0.99	1.18
6	0.062	1.10	0.97	0.97	1.37
7	0.028	1.04	0.98	0.99	1.15
7	0.056	1.09	0.97	0.97	1.29
7	0.083	1.16	0.99	0.93	1.43
9	0.023	1.03	0.98	0.99	1.17
9	0.047	1.08	0.98	0.97	1.31
9	0.062	1.12	0.99	0.95	1.39
12	0.017	1.02	0.99	0.99	1.16
12	0.033	1.06	0.99	0.98	1.28
12	0.050	1.10	1.00	0.95	1.38

The bending moment of the inertia of a circular cross-section with radius r is defined as a reference moment of inertia and labelled I_0. The ratio between I_y and I_0 is also listed in Table 1 for the standardised profiles. The H-profiles are normally slightly more flexible than round profiles.

3. Conclusions

In this paper, an analytical approach was presented to determine the bending stresses and deformations in the hypotrochoidal profile shafts. Valid calculation equations for the area, radii of curvature of the profile contour, and the bending moment of inertia were derived for such profiles. Furthermore, the solutions for bending stresses and deformations were presented. For practical applications, a stress factor was defined for the critical locations on the profile contour.

The analytical results demonstrated very good agreement with both numerical and experimentally determined results.

The stress factors of the bending stresses were determined for all profile geometries standardised according to DIN3689-1, and the values obtained were compiled in a table for practical applications. Based on previous works of the author, the stress factors for torsional stresses were also determined and added to the table. The data allow a reliable and cost-effective calculation of H-profile shafts with a pocket calculator for pure bending as well as torsional loads. This can be very advantageous for SMEs.

Funding: This research was funded by DFG (Deutsche Forschungsgemeinschaft) grant number [DFG ZI 1161/2].

Institutional Review Board Statement: Not applicable.

Informed Consent Statement: Not applicable.

Data Availability Statement: Not applicable.

Conflicts of Interest: The author declares no conflict of interest.

Abbreviations

Formula Symbols:

A	mm^2	Area of profile cross-section
e	mm	Profile eccentricity
e	-	Euler's number
e_{lim}	mm	Profile overlap eccentricity limit
E	MPa	Young's modulus
n	-	Profile periodicity (number of sides)
I_0	mm^4	Corresponding reference moments of inertia for a round cross-section with radius r
$I_{t,ref}$	mm^4	Torsional moment of inertia for reference shaft
I_x, I_y, I_{xy}	mm^4	Surface moments of inertia in the Cartesian coordinate system
$I_{y,ref}$	mm^4	Surface moment of inertia about y-axis for reference shaft
$I_\xi, I_\eta, I_{\xi\eta}$	mm^4	Surface moments of inertia in the rotated coordinate system
l	mm	Length of profile shaft
l_b	mm	Distance to strain gage in experimental test
M_b	Nm	Bending moment
r	mm	Nominal or mean radius
t	-	Profile parameter angle
u_x	mm	Displacement in x direction
x, y, z	mm	Cartesian coordinates

Greek Formula Symbols:

α_{bh}	-	Bending stress factor for profile head
α_{bf}	-	Bending stress factor for profile foot
α_t	-	Torsional stress factor for profile foot

δ_x	mm	deflection
$\varepsilon = e/r$	-	Relative eccentricity
ϕ	-	Rotation angle of the coordinate system
$\lambda = e^{i\theta}$	-	Physical plane unit circle
θ	-	Polar angle
σ_b, σ_z	MPa	Bending stress (z-component of stress vector)
τ_t	MPa	Torsional stress
$\omega(\zeta)$	-	Completed mapping function
$\omega_0(\zeta)$	-	Contour edge mapping function
ζ	-	Complex variable in model plane
ξ,η	-	Coordinates in rotated system

References

1. *DIN 6885-1:1968-08;* Mitnehmerverbindungen ohne Anzug; Paßfedern, Nuten, Hohe Form. DIN-Deutsches Institut für Normung e.V.: Berlin, Germany, 1968.
2. *DIN 3689-1:2021-11;* Welle-Nabe-Verbindung—Hypotrochoidische H-Profile—Teil 1: Geometrie und Maße. DIN-Deutsches Institut für Normung e.V.: Berlin, Germany, 2022.
3. Selzer, M.; Forbrig, F.; Ziaei, M. Hypotrochoidische Welle-Nabe-Verbindungen. In Proceedings of the 9th Fachtagung Welle-Nabe-Verbindungen, Gestaltung Fertigung und Anwendung, Stuttgart, Germany, 26–27 November 2022; Band 2408. pp. 155–169.
4. Gold, P.W. In acht Sekunden zum Polygon: Wirtschaftliches Unrunddrehverfahren zur Herstellung von Polygon-Welle-Nabe-Verbindungen. Antriebstechnik. *Seiten* **2006**, 42–44.
5. Iprotec GmbH, Polygonverbindungen. Bad Salzuflen, Germany. Available online: https://www.iprotec.de (accessed on 13 December 2022).
6. Stenzel, H. Unrunddrehen und Fügen zweiteiliger Getriebezahnräder mit polygonaler Welle-Nabe-Verbindungen, VDI-4. Fachtagung Welle-Nabe-Verbindungen, Gestaltung Fertigung und Anwendung. *Seiten* **2010**, 211–230.
7. Ziaei, M. Westsächsische Hochschule Zwickau, Patentanmeldung: Application of Rolling Processes Using New Reference Profiles for the Production of Trochoidal Inner and Outer Contours. Patent-Nr. DE 10 2019 000 654 A1, 30 July 2020.
8. Ziaei, M. *Torsionsspannungen in Prismatischen, Unrunden Profilwellen mit Trochoidischen Konturen, Forschung im Ingenieurwesen;* Ausgabe 4/2021; Springer: Berlin/Heidelberg, Germany, 2021. Available online: https://www.researchgate.net/publication/355421692_Torsionsspannungen_in_prismatischen_unrunden_Profilwellen_mit_trochoidischen_KonturenTorsional_stresses_and_deformations_in_prismatic_non-circular_profiled_shafts_with_trochoidal_contours (accessed on 14 December 2022).
9. Bronstein, I.N.; Semendjajew, K.A. *Taschenbuch der Mathematik, 7th Auflage;* Verlag Hari Deutsch: Frankfurt am Main, Germany, 2008.
10. Ziaei, M. Bending Stresses and Deformations in Prismatic Profiled Shafts with Noncircular Contours Based on Higher Hybrid Trochoids. *Appl. Mech.* **2022**, 3, 1063–1079. [CrossRef]
11. Zwikker, C. *The Advanced Geometry of Plane Curves and Their Applications, Dover Books on Advanced Mathematics;* Dover Publications: Mineola, NY, USA, 2005.
12. Muskelishvili, N.I. *Some Basic Problems of the Mathematical Theory of Elasticity;* Springer: Dordrecht, The Netherlands, 1977.
13. Sokolnikoff, I.S. *Mathematical Theory of Elasticity;* Robert E. Krieger Publishing Company: Malaba, FL, USA, 1983.
14. *Marc 2020 Manual;* Volume B (Element Library); MSC Software Corporation: Newport Beach, CA, USA, 2020.
15. Schreiter, R. *Numerische Untersuchungen zu Form- und Kerbwirkungszahlen von Hypotrochoidischen Polygonprofilen unter Torsionsbelastung;* Dissertation TU Chemnitz: Chemnitz, Germany, 2022.
16. Ivanshin, P.N.; Shirokova, E.A. Approximate Conformal Mappings and Elasticity Theory. *J. Complex Anal. Hindawi Publ. Corp.* **2016**, *2016*, 4367205. [CrossRef]
17. Ziaei, M. *Analytische Untersuchung Unrunder Profilfamilien und Numerische Optimierung Genormter Polygonprofile für Welle-Nabe-Verbindungen, Habilitationsschrift;* Technische Universität Chemnitz: Chemnitz, Germany, 2002.
18. *Research Report: Entwicklung eines analytischen Berechnungskonzeptes für formschlüssige Welle-Nabe-Verbindungen mit hypotrochoidischen Verbindungen, Abschlussbericht zum DFG-Vorhaben DFG ZI 1161/2 (Westsächsische Hochschule Zwickau) und LE 969/21(TU Chemnitz);* Westsächsische Hochschule Zwickau: Zwickau, Germany, 2020.
19. Lee, K. Mechanical Analysis of Fibers with a Hypotrochoidal Cross Section by Means of Conformal Mapping Function. *Fibers Polym.* **2010**, *11*, 638–641. [CrossRef]
20. Kantorovich, L.V.; Krylov, V.I. *Approximate Methods of Higher Analysis;* Dover Publications: Mineola, NY, USA, 2018.

Disclaimer/Publisher's Note: The statements, opinions and data contained in all publications are solely those of the individual author(s) and contributor(s) and not of MDPI and/or the editor(s). MDPI and/or the editor(s) disclaim responsibility for any injury to people or property resulting from any ideas, methods, instructions or products referred to in the content.

eng

Article

On Long-Range Characteristic Length Scales of Shell Structures

Harri Hakula *

Department of Mathematics and Systems Analysis, Aalto University, Otakaari 1, FI-00076 Espoo, Finland

Abstract: Shell structures have a rich family of boundary layers including internal layers. Each layer has its own characteristic length scale, which depends on the thickness of the shell. Some of these length scales are long, something that is not commonly considered in the literature. In this work, three types of long-range layers are demonstrated over an extensive set of simulations. The observed asymptotic behavior is consistent with theoretical predictions. These layers are shown to also appear on perforated structures underlying the fact these features are properties of the elasticity equations and not dependent on effective material parameters. The simulations are performed using a high-order finite element method implementation of the Naghdi-type dimensionally reduced shell model. Additionally, the effect of the perforations on the first eigenmodes is discussed. One possible model for buckling analysis is outlined.

Keywords: shells; boundary layers; finite element method

Citation: Hakula, H. On Long-Range Characteristic Length Scales of Shell Structures. *Eng* **2023**, *4*, 884–902. https://doi.org/10.3390/eng4010053

Academic Editor: Antonio Gil Bravo

Received: 10 December 2022
Revised: 3 February 2023
Accepted: 1 March 2023
Published: 6 March 2023

Copyright: © 2023 by the author. Licensee MDPI, Basel, Switzerland. This article is an open access article distributed under the terms and conditions of the Creative Commons Attribution (CC BY) license (https://creativecommons.org/licenses/by/4.0/).

1. Introduction

Shell structures and, in particular, thin shells remain challenging for both theoretical and computational structural analysis [1]. One must either use special shell elements or rely on a high-order finite element method, as is performed here. Computing with 3D formulations is still prohibitively expensive. One of the defining features of shells is that every solution of a shell problem can be thought of as a linear combination of features or boundary layers each with its own characteristic length scale, including the so-called smooth component, which typically spans the whole structure. The effects of curvature lead to boundary layers that can also be internal, something that cannot happen in plates, for instance. Moreover, some of the layers can have long, yet parameter-dependent, length scales that have not received much attention in the literature. Indeed, in the first paper introducing modern boundary layer analysis [2], the long-range features on cylindrical shells were omitted since their meaning was not properly understood.

Thin structures are normally modeled as two-dimensional ones via dimension reduction, where the thickness becomes a parameter. For the sake of analysis, the thickness is defined as a dimensionless constant. Once the reduced linear elasticity equations have been obtained, the characteristic length scales can be derived as functions of the parameter. Classical boundary layers have short length scales, and internal boundary layers may have long-range effects along the characteristic curves of such surfaces, but with parameter-dependent widths. The third category of the characteristic length scales, the long-range effects, are the focus of this paper. Every layer is generated by some combination of curvature, kinematic constraints, and loading; in other words, every layer has its own generator. The standard reference for boundary layers of shells is Pitkäranta et al. [3], where every generator of a layer is taken to be either a straight line or a point. This work was later extended via the introduction of curved generators [4,5]. In fact, this extension shows that the collection of boundary layers is not finite.

Shells of revolution are a representative class of thin structures. Let us denote the thickness with d, and the dimensionless thickness with $t = d/L_D$, where L_D is taken to be the diameter of the domain, for example. The practical range in engineering problems is typically $t \in [1/1000, 1/100]$, and already at $t = 1/10$ the dimension reduction is not effective, depending on the model of course. In analysis, t is the convenient parameter, and in the sequel, d and t can be used interchangeably. The long-range layers have characteristic length scales of $L \sim 1/t$ and $L \sim 1/\sqrt{t}$. In addition, there are length scales $\sim \sqrt[n]{t}$, with $n = 1, 2, \ldots$, of which only $n = 1, 2$ are standard boundary layers. For illustrative examples, see Figure 1.

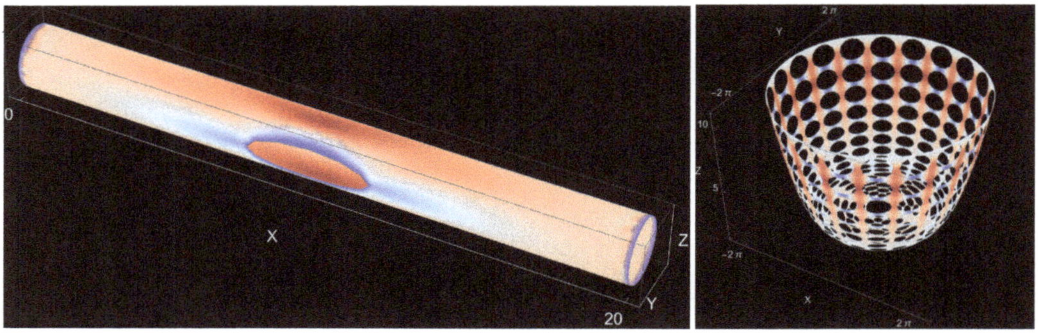

(**a**) (**b**)

Figure 1. Examples of structures subject to long-range layers: transverse deflection profiles. In both cases, the dimensionless thickness $t = 1/100$, all boundaries are kinematically fully constrained, and the loading is the unit pressure. (**a**): Long cylinder with an ellipsoidal hole. (**b**): Circular plate with an elliptic extension. The load is acting only on the extension.

The purpose of this paper is show that many interesting structural responses in shells are due to long-range layers. The main contribution of this paper is to show that these effects also exist in perforated structures. This serves as a useful remainder that even though the internal and long-range boundary layers have wave propagation characteristics, the underlying equations are those of linear elasticity, and phenomena such as wave dispersion do not occur. In the numerical examples below, all three geometry classes, parabolic, hyperbolic, and elliptic, are included. Another contribution of this study is that the results on hyperbolic cases are shown to agree with those on the parabolic ones, as predicted by the theory.

One of the complicating aspects is that the long-range effects are also combined with standard boundary layers, which typically dominate in amplitudes. This is probably one of the reasons why the existing literature is not extensive. As mentioned above, Pitkärantaet al. [3] is the standard reference. In elliptic shells, a curious boundary-layer-like long-range effect can also occur in connection with the so-called sensitivity of the structure. Here, the comprehensive study is by Sanchez-Palencia et al. [6]. In an excellent review article by Pietraszkiewicz and Konopińska [7] in the section on shells of revolution, only standard boundary layers are addressed. Similarly, in the paper by Malliotakis et al. [8] in connection to a wind tower problem, the focus is on the short-range phenomena. This is understandable since the maximal stresses occur due to standard boundary layers. Additionally, as is also evident in the numerical examples below, visualization and precise analysis of the long-range features of the solutions are difficult.

The simulation of thin structures with the finite element method poses its own set of difficulties. A highly influential case study was carried out by Szabo [9], where a complete analysis of a shell structure starting from measurements and ending with error analysis is discussed. The thin shell models have been verified in static and dynamic settings with experiments with reasonable agreement. In particular, one must take numerical locking into account and either use special shell elements or rely on higher-order methods [10]. In this work, the latter approach is adopted and justified using numerical energy convergence observations. One of the open problems in thin shell modeling is the question of buckling. One of the models applicable within the framework used in this paper is due to Niemi [11].

The three examples are motivated by the wind tower example mentioned above [8], interesting modeling problems encountered when long-range effects emerge, [12,13], and recent work on homogenization of perforated shell structures, where the curved generator cases have not been considered [14]. In both [12,13], interesting long-range responses are modeled where the driving force is internal torsion resulting from a bilayer or similar materials. It is an interesting question for future work to see if some connections with the results presented in this work can be found and formulated precisely.

The rest of the paper is structured as follows: In Section 2, the necessary background material is covered. Shell models are introduced in Section 3, and the related layers are discussed in the following Section 4. The set of three examples is presented in Section 5, followed by conclusions in Section 6. An alternative shell model is briefly outlined in Appendix A. Finally, the buckling problem is discussed in the last appendix (Appendix B).

2. Preliminaries

In this section, the necessary background material is introduced. The notation used in the sequel is established here as well.

2.1. Navier's Equations of Elasticity

In this section, the elasticity equations are introduced. For reference, see for instance [15]. Let D be a domain representing a deformable medium subject to a body force f and a surface traction $\mathbf{g}$. The 3D model problem is then to find the displacement field $\mathbf{u} = (u_1, u_2, u_3)$, and the symmetric stress tensor $\sigma = (\sigma_{ij})_{i,j=1}^3$, such that

$$\begin{aligned} \sigma &= \lambda \operatorname{div}(\mathbf{u})\boldsymbol{I} + 2\mu\epsilon(\mathbf{u}), && \text{in } D \\ -\operatorname{div}(\sigma) &= \mathbf{f}, && \text{in } D \\ \mathbf{u} &= \mathbf{0}, && \text{on } \partial D_D \\ \sigma \cdot \mathbf{n} &= \mathbf{g}, && \text{on } \partial D_N \end{aligned} \tag{1}$$

where $\partial D = \partial D_D \cup \partial D_N$ is a partitioned boundary of D. The Lamé constants are

$$\lambda = \frac{E\nu}{(1+\nu)(1-2\nu)}, \qquad \mu = \frac{E}{2(1+\nu)}, \tag{2}$$

with E and ν being Young's modulus and Poisson's ratio, respectively. Further, $\boldsymbol{I}$ is the identity tensor, $\mathbf{n}$ denotes the outward unit normal to ∂D_N, and the strain tensor is

$$\epsilon(\mathbf{u}) = \frac{1}{2}(\nabla\mathbf{u} + \nabla\mathbf{u}^T). \tag{3}$$

The vector-valued tensor divergence is

$$\operatorname{div}(\sigma) = \left(\sum_{j=1}^{3} \frac{\partial \sigma_{ij}}{\partial x_j}\right)_{i=1}^{3}. \tag{4}$$

This formulation assumes a constitutive relation corresponding to linear isotropic elasticity with stresses and strains related by Hooke's generalized law

$$\sigma_v = \mathbf{D}(\lambda, \mu)\epsilon_v, \tag{5}$$

where the constitutive matrix $\mathbf{D}(\lambda, \mu)$ relates the symmetric parts of ϵ and σ.

If the domain D is thin, then one of the dimensions is much smaller than the other two. In standard discretisation of the problem, for instance, with the finite element method, the small dimension, say the thickness, forces the sizes of the elements to be equally small and the simulations become expensive. This could be alleviated with alternative discretisation methods such as isogeometric analysis or high-order finite element method with carefully constructed meshes. Here, the approach taken is to modify the equations via dimension reduction [16].

2.2. Surface Definitions

In this work, the focus is solely on thin shells of revolution. They can formally be characterized as domains in $\mathbb{R}^3$ of type

$$D = \{\mathbf{x} + z\mathbf{n}(\mathbf{x}) \mid \mathbf{x} \in \Gamma, -d/2 < z < d/2\}, \tag{6}$$

where d is the (constant) thickness of the shell, Γ is a (mid)surface of revolution, and $\mathbf{n}(\mathbf{x})$ is the unit normal to Γ. The principal curvature coordinates require only four parameters, the radii of principal curvature R_1, R_2, and the so-called Lamé parameters, A_1, A_2, which relate coordinates changes to arc lengths, to specify the curvature and the metric on Γ. The displacement vector field of the midsurface $\mathbf{u} = \{u, v, w\}$ can be interpreted as projections to directions

$$\mathbf{e}_1 = \frac{1}{A_1}\frac{\partial \Psi}{\partial x_1}, \quad \mathbf{e}_2 = \frac{1}{A_2}\frac{\partial \Psi}{\partial x_2}, \quad \mathbf{e}_3 = \mathbf{e}_1 \times \mathbf{e}_2, \tag{7}$$

where $\Psi(x_1, x_2)$ is a suitable parametrisation of the surface of revolution, $\mathbf{e}_1, \mathbf{e}_2$ are the unit tangent vectors along the principal curvature lines, and $\mathbf{e}_3$ is the unit normal. In other words, $\mathbf{u} = u\,\mathbf{e}_1 + v\,\mathbf{e}_2 + w\,\mathbf{e}_3$.

Profile Functions and Parametrisation

When a plane curve is rotated (in three dimensions) around a line in the plane of the curve, it sweeps out a surface of revolution. Consider a plane curve, the so-called profile function in the xy-plane, $y = \gamma(x)$. Without any loss of generality, in the sequel the surfaces are generated by a curve rotating either around the x-axis or y-axis. This profile function is denoted with $f(x)$ and the resulting surface Γ_f for the case of the x-axis, and for the y-axis with $g(x)$ and the resulting surface Γ_g.

Let $I = [\alpha, \beta] \subset \mathbb{R}$ be a bounded closed interval, and let $f(x) : I \to \mathbb{R}^+$ be a regular function. The shell midsurface Γ_f is parameterised by means of the mapping

$$\begin{aligned} \mathbf{\Psi}_f &: I \times [0, 2\pi] \longrightarrow \mathbb{R}^3 \\ \mathbf{\Psi}_f(x_1, x_2) &= (x_1, f(x_1)\cos x_2, f(x_1)\sin x_2)\,. \end{aligned} \tag{8}$$

For Γ_f

$$A_1(x) = \sqrt{1 + [f'(x)]^2}, \quad A_2(x) = f(x), \tag{9}$$

and

$$R_1(x) = -\frac{A_1(x)^3}{f''(x)}, \quad R_2(x) = A_1(x)A_2(x). \tag{10}$$

Let $J = [\alpha, \beta] \subset \mathbb{R}$ be a bounded closed interval with $\alpha > 0$, and let $g(x) : J \to \mathbb{R}$ be a regular function. In this case, the shell midsurface Γ_g is parameterised by means of the mapping

$$\begin{aligned} \mathbf{\Psi}_g &: J \times [0, 2\pi] \longrightarrow \mathbb{R}^3 \\ \mathbf{\Psi}_g(x_1, x_2) &= (x_1\cos x_2, g(x_1), x_1\sin x_2)\,. \end{aligned} \tag{11}$$

For Γ_g

$$A_1(x) = \sqrt{1 + [g'(x)]^2}, \quad A_2(x) = x, \tag{12}$$

and

$$R_1(x) = \frac{A_1(x)^3}{g''(x)}, \quad R_2(x) = A_1(x)A_2(x). \tag{13}$$

2.3. Perforations

Perforated domains are characterized by the penetration patterns, which in turn depend on the underlying manufacturing processes and the related hole coverage, typically given as a percentage.

The quantity used to characterize perforated sheets of metal is the ligament efficiency η. Let us assume that the holes are ellipses with a, b as the horizontal and perpendicular semiaxis, and the separation of the centers used in the definitions is P_x and P_y, respectively. Following [17–19], one can define both the horizontal and perpendicular ligament efficiency, denoting them as η_x and η_y, respectively. For regular arrays of holes,

$$\eta_x = (P_x - 2a)/P_x, \qquad \eta_y = (P_y - 2b)/P_y, \tag{14}$$

and for triangular arrays, allowing for alternating layers,

$$\eta_x = (P_x - 4a)/P_x, \qquad \eta_y = (P_y - 4b)/P_y. \tag{15}$$

For circular holes, the radius $r = a = b$, of course, and further if the pattern is regular $\eta = \eta_x = \eta_y$. Both pattern types are illustrated in Figure 2. Notice that the triangular pattern in the figure has a tighter packing than that implied by (15).

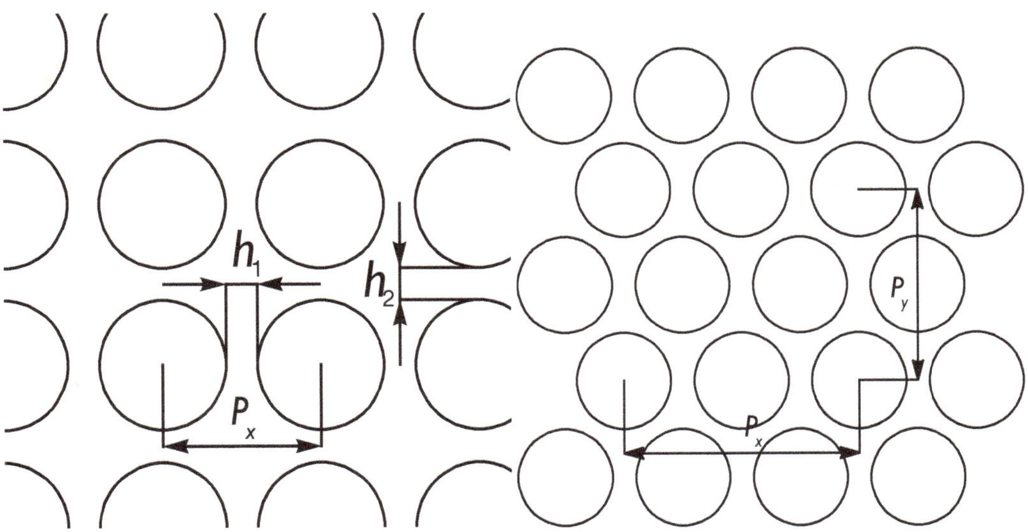

(a) (b)

Figure 2. Penetration patterns. (**a**) Regular pattern. (**b**) Triangular pattern.

2.4. Finite Element Method

All numerical simulations reported here have been computed with two different high-order continuous Galerkin codes in 2D solving the variational formulation on conforming meshes of quadrilateral or triangular elements.

One of the challenges in shell problems is to avoid numerical locking. Instead of using special shell elements [1], one can let the higher-order FEM alleviate the locking and accept that some thickness dependent error amplification or locking factor, $K(t) \geq 1$, is unavoidable. For the hp-FEM solution, one can derive a simple error formulation

$$\text{error} \sim K(t)(h/L_D)^p, \tag{16}$$

where h is the mesh spacing, L_D is the diameter of the domain, and p is the degree of the elements. It is possible that $K(t)$ diverges as t tends to zero, with the worst case being for pure bending problems: $K(t) \sim 1/t$. In (16), the latter part follows from standard approximation theory of the FEM. It is the term $K(t)$ that is shell specific. If the bending part in the energy expressions given below dominates, the energy norm depends linearly on t and hence any energy error estimate has an inverse dependence, that is, $\sim 1/t$. This simple error formula suggests why higher-order methods are advantageous in shell problems: the mesh over-refinement in the "worst" case is $\sim (1/t)^{1/p}$, which for a fixed $t = 1/100$, say, indicates that for $p = 4$ the requirement is moderate in comparison to the case of $p = 1$. This also suggests that convergence in p is a useful measure if the problem is otherwise difficult to analyze exactly. For a more detailed discussion on this and further references, see [20,21].

Implementations

The first solver used in this study is implemented with Mathematica, providing exact geometry handling of the holes via blending functions [22]. The second one is AptoFEM, a parallel code implemented in FORTRAN90 and MPI. Both codes allow for arbitrary order of polynomials to be used in the elements including different orders of polynomials in individual elements in the same mesh [14].

3. Shell Models

If the shell of revolution is defined by the profile function $f(x)$ defined over some interval $x \in I = [x_0, x_1]$, then using the derivatives of the profile function all shell geometries can be classified in terms of Gaussian curvature (see, for instance, [23]). The analysis of shell problems is greatly simplified if the type of curvature is uniform.

1 Parabolic (Zero Gaussian curvature shells). $f''(x) = 0,\ \forall\, x \in I$.
2 Elliptic (Positive Gaussian curvature shells). $f''(x) < 0,\ \forall\, x \in I$.
3 Hyperbolic (Negative Gaussian curvature shells). $f''(x) > 0,\ \forall\, x \in I$.

Dimensionally Reduced Elasticity Equations: Naghdi Model

Consider a shell of (constant) thickness d, the mid-surface ω of which occupies a region to of some smooth surface Γ. This is a three-dimensional body equipped with principal curvature coordinates as defined in Section 2.2 for which the 3D theory of linear elasticity could be considered "exact" for small deformations. One of the classical dimension reduction models is applied here, since these models reveal the nature of shell deformations more explicitly. Such models are often reasonably accurate for thin shells [3]. The displacement field u has five components, u, v, w, θ, and ψ, each of which is a function of two variables on the mid-surface of the shell. The first two components represent the tangential displacements of the mid-surface, w is the transverse deflection, and θ and ψ are dimensionless rotations. The model is similar to the Reissner–Mindlin model for plate bending and is sometimes named after Naghdi. Assume that the shell consists of homogeneous isotropic material with Young modulus E and Poisson ratio ν. Then, the total energy of the shell in our dimension reduction model is expressed as

$$\mathcal{F}(\mathbf{u}) = \frac{1}{2} S(d\, a(\mathbf{u},\mathbf{u}) + d^3\, b(\mathbf{u},\mathbf{u})) - q(\mathbf{u}), \tag{17}$$

where $S = E/(12(1-\nu^2))$ is a scaling factor, q is the external load potential, and $a(\mathbf{u},\mathbf{u})$ and $b(\mathbf{u},\mathbf{u})$ represent the portions of total deformation energy that are stored in membrane and transverse shear deformations and bending deformations, respectively. The latter are quadratic forms independent of d and defined as

$$\begin{aligned}
a(\mathbf{u},\mathbf{u}) &= a_m(\mathbf{u},\mathbf{u}) + a_s(\mathbf{u},\mathbf{u}) \\
&= 12\int_\omega \left[\nu(\beta_{11}(\mathbf{u}) + \beta_{22}(\mathbf{u}))^2 + (1-\nu)\sum_{i,j=1}^{2} \beta_{ij}(\mathbf{u})^2\right] A_1A_2\, d\gamma + \\
&\quad 6(1-\nu)\int_\omega \left[(\rho_1(\mathbf{u})^2 + \rho_2(\mathbf{u}))^2\right] A_1A_2\, d\gamma, \qquad (18)\\
b(\mathbf{u},\mathbf{u}) &= \int_\omega \left[\nu(\kappa_{11}(\mathbf{u}) + \kappa_{22}(\mathbf{u}))^2 + (1-\nu)\sum_{i,j=1}^{2} \kappa_{ij}(\mathbf{u})^2\right] A_1A_2\, d\gamma, \qquad (19)
\end{aligned}$$

where β_{ij}, ρ_i, and κ_{ij} stand for the membrane, transverse shear, and bending strains, respectively. The strain-displacement relations are linear and involve at most first derivatives of the displacement components.

Remark 1. *In the following, we shall omit the constant factor $d\,S$ from the energy expressions. Consequently, all results can be considered to be scaled with a factor $(d\,S)^{-1}$.*

Following [16], the bending strains κ_{ij} are

$$\begin{aligned}
\kappa_{11} &= \frac{1}{A_1}\frac{\partial\theta}{\partial x} + \frac{\psi}{A_1A_2}\frac{\partial A_1}{\partial y}, \\
\kappa_{22} &= \frac{1}{A_2}\frac{\partial\psi}{\partial y} + \frac{\theta}{A_1A_2}\frac{\partial A_2}{\partial x}, \qquad (20)\\
\kappa_{12} &= \kappa_{21} = \frac{1}{2}\left[\frac{1}{A_1}\frac{\partial\psi}{\partial x} + \frac{1}{A_2}\frac{\partial\theta}{\partial y} - \frac{\theta}{A_1A_2}\frac{\partial A_1}{\partial y} - \frac{\psi}{A_1A_2}\frac{\partial A_2}{\partial x}\right. \\
&\quad -\frac{1}{R_1}\left(\frac{1}{A_2}\frac{\partial u}{\partial y} - \frac{v}{A_1A_2}\frac{\partial A_2}{\partial x}\right) \\
&\quad \left. -\frac{1}{R_2}\left(\frac{1}{A_1}\frac{\partial v}{\partial x} - \frac{u}{A_1A_2}\frac{\partial A_1}{\partial y}\right)\right],
\end{aligned}$$

similarly the membrane strains β_{ij}

$$\begin{aligned}
\beta_{11} &= \frac{1}{A_1}\frac{\partial u}{\partial x} + \frac{v}{A_1A_2}\frac{\partial A_1}{\partial y} + \frac{w}{R_1}, \\
\beta_{22} &= \frac{1}{A_2}\frac{\partial v}{\partial y} + \frac{u}{A_1A_2}\frac{\partial A_2}{\partial x} + \frac{w}{R_2}, \qquad (21)\\
\beta_{12} &= \beta_{21} = \frac{1}{2}\left(\frac{1}{A_1}\frac{\partial v}{\partial x} + \frac{1}{A_2}\frac{\partial u}{\partial y} - \frac{u}{A_1A_2}\frac{\partial A_1}{\partial y} - \frac{v}{A_1A_2}\frac{\partial A_2}{\partial x}\right),
\end{aligned}$$

and finally the shear strains ρ_i

$$\begin{aligned}
\rho_1 &= \frac{1}{A_1}\frac{\partial w}{\partial x} - \frac{u}{R_1} - \theta, \qquad (22)\\
\rho_2 &= \frac{1}{A_2}\frac{\partial w}{\partial y} - \frac{v}{R_2} - \psi.
\end{aligned}$$

Remark 2. *When the shell parametrisations defined above are used, all terms of the form $\partial A_i/\partial y$ are identically zero.*

The energy norm $|||\cdot|||$ is defined in a natural way in terms of the deformation energy and taking the scaling into account:

$$\mathcal{E}(\mathbf{u}) := |||\mathbf{u}|||^2 = a(\mathbf{u},\mathbf{u}) + d^2\, b(\mathbf{u},\mathbf{u}). \qquad (23)$$

Similarly for bending, membrane, and shear energies:

$$\mathbf{B}(\mathbf{u}) := d^2\, b(\mathbf{u},\mathbf{u}), \quad \mathbf{M}(\mathbf{u}) := a_m(\mathbf{u},\mathbf{u}), \quad \mathbf{S}(\mathbf{u}) := a_s(\mathbf{u},\mathbf{u}). \tag{24}$$

The load potential has the form $q(\mathbf{v}) = \int_\omega \mathbf{f}(x,y) \cdot \mathbf{v}\, A_1 A_2\; dx\, dy$. If the load acts in the transverse direction of the shell surface, i.e., $\mathbf{f}(x,y) = [0,0,f_w(x,y),0,0]^T$, and $\mathbf{f} \in [L^2(\omega)]^5$ holds, then the variational problem has a unique weak solution $\mathbf{u} \in [H^1(\omega)]^5$. The corresponding result is true in the finite dimensional case, when the finite element method is employed.

In the following discussion, both free vibration and buckling problems will be briefly covered. To this effect, the mass matrix is defined as $\mathbf{M}(t) = t\, M^l + t^3\, M^r$, with M^l (displacements) and M^r (rotations) *independent of t*.

The geometric stiffness matrix used in buckling analysis in the dimensionally reduced case is still without a universally agreed definition (See Appendix B). Here, in the cylindrical case, the geometric stiffness matrix $\mathbf{U}_g$ is taken to be the inner product of the axial derivative of the transverse deflection with itself as suggested in [11]. Formally, $\mathbf{U}_g(\mathbf{v}) = \int_\omega (\partial w/\partial x)^2\, A_1 A_2\; dx\, dy$.

4. Boundary and Internal Layers

The theory of one-dimensional hp-approximation of boundary layers is due to Schwab [24]. Boundary layer functions are of the form $u(x) = \exp(-a\,x/\delta), \quad 0 < x < L$, where $\delta \in (0,1]$ is a small parameter, $a > 0$ is a constant, and L is the characteristic length scale of the problem under consideration. Even though in certain classes of problems it is possible to choose a robust strategy leading to uniform convergence in δ, the distribution of the mesh nodes *depends* on p, and over a range of polynomial degrees $p = 2,\ldots,8$, say, the mesh is different for every p. In 2D, this requires one to allow for the mesh topology to change over the range of polynomial degrees. In this study, the short-range layers have been addressed in the meshes, but optimality in p has not been attempted.

It is useful to define the central concepts in a problem-independent manner.

Definition 1 (Layer Element Width)**.** *For every boundary layer in the problem, one should have an element of width $O(p\,\delta)$ in the direction of the decay of the layer.*

Note, that with c constant, if $c\,p\,\delta \to L$ as p increases, the standard p-method can be interpreted as the limiting method. Boundary layers can also occur within the domains, i.e., be internal layers, or emanate from a point. For our discussion, it is useful to define the concept of boundary layer generators (see [3]).

Definition 2 (Layer Generator)**.** *The subset of the domain from which the boundary layer decays exponentially, is called the layer generator. Formally, the layer generator is of measure zero.*

The layer generators are independent of the length scale of the problem under consideration.

The types of layers that a shell structure can exhibit depend on its geometry, that is, on local curvature. Elliptic, parabolic, and hyperbolic structures each possess a distinctive set of layer deformations. The layer structure is classically assumed to be an exponential solution to the homogeneous Euler equations of the shell problem. In [3], it is shown using the Ansatz $\mathbf{u}(\xi,\eta) = \mathbf{U}e^{\lambda\xi}e^{ik\eta}$ that solutions with $\mathrm{Re}\lambda < 0$ such that the characteristic lengths $L = 1/\mathrm{Re}\lambda \to 0$ are of the form $L \sim t^{1/n}$ where $n \in \{1,2,3,4\}$. Here, ξ is the coordinate orthogonal to the layer generator. From these, the layer with $n = 2$ is present in all geometries, whereas layers with $n = 3$ and $n = 4$ are present only in hyperbolic and parabolic geometries, respectively. The case $n = 1$ arises from a shear deformation and shows up only when a model similar to the model of Reissner and Naghdi is used in analysis.

If the curved generators are included [5], more characteristic lengths can be found. In particular, for elliptic shells with a parabolic curved layer generator, any $n \geq 2$ can be induced. Consider a shell structure generated by a rotation around the y-axis of the profile $g(x) = \alpha(x - x_0)^m, \quad x_0 \leq x \leq x_1$ so that at $x = x_0$ the geometry parameters will vanish, otherwise we have an elliptic shell. The solution to the shell problem under unit pressure is a layer deformation in the scale $L = t^{1/(m+1)}$.

Finally, the long-range layers have the characteristic lengths $L \sim t^{-1}$ and $L \sim t^{-1/2}$, where the first one is the axial torsion boundary layer, and the second is, for instance, induced by kinematic constraints in part of a long cylinder such a T-junction. These layers are referred to as long-range Fourier modes in Section 3.2 of [3]. The layer chart for a long cylinder is given in Figure 3.

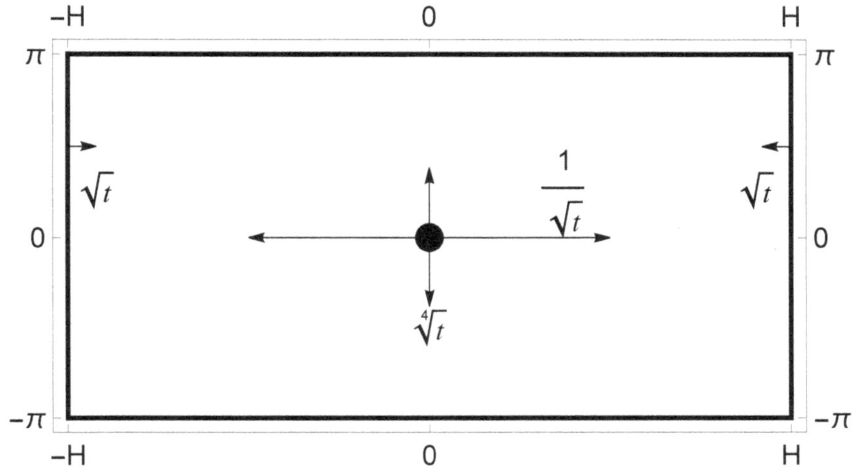

Figure 3. Layer structure: Parabolic long cylinder. The expected length scales are indicated with the arrows. The generators are the boundaries at $x = \pm H$, and the hole in the center. The location of the center does not play a role. If the periodic boundary at $y = \pm\pi$ is free, under torsion there will be a very long layer $\sim 1/t$. The hole also generates a short range axial layer, which is not indicated.

5. Numerical Simulations

The three simulation sets considered here have been tabulated in Table 1. They have been selected to illustrate each of the three main types of long-range layers including perforated variants. The relative effects of the perforations on the fundamental natural frequency are given as ratios of the perforated and reference frequencies. For every realization of the shell geometry, five logarithmically equidistant thicknesses $\in [1/1000, 1/100]$ have been used. All parabolic cases have been computed using both Naghdi and shallow shell models and as expected, the differences are negligible. The hyperbolic and elliptic cases have been simulated with the Naghdi model only. In all cases where short-range layers have been present at the clamped boundaries, the meshes have been adapted a priori to the corresponding length scales. In all structures with perforation patterns, the holes are free; that is, there are no kinematical constraints.

Table 1. The set of simulations with key parameters. The profile functions for the parabolic and hyperbolic cases are $f(x) = 1$ and $f(x) = 1 + (1/2)(x/H)^2$, respectively. For every individual simulation, five logarithmically equidistant thicknesses $\in [1/1000, 1/100]$ have been used. H is the half-width, p the uniform polynomial order, and N is the total number of degrees of freedom. In the non-perforated Slit Shell simulations, the meshes are topologically equivalent. In the perforated cases, the hole coverage percentage is 25%, the triangular pattern is 200 × 10 resulting in 3791 holes, and the regular pattern is 1000 × 10 resulting in 10,000 holes.

Case	Geometry	Perforation	H	p	N
Wind Turbine: Manhole	Parabolic		60	8	197,440
	Parabolic		1000	6	2,127,240
	Hyperbolic		1000	6	2,127,240
Slit Shell: Torsion Effect	Parabolic		100	5	1,907,980
	Parabolic	Triangular	100	5	2,841,675
	Hyperbolic		100	5	1,907,980
	Parabolic		1000	5	1,907,980
	Parabolic	Regular	1000	5	7,126,755
	Hyperbolic		1000	5	1,907,980
Curvature Effect	Mixed	Multipanel		6	490,145

The selected polynomial orders can be justified by considering the energy convergence in p. In Figure 4, two examples of convergence graphs are shown.

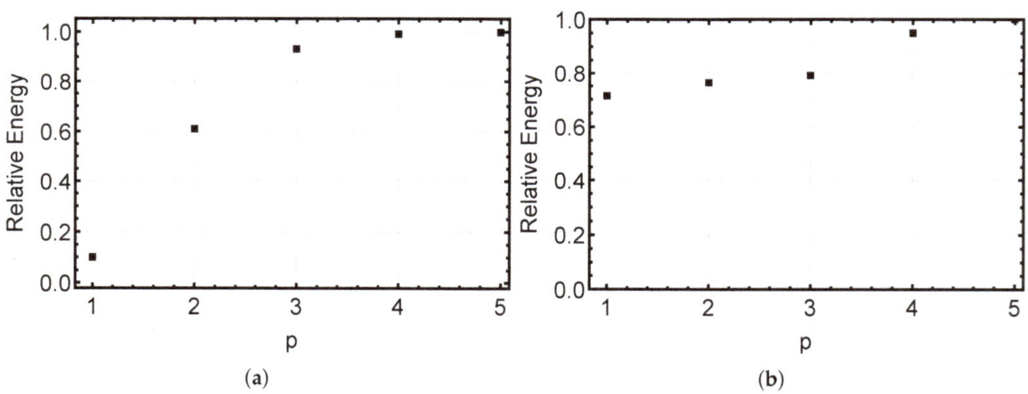

Figure 4. Numerical Locking: p-convergence in energy. Total energy at $p = 5$ is taken as the reference (**a**) Circular plate with an elliptic extension under unit pressure on the extension. (**b**) Long slit parabolic cylinder with a regular perforation pattern under torsion loading. In both cases starting at $p = 4$, the relative error in energy is sufficiently small, justifying the selected polynomial orders.

5.1. Wind Turbine: Manhole

The first simulation concerns the long-range layer on long or tall structures with a kinematically constrained section within the domain. The manhole title is inspired by an example in [8], (Figure 1). The authors discuss the effects of various stiffeners for the manhole of a wind tower. In their FEM analysis figure, the long-range effect is clearly visible but not shown at full length. The construction reported has a dimensional thickness of 3/100, so it falls within the realm of thin structures.

In Figures 1a and 5, the overall solutions are shown for parabolic and hyperbolic shells, respectively. The overall features predicted in the layer chart of Figure 3 are clearly visible in both cases. The effect of the length H of the structure is shown in Figure 6. The long-range extends further in the longer cylinder as expected, since $H = 10 \sim 1/\sqrt{t}$, whereas $H = 100 \gg 1/\sqrt{t}$.

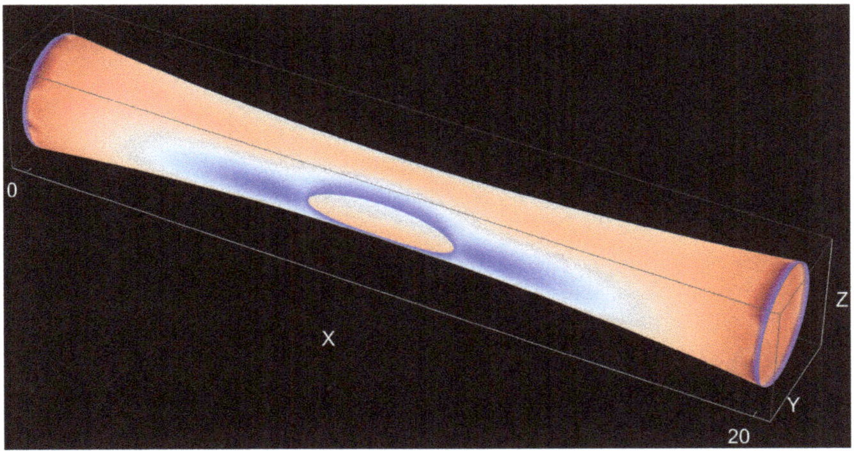

Figure 5. Long hyperbolic shell with an ellipsoidal hole. Transverse deflection profile when the dimensionless thickness $t = 1/100$, all boundaries are kinematically fully constrained, $u = v = w = \theta = \psi = 0$, and the loading is the unit pressure.

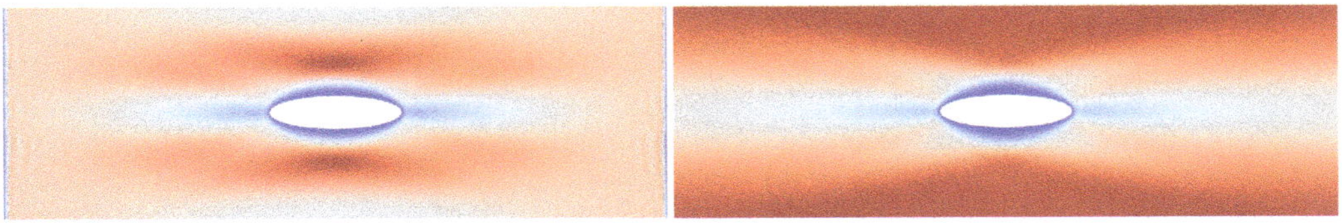

(**a**) (**b**)

Figure 6. Two long cylinders: Transverse deflection profiles. In both cases, the dimensionless thickness $t = 1/100$, all boundaries are kinematically fully constrained, $u = v = w = \theta = \psi = 0$, and the loading is the unit pressure. (**a**) $H = 10$. (**b**) $H = 100$, with the centre section shown.

The asymptotic behavior of the eigenmodes in the free vibration of shells of revolution is known [25]. It is of interest to monitor the effect of the hole to the lowest eigenvalue. Transverse profiles are shown in Figure 7. Since the hyperbolic profile is only mildly hyperbolic and the interval of thicknesses is kept realistic, the profiles appear very similar with the oscillations only slightly more concentrated in the center in the hyperbolic case. Assuming the same material parameters, the eigenvalue amplification due to the hole is smaller in the hyperbolic case. In both cases, the amplification decreases as $t \to 0$. This is due to angular oscillations (wave numbers) increasing as $t \to 0$ and hence the interaction of the hole and the layers becomes weaker.

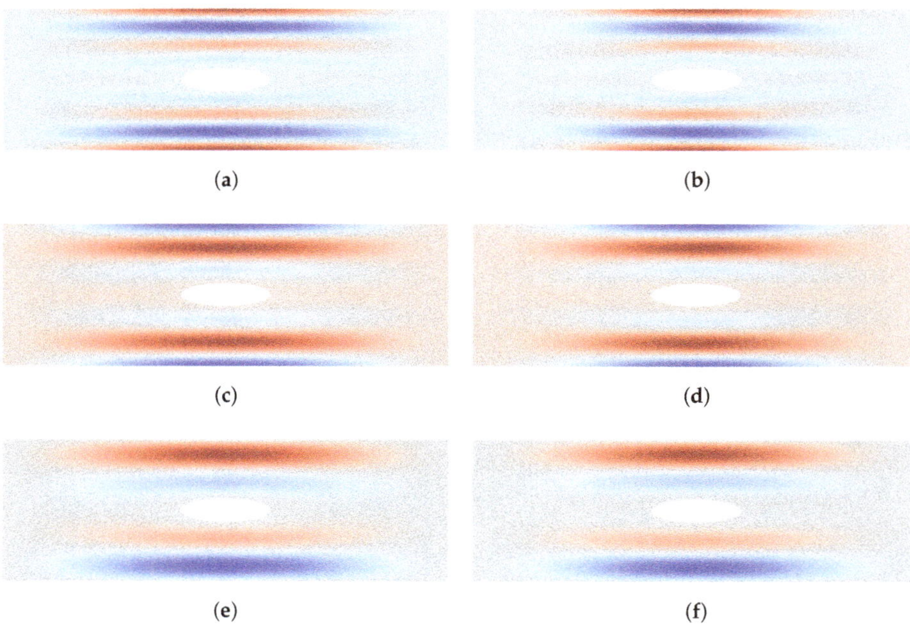

Figure 7. Free vibration: Transverse deflection profiles of first eigenmodes. In both cases, $H = 30$ and all boundaries are kinematically fully constrained, $u = v = w = \theta = \psi = 0$. From top: $t \in \{1/1000, 1/100\sqrt{10}, 1/100\}$. Ratio of the observed eigenvalue over the reference eigenvalues: (**a**,**c**,**e**) (Parabolic) – $\{1.1, 1.3, 1.8\}$. (**b**,**d**,**f**) (Hyperbolic) – $\{1.1, 1.3, 1.7\}$.

Of course, for cylindrical or parabolic shells, there is also the relative long layer of $\sqrt[4]{t}$ in the angular dimension. In Figure 8, this effect is shown by visual comparison of two profiles corresponding to different thicknesses. Since the eigenmodes oscillate in the angular direction as seen in Figure 7, the effect of the hole is very small in the eigenvalues, with the increase in the ratio less than 1%.

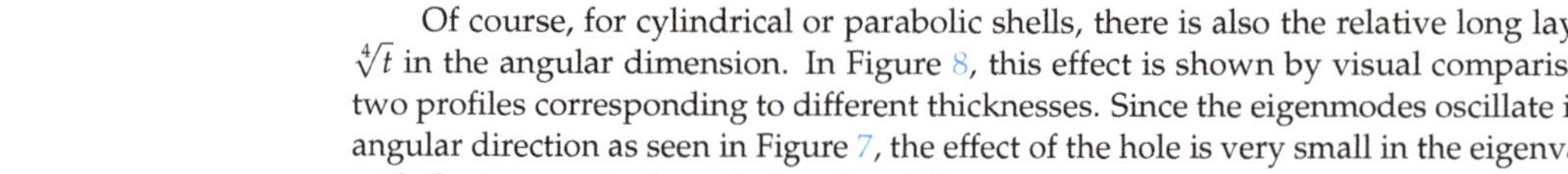

Figure 8. Manhole: Transverse deflection profiles. In both cases, $H = 30$, boundaries are kinematically fully constrained, $u = v = w = \theta = \psi = 0$, and the loading is the unit pressure. (**a**) $t = 1/100$, $w \in [-0.10, 0.24]$. (**b**) $t = 1/1000$, $w \in [-0.56, 0.38]$.

Remark 3. *For hyperbolic shells, the $\sqrt[4]{t}$ layer does not exist. Instead, there is a $\sqrt[3]{t}$ layer along the characteristics of the surface. In Figure 5, the shell is nearly parabolic in the vicinity of the hole and therefore this feature is not visible.*

Confirming the correct length scales is difficult since the long-range effects do not occur in isolation, but in all cases the deflection profile is a linear combination of different characteristic features. Using interpolated representations, the long-range effect is associated with the inflection point of the curve. The two thinnest cases are used to estimate the constant of the layer, and this value is used to predict the corresponding locations in the other cases.

In Figure 9a, another view to the angular layers is given. In addition, in Figure 9b, the agreement with the predicted length scale is illustrated. However, it appears that the clamped end at $x = H$ already affects the overall profile. The estimated constant is $c = 0.85$ leading to model $L \sim c(1/\sqrt{t})$. In Figure 10, two long configurations with different geometries have been considered. The displacement graphs over a set of thicknesses show the stronger short-range effects. As can be seen, in the longer cases, $H = 1000$, the agreement is very good indeed. Notice that the good agreement is on the axial rotation component θ, which means that domain expertise has been necessary to find the right component.

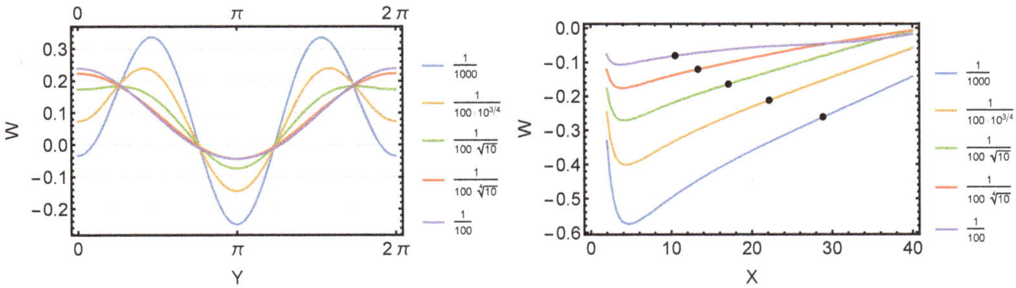

(a) (b)

Figure 9. Manhole: Transverse deflection profiles and predicted characteristic length scales. Case $H = 30$ with ellipsoidal hole at $(1 - H, \pi)$: (**a**) Profiles at $x = 0$. (**b**) Observed characteristic length scales.

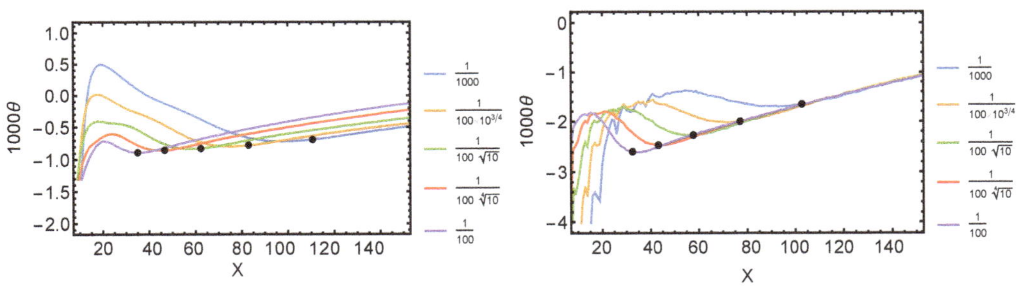

(a) (b)

Figure 10. Long-range layers: Predicted characteristic length scales. In all cases, the inflection points have been computed for the two thinnest cases and the constant has been set with these points. The constants are from left to right: $c_1 = 3.5, c_2 = 3.25$, leading to models $L \sim c_i(1/\sqrt{t})$, $i = 1, 2$. The rest of the points have been selected based on the theoretical prediction. The agreement is surprisingly good. (**a**) Parabolic case, center hole, $H = 1000$, θ-component. (**b**) Hyperbolic case, center hole, $H = 1000$, θ-component.

5.2. Slit Shells: Torsion Effect

The torsion layer $\sim 1/t$ is naturally the most difficult to recover from simulation data. The effect on a slit cylinder is illustrated in Figure 11. The boundary at $x = -H$ is clamped, $u = v = w = \theta = \psi = 0$, but all other boundaries are free. The loading is a unit torsion load acting on the boundary at $x = H$. The technique used in the previous case was not successful here. Indeed, the exact layer could not be found from the data. However, the effect of H can be deduced indirectly by observing the rate of change of curvature of the displacement profile over a set of thicknesses after the loading is scaled by t^3. In Figure 12a, for $H = 100$ the slopes for both geometries, including a parabolic perforated case, the trends are constant. In Figure 12b, for $H = 1000$ the observed trends are somewhat polluted, but the important aspect for this discussion is that as the shell becomes longer, and the torsion effect becomes milder indicating the presence of a long-range layer. Since this phenomenon also exists at $H = 1000 \gg 1/t$ for $t = 1/100$, it is likely that it is due to the corresponding predicted layer.

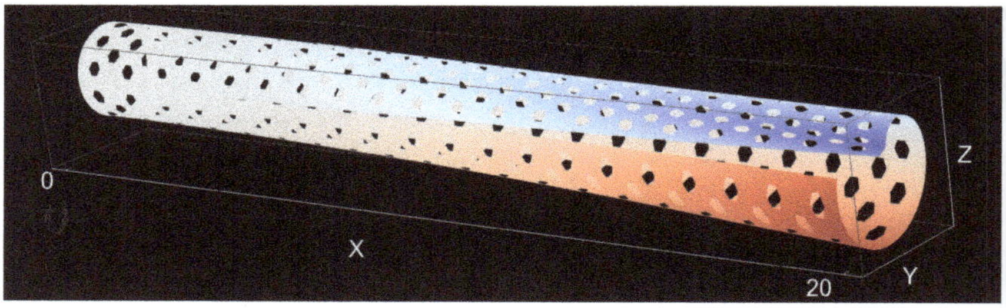

Figure 11. Slit cylinder: Transverse deflection profile. Perforation pattern is triangular with hole coverage of 25%.

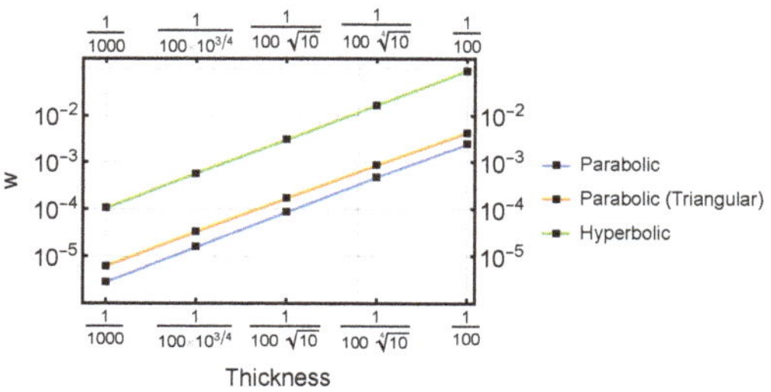

(a)

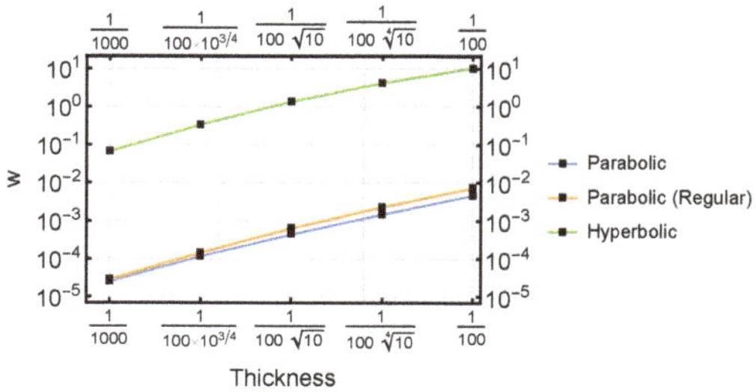

(b)

Figure 12. Slit shell of revolution: Effect of the length of the shell on torsion loading acting on the free end. Parabolic and hyperbolic cases, with a perforated parabolic one as the third option. The torsion layer leads to smaller effect close to the fixed boundary. Shown is the observed transverse deflection at $x = 20$ along the midline of the surface. (**a**) $H = 100$. The slope of the loglog-graphs is =3. (**b**) $H = 1000$. The slope of the loglog-graphs is =2.

Interestingly, the first eigenmode corresponds to the kind of rotation caused by the chosen loading in the static case, and therefore the figure has been omitted. Now, the perforated case is softer (the holes are free), and hence the ratios are less than one, and decreasing as $t \to 0$. For $t = 1/100$, the ratio is 0.64, whereas for $t = 1/1000$, it is 0.42.

5.3. Curvature Effect

The third and final simulation adds two aspects: First, the layer generator is curved, and second, the shell geometry is not of uniform type. The shell of revolution is formed by letting a profile function $g(x) = 1$, $x \in [1, 1+\pi]$, and $= 1 + (x-\pi)^\alpha$, $x \in [\pi, 2\pi]$. In other words, the inner section is a plate and the outer section is an elliptic extension with curvature dependent on α. One realization is shown in Figure 1b. The parameter α determines the length of the layer. Two sets of simulations with $\alpha = 2$ or $= 3$ are computed on the multipanel mesh of Figure 13a. The inner holes are free and therefore as α increases and $t \to 0$ for a fixed loading, the displacement amplitudes increase due to the sensitivity of the problem. However, for a given range of thicknesses, the response of the structure is reasonable (see Figure 13b).

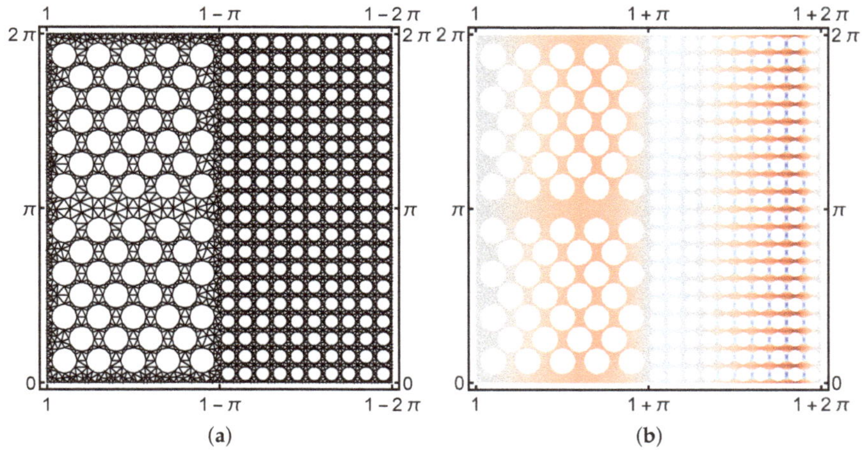

Figure 13. Circular plate with an elliptic extension. Pressure load is acting on the right-hand-side half of the domain. The boundaries at $x = 1$ and $x = 1 + 2\pi$ are clamped, $u = v = w = \theta = \psi = 0$, the $y = 0$ and $y = 2\pi$ boundaries are periodic. The inner circular hole has a radius = 1. (**a**) Multipanel mesh. (**b**) Quadratic extension, w-component (transverse deflection), $t = 1/100$.

For small values of α the recovery of the length scales is successful (see Figure 14). In the previous study on curved generators, the plate was not present, and the structure was not perforated. This clearly indicates that these effects are features of the elasticity equations irrespective of the effective material properties.

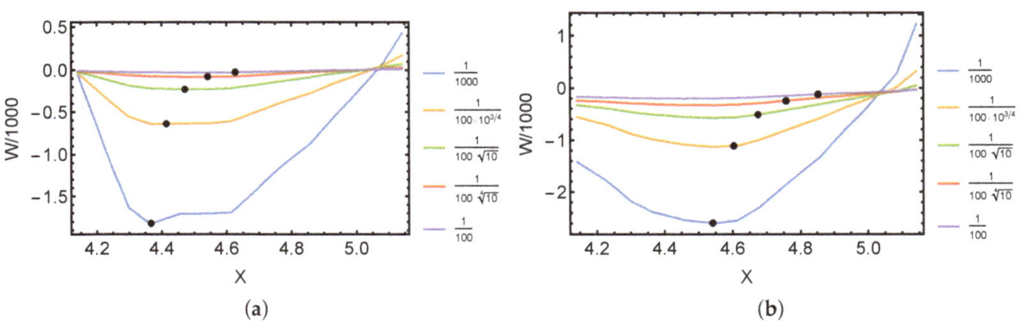

Figure 14. Circular plate with an elliptic extension. Profiles along $y = \pi$. As above, in all cases the inflection points have been computed for the two thinnest cases and the constant has been set with these points. The constant is the same in both cases: $c = 2.25$, leading to model $L \sim c(1/t^{\alpha+1})$. The rest of the points have been selected based on the theoretical prediction. Again, the agreement is very good indeed. (**a**) Quadratic extension, w-component (transverse deflection). (**b**) Cubic extension, w-component (transverse deflection).

The eigenanalysis becomes more involved in this case. As can be seen in Figure 15, the (relative) transverse deflection profiles are almost dramatically different. This is due to the free holes in the elliptic part exhibiting sensitivity [6]. The eigenmode includes strong oscillations in the elliptic part, a phenomenon that does not have any corresponding effect in the reference case where the oscillation is confined to the plate section. It is due to sensitivity that the eigenvalues ratios increase as $t \to 0$. For $t = 1/100$, the ratio is 0.26, whereas for $t = 1/1000$, it is 0.48.

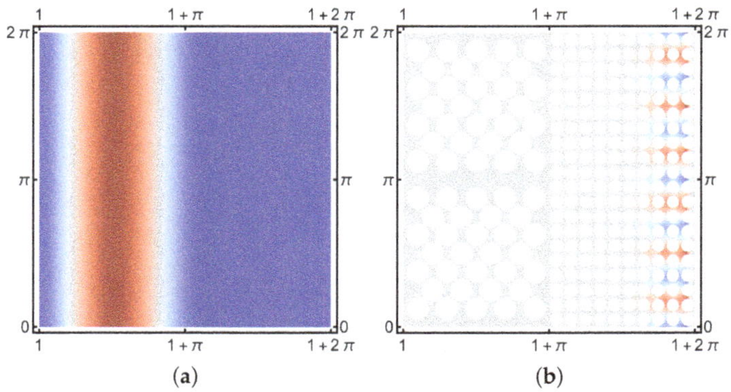

Figure 15. Circular plate with an elliptic extension. First eigenmodes at $t = 1/100$. The boundaries at $x = 1$ and $x = 1 + 2\pi$ are clamped, the $y = 0$ and $y = 2\pi$ boundaries are periodic. The inner circular hole has a radius = 1. (**a**) Reference domain. (**b**) Perforated domain, w-component (transverse deflection).

6. Conclusions

Shell structures have a rich family of boundary layers including internal layers. Each layer has its own characteristic length scale, which depends on the thickness of the shell. The long-range layers do exist, and in certain problem classes play an important role. Interestingly, since they are not well-known, it is possible that in some problems new modeling ideas might be brought forward if only they were recognized in the right contexts.

The simulation of such structures is subject to numerical locking, and high-order finite element methods provide one way to derive reliable solutions. The observed asymptotic behavior is consistent with the theoretical predictions. These layers are shown to also appear on perforated structures underlying the fact these features are properties of the elasticity equations and not dependent on the effective materials.

Funding: This research received no external funding.

Data Availability Statement: Not applicable.

Acknowledgments: I would like to thank the anonymous referees for suggestions that improved the paper considerably.

Conflicts of Interest: The author declares no conflict of interest.

Appendix A. Mathematical Shell Model

In the following, the Naghdi model is simplified by assuming that ω is a domain expressed in the coordinates x and y. The curvature tensor $\{b_{ij}\}$ of the midsurface is assumed to be constant and $a = b_{11}, b = b_{22}$, and $c = b_{12} = b_{21}$. The shell is then called elliptic when $a\,b - c^2 > 0$, parabolic when $a\,b - c^2 = 0$, and hyperbolic when $a\,b - c^2 < 0$. The above assumptions are valid when the shell is shallow, i.e., the midsurface differs only slightly from a plane. In the simplest case, one may set $d\omega = dxdy$ and write the relation between the strain and the displacement fields as

$$\begin{aligned}
&\beta_{11} = \frac{\partial u}{\partial x} + aw, \quad \beta_{22} = \frac{\partial v}{\partial y} + bw, \quad \beta_{12} = \frac{1}{2}\left(\frac{\partial u}{\partial y} + \frac{\partial v}{\partial x}\right) + cw, \\
&\rho_1 = \theta - \frac{\partial w}{\partial x}, \quad \rho_2 = \psi - \frac{\partial w}{\partial y}, \\
&\kappa_{11} = \frac{\partial \theta}{\partial x}, \quad \kappa_{22} = \frac{\partial \psi}{\partial y}, \quad \kappa_{12} = \frac{1}{2}\left(\frac{\partial \theta}{\partial y} + \frac{\partial \psi}{\partial x}\right).
\end{aligned} \tag{A1}$$

This choice of shell model gives us additional flexibility in the design of the numerical simulations since the model admits *non-realizable* shell geometries. This is due to the assumption that the local curvatures are constant at every point of the surface.

Remarkably, for parabolic shells these strains differ from those of the standard Naghdi model only in κ_{12} and ρ_1, when the radius is $= 1$. Naturally, for non-parabolic geometries the differences are much more extensive. Notice that the resulting system has constant coefficients, which simplifies the implementation of the model significantly.

Appendix B. On Buckling Modes

The critical load of the real shell is known to be very sensitive to small geometric imperfections and deviations in boundary conditions, which are difficult to take into account in linear or nonlinear stability theory. As a result, theoretical and experimental results do not agree well in many loading scenarios. In any case, the linear stability theory provides useful information regarding the buckling behavior of thin shells [11].

The first buckling modes are shown in Figure A1. Interestingly, even in the reference case, the lowest mode can exhibit axial oscillations. In the profile of the thinnest configuration with a hole, symmetry appears to be lost. This is due to the extreme ill-conditioning of the problem. Also in contrast with the free vibration, here the eigenvalue ratio between the perforated and reference configurations does not change as $t \to 0$ (see Figure A2). In both cases, the dependence is linear, which is in fact a new result. It should be noted that in simulations with the given buckling model, the observed spectrum is clustered for every fixed thickness. For instance, the relative difference within the first ten modes is less than 1% in the case with a hole, and a fraction higher in the reference case. In fact, it has been proposed that the Shapiro–Lopatinsky conditions are not satisfied in the limit and the coercivity is lost, just as for the sensitive elliptic shells [6].

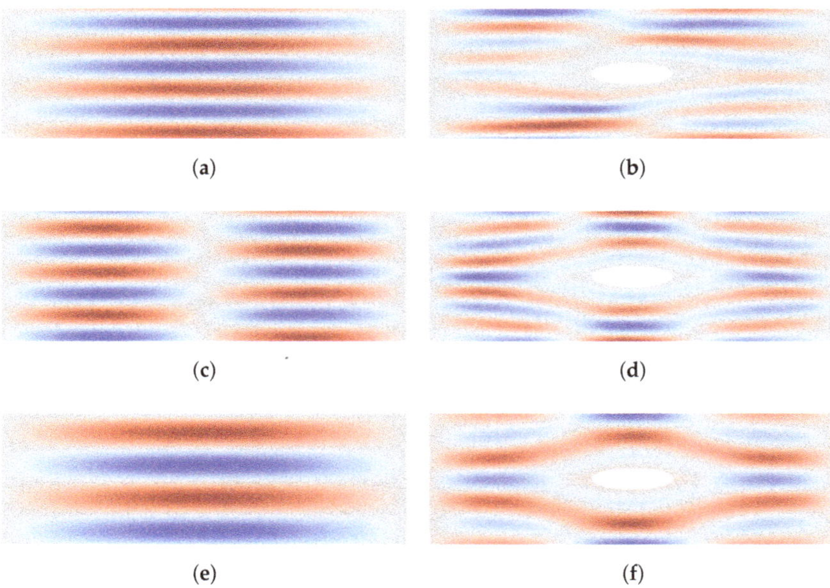

Figure A1. Buckling modes: Transverse deflection profiles of first modes. In both cases, $H = 30$, the hole is kinematically fully constrained, $u = v = w = \theta = \psi = 0$, and the ends have $v = w = \psi = 0$. From top: $t \in \{1/1000, 1/100\sqrt{10}, 1/100\}$. (**a**,**c**,**e**) Parabolic reference. (**b**,**d**,**f**) Parabolic with a hole.

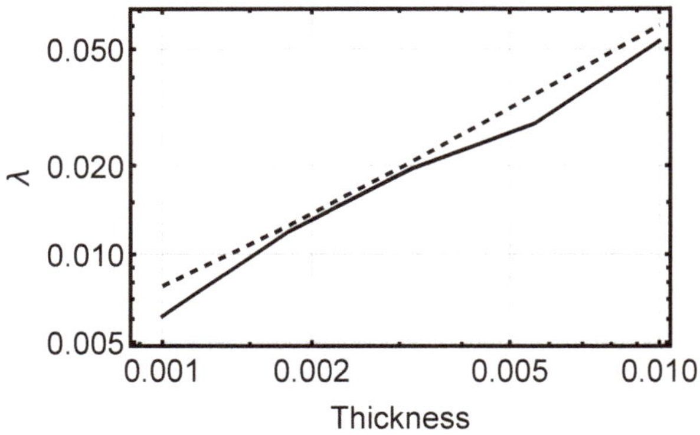

Figure A2. Buckling modes: Linear dependence of the observed smallest λ (eigenvalue corresponding to the critical load) to the thickness. Thick line: Reference. Dashed line: With hole.

This set of simulations simply illustrates the inherent complexity of the buckling problem in this context. Many fundamental concepts remain open starting from the selection of the right model and kinematic constraints.

References

1. Chapelle, D.; Bathe, K.J. *The Finite Element Analysis of Shells*; Springer: Berlin/Heidelberg, Germany, 2003.
2. Pitkäranta, J.; Leino, Y.; Ovaskainen, O.; Piila, J. Shell Deformation states and the Finite Element Method: A Benchmark Study of Cylindrical Shells. *Comput. Methods Appl. Mech. Eng.* **1995**, *128*, 81–121. [CrossRef]
3. Pitkäranta, J.; Matache, A.M.; Schwab, C. Fourier mode analysis of layers in shallow shell deformations. *Comput. Methods Appl. Mech. Eng.* **2001**, *190*, 2943–2975. [CrossRef]
4. Hakula, H.; Havu, V.; Beirao de Veiga, L. Long-Range Boundary Layers in Shells of Revolution. In Proceedings of the 5th International Conference on Computation of Shell and Spatial Structures, Salzburg, Austria, 1–4 June 2005.
5. Hakula, H. Hp-boundary layer mesh sequences with applications to shell problems. *Comput. Math. Appl.* **2014**, *67*, 899–917. [CrossRef]
6. Sanchez-Palencia, E.; Millet, O.; Béchet, F. *Singular Problems in Shell Theory*; Springer: Berlin/Heidelberg, Germany, 2010.
7. Pietraszkiewicz, W..; Konopińska, V.. Junctions in shell structures: A review. *Thin Walled Struct.* **2015**, *95*, 310–334. [CrossRef]
8. Malliotakis, G.; Alevras, P.; Baniotopoulos, C. Recent Advances in Vibration Control Methods for Wind Turbine Towers. *Energies* **2021**, *14*, 7536. [CrossRef]
9. Szabo, B.A.; Muntges, D.E. Procedures for the Verification and Validation of Working Models for Structural Shells. *J. Appl. Mech.* **2005**, *72*, 907–915. [CrossRef]
10. Szabo, B.; Babuska, I. *Finite Element Analysis*; Wiley: Hoboken, NJ, USA, 1991.
11. Niemi, A.H. Numerical buckling analysis of circular cylindrical shells. In Proceedings of MAFELAP 2019, Uxbridge, UK, 18–21 June 2019.
12. Bartels, S.; Bonito, A.; Muliana, A.H.; Nochetto, R.H. Modeling and simulation of thermally actuated bilayer plates. *J. Comput. Phys.* **2018**, *354*, 512–528. [CrossRef]
13. McMillen, T.; Goriely, A. Tendril Perversion in Intrinsically Curved Rods. *J. Nonlinear Sci.* **2002**, *12*, 241–281. [CrossRef]
14. Giani, S.; Hakula, H. On effective material parameters of thin perforated shells under static loading. *Comput. Methods Appl. Mech. Eng.* **2020**, *367*, 113094. [CrossRef]
15. Slaughter, W.S. *The Linearized Theory of Elasticity*; Birkhäuser: Basel, Switzerland, 2002.
16. Malinen, M. On the classical shell model underlying bilinear degenerated shell finite elements: General shell geometry. *Int. J. Numer. Methods Eng.* **2002**, *55*, 629–652. [CrossRef]
17. Forskitt, M.; Moon, J.R.; Brook, P.A. Elastic properties of plates perforated by elliptical holes. *Appl. Math. Model.* **1991**, *15*, 182–190. [CrossRef]
18. Burgemeister, K.; Hansen, C. Calculating Resonance Frequencies of Perforated Panels. *J. Sound Vib.* **1996**, *196*, 387–399. [CrossRef]
19. Jhung, M.J.; Yu, S.O. Study on modal characteristics of perforated shell using effective Young's modulus. *Nucl. Eng. Des.* **2011**, *241*, 2026–2033. [CrossRef]
20. Pitkäranta, J. The problem of membrane locking in finite element analysis of cylindrical shells. *Numer. Math.* **1992**, *61*, 523–542. [CrossRef]
21. Hakula, H.; Leino, Y.; Pitkäranta, J. Scale resolution, locking, and high-order finite element modelling of shells. *Comput. Methods Appl. Mech. Engrg.* **1996**, *133*, 157–182. [CrossRef]

22. Hakula, H.; Tuominen, T. Mathematica implementation of the high order finite element method applied to eigenproblems. *Computing* **2013**, *95*, 277–301. [CrossRef]
23. Do Carmo, M. *Differential Geometry of Curves and Surfaces*; Prentice Hall: Hoboken, NJ, USA, 1976.
24. Schwab, C. *p- and hp-Finite Element Methods*; Oxford University Press: Oxford, UK, 1998.
25. Artioli, E.; da Veiga, L.B.; Hakula, H.; Lovadina, C. On the asymptotic behaviour of shells of revolution in free vibration. *Comput. Mech.* **2009**, *44*, 45–60. [CrossRef]

Disclaimer/Publisher's Note: The statements, opinions and data contained in all publications are solely those of the individual author(s) and contributor(s) and not of MDPI and/or the editor(s). MDPI and/or the editor(s) disclaim responsibility for any injury to people or property resulting from any ideas, methods, instructions or products referred to in the content.

eng

MDPI

Article

A Review of Coastal Protection Using Artificial and Natural Countermeasures—Mangrove Vegetation and Polymers

Deborah Amos * and Shatirah Akib

Department of Civil Engineering, School of Architecture, Design and the Built Environment, Nottingham Trent University, Nottingham NG1 4FQ, UK
* Correspondence: deborah.amos2021@my.ntu.ac.uk

Abstract: Any stretch of coastline requires protection when the rate of erosion exceeds a certain threshold and seasonal coastal drift fluctuations fail to restore balance. Coast erosion can be caused by natural, synthetic, or a combination of the two. Severe storm occurrences, onshore interventions liable for sedimentation, wave action on the coastlines, and rising sea levels caused by climate change are instances of natural factors. The protective methods used to counteract or prevent coastal flooding are categorized as hard and soft engineering techniques. This review paper is based on extensive reviews and analyses of scientific publications. In order to establish a foundation for the selection of appropriate adaptation measures for coastal protection, this research compiles literature on a combination of both natural and artificial models using mangrove trees and polymer-based models' configurations and their efficiency in coastal flooding. Mangrove roots occur naturally and cannot be manipulated unlike artificial model configuration which can be structurally configured with different hydrodynamic properties. Artificial models may lack the real structural features and hydrodynamic resistance of the mangrove root it depicts, and this can reduce its real-life application and accuracy. Further research is required on the integration of hybrid configuration to fully optimize the functionality of mangrove trees for coastal protection.

Keywords: hard engineering techniques; soft engineering techniques; coastal protection; hybrid configuration; hydrodynamic resistance

Citation: Amos, D.; Akib, S. A Review of Coastal Protection Using Artificial and Natural Countermeasures—Mangrove Vegetation and Polymers. *Eng* **2023**, *4*, 941–953. https://doi.org/10.3390/eng4010055

Academic Editor: Antonio Gil Bravo

Received: 2 November 2022
Revised: 13 February 2023
Accepted: 13 February 2023
Published: 8 March 2023

Copyright: © 2023 by the authors. Licensee MDPI, Basel, Switzerland. This article is an open access article distributed under the terms and conditions of the Creative Commons Attribution (CC BY) license (https://creativecommons.org/licenses/by/4.0/).

1. Introduction

In the coastal region, dry land and a maritime environment (water and submerged land) coexist in a zone where terrestrial functions and land uses directly affect the marine environment, and vice versa. Physiological factors such as tides, waves, nearshore eddies, sand movement, and rivers impact coastlines. In several coastal cities worldwide, coastlines cover ecosystems and habitats that generate goods and services for the local population. Coastal areas also serve as the origin or backbone of the national economy [1].

According to Zanuttigh, erosion and flooding presently pose serious hazards to coastal communities, so developing defense mechanisms capable of dealing with the increasing sea level and more frequent storms caused by climate change is a significant challenge [2]. Different techniques are used to protect coastlines against erosion, including hard engineering and soft engineering. In hard engineering, solid structures are used to withstand erosion pressures, such as sieves, dikes, embankments, piers and revetements, and breakwaters. The use of soft engineering methods of coastal protection involves taking into consideration all aspects of preservation, including environmental, sociological, and economic aspects, and utilizing smaller structures made of natural materials. Currently, many parts of the world prefer natural coastal defenses that employ vegetation such as mangroves [3–6].

Mangroves are vegetation formations that develop on alluvial soils in coastal and estuarine locations which are frequently inundated by ocean tides. Researchers have extensively studied the performance of mangrove forests in reducing waves caused by erosion, including [7–9] who conducted laboratory experiments on mangroves as coastal protection. Mangrove trees have been proven and used in several locations as a solid structure capable of shielding coastlines against erosion. For decades, this has led to problems in establishing natural coastal protection, for example, mangrove-seedling-trees being destroyed by waves or tides before they have a chance to grow firmly, which requires at least two years of plantation. Planting them requires temporary structures, according to Verhagen [10]. As a result of this challenge, a natural coastal protection system combining natural and temporary artificial structures is recommended [11].

Mangroves grow in tidal zones along estuaries and coastal areas. While considering mangrove regeneration, it is crucial to consider the appropriate habitat and planting strategy. Their species are selected based on the existing species in the surrounding region as well as their access to seed. Yuanita et al. conducted a physical modelling experiment on different configurations using four different types of model settings without mangroves and with the presence of mangroves. A modelled mangrove seedling was carried out in a wave flume made of iron bars [12].

Several studies currently indicate that floods attenuate differently but they fail to address the role of major factors such as slope bathymetry, forest area, forest channelization, plant density, flood amplitudes and durations, etc. in determining those variations. More research is needed to understand how forest and storm features impact flood attenuation rates in mangrove forests so that informed decisions can be made about mangrove management. Natural resources necessary for human survival and growth have historically been found in coastal areas [13]. Today, coastal areas remain attractive due to their abundant ecological benefits. The majority of big cities being located near coastal regions, such as New York, Tokyo, Shanghai, and London, as argued by Nicholls [14], and the population density in coastal regions being three times the global mean, are indicative of society's desire to live near the shore [15]. The socioeconomic status of coastal communities in the UK makes this particularly important due to the UK's diverse coastal areas.

Natural erosion and flooding caused by coastal storms, such as flooding and coastal erosion threats, are becoming more common as a consequence of climate change due to tidal marshes and mangrove [16]. The researchers argue, however, that tidal wetlands are not able to reduce all risks equally, and that hazard reduction is governed by specific conditions. As a result of severe weather conditions, wetland qualities, and relatively large coastal terrain geometries, long-period severe storms that raise ocean levels by several metres for about a day are less effectively attenuated. Although storm damage to vegetation (especially mangrove trees) is often severe, and recovery can take years, wetlands generally assist in reducing erosion.

Slinger and Vreugdenhil demonstrated the importance of nature-based solutions for coastal management using a critical reflection technique centered on the design process. They distinguish four axes in attempting to determine the extent to which a hydraulic infrastructure forms a nature-based solution: the degree of inclusion of ecological knowledge; the extent to which the full infrastructural lifecycle is addressed; the complexity of the actor arena considered; and the resulting form of the infrastructural artefact. They classified traditional and new sea defense facilities on the North and South Holland coasts along the axes indicating how nature-based newly implemented solutions are and how broadly society values and stakeholders are included in the design process [17].

2. Coastal Engineering Protection

The coastal zone is a sensitive area in which the balance could be disturbed by a variety of factors; therefore, engineers, planners, and government agencies must pay close attention to detail before proceeding with any engineering activities along the coastline. Coastal behavior is largely site-specific, which means that a host of different factors must be considered closely. It is important to ensure that activities near beaches are ecologically sound, particularly in metropolitan areas. In coastal protection, measures are classified as hard (gabions, seawalls, offshore detached breakwaters) and soft (artificial nourishment of beaches, bio shields/vegetation, dune stabilization, geosynthetic application) as discussed previously.

2.1. Hard Engineering Techniques

Typically, hard engineering consists of the erection of gravity infrastructure made up of dunes, concrete structures, or rubble with a trapezoidal cross-section that is designed to withstand the waves along the shoreline. When structures are built along the coast, they are often irreparably damaged. Many of the projects are usually undertaken to provide a quick solution to erosion concerns; they are most effective when they are meticulously constructed with a comprehensive understanding of the wave geography, the local bathymetry, and the sediment properties. Hard engineering structures along the coasts are groins, seawalls, breakwaters, and offshore breakwaters (emerged and submerged). Hard engineering approaches have strong impacts on the environment and are expensive to implement and maintain.

2.1.1. Seawall

This structure prevents erosion immediately along a coastal stretch, but it may not contribute to or expand beach width. It may be necessary, however, in many circumstances to regularly repair seawalls, especially those constructed of rubble mounds. Figure 1 shows a cross-section of the seawall harbor. There are several practical obstacles in transporting tonnes of rubble mound to the beach, as well as in continually fabricating concrete structures to drop along the coastlines. Hard methods and gravity systems can be efficient if the local soil structure are sustainable and construction materials are easily obtained at the construction site location. Gravity systems and hard methods can be efficient if the local soil structure is sustainable and construction materials are easily obtained on the construction site. The disadvantage of a seawall is that waves can erode the wall, defeating its purpose.

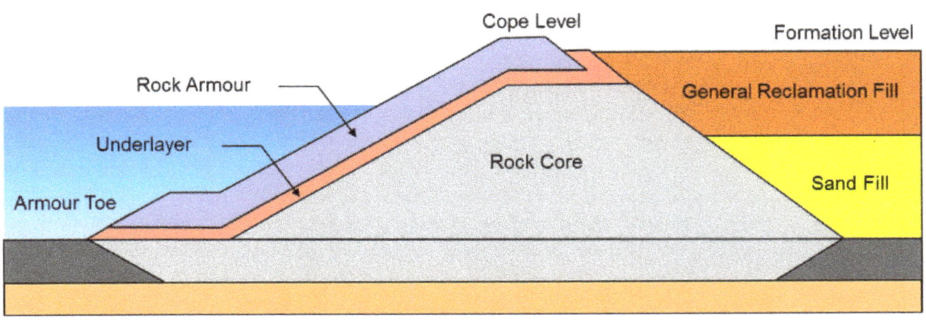

Figure 1. Cross Section of a Seawall. Adapted with permission from Ref. [16], 2013, Firth et al.". More details on "Copyright and Licensing" are available via the following link: https://www.mdpi.com/ethics#10 (accessed on 12 February 2023).

2.1.2. Gabion

According to [16], the utilization of gabion boxes as submarine reefs could be considered a soft engineering solution to counteract coastal flooding since they contribute to fostering ocean life around them. Sundar and Murali went into detail about the use of gabions around the Kerala coast. Gabion boxes are considered a hard engineering solution when used as an alternative to rubble or concrete armour layers in traditional shore-linked structures. The gabion boxes were originally used to repair a damaged seawall cross-section. Although gabions are a hard engineering structure, they are not very attractive and effective [17].

2.1.3. Offshore Detached Breakwater

Breakwaters that are disconnected offshore generate areas of low energy on their leeside, allowing for the creation of salient and, eventually, Tombolos, the details of which are described by Sukanya [18]. The cost and time involved in the construction of offshore disconnected breakwaters have prevented them from being implemented over impacted portions of Indian coastline. Waves are deflected by the breakwaters' ends, creating a quiet zone between them. Their offshore run parallel to the eroding shoreline and contribute to the formation of salient in time, which in turn leads to the formation of tombolo which enlarges the beach. Furthermore, it can also be built as part of wave energy conversion (WEC) systems, such as an oscillating water column [19].

2.2. *Soft Engineering Techniques*

It is a well-designed measure, which has little or no impact on the coastal environment. Opposed to hard measures, this is a lengthier process. Metrics like these require extensive knowledge. In this term, artificial beach nourishment and natural vegetation are two of the most common solutions. Over time, geo-synthetics were increasingly used in coastal protection measures, where polymer-based synthetic fibers were utilized for drainage, separation, filtration, and retention [20]. The term "soft structures" refers to structures completely or partially composed of geo-synthetic materials, such as seawalls, underwater breakwaters, and submarine reefs.

2.2.1. Coral Reef

Its' unique structures—some emerging from deep levels to the surface of the ocean, and in many cases extending parallel to coasts for tens or hundreds of kilometers—place them on the front line of coastal protection. The structural geometry and ruggedness of reef formations determine their impact on currents and waves. This complicated structure is a result of the biotic proliferation of habitat-forming organisms, particularly hard corals, and coralline algae. In addition to reducing coastline flooding, reef roughness has been found to have a substantial impact on reducing massive energy flows from underlying seas into the reef structure, greatly slowing the action of waves [21,22].

In tropical regions, coral reefs play an extremely important role in dispersing wave energy. On the other hand, fragmented reef patches and channels may be able to enhance or direct tidal energy locally [23,24]. The impact of storms on habitat and the kind of coastal protection provided by reefs must also be acknowledged [25].

Sea level rise also poses a critical threat to reef structures, including beaches and islands that are connected. As evidenced by geological data from the Great Barrier Reef, coral reefs may grow rapidly [26], although such growth is dependent on reef stability. There are many regions where land has formed from coral reef deposits that are sculpted into beaches and islands by storms and sometimes boosted by windblown sediments [27]. A massive analysis of Pacific islands found that, while some islands are shrinking in area and many have dynamic borders, all of these mechanisms could be adequate to allow sustained island expansion or maintenance under some conditions of rising sea levels: despite the slight rise in sea levels that has occurred to date, the total area of coral islands appears to have expanded [28], although coastal development and climate change, particularly ocean acidification, may alter such processes.

Psychologically, sea level rise, as well as the possibility of sea level rise associated with changes in island sediment migration, may pose serious risks even if landmasses are not substantially reduced [27,29,30]. As with coastal wetlands, reef formations have varying impacts throughout space on coastal protection. The primary sources of variability are listed in Table 1. A better understanding of these causes and their measurement is crucial to fully analyze how well a reef protects.

Table 1. Significant variation determinants in the coastal protection function of coastal wetlands and coral reefs. Coral reef data are based on [21,31–36]. Wetlands data are based on [4,35,37–39].

	Coral Reef	Coastal Wetlands
Ecosystem determinants	• Prevailing tides • Water depth • Exposure • Distance to shore • Slope • Wave Characteristics • Bathymetric • Topography	• Bathymetric • Topography • Wave Characteristics • Drainage System • Presence and frequency of disturbances • Slope • Distance from sediment source • Exposure • Prevailing tides • Soil characteristics • Distance to other ecosystems • Adjacent land use • Distance to shore • Water depth over plants
Abiotic Determinants	• Presence and proximity of other ecosystems (e.g., seagrass) • Reef width • Levels of bioerosion • Reef profile • Roughness • Dominant species (corals and calcareous algae)-Skeletal morphology, growth rates, disease resistance • Meso-scale structure e channels, fragmentation • Reef surface depth • Resistance and resilience (capacity to survive or recover from impacts)	• Fragmentation • Habitat width • Vegetation structure, salt marshes: plant height, vegetation stiffness • Plant density • Vegetation structure, mangroves: canopy height, aerial root physiognomy, age class distribution, sub-canopy elements • Dominant species • Resistance and resilience (capacity to survive or recover from impacts)

2.2.2. Mangrove Forest

Mangrove trees have been proven and used in several locations as a solid structure capable of shielding coastlines against erosion. This has caused problems for decades in establishing natural coastal protection, for example mangrove-seedling-trees being destroyed by waves or tides before they have a chance to grow firmly, which requires at least two years of plantation. Planting them requires temporary structures, according to Verhagen [10]. As a result of this challenge, a natural coastal protection system combining natural and temporary artificial structures is recommended [12]. Mangrove species are selected based on the existing species in the surrounding region as well as their access to seed. Figure 2 illustrates a natural coastal protection system.

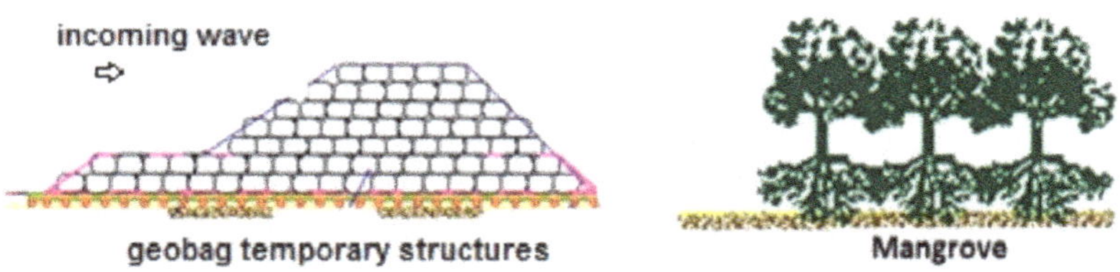

Figure 2. Temporary Structures and Natural Coastal Protection [12]. Reprinted/adapted with permission from Ref. [12]. 2019, Yuanita et al. More details on "Copyright and Licensing" are available via the following link: https://www.mdpi.com/ethics#10 (accessed on 12 February 2023).

A mangrove ecosystem is a tropical or subtropical wetland forest located between the land and the ocean composed of saltwater-adapted trees, shrubs, palms, and ferns. As mangroves grow at or above mean sea level, floods vary from near-constant to irregular [40–44]. Giri et al. estimate that mangrove ecosystems cover 152,400 square kilometres worldwide, distributed across 123 nations, and account for 30–35% of tropical wetland forests [43–45]. They are not one morphological group, but rather a variety of plant species with special adaptations that allow them to survive in the severe intertidal environment [40,42,46].

Traits and Adaptation

With both physiological and morphological adaptations, mangrove plant species are uniquely adapted to frequently waterlogged, salines, and turbulence intertidal environments, including:

- Extensive aerial rooting systems.
- Mechanisms for salt exclusion, tolerance, or secretion.
- Conservative resource-capture and growth strategies, including investments in buoyant, viviparous propagules for several species [47].

Rhizophora species have tall lateral prop roots (or stilt roots). In some cases, as well as in *Avicennia* spp. [e.g., *officinalis*]), shallow but far-reaching aerial roots producing surface-penetrating pneumatophores in *Avicennia*, *Laguncularia*, *Lumnitzera*, *Sonneratia*, and *Xylocarpus* spp., surface-penetrating knee roots in *Bruguiera*, *Ceriops* and *Xylocarpus* spp., plank roots in *Camptostemon* and *Xylocarpus* spp., and buttress-forming stems in *Heritiera* and *Kandelia* spp. assist in stabilising mangrove stems (Figure 3; [41,43,46,47]). In the saline intertidal zone, high root:shoot ratios are a key factor for absorbing water [47], as well as tolerance of strong intertidal disturbances [48]. The presence of lenticels allows root aeration in anaerobic, water-logged sediments with surface-penetrating aerial roots [46,47]. Mangrove roots such as *Aegialitis*, *Aegiceras*, *Avicennia*, *Bruguiera*, *Ceriops*, *Excoecaria*, *Osbornia*, *Rhizophora*, and *Xylocarpus* spp. are also capable of excluding salt from tissues by ultrafiltration; other species actively secrete salt from tissues such as *Acanthus*, *Aegialitis*, *Aegiceras*, *Avicennia*, *Laguncularia* and *Sonneratia* species or from senescent leaves such as *Excoecaria* and *Xylocarpus* spp. [46,47].

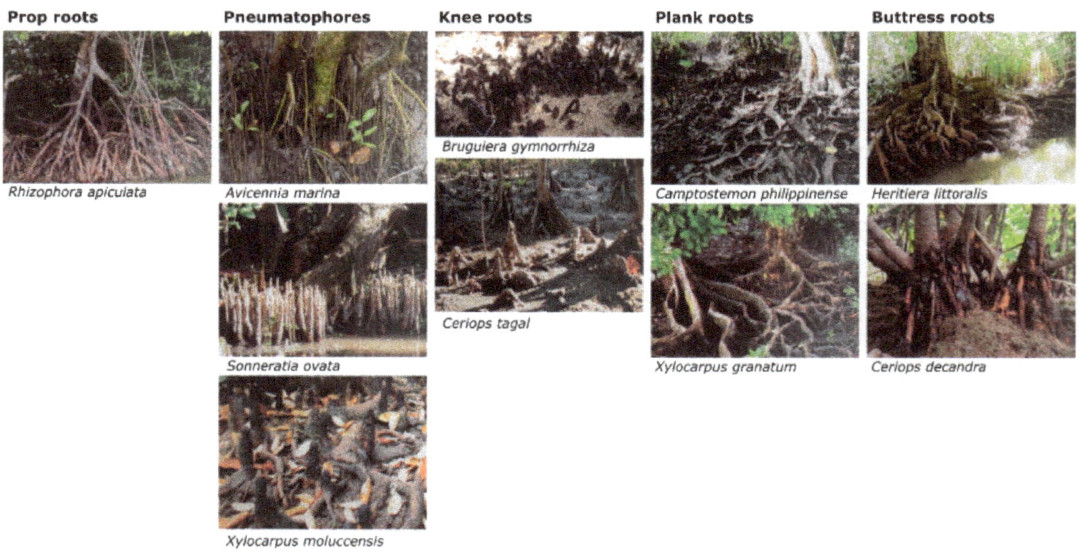

Figure 3. Mangrove Species. Reprinted/adapted with permission from Ref. [41], 2010, Polidoro et al. More details on "Copyright and Licensing" are available via the following link: https://www.mdpi.com/ethics#10 (accessed on 12 February 2023).

Mangrove species make significant investments in leaf production, often producing lengthy, fleshy leaves with tough outer layers of the epidermis and specialised salt excretory glands that reduce transpiration losses at the expense of reduced number of leaves and photosynthetic activity [41,46,49]. *Rhizophora apiculata* mangrove seedlings can be planted on the shore when they are more than 30 cm tall and have four leaves. Seedlings of *Rhizophora mucronata* can be planted when they are at least 55 cm tall and have at least four–six leaves. As mangroves cannot thrive in either wet or dry conditions, they should be planted in places where both wet and dry conditions exist daily.

Experimental Models of Coastal Protection

Models are conducted using a physical modelling experiment on different configurations using four distinct types of model settings without mangroves and with the presence of mangroves [12]. A modelled mangrove seedling was carried out in a wave flume made of iron bars. Different types of configurations are illustrated in Figure 4.

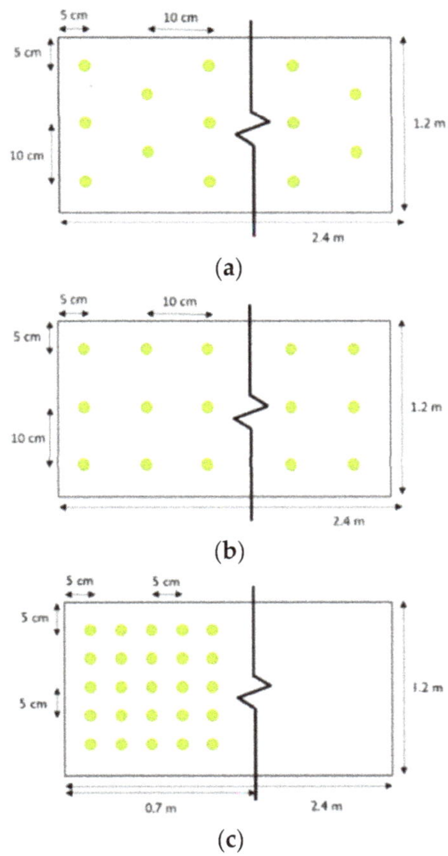

Figure 4. Types of Configuration (**a**) staggered arrangement 10 cm; (**b**) tandem arrangement 10 cm; (**c**) tandem arrangement 5 cm. Reprinted/adapted with permission from Ref. [12]. 2019, Yuanita et al. More details on "Copyright and Licensing" are available via the following link: https://www.mdpi.com/ethics#10 (accessed on 12 February 2023).

In this study, the influence of the mangroves model was examined using the wave transmission coefficient (K_t). A transmission coefficient is the ratio of the transmitted wave height (H_t) to the starting wave height (H_i). To determine the transmitted wave height (H_t), data from wave gauge CH_3 was used, while the initial wave height (H_i) was determined by data from wave gauge CH_2.

$$K_t = \frac{Transmiited\ Wave\ Height}{Incident\ Wave\ Height} = \frac{H_{t\ (CH_3)}}{H_{i\ (CH_2)}} \tag{1}$$

The research objective was to examine the wave height reduction with different mangrove densities, and to investigate the effect of mangrove seedling tree patterns on wave attenuation. The experimental testing was carried out in a narrow wave flume with a mangrove model as the primary natural barrier and geotextile geo-bag models as a temporary constructed construction. During this laboratory experiment, several wave scenarios were established. The study focused on the wave propagation findings over mangrove seedling trees in order to discover the most effective configuration of mangrove tree planting against wave. The results revealed that the wave height reduction in areas with mangroves was twice as big as that in bare land [12].

During the research it was also discovered that the variation in wave attenuation comparing tandem and staggered tree configurations was 20% lesser and that the temporary structure considerably reduces wave height and protects the growth of mangrove seedlings against wave action.

Safari et al. in their study computed the transmission coefficient (K_t) as the ratio of the residual wave height after models to the incident wave height before models in Equation (2) [50,51]. To overcome the limitations of the previously described armour blocs, Hogue et al. investigated the newly designed armour unit, called 'The Starbloc®,' which is made up of a centralized hexagonal core, three legs, and two noses. Its structural characteristics facilitate simplified mobility, much better positioning, and much better hydraulic stability [52].

$$K_t = \frac{Wave\ Height\ after\ Models}{Wave\ Height\ before\ Models} = \frac{H_{aft}}{H_{bfr}} \tag{2}$$

An experimental investigation of the efficiency of artificial Xbloc walls made of hybrid polymer and mangrove root models for water wave defense was conducted by Safari et al., as shown in Figure 5, Ref. [51].

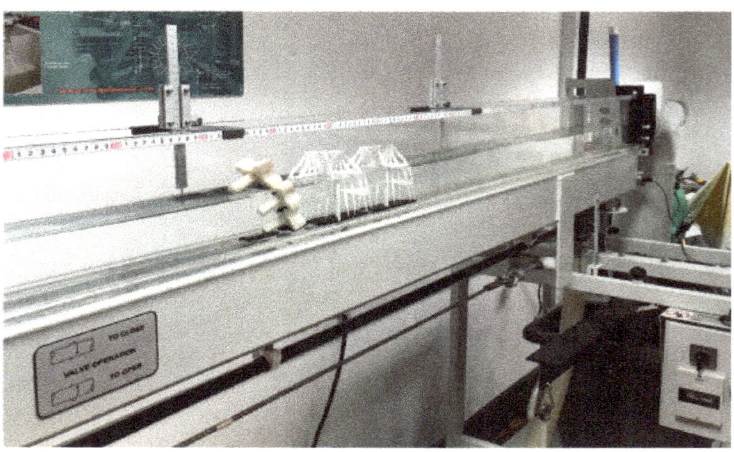

Figure 5. Hybrid Configuration using one Xbloc wall with two mangrove roots in a 5 m flume tank. Reprinted/adapted with permission from Ref. [51]. 2018, Safari et al. More details on "Copyright and Licensing" are available via the following link: https://www.mdpi.com/ethics#10 (accessed on 12 February 2023).

Three Xbloc pieces were placed on each other and bonded with water-resistant adhesive to form one Xbloc wall. Software such as SolidWorks and AutoCAD were used to create fake models, which were 3D printed, laser cut, and superglued. The test was carried out using a variety of single and multiple Xbloc barriers and mangrove root simulations. For six alternative model setups, changes in wavelength, height, celerity, and period were found. The results showed that the celerity, height, and wavelength were successfully reduced, as well as the wave period being lengthened (one cycle time).

In the research carried out, it was discovered that the hybrid configuration of one Xbloc wall and two mangrove roots gave the best protection, lowering the wavelength, celerity, and height by 5.50%, 26.46%, and 58.97%, respectively, and delaying the wave duration by 28.34%. The configuration with only one set of mangrove roots model had the lowest attenuation. As a result, wave reduction utilizing the hybrid action of artificial polymer made Xbloc walls and mangrove roots was superior since it permitted wave energy dissipation to a larger extent than using just Xbloc walls or mangrove roots alone.

As shown in Equation (3), Zwicht computed the transmission coefficient in consideration of wave height and wave energy [53]. Their studies specify the reflection and dissipation coefficients as two additional wave attenuation analysis factors. Their linking method involves the energy balance among the three factors.

$$K_t = \frac{Transmitted\ Wave\ Energy\ (after\ forests)}{Incident\ Wave\ Energy\ (before\ forests)} = \frac{E_{m0,t}}{E_{m0,i}} \quad (3)$$

Hogue et al. investigated the uneven wave attenuation performance of mangrove forests in terms of wave dissipation, reflection, and transmission coefficients. The experiment was carried out in a Twin Wave Flume (TWF), with the bigger flume containing quantified *Rhizopora* sp. mangrove trees and the smaller flume not. *Rhizophira* sp. was extremely efficient in minimising tsunami-induced flow due to the complexity and thickness of its rhizome. The wave energy diminished exponentially throughout the flume forest area, and the amount of the energy dissipated decreased from the front of the vegetation to the end having more wave attenuation at the mangrove forest Ref. [52].

Artificial coastal protection measures were examined by Zwicht, who analyzed the effect of concrete unit weight on the hydraulic stability along with our ability to establish the appropriate computational model of the stability number (Ns) [53]. Based on the model testing, it was evident that as the specific weight increases, so does the hydraulic stability; however, when factoring in the impact of varied gradients, relevant data were obtained. For gradients of 2:3 and greater, stability was observed to be higher than predicted from the previous Ns equation, whereas stability was lower for gradients of 1:2. During coastal protection, the stability of armour bloc units depends on their structure, packing density, and deployment pattern (random or organized). Acropode® and Xbloc®, which are single layer interlocking armour block units, can be damaged by oscillations. As a result of a weak foundation or inadequate interconnections, blocks wobble during this phase of destruction, causing variations in their optimum state.

A laboratory experiment of wave attenuation through cylinder arrays, mimicking wave attenuation processes through a coastal man-grove forest, was conducted in a flume of the Fluid Mechanics Laboratory at Delft University of Technology by Phan et al. The effective length, height, and width of the flume is 40 m, 1 m, and 0.8 m, respectively. Numerical modeling was constructed based on SWASH model using Morrison's equation shown in Equation (4) [54].

$$F_x = \frac{1}{2}\rho C_D h_v b_v N_v U\,|U| \quad (4)$$

The physical model was constructed in a way that the numerical results can be directly compared with the experimental results. A wide variety of wave characteristics, such as regular, irregular, broken, and non-broken waves, were used in the experiment to obtain additional information. The findings support the idea that vegetation can reduce wave heights. Furthermore, the vegetation influenced the set-down of the waves rather than the set-up of the waves. Data from the experiment were used to assess the effect of wave nonlinearity on wave reduction techniques.

Maza et al. investigated the physical processes involved in flow-mangrove interaction, wave attenuation, and drag forces along a 1:6 scale fringe Rhizophora mangrove forest. A 26 m long forest composed of 135 models built reproducing mature Rhizophora mangrove trees with 24 prop roots were used for the experiment. Using both experimental and numerical approach, it was observed that water depth, the accompanying mangrove frontal area, as well as wave height were shown to be the major variables causing wave attenuation for short waves. Wave shoaling was caused by the forest's seaward slope, which increases the wave steepness. Therefore, the pressures imposed on the mangroves began to rise after 3–4 m. Wave decay models that match wave heights well produce smaller pressures farther into the forest [55].

3. Conclusions

Models are critical for forecasting and monitoring mangrove functioning and sustainability. Classification techniques are necessary to characterize mangroves for use in coastal flood risk mitigation. Secondly, experimental and numerical mangrove models may be used to replicate severe flooding circumstances (functionality) and anticipate long term development (persistence) in order to analyze the impacts of climatic and human induced alteration. While mangrove model configuration has been extensively used, the creation of experimental and numerical methods with predictive validity is an ongoing area of study.

Globally, coastal areas suffer endemic problems of human induced problems associated with increase in population growth while dealing with the effect of naturally occurring climate change and increased susceptibility to coastal flooding. Mangrove forests can aid flood mitigation and help adapt to climate change. Mangroves are suitable for minimizing coastal flooding when combined with artificial structures. Many researchers are experimenting with different methods of coastal protection measures using a combination of hard and soft engineering structures as hybrid coastal defense strategies. In order to reduce coastal flooding using mangrove forests, there is need to study, analyze, and simulate the essential processes, patterns, and limitations to mangrove efficiencies.

This review provides an overview of the existing literature on experimental modeling and numerical approaches for the effective use of mangrove trees and artificial polymers in coastal protection. Mangrove roots occur naturally and cannot be manipulated unlike artificial model configuration which can be structurally configured with different hydrodynamic properties. Artificial models may lack the real structural features and hydrodynamic resistance of the mangrove root it depicts, and this can reduce its real-life application and accuracy.

4. Innovation and Future Research Direction

This research is limited to finding the influence of using natural and artificial countermeasures considering different reviews of past literatures on the use of hybrid polymer and mangrove trees. The study is to examine the effectiveness of using the combined polymer and mangrove roots in comparison with each model being used separately for coastal protection. This study recommends the following:

- The artificial models may lack the actual structural features and hydrodynamic resistance of the natural mangrove tree species it depicts, reducing accuracy when used in real-world applications. Further research should be undertaken to model the real-life properties of mangroves so that greater adaptability and resistance can be validated to real life applications.
- The use of digital devices should be adopted for future research to reduce human errors when taking the reading during the process of collecting data.
- The application of Artificial Intelligence and Machine Learning could be applied to predict the future wave reduction in using structural measure and nature based solution

Author Contributions: Conceptualization, D.A. and S.A.; methodology D.A. and S.A.; writing—original draft preparation, D.A.; writing—review and editing, D.A. and S.A.; supervision, S.A. All authors have read and agreed to the published version of the manuscript.

Funding: This research received no external funding.

Institutional Review Board Statement: Not applicable.

Informed Consent Statement: Not applicable.

Data Availability Statement: Not applicable.

Conflicts of Interest: The authors declare no conflict of interest.

References

1. Ketchum, B.H. *The Water's Edge: Critical Problems of the Coastal Zone*; MIT Press: Cambridge, MA, USA, 1972; p. 414.
2. Zanuttigh, B. Coastal flood protection: What perspective in a changing climate? The THESEUS approach. *Environ. Sci. Policy* **2011**, *14*, 845–863. [CrossRef]
3. Bao, T.Q. Effect of mangrove forest structures on wave attenuation in coastal Vietnam. *Oceanologia* **2011**, *53*, 807–818. [CrossRef]
4. Gedan, K.B.; Kirwan, M.L.; Wolanski, E.; Barbier, E.B.; Silliman, B. The present and future role of coastal wetland vegetation in protecting shorelines: Answering recent challenges to the paradigm. *Clim. Chang.* **2010**, *106*, 7–29. [CrossRef]
5. Parvathy, K.G.; Bhaskaran, P.K. Wave attenuation in presence of mangroves: A sensitivity study for varyingbottom slopes. *J. Ocean. Clim. Sci. Technol. Impacts* **2017**, *8*, 126–134.
6. Othman, M.A. Value of mangroves in coastal protection. *Hydrobiology* **1994**, *258*, 277–282. [CrossRef]
7. Husrin, S.; Strusinska, A.; Oumeraci, H. Experimental study on tsunami attenuation by mangrove forest. *J. Earth Planets Space Vol.* **2012**, *64*, 973–989. [CrossRef]
8. Strusińska-Correia, A.; Husrin, S.; Oumeraci, H. Attenuation of solitary wave by parametized flexible mangrove models. *Coast. Eng. Proc.* **2014**, *1*, 1–34.
9. Hashim, A.M.; Catherine, S.M.P. A Laboratory Study on Wave Reduction by Mangrove Forests. *APCBEE Procedia* **2013**, *5*, 27–32. [CrossRef]
10. Verhagen, H. The Use of Mangroves in Coastal Protection. In Proceedings of the COPEDEC 2012: The 8th International Conference on Coastal and Port Engineering in Developing Countries, Chennai, India, 20–24 February 2012; pp. 20–24.
11. Yuanita, N.; Kurniawan, A.; Paramashanti, P.; Laksmi, A.A. Natural Coastal Protection System Preliminary Design. *J. Subsea Offshore-Sci. Eng.* **2018**, *14*, 1–5.
12. Yuanita, N.; Kurniawan, A.; Setiawan, H.; Hasan, F.; Khasanah, M. Physical model of natural coastal protection system: Wave transmission over mangrove seedling trees. *J. Coast. Res.* **2019**, *91*, 176–180. [CrossRef]
13. Özyurt, G.; Ergin, A.Y.Ş.E.N. Application of Sea Level Rise Vulnerability Assessment Model to Selected Coastal Areas of Turkey. *J. Coast. Res.* **2019**, *56*, 248–251.
14. Nicholls, R. Coastal megacities and climate change. *GeoJournal* **1995**, *37*, 369–379. [CrossRef]
15. Small, C.; Nicholls, R.J. A Global Analysis of Human Settlement in Coastal Zones. *J. Coast. Res.* **2003**, *19*, 584–599.
16. Firth, L.B.; Thompson, R.C.; Bohn, K.; Abbiati, M.; Airoldi, L.; Bouma, T.; Bozzeda, F.; Ceccherelli, V.; Colangelo, M.A.; Evans, A.; et al. Between a rock and a hard place: Environmental and engineering considerations when designing coastal defence structures. *Coast. Eng.* **2014**, *87*, 122–135. [CrossRef]
17. Sundar, V.; Murali, K. *Planning of Coastal Protection Measures along Kerala Coast: Paper presented at the State Government of Kerala by IIT Madras*; Department of Ocean Engineerimg, Indian Institute of Technology: Madras, Chennai, India, 2007.
18. Sukanya, R.; Sundar, V.; Sannasiraj, S.A. Geo-Technical Stability and Sensitivity Analysis of Geo-Synthetic Seawall at Pallana Beach, Kerala, India. In *Proceedings of the Fifth International Conference in Ocean Engineering (ICOE2019)*; Springer: Singapore, 2021; pp. 15–26. [CrossRef]
19. Pilarczyk, K.W.; Zeidler, R.B. *Offshore Breakwaters and Shore Evolution Control*; Rotterdam, A.A., Ed.; Balkema Publishers: Leiden, The Netherlands, 1996.
20. Phan, K.L.; Stive, M.J.; Zijlema, M.; Truong, H.S.; Aarninkhof, S.G. The effects of wave non-linearity on wave attenuation by vegetation. *Coast. Eng.* **2019**, *147*, 63–74. [CrossRef]
21. Kench, P.S.; Brander, R.W. Wave processes on coral reef Flats: Implications for reef geomorphology using Australian case studies. *J. Coast. Res.* **2006**, *22*, 209–223. [CrossRef]
22. Monismith, S.G. Hydrodynamics of Coral Reefs. *Annu. Rev. Fluid Mech.* **2007**, *39*, 37–55. [CrossRef]
23. Cochard, R.; Ranamukhaarachchi, S.L.; Shivakoti, G.P.; Shipin, O.V.; Edwards, P.J.; Seeland, K.T. The 2004 tsunami in Aceh and Southern Thailand: A review on coastal ecosystems, wave hazards and vulnerability. *Perspect. Plant Ecol. Evol. Syst.* **2008**, *10*, 3–40. [CrossRef]

24. Fernando, H.; Samarawickrama, S.; Balasubramanian, S.; Hettiarachchi, S.; Voropayev, S. Effects of porous barriers such as coral reefs on coastal wave propagation. *J. Hydro-Environ. Res.* **2008**, *1*, 187–194. [CrossRef]
25. Woodley, J.D. The incidence of hurricanes on the north coast of Jamaica since 1870: Are the classic reef descriptions atypical? *Hydrobiologia* **1992**, *247*, 133–138. [CrossRef]
26. Perry, C.T.; Smithers, S.G. Cycles of coral reef 'turn-on', rapid growth and 'turn-off' over the past 8500 years: A context for understanding modern ecological states and trajectories. *Glob. Change Biol.* **2011**, *17*, 76–86. [CrossRef]
27. Woodroffe, C.D. Reef-island topography and the vulnerability of atolls to sea-level rise. *Glob. Planet. Chang.* **2008**, *62*, 77–96. [CrossRef]
28. Webb, A.P.; Kench, P.S. The dynamic response of reef islands to sea level rise: Evidence from multi-decadal analysis of island change in the Central Pacific. *Glob. Planet. Chang.* **2010**, *72*, 234–246. [CrossRef]
29. Briguglio, L. The Vulnerability Index and small island developing states: A review of conceptual and methodological issues. In *Paper Prepared for the AIMS View of Conceptual and Methodological Issues, Praia, Cape Verde*; UNESCO: Paris, France, 2003; pp. 1–5.
30. Moore, W.S. The subterranean estuary: A reaction zone of ground water and sea water. *Mar. Chem.* **1999**, *65*, 111–125. [CrossRef]
31. Brander, R.W.; Kench, P.S.; Hart, D. Spatial and temporal variations in wave characteristics across a reef platform, Warraber Island, Torres Strait, Australia. *Mar. Geol.* **2004**, *207*, 169–184. [CrossRef]
32. Gourlay, M. Wave set-up on coral reefs. 1. Set-up and wave-generated flow on an idealised two dimensional horizontal reef. *Coast. Eng.* **1996**, *27*, 161–193. [CrossRef]
33. Gourlay, M. Wave set-up on coral reefs. 2. set-up on reefs with various profiles. *Coast. Eng.* **1996**, *28*, 17–55. [CrossRef]
34. Lacambra, C.; Spencer, T.; Moeller, I. Tropical Coastal Ecosystems as Coastal Defences. In *The Role of Environmental Management and Eco-Engineering in Disaster Risk Reduction and Climate Change Adaptation*; ProAct Network: Genolier, Switzerland, 2008.
35. Sheppard, C.; Dixon, D.J.; Gourlay, M.; Sheppard, A.; Payet, R. Coral mortality increases wave energy reaching shores protected by reef flats: Examples from the Seychelles. *Estuar. Coast. Shelf Sci.* **2005**, *64*, 223–234. [CrossRef]
36. McIvor, A.L.; Möller, I.; Spencer, T.; Spalding, M. Reduction of Wind and Swell Waves by Mangroves. In *Natural Coastal Protection Series: Report 1. The Nature Conservancy*; University of Cambridge: Cambridge, UK; Wetlands International: Cambridge, UK, 2012.
37. Shepard, C.C.; Crain, C.M.; Beck, M.W. The protective role of coastal marshes: A systematic Review and Meta-analysis. *PLoS ONE* **2011**, *6*, e27374. [CrossRef]
38. Duke, N. Mangrove floristics and biogeography. In *Tropical Mangrove Ecosystems*; American Geophysical Union: Washington, DC, USA, 1992.
39. Primavera, J.; Sadaba, R.; Lebata, M.; Hazel, J.; Altamirano, J. *Handbook of Mangrove in the Philippines—Panay*; SEAFDEC Aquaculture Department: Iloilo, Philippines, 2004.
40. Polidoro, B.A.; Carpenter, K.E.; Collins, L.; Duke, N.C.; Ellison, A.M.; Ellison, J.C.; Farnsworth, E.J.; Fernando, E.S.; Kathiresan, K.; Koedam, N.E.; et al. The loss of species: Mangrove extinction risk and geographic areas of global concern. *PLoS ONE* **2010**, *5*, 10095. [CrossRef]
41. Spalding, M.; Kainuma, M.; Collins, L. *World Atlas of Mangroves*; Earthscan: London, UK, 2010.
42. Giri, C.; Ochieng, E.; Tieszen, L.L.; Zhu, Z.; Singh, A.; Loveland, T.; Masek, J.; Duke, N.C. Status and distribution of mangrove forests of the world using earth observation satellite data. *Glob. Ecol. Biogeogr.* **2011**, *20*, 154–159. [CrossRef]
43. FAO. *The World's Mangroves 1980–2005*; FAO: Rome, Italy, 2007.
44. Hogarth, P. *The Biology of Mangroves and Seagrasses*; Oxford University Press: Oxford, UK, 2007.
45. Tomlinson, P. *The Botany of Mangroves*; Cambridge University Press: Cambridge, UK, 1986.
46. Alongi, D. Mangrove forests: Resilience, protection from tsunamis, and responses to global climate change. *Estuar. Coast. Shelf Sci.* **2008**, *76*, 1–13. [CrossRef]
47. Balun, L. Functional Diversity in the Hyper-Diverse Mangrove Communities in Papua ew Guinea. Ph.D. Thesis, University of Tennessee, Knoxville, TN, USA, 2011.
48. Primavera, J. Field Guide to Philippines Mangroves. Zoological Society of London. 2009. Available online: https://www.zsl.org/sites/default/files/media/2015-06/Field% (accessed on 8 October 2022).
49. Sabari, A.A.; Oates, A.R.; Akib, S. Experimental Investigation of Wave Attenuation Using a Hybrid of Polymer-Made Artificial Xbloc Wall and Mangrove Root Models. *Eng* **2021**, *2*, 229–248. [CrossRef]
50. Safari, I.; Mouaze, D.; Ropert, F.; Haquin, S.; Ezersky, A. Hydraulic stability and wave overtopping of Starbloc®armored moundbreakwaters. *Ocean. Eng.* **2018**, *151*, 268–275. [CrossRef]
51. Hoque, A.; Husrin, S.; Oumeraci, H. Laboratory studies of wave attenuation by coastal forest under storm surge. *Coast. Eng. J.* **2018**, *60*, 225–238. [CrossRef]
52. Van Zwicht, B. Effect of the Concrete Density on the Stability of Xbloc Armour Unit. Master's Thesis, Delft University, Delft, The Netherlands, 2009. Hydraulic Engineering Section.
53. Sundar, V.; Sannasiraj, S.A.; Babu, S.R. Sustainable hard and soft measures for coastal protection—Case studies along the Indian Coast. *Mar. Georesources Geotechnol.* **2022**, *40*, 600–615. [CrossRef]

4. Maza, M.; Lara, J.L.; Losada, I.J. Experimental analysis of wave attenuation and drag forces in a realistic fringe *Rhizophora* mangrove forest. *Adv. Water Resour.* **2019**, *131*, 103376. [CrossRef]
5. Temmerman, S.; Horstman, E.M.; Krauss, K.W.; Mullarney, J.C.; Pelckmans, I.; Schoutens, K. Marshes and Mangroves as Nature-Based Coastal Storm Buffers. *Annu. Rev. Mar. Sci.* **2022**, *15*, 95–118. [CrossRef]

Disclaimer/Publisher's Note: The statements, opinions and data contained in all publications are solely those of the individual author(s) and contributor(s) and not of MDPI and/or the editor(s). MDPI and/or the editor(s) disclaim responsibility for any injury to people or property resulting from any ideas, methods, instructions or products referred to in the content.

eng

Communication

Measuring the Adoption of Drones: A Case Study of the United States Agricultural Aircraft Sector

Roberto Rodriguez III

Spatial Data Analysis and Visualization Laboratory, University of Hawaii at Hilo, Hilo, HI 96720, USA; roberto6@hawaii.edu

Abstract: Unmanned aircraft systems (UAS), commonly referred to as drones, are an emerging technology that has changed the way many industries conduct business. Precision agriculture is one industry that has consistently been predicted to be a major locus of innovation for UAS. However, this has not been the case globally. The agricultural aircraft sector in the United States is used as a case study here to consider different metrics to evaluate UAS adoption, including a proposed metric, the normalized UAS adoption index. In aggregate, UAS operators only make up 5% of the number of agricultural aircraft operators. However, the annual number of new UAS operators exceeded that of manned aircraft operators in 2022. When used on a state-by-state basis, the normalized UAS adoption index shows that there are regional differences in UAS adoption with western and eastern states having higher UAS adoption rates while central states have significantly lower UAS adoption rates. This has implications for UAS operators, manufacturers, and regulators as this industry continues to develop at a rapid pace.

Keywords: unmanned aircraft system; UAS; unmanned aerial vehicle; UAV; drone; agriculture

Citation: Rodriguez, R., III. Measuring the Adoption of Drones: A Case Study of the United States Agricultural Aircraft Sector. *Eng* **2023**, *4*, 977–983. https://doi.org/10.3390/eng4010058

Academic Editor: Antonio Gil Bravo

Received: 31 December 2022
Revised: 27 February 2023
Accepted: 16 March 2023
Published: 17 March 2023

Copyright: © 2023 by the author. Licensee MDPI, Basel, Switzerland. This article is an open access article distributed under the terms and conditions of the Creative Commons Attribution (CC BY) license (https://creativecommons.org/licenses/by/4.0/).

1. Introduction

Unmanned aircraft systems (UAS), also referred to as unmanned aerial vehicles (UAV) and drones, have made great strides globally as regulatory frameworks have gradually accommodated this growing sector. Precision agriculture is frequently projected to be the most significant industry to benefit from these new tools [1–3]. However, these optimistic predictions have not been achieved [4,5]. In many cases throughout the world, regulatory hurdles remain in the United States [6], Europe [7,8], India [9], and Africa [10]. Meanwhile, UAS have been at the forefront of aerial application in Japan [11], China [12], and Korea [13].

Traditional aerial applications of plant protection products have been at the core of developments in UAS [14–17]. Additionally, some more specific applications in agriculture and forestry have crossed over from manned aircraft, including insect sampling [18,19], encapsulated herbicide applications [20], and aerial ignition [21]. Novel applications that are unique to this platform include vegetation sampling [22–24] and cattle herding [25]. While these new developments have introduced more cases for UAS, the implementation has proven difficult to measure.

Recent bibliometric studies on agricultural UAS have shown an increasing trend [26–28]. However, these studies are biased toward research and remote sensing. This study seeks to measure the implementation of UAS by industry—specifically, agricultural aircraft operators. Several metrics for the assessment of technology adoption have been developed, which are typically based on the percentages of users [29]. For agricultural technology, the Agricultural Technology Adoption Index uses an area that the technology is operated within as the base metric [30]. While the area is an effective base measurement, this data is not publicly available for applied areas of different aerial application technology.

In this study, the adoption of UAS compared to existing manned aviation is considered the use of a number of agricultural aircraft operators in the USA as a case study: Section 2 describes the materials and methods, including operator data acquisition and analysis; Section 3 describes the results of aggregated data analysis and annual trends; Section 4 discusses the experimental results including implications for regulators; and Section 5 concludes the work.

2. Materials and Methods

Data on agricultural aircraft operators was downloaded from the FAA databases [31] on 24 November 2022. Agricultural aircraft operators are regulated under Title 14 of the Code of Federal Regulations Part 137. The data was partitioned first by operator type (i.e., Part 137) and then by aircraft operated to separate those operators who had UAS listed on the operator certificates and those who did not. Data were then aggregated by year and by state for temporal and spatial analyses. Agricultural aircraft operators utilizing UAS who had certificates prior to the introduction of Part 107, i.e., operated manned aircraft and added UAS to their existing certificate, were aggregated together during the temporal analysis. The number of farms and the average farm size on a state basis, based on a 2021 USDA report [32], were also incorporated into the analysis. ANOVA was performed using R [33] to identify significant factors correlating with the number of agricultural aircraft operators using UAS.

To illustrate the adoption of UAS, an additional metric, the normalized UAS adoption index, I, is defined as

$$I = \frac{n_{\mathrm{UAS},x}}{n_{\mathrm{UAS}}} - \frac{n_{\mathrm{M},x}}{n_{\mathrm{M}}}$$

where n_{UAS} is the total number of agricultural aircraft operators using UAS, n_{M} is the total number of agricultural aircraft operators only using manned aircraft, and the subscript x denotes the quantity at the individual state level. The normalized UAS adoption index was calculated for each state and regional trends were analyzed qualitatively.

3. Results

Following the initial introduction of the Part 137 operator certificate in 1967, there has been a steady increase in the number of operators (Figure 1). We see a similar pattern for UAS agricultural aircraft operators following the introduction of 14 CFR 107 in 2016 and the standardized exemption for agricultural UAS operations. At the end of the study period, there were 1767 Part 137 operators, of which 93 (5%) made use of UAS.

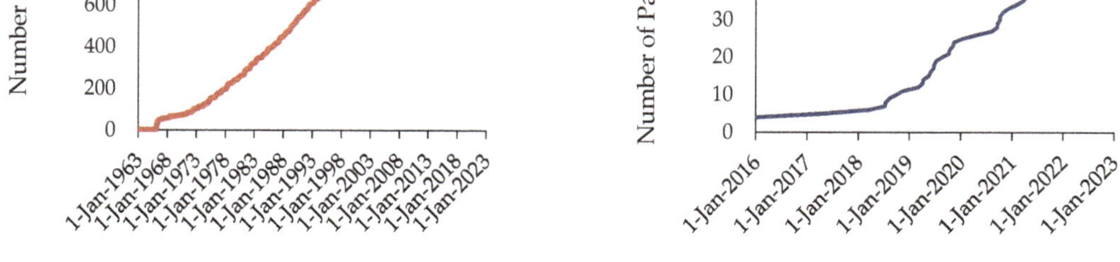

Figure 1. The total number of manned ((**A**), red) and unmanned ((**B**), blue) agricultural aircraft operators in the United States.

Focusing on manned agricultural aircraft operators, we see that the annual increase in the number of operators was relatively stable after 1986 until 2008 (Figure 2). The years from 2009 to 2020 saw an increase in the annual increase in operators, with a sharp increase starting in 2014. This increase is likely due to changes in the certification process following the FAA Modernization and Reform Act of 2012 [34], and in response to an audit report by the Inspector General [35]. Since 2021, the rate has fallen back to average levels of 29.3 new operators per year, likely due to complications in the certification process caused by COVID-19, e.g., restrictions on travel and meetings preventing in-person knowledge and skill tests.

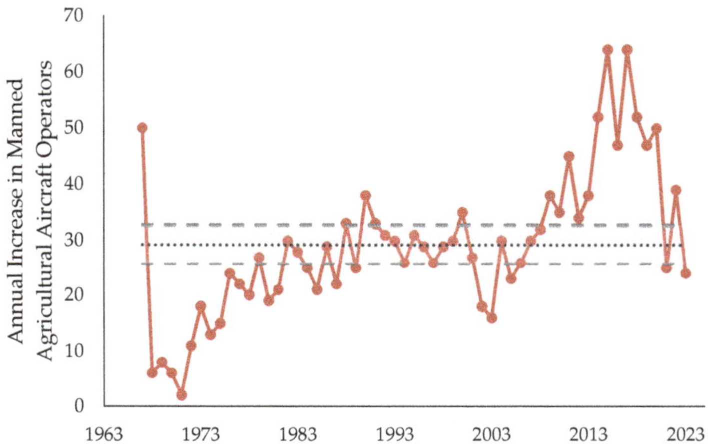

Figure 2. The annual increase in manned agricultural aircraft operators. The average annual increase is 29.3 (black dashed) with a 95% confidence interval of 25.8–32.7 (dashed grey).

Comparing the rate of new operators per year, we see that the addition of UAS operators had a significantly slower start than manned agricultural aircraft operators, which did not have the initial spike that manned agricultural aircraft operators experienced (Figure 3). However, the rate of new additions has rapidly climbed and in 2022, for the first time, the addition of new UAS agricultural aircraft operators has exceeded that of manned agricultural aircraft operators and also the average annual rate of new manned agricultural aircraft operators.

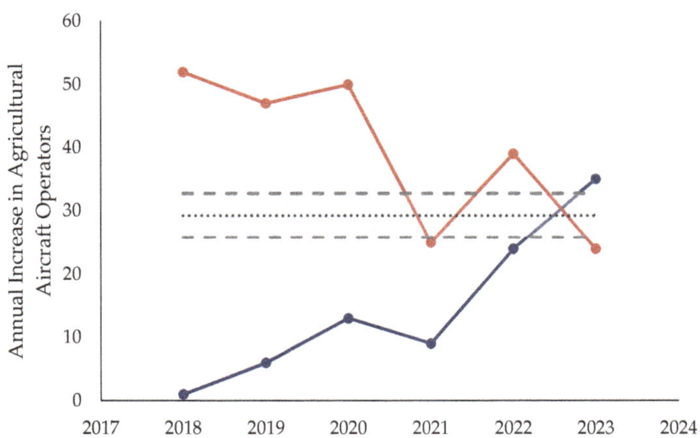

Figure 3. The annual increase in manned (red) and unmanned (blue) agricultural aircraft operators and the average annual increase in manned operators since 1967, with a 95% confidence interval.

The ANOVA analysis showed that the number of farms in a state and the number of manned agricultural aircraft operators were both significant factors in determining the number of UAS agricultural aircraft operators (Table 1). On a state-by-state basis, the number of UAS agricultural aircraft operators is only weakly correlated with the number of manned agricultural aircraft operators (Figure 4A). This indicates that UAS are not simply replacing a portion of the existing aerial application market. Over a third of states do not have a UAS agricultural aircraft operator, and yet half have two or more, which indicates a regional bias (Figure 4B). The primary factor in this regional bias is the number of farms in a particular state. The normalized UAS adoption index (Figure 5) further illustrates this regional bias with states with relatively high adoption indices concentrated together. A positive index value indicates that the rate of increase in the number of agricultural aircraft operators using UAS exceeds that of operators using only manned aircraft while a negative index value indicates the opposite.

Table 1. Results of statistical analysis of factors affecting number of UAS agricultural aircraft operators. Only significant results, number of farms, and number of manned agricultural aircraft operators are shown.

Variable	Sum of Squares	Degrees of Freedom	F	p	η^2
Number of Farms	127.218	1	62.768	<0.001	0.57
Manned Agricultural Aircraft Operators	9.521	1	4.698	0.03	0.09
Residual	95.26	47			

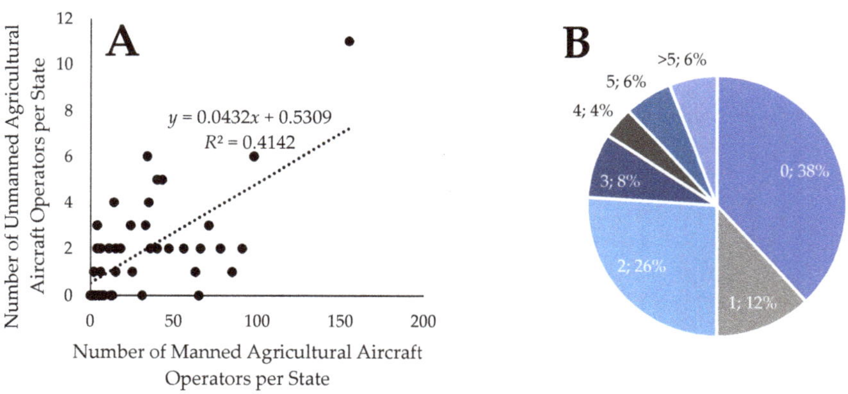

Figure 4. (**A**) Number of unmanned agricultural aircraft operators vs number of manned agricultural aircraft operators. (**B**) Number of unmanned agricultural aircraft operators by state.

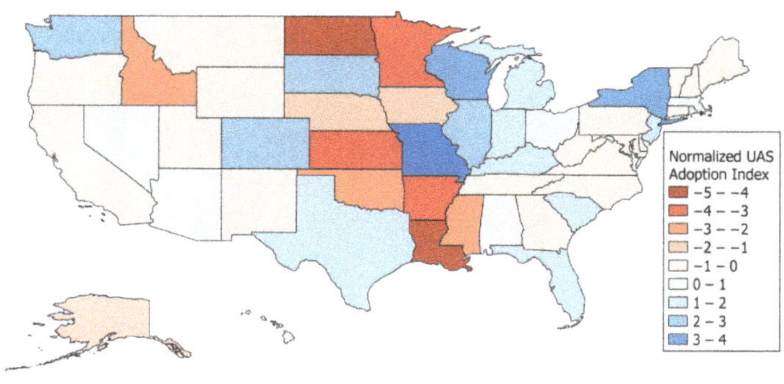

Figure 5. Normalized UAS adoption index across the United States indicates significant regional bias in the adoption of UAS for aerial application.

4. Discussion

While the total number of agricultural aircraft operators utilizing UAS is relatively low compared to that of manned agricultural aircraft operators, the rapid increase in the annual rate of new agricultural aircraft operators using UAS, which has now overtaken the rate of new manned agricultural aircraft operators, provides a strong argument that the industry has finally started taking these new tools seriously. This also has ramifications for the FAA as the agency must now account for double the amount of new Part 137 applicants, with roughly half being potential UAS agricultural aircraft operators. The relatively high number of states without an agricultural aircraft operator with UAS also impacts local Flight Standards Districts Offices (FSDO), with many having no experience with these new aircraft, which will lead to difficulties performing oversight and inspections of these new agricultural aircraft operators.

Based on the individual state analysis, there is still some regional bias to operators, as has been previously noted [6]. In particular, western and eastern states have higher UAS adoption rates while central states have significantly lower UAS adoption rates. The primary determining factor is the number of farms in a particular state, with the number of manned agricultural aircraft operators having a smaller effect size. Additional regional variability may be due to the types of crops in these areas and continuing regulatory barriers. Manufacturers may use this information when considering customer service locations, e.g., for repairs of UAS.

The normalized UAS adoption index as a metric was able to capture this regional bias. To analyze individual factors such as regulation and crops, alternative geographic boundaries could be used in place of state boundaries, such as Flight Standards Districts. The normalized UAS adoption index could be further applied within the United States to analyze other types of operator, such as Part 135 air carrier operators, or applied on a global scale to analyze the adoption across different nations in order to understand how variations in regulations have helped or hurt the adoption of UAS.

The limitations of this study include the use of the headquarters' location listed on the certificate and the lack of applied area data in the calculation of adoption rates. The service area of an agricultural aircraft operator can extend beyond the state that the base of operations is located in. In particular, adjacent states typically accept the pesticide applicator license based on reciprocity. Due to the limited payload capacity of UAS, the applied area per flight is typically much lower than that of manned aircraft. This would result in a bias in area calculations, as manned aircraft are currently a more economical alternative to UAS.

5. Conclusions

Based on aggregate numbers of operators in the USA, UAS still have a long way to go in comparison to manned agricultural aircraft operators in agricultural operations, with only 5% of agricultural aircraft operators using UAS. However, in terms of new operators being added to the sector, UAS are now leading the charge. The normalized UAS adoption index, a proposed metric to evaluate the introduction of UAS into a sector, applied on a state-by-state basis, indicates a strong regional bias in the distribution of these operators. This index may be applied to other operator types and other geographic boundaries to determine factors that may be impacting UAS utilization.

Funding: This research received no external funding.

Institutional Review Board Statement: Not applicable.

Informed Consent Statement: Not applicable.

Data Availability Statement: All underlying data is made publicly available at https://av-info.faa.gov/dd_sublevel.asp?Folder=\AIROPERATORS (accessed on 24 November 2022).

Conflicts of Interest: The author declares no conflict of interest.

References

1. Wargo, C.; Snipes, C.; Roy, A.; Kerczewski, R. UAS Industry Growth: Forecasting Impact on Regional Infrastructure, Environment, and Economy. In Proceedings of the 2016 IEEE/AIAA 35th Digital Avionics Systems Conference (DASC), Sacramento, CA, USA, 25–29 September 2016; pp. 1–5.
2. Doering, C. Growing Use of Drones Poised to Transform Agriculture. *USA Today* **2014**, 23.
3. AUVSI. *The Economic Impact of Unmanned Aircraft Systems Integration in the United States*; Association for Unmanned Vehicle Systems International (AUVSI) Economic Report; AUVSI: Denver, CO, USA, 2013.
4. Hunt, E.R.; Daughtry, C.S.T. What Good Are Unmanned Aircraft Systems for Agricultural Remote Sensing and Precision Agriculture? *Int. J. Remote Sens.* **2018**, *39*, 5345–5376. [CrossRef]
5. Freeman, P.K.; Freeland, R.S. Agricultural UAVs in the U.S.: Potential, Policy, and Hype. *Remote Sens. Appl. Soc. Environ.* **2015**, *2*, 35–43. [CrossRef]
6. Rodriguez, R. Perspective: Agricultural Aerial Application with Unmanned Aircraft Systems: Current Regulatory Framework and Analysis of Operators in the United States. *Trans. ASABE* **2021**, *64*, 1475–1481. [CrossRef]
7. Lowenberg-DeBoer, J.; Behrendt, K.; Ehlers, M.-H.; Dillon, C.; Gabriel, A.; Huang, I.Y.; Kumwenda, I.; Mark, T.; Meyer-Aurich, A.; Milics, G.; et al. Lessons to Be Learned in Adoption of Autonomous Equipment for Field Crops. *Appl. Econ. Perspect. Policy* **2022**, *44*, 848–864. [CrossRef]
8. Reger, M.; Bauerdick, J.; Bernhardt, H. Drones in Agriculture: Current and Future Legal Status in Germany, the EU, the USA and Japan. *Landtechnik* **2018**, *73*, 62–80.
9. Srivastava, S.; Gupta, S.; Dikshit, O.; Nair, S. A Review of UAV Regulations and Policies in India. In Proceedings of the UASG 2019, Roorkee, India, 6–7 April 2019; Jain, K., Khoshelham, K., Zhu, X., Tiwari, A., Eds.; Springer International Publishing: Cham, Switzerland, 2020; pp. 315–325.
10. Ayamga, M.; Tekinerdogan, B.; Kassahun, A. Exploring the Challenges Posed by Regulations for the Use of Drones in Agriculture in the African Context. *Land* **2021**, *10*, 164. [CrossRef]
11. Sheets, K.D. The Japanese Impact on Global Drone Policy and Law: Why a Laggard United States and Other Nations Should Look to Japan in the Context of Drone Usage. *Ind. J. Glob. Legal Stud.* **2018**, *25*, 513. [CrossRef]
12. Yang, S.; Yang, X.; Mo, J. The Application of Unmanned Aircraft Systems to Plant Protection in China. *Precis. Agric.* **2018**, *19*, 278–292. [CrossRef]
13. Xiongkui, H.; Bonds, J.; Herbst, A.; Langenakens, J. Recent Development of Unmanned Aerial Vehicle for Plant Protection in East Asia. *Int. J. Agric. Biol. Eng.* **2017**, *10*, 18–30.
14. Wang, G.; Lan, Y.; Qi, H.; Chen, P.; Hewitt, A.; Han, Y. Field Evaluation of an Unmanned Aerial Vehicle (UAV) Sprayer: Effect of Spray Volume on Deposition and the Control of Pests and Disease in Wheat. *Pest Manag. Sci.* **2019**, *75*, 1546–1555. [CrossRef]
15. Woldt, W.; Martin, D.; Lahteef, M.; Kruger, G.; Wright, R.; McMechan, J.; Proctor, C.; Jackson-Ziems, T. Field Evaluation of Commercially Available Small Unmanned Aircraft Crop Spray Systems. In Proceedings of the 2018 ASABE Annual International Meeting, Dearborn, MI, USA, 31 July 2018; American Society of Agricultural and Biological Engineers: St. Joseph, MI, USA, 2018; p. 1.
16. Martin, D.E.; Rodriguez, R.; Woller, D.A.; Reuter, K.C.; Black, L.R.; Latheef, M.A.; Taylor, M.; López Colón, K.M. Insecticidal Management of Rangeland Grasshoppers Using a Remotely Piloted Aerial Application System. *Drones* **2022**, *6*, 239. [CrossRef]
17. Chen, H.; Lan, Y.; Fritz, B.K.; Hoffmann, W.C.; Liu, S. Review of Agricultural Spraying Technologies for Plant Protection Using Unmanned Aerial Vehicle (UAV). *Int. J. Agric. Biol. Eng.* **2021**, *14*, 38–49. [CrossRef]
18. Mulero-Pázmány, M.; Martínez-de Dios, J.R.; Popa-Lisseanu, A.G.; Gray, R.J.; Alarcón, F.; Sánchez-Bedoya, C.A.; Viguria, A.; Ibáñez, C.; Negro, J.J.; Ollero, A.; et al. Development of a Fixed-Wing Drone System for Aerial Insect Sampling. *Drones* **2022**, *6*, 189. [CrossRef]
19. Kakutani, K.; Matsuda, Y.; Nonomura, T.; Takikawa, Y.; Osamura, K.; Toyoda, H. Remote-Controlled Monitoring of Flying Pests with an Electrostatic Insect Capturing Apparatus Carried by an Unmanned Aerial Vehicle. *Agriculture* **2021**, *11*, 176. [CrossRef]
20. Rodriguez III, R.; Leary, J.J.K.; Jenkins, D.M. Herbicide Ballistic Technology for Unmanned Aircraft Systems. *Robotics* **2022**, *11*, 22. [CrossRef]
21. Lawrence, B.; Mundorff, K.; Keith, E. The Impact of UAS Aerial Ignition on Prescribed Fire: A Case Study in Multiple Ecoregions of Texas and Louisiana. *Fire Ecol.* **2022**, *19*, 11. [CrossRef]
22. Perroy, R.L.; Meier, P.; Collier, E.; Hughes, M.A.; Brill, E.; Sullivan, T.; Baur, T.; Buchmann, N.; Keith, L.M. Aerial Branch Sampling to Detect Forest Pathogens. *Drones* **2022**, *6*, 275. [CrossRef]
23. Krisanski, S.; Taskhiri, M.S.; Montgomery, J.; Turner, P. Design and Testing of a Novel Unoccupied Aircraft System for the Collection of Forest Canopy Samples. *Forests* **2022**, *13*, 153. [CrossRef]
24. Charron, G.; Robichaud-Courteau, T.; La Vigne, H.; Weintraub, S.; Hill, A.; Justice, D.; Bélanger, N.; Lussier Desbiens, A. The DeLeaves: A UAV Device for Efficient Tree Canopy Sampling. *J. Unmanned Veh. Syst.* **2020**, *8*, 245–264. [CrossRef]
25. Li, X.; Huang, H.; Savkin, A.V.; Zhang, J. Robotic Herding of Farm Animals Using a Network of Barking Aerial Drones. *Drones* **2022**, *6*, 29. [CrossRef]
26. Singh, A.P.; Yerudkar, A.; Mariani, V.; Iannelli, L.; Glielmo, L. A Bibliometric Review of the Use of Unmanned Aerial Vehicles in Precision Agriculture and Precision Viticulture for Sensing Applications. *Remote Sens.* **2022**, *14*, 1604. [CrossRef]
27. Rejeb, A.; Abdollahi, A.; Rejeb, K.; Treiblmaier, H. Drones in Agriculture: A Review and Bibliometric Analysis. *Comput. Electron. Agric.* **2022**, *198*, 107017. [CrossRef]

8. Raparelli, E.; Bajocco, S. A Bibliometric Analysis on the Use of Unmanned Aerial Vehicles in Agricultural and Forestry Studies. *Int. J. Remote Sens.* **2019**, *40*, 9070–9083. [CrossRef]
9. Skare, M.; Riberio Soriano, D. How Globalization Is Changing Digital Technology Adoption: An International Perspective. *J. Innov. Knowl.* **2021**, *6*, 222–233. [CrossRef]
0. Jain, R.; Arora, A.; Raju, S.S. A Novel Adoption Index of Selected Agricultural Technologies: Linkages with Infrastructure and Productivity. *Agric. Econ. Res. Rev.* **2009**, *22*, 109–120.
1. FAA Data Downloads: Air Operators. Available online: av-info.gov/dd_sublevel.asp?Folder=\AIROPERATORS (accessed on 24 November 2022).
2. National Agricultural Statistics Service. *Farms and Land in Farms 2021 Summary*; United States Department of Agriculture: Washington, DC, USA, 2022.
3. R Core Team. *R: A Language and Environment for Statistical Computing*; R Core Team: Vienna, Austria, 2019.
4. United States House of Representatives. *FAA Modernization and Reform Act of 2012*; United States House of Representatives: Washington, DC, USA, 2012.
5. Federal Aviation Administration. *Weak Processes Have Led to a Backlog of Flight Standards Certification Applications*; Federal Aviation Administration: Washington, DC, USA, 2014.

Disclaimer/Publisher's Note: The statements, opinions and data contained in all publications are solely those of the individual author(s) and contributor(s) and not of MDPI and/or the editor(s). MDPI and/or the editor(s) disclaim responsibility for any injury to people or property resulting from any ideas, methods, instructions or products referred to in the content.

MDPI
St. Alban-Anlage 66
4052 Basel
Switzerland
Tel. +41 61 683 77 34
Fax +41 61 302 89 18
www.mdpi.com

Eng-Advances in Engineering Editorial Office
E-mail: eng@mdpi.com
www.mdpi.com/journal/eng

www.ingramcontent.com/pod-product-compliance
Lightning Source LLC
Chambersburg PA
CBHW041932260926
51850CB00018B/90
* 9 7 8 3 0 3 6 5 7 5 3 0 8 *